HEALTH MONITORING OF BRIDGES

This edition first published 2009

Registered office
John Wiley & Sons Ltd, The Atrium, Southern Gate, Chichester, West Sussex, PO19 8SQ, United Kingdom

For details of our global editorial offices, for customer services and for information about how to apply for permission to reuse the copyright material in this book please see our website at www.wiley.com.

Library of Congress Cataloging-in-Publication Data

Wenzel, Helmut.
Health monitoring of bridges / By Helmut Wenzel.
p. cm.
Includes bibliographical references and index.
ISBN 978-0-470-03173-5 (cloth)
1. Bridges–Inspection. 2. Bridge failures–Prevention. I. Title.
TG305.W46 2009
624.2028'7–dc22

2008043726

ISBN: 9780470031735

A catalogue record for this book is available from the British Library.

Set in 9/11 pt Times by Thomson Digital, Noida, India

HEALTH MONITORING OF BRIDGES

Helmut Wenzel

VCE Holding GmbH,
Vienna, Austria

WILEY

A John Wiley and Sons, Ltd, Publication

Contents

Figures

Tables

Foreword

Bridges are the flagships of our transportation infrastructure, on which society heavily depends. Operation and maintenance have become more and more complex with the increased age of our bridge stock. Structural Health Monitoring, as part of lifecycle management procedures, experienced a growth in importance recently. To maintain and improve the high quality and high level of service to the public it is essential to know the lifecycle performances of structures to ensure long service life and durability.

Structural Health Monitoring of bridges comprises too many approaches and aspects to be completely covered in one publication. This book therefore concentrates on the current practice and methodologies of dynamic monitoring. The theorem that the health of a structure is expressed in its dynamic characteristic is exploited.

Other sectors like mechanical engineering or aeronautics are operating in conditions where the properties of their structures are well known and are operating under controlled conditions. Civil engineering has to account for numerous non-linearities as well as dominating environmental factors that are able to hide useful information in the records. This leads to the situation that the results carry a portion of uncertainty that we have to deal with as effectively as possible. It is most likely that one or the other approach will be overruled by future research work and the methodologies considerably improved. It therefore is of highest importance that the raw data of any monitoring campaign are stored properly in order to apply new algorithms in the future, or to enable qualitative comparison between subsequent measurements.

It further has to be mentioned that bridge management approaches are dominated by political factors or incidents like bridge collapses. This might hinder the best possible exploitation of the methodologies. It further has to be recognized that currently a search for proper bridge management procedures is underway, which might lead to adaptation of the described approach.

The vision for structural health monitoring of bridges is an integrated decision support system, web based and featuring a most user friendly surface. It contains the following elements:

- A display embedded in a GIS environment reporting the status of any structure in a network
- A database with web interface
- Permanent and mobile monitoring units
- Data handling, transfer and cleaning routines
- A knowledge and history base for statistical comparison
- A database on dynamic bridge simulation including automatic model update routines
- A case based reasoning system to compute the proposals for decision making
- Interfaces to existing bridge databases and relevant codes and standards

The key to success is high quality data combined with realistic identified models and deterioration laws quantitatively supported by monitoring. The output can either be a reliability index, a safety level, a graphic symbol or any other output value as desired by the bridge owners. Due to the complexity of the subject there will be limitations to this approach that can be narrowed in combination with engineering judgment. There is a long way ahead of us before the computerized SHM systems will be superior to the experience of a senior bridge engineer. It is doubtful that the human input can be replaced in civil engineering entirely and a fruitful combination is proposed. The proposed SHM is to be seen as a tool and support for the bridge engineer as well as an indicator to the bridge operators when critical situations are developing and a human expert ought to be called.

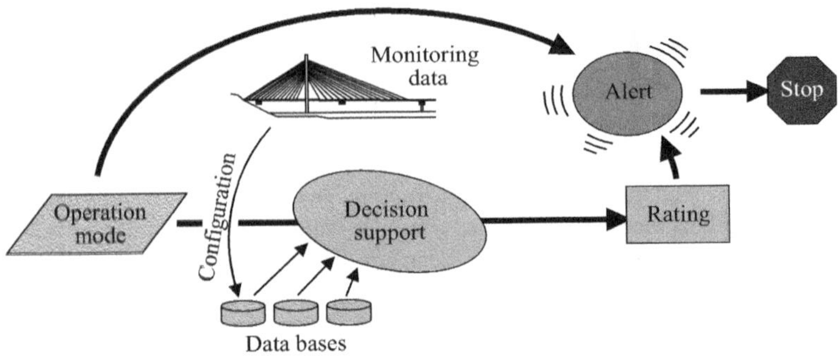

Figure 1 System architecture of VCDECIS

Structural Health Monitoring in practice requires a structured approach. Too often only fragments of the complete procedure are applied. The best and most satisfactory results will be achieved if all issues behind the following 12 activities are addressed.

1. SHM concept (clear objectives!) and design
2. Optimization and cost benefit analysis
3. Hardware
4. Software
5. Communication and web interface
6. Commissioning and start-up
7. Reporting structure
8. Periodic reporting
9. Analyses and expertise
10. Thresholds and warning levels
11. Periodic maintenance
12. System upgrade

Not all issues are necessarily addressed by the same specialist. Often teamwork is the way to succeed.

List of Contributors

The following persons and institutions respectively provided the author with major contributions:

Günther Achs, Vienna Consulting Engineers
A. Emin Aktan, Drexel Intelligent Infrastructure and Transportation
P. Barras, CETE du Sud Ouest, Laboratoire Régional des Ponts et Chaussées
Andrea Bergamini, Swiss Federal Laboratories for Materials Testing and Research
T. Bolle, Infokom
James Brownjohn, University of Plymouth
L.K. Cheng, TNO TPD
Guido DeRoeck, Katholieke Universiteit Leuven
Rainer Flesch, Arsenal Research GmbH
C.-P. Fritzen, University of Siegen
Peter Furtner, Vienna Consulting Engineers
Georg Gutenbrunner, Vienna Consulting Engineers
Karim Hariri, University of Technology at Braunschweig
Olaf Huth, Swiss Federal Laboratories for Materials Testing and Research
Daniele Inaudi, SMARTEC SA
Raid Karoumi, Royal Institute of Technology
Anne S. Kiremidjian, Stanford University
J.M. Ko, Hong Kong Polytechnic University
S. Lesoille, Laboratoire Central des Ponts et Chaussées
Andrea Mordini, Vienna Consulting Engineers
Bart Peeters, LMS International
Udo Peil, University of Technology at Braunschweig
Rolf Rohrmann, Federal Institute for Materials Research and Testing
Dominique Siegert, Laboratoire Central des Ponts et Chaussées
Hiroshi Tanaka, University of Ottawa
Robert Veit-Egerer, Vienna Consulting Engineers
Johan Wiberg, Royal Institute of Technology

GEOCISA.
Geodetic and Geophysical Research Institute of the Hungarian Academy of Sciences,
RAMBØLL.
Swedish National Road Administration,
TRL Limited.

Preface

The demand for structural health monitoring of bridges was recognized after the Silver Creek Collapse in the United States (1967) and the collapse of the Reichsbrücke in Vienna (1976). Monitoring in civil engineering became feasible with the development of PCs and suitable hardware in the beginning of 1990. This subject is now discussed deeply in the bridge engineering community. In respective conferences and symposia prominent spaces are reserved and Stanford University organizes the bi-annual international workshop on Structural Health Monitoring where the community meets and discusses progress and the roadmap for future development.

During the Stanford workshop in 2005 the idea to write books on structural health monitoring in the various sectors was widely discussed. This book covers bridges, which are the flagships of civil engineering. This sector is particular challenging as every bridge is a prototype and the usual risk paradigms applied to other sectors do not bring the desired success. To show the variety of applications a large number of practical cases were collected in a European research project (SAMCO) which is presented here, together with the theoretical theses and a helpful compilation of terms and formulations.

I would like to mention that this sector is still under development and progressing quickly, which might make some of the statements outdated or obsolete. Nevertheless it is intended to provide a basis for the necessary future development toward integrated decision support systems for our bridges.

Acknowledgments

The advice, support and understanding of fellows, colleagues and family members is required to write a book. The author is well aware that without this help it would not have been possible to do that. The book represents a summary of actual bridge research and assessment work performed during the past years. I would like to particularly acknowledge the opportunities given to me by our clients European Commission (Mr. Katalagarianakis), ASFINAG (Mr. Fink, Mr. Ritzberger), NÖ-LR (Mr. Talmann), ÖBB (Mr. Presle) and BMVIT (Mr. Breyer and Mrs. Eichinger).

Many thanks go to my fellows in the company's research and development department, namely Mr. Peter Furtner, Mr. Robert Veit-Egerer, Mr. Martin Stöger, Mr. Ernst Forstner, Mr. Andrea Mordini and Mrs. Bianca Mick. An extensive contribution has been received by Prof. Hiroshi Tanaka who compiled Chapter 12. Valuable information and fruitful discussion with the global experts in this sector gave plenty of inspiration and targeted advice. Yozo Fujino (University of Tokyo), Emin Aktan (Drexel University), Dan Frangopol (Lehigh University), Anne Kiremidjian (Stanford University) and Aftab Mufti (ISIS Network Canada) are to be named.

This book would not have been possible without the research projects carried out recently. Particular the projects in the 5th and 6th framework program of the European Commission, namely SAMCO (G1RD-CT-2001-05040) and SAFE PIPES (NMP-CT-2005-013898) enabled me to produce a wide portion of the work presented. National projects like BRIMOS and HOT SPOT (supported by BMVIT and FFG) enabled the development of practical applications.

The author further would like to acknowledge the contributions of numerous owners of bridges, research fellows and colleagues in practical work. Without this support and real applications the progress would not have been half as good. Finally I would like to say thank you to all who I was not able to mention here.

1

Introduction and Motivation

Bridges are the flagships of civil engineering. They attract the greatest attention within the engineering community. This is due to their small safety margins and their great exposure to the public. Early bridges were the backbone of powerful empires from China to Rome and the Incas in America. Currently the transportation infrastructure is directly related to the economic success of a nation. Bridges are admired for their function but also primarily for their esthetic impact. Imagine New York without her bridges, Japan without the Honshu Shikoku project or Europe without the Greatbelt Link. This book will contribute to the preservation and maintenance of these important elements of modern society.

1.1 Health Monitoring

The global higher transportation network operates about 2.5 million bridges. Current bridge management systems rate them using various methodologies and approaches. This results in very inhomogeneous statistics. The US Federal Highway Agency (FHWA) stated in 2005 that 28% of their 595 000 bridges are rated as being deficient, with only a portion of these (about 15%) being deficient for structural reasons. In Europe this figure varies around 10% being structurally deficient. No figures are available for the Asian networks. Nevertheless if we consider an average of 10% structural deficiency, we are looking at 250 000 bridges that definitely require structural health diagnosis, improvement and monitoring. As structural health monitoring (SHM) should be used in a preventive capacity before bridges become deficient, this considerably increase the number of its applications above the global estimate of 10% that are structurally deficient.

Structural health monitoring is the implementation of a damage identification strategy to the civil engineering infrastructure. Damage is defined as changes to the material and/or geometric properties of these systems, including changes to the boundary conditions and system connectivity. Damage affects the current or future performance of these systems.

The damage identification process is generally structured into the following levels:

- Damage detection, where the presence of damage is identified.
- Damage location, where the location of the damage is determined.
- Damage typification, where the type of damage is determined.
- Damage extent, where the severity of damage is assessed.

An extensive literature has developed on SHM over the past 20 years. This field has matured to a point where several accepted general principles have emerged. Nevertheless these principles are still challenged and further developed by various groups of interest. Strategies in mechanical engineering or aerospace adopt different approaches, but nevertheless the civil engineering community can considerably benefit from them. Separate approaches are necessary to consider that civil engineering structures are each a prototype.

1.2 Client Requirements and Motivation

The construction sector is conservative. The implementation of new technologies needs a clear requirement and motivation to be accepted by owners and operators. It has been recognized that the current practice does not satisfy the needs of shrinking budgets and aging structures. Nevertheless they satisfy valid codes and standards. Before a breakthrough in implementation of new technologies can happen the requirements and motivation have to be clearly understood and argued against potential clients.

Three main drivers might be approached in the promotion of SHM. The motivation to apply and order services based on the new technologies can be:

- Responsibility driven, which means the new methods become standard applications supported by codes, standards and guidelines.
- Economically driven, such as situations where a ranking of structures to be rehabilitated is necessary because of insufficient budget available or the need to use a structure for a certain time period longer than designed.
- Curiosity-driven motivations comprise those cases where clients would like to know more about their important and complicated structures. Results can also lead to better planning for future structures.

From the above-mentioned motivations the following requirements can be derived. These are typically services requested from the technology providers:

- A certificate that a structure satisfies the requirements from codes, standards and guidelines comprises a main business opportunity. Many recommendations already consider the increase of maintenance periods so that measurements can be taken. The provision of such certificates by engineers is common practice in Europe. Other parts of the world do not apply this system. It has led to an impressive evolution of bridge technology in Europe, which has been exported worldwide. It creates an environment for quality construction.
- The transfer of liabilities and responsibilities for structures in terms of technical and operational matters takes place with the huge privatization drive we can observe currently. Clients are systematically transferring the stock of structures into private hands. The new players involved are open to new applications that are able to support innovative and economic maintenance strategies.
- Special structures require special attention. The necessary top expertise cannot always be available with every owner or operator. The top experts for each region will be required to offer the newest technologies for their work.
- A shortage in the capacity of personnel to carry out routine maintenance and assessment works at the bridge stock also leads to new opportunities. As these services are normally tendered, new technologies might have an economic and technology edge.
- In case of emergency or accidents the generation of a secure situation is desired by affected owners. Any assessment based on the results of measurements is more likely to be accepted than subjective assessment by the expert. The clients want to sleep well because somebody else is permanently watching and assessing their structures.
- Ad hoc assessment in case of doubt or emergency also comprises this application area. The subjective conventional assessment produces too many negative scores on structures, and doubts are raised. A quantitative assessment is desired.

- The optimization of maintenance concepts requires input on which this process can be performed. The more data are available, the better the organization will be and the better the available maintenance concepts. A reduction of the remaining risks helps to make decisions with lower safety margins.
- The determination of priorities, through a quantification based on measurements, helps to satisfy the growing demand in combination with shrinking budgets. This assessment can come up with better scores, minimizing the number of structures requiring immediate intervention. Decision support for investment planning can be offered on the basis of the above-mentioned services. Every new measurement improves the database and as such improves the quality of the results and supports the necessary decision making.
- Life-cycle cost determination helps to increase the periods when budgetary planning is necessary. The demand for retrofit and maintenance can be estimated over the whole life period of a structure or even of a fleet of structures.
- The direct link of structural performance to operation of a structure can be established. Very often information about an optimal speed or frequency in the traffic can be determined that can be used by the operation personnel of a transportation infrastructure and communicated to the drivers through telematic devices.
- Hot spot identification technologies are very often requested in case the weakest point of the system or a significant accumulation of incidents is observed. Clients would like to know where to look first and what the background of certain phenomena could be.
- The prediction of structural performance for future loading scenarios is a further specific item requested. When a nonlinear behavior can be expected, special expertise becomes necessary.
- Fleet observation is desired to improve the quality of assessment when the number of structures is huge. For this the conception has to be subdivided into stages depending on the depth of information required.

The selection of a suitable observation concept has to be based on mainly external factors. These are the number of structures to be observed in combination with the budget available. For this purpose it is necessary to offer services on increasing quality levels. The levels can be subdivided into spot, periodic, permanent and online assessment campaigns at structures. The respective features are:

- A spot observation should comprise a very quick measurement campaign with a few simple to handle sensors only. It should provide information on the general condition of a structure in order to create a ranking.
- Periodic assessment means a measurement campaign on a structure, which is repeated after a specified period of time, to generate information on the performance over time. This spot information might comprise rather long periods.
- Permanent observation and assessment of structures becomes necessary when certain limits are passed. This observation allows a very detailed assessment based on permanent recordings and can help to implement quick decision making.
- Online observation and assessment allows warning through electronic media, be it through a short message service (SMS) in the simple case or an online status through the internet. Decisions might be taken by the computer based on the measurement data. These alert systems will only be applied at extremely critical structures.

In general it has to be stated that clients need and desire support of their work and not issues that make it more complicated. In this respect also the procedures have to be carefully watched and permanently improved. The information policy also plays a major role in the client–consultant relationship. The new methodologies are rather complex and require a deep understanding of structural dynamics, physics and measurement techniques. Due to the fact that this expertise is rarely available at the owner's engineering department, the fear to be exposed to unknown black box applications has to be taken from their shoulders.

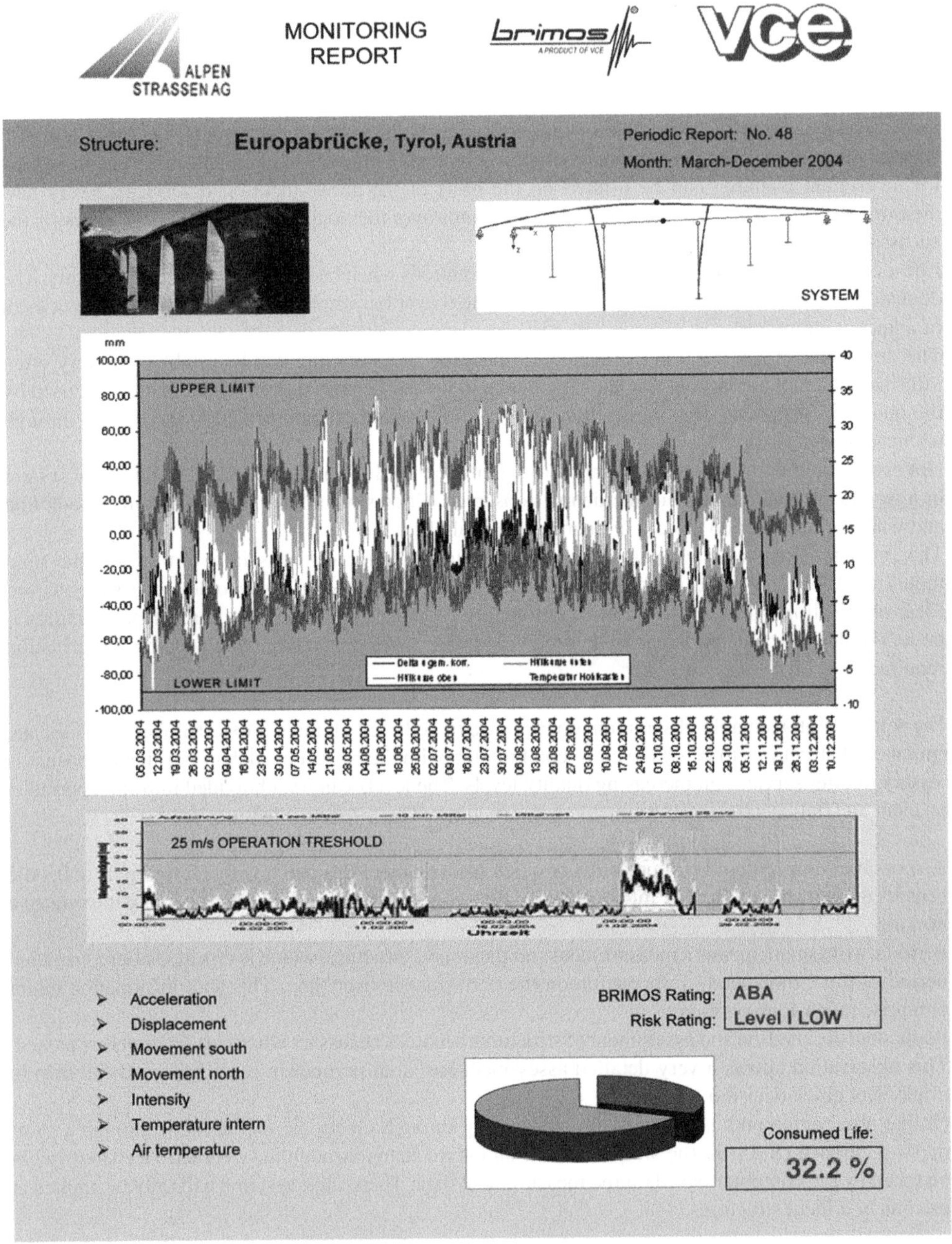

Figure 1.1 Periodic SHM report of a bridge

On the other hand they are spending considerable amounts of money and would like to be informed frequently about progress and results. Therefore we have to ensure that the technology part is in good and competent hands and that they will receive the necessary information they desire. The best success has been achieved with very simple reporting techniques. A periodic report received by email comprising single page information is preferred. The example shown in Figure 1.1 provides such a typical weekly report. The main information is provided in a single window, where upper and lower normalized thresholds are given and the measurement results within this period are placed within these thresholds. With one look at this graph the personnel can immediately see whether any of the thresholds have been exceeded. The client is satisfied because all indicators are positive and the ordered observation is permanently working.

The periodic report should provide on this single page the following information:

- A photograph and a system plot of the structure under observation for easy and quick identification.
- A window with the periodic results placed within the relevant thresholds over the observation period.
- Eventually a second window with special information required by the client, such as wind speed information or any other quantity desired.
- Finally a rating should be provided that is based on the measurements taken in the reporting period. This rating should enable the client to immediately see whether any changes have happened.
- Eventually the specification of a remaining life capacity can be provided if the necessary data are recorded.

Besides this one-page record for the client a scientific report should be generated by the system for the expert. This will enable a quick assessment of all the single measurements in order to acquire the necessary expertise or learn from the performance. Every year on average the system should be calibrated with the information gained. This might also comprise a change in the rating and will update the remaining life capacity based on existing knowledge.

2

Bridge Management and Health Monitoring

Any bridge management policy is a multidisciplinary task where subjects from a number of fields like structural engineering, computer science as well as economics have to be involved. The currently operated bridge management systems (BMSs) are very efficient tools where the mentioned fields are combined with the final objective of optimization of maintenance budgets within a stock of existing bridges. All aspects (structural, computer science, economic, social and user costs) are of vital relevance for the good performance of any BMS.

The condition of bridges is normally assessed through a condition rating or condition index, as a ranking of a bridge condition in comparison to others of a bridge stock. The condition index should be an indicator for:

- A ranking of bridges in a bridge stock from the most deteriorated to the best.
- Assessment of the capacity of a bridge to determine eventual capacity reduction factors.
- Identification of trends in the deterioration process that can lead to an estimation of the expected serving life using the condition rating extrapolated to successive time intervals.

All relevant data are normally stored in databases, which could be used for other purposes such as an information source for maintenance works. Such web-based services are slowly developing but will conquer the market soon.

Currently bridge owners use one of the commercially available systems such as PONTIS or have developed their archives into BMSs. The characteristics of these are given by Casas (2006):

- All administrations have a systematic inspection procedure (normally divided into three levels), but not all use the results of the inspection in a comprehensive and objective way to derive a condition rating, either numerically (in the range 1–10, 1:100) or grammatically (poor, fair, acceptable, good, . . .).
- At the present time, all administrations and transportation agencies use visual inspection as the main source of relevant data to carry out the condition assessment.
- Only in a reduced number of countries does the code or guideline for the load capacity assessment use directly the final result of the condition assessment in the capacity rating process. A good example here would be the recently developed Danish BMS DANPRO+.
- There are basically two approaches to evaluation of the condition of the whole structure based on the condition assessment of its element. The first is based on a commutative condition rating, where the

Health Monitoring of Bridges Helmut Wenzel

most severe damage on each element is determined for each span if there is a superstructure and for each part of the substructure, the carriageway and accessories. The final result is the condition rating for the structure. The second method uses the highest (or lowest, depending on the measuring scale) condition rating of the bridge components as the condition rating for the whole structure itself.

- Each administration uses different condition rating techniques. Such a situation may derive from the fact that the same bridge, assessed by two engineers from different countries, can be rated with different grades.
- A clear division exists between methods that are purely subjective, those based on simple scoring (by assigning a number of deficiency points to the inspected structural member, in compliance with the rules adapted for the classification and evaluation of damage) and those where the final condition rating is obtained by calculation (where the weighting of a set of selected essential damage types is done based on the report by the inspector).
- Most methods divide the whole bridge into several parts or components and these components into elements. In such cases there is no clear or objective indication of how to pass from the condition of the individual element to the global condition rating of the bridge, and this is made by the engineering judgment of the inspector.

Also important by itself, and even more important than the condition index for a reliable management of the bridge stock, is the calculation or modeling of the condition index deterioration with time. Numerous models have been developed that describe the transition of a variety of condition states over time. Many of the models are based on a linear deterioration of condition states where the condition rating at any time can be computed and a deterioration rate can be expressed in terms of condition rating loss per year. Hearn (1995) compiled an extensive list of these models. Some are based on data while others rely on expert opinion. Actually most of the existing BMSs use the Markov chain approach to model condition deterioration.

Future trends in this area are the incorporation of quantitative monitoring results with the qualitative expert's opinion and the results of visual inspection by fuzzy logic or neural networks. These trends are directly linked to the adoption of a probabilistic framework for the assessment of bridge condition. Frangopol and Das (1999) have claimed the limitations of the condition index to characterize discrete condition states and the Markovian approach for deterioration modeling, among others, in the lifecycle cost management of existing bridges. Much effort has been done in the formulation of deterioration models, mainly in the case of chlorite-induced corrosion of concrete bridges, where several analytical, semi-empirical and empirical methods exist. However at the moment nonconclusive models still exist, as very often the predicted values differ from the experimental results.

The use of nondestructive testing (NDT), monitoring and health monitoring in bridge condition assessment as an alternative and supplement to the actual procedure basedon visual inspection is the area of most increased activity. The main topics to be solved are the following: how to deal with the measuring errors, uncertainty and noise in relation to the different sensors and measuring techniques, and how to integrate the data from the point in time of NDT and from continuous monitoring in the assessment of the condition rate. It seems that using model updating techniques is the best way, as demonstrated by Faber and Sorensen (2002), Enright and Frangopol (1999) and Rafiq (2005).

In the USA a unified condition assessment procedure has existed for a long time. In Europe there are currently different condition assessment methods in each country. Harmonization of these methods is needed in the near future.

2.1 Bridge Management Philosophy

The two basic values of budget and function are what bridge management philosophies are based on. Aesthetics can be another important driver, but applies mainly to the few flagship structures. The engineering community has accepted the principles of preventive maintenance, meaning that investment into structural health in time keeps the safety level of a structure constant. For high quality structures this requires an average spending of 0.8–1% of the replacement value annually. Another valid strategy is to let bridges deteriorate until they reach the safety limit and are replaced. This strategy is applied when insufficient funds are available for preventive maintenance, and functional requirements very often ask for replacement anyway. In fact more structures are functionally rather than structurally deficient. This strategy requires less money overall, but carries the burden of aesthetic impacts. Typical examples are the bridges of New York where the lack of funding has left no choice on strategy. Both philosophies require SHM. In preventive maintenance it is required to detect the point of intervention, where rehabilitation shows the highest effect. The functionally driven philosophy requires SHM in order to detect when safety limits are reached.

2.2 Structural Health Monitoring

2.2.1 Definitions

Structural health monitoring is the implementation of a damage identification strategy to the civil engineering infrastructure. Damage is defined as changes to the material and/or geometric properties of these systems, including changes to the boundary conditions and system connectivity. Damage affects the current or future performance of these systems.

The damage identification process is generally structured into the following levels:

- Damage detection, where the presence of damage is identified.
- Damage location, where the location of the damage is determined.
- Damage typification, where the type of damage is determined.
- Damage extent, where the severity of damage is assessed.

Extensive literature has developed on SHM over the last 20 years. This field has matured to a point where several accepted general principles have emerged. Nevertheless these principles are still challenged and further developed by various groups of interest. Strategies in mechanical engineering or aerospace adopt different approaches, nevertheless the civil engineering community can considerably benefit from these efforts. Separate approaches are necessary to consider that civil engineering structures are each a prototype.

2.2.2 Structural Health Monitoring Axioms

At the Stanford SHM Workshop in 2005 Farrar et al. have specified axioms for structural health monitoring, which are an attempt to formulate common rules and understanding to support the "fundamental truth" that has been argued by the community. These axioms do not represent operators for SHM. In order to generate methodologies it will be necessary to add a group of algorithms that carry the SHM practitioner from data to a decision. The discipline of statistical pattern recognition is proposed for this approach. The axioms formulated are:

Axiom 1. The assessment of damage requires a comparison between two system states.

Axiom 2. Identifying the existence and location of damage can be done in an unsupervised learning mode, but identifying the type of damage present and the damage severity can only be done in a supervised learning mode.

Axiom 3. Without intelligent feature extraction, the more sensitive a measurement is to damage, the more sensitive it is to changing operational and environmental conditions.

Axiom 4. There is a trade-off between the sensitivity to damage of an algorithm and its noise rejection capability.

Axiom 5. The size of damage that can be detected from changes in system dynamics is inversely proportional to the frequency range of excitation.

In the following these axioms shall be looked at with the knowledge of civil engineering SHM and commented accordingly (see Section 4.5).

2.2.3 Condition Assessment

Knowledge about the condition of a bridge or its elements is the most essential information required. Structural health monitoring provides the opportunity to quantify the condition and to provide the basis for decisions. The methodology and tools are described in the following chapters. Here a selection of the most essential condition assessment cases is provided as an introduction into SHM. Applications are given in Chapter 9.

In civil engineering the procedure and tools are best developed for bridges. Some kind of SHM always existed in this sector. Figure 2.1 shows how these procedures have developed from simple inspection routines to highly sophisticated monitoring campaigns.

The extent of monitoring is mainly dependent on the required results. Currently five levels are used in order to determine the depths of investigation. These are:

Level 1: Rating. This represents the conventional assessment of the structure starting with a visual field inspection that provides a subjective impression of the condition of the structure. Some preliminary analytical investigation is performed in order to provide a rating as a basis for decisions. This would be the typical application of a bridge management system such as PONTIS or DANBRO. Many bridge owners use databases to store the results.

Level 2: Condition assessment. A rough visual field inspection has to be an element of any SHM campaign. After that a decision has to be made whether the conventional approach is satisfactory or an extended or even sophisticated additional approach is taken. This determines the type and quantity of instrumentation. For condition assessment a simple instrumentation is sufficient and a simple decision support system will provide the necessary additional information. Storage and treatment of data should also be done in the existing database. A link to existing conventional tools is available. The monitoring can be performed at single spots only.

Level 3: Performance assessment. This intermediate level uses the same procedure as described for Level 2. The level of assessment and performance elaboration in the decision support process is considerably higher as additional information such as mode shape is measured and elaborated. This provides additional indicators for the assessment and will demonstrate the performance of the structure. It obviously requires a denser instrumentation and synchronous monitoring.

Level 4: Detail assessment and rating. The next step will be to establish an analytical model representing the structure. The model will be compared with the monitoring results. If that identification is simple, a step back towards Level 3 might be taken. If phenomena are detected that cannot be explained from the records, further steps have to be taken to clear the situation. The most obvious thing is to introduce a permanent record over some period of time to capture the necessary phenomena valid for this specific case. Load testing also has been proven successful to establish performance parameters. With these results a simple model update can be performed to assess the results and provide a rating. Extensive monitoring is required. The records should cover at least 24 hours, but preferably much longer to capture environment and traffic situations.

Level 5: Lifetime prediction. For a serious lifetime prediction the records available have to be long enough to cover at least three cycles relevant for the structure. This is normally in the order of three

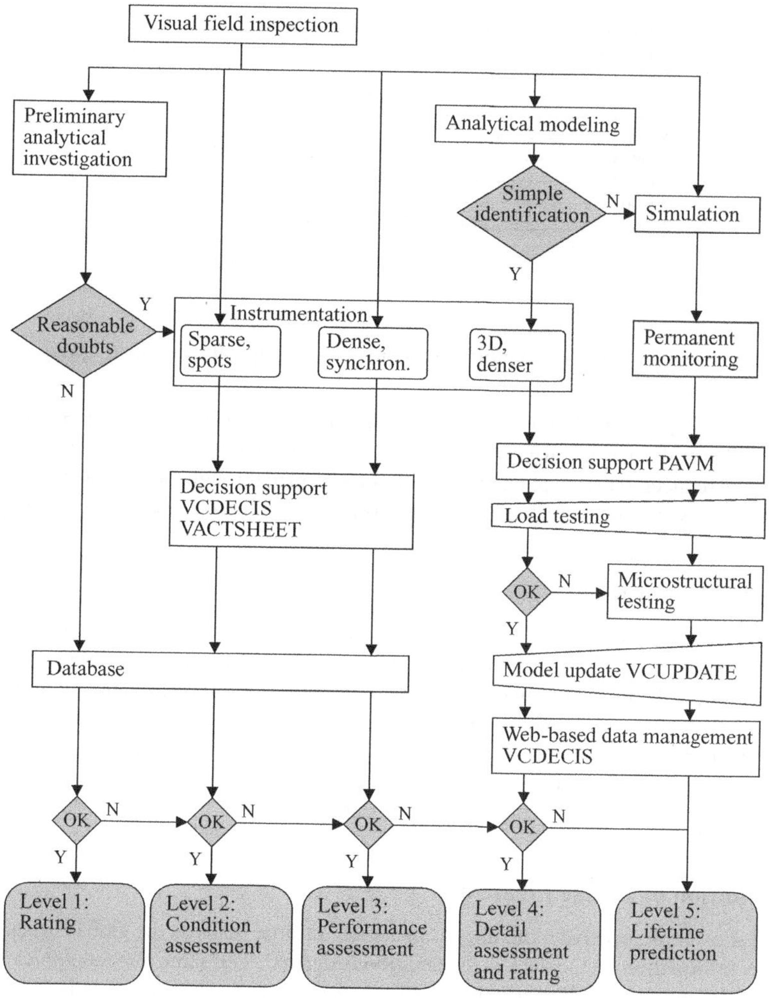

Figure 2.1 Typical hierarchical concept for the SHM procedure for bridges

years. Simulation should be run from the analytical model in order to achieve a theoretical performance for comparison. To handle the large quantity of data, special software for decision support is required. Load testing will be targeted and extensive. In addition, microstructural testing might be useful in order to look into the performance of single elements of a structure. The update process will be extensive and consider several conditions of the structure. This includes particularly the loaded and unloaded cases and all the non linearities involved. In cases of reasonable doubt, this monitoring system should be operated online and web based with a warning computed by decision support. The final lifetime prediction can then be performed as described in Section 7.3.

The costs related to these procedures are given in Figure 2.2. These costs mainly depend on the extent of the monitoring campaign and the number of man-hours to be invested in modeling, simulation and update procedures. The figures provided in the graph are based on 2006 prices for a typical three-span

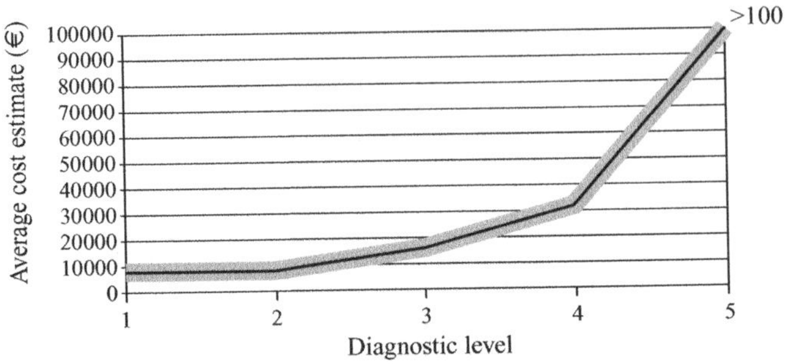

Figure 2.2 Cost development of monitoring campaigns for a typical three-span bridge (€, base 2006)

bridge with an average length of 150 m. The price can also be influenced by the number of spans, by the type of structure and also particularly by the conditions for the monitoring campaign. It is expected that the prices will be reduced rather than increased. This can happen through the introduction of time-saving modeling procedures and sophisticated monitoring software, but these are still to be developed.

2.2.3.1 Global Condition

Many bridges that have passed their critical age, which have been designed following now inadequate loading standards, might have been damaged by deterioration or modern overloading. The question of whether the bridge still fulfills its function is often asked by the owners. A simple ambient vibration monitoring (AVM) campaign provides reliable information on the condition, as demonstrated by and Wenzel and Pichler (2005). The principle is to compare the actual behavior measured on site with a theoretical model that represents the designer's concept. The methodology is based on function only and does not provide information on eventually existing damage.

2.2.3.2 Condition of Bridge Elements

Deck slabs, for example, receive more direct impact from traffic loads as the global structure, therefore in some cases information on the deck slab condition is desired. The same methodologies as mentioned before can also be applied locally. The behavior of a deck slab as a structural system in itself is monitored and compared to the desired behavior. Weak points of the structure can easily be filtered out by this method. Details can be found in Chapter 7 and in Wenzel and Pichler (2005).

2.2.3.3 Cable and Hangers

Cables and hangers are bridge elements with very different characteristics. Axial forces dominate its function, but bending often cannot be neglected. Special assessment routines have been developed for them. As they are prone to aerodynamic or parametric excitation it is often more difficult to understand their behavior. If the behavior is not yet well understood, condition assessment becomes even more difficult. Details and examples are provided in Section 7.1.2.

2.2.3.4 Nonstructural Elements

In some cases nonstructural elements are the reason for deficiencies of structures. A typical case is the condition of the asphalt pavement of a bridge, which when damaged influences the structural response considerably. Other elements such as guard rails, hand rails, masts or lighting systems contribute to the structural response considerably. On the other hand, if properly separated, information on the condition of these nonstructural elements can also be obtained. Recent failures of lighting poles have highlighted such a requirement.

Commercial aspects very often dominate the procedure. Science has offered a large variety of tools that can be applied.

2.2.4 *Codes, Standards and Legal Aspects*

A most useful standard on SHM, particularly for bridges, has been elaborated in the Structural Assessment Monitoring and Control (SAMCO) network. Due to its size it cannot be repeated here.

2.2.5 *Tools*

The tools for SHM are best described in the comprehensive *Encyclopedia of Structural Health Monitoring* published by Wiley. It comprehensively covers all fields of SHM.

2.3 Examples of Bridge Management Systems

Information on global bridge management systems (BMSs) has been collected upto the year 2006. The following comprises examples of such systems, including numbers and costs that were valid at that time. It shall be seen as a source of information on existing approaches globally. Some up-to-date information on these fleets of bridges can be found under the web links provided.

2.3.1 *Introduction*

Bridges are large and expensive structures that are of great importance to our economy and society, but they are often exposed to hard environmental and weather conditions. Therefore, bridge engineering and monitoring can ensure their capacity to resist these conditions and thus negative impacts on our economy and society can be prevented.

As a bridge reaches the end of its service life, which can be the result of structural damage and/or material degradation, it should have reached a minimum acceptable performance level. In order to determine this level many factors have to be taken into consideration. A BMS can determine the right time for improvements on a bridge and it can improve the overall condition of an agency's network of bridges early enough.

A BMS is a decision support tool that consists of inventory (data regarding the characteristics and condition of the bridge), inspection (examinations of the bridge) and recommendations (regarding the maintenance and improvement of the bridge). It is also capable of prioritizing the allocation of funds. A BMS is important for every stage of a bridge's life and consists of the following components:

- data storage;
- cost and deterioration models;
- optimization and analysis models;
- updating functions.

2.3.2 *Structure of BMSs*

According to the "good practice" guidelines of the American Association for State Highway and Transportation Officials (AASHTO), the basic four modules of a BMS are:

- A database containing all the data required to carry out the management functions and that are going to be used as input for modeling.
- Deterioration modules for predicting the conditions and the deterioration rate of the bridge's components at a given time in the future.
- Cost modules in order to identify the required funding for the proposed maintenance strategy, the rehabilitation and replacement (R&R) procedure and the financial impact as a result of poor bridge conditions.
- Optimization algorithms in order to analyze alternative funding scenarios and select the most cost-effective R&R strategy.

According to the existing literature, a BMS can be divided into four basic modules that are analysed below: database module, rehabilitation and replacement module, maintenance module and project-level module. Two supplementary modules, the modeling and inspection modules, which act as input sources for the previous modules, are also presented.

2.3.2.1 Database Module

The database forms the heart of a BMS and it contains information on every bridge in the network. Since all the functions of the BMS originate from the database, the quality of the system is closely related to the quality of the database. However, the quality of the database is not dependent on the amount of data it contains but on the significance of these data. The data of a BMS database vary according to the need and the aims of the user but they can generally be divided into the following categories: inventory variables, bridge condition variables, bridge appraisal and proposed improvement variables, and historical variables.

The inventory variables include the bridge location, type and classification, whereas the field inspection variables concern the condition of the bridge. The bridge appraisal and proposed improvement variables derive mainly from other modules, apart from variables such as the cost of improvements or the condition of the bridge, which are provided by the user. Unlike the inventory variables, the bridge appraisal and proposed improvement variables may change as they will be generated by the system. Following an improvement action, historical variables regarding this action and associated costs have to be entered into the system.

It is understood that, due to the size and amount of data, high quality database management software is required. Moreover, a relational database scheme has to be designed. This can happen by designing an entity-relationship model and then convert it to the corresponding relations or tables, or by designing a single relation scheme (universal relation).

2.3.2.2 Rehabilitation and Replacement (R&R) Module

This module sets the priorities for the allocation of funds. As the rehabilitation of bridges and the replacement of their components can be particularly costly, many factors have to be analyzed in order to set the priorities of these actions. These R&R actions can belong in one of the following categories: sufficiency rating, level-of-service deficiency rating, incremental benefit/cost analysis and mathematical programming.

2.3.2.3 Maintenance Module

This module will handle the bridges of the network that, due to their satisfactory condition, do not require any R&R action. The module does not use a complicated optimization procedure but a simple ranking method. This happens because the cost of maintenance is lower than that of R&R and because the amount of maintenance activities can be very high. Therefore, all the maintenance activities should be coded and a standardized listing of these activities should be developed. The maintenance actions of the listing should be observed and recorded during the inspection phase.

2.3.2.4 Project-Level Module

This module examines the various bridge improvement alternatives by undertaking cost/benefit analyses. The input information for this module is provided from the database and modeling module and the results can be used at network level. Initial cost analysis, life-cycle cost analysis and incremental benefit/cost analysis are the methods that are usually used in the project-level module.

2.3.2.5 Modeling Module

The modeling module is used to predict future condition ratings, future load capacities, cost and effectiveness of the R&R activities and user cost parameters. It applies statistical analysis to the historical data of the database and it calibrates prediction models. By analyzing the historical data the module establishes performance standards and unit costs for the R&R activities. It can also predict user cost parameters through statistical analysis. For example, it can predict the increase rate of average daily traffic (ADT) for various road classifications and the percentage of heavy-vehicle traffic. Moreover by analyzing accident data, the relations between the accident rate and the characteristics of the deck can be investigated.

2.3.2.6 Inspection Module

At this stage the inspection phase is initiated. All the information regarding these inspections (inspector's data, reports and observation, dates, types and frequency of inspections, etc.) are obtained from this module.

Inspection Types

There are various types of inspections, each one being appropriate for different stages of the useful life of the bridge and each one having different frequency, level of intensity, detail and degree of testing. For an appropriate inspection level the Bridge Authority needs to undertake the following types of inspections: initial, routine, in-depth, damage and special.

Inspection Frequency

Inspection frequency should not exceed two years unless past reports and performance history justifies otherwise.

Inspection Tools and Special Equipment

Inspection equipment and tools may vary from basic (camera, ruler, chipping hammer, paint scraper, wire brush, venire, 20-m tape, flashlight, field-marking crayon) to special (rigging, staging, workboats, snoopers, aerial buckets, traffic protection devices, and under-bridge inspection equipment).

Inspection Staff

The inspection team should consist of people with knowledge and experience in structural behavior and design of bridges, materials and its behavior through time, weathering, etc., typical construction practices and telltale signs of their proper or improper execution. Collaboration with members of the original design team of the bridge would be beneficial.

2.3.3 Recent Examples and Information

A good practical overview can be obtained during the annual IABSE (International Association of Bridge and Structural Engineering) symposia and conferences. These events provide a practical overview on real cases and the solutions chosen. The following information has been obtained from the 2006 IABSE event in Copenhagen. It contains some useful figures on this subject.

Bridge Management System DANBRO+

The Danish road authority has designed a new BMS, which was released in December 2006 (Bjerrum et al. 2006). There have been ten years of development. The system comprises 41 big bridges with a total of 240 000 m^2 and a replacement value of 1.8 billion €. Their budget for maintenance is € 15 million annually. The new system will contain all the necessary information to perform the maintenance jobs and even ask questions.

The costs for the system design and implementation have been:

- € 130 000 for the definition of the requirements and the conception;
- € 980 000 for the software;
- € 540 000 for testing, project management and implementation.

This makes a total of € 1.65 million for DANBRO+ and it is considered the most appropriate BMS that exists.

Data of the Great Belt Bridge (Laursen et al. 2006)

This bridge was opened in 1998 at a cost of € 2.85 billion. It is 19 km long and carries four lanes. Before the bridge opening 8100 vehicles per day crossed the straight. Just after opening, this increased to 18 000 vehicles per day and is currently standing at 24 500 vehicles per day (3×). They recently invested 60 million Danish krona (approximately € 8 million) for inspection equipment to be able to inspect each and every part of the structure.

The main work of maintenance is carried out by subcontractors, who partly have to work on standby. In practice a problem can occur during a change in subcontractor, however this is the most economic solution.

At the Øresund Bridge close by the number of vehicles at opening has been 5500 per day in the year 2000, which increased to 11 000 vehicles per day for the year 2005, which is double the capacity.

The Netherlands (Van Beek and Djorai 2006)

The Netherlands have a higher road network of 3250 km with more than 4000 bridges of total length 227 km. The value is € 25 billion. They are managed by the Rikswaterstat.

Croatia (Radic et al. 2006)

Croatia has a number of old major bridges to manage. The Arch Bridge at Krk with a span of 390 m has been constructed out of pre-cast elements (six per section) that have been assembled by a cable crane. This bridge is in very bad condition and has to be retrofitted. The deck plate is only 13 cm thick. The columns, with heights of up 70 m, have walls of only 30 cm thickness. This bridge and four other arch bridges comprise the most difficult rehabilitation cases.

The new cable-stayed bridge at Dubrovnik had an interesting story of cable vibrations. Besides the usual rain/wind vibrations at 10.20–11.10 m/s (about 15 incidents in five years), another phenomenon has been observed that is related to heavy wind with snowfall at wind speeds of 19.6–24.3 m/s. All cables vibrated violently. On photographs snow accumulation on the windward side has been observed. The first incident happened in March 2005 and lasted two hours. The same incident occurred again in March 2006 for about six hours. The amplitudes were more than 2.5 m at cables of 200 m length. Almost all cables have been affected. The angle of attack in the horizontal plane was 34°. The cables reacted so violently that they hit lighting poles on the bridge, causing considerable damage. The cables are of the

Dywidag type with a diameter of 22.5 cm and 720 t maximum. The angle of the longest cable is 22°. Another observation was that the bolts that fixed the sheath tubes with the structure were pulled out and fell off, particularly on the pylon side. As a solution, mechanical dampers have been installed. It will be checked whether they are effective.

Serbia (Jensen et al. 2006)

The bridges in Novisad, which were damaged by airplane attack in the Balkan war in 1999, have been reconstructed. An interesting fact is that in the last 100 years there have been nine bridges across the Danube in Novisad, out of which only one survived and eight have been destroyed by war action. The cable-stayed bridge at Novisad showed a slow collapse; it took about one hour to go down. Out of the 12 missiles fired at the bridges, only six were successful. The remaining six have to be removed with difficulty.

Canada (Buckland and Matson 2006)

Peter Buckland of Buckland & Taylor from Vancouver reported on the rehabilitation of major bridges in Canada and the USA. His credo is "Analysis last – think first." Very impressive solutions have been presented. Exchange of whole segments of structures under most restrained conditions have been presented. At the suspension bridge at Lions Gate in Vancouver, 12-m long sections of the deck structure have been removed and replaced within six hours during the night.

2.4 Protection of Bridges against Man-Made and Natural Hazards

After the dramatic events of September 11, 2001, and the follow-up events in London and Madrid, the need to protect our transportation infrastructure has been clearly demonstrated. As bridges are the most vulnerable elements, this subject has to receive high attention. Monitoring can help to highlight eventual activities, but mainly to identify the weakest elements that can be protected by strengthening or retrofit.

In many cases similar structural response for both the seismic and the blast load case has been observed, (e.g. F. Seible, University of California, San Diego). For example, both can result in progressive structural collapse, and mitigation measures (such as design for redundancy and ductility, as well as retrofit measures to increase concrete confinement) apply to both seismic and blast conditions. The vulnerability of these structures under a range of potential threat scenarios must be quantified. New technologies are to be developed and tested. Computational seismic and blast physics models have to be improved and validated for damage prediction and the determination of effective mitigation strategies. It is recommended that the same engineering team be used to deal with seismic questions in the assessment of eventual blast effects and related matters.

Further Reading

Bjerrum J, Larsen E, Bak H and Jensen F (2006) Internet-based management of major bridges and tunnels using DANPRO+. *Conference on Operation, Maintenance and Rehabilitation of Large Infrastructure Projects, Bridges and Tunnels*, Copenhagen.

Buckland P and Matson D (2006) Increasing the load capacity of major bridges. *Conference on Operation, Maintenance and Rehabilitation of Large Infrastructure Projects, Bridges and Tunnels*, Copenhagen.

Casas JR (2006) Bridge management: Actual and future trends In *Bridge Maintenance, Safety, Management, Life-Cycle Performance and Cost* (ed. Cruz PJ, Frangopol DM and Neves LC), pp. 21–30. Taylor and Frances, London.

Enright M and Frangopol D (1999) Condition prediction of deteriorating concrete bridges using Bayesian updating. *Journal of Structural Engineering* **125**(10), 1118–1125.

Faber M and Sorensen J (2002) Indicators for inspection and maintenance planning of concrete structures. *Structural Safety* **24**, 377–396.

Farrar C, Worden K, Mansin G and Park G (2005) Fundamental Axioms of Structural Health Monitoring. *Proceedings of 5th International Workshop on Structural Health Monitoring*. Stanford, CA.

Frangopol D and Das P (1999) Management of bridge stocks based on future reliability and maintenance costs. In *Current and Future Trends in Bridge Design, Construction and Maintenance* Thomas Telford. (ed. Das PC, Frangopol DM and Nowak AS), pp. 45–58.

Hearn G, Frangopol D and Szanyi T (1995) *Report on Bridge Management Practices in the United States*. Technical Report, University of Colorado, Boulder.

Jensen J, Mogensen V, Sorensen O, Dunica S and Bojovic A (2006) Danube clearance and sloboda bridge reconstruction in Novi Sad. *Conference on Operation, Maintenance and Rehabilitation of Large Infrastructure Projects, Bridges and Tunnels*, Copenhagen.

Laursen E, Knudsen A and Andersen M (2006) Management of the concrete structures of the Great Belt Link. *Conference on Operation, Maintenance and Rehabilitation of Large Infrastructure Projects, Bridges and Tunnels*, Copenhagen.

Radic J, Bleiziffer J and Tkalcic D (2006) Rehabilitation of large arch bridges. *Conference on Operation, Maintenance and Rehabilitation of Large Infrastructure Projects, Bridges and Tunnels*, Copenhagen.

Rafiq M (2005) *Health monitoring in proactive reliability management of deteriorating concrete bridges*. PhD thesis, School of Civil Engineering, University of Surrey.

Van Beek T and Djorai B (2006) Inspection and maintenance products for civil structures. *Conference on Operation, Maintenance and Rehabilitation of Large Infrastructure Projects, Bridges and Tunnels*, Copenhagen.

Wenzel H and Pichler D (2005) *Ambient Vibration Monitoring*. Wiley, Chichester.

3

Bridge Rating and Risk Assessment

Individual rating systems exist at most of the clients' databases. Commercial programs such as BRIDGIT and PONTIS are widely used in the USA. They create a score based on input provided by bridge inspectors, which is used for a rating. The unsatisfactory situation that up to 30% of the bridge stock is rated deficient comes from the lack of quantitative assessment tools. Inspection ratings are very subjective and reflect the reluctance to take any risk by the inspectors.

In order to achieve better and more realistic ratings the current practice has to be complemented with measured and computed values. Better results are achieved if the rating is based on three approaches, namely:

- the existing conventional inspection;
- the computation of a rating out of measured values from bridge tests;
- the rating arising from comparison of measurements with computed models.

The final target has to be to reduce the number of deficient-rated bridges from the current 30% of the bridge stock to the justified 10–15%, which will enable a ranking of structures in order to target the rehabilitation effort.

The following need to be distinguished:

- Cases where immediate action is required. These cases are rare and comprise less than 0.3% of the bridges concerned.
- Cases where intervention within reasonable timeframes is required. This group comprises probably 4–5% of the bridge stock involved.
- Cases where long-term action is required. This group completes the cases up to the mentioned 10–15% of the bridge stock.
- Cases where no action is required. This group represents the majority of cases where a previous deficient rating can be upgraded to "normal condition".

3.1 Inspection Rating

As one of the starting points for bridge management the collapse of the Silver Bridge in the USA in December 1967 can be quoted. The NBI (National Bridge Inventory) program began in 1971 and represents the establishment of a national bridge inspection program. It has been focused on the elimination of deficient bridges. A rating is provided through be annual inspections of the 597 000 bridges of the Federal Highway Administration (FHWA). About 116 fields of data are collected. The inspectors provide the condition ratings. The appraisal ratings are determined by rules introduced into the procedure. Software packages such as PONTIS have been developed and are widely used.

The eligibility of bridges for the Highway Bridge Repair and Rehabilitation Program (HBRRP) is determined by deficiency. The current guideline (issued 1995) by the department of transportation (report no. FHWA-PO-96-501) is the "Recording and Coding Guide for the Structure Inventory and Appraisal of the Nations Bridges".

The NBI ratings follow a numeric code from 0 to 9. Condition ratings are performed for superstructure, substructure, deck and culverts. Appraisal ratings are provided for waterway adequacy (frequency of overtopping), structural evaluation (load rating), approach alignment (speed reduction), deck geometry (roadway widths), and unclearancies (vertical and lateral).

The NBI condition and appraisal ratings are as follows:

Condition rating	Appraisal rating
9 Excellent condition	9 Superior to present desirable criteria
8 Very good condition	8 Equal to present desirable criteria
7 Good condition	7 Better than present minimum criteria
6 Satisfactory condition	6 Equal to present minimum criteria
5 Fair condition	5 Somewhat better than minimum adequacy to tolerate being left in place as is
4 Poor condition	4 Needs minimum tolerable limits to be left in place as is
3 Serious condition	3 Basically intolerable requiring high priority of corrective action
2 Critical condition	2 Basically intolerable requiring high priority of replacement
1 Imminent failure condition	1 Not used
0 Failed condition	0 Bridge closed

Figures 3.1 and 3.2 show the bridge rating summary of 2007 for the USA. They clearly show that the majority of deficiencies are concentrated on the topics deck widths (geometry) and load. Figure 3.2 gives a more detailed look into the deficiency summary (ratings 0–4). The huge demand on this subject becomes visible here. The reliability of the condition rating has been questioned recently. In a blind test only 4% of the inspectors have found hidden damages in a bridge selected by the FHWA. Better and more reliable methodologies become necessary. The methods used for load rating, namely based on load factor

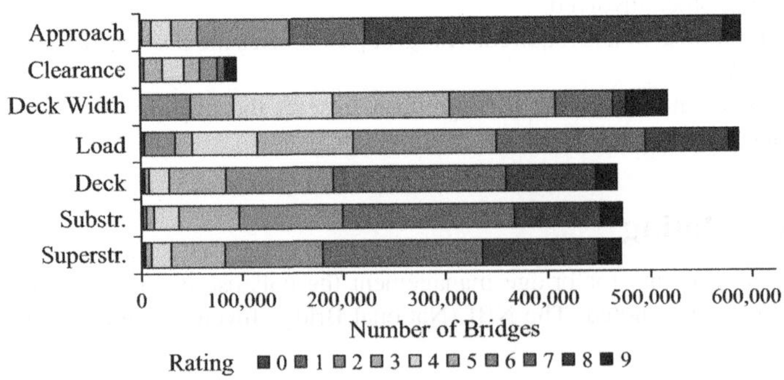

Figure 3.1 Rating of the entire FHWA bridge stock

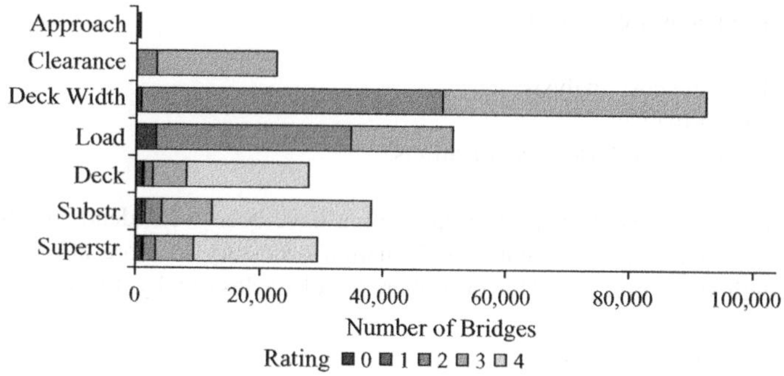

Figure 3.2 Detailed rating of the entire FHWA bridge stock in classes 0–4 (deficient)

calculation and visual inspection, have produced a total of 23 817 bridges rated as deficient. The number of bridges tested is negligible. This can be seen as one of the shortcomings of the NBI program:

- the NBI program is adequate for administration of the national HBRRP program;
- it is inadequate for bridge performance measurement;
- most states augment NBI data;
- condition ratings are based on visual inspections that are not reliable and fragmented;
- the program is not adequate for owner-level bridge management.

On element-level inspection the PONTIS program has provided a significant advance on:

- element-level inspections;
- more discretized condition state data;
- more quantitative condition state data;
- provides network (population)-level decision support.

The limitations of element inspections are:

- the condition states are still based solely upon visual inspection;
- invisible deterioration, damage or distress is not detected or measured;
- operational performance is not measured;
- vulnerability and reliability are not adequately considered.

This opens a wide field for SHM activities with particular detection and measurement needs on damage: impact, overload, scour, seismic, fracture, settlement, foundation, inoperative bearings, movement, lack of movement and cracking. On deterioration the following are noted: corrosion, fatigue, water absorbtion, loss of pre-stress, unintended structural behaviour and soil stiffness.

For performance measurement the needs are:

for operation	for service
Average daily traffic (ADT)	Congestion
Freight tonnage	Accidents
Stress	Availability
Strain	Load capacity
Deflection	
Displacement	

The additional data needs are defined as:

- data to support lifecycle cost analysis;
- data to measure performance;
- data to support performance-based specifications.

The shortcomings of the visual inspection and rating have been well identified. Technology exists and can be used to provide the quantitative information necessary to more effectively manage civil infrastructure throughout its lifecycle. However, there are obstacles to why it is not being used:

- specific data needs are not defined;
- standards do not exist;
- each bridge is unique;
- systems are still expensive;
- systems are unproven;
- bridge owners are not convinced of value;
- not required by law.

To alleviate these shortcomings the FHWA has introduced the Long-Term Bridge Performance program (LTBP), a new program of research and development to address these needs. The LTBP program aims to:

- Select representative samples of bridges.
- Perform a program of detailed periodic quantitative inspections and evaluations.
- Be performed on a long-term basis (at least 20 years).
- Create a subset of instrumented "smart" bridges to develop, refine and standardize SHM systems and to monitor operational performance.
- Conduct forensic autopsies of decommissioned bridges on a large scale.
- Perform forensic studies of bridge failures.

The data to be collected are defined as:

- quantitative data on bridge condition;
- quantitative data on deterioration;
- quantitative data on operational performance of bridges;
- quantitative lifecycle data;
- forensic data on bridge components;
- forensic data on bridge failures;
- statistically valid data to support a probabilistic reliability-based approach to asset management.

The LTBP has been authorized and the FHWA is implementing it in 2007. The role of SHM within it is defined as:

- to detect and report any unsafe conditions;
- to detect and report unusual or unexpected events or behavior;
- to measure operational performance;
- to support data given by performance-based asset management.

It is expected that a new world standard can be developed for this subject within the lifetime of this ambitious program.

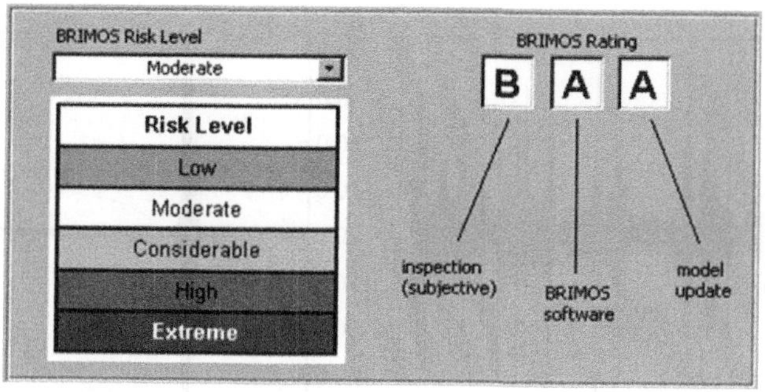

Figure 3.3 The BRIMOS®rating. The categories are defined as follows: (A) good condition; (B) good condition with local damage; (C) problematic condition

3.2 The BRIMOS® Rating

The BRIMOS® (Bridge Monitoring System) rating is a classification based on several research projects started in 1995. About 1000 structures have been assessed and the experience has been incorporated into the assessment procedure. It is based on the vibrational signature of a structure, which is obtained by a measurement campaign. Depending on the extent of this campaign various properties can be computed, which are combined to form the BRIMOS rating. This classification (Figure 3.3) allows a fast identification of the structure's integrity as well as the corresponding risk level based on measured dynamic parameters (eigenfrequencies, mode shapes, damping pattern in the lengthwise direction, vibration intensity and static as well as dynamic vertical displacements), visual inspection, finite element model update and reference data (BRIMOS database and BRIMOS knowledgebase). The result is a factor, related to a predefined risk level.

The components of the rating system are described in the following sections.

3.2.1 Quality of the Signals

The results of the assessment are dependent on the quality of the signals. This is mainly dependent on the measurement campaign where the quality of hardware and the extent of recording can provide considerable differences. Therefore the BRIMOS rating contains an element (weighting factor) that assesses the quality of the signals in order to avoid poor campaigns that missed the real action achieving too good ratings. The single items to be considered here are:

1. **Length of the records.** Recent studies have shown that the minimum record length should be approximately 44 min, which represents two records of 660 000 points at a sampling rate of 500 Hz. It can be assumed that during this period a reasonable representation of the ambient vibration behavior (Figure 3.4) is recorded. Nevertheless this cannot be sufficient for the recording of representative events on a structure if there is little traffic. Longer records will give a better rating and shorter records will have to receive a considerable penalty for the risk involved. (Note: It has been found that the length of records can have significant influence on the results. To make campaigns comparable it will therefore be necessary to compare records of equal length. This is not necessarily dependent on the sampling rate.)
2. **Sampling rate.** The previous specified sampling rate of 100 Hz should be applied as a minimum. Recent research suggests that higher sampling rates (Figure 3.5) might be able to improve the results. If storage and data transfer are not guiding factors it is recommended to sample with 200 or even 500 Hz. Nevertheless the record length of 11 min should be maintained. Any record with a lower

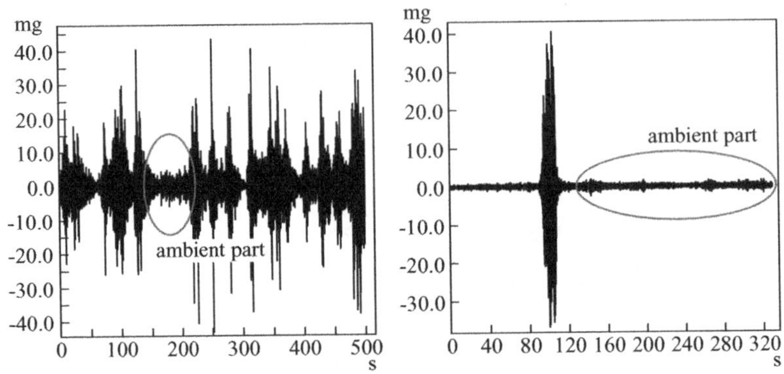

Figure 3.4 Lively and ambient record

sampling rate should receive a penalty in the rating. Reasons to use lower sampling rates can be the wish to limit the number of data to be transferred for energy supply reasons, e.g. when wireless modes are applied. It should be tried in future to eliminate data transfer as a limiting factor.

3. **Energy content.** Each record represents the history of events that happened during the monitoring period. The evaluation procedure to be used depends on this value. Pattern recognition methodologies might be used to identify events and help to choose the right approach. Furthermore it has to be considered that many of the dynamic properties such as damping (Figure 3.6) are also dependent on the amplitudes and thus on the energy content (see Section 4.4 on damping).
4. **Ambient sections in the signal.** In order to obtain the basic natural frequencies it might be necessary to have sufficient ambient sections (Figure 3.7) in the signal. If too much traffic is recorded, the bridge is always excited and additionally loaded by the traffic, which does not allow the development of

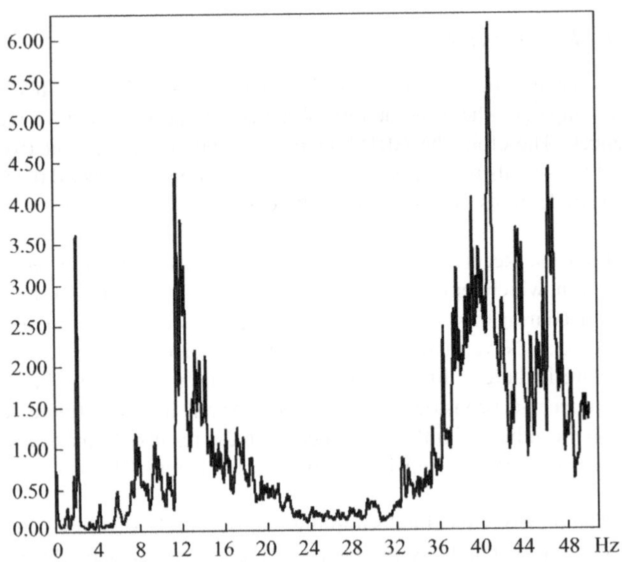

Figure 3.5 Record with low and high frequency content

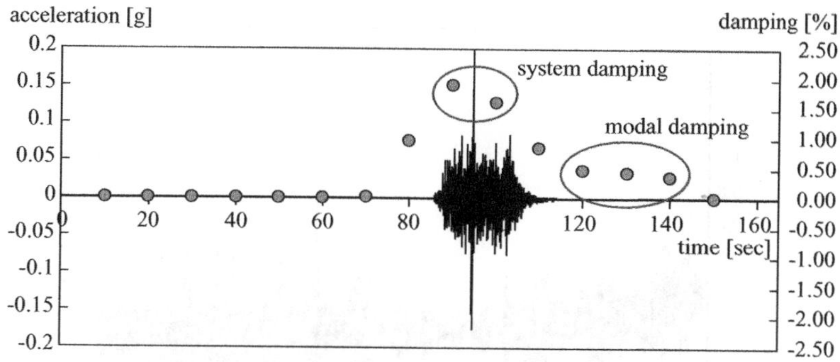

Figure 3.6 Damping related to energy content

a free vibration scenario. This assessment will determine how much compensation is necessary to obtain correct values.

5. **Drift, spikes and noise.** Such phenomena cannot be completely avoided during monitoring. Disturbance from passing trains, the illumination of a bridge or other electric sources is quite common. In many cases the monitoring system on the bridge cannot be properly earthed in order to avoid such false records. The software should enable such unwanted signals to be identified. In most cases the characteristic of the signal is quite unique and can be detected by pattern recognition procedures. Electrical shocks are represented by extremely sharp rises, which means that the difference between two measurement points is orders of magnitude greater than normal signals. A drift in the signal can

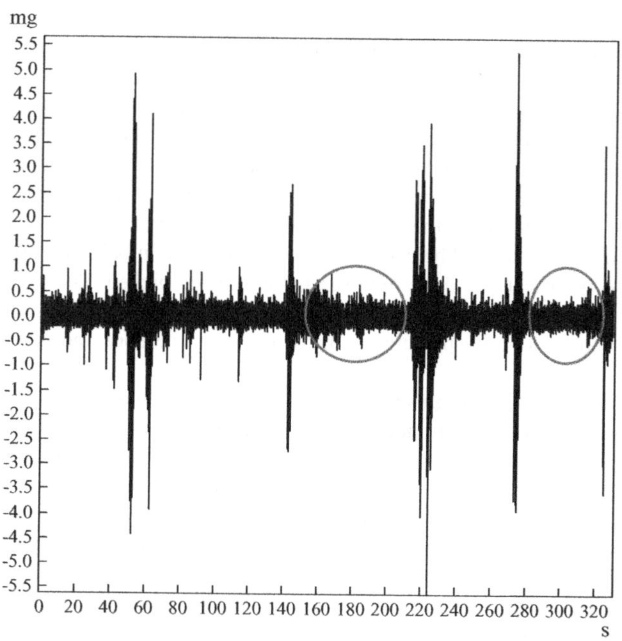

Figure 3.7 Ambient sections in the signal

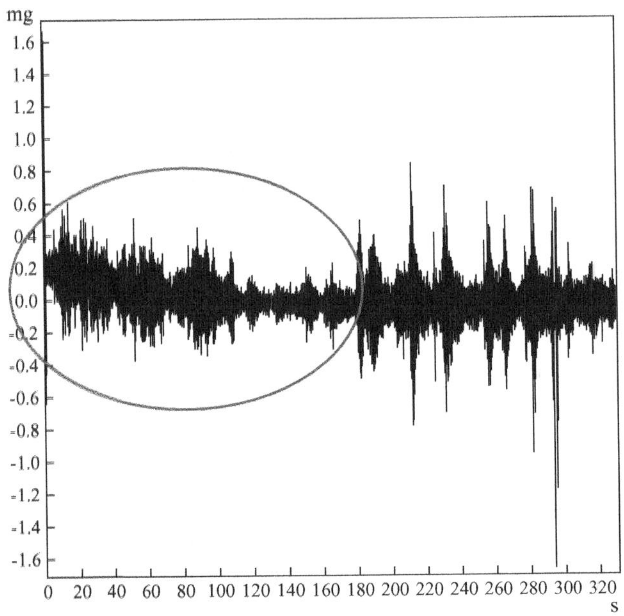

Figure 3.8 Drift at the beginning of the signal

indicate that the sensor is inclined due to bridge displacement or an influence from electric power. These phenomena also can be detected by pattern recognition methods. The background noise also should be determined in order to assess the quality of the measurement campaign (Figure 3.8).

6. **External input.** Bridges are normally located in areas where other activities are common. This might provide a considerable input from external sources. Particular cases have been underground lines, which provide vibrations on structures without being seen. Furthermore, passing trains at considerable distances also transfer vibrations to a structure, and the application of rotational machinery has been recorded. Such external input can be detected by the provision of an external sensor outside of the system. It can be subtracted from the signal if there is a distinct input (Figure 3.9).
7. **Calibration.** The sensors should be calibrated before a major campaign is carried out to ensure that they are all working properly and the right sensors have been picked. The results of this calibration can be inserted as a calibration factor into the calculations. In most cases an internal calibration is sufficient, which is done by placing all the sensors on one spot and adjusting the amplitudes (Figure 3.10).
8. **Sensor types.** The types of sensors used also provide considerable differences in the assessment. For very low frequencies geophones provide the best results, but they are not reliable if higher frequencies are obtained. Piezoelectric sensors have a very high background noise and therefore very often do not pick up low frequencies. Therefore, information on which sensor has been used should be carefully recorded to allow reasonable assessment of the phenomena found in the signals (Figure 3.11).
9. **Monitoring system type.** There are major differences between the monitoring systems used. When we are looking for the new wireless systems we have to deal with the jitter, which represents the different time stamp of each record influenced by the operation system. This phenomenon can considerably reduce the quality of assessment of eigenforms or any other value where the phase is important. The rating has to consider whether to use a highly developed monitoring system with multiple sensors recorded through cables or a low cost solution.

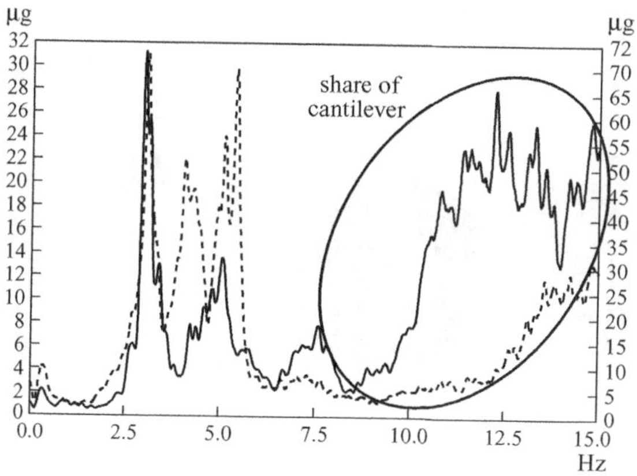

Figure 3.9 Spectrum of the cantilever (continuous line) and box girder (dashed line)

10. **Monitoring conditions.** Each campaign is carried out under distinct monitoring conditions. These might be favorable or, more likely, unfavorable. The conditions have to be recorded and assessed. A characteristic of the conditions has to be provided in the measurement record (Figures 3.12 and 3.13) and stored in the database for subsequent assessment.
11. **Monitoring layout.** The layout of the monitoring campaign also is a major factor for assessment. Depending on the costs to be considered, very limited campaigns are generally performed. The limitations therefore have to be clearly considered in the assessment. Advanced systems such as BRIMOS provide an optimal sensor layout for each structure where the geometry is known. It has to be assessed how much of this optimum layout has been achieved during the campaign (Figure 3.13).

Any other observation of the monitoring personnel has to be recorded and considered, therefore this list could be extended for extraordinary structures or conditions.

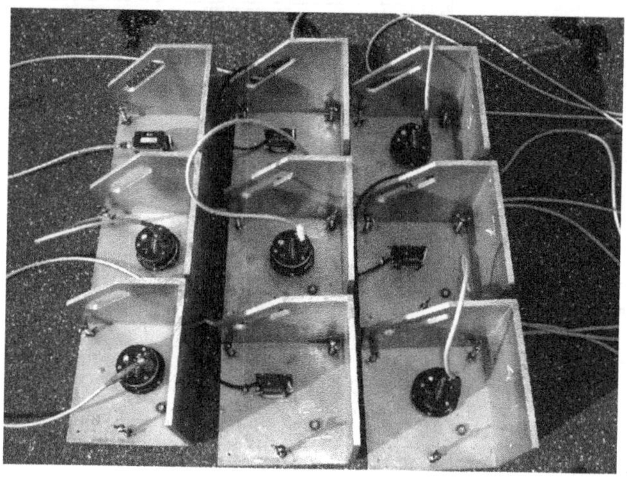

Figure 3.10 Calibration of the sensors before starting the measurements

Figure 3.11 Different sensor types (from left to right): Brüel Kjaer 4514, Kistler 8393A2 and EpiSensor FBAES-T accelerometers, and Lennartz LE-3D-5S seismometer

3.2.2 Signal Conditioning

Signals can be improved and assessed by various signal conditioning methodologies. A simple assessment tool is the autocorrelation function. It should not be forgotten that raw signals have to be stored but only conditioned signals are used for assessment. The conditioning routines might vary or improve the signal and allow better subsequent assessment if needed.

It has to be expressed that quite often the signals found are so disturbed that they influence the procedure. Such signals should be dumped and not conditioned. The best procedure is to delete them in order to avoid false assessment by subsequent campaigns.

vce

brimos
A PRODUCT OF VCE
Measurement record

Object:	**Aschach Bridge Brandstatt**		
Object Short Term: **ASCHACH** Ref. No. 00 / 076	Measurement team: Lutz, Stöger	Measurement on: **10.05.00** Time from: 13:40 to: 16:00	
Number of spans: 1 Cross section: box girder Building material: concrete		Device: ☑ µMUSYCS ☐ other	
file: Aschach_layout.xls printed: 22.06.2001		☑ First measurement ☐ Follow-up measurement	

Comments:			
Meteorological data:	Cloudiness: ☐ very cloudy ☐ partly cloudy ☐ slightly cloudy ☐ sunny intervals ☑ sunny	Wind speed: Wind direction: Precipitation: ☐ heavy rain ☐ slight rain ☑ no rain	Air temperature: 28 °C
Traffic volume:	☐ Pedestrian ☐ Cyclist	☑ Car traffic light ☐ heavy ☑ Truck traffic light ☐ heavy	Miscellaneous items:
Measurement parameters:	**µMUSYCS** Active channels ☑ C1-C3 ☑ C7 ☑ C9 ☑ C11-C13 ☑ C17 ☑ C4-C6 ☑ C8 ☑ C10 ☐ C14-C16 ☐ Laser Scanning time: ms; Test results; Modulation; Filter; dB		

Figure 3.12 Measurement record

Figure 3.13 Heavy transport

As the combination of records is rather difficult it is advisable to record longer periods and to subsequently cut them into the necessary pieces. This was previously limited by the available buffer storage facilities but should not be a problem any more. Also the size of files is not a limiting factor.

Pattern recognition routines can search the available records for ambient sections and for distinct events (Figure 3.15), from which the necessary data pool for assessment can be created.

3.2.3 Damping Values

Damping has been found to be the most promising value for damage and condition detection. Nevertheless with ongoing research the subject has become more complex than before (see Figure 3.16), with different terms being used for the same phenomenon and difficulty finding the relevant definition for which kind of damping to use where. For these reasons a separate section on damping has been included in this book (Section 4.4). For the BRIMOS rating the following values are considered.

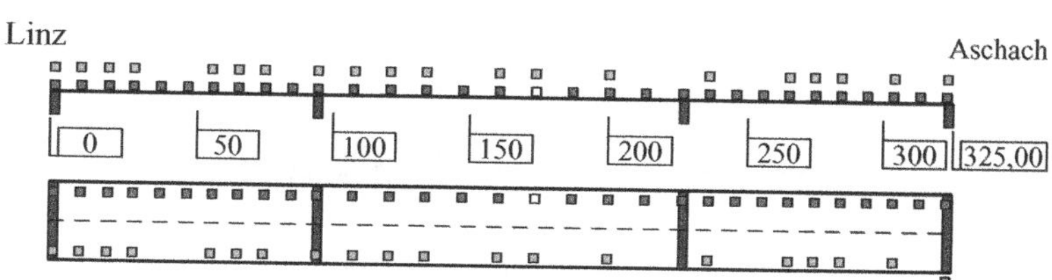

Figure 3.14 Example of a sensor layout

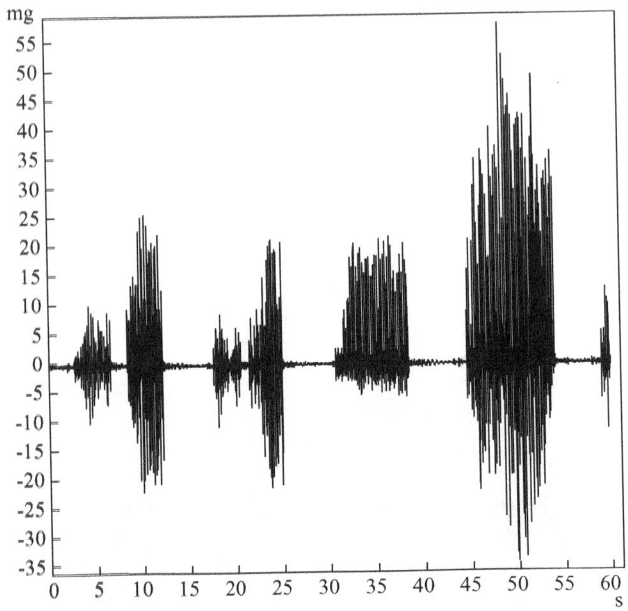

Figure 3.15 Unusual signal

Random Decrement Technique Value on the Representative File

The various values of damping are assessed separately. This requires considerable computation effort as described in Section 4.4. The considered values are:

Modal Damping

The values for modal damping (Figure 3.17) indicate how vulnerable a structure is towards resonance. In most of the cases it is sufficient to determine the damping value for the dominant frequency. It can then be compared to any possible excitation frequency to assess the danger of resonance. Very often we observe resonance phenomena in higher frequencies that relate to the behavior of structural members

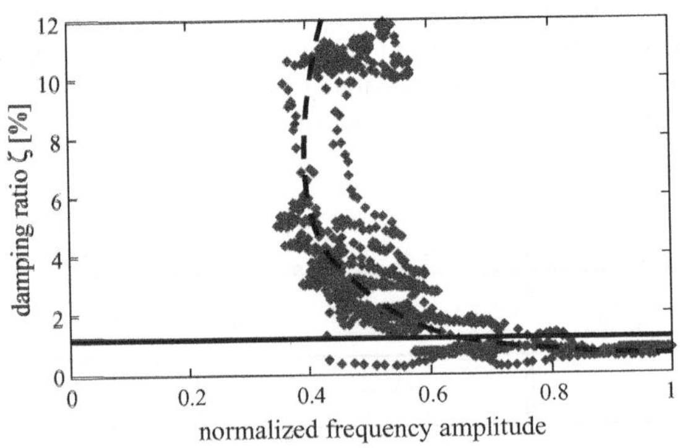

Figure 3.16 Damping estimation for different windows of a certain file (cf. Figure 4.78)

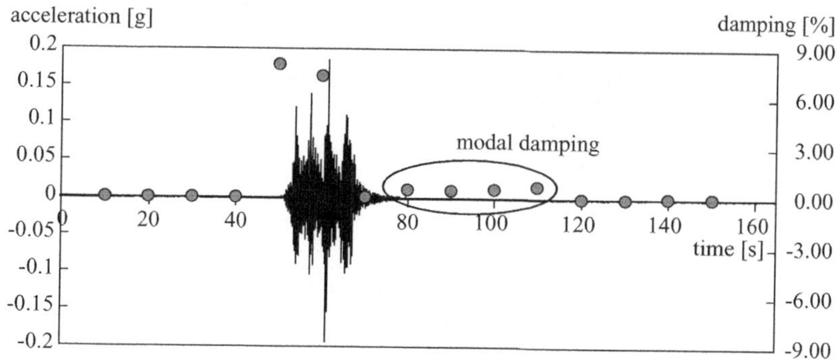

Figure 3.17 Modal damping portion in signal

such as cantilever slabs or cross beams. In this case these frequencies have to be looked at too. Currently the dominant frequency is determined by the highest amplitude in the spectrum. If this represents a local mode it is advisable to look at the global modes separately. Furthermore it can be studied whether there are other single frequencies represented that show unusual damping characteristics.

The recommended way to determine modal damping is to cut the record into separate windows, take those that reach amplitudes over 60% of the greatest amplitude and compute the damping value as a mean. The results are normally margined below the random decrement technique (RDT) value over the whole length of the file.

There are no clear allowable or recommended damping values in any of the standards or recommendations but there is a consensus that 1.5% damping is sufficient to suppress any unwanted vibration.

System Damping

This develops out of modal damping when the amplitudes are higher. Additional sources of damping are activated. This might be the energy consumption of any friction of bearings, expansion joints, bending action of piers, behavior of the bridge outfitting and also the structure–vehicle interaction. This damping value is nonlinear and increases with amplitude.

System damping (Figure 3.18) can be determined when damping values are computed out of signals with high energy input, such as the passage of a train or a heavy truck. The selection of the window has to be carefully done and the method of computation also influences the results. Maximum likelihood methods are to be applied instead of random decrement techniques.

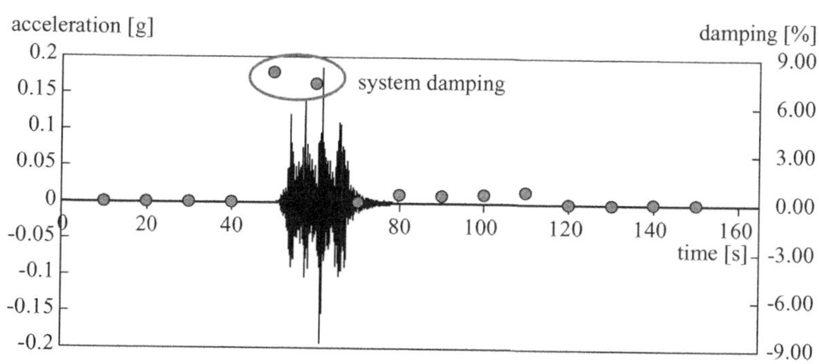

Figure 3.18 System damping portion in signal

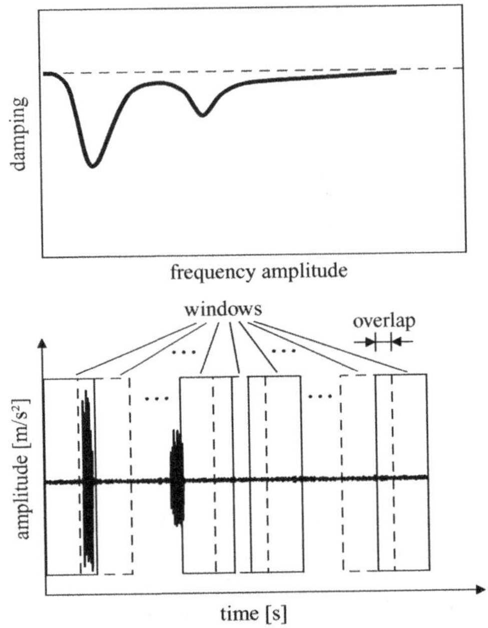

Figure 3.19 Change of damping during train passage events

There is a common understanding that system damping can be assumed to be around 5%. Recent studies show that for traffic excitation and particular railway bridges these values might well be considerably higher. Nevertheless there are cases where the excitation frequency meets with a dominant or subdominant structural frequency. In such cases damping can become negative and have to be clearly identified in the retransformed time domain signal.

Impact Damping

This value is closely related to the previously described system damping. Here we understand a single event and look at the structural behavior after that. It can be determined by applying curve-fitting methodologies to the primary time domain signal. It is also a nonlinear function.

It is difficult to assess how representative a single event is for the assessment of the structural behavior. For railway bridges this can be repeated easily by using more train passages (Figure 3.19). For road bridges it might be difficult to record a number of similar events. This kind of assessment will be better the longer the records are. Best results are only achieved from permanent monitoring over a considerable length of time.

There are no codes or guidelines available to assess this value. Recommendations are to be developed in order to standardize this procedure. In many cases it might not be possible to apply this character because of the lack of any representative event.

Half-power Bandwidth

The easiest and quickest way to determine damping is to compute the half-power bandwidth. This is the bandwidth you obtain when you cut the spectrum at the half-power level, and is defined by the amplitude divided by the square root of 2 (Figure 3.20). This method gives a rough indication of the magnitude of damping but contains many uncertainties due to unknown excitation and other structural features that influence this value. But for a rough quantification or comparison of structures it can be useful, particularly for the assessment of structural members such as cables.

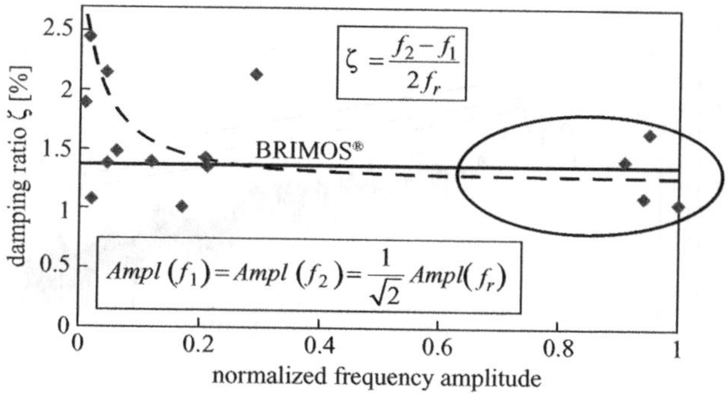

Figure 3.20 Amplitude-dependent damping by half-power bandwidth

As described, this value is computed for the various frequencies found in the spectrum by determination of the bandwidth at half-power level. This can easily be automated. It is also advisable to carry out this computation after several attempts of windowing the signal in order to sort out the disturbed cases.

Aerodynamic Damping

On major structures another damping value becomes interesting, which is the aerodynamic damping generated from the action of the bridge. This value only plays a major role in slender and large structures, and also for structural elements that fit this character.

There are formulas that enable the calculation of aerodynamic damping, as explained in Chapter 12. If better values are required, wind tunnel tests might be conducted to measure these values.

Damping Functions

It is desirable to find out which action on the structure triggers which dynamic behavior. The different behaviors described above might be combined to form a damping function. A fundamental question not yet answered by research and development is the realistic collaboration of the various modes under distinct loads (Figure 3.21). It is easy to determine the damping of a single frequency of a cable, but how are these single values related to each other and how do they combine to form a cable damping value? The currently used approach of applying filters to the signal very much depends on the selection of the filter. Reasonable results can be achieved when such a filter signal is computed and the single damping values are interrelated in a damping function.

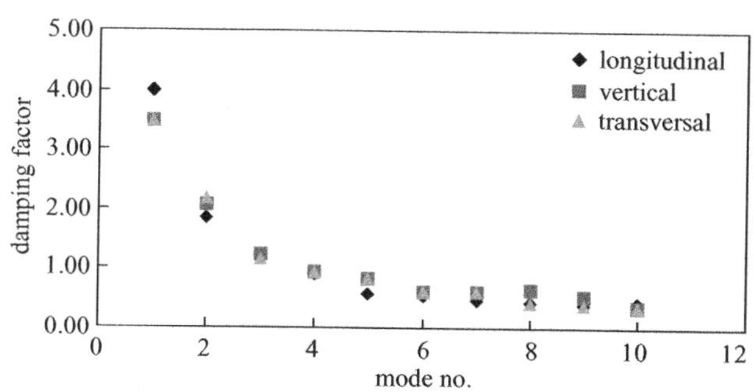

Figure 3.21 Dependence of damping on mode number

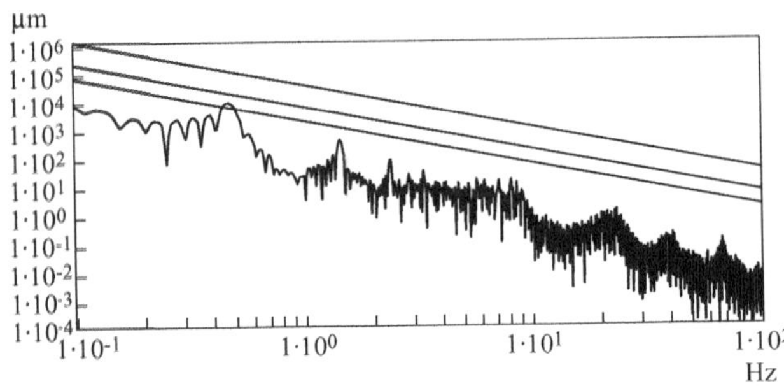

Figure 3.22 Vibration intensity plot

This problem is under discussion in the various engineering societies and there might be recommendations developed in the near future.

3.2.4 Vibration Intensity

A most important value for the assessment of a structure is the vibration intensity within the various modes. The classification of this vibration into critical or noncritical values can indicate whether the structure might be subject to local or global damage from fatigue or related material problems. The values applied for classification in Figure 3.22 have turned out to be conservative. In real tests they have been justified when it comes to damaging scenarios through real impact, such as using a hammer.

In practice, structures with high vibration intensities have a history of local problems particular with the outfitting. Expansion joints and bearings are subject to exchange more frequently than under normal conditions.

The computation of vibration intensities can also be used to assess the impact of single events on a structure that might lead to restrictions or special conditions for use.

3.2.5 Eigenform

If synchronous measurement is performed, determination of the corresponding eigenform is easy within today's program systems because it automatically computes a plot of the related eigenform (Figure 3.23). This plot can be assessed to find out whether it is the expected form or shows unexpected features. This relates particularly to any displacement at areas where a bearing should avoid such an action. Identification of faulty foundations or piles thus becomes feasible.

In most applications it is not possible to compute an eigenform because of missing synchronous measurements. In this case only single points can be looked at by double integration.

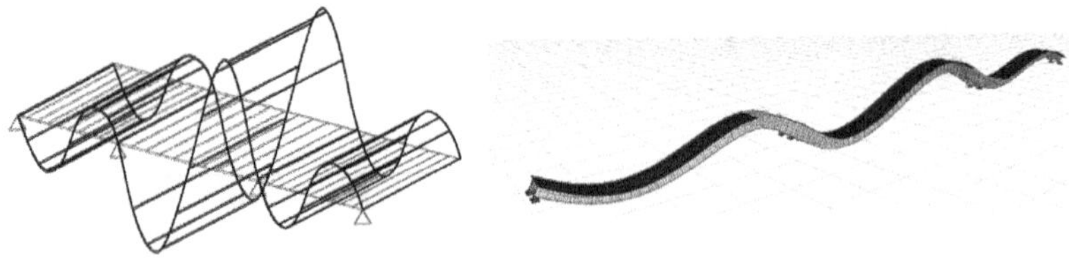

Figure 3.23 Second eigenforms, measured (left) versus computed (right)

3.2.6 Redundancy

For the assessment of a structure it is also important to have a clear indication of the redundancy in case of local failures. This is mainly based on assessment of whether the failure of a single structural member will cause collapse or not. A single span beam is more vulnerable than a continuous beam. Frames often offer additional capacities. Pre-cast element structures are more vulnerable than cast-in-place structures. There are no clear methodologies available to assess this redundancy from monitoring results. Here an engineering assessment has to replace the measured quantities.

3.2.7 Type of Structure

As explained above, the type of structure plays a major role in the rating. Single span beams are most vulnerable compared to the additional capacity provided by a continuous beam or a grid of girders (Figure 3.24). From experience it is known that a specific number of structures are more vulnerable and develop problems more often. This experience also can be used to improve the rating. Under normal circumstances the type of structure would go into the first category of ratings, namely the inspection-based methodologies. Nevertheless in many cases nowadays there is no inspection carried out and only a quick measurement is taken. In these cases it is advisable to include an element to assess the type of structure into the rating.

3.2.8 Environmental Conditions

It is obvious that the environmental conditions determine the deterioration rates of structures, therefore these conditions should be evaluated in the rating procedure. There are a number of options:

- The hydrological regime can be assessed by using local weather data. The quantity of rainfall and its intensity will influence the structural integrity.
- The elevation of a structure indicates the length of the winter period and the duration of cold temperatures. Such an assessment can be made from meteorological records.
- A most important factor for resistance is the use of salt and its role as a corrosion agent that may lead to structural deterioration.
- The exposure of a structure to wind also should be considered. When structures are situated in rather windy areas hazardous local phenomena can develop.
- Typical moisture contents of the surrounding air might be considered in the assessment. These values should be determined on site by a measurement campaign.
- The exposure of a structure might be determined by the assessment of satellite images, which provide information on the local conditions.

Figure 3.24 Continuous beams versus single span pre-cast beams

	Risk Level	*Why*	*What to do*
green	Low	Info	Regular operation
	Moderate	Info	Long term action
yellow	Considerable	Show development	Mid-term action
	High	Demonstrate risk	Immediate action
red	Extreme	Show fault	Automatic alert, ACTION

Figure 3.25 Rating system

- The solar radiation exposure can be computed when the structure is geo-referenced in a geographic information system (GIS).

3.2.9 Computation of the Rating Value

The best representation of the rating can be through values out of which thresholds are computed. The basis is the RDT value over the total record length. This value is expressed as a percentage. For all the other items additions or subtractions to this percentage are provided. As single values the following are proposed:

- for the quality of signals, from −0.10 to 0.10;
- for modal damping, from 0 to 0.10;
- for system damping, from 0 to 0.20;
- for event damping, from 0 to 0.05;
- for aerodynamic damping, from 0 to 0.05;
- for the damping functions, from 0 to 0.10;
- for vibration intensity, from 0 to 0.20;
- for eigenforms, from 0 to 0.10;
- for redundancy, from 0 to 0.30;
- for the type of structure, from 0 to 0.10;
- for environmental conditions, from 0 to 0.10.

All these values are added to a final value that is compared to the thresholds in the typical traffic light diagram of Figure 3.25. All results in the green area represent very good conditions. All results in the yellow area represent good conditions but an indication of local problems to be studied in more detail or to be treated again after a certain period. All figures in the red area are to be treated immediately in order to improve the rating by additional input or by taking action such as closing of a structure or limiting loads.

3.3 Probabilistic Approach in SHM

As discussed in the previous chapter there are many factors that influence the health assessment of a bridge. The most appropriate way to come to a decision out of multiple input is a probabilistic approach. Out of the various components of the condition rating, a decision tree has to be constructed. For straightforward structures a deterministic solution works satisfactorily, but with increased complexity useful results need the introduction of probabilities.

Much has been published on this subject. It is a domain made for cases like SHM but is rarely used. Bridge owners are attracted by the specific information a deterministic approach provides without the inherent uncertainties. Ratings by inspectors are subjective, creating the known problems with large bridge stocks. It is a major challenge for the engineering community to satisfactorily solve this problem and offer solutions on a probabilistic basis that satisfy the owners.

Analysis Approach:

The most sophisticated and detailed analysis approach can be driven by a probabilistic seismic hazard analysis (PSHA) under consideration of all uncertainties (epistemic and aleatory). Recent probabilistic seismic hazard analyses have only regarded the source and attenuation effects without consideration of the local site parameters. In the present case of wide span bridge structures the different foundation conditions of the piles can vary in a proper manner. Therefore the seismic hazard analysis should be adjusted to take into account not only the global conditions but also the local conditions of the subsoil. To have these information measurements on site is indispensable.

1. Analysis system:
   - source models (definition of the sources)
     - database of focal mechanism (focal depth, strike, dip, rake, magnitude);
     - parametric study on the focal mechanism (which mechanism leads to the ultimate maximum earthquake?);
   - attenuation modeling
     - existing attenuation laws (e.g. Ambraseys, Campbell, Boore, etc.);
     - local attenuation law (regression analysis according to local earthquake events – depending on the outline and reliability of the earthquake database);
   - site modeling – structural models
     - finite element model of the bridge section;
     - local soil model for each foundation of the bridge piers, extracted and updated by seismic system identification (SEISMID) measurements;
     - local radiation conditions (array measurements to define the dominant propagation direction).

The solution of a PSHA is done by means of a logic tree (decision tree structure of epistemic uncertainties; Figure 3.26) (e.g. epistemic uncertainties arise if the scientific understanding is imperfect; the opposite of this are aleatory uncertainties, e.g. subjective input in the regression analysis itself). Thereby each path of the logic tree represents one possible combination of parameters.

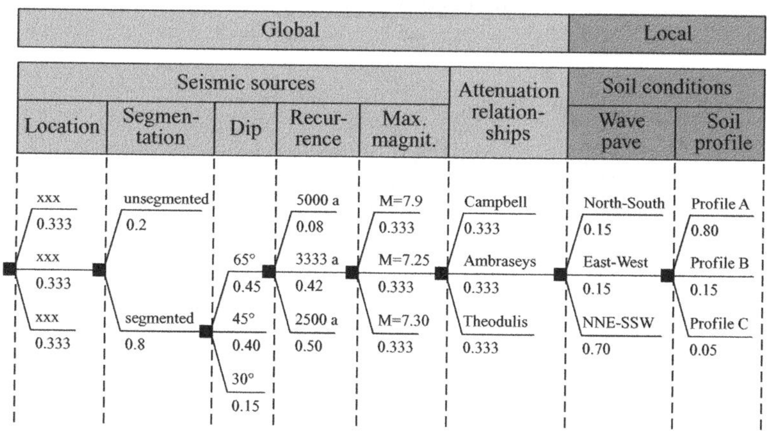

Figure 3.26 Probabilistic seismic hazard analysis – logic tree

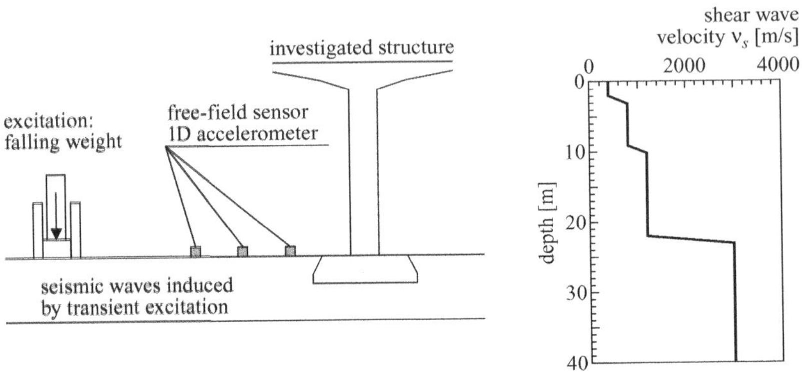

Figure 3.27 Test setup for v_s measurements (left) and typical profile of the shear wave velocity (right)

2. Seismic system identification measurements:
   To receive the necessary information about the soil conditions on site the SEISMID technique is applied. The measurements can be subdivided into two different methods:
   - measurement of the average shear wave velocity (Figure 3.27);
   - investigation of the predominant wave path (Figure 3.28): according to the different arrangement of subsoil layers, it is possible that there are local varying wave paths.

Results

- Seismic hazard curve with emphasis on local conditions (horizontal acceleration plotted against return period for each bridge foundation) with respect to uncertainties.
- Uniform Seismic Hazard Spectra.
- Fundamental local soil parameters as information for detailed numerical studies.

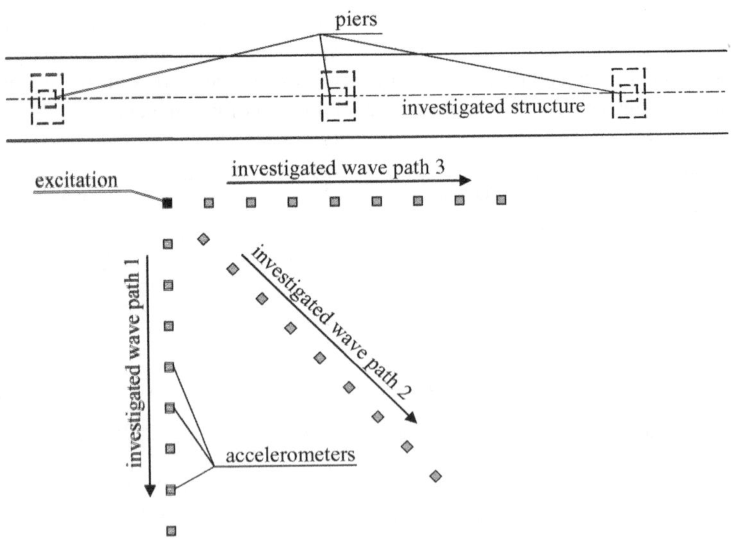

Figure 3.28 Investigation of the predominant wave path by instrumentation of three different directions with dynamic excitation

3.4 Risks from Natural Hazards

A connection from climate change to bridge performance has been indicated by various researchers. Nevertheless, the fact that an increasing density of population pushes our structures into less favorable locations contributes much more to this issue. It simply has not been necessary to build bridges in dangerous areas before.

On the other hand we experience rainfall in unknown intensities. This not only leads to problems with the bridge dewatering system, but also becomes responsible for landslides in previously not exposed locations. Another aspect is that the assessment of the earthquake hazard has been changed considerably after some measured data became available from recent earthquakes. It has been found that the demand on our structures from earthquakes is about double what the codes request us to apply. This provides the opportunity for a general reassessment of our bridge stock, which can usefully be combined with SHM activities. Due to this fact the subject of seismic assessment is given wider scope in the following sections.

The subject of extreme temperatures should not be forgotten. Several monitoring campaigns have indicated that the current codes for such cases are not sufficient. The amendment of codes to cover all these findings is underway and will be combined with routine SHM activities.

3.4.1 Earthquakes

Earthquakes have received growing attention recently. Since the earthquakes of Northridge in 1994 and Kobe in 1995 our knowledge has increased and our assessment of the hazard has changed. Actual data from the Kocaeli earthquake in Turkey in 1999 and Chi Chi in Taiwan in 1999 have confirmed that earthquake loads have been underestimated by approximately a factor of two. This creates the situation that earthquake loads are the governing load cases in bridge design and assessment, which is why this subject goes closely along with SHM. Both campaigns, namely the health assessment as well as the earthquake assessment, might usefully go hand in hand. This is a good occasion to carry out useful action on bridge retrofitting, from which the bridge resistance and therefore its lifetime can benefit considerably.

For bridges there are two main subjects of concern:

- an increased displacement demand, which might lead to bearing failures and consequential collapse;
- excessive ground shaking from local site effects.

For the first phenomenon, so-called microzonation methodologies such as horizontal/vertical (H/V) ratio or seismic wave dispersion analysis are briefly described here. For the second phenomenon the identification of eventual local sources through remote sensing has proven to be most successful. A brief description is enclosed here.

Recent earthquake scenarios in densely populated areas led scientists' attention to sophisticated seismic hazard prevention. Especially in urban areas, which are located predominantly on a soft soil layer, waves of moderate seismic events can be amplified greatly. The enhancement of seismic exposure for local building structures, the so-called site effect, led to an ascending request for a precise investigation of the realistic seismic vulnerability.

Additionally, earthquake hazard prediction has changed dramatically in later years. According to recent studies (Hinsch and Decker 2003) scientists of the University of Vienna found that different seismic slip rates of various sectors of the Vienna Basin Transform Fault (VBTF) indicate locked fault elements. They estimate that a strong-motion earthquake (at least magnitude $M = 6.1$) can be expected to normalize the slip rate of the slowest sector of the VBTF to the rate of the fastest sector.

With regard to the economic and social importance (about 2.4 million inhabitants producing 45% of the Austrian GDP) of the Vienna Basin area it will be strongly recommended to determine possible site effects and, furthermore, enforce the proposed microzonation of that area to enhance the standard of seismic hazard analysis.

Additional to the increase in safety assessment, the proposed microzonation studies will lead to economic advantages regarding the conversion of national building regulations to the European Standards. Compared with the currently available national regulation ÖNORM B 4015 (4015 OB 2002) in the revised version of Eurocode 8 (1998-1 EDE 2005) the subsoil classification scheme is more advanced and adjusted to local soil properties. In particular five different subsoil classes (A–E) are defined in Eurocode 8. Their distinguishing coefficient is the average shear wave velocity $v_{s,30}$ dominating in the first 30 m of the subsoil layers. It therefore will be necessary to have detailed information about the local site conditions, which currently is an exception or is only available for relevant building structures. It therefore will be one of the primary objectives of the proposed method to obtain as much information as possible from the in situ measurements, to enhance the basic principles and simplify the application of Eurocode 8.

3.4.1.1 Preview

According to building regulations the probability of earthquake events in Austria is comparable to those in neighboring countries, similar to that in Switzerland and substantially higher than that in Germany. It is also noticeable that the highly populated areas in Austria (Vienna, Graz and Innsbruck) are located in the more endangered areas compared with the areas of high population density in Germany (Berlin, Munich, Hamburg) and Switzerland (Zürich, Bern) (see Figure 3.29).

Nevertheless, contrary to neighbouring countries, in Austria no microzonation study exists. In recent years extensive earthquake hazard analysis was accomplished in Germany, concentrated on the area of Cologne. Managed by the GeoForschungsZentrum Potsdam several tests were performed based on the H/V ratio and a critical frequency ratio between the natural frequency of the building structure and the subsoil, which led to a local earthquake hazard map (Merz and Apel 2004).

In Switzerland highly sophisticated research has been done particularly in the area of Basle. Numerous regulations and reports are available and indicate the high quality standard (Kind et al. 2002; für Wasser und Geologie BB 2004).

Supplementary to national research work, several Europe-wide projects (SESAME, RISK-UE, SEISMOCARE) have been accomplished and achieved an international standard in earthquake hazard assessment. The objective of this report is to develop a method for Austria, according to the existing European research work, to close the technological gap in seismic hazard prevention. The main advantages and highly sophisticated technologies of these projects should be combined in the SEISMID technology. Therefore a short description of the individual parts and work packages of these projects is given.

Research Project SESAME

The Europe-wide (Belgium, France, Germany, Greece, Italy, Norway, Portugal, Slovakia and Switzerland) SESAME project (Site Effects Assessment using Ambient Excitations; (Research Project SE 2004)) had the aim of estimating the site effect with special attention on urban areas and low cost techniques. One of the objectives of the project was to investigate the reliability of the H/V and the array technique. During the project the influences of the experimental conditions on the results were studied and highlighted (Duval et al. 2004). In an additional report (Research Project WP12 SE 2006) user guidelines for the H/V spectral ratio technique on ambient vibrations are defined.

The very sophisticated conclusions of these investigations are a helpful input for experimental measurements of ambient noise as well as for numerical data evaluation of the SEISMID project.

Research Project RISK-UE

RISK-UE (Research Project RUE 2004) was a European project performed from 2001 until 2004, in which methodologies for earthquake-risk scenarios for several cities (Barcelona, Bitola, Bucharest, Catania, Nice, Sofia and Thessaloniki), including both current and historical buildings, were developed.

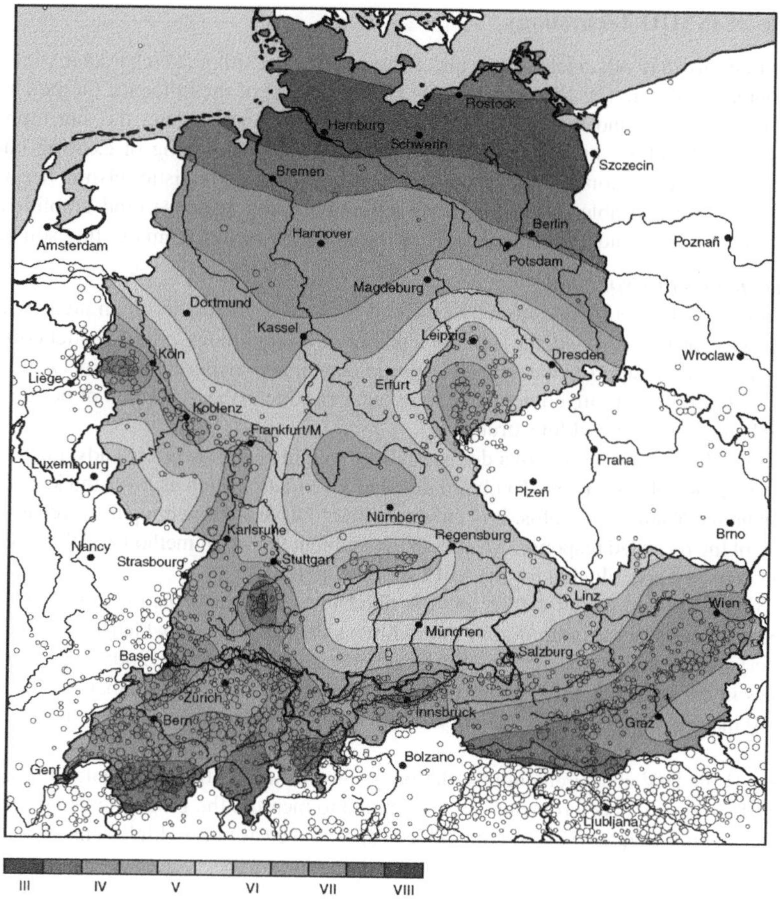

Figure 3.29 Assessment of earthquake hazard exposure for Germany, Austria and Switzerland, according to Grünthal et al. (1998)

The project was based on seismic hazard assessment and stock-taking of buildings in order to identify the critical issues of urban systems. In the Report of Work Package 2 (Research Project WP2 RUE 2003) deterministic seismic hazard evaluations for urban areas were developed, resulting in attenuation relationships for European regions. With the application of slight modifications these approaches will be useful for the Vienna Basin Transfer Fault, in order to obtain the effective ground motion forces for the urban areas in Vienna.

Research Project SEISMOCARE

The SEISMOCARE European project (Research Project 1998) deals with computer-aided reduction of seismic risk with application in existing cities. Its main aim was to develop an integrated methodology for earthquake hazard assessment and to produce a software package for reliable predictions of losses due to earthquakes.

The SEISMOCARE research work will be beneficial for the implementation of GIS in the SEISMID project and should be an initiation for developing software packages for integrated seismic hazard prevention.

3.4.1.2 The SEISMID Technology

Local site conditions may adversely affect all of the important earthquake characteristics (amplitude, frequency content, duration; see (Kramer 1996)). The dimension of the influence on the characteristics depends on the geometry and material properties of the building structure, the site topography and the input motion. Therefore, numerous in situ measurements, stock-taking of existing buildings and adjusted earthquake attenuation relationships are required to obtain a realistic seismic hazard mapping. Additionally it is indispensable to find correlations between site response (and amplification factor, respectively) and the soil conditions to achieve the amplification factors from existing soil profiles.

In Situ Measurement Procedure

Ambient vibration testing is a very attractive method for measurements in urban areas with moderate seismicity because, on the one hand, the low ambient vibration amplitudes indicate better correlation with moderate seismic events (approximately linear-elastic constitutive material behavior) and, on the other hand, seismic cross-hole tests are complicated in urban areas due to the disruption of building occupants and induced damages on the building structure.

However, forced vibration tests with a defined falling weight have exceptional advantages concerning the parameters of the subsoil layers. To obtain detailed information such as layer thickness, shear wave velocity or dynamic Young's modulus, it is indispensable to have a defined input source to get a correct interpretation of the recorded response. Thus both the ambient vibration method and the forced vibration technique seem to be very valuable for the SEISMID technology.

In Figure 3.30 the forced vibration tests are described whereby the soil parameters can be determined. The refracted seismic waves are recorded by three-dimensional geophones, which are located in arrays on the subsoil surface. According to the theoretical fundamentals (Knödel et al. 2005) the distance between the geophones in the array has to be chosen in combination with the wavenumbers of the input signal. In Figure 3.30 the seismic impedance is denoted by I as the product of the wave velocity v and the soil density ρ.

In Figure 3.31 the H/V measurement installation is shown. Around the investigated building structure several three-dimensional accelerometers are assembled to measure the ground response of the ambient excitation. The natural frequencies (and the mode shapes in the case of multiple sensors) of the building structure are specified with internal three-dimensional accelerometers.

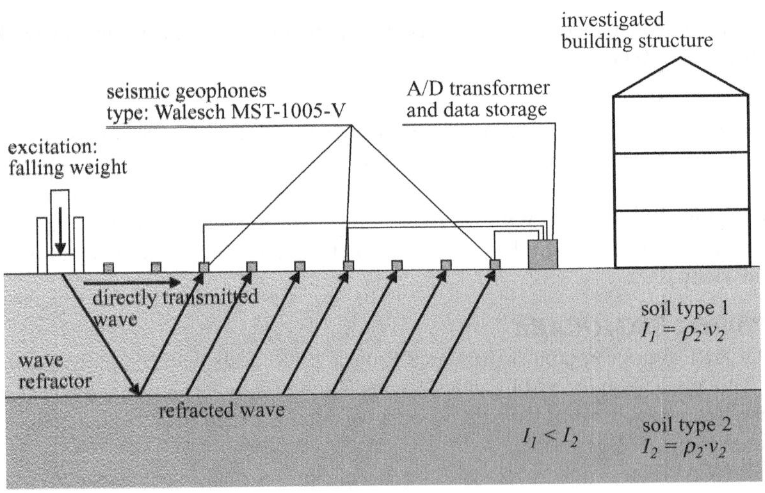

Figure 3.30 Seismic refraction method with proposed falling weight excitation

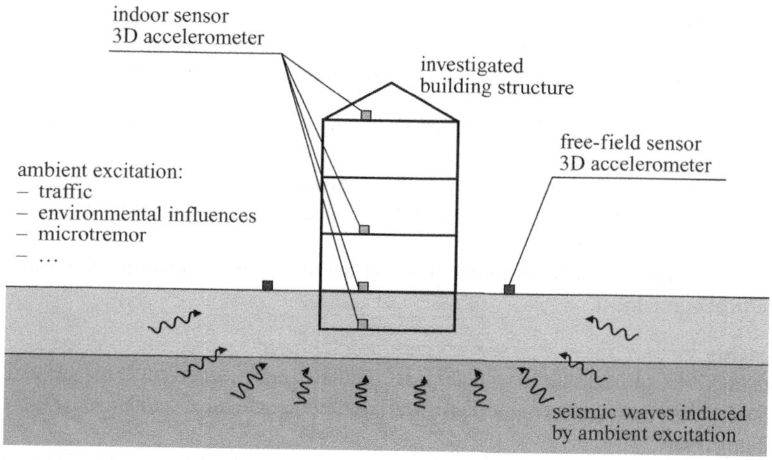

Figure 3.31 Instrumentation for ambient H/V ratio measurements

The H/V Ratio Technique

To estimate the local amplification factor of seismic waves it is preferable to use the H/V ratio method according to Nakamura (1989) due to the fact that it is very simple and user friendly. The method is convenient for almost every region in Europe because it was successfully accomplished in countries with moderate seismicity where no strong motion data were available (see Section 3.4.2). Therefore the input for the H/V computation can also consist of transient or even ambient excited signals.

Nevertheless in the case of the amplification factor it is more beneficial to use ambient excitation because the large number of samples needed for statistical evaluation would, for transient excitation, lead to unacceptable disruption for the occupants. The ambient vibration technology for seismic microzonation was effectively applied in former projects, see the SESAME project (Research Project SE 2004) or the Scientific Report for the DNFK (Bormann et al. 2004).

With the proposed method (Steimen et al. 2005) the following local soil parameters can be defined:

- Basic natural frequency f_0 of the subsoil.
- With cognition of the basic natural frequency f_0, either the shear wave velocity v_s or the layer thickness can be calculated if one or both values are known. Because of the fact that by means of the proposed method neither the shear wave velocity v_s nor the layer thickness can be defined exactly, additional measurements have to be arranged (see Section 3.3).
- Estimation of the amplification factor. The peak of the H/V spectra may not be transferred directly into an amplification factor. In several publications this value is interpreted as a lower limit value for the amplification of shear waves.

The computation and furthermore the numerical implementation of the H/V procedure can be described according to the SESAME project (Research Project SE 2004) as follows (see Figure 3.30):

Pre-Processing

- Three-dimensional input (the accelerometer in northward direction to get North–South, East–West and vertical components).
- Windowing of the signal (in our case only the ambient parts are of interest, but for transient excitation only the transient parts of the time response are of interest).
- Offset removal.

Main Data Processing

The three different components of the signal were considered separately. The main data processing is repeated for every input signal (*n* steps according to the numbers of preliminary separated windows):

- Fast Fourier transformation (FFT) is applied to obtain the several spectral amplitudes of the three components.
- Smoothing of the three spectral amplitudes with a bandwidth factor of six.
- Afterwards, the resulting horizontal component is evaluated on the quadratic average of the North–South and East–West component.
- The resulting H/V value is the logarithmized (on base ten) ratio of the resulting horizontal component and the vertical component.

Post-Processing

- The output of the data processing (*n* signals) is accumulated and averaged over the number of windows.
- To consider the experimental and numerical uncertainties, the statistical standard deviation is calculated (in logarithmic scale).
- Finally the averaged H/V ratio and the standard deviation are set back to linear scale (see Figure 3.32).

Shear Wave Velocity v_s

The shear wave velocity v_s and the thickness of the subsoil layers are essential parameters for the accurate determination of site effects as well as the economic seismic design according to Eurocode 8. Depending on the local existing soil layers, several seismic methods can be used. If the soil impedance I (denoted as the product of the wave velocity v and the soil density ρ) of the upper layer I_1 is smaller than the soil impedance of the subjacent layer I_2, the method of seismic refraction can be used (see Figure 3.30). The seismic waves will be excited with a defined falling weight and registered with a couple of three-dimensional geophones arranged in arrays on the subsoil surface at defined distances (see Section 3.4.1.2).

In other publications (Havenith et al. 2006) different methodologies are used to measure the shear wave velocity. The main advantage of the proposed forced vibration tests is the very high quality of the measured data. On the other hand, with additional accelerometers inside the buildings, very important information about the dynamic soil–structure interaction can be found. Due to the fact that full-scale tests are very rare, such information is of great relevance and can be used to verify numerical simulations of dynamic soil–structure interaction analysis.

Vulnerability of Existing Building Structures

The seismic vulnerability of existing building structures is one of the most important steps in an integrated seismic hazard analysis and microzonation study. In particular the evaluation of existing buildings should be as detailed as necessary and as precise and elementary as possible. One of the most important publications in this area of research was done for the city of Basle (Lang 2002). The general concept of the method based on nonlinear static procedures is to develop a vulnerability analysis and calculate the capacity and seismic demand of the building structures.

This methodology can be almost completely adapted for residential buildings of the microzonation study in Vienna, because of the similar topology of building structures. Other types of structures, such as bridges, infrastructure facilities or high rise buildings, require a more precise vulnerability analysis.

3.4.1.3 Application

The SEISMID technology should lead to an integrated seismic hazard analysis method that can be used for different applications. One of the most important will be the development of a microzonation map. Furthermore the technology can be used to drive detailed analysis for important building structures with unsteady layouts. It therefore can be used to obtain detailed soil parameters such as the shear wave

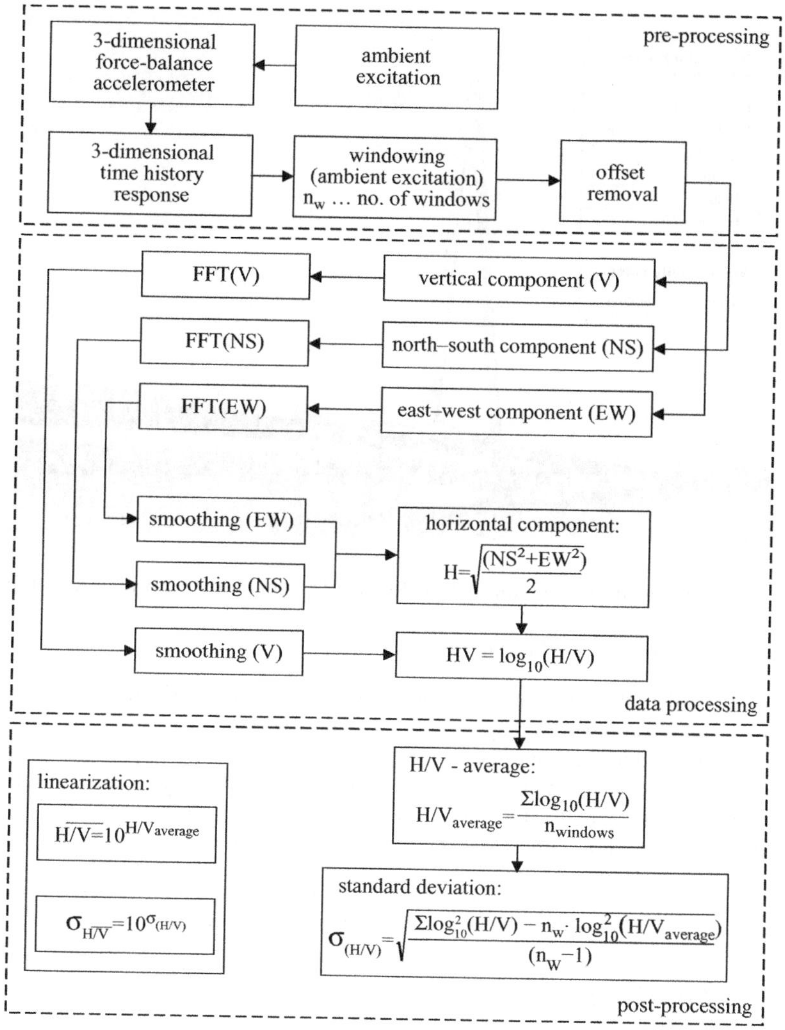

Figure 3.32 Flowchart of processing the average H/V ratio ($\overline{H/V}$) and standard deviation ($\sigma_{\overline{H/V}}$)

velocity v_s, dynamic Young's modulus E_{dyn} and others as input for numerical simulations. Bridges are one of the most important and highly vulnerable building structures and therefore deserve to be protected. In Section 3.4.2 the proposed application of the SEISMID technology on bridges will be described.

Bridges

In many cities bridges are the vulnerable chain link of lifeline infrastructure and often connect several districts. They are therefore one of the most important building structures but unfortunately also one of the most vulnerable (see the Vienna Donaustadtbrücke in Figure 3.33).

Supplementary to the basic investigations mentioned in Section 3.4.1, bridges require a more precise investigation because their supporting elements are often more exposed and exploited than those of residential buildings. It is obvious that they are often based on different subsoil, resulting in different amplification factors for each bearing.

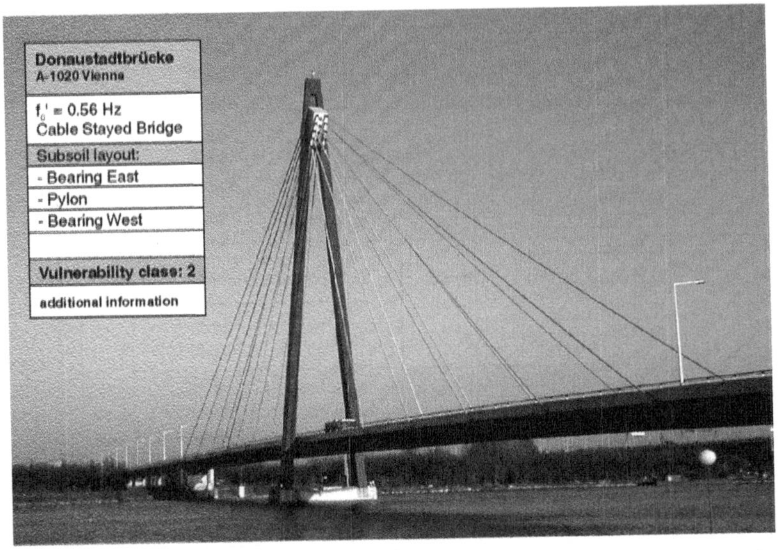

Figure 3.33 Example of implementation of the SEISMID technology for bridge structures in Vienna

Additional investigations should be concentrated on:

- brief knowledge of the subsoil parameters at each foundation based on verified measurements (borehole data should be used to compare and update the in situ measurements);
- natural frequencies and mode shapes of the bridge deck;
- natural frequencies and mode shapes of supplementary bearing structures (cables, pylon, piers, . . .);
- amplification factor for each investigated foundation;
- bearing capacity of the whole structure and several parts (measurements or numerical simulations).

Careful attention has to be paid to the carrying of horizontal reaction forces caused by seismic waves because most bridges are very vulnerable against their horizontal direction.

3.4.2 The SEISMID Approach

Ambient vibration testing is a very attractive method for measurements in urban areas with moderate seismicity because, on the one hand, the low ambient vibration amplitudes indicate better correlation with moderate seismic events and, on the other hand, seismic cross-hole tests are complicated in urban areas due to the disruption of building occupants and induced damage on the building structure.

However forced vibration tests with a defined falling weight have exceptional advantages concerning the parameters of the subsoil layers. To obtain detailed information (e.g. layer thickness, shear wave velocity or dynamic Young's modulus) it is indispensable to have a defined input source to get a correct interpretation of the recorded response.

3.4.2.1 Measurement Procedure

Thus both the ambient vibration method and the forced vibration technique seem to be very valuable for the SEISMID technology. In Figure 3.34 the forced vibration tests are described whereby the soil parameters can be determined. The refracted seismic waves are recorded by three-dimensional geophones that are located in arrays on the subsoil surface. According to the theoretical fundamentals (Knödel et al. 2005)

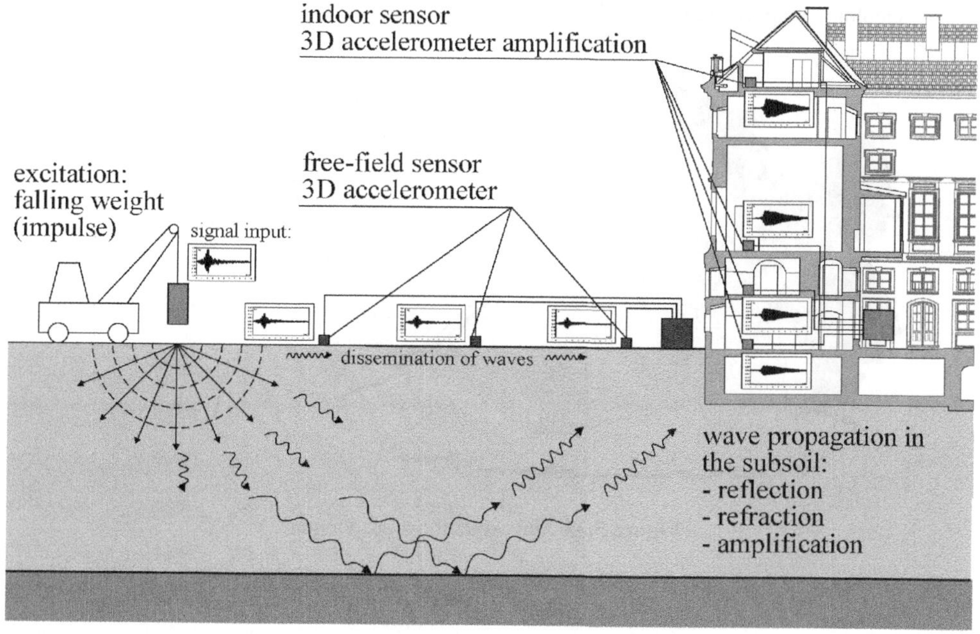

Figure 3.34 Experimental setup for impulsive excitation

the distance between the geophones in the array has to be chosen in combination with the wavenumbers of the input signal.

3.4.2.2 Vulnerability of Structures

One of the most important steps in an integrated seismic hazard analysis and microzonation is the seismic vulnerability of existing building structures. The evaluation of existing buildings should be as detailed as necessary and as precise and elementary as possible. A general concept of the method, based on nonlinear static procedures to develop a vulnerability analysis and calculate the capacity and seismic demand of the building structures, had been done for Basel (Lang 2002).

This methodology can be almost completely adapted for residential buildings of the microzonation study in Vienna, because of the similar topology of building structures. A more precise vulnerability analysis for other types of structures (bridges, infrastructure facilities or high rise buildings) is required.

3.4.3 Remote Sensing and Geographic Information Systems

Numerous studies have demonstrated the value and the use of GIS-integrated datasets as satellite radar data (ERS, SRTM and LANDSAT TM data), geomorphologic and geologic field data and seismotectonic data from different earthquake prone areas (Ponte and Moccia 1995; Cornet et al. 1997; Theilen-Willige 1999b) for seismic microzonation research.

The GIS techniques are used for regional analysis and prediction. Digital datasets include: an inventory of seismic records; geological mapping; extensive geotechnical data on rock properties; high-resolution digital elevation data; and suitable high-resolution remote sensing data. Both space and airborne remote sensing systems now have resolutions that permit detailed geomorphologic and geologic mapping to be

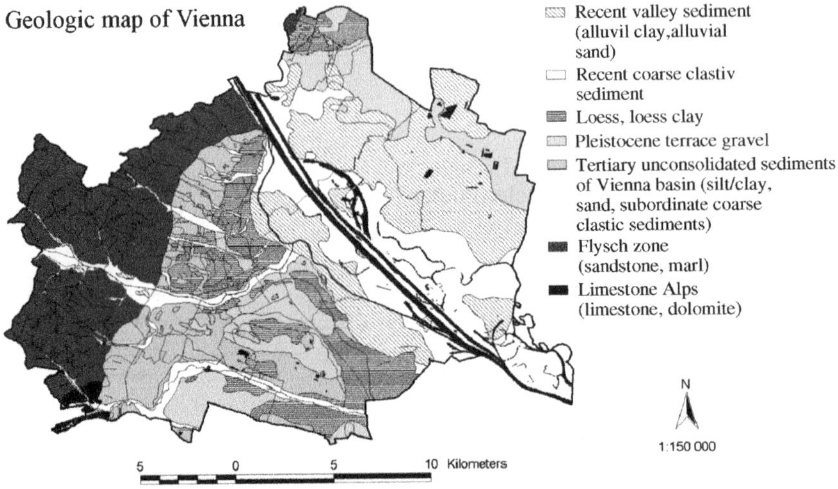

Figure 3.35 Investigation area, Vienna

conducted. This mapping procedure is used to produce hazard risk maps as a basis for urban development planning in Austria, especially for the area of Vienna (Figure 3.35).

Remote sensing data can be used to map factors that are related to the occurrence of higher earthquake shock and/or earthquake-induced secondary effects: factors such as lithology (unconsolidated sedimentary covers), faults, steep slopes, vegetation and land use. Special attention is focused on precise mapping of traces of faults on satellite imageries, predominantly on areas with distinct expressed lineaments, as well as on areas with intersecting/overlapping lineaments and on areas with unconsolidated sedimentary covers. The lineament analysis based on satellite imageries can help to delineate local fracture systems and faults that might influence seismic wave propagation and influence the intensity of seismic shock. Merging lineament maps with isoseismal maps, for example, contributes to a better knowledge of subsurface structural influence on seismic shock intensity and of potentially earthquake-induced secondary effects (e.g. landslides or soil liquefaction).

3.4.4 Objectives and Approach

One objective is the development of a standardized and common operational Information System focused on the monitoring of an infrastructure integrating different remote sensing, geophysical and geotechnical data in order to support national and local authorities in Austria.

The aim of this study is to investigate the potential of remote sensing and GIS techniques to improve the understanding of the subsurface conditions influencing damage intensity in the Vienna area. The study is an attempt to integrate various datasets (such as LANDSAT ETM data and satellite radar data SRTM), aerial photographs and seismotectonic data from Austria to obtain a better general understanding of the processes influencing the damage intensity of stronger earthquakes, including primary and secondary effects such as ground motion, liquefaction potential and landslide susceptibility, and to improve hazard maps. One focus of this research is to investigate the use of satellite imageries for the detection of the tectonic pattern and of earthquake-induced faults and fracture patterns at the Earth's surface (Figure 3.36).

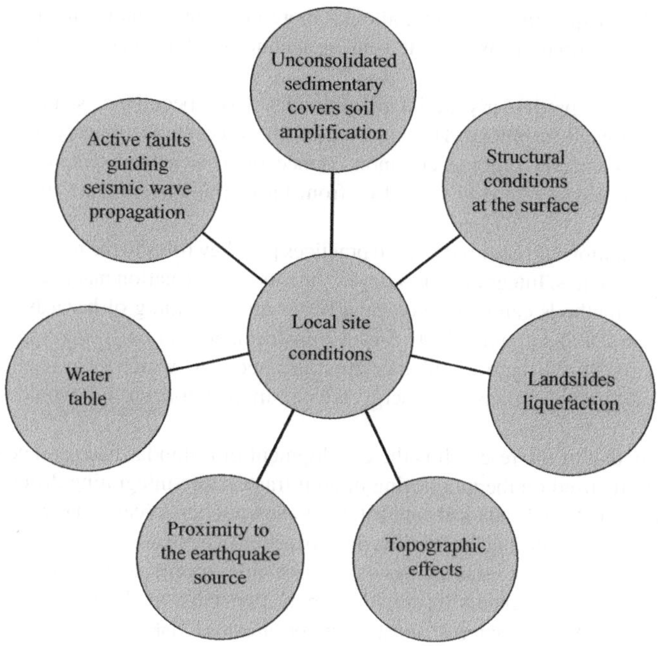

Figure 3.36 Visualization of earthquake damage influencing factors

3.4.4.1 Evaluation of Maps and Other Data

Results of evaluations of remote sensing image data had been compared and combined with available topographic, geological and geophysical data, especially with the maps of known faults and fracture zones. Digital terrain models (DTM) were used to investigate the influence of morphology on macroseismic intensity. Hillshade and slope gradient maps derived by DEM data provide a database for hazard mapping.

The LANDSAT ETM satellite data were merged with DTM data. Compared with two-dimensional images such as the available satellite pictures, DTM images have the advantage of representing the vertical extension of the earth's surface by giving height values for every pixel. With digital image processing techniques, customized perspective views can be generated to meet specific requirements for risk mapping. An exaggeration of the topography often is necessary to distinguish fault segmentation associated with slip deficits and low relief. Fault segment boundaries may be identified.

Digital elevation data are available at different resolution and quality: GTOPO30 data (1 km ground resolution) provided by the US Geological Survey and SRTM data provided by the Shuttle Radar Topography Mission (90 m resolution). Available geological and geotechnical data will be collected, interpreted and mapped. It is examined to what extent, by overlays and linkages of the data among themselves, additional information can be won and relationships can be represented in a more descriptive way.

3.4.4.2 Evaluation of Remote Sensing Data and Image Enhancement by GIS

Innovations in GIS technology are increasingly accepted tools for the presentation of earthquake hazard vulnerabilities and risks. Over the past few decades, earth observation technology has proved to be an increasingly powerful tool to monitor and assess the Earth's surface on a regular basis. Earth observation satellites such as NOAA, LANDSAT, SPOT, ERS, or ENVISAT, with increasing capabilities in terms of spatial, temporal and spectral resolution, allow a more efficient, reliable and affordable monitoring of the environment over time. Thus, remote sensing technology has become a fundamental input for GIS.

Remote sensing data, particularly aerial photographs, have become a standard tool for providing an earthquake damage inventory. The application of stereoscopic imagery, satellite imagery and RADAR imagery has not yet become common. Hence there is a huge potential for forecasting in this area. Image data, either remotely sensed or from terrestrial systems, are innovative tools in this area.

Exchange of information and communication practices play key roles in the realization of effective disaster risk reduction activities. Integrating new developments in information management with established and more traditional methods can help to create a better understanding of hazards and risks. Effective information management and communication is also instrumental for early warning systems and effective mitigation efforts. It will help to enhance Austria's security, which is in itself a precondition of numerous community policies (transport, energy, telecommunication, etc.). It would foster cross-border cooperation.

The primary objective of the research is the development of a standardized and common operational Information System focused on the monitoring of an infrastructure integrating different remote sensing data in order to support national and local authorities, as shown schematically in Figure 3.37. Reaching the objective of the request, considering the improvement of security measurements, requires the development of reliable and cost-effective information systems being able to provide national and local authorities with accurate, timely and continuous information about the status of the infrastructure and processes influencing their safety. Special interest must focus on areas of risk where urgent measurements are required to be implemented.

The remote sensing data will be used for generating an image-based GIS (see Figure 3.38). The various datasets (such as LANDSAT TM and satellite radar topography (SRTM) data, and topographic, geological and geophysical data) from the investigation area are integrated as layers into GIS using the software

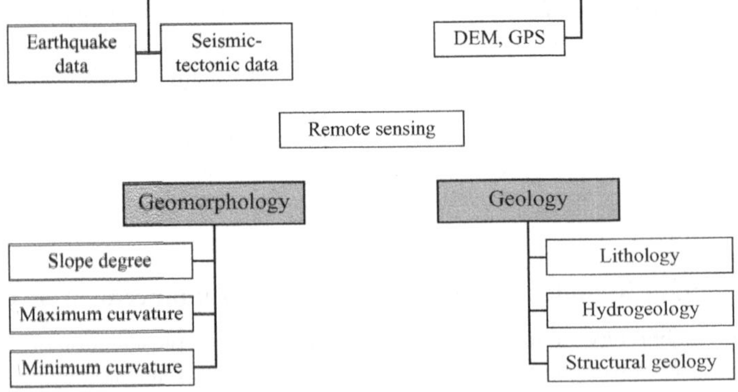

Figure 3.37 Remote sensing and GIS contribution to a GIS database

Figure 3.38 Maps based on SRTM: combining LANDSAT and SRTM data in order to detect sites of higher damage risk

ArcView GIS 3.3 with the extensions Spatial Analyst and 3D-Analyst and software ArcGIS 9.0 of ESRI. The Hydro-tools of ArcView are used as well.

3.4.4.3 Digital Image Processing and Evaluation

The technology of actual earthquake prediction, to enable sounding of warning alarms beforehand to save people and resources, is still in its infancy. However, seismic risk analysis can be carried out for designing structures such as bridges, buildings, etc. Local site conditions influencing damage intensity can be mapped. Remote sensing and GIS can provide valuable inputs to this aspect, for example for active fault mapping.

For enhancing the satellite data of the test sites, digital image processing will be carried out. Various image enhancement tools delivered by ENVI Software/ CREASO were tested, for example for finding the best suited Red, Green, Blue (RGB) combinations or contrast stretching parameters for geologic-tectonic evaluation purposes. Image classification and color-coding variations of the different remote sensing data will be tested in order to enhance the detectability of geologic structures as fault zones in the subsurface that might influence the security of infrastructural facilities, for example during earthquakes or landslides.

The RGB principle is reviewed briefly: Three images from the different LANDSAT bands to be used as end-members in a triplet are projected, one image through one primary color each, i.e. one image is coded in blue, the second in green and the third in red. In this way, each image is given a particular false color (Gupta 1991). In principle, any image can be coded in any color.

The LANDSAT data were merged with SRTM-derived maps (Figure 3.38). Compared with two-dimensional images such as the available satellite pictures, digital elevation models (DEM) have the advantage of representing the vertical extension of the earth's surface by giving height values for every pixel. With digital image processing techniques, for example, customized perspective views can be generated to meet specific requirements for risk mapping. For a geomorphologic overview, SRTM data terrain parameters were extracted from a DEM as shaded relief, aspect and slope degree (magnitude of maximum gradient, steepest slope angle), minimum and maximum curvature or plan convexity maps representing surface parameters. Digital processing of elevation data may also provide evidence for neotectonic activity. As a complementary tool Google Earth software was used to benefit from the three-dimensional (3D) imageries of the various investigation areas (http://earth.google.com/).

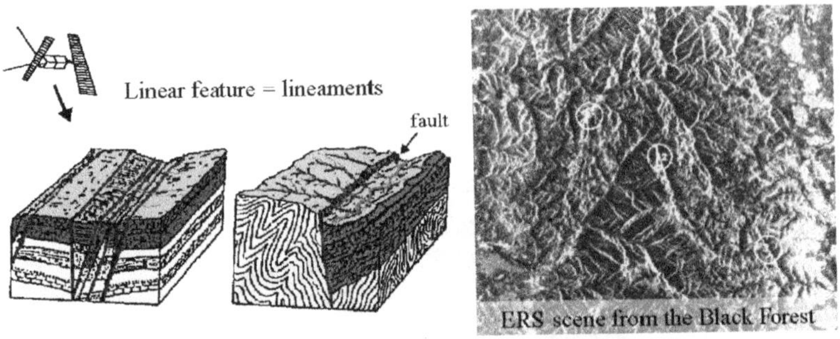

Figure 3.39 Lineament mapping

3.4.4.4 Lineament Analysis

Detailed lineament analysis helps to detect areas prone to slope failure. The movements of slopes are structurally controlled by surfaces of planes of weakness, such as faults, joints and bedding planes. A careful search to locate areas with close spacing of faults and joints, especially where they overlap and intersect, helps find evidence of possible continued movements and of potential take-off domains.

A very important component of this part of the project is the methodology of lineament analysis (mapping of linear features visible on the imageries) based on satellite imageries for providing information of the geologic structure of the subsurface (Figure 3.39). A combined evaluation of structural field data, seismotectonic data and lineament analysis based on satellite imagenes will be carried out. The following linear and curvilinear features and risk sites will be identified and mapped in the investigation areas:

- lineaments;
- probable fault zones;
- structural features;
- probable sites of slope failure, liquefaction, subsidence.

As risk areas will be mapped regions with higher risk of earthquake ground shaking, earthquake-induced secondary effects such as landslides or liquefaction, or stronger ground motion due to seismic wave amplification (Figures 3.40 and 3.41).

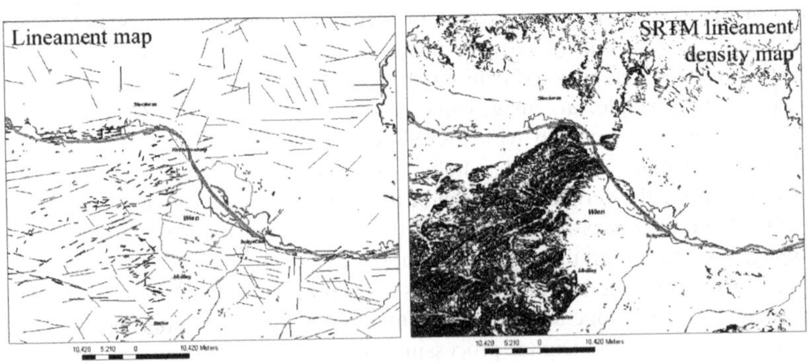

Figure 3.40 Risk factors

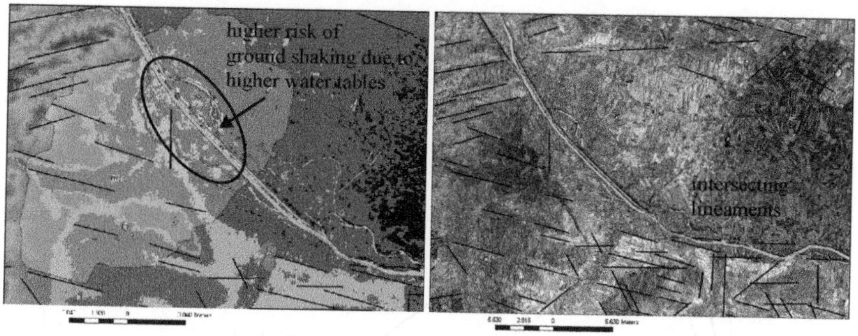

Figure 3.41 Potential risk of higher ground shaking for the Vienna Basin

3.4.4.5 Geomorphologic Setting

Part of a GIS-integrated standard investigation should be the analysis of earthquake density and distribution and the spatial relationship of areas with higher earthquake occurrence to landslide occurrence. The concentration of larger fault zones and intersecting fault and fracture zones is another important factor influencing slope instability that can be visualized as a layer in a GIS.

The investigation of spatial relationships between the occurrence, distribution and intensity of landslides and the position of fault and fracture zones can be carried out in a standard manner using the geoprocessing wizard tools of GIS software.

3.4.5 Conclusions

ENVISAT and LANDSAT imageries were used to create maps within an "Earthquake Hazard Information System" as important layers in a GIS database in order to perform user-defined computations of earthquake hazard maps as required by (Erdik and Swift-Avci 1997). ENVISAT ASAR are of high importance and value concerning the detection and mapping of areas prone to earthquake damage. Lineament analysis based on ENVISAT ASAR radar-imageries can help to delineate those local fracture systems and faults that might influence seismic wave propagation and thus influence the intensity of seismic shock, as larger outcropping faults can cause a reduction of seismic velocities or multiple reflections amplifying vibration intensity and duration.

A GIS-based satellite image interpretation technology gives the opportunity to monitor the current condition and therefore offers reliable actual information. On a local scale, a knowledge-based system that uses field data gained through permanent monitoring and a knowledge base extracted from human experience (know-how) and also history data is ideal for early warning. The knowledge-based system uses a factor matrix with influence factors for mass movements and take-off domains.

By providing up-to-date information and integrating the results with traditional earthquake hazard assessment studies, coherent and secure information is provided. By these means the information can be used for early-warning, for decision support in disaster management.

3.4.6 Attenuation Functions

The determination of useful attenuation functions for a specific bridge site represents a science in itself. Depending on the distance to the source, a different peak ground acceleration (PGA) has to be expected. This normally is not the task of the bridge designer or the bridge owner. Regional studies are recommended that can come up with a useful PGA map for application in the design. It is obvious that this map covers

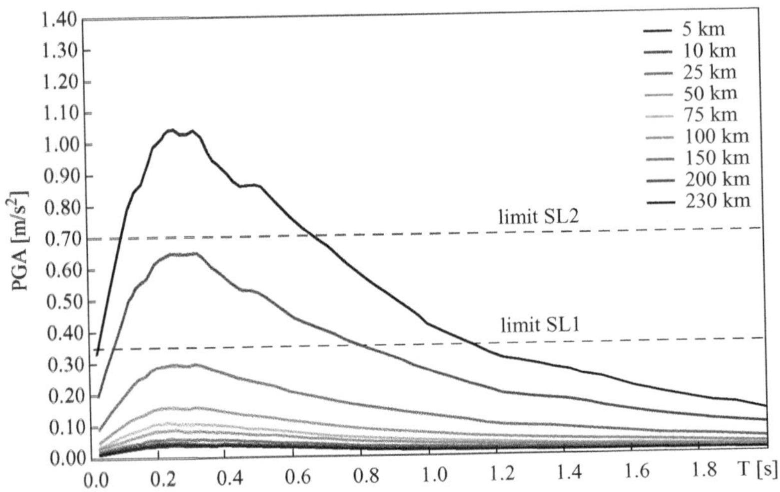

Figure 3.42 Limits of peak ground acceleration used for the design of the nuclear power plant Temelin; spectral attenuation relationship according to Ambraseys et al. (1996)

the worst case and does not consider eventual reductions due to good local conditions. If the safety cannot be demonstrated by applying the local map, attenuation functions could be applied in detail to result in more favorable input values for assessment of the structure. Figure 3.42 shows a typical attenuation function computed on Ambraseys et al. (1996) functions.

3.4.7 Landslides

Landslides become particularly important in the bridge management process, when slow movements change the structural system or impose heavy external loads on the bridge. To detect these movements several approaches have been taken:

- By using remote sensing data, landslides on a global scale can be detected and monitored. INSAR methodologies can quantify even small movements (down to 1 mm).
- A detected moving slope should then be instrumented by conventional systems such as total stations or a GPS network. The level of observation should be determined individually.
- When the movement becomes critical, instruments such as inclinometers and piezometers are buried in the slope to provide the basis for an early warning.

Attention has to be paid so that this information is properly used in the bridge management process. In practice different departments of the road management can be responsible for the structure or for natural hazards.

3.4.8 Fire

Fire below a bridge is an issue of bridge health monitoring. The extent of damage mainly depends on the temperature the structure has been exposed to. Furthermore, the time of exposure is critical. Nevertheless fires are exceptional cases and cannot be covered by the ordinary maintenance program. Special procedures should be implemented if a fire is detected.

3.4.9 Flood Scour

Scours are the predominant reason for bridge collapses worldwide. Various attempts have been made for the monitoring of scours. Technically it is not too challenging. The actual water depths can easily be monitored by an echo sounder and several proposals for contact sensors (which sound an alarm if exposed) have been seen. However no system has been found in practice because of the discrepancy between the life expectation of such a system and the life expectation for a bridge. Another reason is the huge number of piers affected.

3.5 Vehicle and Ship Impact

In the early morning of December 17, 2005 a ship loaded with more than 3500 tons of iron ore traveling up the Danube River crashed into the center pier of the railway bridge in Krems, Austria. The pier was sheared off above the water surface and moved 2.17 m upstream (Figure 3.43).

The temporary stability of the damaged pier was checked by dynamic measurements. Furthermore a monitoring and alarm system based on vibration monitoring was installed on the bridge to observe the behavior of the damaged bridge for two months until the lifting of the superstructure. Because the shipping traffic under the bridge should not be interrupted, the system was designed to warn ships, as well as workers on the structure, early enough in advance to a collapse.

The following pages describe the layout and function of the monitoring and alarm system, including real-time data analysis and the results for the whole monitoring period.

3.5.1 Introduction and Scope of Work

The submerged part of the pier stub was checked by divers and they noticed numerous cracks. Due to the check results, failure mechanisms were assumed: failure of the substructure, which is announced by an inclination; and brittle fracture of the remaining pier shaft due to the loads from the structure.

For a better understanding of the situation, measurements were performed to be able to check the stability of the damaged pier. A monitoring system based on vibration and inclination measurements was installed for permanent monitoring of the damaged pier and the supported superstructures in order to be

Figure 3.43 Damaged pier

able to detect critical situations at an early stage. For such permanent monitoring two different situations had to be taken into consideration:

Nonworking Phases
During the nonworking phases the bridge was monitored in a completely automated way without any personnel. If previously defined threshold values were exceeded, a defined group of persons was automatically alarmed via SMS.

Securing Works
The automated alarm system was deactivated during the working phases. In order to be able to identify critical situations on time and to warn the workers on time, the measuring devices were supervised by engineers on site.

3.5.2 Stability Check of Damaged Pier

3.5.2.1 Measurements

The measurements on the stability check were carried out on December 21, 2005. Two mobile measurement devices (BRIMOS Recorder 800) were used. Each of these units has two highly sensitive 3D acceleration sensors for vibration measurement at structures, especially under ambient conditions.

Measurements were performed at damaged pier 5, undamaged pier 4 and at both superstructures supported by pier 5. During the measurements the BRIMOS recorder was always located at the pier and the external sensor at the superstructure.

3.5.2.2 Evaluation of the Measurement Data

The evaluation of all measurement data was performed by means of BRIMOS 9.03 software. The evaluation procedure took place automatically in several steps, which are briefly described below:

- Conversion of the acceleration signals of *V* into *g*, depending on the respective sensor.
- Establishment of a signal window, where the *x*-axis describes the length of the file and the *y*-axis is scaled automatically.
- Elimination of interfering signals or of signal areas disturbed by exterior influences.
- Evaluation of the vibration displacements by numeric double integration.
- Evaluation of single-channel spectra and of ANPSD (averaged normalized power spectra density).
- Determination of the damping values by means of the random decrement technique (RDT) for all eigenfrequencies and sensors.

The operations carried out in the individual evaluation steps are not commented on in detail here. For more details, please refer to (Wenzel and Pichler 2005).

3.5.2.3 Results of the Stability Check

The dynamic parameters for damaged pier 5 and undamaged pier 4 were determined by evaluation of the measuring data in the frequency and time range. From a comparison of the values, conclusions could be drawn on the condition of the damaged pier.

The great mass of the superstructures influenced the dynamic behavior of both piers. For both piers the same eigenfrequencies were identified in cross-direction (normal to the flowing direction). This can be attributed to a reciprocal transmission of the vibration behavior of the piers by the superstructures. This behavior indicates that the damaged pier was held in position by the supported superstructures because of constraining forces introduced by the displacement and distortion of the structures during the ship crash. The test also assessed the monolithic behavior of the displaced pier shaft. From the measuring results no signs of an imminent brittle fracture can be seen.

3.5.3 Permanent Monitoring and Alarm System

The permanent monitoring unit with alarm system was installed by the VCE measuring team on December 22, 2005. The measuring system was in permanent operation up to its removal on February 16, 2006. The system is based on the assumption that critical situations are recognized due to deformations of the pier or the structures.

3.5.3.1 System Description and Instrumentation

The permanent monitoring system is based on the BRIMOS technology in its composition and its function. Four highly sensitive acceleration sensors are used, which enabled very precise inclination measurements with an accuracy of less than 0.001°. The sensors were installed at the following points:

- Damaged pier, upstream (Figure 3.44).
- Damaged pier, downstream.
- At both superstructures supported by pier 5 at a distance of approx. 20 m of the pier.

The measurement unit, consisting of uninterruptible power supply, cabling, data logger, PC and modem, was accommodated in a heated container at the left shore directly beside the bridge.

After installation the proper function of the system was tested by manual inclination of the sensors. Figure 3.45 shows the results of this test: the raw signal; the inclination and the trigger criterion with the threshold values.

Figure 3.44 Sensor on the damaged pier

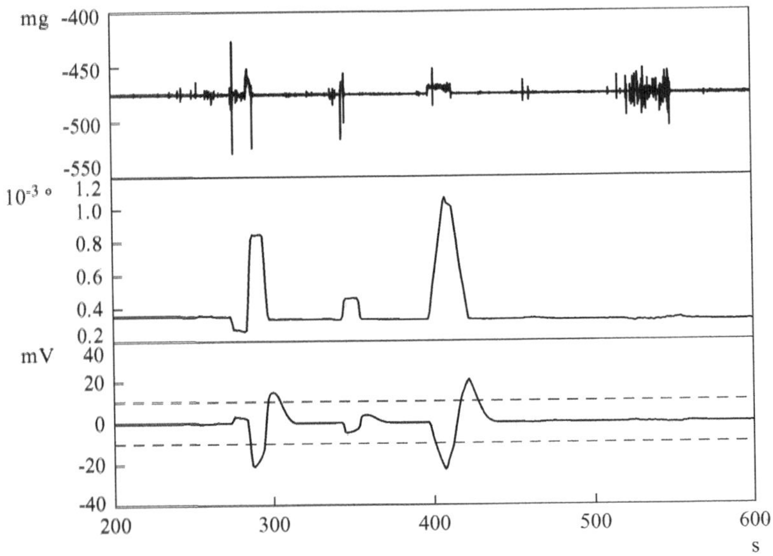

Figure 3.45 Test of the monitoring system

3.5.3.2 Evaluation and Alarming

The evaluation of all measuring data and the checking for fixed threshold values was performed by software specially created for this purpose on site in real time. The permissible deformation was established by the soil expert on site at 1 cm. Due to the geometry, a threshold value of 0.1 was determined for alarming.

The "normal" inclination curve over the day due to temperature fluctuations and solar radiation had to be considered. From the observation over two days, from December 22 to 23, 2005, this inclination curve caused by temperature expansions was determined. For monitoring and alarming, two different situations had to be taken into consideration:

1. **Nonworking phases.** During the nonworking phases the bridge was monitored in a completely automated way without any personnel. If previously defined threshold values were exceeded, a defined group of persons was automatically alarmed via SMS.
2. **Securing works.** During the securing works, interruption of the highly sensitive acceleration sensors was a danger. This could have led to the threshold values being exceeded and the triggering of false alarms. Therefore the automated alarm system was deactivated during the working phases. In order to be able to identify critical situations on time and to warn the project engineers on time, the measuring devices were monitored by engineers on site.

For the monitoring system a permanent function control was established. For any malfunction, such as a power failure, an automatic alarming of an available measurement engineer was planned via SMS. For the bridging of short-term power failures an uninterruptible power supply (UPS) was installed.

3.5.3.3 Results of the Permanent Monitoring

All recorded measuring data were saved on site and subsequently evaluated and graphically processed. Figure 3.46 shows typical results of permanent monitoring. The influence of insolation on temperature expansion and the subsequent pier inclination can be clearly recognized. In Figure 3.47 the slight

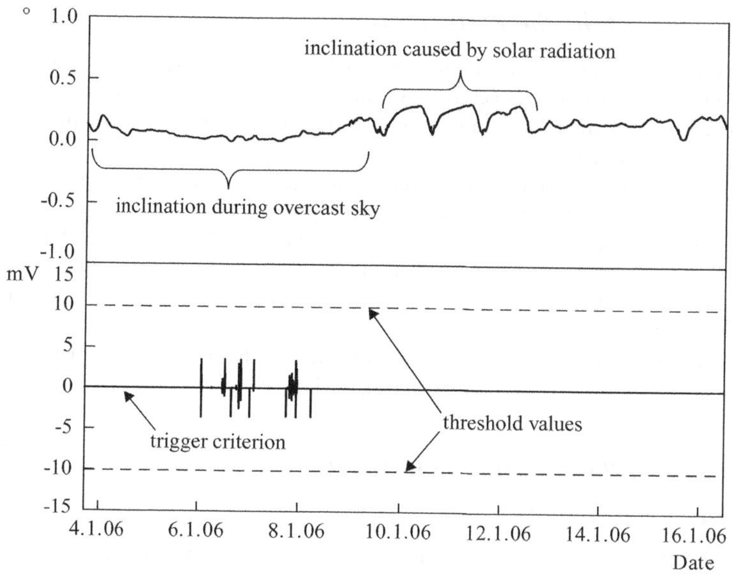

Figure 3.46 Typical monitoring results from the damaged pier

movement caused by loosening of the rail fastenings and cutting of the rails can be seen. On January 31 a sensor was knocked over during the cutting of the cable trough. This led to all threshold values being exceeded. The slight change in position of the sensor during the resetting was expressed by a displacement in the measurement record.

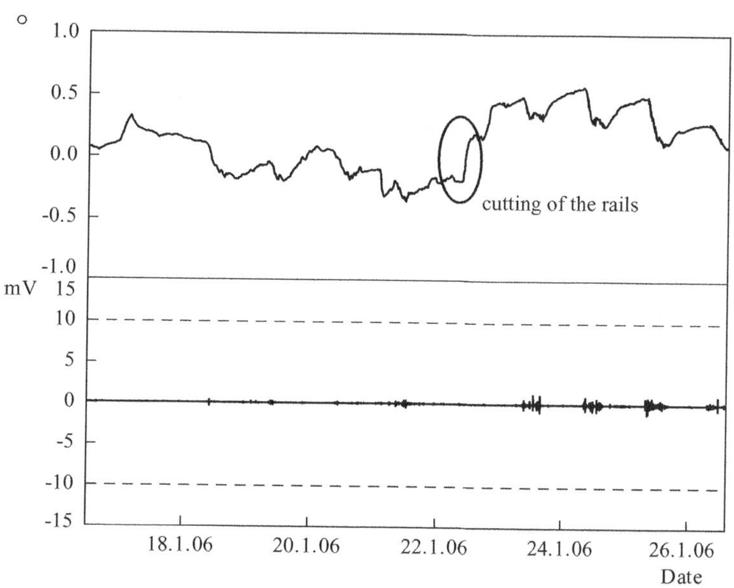

Figure 3.47 Inclination during cutting of the rails

3.5.4 Assessment of the Superstructures

As explained in detail earlier, statements on the actual load-bearing behavior can be made by means of determination of the dynamic characteristic on the basis of vibration measurements under ambient conditions. Forecasts on the future development and residual life expectancy of structures become possible thanks to periodic measurements.

The monitoring and alarm system could be used for determination of the dynamic characteristic of the railway bridge over the Danube in Krems. Two highly sensitive 3D acceleration sensors of the measurement unit were placed at the two superstructures supported by the damaged pier at a distance of approx. 20 m from the pier. At both sensor positions the dynamic characteristic was determined based on the measurement data. The frequency spectra with the respective structural eigenfrequencies reflect the condition of the structure.

Any changes in the load-bearing behavior are documented by a check after the remounting of the superstructures. This check can also serve as a basis for later condition assessments.

3.5.5 Summary of the Measurement Results

3.5.5.1 Results of the Stability Check

On December 21, 2005 the vibration characteristic of damaged pier 5 and undamaged pier 4 was measured by means of a mobile measurement unit. Evaluation of the measurement data in the frequency and time range and a comparison of the results enabled conclusions to be made on the stability of the damaged pier.

The monolithic behavior of the displaced pier shaft was assessed by the check. From the measurement results no signs of any brittle fracture could be recognized.

3.5.5.2 Results of the Permanent Monitoring

The permanent monitoring unit with alarm system was installed by the VCE measurement team on December 22, 2005. The measuring system was in permanent operation up to its removal on February 16, 2006.

The system is based on the assumption that critical situations are previously recognized due to deformations of the pier or the structures. Therefore damaged pier 5 and both structures supported by this pier were instrumented with highly sensitive sensors. From observation over two days, December 22 and 23, 2005, the "normal" inclination curve was determined over the day and subsequently threshold values were determined. If threshold values were to be exceeded, an automatic triggering of an alarm was established. Basically two situations could be distinguished:

1. Monitoring during the nonworking phases up to the lifting of the structures.
2. Execution of securing and dismantling works at the structure.

In the whole monitoring period the threshold values were never exceeded during the nonworking phases. During the works at the structure, SMS alarming was deactivated. The monitoring of the works and the observation of the monitoring system was carried out by personnel on site.

3.5.5.3 Results of the Measurements at the Superstructures

Two 3D acceleration sensors were mounted at the superstructures during the whole monitoring period. From the recorded measurement data the dynamic characteristic of the superstructures was established.

The latter reflects the condition of the structure. Changes in the load-bearing behavior can be documented by checks after the remounting of the structures.

3.5.6 Final Comments

The monitoring system with SMS has proved a success in the configuration used. The measurement unit worked faultlessly and without any failures during the whole monitoring period. The additional presence of measurement personnel during the working phases at the structure (securing works, dismantling, etc.) has proved appropriate and necessary as the trigger level for alarming was repeatedly exceeded by the works. The system with the sensors installed directly at the pier and at the structures supplied accurate and reliable values. In addition the measurement data recorded on the structures reflected the current condition of the structures before lifting. Any changes can be documented by a check after remounting.

3.6 Man-Made Hazards

While seismic hazard mitigation for bridges has developed over many years into an established science, blast mitigation research for bridge structures has just commenced. CALTRANS (California Department of Transportation) and the University of California at San Diego (Powel Structural Research Laboratories) have performed a series of bridge blast performance characterization tests on reinforced concrete bridge decks. Another test series has been performed on orthotropic steel bridge decks. The results of these two tests are representative of large, heavily utilized bridge structures in California as well as many other locations around the world.

Testing has shown that retrofit technologies developed for seismic mitigation have proven very effective in mitigation blast damage, in particular those using carbon-reinforced fiber polymer wrapping. Comparison between seismic and blast loading is providing an insight into what other technologies are successful in mitigating multiple structural hazards. Considerable research and development work is ongoing in the USA on this subject and new knowledge will appear soon. The tests performed by the University of California, at San Diego have shown that it takes a considerable mass of explosives to introduce critical damage to a bridge structure. A bomb of 10 kg of TNT exploded on the top of a bridge only causes some spalling of concrete at the bottom of the deck plate. The upper side was almost unharmed. A bomb of 25 kg of TNT punched a hole into the bridge deck of diameter approximately 80 cm. Only a bomb size of 100 kg of TNT causes partial collapse of the bridge structure. The main difference to seismic loading is time. Earthquakes have a duration of 10–60 s, whereas a blast lasts only 2 ms. A blast test to cellular steel structures also showed that only parts of the structure are damaged, which does not easily lead to progressive collapse.

Blast mitigation research to date has shown that there are many similarities to earthquakes in the hazard and structural response. On the load side, both blast and earthquake loads are largely unknown in terms of location, magnitude, intensity, type, etc., and both have characteristics of rapid attenuation with distance from the source mechanism. In terms of consequences, both actions can result in progressive structural collapse, requiring redundant structural systems for mitigation. The critical local structural response, i.e., the performance of individual structural elements or members, is dominated by brittle shear failures in both cases and requires shear strength and ductility to prevent local failures. Finally the load–structure interaction can modify the force and deformation response significantly. Thus, based on the above-outlined similarities, a unified design and/or retrofit approach to multihazard mitigation seems reasonable. Table 3.1, taken from (Seible et al. 2006), presents a summary of the similarities between seismic and blast loads.

On the other hand, there are also significant differences in these two extreme event scenarios, such as the load duration and associated strain rate effects and the extent of initial damage or excitation. In

Table 3.1 Similarities between seismic and blast hazards

Seismic	Blast
Input loading history unknown (location, intensity, duration, frequency content, fling)	Input loading history unknown (standoff, charge weight, type modifications)
Attenuation of input	Attenuation of pressure
Strength required for shear	Strength required for shear
Deformation capacity required for flexure (ductility)	Deformation capacity required for flexure (ductility)

Table 3.2 Differences between seismic and blast hazards

Seismic	Blast
Entire system is affected	Local subassemblages are affected
Bridge response 1–2 s	Bridge component response 0.1–1 s
Strain rate effects can be neglected	Strain rate effects are significant
Duration 10–45 s	Duration 1–2 ms
	Containment required against fragmentation and punching shear

addition to strain rate effects, the blast load case also has a shattering or fragmentation effect on brittle materials for close-in charges, and mitigation measures require containment considerations, something not paramount in seismic hazard mitigation.

Table 3.2 presents a summary of the differences between seismic and blast hazards. Both the similarities and the differences should be further explored to form the basis for multihazard mitigation in bridge structures.

Further Reading

1998-1 EOE (2005) *Auslegung von Bauwerken gegen Erdbeben - Teil 1: Grundlagen, Erdbebeneinwirkungen und Regeln für Hochbauten.* Technical report, Österreichsiches Normungsinstitut.

4015 OB (2002) *Belastungsannahmen im Bauwesen - Außergewöhnliche Einwirkungen - Erdbebeneinwirkungen - Grundlagen und Berechungsverfahren.* Technical report, Österreichisches Normungsinstitut.

Ambraseys NN, Simpson KA and Bommer JJ (1996) Prediction of horizontal response spectra in Europe. *Journal of Earthquake Engineering and Structural Dynamics* **25**, 371–400.

Boore DM, Joyner WB and Fumal TE (1997) Equations for estimating horizontal response spectra and peak acceleration form western North America earthquakes: a summary of recent work. *Journal of Seismological Research Letters* **68**(1), 128–153.

Bormann P, Parolai S and Milkereit C (2004) *Erdbebenmikrozonierung zur Kartierung standortspezifischer Erschütterungsübertragung.* Technical Report ISSN 1610-0956, TP B2.1 Scientific Technical Report for DFNK (Deutsches Forschungsnetz Naturkatastrophen).

Conrad O (1998) *Ableitung hydrologisch relevanter Reliefparameter aus einem digitalen Geländemodell (am Beispiel des Einzugsgebietes Linnengrund / Kaufunger Wald).* Master's thesis, Institut für Geographie, Georg-August-Universität zu Göttingen.

Cornet F, Helm J, Poitrenaud H and Etchecopar A (1997) Seismic and aseismic slips induced by large scale fluid injections. *Pure and Apllied Geophysics* **150**, 563–583.

Crozier MJ (1986) *Landslides – Causes, Consequences and Environment.* Croom Helm, London.

Douglas J (2001) *A Comprehensive Worldwide Summary of Strong-Motion Attenuation Relationships for Peak Ground Acceleration and Spectral Ordinates*. Engineering Seismology and Earthquake Engineering ESEE Report No. 01-1, Department of Civil and Environmental Engineering, Imperial College, London.

Duval AM, Chatelain JL, Guilier B and the SESAME WP02 Team (2004) Influence of experimental conditions on H/V determination using ambient vibrations (noise). *Proceedings to ICSDEE & ICEGE 2004*, Berkeley, CA (ed.).

Erdik M and Swift-Avci J (1997) Utilitizing GIS for earthquake damage scenario development. In *Radar-Interferometrie zur Messung der Erdkrustendynamik* (ed. Massonet D). *Spetrum der Wissenschaft* **9**, 56–65.

Fabbri A and Chung C (2001) Spatial support in landslide hazard predictions based on map overlays. *Proceedings of International Association for Mathematical Geology Annual Meeting*, Cancun, Mexico.

Fäh D and Bachmann H (2000) Earthquake scenarios for Switzerland. *Proceedings of the EuroConference on Global Change and Catastrophe Risk Management: Earthquake Risks in Europe*, Laxenburg, Austria.

Franzke HJ, Werner W and Wetzel HU (2003) Die Anwendung von Satellitenbilddaten zur tektonischen Strukturanalyse des Schwarzwaldkristallins und des angrenzenden Oberrheingrabens. *Jahresheft des Landesamts für Geologie, Rohstoffe und Bergbau, Baden-Würtemberg* **39**, 25–54.

Für Wasser und Geologie BB (2004) *Verfahren zur Erstellung und Verwendung von Mikrozonierungsstudien in der Schweiz*. Technical Report Bestellnummer 804.806.d.

Geyer O, Schober T and Geyer M (2003) *Sammlung geologischer Führer 94 - Die Hochrhein-Regionen zwischen Bodensee und Basel* xi edition edn. Gebr. Bornträger, Berlin - Stuttgart.

Giardini D, Wiemers S, Fäh D and Deichmann N (2004) *Seismic Hazard Assessment of Switzerland*. Technical Report, Swiss Seismological Service, ETH Zürich.

Grünthal G (1998) *European Macroseismic Scale 1998 (EMS-98)*. European Seismological Commission, Subcommission on Engineering Seismology, Working Group Macroseismic Scales, Luxembourg.

Grünthal G, Mayer-Rosa D and Lenhardt W (1998) Abschätzung der Erdbebengefährdung für die D-A-CH-Staaten Deutschland, Österreich, Schweiz. *Bautechnik* **75**(10), 753–767.

Gupta R (1991) *Remote Sensing Geology*, 1st edn. Springer Verlag, Heidelberg.

Havenith H, Roten D, Fäh D and Giardini D (2006) *Wave Velocities from the Measurement of Ambient Vibrations on a Small Scale Seismic Array*. Technical Report, http://www.seismo.ethz.ch/hazard/risk/flyers/WaveVelocities.html.

Hinsch R and Decker K (2003) Do seismic slip deficits indicate an underestimated earthquake potential along the Vienna Basin Transfer Fault System?. *Terra Nova* **15**(5), 343–349.

Idriss IM (1993) *Procedures for Selecting Earthquake Ground Motions at Rock Sites*. Technical Report NIST GCR 93-625, National Institute of Standards and Technology.

Káarnik V (1996) *Seismicity of Europe and the Mediterranean*. Academy of Sciences of the Czech Republic StudiaGeo and Geophysical Institute Praha.

Kind F, Fäh D, Zechner E, Huggenberger P and Giardini D (2002) Seismic zonation from a 3D seismic velocity reference model of the area of Basle, Switzerland. *EGS XXVII General Assembly, Abstract 720*, Nice.

Knödel K, Krummel H and Lange G (2005) *Handbuch zur Erkundung des Untergrunds von Deponien und Altlasten - Band 3: Geophysik*. Springer-Verlag, Berlin.

Kramer S (1996) *Geotechnical Earthquake Engineering*. Practice Hall.

Landesamt für Geologie, Rohstoffe und Bergbau BW (1998) Geowissenschaftliche Übersichtskarten von Baden-Würtemberg, 1:350,000, 20 landesweite Karten für Planung, Wirtschaft und Umwelt, CD.

Lang K (2002) *Seismic vulnerability of existing buildings*. Dissertation No. 14446, ETH Zürich.

Lee S and Digna GE (2005) Landslide susceptibility mapping using probability and statistics models in Baguio City. *31st International Symposium on Remote Sensing of Environment*, Saint Petersburg, Russia.

Merz B and Apel H (2004) *Risiken durch Naturgefahren in Deutschland*. Technical Report, Abschlussbericht des BMBF-Verbundprojektes Deutsches Forschungsnetz Naturkatastrophen (DFNK); Geoforschungszentrum Potsdam.

Mälzer A, Rösch H, Misselwitz I, Ebert M and Moosmann D (1988) *Höhenänderungen in der Nordschweiz und im Südschwarzwald bis zum Bodensee*. Technical Report NTB 88-05, NAGRA, CH-Baden.

Nakamura Y (1989) A method for dynamic characteristics estimation of subsurface using microtremor on the ground surface. *Quarterly Report of the Railway Technical Research Institute* **30**(1), 25–33.

Neuhäuser B (2005) *GIS-Gestützte, Probabilistische Beurteilung der Gefährdung durch Massenbewegungen - Einsatz von Geoinformationssystemen (GIS) zur Multikriterien Beurteilung der Rutschanfälligkeit dargestellt am Beispiel der Schwäbischen Alp*. Master's thesis, Institut für Geographie und Angewandte Geoinformatik, der Universität Salzburg.

Plate EJ and Merz B (2001) *Naturkatastrophen. Ursachen-Auswirkungen-Vorsorge.*. E. Schweizerbardt'sche Verlagsbuchhandlung (Nägele u. Obermiller), Stuttgart.

Ponte S and Moccia A (1995) Validating a spaceborne SAR simulator by using SIR-C/X-SAR data. *46th International Astronautical Federation (IAF) Congress*, number IAF-95-B.6.03., Oslo, Norway.

Popescu M (1994) A suggested method for reporting landslide causes. *Bulletin of the International Association of Engineering Geology* **50**, 71–74.

Popescu ME 2002 Landslide Causal Factors and Landslide Remediatial Options *Proceedings of the 3rd International Conference on Landslide, slope stability and safety of Infrastructures*, Singapore.

Research project RISK-UE (2004) *An Advanced Approach to Earthquake Risk Scenarios with Applications to Different European Towns*. Technical Report, Project No. EVK4-CT-2000-00014, Energy Environment and Sustainable Development, European Union.

Research project SE (1998) *Computer Aided Reduction of Seismic Risk with Application in Existing Cities, Town Planning and Construction*. Final Report, Project No. EVN4-CT97-0588, SEISMOCARE.

Research Project SE (2004) *Site Effects Assessment using Ambient Excitations*. Final Report, Project No. EVG1-CT-2000-00026, SESAME, European Union.

Research Project WP12 SE (2006) *Guidelines for the Implementation of the H/V Spectral Ratio Technique on Ambient Vibrations – Measurements, Processing and Interpretation*. Deliverable d23.12, Project No. EVG1-CT-2000-00026, SESAME, European Union.

Research Project WP2 RISK-UE (2003) *An Advanced Approach to Earthquake Risk Scenarios with Applications to Different European Towns – WP2 Basis of a Handbook of Earthquake Ground Motion Scenarios*. Technical Report, Project No. EVK4-CT-2000-00014, Energy, Environment and Sustainable Development, European Union.

Rinaldis D, Berardi R, Theodulidus N and Margaris B (1998) Empirical prediction models based on a joint Italian & Greek strong-motion database: I, peak ground acceleration and velocity. *Proceedings of Eleventh European Conference on Earthquake Engineering*.

Ruff M (2005) *GIS-gestützte Risikoanalyse für Rutschungen und Felsstürze in den Ostalpen (Vorarlberg, Österreich)*. PhD thesis Fakultät Bau-, Geo- und Umweltwissenschaften, Universität Karlsruhe.

Sabetta F and Pugliese A (1987) Attenuation of peak horizontal acceleration and velocity from Italian strong-motion records. *Bulletin of the Seismological Society of America* **77**(5), 1941–1513.

Savvaidis P, Theilen-Willige B and Neuhäuser B (2006) *LANDSLIDE ALERT SYSTEM: Guidelines for the Standardization of Landslide – GIS Layer Structure*. Final Report of Integrated Optimization of Landslide Alert Systems (OASYS) EVGG1-2001-00061, VCE Holding GmbH, Vienna.

Schneider G (2004) *Erdbeben - Eine Einfhürung für Geowissenschaftler und Bauingenieure*. Spektrum Akademischer Verlag, Elsevier, München.

Seible F, Hegemier G, Wolfson J, Conway R, Arnett K and Baum J (2006) Protection of our bridge infrastructure against manmade and natural hazards In *Bridge Maintenance, Safety, Management, Life-Cycle Performance and Cost* (ed. Cruz PJ, Frangopol DM and Neves LC), pp. 13–20. Taylor and Francis London.

Sieberg A and Lais R (1925) Das mitteleuropäische Erdbeben vom 16.11.1911, Bearbeitung der makroseismischen Beobachtungen. *Veröffentlichungen der Reichsanstalt für Erdbebenforschung* **H.4**, Jena.

Smit P (1998) Strong-motion attenuation model for central Europe. *Proceedings of the Eleventh European Conference on Earthquake Engineering*.

Sponheuer W (1960) Methoden zur Herdtiefenbestimmung in der Makroseismik. *Freiberger Forschungsheft C88*. Akademie-Verlag, Berlin.

Steimen S, Wössner J and Fäh D (2005) *Geophysikalischer Feldkurs - Natürliche Bodenunruhe als Instrument der seismischen Mikrozonierung*. Technical Report, Lecture Notes, Geophysikalischer Grundkurs - ETH Zürich.

Teledata GGs (2003) *ALPSLOPE Monitoring of Unstable Slopes in the Italian Alps Based on Remote Sensing*. Executive Summary of DUP – Small Service Project, ESA//ESRIIN Contract No. 15646//01//II-LG, Teledata GeoConsult GmbH-srl, Bozen, Italia.

Theilen-Willige B (1999a) *Erdbebengefährdung im Bodenseegebiet - Abschlussbericht für das XEP-Projekt (Erstellung beispielhafter X-SAR-Produkte) des DFD/DLR*. Technical Report, Oberpfaffenhofen, CD-ROM.

Theilen-Willige B (1999b) *Erdbebengefährdung im Bodenseegebiet - Fernerkundungsmethoden bei der Erfassung von untergrundbedingten Effekten bei Erdbeben im westlichen Bodenseegebiet*. PhD thesis Technische Universität Berlin.

Theilen-Willige B (2000) Seismic hazard zoning based on evaluations of remote sensing data (LANDSAT TM-/SIR-C/X-SAR-Radar) of the Lake Constance area/ southwest Germany in comparision with field check. *Zeitschrift Photogrammetrie - Fernerkundung - Geoinformation (PFG)* **1**, 19–32.

Theilen-Willige B (2002) Beitrag zur Fernerkundung zur Erdbebenvorsorge - Fernerkundungsmethoden bei der Erfassung von durch Erdbeben und durch Erdbebenfolgeschäden gefährdeten Bereichen. In *Proceedings zum Symposium Naturkatastrophen in Mittelgebirgsregionen* (ed. Fiedler F), pp. 245–270. Verlag für Wissenschaft und Forschung GmbH, VWF, Berlin.

Theilen-Willige B (2005) *Remote Sensing and GIS Contribution to Seismic Microzonation Studies in NE-Austria - Vienna Area.* SEISMID Project Report.

Theilen-Willige B and Neuhäuser B (2006) *Remote Sensing and GIS Contribution, GIS Integrated Geologic Evaluations of Remote Sensing – Data from the Test Sites for the Detection and Monitoring of Landslides.* Final Report of Integrated Optimization of Landslide Alert Systems (OASYS) EVG1-2001-00061, VCE Holding GmbH, Vienna.

Travasarou T, Bray JD and Abrahamson NA (2003) Empirical attenuation relationships for Arias Intensity. *Journal of Earthquake Engineering & Structural Dynamics* **32**(7), 1133–1155.

Wenzel H and Pichler D (2005) *Ambient Vibration Monitoring*. J. Wiley & Sons Ltd., Chichester.

Wood J (2004) *The geomorphological characterisation of digital elevation models*. PhD thesis, University of Leicester.

4

Damage Detection and Assessment

The various disciplines require different approaches in SHM. The first successful attempts have been taken by the mechanical engineers using analysis of vibrations for the prediction of damage of machinery. It is successful because the process is rather stationary and the options for change are only few. Another sector, namely the automobile industry, discovered the advantages of diagnosis systems, which nowadays are available in almost any new car. They benefit from the huge numbers produced with always the same objectives asked for, which allows a very fine-tuned process. In aerospace, SHM profits from the very well-defined material properties and geometries. Again the objectives are very similar from case to case and stable routines can be developed. The practice is focused on the detection of crack and fatigue indicators in aircraft bodies.

Bridges are individual structures that have very little in common with each other. Almost any new bridge is a prototype. The combinations of facts, use, properties, boundary conditions and geometry create a huge number of unknowns, therefore a uniform monitoring process is not feasible. On the other hand, in structural engineering the safety margins are higher to cover the unknowns. This allows methodologies that which can accept a certain limit of uncertainties. Monitoring-based assessment will always produce better results than conventional methodologies. The bridge engineer's mind is still guided too much by the idea of a perfect theoretical solution supported by today's computer programs. He therefore also asks this perfect solution from any assessment process. Considerable educational effort is necessary to support the requirements of realistic damage detection and assessment in bridge engineering.

There is a major difference between civil engineering and all the other disciplines. In mechanical engineering, automotive or aeronautics monitoring systems are designed to work permanently over the lifetime of a structure or component. In civil engineering this is not feasible financially and also we have to live with the fact that most of our structures exist already with a widely unknown history. Therefore the development has to go into monitoring campaigns first and only then can we think about permanent monitoring at those structures where a realistic problem has been detected. These facts explain the dominance of vibration-based methodologies in the SHM of bridges. In most cases the starting point is an isolated Ambient Vibration Monitoring campaign in order to identify the basic quality or eventual deviation from expectations. After system identification an eventual permanent system can be designed, which then uses the wide range of sensors and methods that are available now.

Health Monitoring of Bridges Helmut Wenzel

Figure 4.1 Europabrücke – overview

4.1 Weak Point Detection and Fatigue Assessment

Bridges are aging and traffic is growing, which creates a demand for accurate fatigue life assessment. The Europabrücke – a well-known Austrian steel bridge near Innsbruck, opened in 1963 – is one of the main alpine north–south routes for urban and freight traffic (Figure 4.1). It represents a bridge generation where bridge designers placed priority on building material economization. A long-term preoccupation of VCE with BRIMOS on the Europabrücke (since 1997) with regard to fatigue problems and possible damage led to the installation of a permanent monitoring system in 2003.

Since that time a lot of investigations and additional measurements have been devoted to innovative, mainly monitoring-based fatigue assessment. As lifetime predictions in modern standards depend on lots of assumptions, the emphasis is to replace those premises – referring to loading – by measurements. In that context the present work is focused on three levels:

Level I. Global behavior – primary load-bearing members (traffic loading observation mainly based on laser-supported global deflection measurement).

Level II. Cross-sectional behavior – dynamic truck-weight classification system, which utilizes laser-calibrated accelerometers reproducing vertical cantilever deformations based on pattern recognition.

Level III. Local behaviour – e.g. the bottom and top joints of the bridge's torsional bracings (verification with additional strain gauge measurements).

In each of these levels of analysis the consumption of the structure's overall capacity per year is to be determined by analyzing the relation between the randomly induced traffic loading (vehicles per day) and the fatigue-relevant, dynamic response of the structure, exclusively caused by freight traffic. An indispensable requirement is to reduce the permanent monitoring system's data by Rainflow Counting, describing the remaining fatigue-relevant recurring response cycles in different categories of intensity and occurrence. As the present lifetime calculations are performed in terms of stresses by means of damage accumulation, global and local Finite Element Analysis is necessary for the transition of measuring data. Detailed knowledge about the progression of the prevailing traffic from the very beginning up to the present and the implementation of published future trend studies for the next ten years can be used for an extrapolation of the measured impact for the whole lifetime. As this research work tries to encourage in situ measurements instead of "design situations," it also aspires to analyze the consequence of statistical scatter in each level of impact as well as for fatigue resistance.

4.1.1 Introduction

Previous measurements at the Europabrücke matched very well with the comparative analytical calculations, but they also exhibited a remarkable level of intensity of the loading impact. Currently the

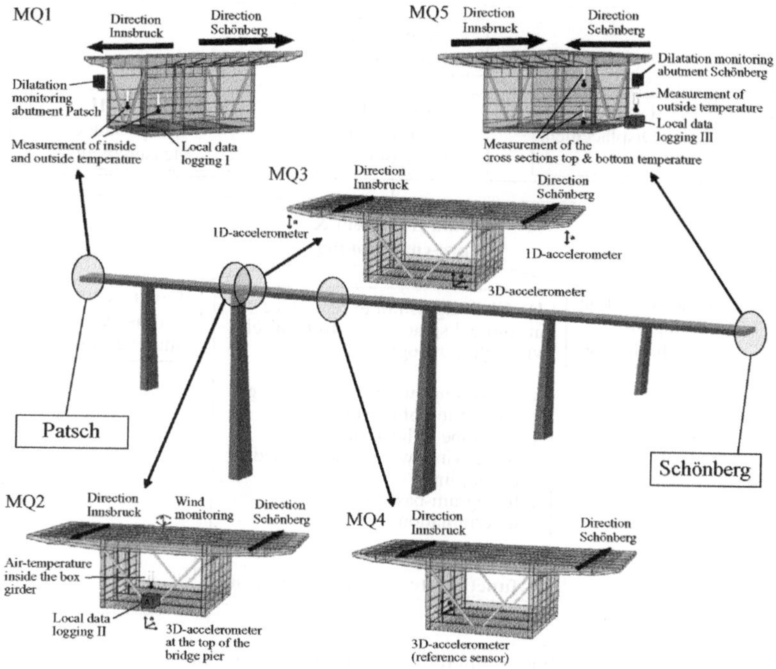

Figure 4.2 The permanent monitoring system and location of measurement points on the bridge

bridge is stressed by more than 30 000 motor vehicles per day (approximately 20% freight traffic). The superstructure is represented by a steel box girder (width = 10 m; variable height along the bridge length 4.70 – 7.70 m) and an orthotropic deck and bottom plate. This motorway bridge with six spans differing in length (longest span 198 m, supported by piers with an elevation of 190 m) and a total length of 657 m comprises six lanes, three for each direction distributed on a width of almost 25 m. To reach the already defined goals, a permanent monitoring system has been developed in a stepwise manner (see Figure 4.2). It consists of 24 measuring channels (sampling rate 100 Hz) representing the main span's, the pier's and the cantilever's accelerations, the abutment's dilatation, wind speed and direction and temperatures at several locations.

4.1.2 Methodology

The flowchart in Figure 4.3 gives an overview of the intended analysis. Even if the present methodology represents a tailor-made approach, it can be modified and transferred to other bridge structures.

First, parts of some generally applied methodology are mentioned, before further specifications are discussed.

4.1.2.1 Finite-Element-Based Stress Determination

As the present lifetime calculations are performed in terms of stresses (Stress-Life Approach) Finite Element (FE) analysis is necessary for the transition of measurement data. In the course of using shell elements for the analysis of welded components, two different methods of fatigue life prediction – the nominal stress approach and structural or geometric (hot spot) stress approach – are used and compared.

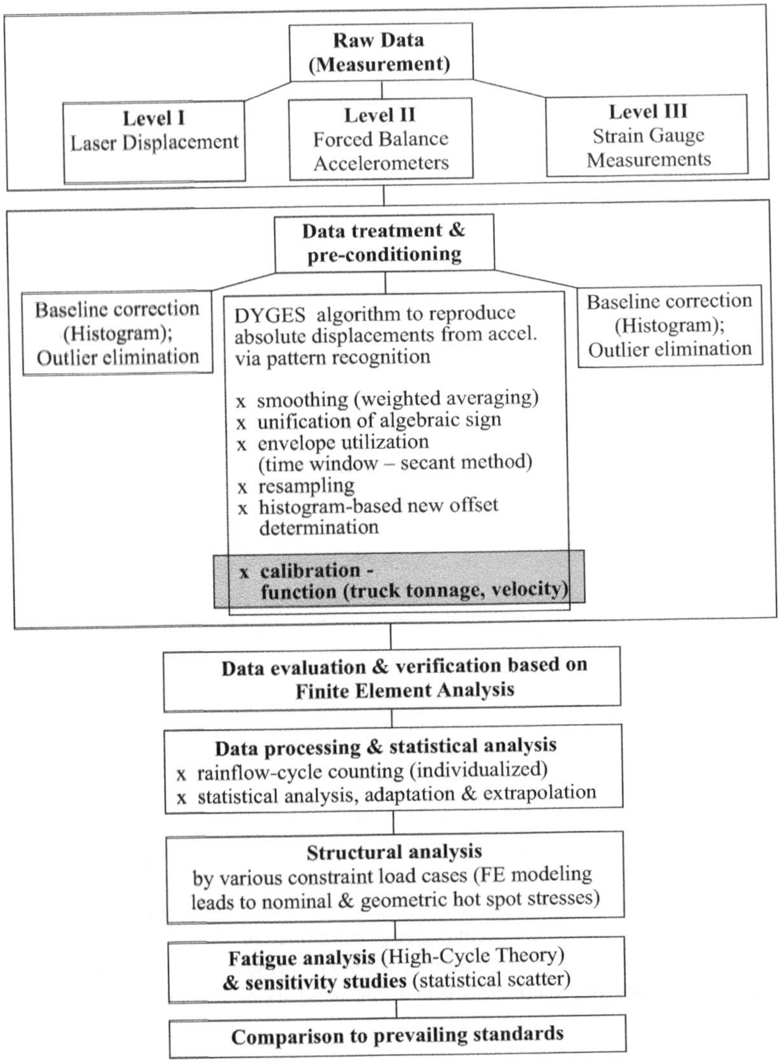

Figure 4.3 Detailed flowchart of the methodology

The nominal stress approach – for nonwelded or welded details in areas of potential crack occurrence – deals with stresses derived from simple beam models or from coarse-mesh FE models. The full-scale test-based curves take into account residual stresses, welding profile and imperfections in the material due to manufacturing. Stress concentrations resulting from the gross shape of the structure are also included in the nominal stress (Simonsen 2001). Possible imperfections of the structural members are considered by stress concentration factors.

Structural or geometric (hot spot) stresses consist of membrane- and shell-bending stress components (Figure 4.4). They include nominal stresses (stresses from structural discontinuities and presence of

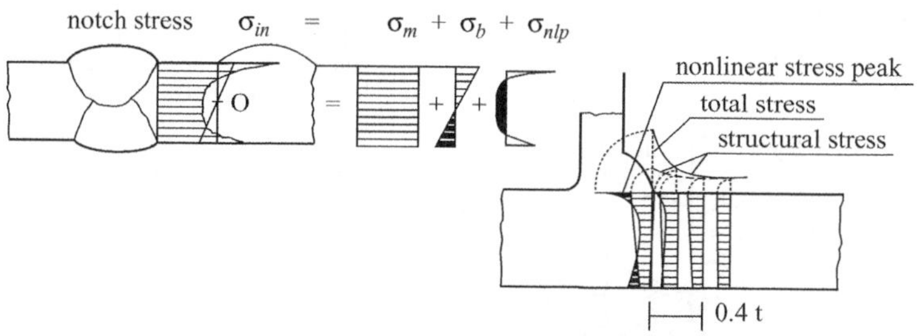

Figure 4.4 Variation in the through-thickness stress distribution approaching the weld toe (Niemi 2000)

attachments) but they do not include stresses occurring from the presence of welds (Simonsen 2001). As singularities at the weld toe are difficult to represent, FE modelling cannot directly give the actual peak stress at the weld toe. Therefore various types of stress extrapolation methods have been developed to overcome this problem (Figure 4.4).

At the beginning the structure is modeled with a coarse shell element mesh, therefore a refinement in the relevant hot-spot areas is necessary. The Hot Spot Designer's Guide published by Prof. Niemi (2000) defines certain directives that which are of crucial importance. At a distance of 0.4 t from the weld toe, the nonlinear component has practically vanished and the distribution is almost linear (Niemi 2000). Fatigue analysis of welded assemblies should be performed using the maximum principle stress range within ±60° of the normal to the weld toe or the axial stress component perpendicular to the weld toe. Depending the analyzed structural detail and the FE software used, an appropriate meshing procedure is developed. Then a linear or quadratic extrapolation can be applied to derive the hot-spot stresses on the plate surface using stress values from the calculated points (mesh) in certain distances to the weld toe (e.g. Figure 4.5).

A comparative lifetime calculation with nominal stresses is also to be carried out. Based on the suggestion of the authors of *Ermittlung von Dauerschwingfestigkeitskennwerten für die Bemessung von geschweißten Bauteilverbindungen auf der Grundlage örtlicher Strukturbeanspruchungen* (2002) these nominal stresses are identified at distance points twice the plate thickness away from the weld toe. It is necessary to bear in mind that fatigue criteria originally were defined for uniaxial loading. In cases of complex loading (multi-axial, nonproportional) a fatigue criterion needs to be defined in order to apply fatigue curves obtained under uniaxial loading (des Industries Mechaniques CT 2001). Even if numerous criteria have been proposed for understanding multi-axial fatigue, these criteria reflect only the behavior of unwelded zones. In the current state of knowledge, the best criterion for the present work to describe the cracking of welded zones is the principal stress range, as long as the directions of the principal stress do not vary in cyclic loading cases.

Randomly influenced load–time histories lead to varying mean stresses. For nonwelded details the endurance is reduced as the mean stress becomes more tensile. In these cases empirical corrections (Goodman and Gerber theory as an alternative) are applied. Cracks in compression zones tend to arrest but are typically not structurally significant. Thus members or connections for which the stress cycle is at least partially in tension are currently required to be observed under consideration of fatigue tasks.

For welded details the endurance is usually not reduced. Due to shrinkage effects, which are locked into the weld regions at fabrication, the weld's residual stresses often reach the tensile yield region of the material. This is the reason why fatigue analysis of welded assemblies normally has to be performed using the full stress range regardless of the mean stress during the cycle. The crack cannot distinguish between applied and residual stresses. Thus, for the purpose of analysis, the *S–N* (stress–number [of

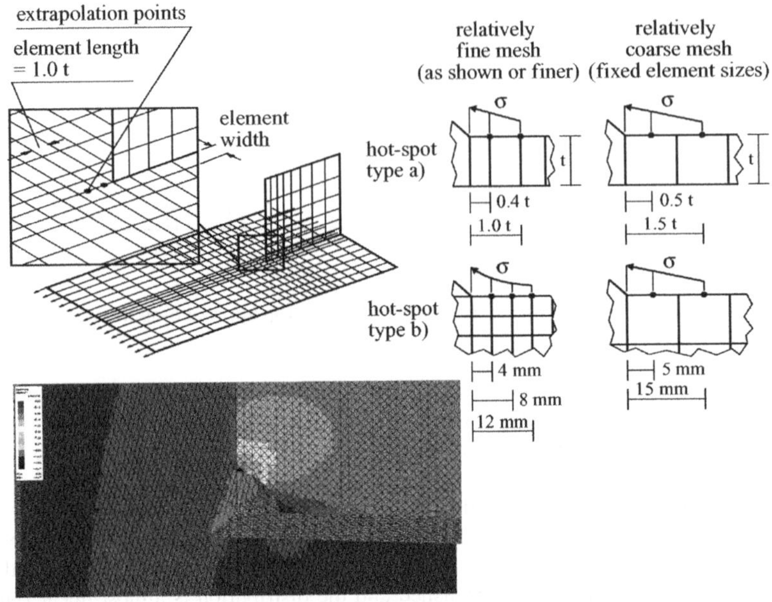

Figure 4.5 Meshing and structural (hot spot) stress evaluation using FE-based surface stress extrapolation (Hobbacher 2003)

cycles]; see Figure 4.6) curve always assumes the worst, i.e. that the maximum stress in the cycle is at yield point in tension. It is particularly important to exhibit this fact as it means that fatigue cracks can grow in parts of members that are nominally "in compression" (Leuven undated). This means that the real mean stress occurring in welded details remains independent of the applied mean stress – its effect is included in the fatigue curves for welded assemblies (des Industries Mechaniques CT 2001). For the same reason dead load effects are not significant when analyzing the fatigue lifetime of welded details.

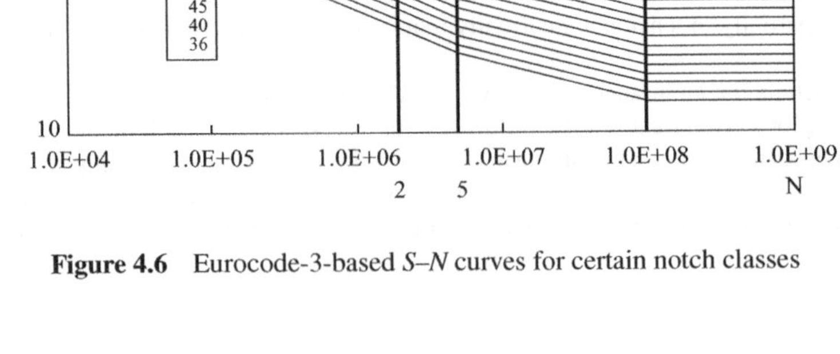

Figure 4.6 Eurocode-3-based *S–N* curves for certain notch classes

4.1.2.2 Stress-Life Approach and Damage Accumulation

Fatigue analysis by comparing the number of loading cycles n_i on a certain stress-range level $\Delta\sigma$ to an allowable number of cycles N has become a matter of course in civil engineering. Figure 4.6 shows Eurocode-3-based S–N curves according to fatigue tests of diverse constructional elements. They already include the impact of local notches due to welding, stress orientation, residual stresses, metallurgical conditions, etc. Each notch class is defined by the detail category $\Delta\sigma_c$ (belonging to two million cycles), the constant amplitude fatigue limit $\Delta\sigma_D$ (belonging to five million cycles) and the cut-off limit $\Delta\sigma_L$ (belonging to 100 million cycles). Stress ranges smaller than $\Delta\sigma_L$ are of no fatigue significance.

Cases of nonperiodic loading as in the present situation demand the well-known Modified Damage Accumulation Concept by Palmgren-Miner (Ramberger 1998):

$$D_i = \frac{n_i}{N_i}, \quad D_1 = \sum_{i=1}^{j} \frac{n_i}{N_i}, \quad D_2 = \sum_{i=j+1}^{z} \frac{n_i}{N_i}, \quad D = \sum_{i=1}^{z} D_i = D_1 + D_2. \tag{1}$$

All partial damages D_i of stress ranges corresponding to $m = 3$ are summarized in D_1, those corresponding to $m = 5$ are reflected in D_2, while stress ranges smaller than $\Delta\sigma_L$ are of no fatigue significance. The nonperiodic loading leads to the necessity of modeling Wöhler curves with two different slopes before reaching the cut-off limit. The decrease of slope between two and five million cycles complies with the modified Miner rule (Haibach 2002) and represents the reduced fatigue limit due to gradual damage in comparison to earlier stated cut-off limits.

Basically the fatigue process can be subdivided into two phases: initiation (i.e. development and early growth) and propagation (i.e. growth of a crack to failure). For the present research work the stress-life approach, dealing with structural (hot spot) Wöhler curves given for $N \geq 10^4$, is applied ("crack initiation" with service life to form "small cracks"). The fatigue threat – exclusively caused by truck traffic – demands High Cycle Fatigue Theory as it is best suited to long-life applications. It does not distinguish between initiation and propagation phases, but deals with the total lifetime. Linear elastic material behavior can be assumed, as the structural hot-spot stress range should not exceed twice the yield strength of the material (Hobbacher 2003). Stresses and strains are assumed to remain elastic. Failure is predicted if $D \geq 1$ (inception of a visible crack but the crack growth itself is not accepted). The Eurocode 3 Annex A2 (*Fatigue Strength of Steel Structures* 2002) includes suitable provisions for which of the classified S–N curves (Figure 4.6) can also be used when applying the geometric stress approach as an alternative to the curves published in Haibach (2002).

These standardized stress-life curves are based on functions derived from experimentally obtained mean values m_x. This means that these data are scattered. Fatigue – a serviceability limit state – is therefore defined as exceedance of 5% probability of failure x_5, calculated with Gaussian distribution. As this research work tries to encourage in situ measurements instead of "design situations," it is aspired to analyze the consequence of statistical scatter on fatigue resistance.

Many methods for the statistical evaluation of fatigue resistance are available. The authors have chosen an approach according to Spaethe (1992), who recommends presupposing a log-normal distribution for stresses at constant fatigue life, with a certain variance V_x increasing with lower notch classes. As fixed variances from standard values are used, no confidence level has to be considered (Spaethe 1992). The characteristic values x_p, which vary statistically at a specified fractile, are determined in two steps (see Equations 2 and 3 respectively):

$$x_p = m_x \pm k_p\sigma_x = m_x(1 \pm k_p V_x) \tag{2}$$

$$x_5 = m_x - k_5\sigma_x = m_x(1 - k_5 V_x) \Rightarrow m_x = \frac{x_5}{(1 - k_p V_x)} \tag{3}$$

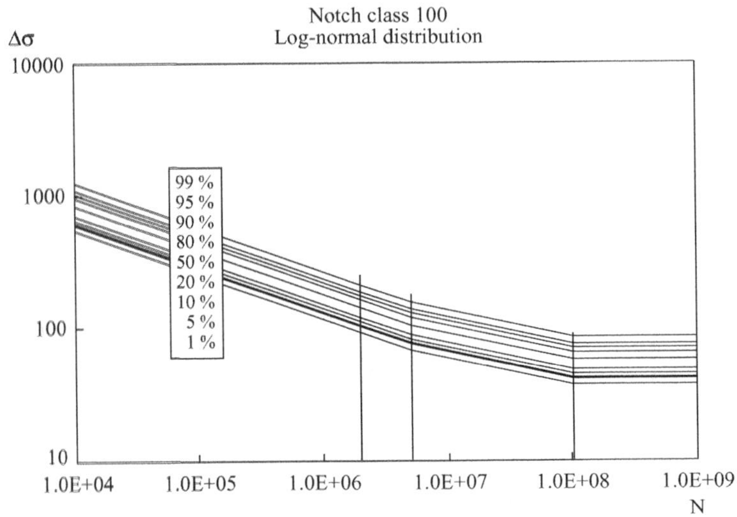

Figure 4.7 Statistical scatter of *S–N* curves

The use of a positive sign in Equation (2) leads to characteristic values from 50 to 100%, whereas a negative sign is used to calculate characteristic values x_p between 0 and 50% probability of failure. k_p represents the standardized, so-called fractile factor, and σ_x is the standard deviation.

$$\left.\begin{aligned} \sigma_u &= \sqrt{\ln(1+V_x^2)} \\ m_u &= \ln m_x - \tfrac{\sigma_u^2}{2} = \ln\left(\frac{m_x}{\sqrt{1+V_x^2}}\right) \end{aligned}\right\} \Rightarrow x_p = \exp(m_u \pm k_p \sigma_u) \tag{4}$$

After the reference value m_x has been derived, Equation (4) describes the transition from Gaussian distribution to log-normal-distributed characteristic values of x_p for an appropriate description of scattered Wöhler curves (Figure 4.7):

4.1.2.3 Rainflow Algorithm

The elapsed time of the structure's response due to randomly induced traffic loading is recorded by high precision sensor data. An indispensable requirement is to reduce the enormous amount of information from the permanent monitoring system to a few statistical parameters for further assessment. The Rainflow Counting Method reduces the sensor data's complete loading time-history represented by random sequences of peaks and valleys to a set of fatigue-relevant recurring response cycles in different categories of intensity and occurrence and has become state of the art in fatigue analysis related to nonperiodic loading.

The analyzed random time series may be considered as matching pairs of reversals: the reversal from the maximum to the minimum of the assessed signal and the reversal from the minimum to the maximum, with all the other reversals effectively interrupting these two. This phenomenon is often called the "material memory." A material subjected to a sequence of reversing loads apparently interprets each closed cycle (matching pair of reversals) as a temporary interruption of a larger strain range, and remembers which complementary hysteresis part applies for this larger event.

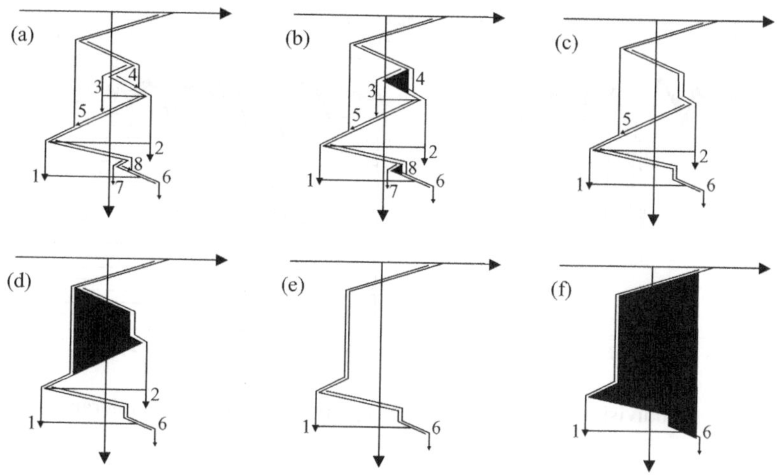

Figure 4.8 Rainflow counting – example

The algorithm used by the author is based on Naubereit and Weihart (1999) but includes certain specifications, which are listed in the following:

- The assessment starts by determining the range of evaluation for every analyzed signal, depending on its maximum absolute value. This range is defined from this maximum value to the same value with the opposite algebraic sign. In this way the look of the correlation matrix indicates its qualitative plausibility automatically. Before the counting procedure starts the whole signal is subdivided into constant sections.
- The explanation continues with an exemplary displacement time-history, which has to be analyzed. After the signal is rotated 90° in the clockwise direction, an imaginary flow of water is initiated in every peak and valley. For a better understanding we focus on a short sequence in Figure 4.8. We follow a certain flow (specified as number 2) until it experiences a drop.
- If the observed flow intersects a second flow originating from a peak or valley of a smaller absolute value than the origin of the first flow, then a cycle can be counted. The cycle ranges between the values of the point of intersection and the origin of the second flow (Figure 4.8b – closed loops 3, 4 and 7, 8 are identified and saved).
- If the second flow starts at a peak or valley of a larger absolute value than the origin of the first flow, a cycle can also be counted. In this case however the cycle ranges between the values of the origin of the first flow and the drop-off point (Figures 4.8d and 4.8f – closed loops 2, 5 and finally 1, 6 are identified and saved).
- Once a cycle is counted, its data points are imaginarily removed from the graph and the counting process continues.
- All hystereses within a certain subdiagonal have the same amplitude. Fatigue relevance increases when the analyzed diagonals are further from the main diagonal.
- Pairs of reversals are stored until a complementary reversal is identified. Finally the residual of the remaining reversals is registered in the Rainflow Matrix in the same manner as closed cycles. The intention is to counteract a possible impreciseness of the counting procedure, which strongly depends on the stated subdivision into constant sections.
- Existing cycles that do not exceed the predefined section margin are ignored.

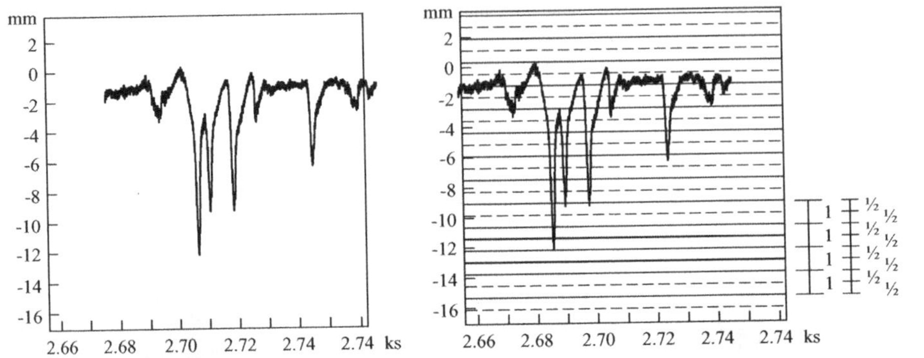

Figure 4.9 Subdivision of the displacement signal into sections with tolerance envelopes

- Each section limit (continuous lines) has an imaginary envelope represented by half of the class size (Figure 4.9). This means that local extreme values within this "tolerance envelope" (dashed lines) are assigned to its corresponding section limit.
- Optionally – depending on the analysed signal and the required accuracy of evaluation – 20 × 20 matrices (20 sections) or 30 × 30 matrices are typically used.
- Summation within a matrice's row provides local minima values per section (Figure 4.45).
- Rainflow Matrices for a certain timeframe contain no more information of order or variation in time.

As a result we obtain a counting matrix that shows the frequency of occurrence of closed cycles n_i from one certain level of displacement to another. By means of Rainflow counting the traffic loading becomes anonymous – its results are not directly comparable to conventional traffic counting anymore. Figure 4.30 shows a possible result of the whole procedure corresponding with Figure 4.29.

A transition from the Rainflow Matrix to a so-called Damage Matrix is necessary to follow up the consequence of fatigue for each analyzed detail (Veit and Wenzel 2006). Depending on the measured timeframe the derived Rainflow Matrix is extrapolated to a period of one year. After having assigned the area of interest to a certain notch class (*S–N* curve), the significance of all elements of the matrix in terms of stresses can be determined by its partial damages D_i, accumulating to a total damage D per year. Thus, the unit of the vertical axis of the Damage Matrix, originally defined in terms of frequency of occurrence, is replaced by damage rate r (Equation 5):

$$r = 100\frac{D_i}{D} \tag{5}$$

To get a better understanding of every element's damage relevance r, Equation (5) is used to transform the structural cycle-counting matrices (Figure 4.45) into damage assessment or damage quantification matrices (Figure 4.48).

Adaptation of Rainflow Matrices

The derived counting matrices have to be adapted for reasons of necessary extrapolation and superposition. The Europabrücke was subsequently equipped with the described permanent monitoring system. Even if this facility provides performance data over a long period of time, this timeframe is still relatively small compared to the bridge's overall lifetime. This is the reason why the derived counting matrices have to be adapted to time segments outside the monitoring period. Following this intention the changing operational conditions in terms of varying traffic volume, as well as varying effective tonnage, can be taken into consideration.

Studies such as the one by Drešler et al. (1996) have shown that an extrapolation of Rainflow Matrices – using detailed knowledge about the progression of traffic volume – by uniform scaling of the number of occurrences is not adequate for lifetime estimations. The authors, advanced approach considers a method developed for use in the Rainflow domain based on Drešler et al. (1996), which covers the remaining scatter of prevailing loading spectra – assuming that the physical process corresponding to the extrapolated matrix is not of a significantly different type than the process generating the initial matrix.

The concept for adapting counting matrices uses nonparametric density estimators. A certain set of measured discrete points is described by a function $f(x)$ that depends on data values Y_i measured at points x_i $(i = 1, \ldots, N)$ with a random error ε_i:

$$Y_i = f(x_i) + \varepsilon_i \tag{6}$$

where $f(x)$ is estimated by the mean value and weighted by the function's smoothness:

$$\overline{f(x)} = \frac{1}{N}\sum_{j=1}^{N} w_j(x) Y_j(x_i) \tag{7}$$

In one-dimensional cases, the Gaussian kernel estimator (Equation 8) would be used:

$$k_\sigma(u) = \frac{1}{2\pi\sigma} e^{-\frac{u^2}{2\sigma^2}} \tag{8}$$

However, Rainflow histograms have an arbitrary shape, which can only be described by a nonparametric approach. Equation (9) is introduced for two-dimensional purposes to transform the discrete Rainflow Matrix into a smooth function that is more readily accessible:

$$k_\sigma(u, v) = \frac{1}{2\pi\sqrt{\Sigma}} e^{-\frac{1}{2}[u,v]\Sigma^{-1}\begin{bmatrix} u \\ v \end{bmatrix}} \tag{9}$$

where $k_\sigma(u, v)$ describes the adaptive, two-variant density estimator (normal distribution, mean= 0; covariance matrix Σ), which considers the mechanical background of the Rainflow Matrices ("information from a certain level to another one") as well as the characteristics of fatigue relevance within a counting matrix:

$$\Sigma = \frac{1}{2}\sigma^2 \begin{bmatrix} \lambda^2 + 1 & \lambda^2 - 1 \\ \lambda^2 - 1 & \lambda^2 + 1 \end{bmatrix} \tag{10}$$

The choice of Σ reflects the special need that entries of the same subdiagonal should have a higher influence than those of other subdiagonals. As visible in Figure 4.10, the contour lines of the kernel estimator finally describe ellipses. Their principal axes are parallel to the subdiagonals and λ provides the ratio between the large and the small principal axis (Drešler et al. 1996). σ^2 is going to be chosen individually for each entry to guarantee that there are enough values to perform a statistically good averaging.

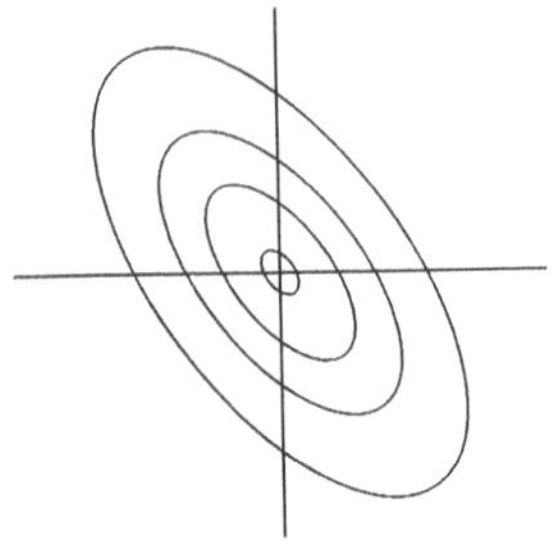

Figure 4.10 Some contour lines for $\lambda = 2$ (Dreŝler et al. 1996)

Rainflow Matrices – derived by an equal subdivision of the intensity of impact as well as by an equal assignment of entries – can be superposed linearly. Such Rainflow Matrices correspond to several loading histories, stringed together. This enables the authors to consider changing operational conditions, such as the variation of the loading level (notional truck weight), simply by rescaling the subdivision of the Damage Matrices.

4.1.3 Weak Point Determination

4.1.3.1 Objectives

In the course of previous investigations – and parallel to visual inspections – the steel bridge's torsional bracings and their joints turned out to be the main subject of interest (Figure 4.11), as their response is most sensitive to the global, freight-traffic-dominated impact. Over the years periodic bridge inspections were undertaken. With regard to the bracings it can be stated that the following type of deficiencies were most commonly identified (Figure 4.12): crack formation in the corrosion protection coat, loosening bolts and, more seldom, fatigue cracks, buckling and fracture of certain connecting plates.

The present section shows the results of an ambitious measurement campaign. All 144 torsional bracing beams – belonging to both driving directions – were monitored with accelerometers to assess their structural integrity (Figure 4.13). An effective and stable approach was chosen by comparing calculated, expected values of structural stiffness with the measured ones to localize weak points or weak areas within the whole structure as a basis for continuative performance analysis. The conception shown assures the determination and observation of slowly progressing processes in the structure, which might lead to local damage or to deterioration of the structure's operational integrity.

Figure 4.11 Design of the torsional bracing's bottom and top joints – Europabrücke

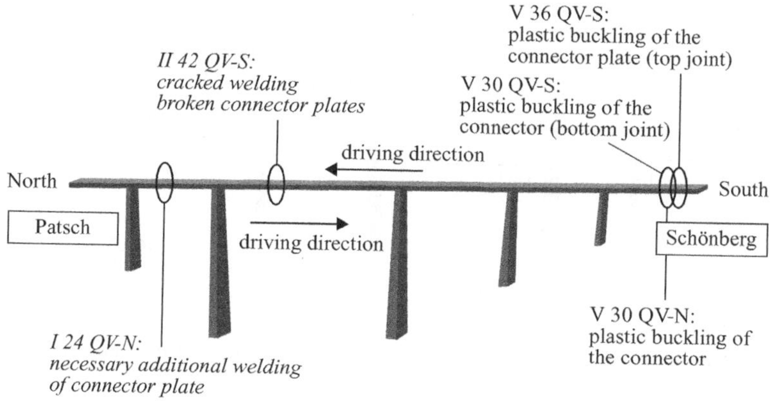

Figure 4.12 Deficiencies since 1983

The described monitoring investigation of all diagonal bracings was undertaken in order to determine the condition of maintenance of these structural members (their integrity) as well as the load-bearing capacity by means of BRIMOS. Along with the conventional bridge assessment this investigation supports the determination and location of potential problem zones based on the measured structure's vibration behavior. On the one hand the relevance of visible damage (cracks, buckled joint plates) was quantified (Figure 4.14). On the other hand locations or areas without visual indicators but already giving notice of potential local failure were identified and assessed by means of modal analysis.

The present investigation is to be understood as an initial measurement. Possible upcoming measurements are to be referred to this initial one and possible changes of the structure's operational integrity in the course of time can be quantified with this approach. The present monitoring campaign was also combined with an accompanying, visual inspection of the structure to document obvious damages or irregularities. Besides a few smaller fatigue cracks, two of the diagonal bracings are to be exhibited. In both cases there was evidence of explicit damage – driven by torsional overstressing – caused by freight traffic. According to the bridge's nomenclature the torsional bracing V 24 QV-N has been overstressed by dynamic torsional loading with remarkable axial force cycles, which led to a distinctive fatigue crack, growing progressively within the connection plate between the diagonal bracing and the orthotropic bottom plate (Figure 4.15).

The torsional bracing IV 30 QV-S has been excessively loaded by static torsional impact with remarkable axial forces, which led to a distinctive plastic buckling, weakening the connection plate

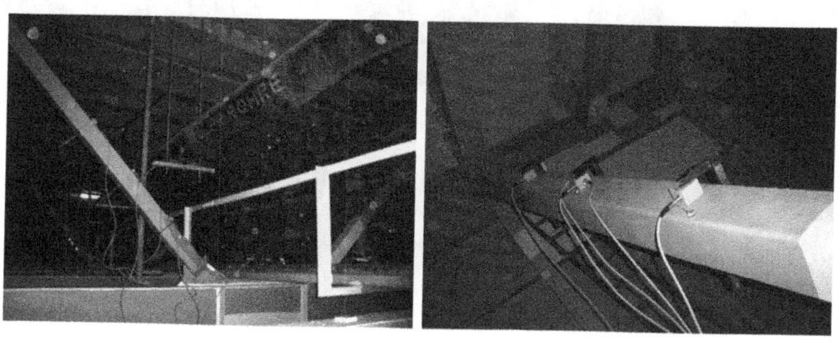

Figure 4.13 Accelerometer-based vibration monitoring of the torsional bracings

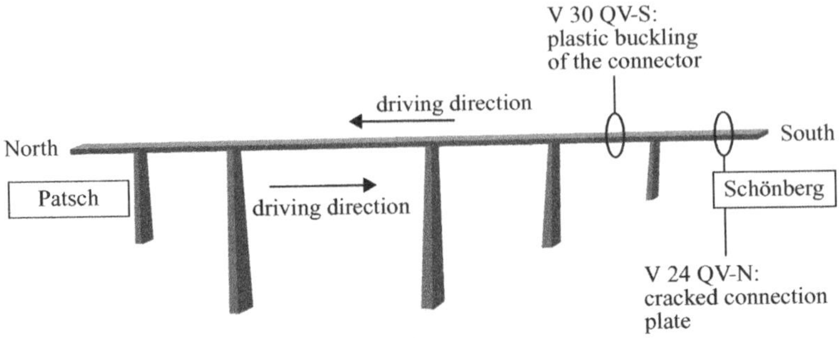

Figure 4.14 Deficiencies revealed in the course of the measurements in 2006

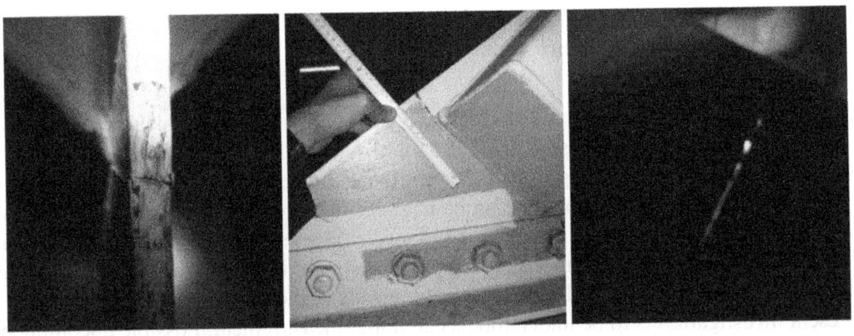

Figure 4.15 Fatigue crack (13 cm long), weakening the connection plate between the torsional bracing V 24 QV-N and the orthotropic bottom plate

between the diagonal bracing and the orthotropic bottom plate (Figure 4.16). Even if there is no evidence of this certain damage within the documentation of the periodic bridge inspections, it is possible that this buckling has its origin in a constraint stressing in the course of the mounting workings.

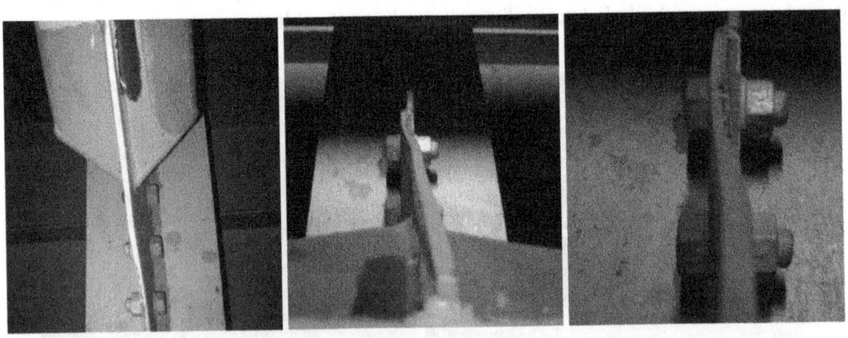

Figure 4.16 Plastic buckling, weakening the connection plate between the torsional bracing IV 30 QV-S and the orthotropic bottom plate

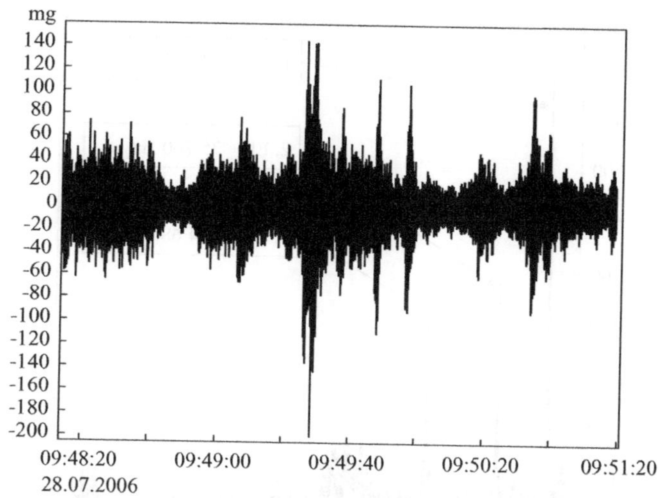

Figure 4.17 Effective vertical accelerations under ambient conditions at bracing II 42 QV-N

4.1.3.2 Methodology – Determination of Expected Values for the Bracing's Effective Stiffness

The measurements were made under ambient conditions (Figure 4.17) on the one hand (environmentally excited vibrations) and under normal site traffic, including the dominating affection by passages of heavy vehicles, on the other hand. The remarkable dynamic response of the bracings in the range of 0.1–0.2 g, which occurs quite frequently, is to be pointed out. The frequency spectra – representing the effective dynamic stiffness of the structural members – show a distinctive dynamic characteristic in all three analyzed dimensions. The entire investigation is focused on the diagonal's vertical direction (perpendicular to the traffic's driving direction).

The extracted eigenfrequencies (Figure 4.18) are assumed to be a stable and effective indicator for probable anomalies. In the present case the lower frequencies (the natural as well as the following two) are suitable to assess the integrity and functionality of the boundary conditions of the torsional bracings. As already stated, the design of the bracing's joints turned out to be the main subject of interest, being most sensitive to the global, freight-traffic-dominated impact (Figures 4.11, 4.15 and 4.16).

The eigenfrequency f_i is defined as a function that depends on the cross-section's bending stiffness EI, the structural member's length L, the mass per meter m and the type of boundary conditions (λ). Utilization of the present equation without considering axial forces is permissible, as the acting forces in the present case are relatively small – furthermore, probable changes of the effective axial forces are occurring in a quite short timeframe, which minimizes their affect on the dynamic response in the course of frequency analysis.

It can also be shown that the relation between the diagonal's free vibration length and the geometrical length is constant, which also eliminates its influence on the judgment of the results of the present methodology. The project documentation delivers all input parameters to derive the expected values for eigenfrequencies based on Blevins (1979):

$$f_i = \frac{\lambda_i^2}{2\pi L^2}\sqrt{\frac{EI}{m}} \tag{11}$$

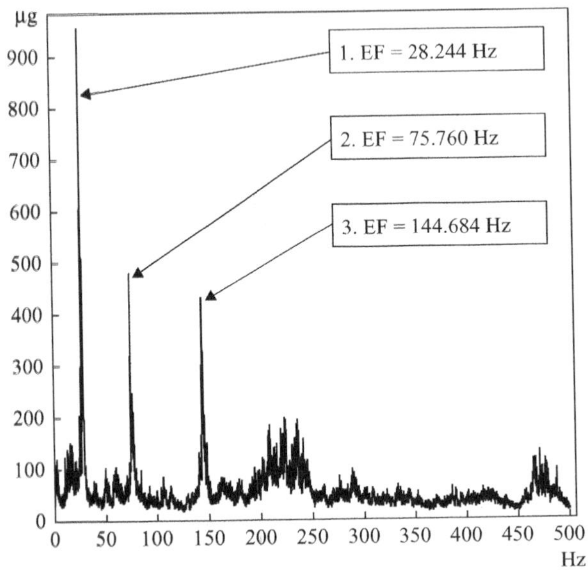

Figure 4.18 Smoothed frequency spectrum (vertical) under ambient conditions at bracing II 42 QV-N

Figure 4.19 shows the progression of derived expected values for f_1 assuming different types of boundary conditions (clamped on both ends, clamped–hinged and an approximately determined rotational spring on both ends). The ground view of the bridge deck has a certain flexion shape in both abutment areas and a corresponding lateral inclination, leading to different lengths of the torsional bracings in the bridge's lengthwise direction for the uphill and the downhill driving direction (Figures 4.19 and 4.20).

Of course the differing length of the diagonals due to the varying steel box-girder's height (decreasing from the main span to both abutments) is also considered. As Equation (11) demands squared values of the structural member's lengths, parabolic functions represent the expected eigenfrequencies in the bridge's lengthwise direction. Consideration of different supporting conditions is implemented into the constant parameter λ. The bigger its value, the more the three derived parabolic functions seem to drift

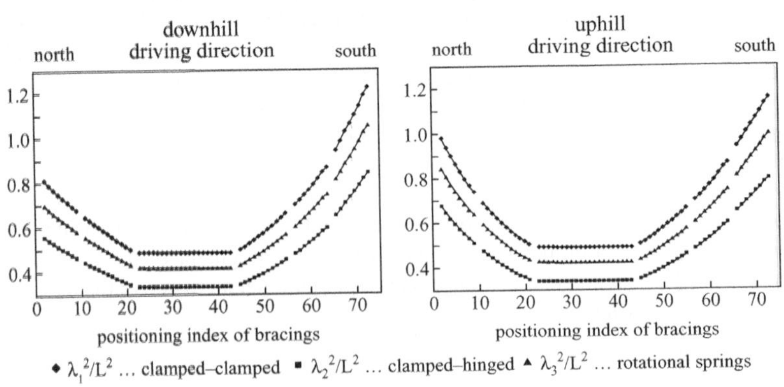

Figure 4.19 Pattern of the functions of the expected values of f_1 for varying boundary conditions

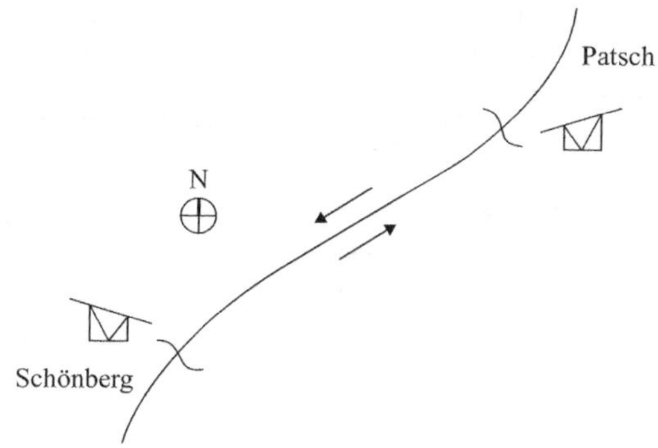

Figure 4.20 Schematic ground view of the flexion shape of the bridge deck – the lateral inclination demands varying lengths of the torsional bracings in the bridge's lengthwise direction

away from each other, with increasing values for expected eigenfrequencies at the same time due to the nonlinearity of variation of the diagonal lengths.

The methodology developed delivers a standardized kind of comparison. The utilization of Equation (11) – considering the varying length of the torsional bracings in the uphill and the downhill driving direction – automatically ensures that each expected value of the structural member's dynamic response includes the same uncertainty of modeling.

The present approach intends to provide a horizontal progression of the calculated ratio between measured and expected values for eigenfrequencies under regular conditions. Probable deviations from this trend line are much better recognizable than from the progression of the measured values. This fundamental assumption of the methodology is confirmed when the progression of the calculated ratio between each of the three functions for expected values (Figure 4.19) is compared to the other ones (Figure 4.21). Analogous to the comparison of the individual functions of expected values to each other (Figure 4.21), the comparison between the measured and the expected values of the bracing's effective stiffness was done.

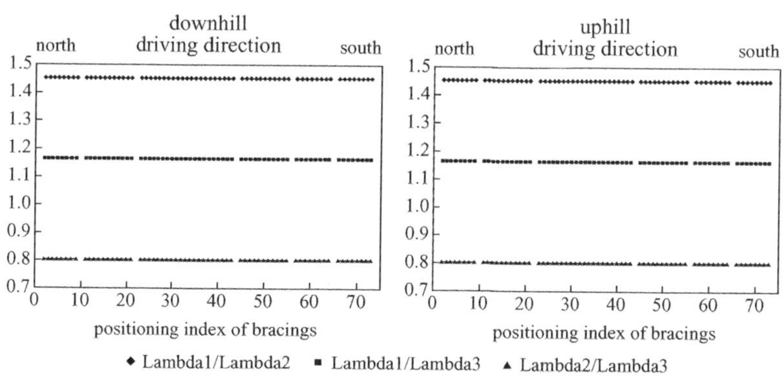

Figure 4.21 Pattern of the ratio of calculated functions for the three expected values of f_1 relative to each other for varying boundary conditions

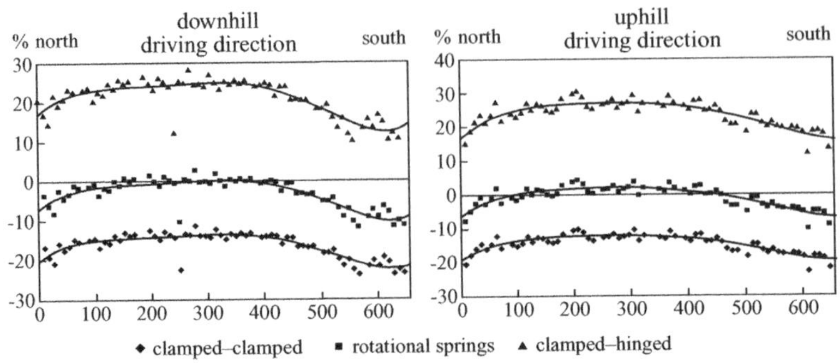

Figure 4.22 First eigenfrequency: trend of deviation of measurement results from expected calculated values in the bridge's lengthwise direction for varying boundary conditions

4.1.3.3 Results of the Assessment

When analyzing Figure 4.22 – focusing on the natural frequencies – it is obvious that firstly there is a systematic, widely constant deviation of the monitoring results from the expected, calculated ones (horizontal progression). From this progression, singular as well as sectional distinctive parts – leaving the horizontal trend line – are isolated, which is an intention of the chosen methodology. In these cases it is remarkable that the deviations – normally occurring nearer to the analytical model "clamped on both ends" – are then located nearer to the analytical model "clamped on one end and hinged on the other end." In the following, three main reasons (indicating clear evidence of systematic structural weakness) are explained:

- Up to now the investigation has been showing a consistent trend, confirmed by each of the first three observed eigenfrequencies. The trend line representing the deviation between measured and expected values has a horizontal progression in the central part of the bridge (bridge spans I, II and III) while the outer spans B, IV and V (downhill driving direction) and the outer spans IV and V (uphill driving direction) already indicate a degradation of the torsional bracing's integrity over the entire span length. The reason might be the ability of the central spans to redistribute internal forces and restraint, while the outer spans have this possibility only in a limited manner – which forces their torsional resistance to be activated regularly. This has an impact on the torsional bracing's joints, as an effective degradation of the integrity of their boundary conditions is clearly identifiable with a decrease of the natural frequency. These statements are particularly confirmed by the fact that in the past fatigue damage of the diagonal bracings has mostly occurred in the outer bridge spans.
- Figure 4.22 also shows that the assessment of deviations of measured natural frequencies f_1 to the expected values can be related to every type of boundary condition (clamped on both ends; clamped–hinged and an approximately determined rotational spring on both ends). The flexion shape of the obtained trend line remains, which indicates that the observed phenomenon is insensitive with regard to the type of supporting conditions.
- By comparing the left and right parts of Figure 4.22 it becomes clear that the diagnosis of reduced structural integrity in the outer bridge spans is particularly evident in the downhill driving direction. The condition of the torsional bracings belonging to the uphill driving direction seems to be much less affected even if they show the same degradation phenomenon next to the southern abutment. The reason is most likely the heavy freight traffic impact suddenly entering the bridge structure and moving downwards, even if finally the whole assembly's torsional resistance (consisting of two bracings) is activated. Again these statements are particularly confirmed by observations during visual inspections,

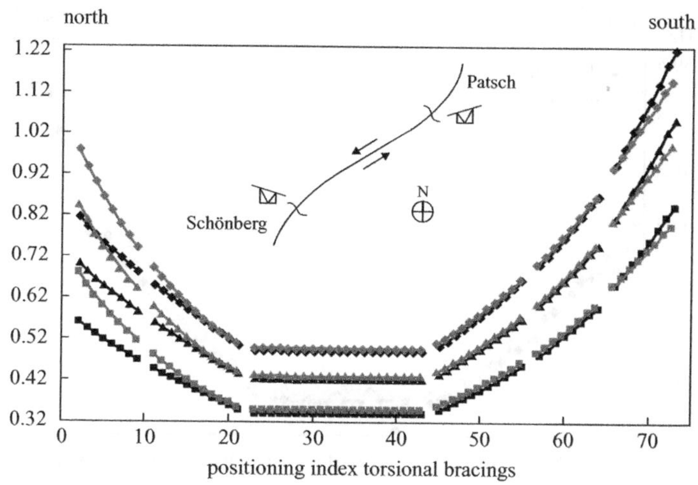

Figure 4.23 Comparison of certain functions of the three expected values of f_1 relative to each other for varying boundary conditions

as in the past fatigue damage of the diagonal bracings has mostly occurred in the downhill driving direction.

- Figure 4.23 shows the predetermined progression of expected values depending on the varying lengths of the torsional bracings and different boundary conditions. It has already been stated that the differing length of the diagonals due to the varying steel box-girder height (decreasing from the main span to both abutments) is considered. Additionally the bridge deck's ground view has a certain flexion shape in both abutment areas and a corresponding lateral inclination, leading to different lengths of the torsional bracings in the bridge's lengthwise direction for the uphill and the downhill driving direction (Figures 4.19 and 4.20). For bridges with a satisfactory integrity of its torsional bracings, these geometrical properties demand that bracings belonging to the downhill driving direction and getting closer to the southern abutment are becoming successively shorter. The corresponding function describing the expected eigenfrequency values for the downhill direction includes increased expected frequency values in comparison to those values belonging to the uphill driving direction. For bracings getting closer to the northern abutment these conditions are assumed to be the other way round.

  When observing and comparing the measured eigenfrequencies - belonging to both driving directions (Figures 4.24 and 4.25), it becomes evident that the necessary assumptions described in the previous paragraph with regard to the outer bridge spans are not confirmed, especially not in the most critical area – the downhill driving direction next to the southern abutment. The necessary exceedance of the uphill values by the downhill values does not appear. The values corresponding to the downhill driving direction's function are mostly in the same order as those from the uphill direction or even smaller. At the northern abutment a distinctive decline of the measured function to the expected parabolic progression – again belonging to the downhill driving direction – can be observed.

4.1.3.4 Conclusions

In the following, all fundamental observations and findings are summarized for reasons of clarification. Again it can be stated that there are distinctive deviations from the measured to the expected functions for the outer bridge spans of the downhill driving direction - while the uphill driving direction is affected by the same set of problems mainly in the area near the Schönberg abutment.

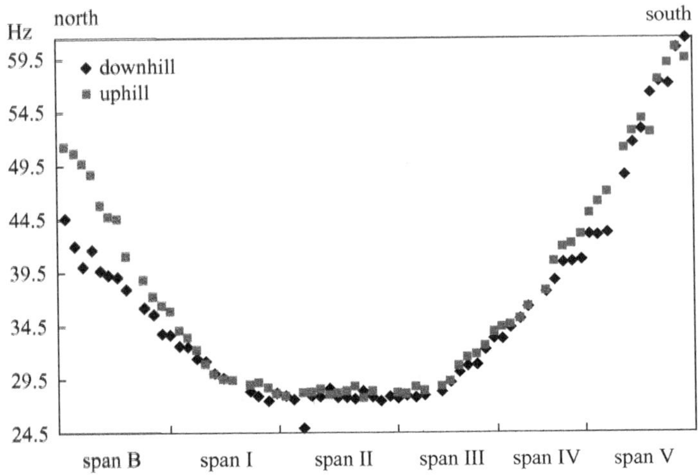

Figure 4.24 Measured values of the structural member's response (f_1)

All of these statements lead to a determination of weak points and weak areas, which are highlighted in Figure 4.26. Additional to the areas being identified as quite noticeable, there are single bracings whose loss of integrity seems to be in the range of the outer bridge span areas. Furthermore the diagnosis revealed even more diagonal bracings, indicating that there is already possible local damage or further deterioration of the structural member's operational integrity.

4.1.4 Providing, Conditioning and Implementing the Loading Impact

Fatigue is understood as a serviceability limit state for bridges since the occurring fatigue cracks often do not directly cause structural failure. Phenomena such as redundancy and ductility have usually prevented steel bridges from catastrophic collapse. In long span bridges the load on the primary superstructure is dominated by the dead load. As the fluctuating live load part usually is relatively small, fatigue is of secondary importance. The deck, stringers and transversal girders are mainly subjected to live load and

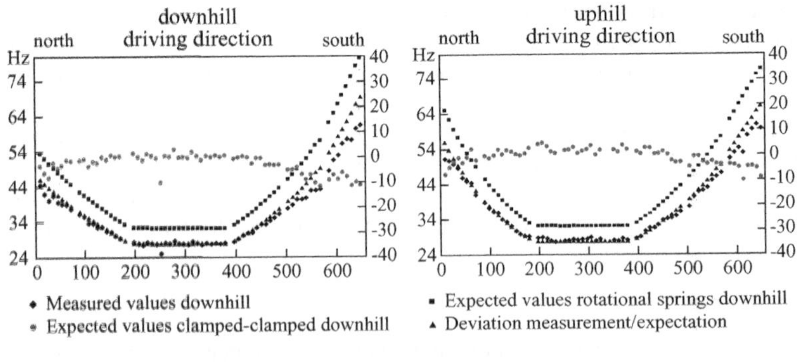

Figure 4.25 Functions of the expected values of f_1 for varying boundary conditions, measured values of the structural member's response (f_1) and trend of deviation of the measurement's results to calculated expected values in the bridge's lengthwise direction

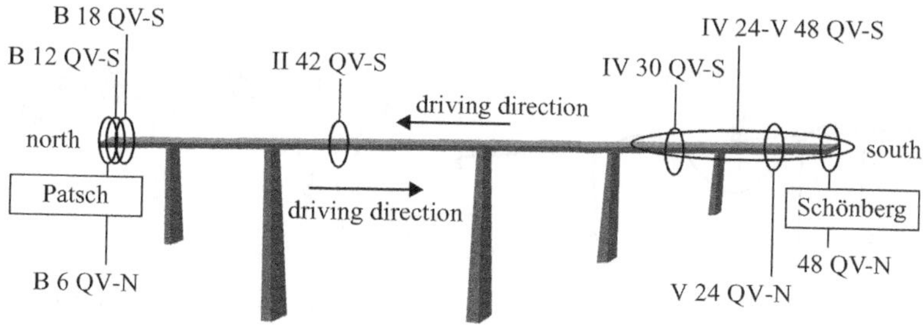

Figure 4.26 Monitoring-based determination of weak points and weak areas for both driving directions

therefore they may be controlled by fatigue. To ensure a unitary and integral approach, the discussion will be focused on three levels, where associated hot spots are identified for continuative analysis with regard to different levels of relevance (isolation of remaining, fatigue-relevant loading cycles n_i from randomly occurring traffic).

4.1.4.1 Level I – Global Behavior

Assessment of Global Impact

The following section deals with measurement of vertical displacement of the bridge's main span. The aim was to obtain measured impact data for the global level of analysis. Figures 4.27 and 4.28 show schematically how this was realized. The measuring system consists of a laser-transmitter unit (stationary

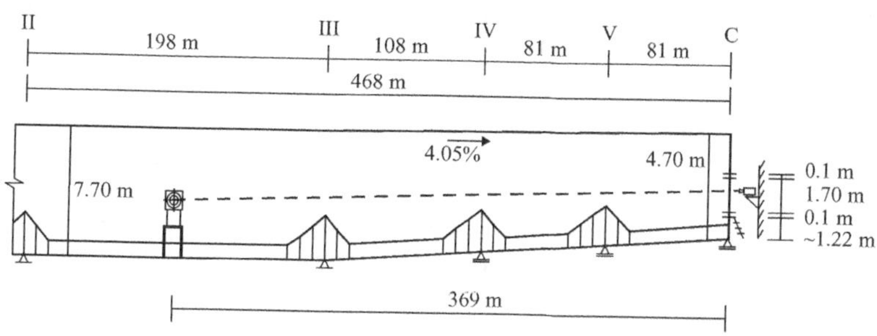

Figure 4.27 Schematic representation of measurement of the vertical displacement of the main span

Figure 4.28 Optoelectric receiver unit (left) and laser-transmitter unit (right)

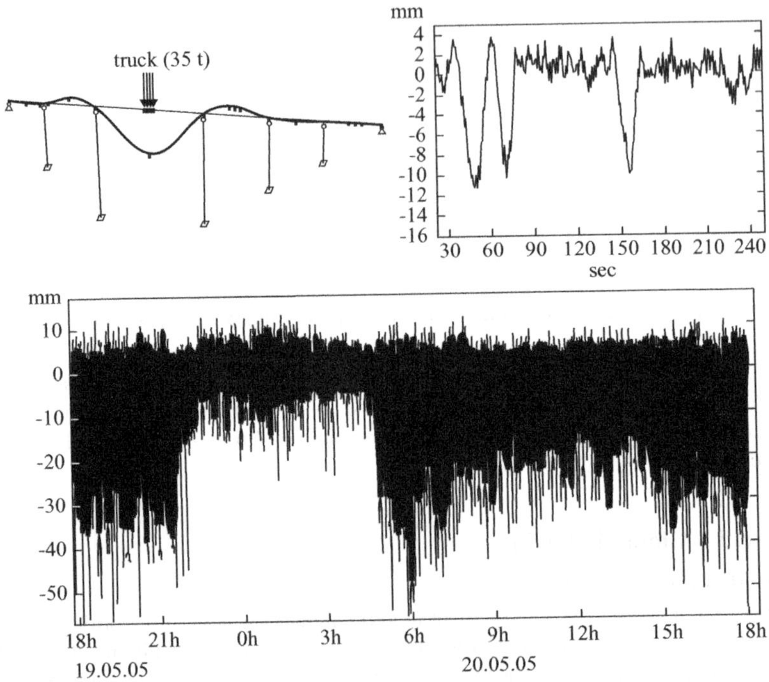

Figure 4.29 The vertical response of the main span due to traffic loading (19.05.2005,17:50–20.05.2005,18:05)

and located at the bridge's abutment) and the receiver unit at a distance of 369 m. The laser transmitter sends out a narrow, point-type laser beam in a fixed direction. The optoelectric receiver measures the position of the optical center of gravity of the laser beam's hit point on the optical screen. Via a base frame, the receiver is tied positively with the orthotropic bottom plate. Regarding the cross-section's contour accuracy, the span's global vertical displacement is measured in a sufficiently precise manner.

In Figure 4.29 the result of a 24-h measurement campaign is visualized. The total (static and dynamic) displacement of the span was recognized to be in the range of −12 and 56 mm. The verification of plausibility can easily be done by analyzing the signal during the night hours (22:00–06:00), where regular freight traffic is not allowed to cross the bridge structure. The exceptional truck passages occurring during this time window show vertical displacements in the range of −10 and 18 mm. Static loading by a truck with 35 tons would result in an approximate displacement of 14 mm (based on global structural analysis).

In addition to quantitative plausibility, the qualitative plausibility was also checked. For this reason some isolated truck passages could be verified directly in the signal belonging to the night hours again. Such passages lasted for 20 s, which corresponds really well with the fact that a truck – maintaining the prevailing speed limit of 40 km/h – needs approximately 20 s to pass the 198-m long main span.

With regard to continuative global stress analysis it was demonstrated that the desired measurement data were obtained under representative conditions. It needs to be emphasized that a series of measurements had to be created on several working days by repeating such a measurement campaign to cover the randomness of this individual process.

Figure 4.30 shows the remaining response cycles of the bridge's main span due to the occurring traffic related in Figure 4.29. The fatigue analysis itself is going to be performed in terms of stress.

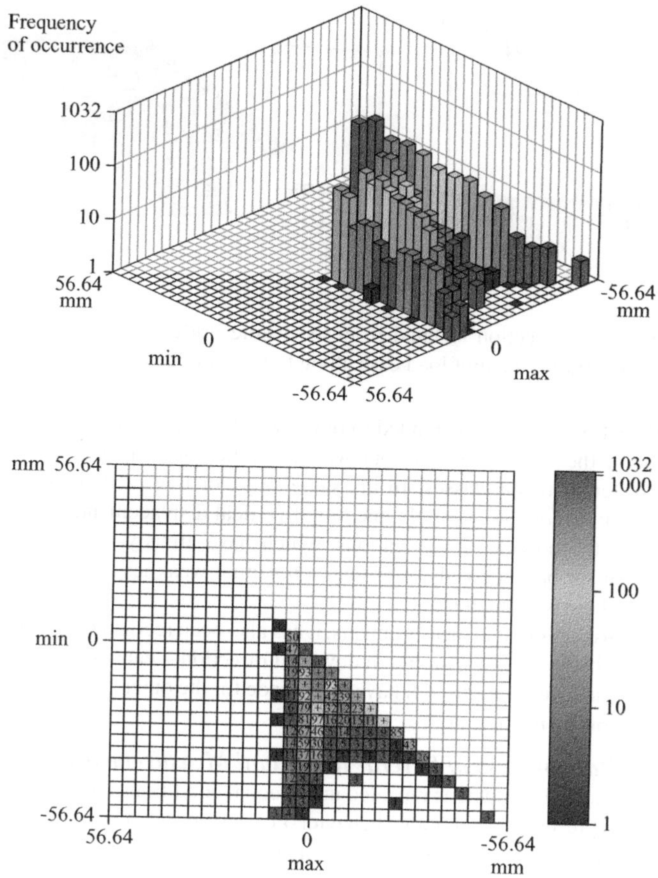

Figure 4.30 Rainflow counting of global bridge displacement

Firstly, the chosen approach demands some global structural analysis. The obtained displacement cycles are implemented into a global framework model of the whole bridge by means of a single constraint load case (Figure 4.31, left) with RSTAB software. System identification performed with this software and compared to measurements afterwards showed that global behavior is represented quite well. On the other hand fundamental phenomena such as the effective width, which is of eminent importance for accurate stress analysis, are disregarded with RSTAB. To overcome this problem, the following mechanical solution was developed.

As already shown, the measured displacement cycles are implemented into a global framework model of the whole bridge by means of a single constraint load case. Subsequently a 63-m long bridge segment – modeled with shell elements (RFEM) – is prepared (Figure 4.31, right), which can be individualized in its geometrical properties depending on the determined area of interest. The transition from the global (framework-based) state variables to the local (FE (shell element) based) ones is realized through assignment of displacements and cross-sectional rotations from the global bridge model along the outer edges of the modeled bridge segment (Figures 4.31 and 4.32).

The applied hypothesis for the implementation of measured, global displacement data on the global bridge model via unit load case (restraint at the sensor position) has to be confirmed. For that reason

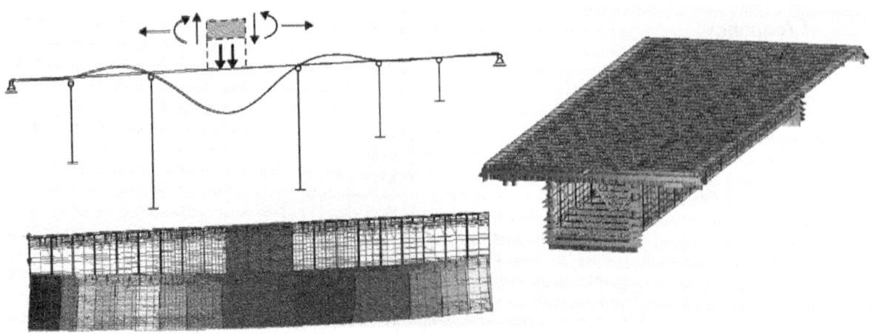

Figure 4.31 A 63-m long section isolated from the whole bridge structure and modeled with shell elements under the constraint of a unit load case along the outer edges

further fatigue calculations are to be extended to other possible loading configurations – with the same displacement value at the sensor position, but with other flexion shapes – leading to local analysis corresponding to different notch classes. In the same way, loading configurations related to explicit torsional impact, and in accordance with the measured vertical displacement, must be determined. This need is motivated by the previous investigations, which have already revealed that the box girder's torsional bracings (square bar diagonals) represent a potential weakness and thus they are going to be one of the main global areas of interest.

During numerous inspections and measurements, it has also been recognized that the impact of trucks – suddenly entering the bridge's first span in the downhill direction – causes noticeable global vibration. Therefore global fatigue investigations will have to be made for the outer spans, following the procedural principles described for the main span.

Evaluation of the Randomness of Several Individual Measurements

Several global measurement campaigns repeated under equal conditions indicate a random sample from the universal set that represents global bridge behavior. For reasons of complexity and economization this laser-supported measurement configuration is not capable of being a part of the permanent monitoring system. To cover the randomness of these several individual measurements, an approach from Bronstein

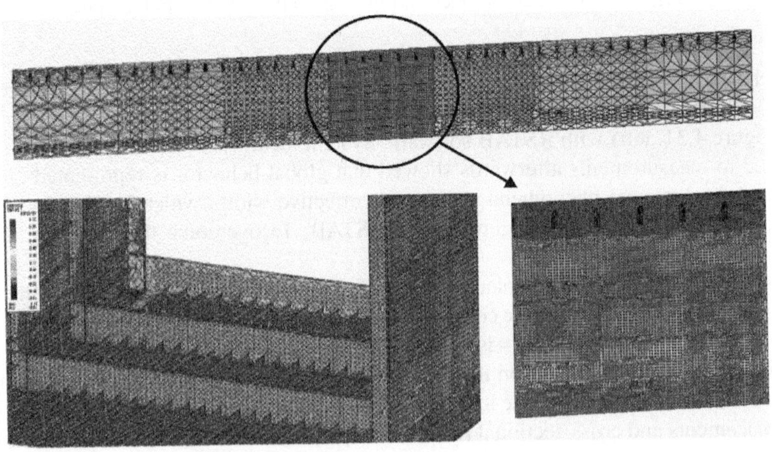

Figure 4.32 Analysis of principal stresses: overview and detail for the orthotropic bottom plate and the box girder itself, corresponding to Figure 4.31

et al. (2001) is chosen. Ultimately the confidence interval for the measured mean value with an unknown variance is derived. Following the hypothesis in Bronstein et al. (2001), the random variables obtained from such individual measurements follow a Gaussian distribution regardless of the universal set's distribution function. Thus confidence intervals for the obtained measurement data are established by means of an accepted, predetermined probability of error $\alpha/2$ (Equation 12):

$$\mu = \overline{x} \pm \frac{s}{\sqrt{n}} t_{\alpha/2;n-1} \tag{12}$$

where μ represents the modified reference value (e.g. accumulated damage D – but could also be the Rainflow procedure's section values leading to damage D) belonging to individual measurements, $\overline{x}$ is the measurement-based mean value, s is the measurement-based standard deviation, n is the amount of performed measurements and $t_{\alpha/2;n-1}$ is the quantile corresponding to a t-distribution.

4.1.4.2 Level II – Cross-Sectional Behavior

The bridge's 7.5-m long cantilevers are primarily heavily loaded by trucks, as those with a permissible maximum weight of more than 7.5 t are allowed to use only the first lane. This emphasizes the importance of this level; therefore it was decided to incorporate the cantilever itself into the permanent monitoring system.

For permanent monitoring of the cantilever vibration, so-called forced balance accelerometers (FBA11, sampling rate 100 Hz) are used. Those transducers register the alteration of electrical capacity (by means of voltage) based on the variation of their position due to a springy mass – influenced by acting accelerations. According to Figure 4.33, three one-dimensional sensors were installed in the region of bridge pier II to describe the loading behavior in the transverse direction. Two of them are located at a defined distance of 15 m – in the same direction of traffic – to verify recurring truck passages and their related velocity without any disturbance of traffic by smoothing the signal with various time intervals (weighted averaging) and subsequent automatic "peak-picking." This procedure is comparable with the utilization of digital filters and widely eliminates the influence of interferences in the higher frequency range. Nevertheless it was found that smoothing maintains the whole energy impact much better than digital filters, especially for trucks passing the measurement section in quick succession.

Consequently the structural member's response – the sensor signal of acceleration - has to be transformed into a signal of absolute displacement, as this is the decisive variable of interest for further stress-based fatigue analysis. This state variable typically cannot be derived mathematically from transducer accelerations. Elimination of the acceleration's offset and double integration lead to vibration about the neutral axis in terms of relative displacements (Wenzel and Pichler 2005). For that reason the so-called DYGES algorithm (the German acronym for Dynamic Weight Registration System) was developed, which enables the authors for the first time to generate absolute displacements from relative accelerations by means of digital signal processing. This pattern recognition-based procedure (see

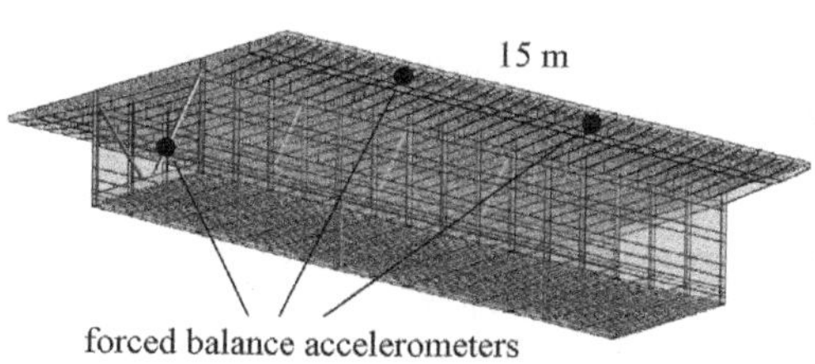

Figure 4.33 Vehicle recognition by acceleration sensors placed at the bridge cantilevers

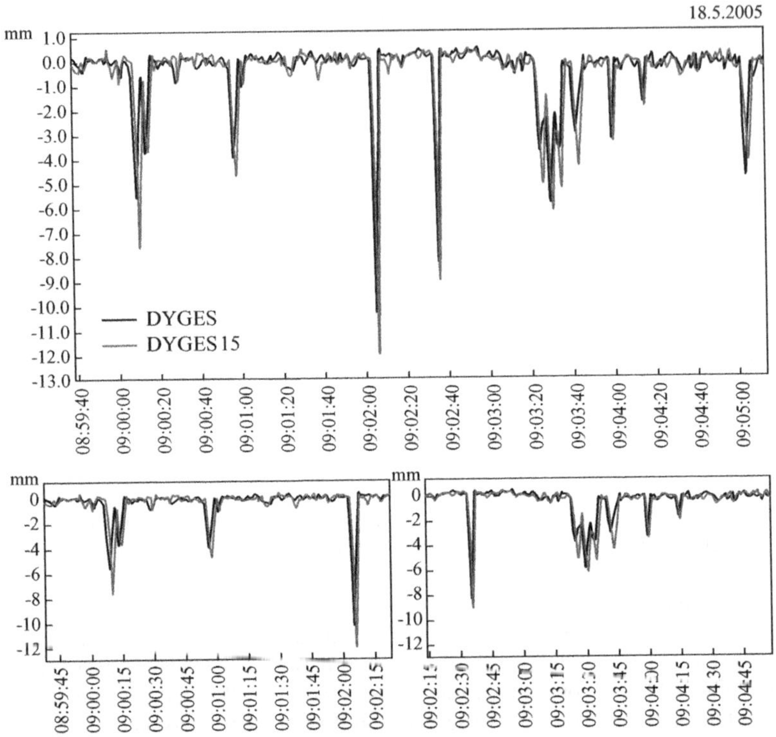

Figure 4.34 Reproduced absolute vertical cantilever displacements based on accelerometer measurements before the input of calibration (DYGES progression – specified in flowchart of Figure 4.38)

Figure 4.38) mainly consists of weighted averaging, unification of the signal's algebraic sign (squaring), an envelope function (based on the time window secant method) and a histogram-based new offset determination.

Figure 4.34 compares the reproduced vertical displacement patterns based on accelerometer measurements DYGES and DYGES15, and gives an insight into how single and repeated truck crossings can be identified and isolated (sequence of local extrema) for further assessment. The influence of prevailing velocities and motor vehicle tonnage has still to be considered (see Figure 4.44). In-succession velocity (peak picking) and deflection intensity can be determined for further calibration purposes based on the identification of the occurring, identical truck crossings – extracted from the two successively located accelerometers (Patsch and Patsch15). It is to be specified that the pattern of DYGES15 is mainly used for velocity identification. The reproduction of a "quasi-laser" progression is exclusively based on the DYGES pattern itself as the laser calibration in terms of vertical displacements could only be performed for the first of the two successively located cantilever accelerometers. Furthermore it is to be emphasized that this approach of deriving cantilever displacements due to traffic loading is a structure-specific approximation, which had to be optimized by the calibration of truck crossings for consistent analysis. This information was obtained in the course of an additional measurement campaign in May 2005, which was conducted with the following aspects: trucks with varying, well-known loads (a fully loaded truck and a truck without any cargo) passed the measurement section with stepwise increased velocities (from 20 to 60 km/h). This process was repeated with simultaneous measurement of vertical deflection (supported by laser) and acceleration (Figure 4.35). The essential sources of interest, which had necessarily

Figure 4.35 Laser transmitter unit (left), optoelectric receiver unit (middle) and fully loaded (42.8 t) calibration truck (right)

to be implemented into the DYGES procedure, were the dynamic amplification due to the influence of transported weight as well as due to the influence of the truck's velocity.

The calibration-based part of the algorithm (see Figure 4.3) is specified in Figure 4.36. The measurement section is located directly above bridge pier II, thus the analytical calculation (for verification

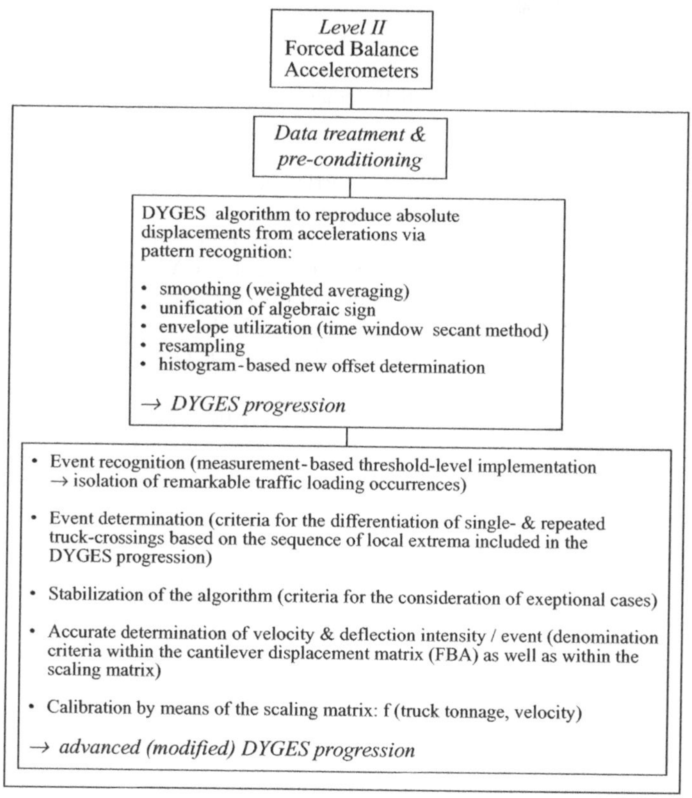

Figure 4.36 Flowchart of the calibration approach for the cantilever's acceleration sensors (included in the procedure outlined in Figure 4.3)

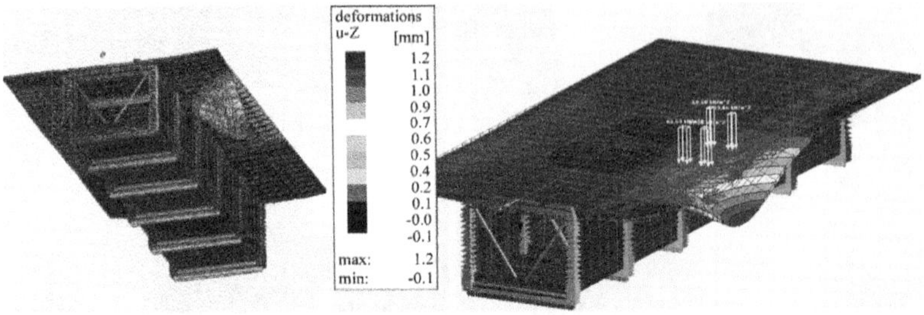

Figure 4.37 Bridge segment modeled with shell elements and stressed by an unloaded truck (8.7 t)

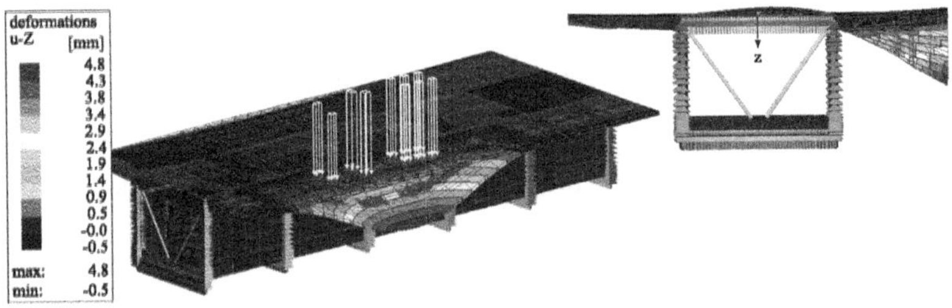

Figure 4.38 Bridge segment modeled with shell elements and stressed by a heavily loaded truck (42.8 t)

purposes) with the same (static) truck loading was carried out on a bridge segment modeled with shell elements and prevention of global deformation (Figures 4.37 and 4.38).

The cantilever deformations achieved for the unloaded as well as for the loaded truck show very good accordance with the measured values obtained (Figure 4.39). When analyzing Figures 4.37 and 4.38, one decisive assumption of the DYGES algorithm is confirmed analytically. The cantilever's response explicitly stressed by truck loading does not interact with the cantilever belonging to the opposite driving direction. This fact supports the observation already made during extensive measurement data analysis.

Figure 4.39 shows a very remarkable result of the laser-supported calibration measurement campaign. The pattern was obtained by weighting both truck-crossing repetition procedures. The functions, which

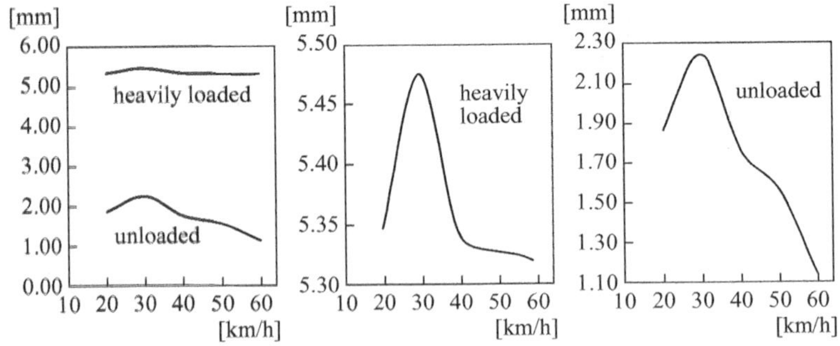

Figure 4.39 Amplification pattern (laser-supported absolute displacements) of the cantilever

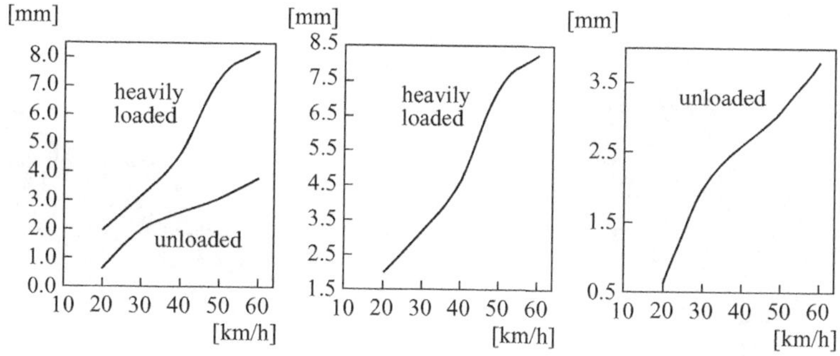

Figure 4.40 Amplification pattern (accelerometer-supported noncalibrated absolute displacements) of the cantilever

can be understood as an amplification factor pattern, have similar characteristics for heavy loaded trucks as for unloaded ones, even if the latter is more distinctive.

Long-term monitoring-based fatigue investigation (High Cycle Fatigue) assumes that stresses and strains remain elastic. For that reason these two functions can be connected via generatrix (linear interpolation), which enables the information for an automatic registration of all truck weights occurring under characteristic operational conditions to be used.

Figure 4.40 provides the complementary information to the results presented in Figure 4.39 for simultaneously performed accelerometer-supported cantilever measurements. The patterns were again obtained by weighting the DYGES-based results of both truck-crossing repetition procedures (Figure 4.3). The functions, which can be understood as an amplification-factor pattern, again have a distinctive trend line for heavily loaded trucks as well as for unloaded ones, even if each flexion shape is an individual one in detail.

Successively these two functions are connected via generatrix (linear interpolation), which enables the information for an automatic registration of all truck weights occurring under characteristic operational conditions to be used. In particular, comparison of both derived areal functions (Figures 4.41 and 4.42) strongly supports the need to implement a calibration sequence into the developed DYGES algorithm.

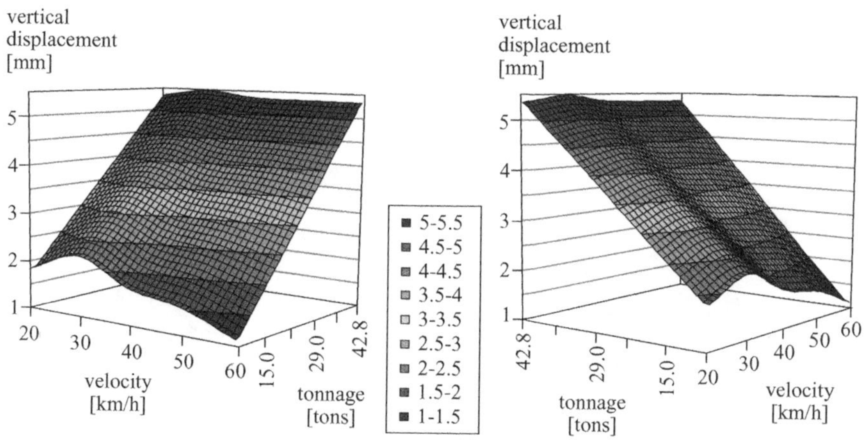

Figure 4.41 Areal function for absolute vertical cantilever displacements (laser) with varying weight and velocity

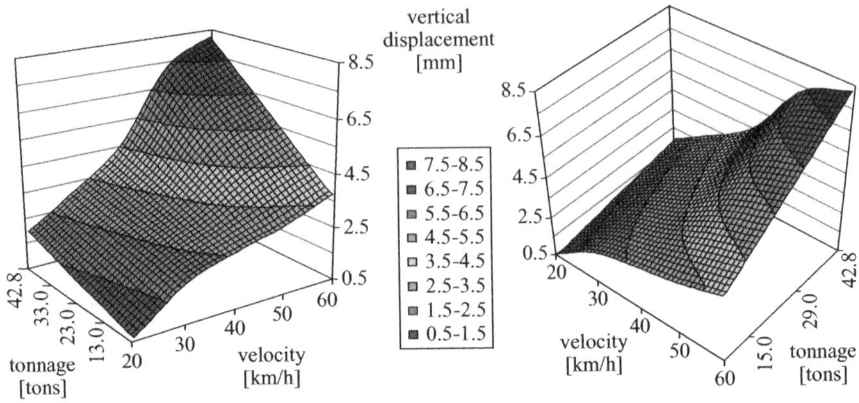

Figure 4.42 Areal function for reproduced absolute vertical cantilever displacements (accelerometer) with varying weight and velocity

Thus a scaling matrix is derived from both matrices, which fulfils this requirement to calibrate the reproduced, accelerometer-based absolute displacements (Figure 4.43).

The scaling of a registered event itself is accomplished by means of experimentally determined threshold levels and by using sequences of local extrema included in the DYGES progression. Each node within the time-history of these isolated events is multiplied by the same appropriate scaling factor before being implemented into the initial DYGES function again (Figures 4.43 and 4.44). Further details are included in Figure 4.36. Results from the current state of the transformation algorithm are typically visualized in Figure 4.44 – the influence of prevailing velocities and motor vehicle tonnage has already been considered. Of course the output of the calibration measurement campaign is limited in terms of the registered truck's weight and velocity. As each occurring case needs to be assessed, those cases that would be related with weights smaller than 8.7 tons are stated with this value (conservative approach in accordance with the Scaling Matrix). Regarding the truck weight it can be assumed that values lower than 8.7 tons (related to approximately 1.5 mm cantilever deformation) do not exceed the Wöhler curve's cut-off limits. Those cases that would be related to weights higher than 42.8 tons are stated with this certain value (nonconservative approach in accordance with the Scaling Matrix). Values higher than 42.8 tons should

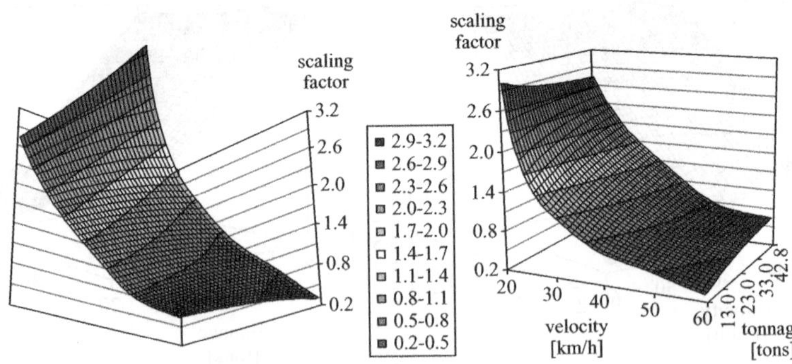

Figure 4.43 Scaling Matrix to calibrate reproduced absolute vertical cantilever displacements (accelerometer and laser) with varying weight and velocity

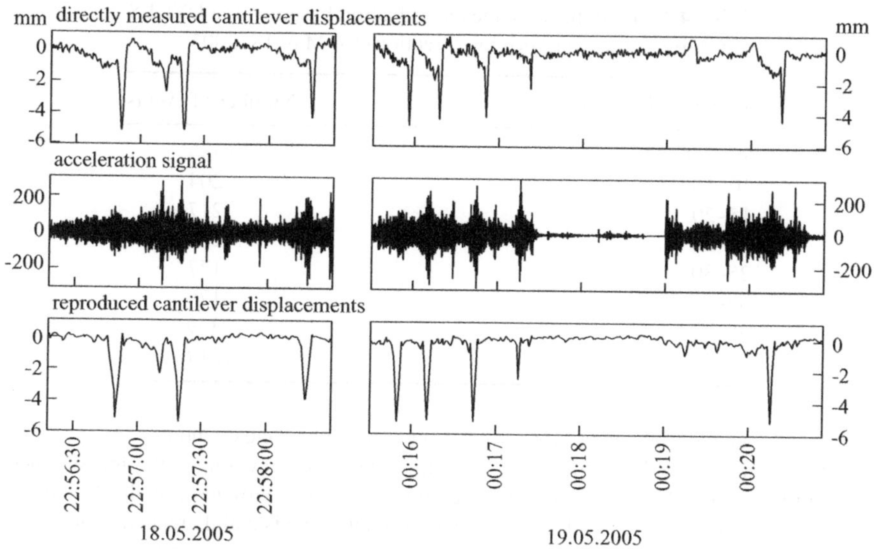

Figure 4.44 Reproduced cantilever displacements versus directly measured ones and acceleration signal

not be exceeded remarkably, as extended observation by the bridge owner confirmed compliance with the permissible maximum weight (44 tons).

Regarding the truck velocity it can be assumed that cases that would be related to velocities lower than 20 km/h are stated with this certain value (nonconservative approach in accordance with the Scaling Matrix) even if such cases should be negligible. Cases that are going to be related to velocities higher than 60 km/h are also stated with this certain value (conservative approach in accordance with the Scaling Matrix). It needs to be kept in mind that values higher than 60 km/h are not possible during the daytime. This assumption is based on an existing speed limit (40 km/h) but it is also a logical conclusion of the fact that the prevailing intensity of traffic flow prevents values higher than 60 km/h.

Long-term monitoring – leading to an automatic registration of truck volume, truck velocity and truck weight – will state more precisely how often exceedance of 60 km/h will occur during the night time. In any case Figures 4.42 and 4.43 clearly show that the usage of the upper-bound values of the Scaling Matrix at least leads to conservative solutions for exceedance cases of velocity even if the exact information for rescaling is not available.

Exact results of the performed traffic analysis (Tables 4.1 and 4.2) will strongly support the observance of randomness of the output information and its relevance for fatigue analysis based on statistical studies. Following the described principles the cantilever's vertical deformations are reproduced qualitatively (progression) and quantitatively with sufficient accuracy for further fatigue assessment.

A detailed observation of Figure 4.44 shows two main perceptible deviations between the original (top) and the reproduced (bottom) qualitative displacement pattern. With regard to continuative fatigue analysis their influence on the obtained data can be neglected and is explained as follows:

1. Existing cycles – mainly occurring during unloaded vibration about the neutral axis – that do not exceed the predefined correlation matrix's section margin are ignored.
2. The imaginary flow of water, which is initiated in every peak and valley in the course of Rainflow counting, is followed until it experiences a drop. Referring to the principles of the applied Rainflow algorithm (Section 4.1.2.3), there is no discrepancy when small pattern-sequences are not available anymore, as long as all occurring local extreme values are made available for further analysis.

Table 4.1 Example tonnage classification based on the DYGES algorithm (downward driving direction) for May 18, 2005

Tonnage classification	Number of events
8.7–10	1386
10–15	204
15–20	217
20–25	237
25–30	177
30–35	176
35–42.8	339
>42.8	1319

Neither case is a source of uncertainty for further fatigue-relevant cycle counting.

Finally it is to be specified that the extracted output, which is drawn upon the measurement-based DYGES algorithm and conditioned for further structural analysis, leads to cantilever displacements as well as to a freight traffic classification (volume / tonnage / velocity) and is mainly related to one acceleration sensor (Patsch). Events that might possibly be identified at the other accelerometer located successively 15 m away from the first one (Patsch15) but not confirmed at Patsch have to be ignored, as well as in the opposite situation. Furthermore an additional outlier sensitivity criterion was implemented to enable a stable recognition of vehicle registration by means of the two corresponding cantilever accelerometers. This also demanded time-frame-equivalent velocities to be established from which certain minimum value (10 km/h) up to which certain maximum value (115 km/h) identified loading events can be linked to each other (Figure 4.36).

Figure 4.45 shows the result of Rainflow counting for a monitoring period of one month based on the present stage of development of the introduced reproduction procedure for cantilever displacements (DYGES). The influence of all levels of loading is obvious but is eliminated via a fatigue-relevant transition to a corresponding Damage Matrix (Figure 4.46), which is of eminent importance for further FE calculations.

In addition to the methodology's output in the Rainflow domain, which has become a matter of course in performance calculation nowadays, the introduced DYGES algorithm also enables the authors to extract more conventional classification data (Tables 4.1 and 4.2) in terms of traffic volume and individual truck

Table 4.2 Example velocity classification based on the DYGES algorithm (downward driving direction) for May 18, 2005

Velocity classification (km/h)	Number of events
<20	315
20–25	100
25–30	129
30–35	236
35–40	795
40–45	1067
45–50	418
50–55	134
55–60	88
>60	773

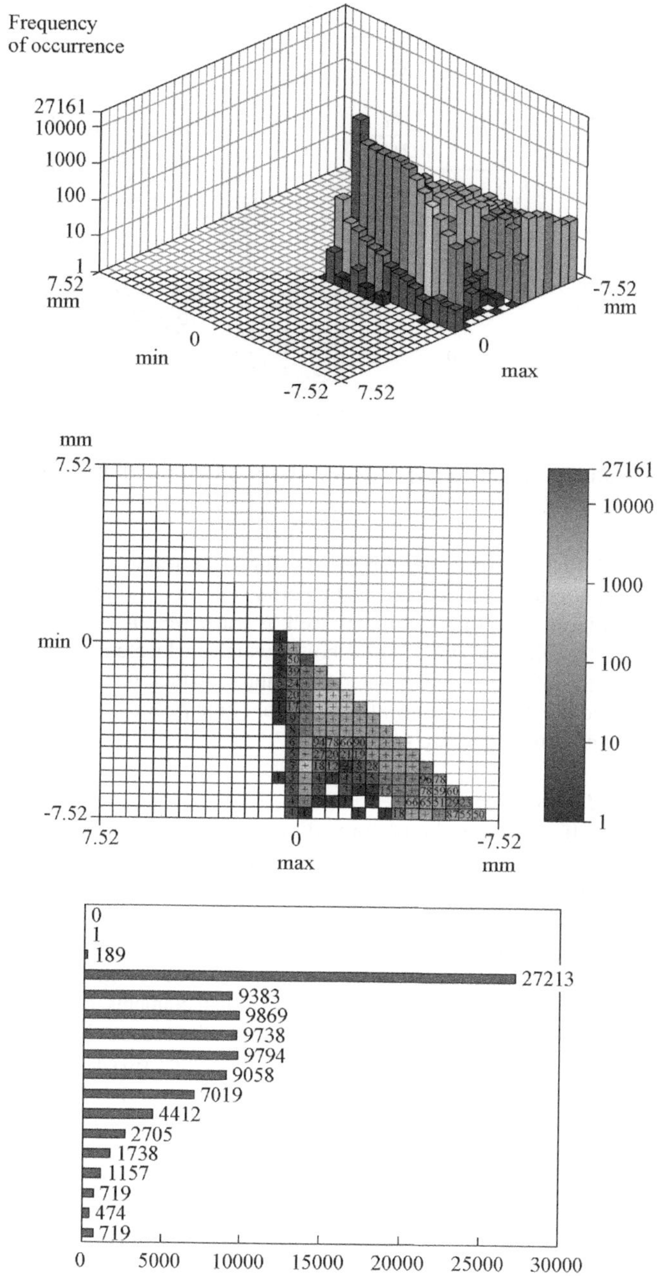

Figure 4.45 The axonometric projection of the rainflow matrix, its ground view and the appropriate level-crossing histogram for a representative month (July 2004)

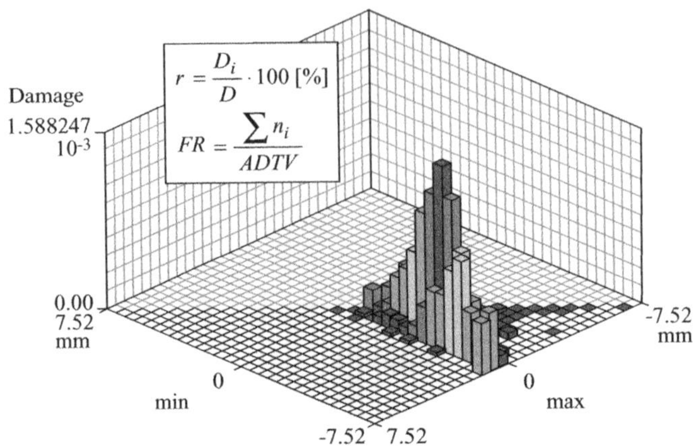

Figure 4.46 Damage Matrix corresponding to the Rainflow Counting Matrix (Figure 4.45)

weight for any aspired time-segment. The fatigue-causing truck traffic is registered, classified via in situ procedures that were established as part of the permanent monitoring system. This heavy freight traffic classification in terms of traffic volume as well as truck weight was successfully validated by comparison with officially collected data at the tollbooth, which are located directly at the Europabrücke. This tollbooth provides the information on counted truck units and their number of axes. Additionally this validation was supported by video recordings for individual time periods, which lead to a final approval of the DYGES algorithm's output (Figure 4.47).

Fatigue Analysis in Terms of Stresses

The fatigue analysis itself is going to be performed in terms of stresses, which means that comparative lifetime calculations with nominal as well as structural stresses are going to be carried out. The fatigue threat exclusively caused by truck traffic leads to the application of High Cycle Fatigue Theory because the stress-life approach is most suitable for long-life applications in civil engineering. The displacement cycles obtained via Rainflow analysis from the reproduced cantilever displacement function (DYGES algorithm) are implemented into bridge segments, which can be individualized in their geometry depending on the determined area of interest (modeled with shell elements, e.g. Figure 4.5). Stresses and strains are assumed to remain elastic. This facilitates the calculations, as the monitoring-based impact data are implemented into the FE models by means of a single static load case (restraint by a vertical unit displacement along the cantilever's outer edge in the area of the sensor position) that is to be calculated and related to all other occurring displacement cycles afterwards.

Figure 4.33 showed a bridge structure segment where the cantilever acceleration sensors are installed. Shell elements are going to be used in the course of FE analysis to model the middle plans of the plates. Plate thickness is given as a property of the elements. The structure will be modeled with a coarse mesh and therefore a refinement at the relevant hot-spot areas is necessary.

The author would like to point out the following conclusions as a result of detailed investigation for this level of fatigue analysis: The appropriate model of the hot-spot areas includes corner arcs for welds as well as additional stiffening plates for connector units. The weld geometry itself is neglected. The location of the weld toes at structural intersections represents a very conservative solution for the ongoing stress extrapolation. The physically imprecisely modeled overlapping area can alternatively lead to the consideration of an offset (half of the corresponding plate thickness) before the relevant values for stress extrapolation are determined (Figure 4.48).

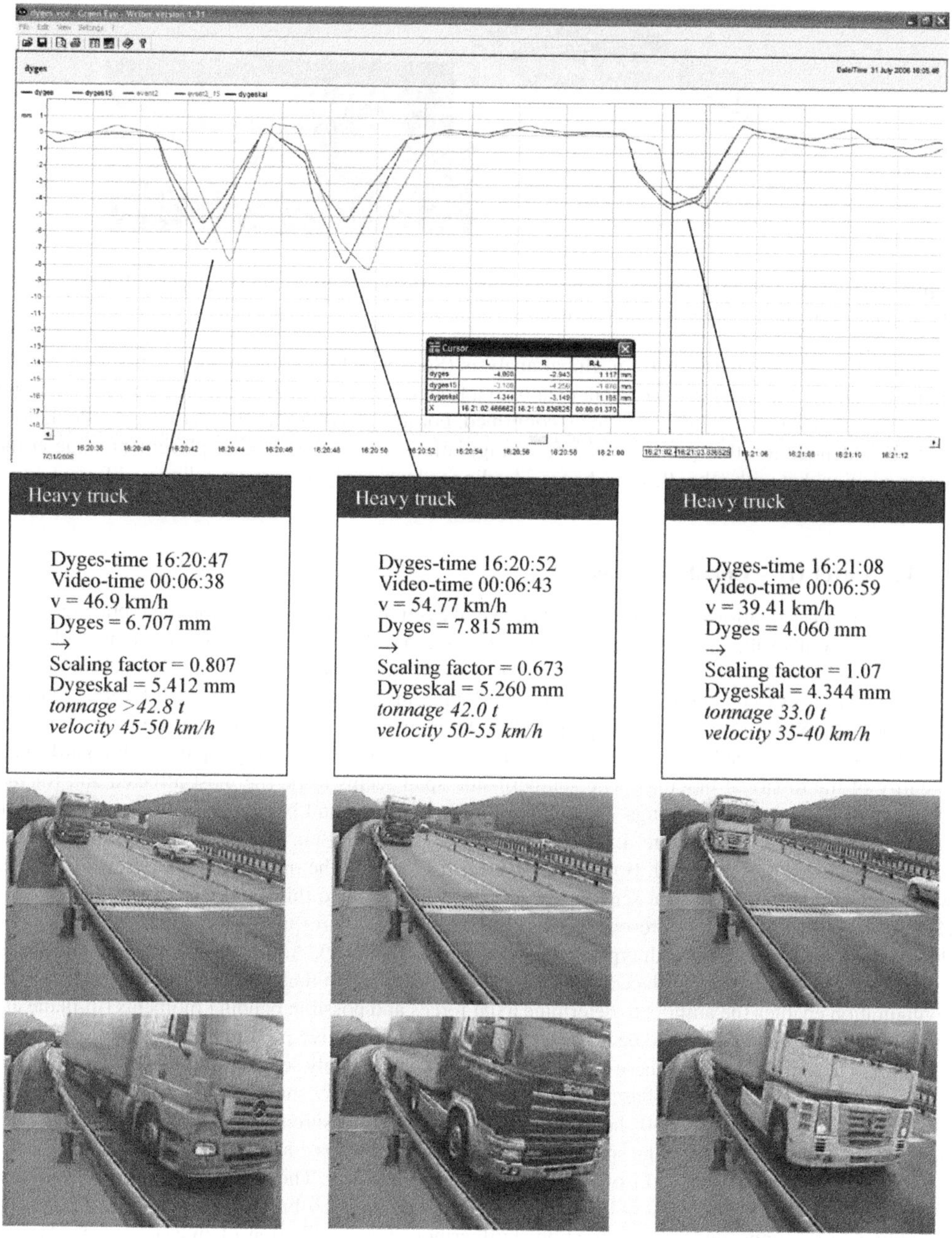

Figure 4.47 Video-supported validation of freight traffic classification based on permanent monitoring

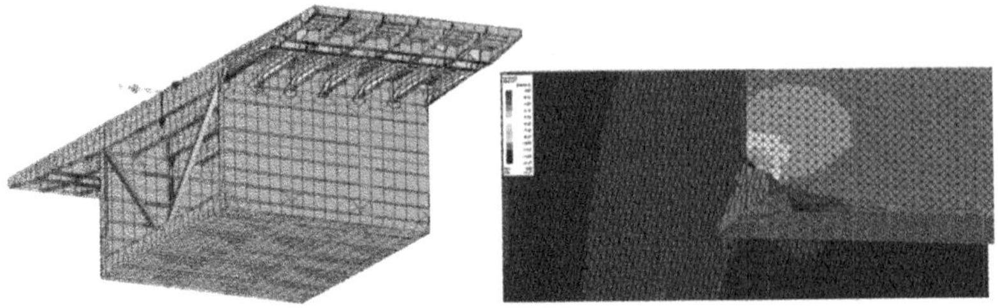

Figure 4.48 Finite element model of a bridge segment stressed with the unit load case along the nodes of the cantilever's outer edge (nominal stress)

Similar to the approach shown for global analysis (Section 4.1.4.1), the applied hypotheses for the implementation of derived displacement data onto a bridge segment model via unit load case (restraint along the cantilever's outer edge) has to be confirmed. For that reason again, further fatigue calculations are to be extended to other possible loading configurations – with the same displacement value at the sensor position, but with other flexion shapes – leading to local analysis corresponding to different notch classes.

4.1.4.3 Level III – Local Systems

The previous investigations, especially those shown in Section 4.1, demand the torsional bracings's bottom and top joints subjected to fatigue threat to be assessed. Comprehensive observations lead to the fact that these details are the most endangered ones in the course of analyzing the Europabrücke. This research work tries to encourage the target of developing innovative methods in each level of analysis. Therefore the following concept was prepared in the course of a one- week measurement campaign. Section 4.1.3.3 has already indicated effective degradation at the outer bridge spans with regard to the integrity of the torsional bracings. The southern side span seems to be the most affected and for that reason two corresponding bracings – belonging to this side span and being next to the ones where the remarkable fatigue crack (Figure 4.15) was identified – were chosen. An aerial measurement grid of nine strain gauges was prepared. Both joints at the top as well as the ones at the bottom were assessed with single strain gauges. These sensors are placed twice the plate thickness away from the weld toe (Figure 4.49), making them represent nominal (principal) stresses for uniaxial loading. For that reason these values are comparable with typical stress-life curves (Wöhler). Additionally both bracings are assessed with four strain gauges, placed approximately in the middle of these structural members. The latter configuration enables the authors to determine axial forces and possible bending moments (in-plane and out of plane) accurately. It has to be emphasized, that the continuative performance calculations are not intended to be mainly based on these measurements. They should only serve as a source of validation and calibration for the prepared FE-based hot-spot analysis. Figure 4.49 shows the described configuration of strain gauges, and Figure 4.50 shows exemplary results of measured strains. While observing a few daily cycles of loading impact, the spectrum of effective axial forces could be determined (Figure 4.51).

First, these strains are verified from the qualitative point of view. The strain functions obtained show an exact correlation of the sensors installed on the directly stressed bracing with those at the bracing belonging to the opposite driving direction. Furthermore the distribution of internal forces showing decreasing values from the source of impact – the directly stressed top joint – to the top joint on the opposite side is very distinctive.

For a quantitative verification of the measurement-based effective forces acting on the torsional bracings due to freight traffic, approximative FE Analysis was performed. Responding reference values in the range 55–60 kN were determined, which confirms the consistency of the measured values. This represents a good basis for detailed hot-spot analysis on local FE models of the top and bottom joints.

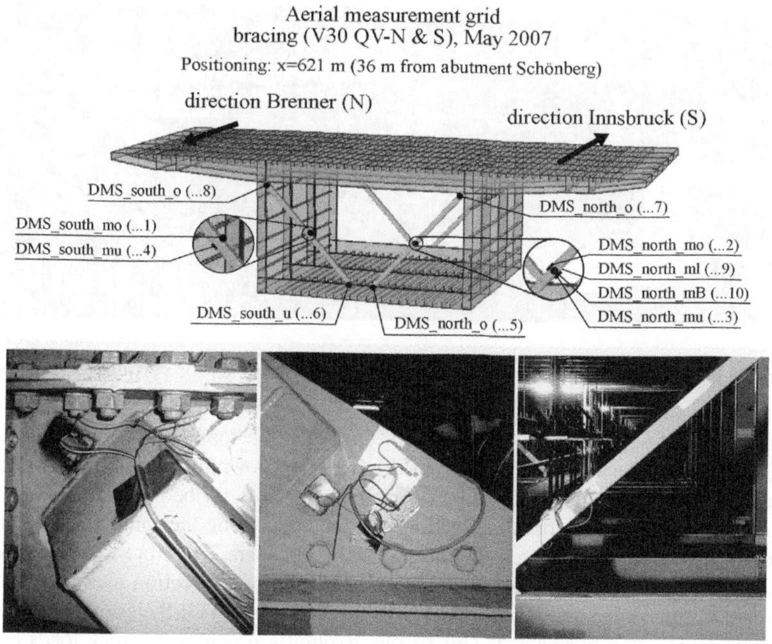

Figure 4.49 Strain gauge sensors installed along an assembly of two corresponding torsional bracings

4.1.5 Tailor-Made Loading Model for Performance Prediction

A step-by-step methodology has been prepared (Figure 4.3), that uses the isolated hot-spot areas for detailed fatigue analysis to determine how grave the present situation is with regard to the remaining service lifetime of the relevant structural members analyzed with the monitoring-based impact. In the following an insight is given into how the results of the mentioned one-week measurement campaign are extracted and implemented into the developed methodology for performance prediction.

First, accompanying video recordings were undertaken to eliminate possible uncertainties and to provide complementary information. Extensive data mining leads to frequency distributions including

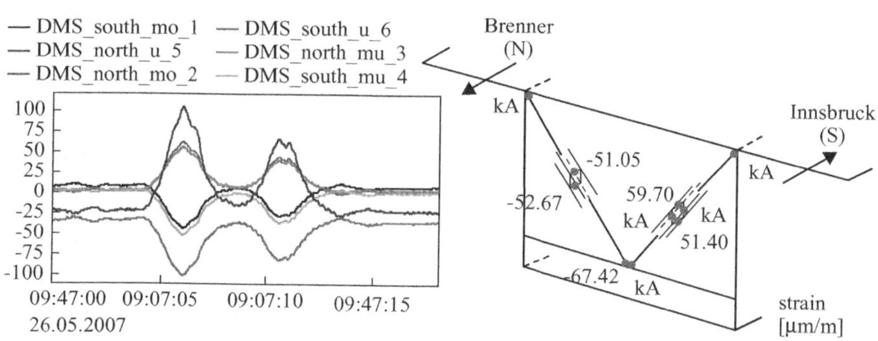

Figure 4.50 Exemplary response of both bracings in terms of measured strain due to a certain truck loading impact

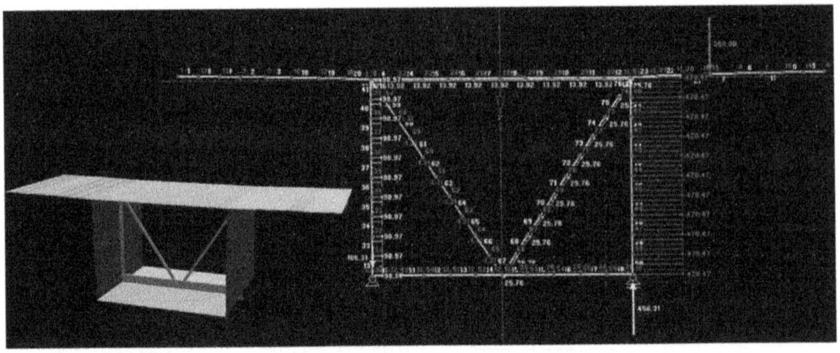

Figure 4.51 Effective axial forces by means of FE analysis for verification purposes

different kinds of loading scenarios (Figure 4.52). One of these loading configurations – two trucks simultaneously stressing the bridge's side span by driving in the downhill direction – was chosen to show its consequence in all of the three levels of performance analysis described (Figure 4.53).

Global response in terms of vertical deflection was recorded using the same assessment conception demonstrated in Figure 4.27 – adopted for the southern side span. Therefore the global response due to the chosen scenario can be observed via laser-supported global deflection measurement. In Section 4.1.4.2 the main feature of the permanent monitoring system – the DYGES algorithm (Dynamic Freight Traffic Classification) – was introduced. This feature explicitly provides the truck loads corresponding to the chosen global bridge span deflection.

Finally the effective axial forces and bending moments acting on the bracings as well as the stress cycles can be extracted from the monitored strain data recorded at the directly stressed bracings (Figures 4.49 and 4.50).

In the following it is explained how the established tailor-made loading model (Figure 4.52) is utilized for performance analysis of Level I (Global Impact): Monitoring-based impact data for a certain representative time period (one day, one week, e.g. Figure 4.54) are assessed using Rainflow analysis. To express the damage rate for this loading function each of the observed, typically occurring loading

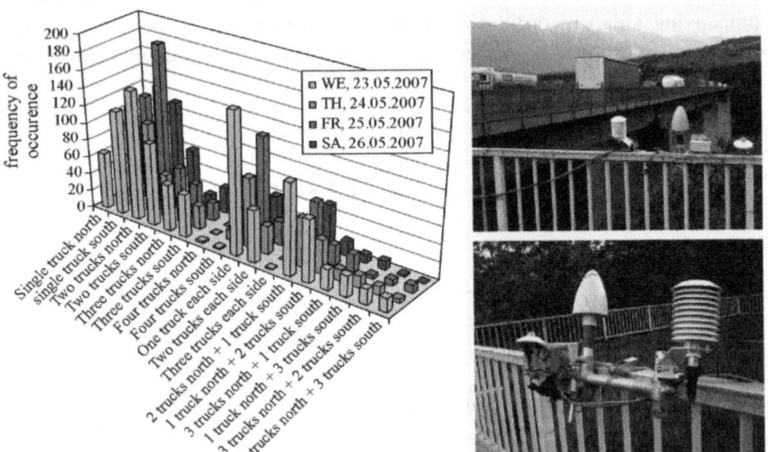

Figure 4.52 Video-supported classification of loading scenarios and resulting frequency distribution

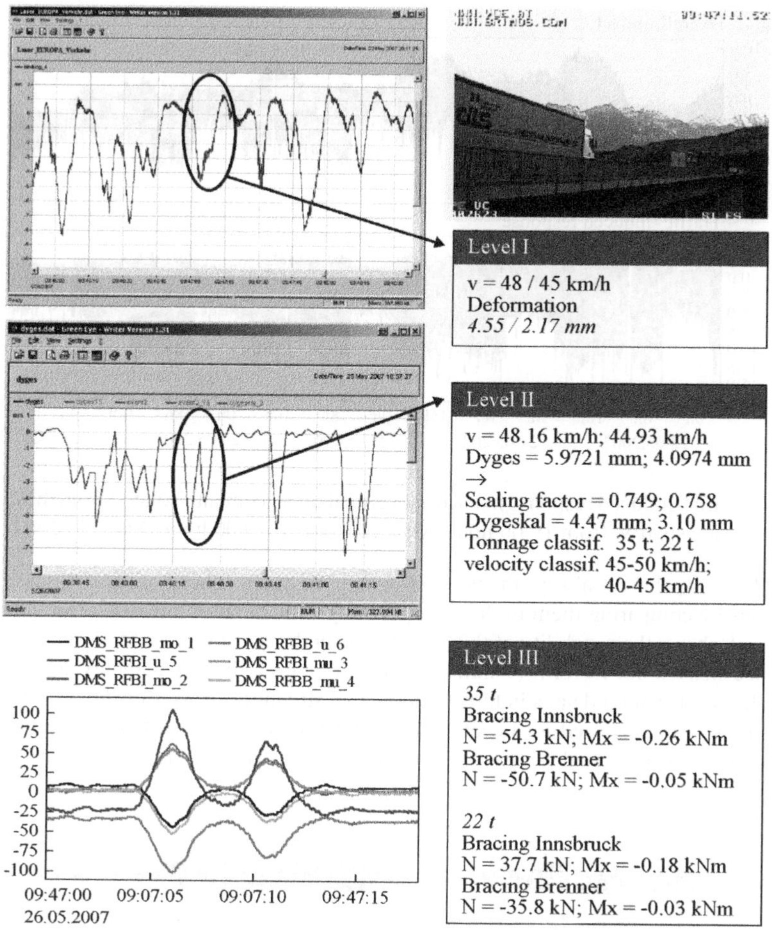

Figure 4.53 Randomly selected loading scenario and the structural response, expressed by means of the three-level approach

scenarios (Figure 4.52) are implemented one by one into a structural FE model using certain truck loads (DYGES). This leads to a corresponding structural stress cycle $\Delta\sigma$. With truck traffic causing High Cycle Fatigue, linear elastic material behavior can be assumed. Stresses and strains are assumed to remain elastic. This facilitates the calculations, as a single load case due to the chosen scenario is to be calculated in terms of stresses, while all other truck loads are automatically included within the Rainflow Matrix. The Rainflow Matrix, the calculated stress cycle value and the stress-life curve corresponding to the analysed relevant structural detail lead to a Damage Matrix (Section 4.1.2.3). This assumes that only the chosen loading scenario is occurring. Based on the frequency distribution of loading scenarios (Figure 4.52) the chosen entry gets its weighting within all other relevant loading cases to quantify the consumed loading capacity relative to a certain observation period. The same procedure is repeated for the other entries of the observed loading scenario distribution, leading to an accumulated damage per determined observation period.

In the course of the FE calculations, following the consequence for the fatigue resistance of structural details relevant for Level I, a parallel determination for those structural details relevant for Level III can

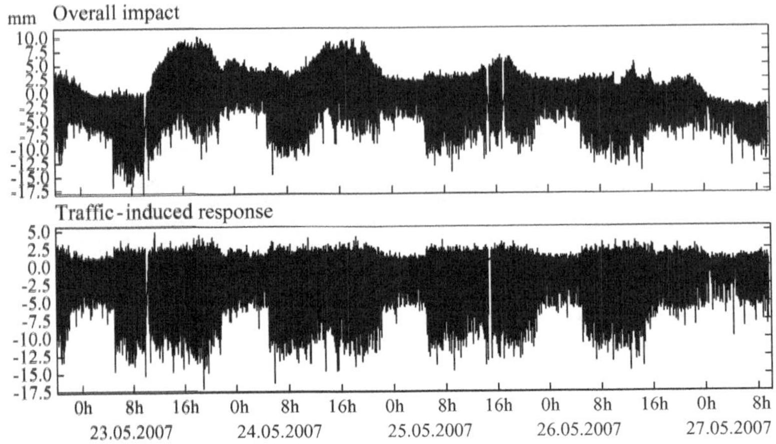

Figure 4.54 Traffic-induced response in terms of vertical displacement, isolated from the overall impact for the bridge's side span V, from 19:00 hours May 22, 2007 to 10:00 hours May 27, 2007

be done. Additionally there is always the possibility to make a cross-check of the stresses obtained from the calculations by comparing them to the ones given by explicit strain measurement at representative positions, which shows the capability of this integral methodology (Figure 4.55).

Performance analysis for Level II (Cantilever Impact) seems to be less complex, as the consequence on fatigue-relevant structural details in terms of principal structural stresses occurs perpendicular to the stress cycles previously analysed for Level I. The DYGES-based truck loading impact (Figure 4.56) is implemented into structural models again by means of a deformation constraint load case affecting the cantilever's outer edges.

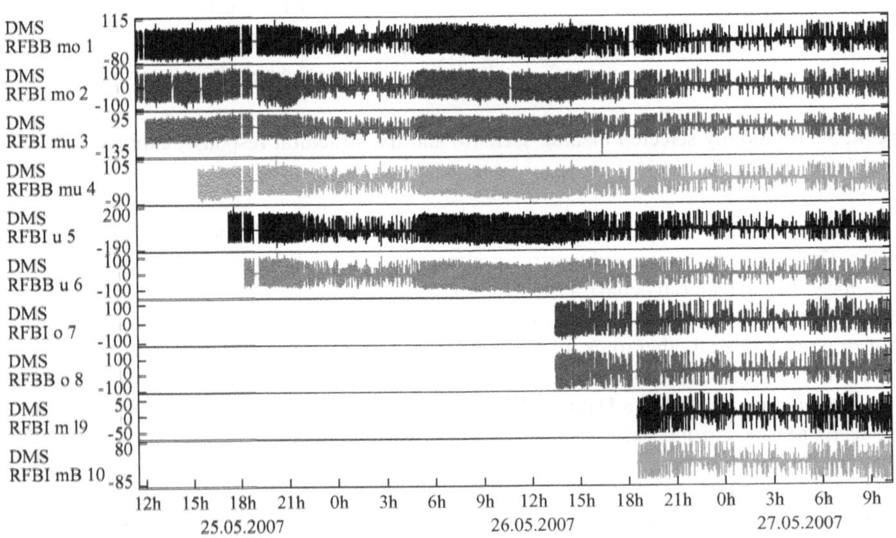

Figure 4.55 Successively added strain gauges from 11:00 hours May 25, 2007 to 11:00 hours May 27, 2007

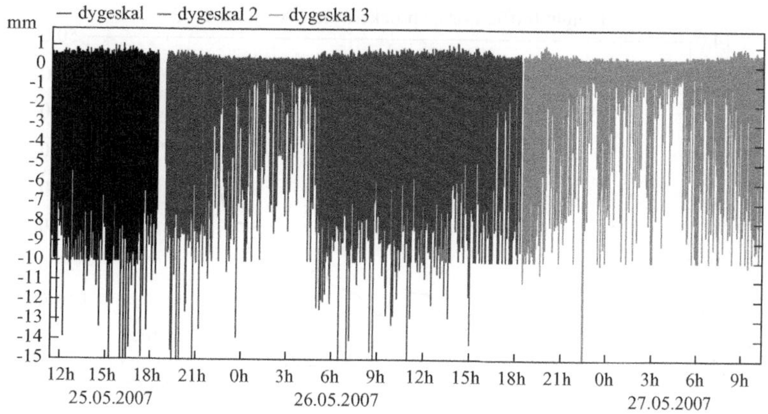

Figure 4.56 Corresponding pattern of the dynamic weight registration based on cantilever acceleration measurements (reproduced cantilever deformation via pattern recognition) – a feature implemented into permanent monitoring: 11:00 hours May 25, 2007 to 11:00 hours May 27, 2007

4.1.6 Remaining Service Lifetime by Means of Existing Traffic Data

The demonstrated damage calculations refer to a sufficient amount of workdays and weekend-days to enable a representative extraction of a damage rate per week, which consequently can be extrapolated to enlarged time periods. Thus the assessment of an analyzed detail via the Damage Matrix calculated for a measurement period of a whole year (i.e. "Damage-per-year effect") is possible to determine the consumption of the overall loading capacity per year. Detailed knowledge about the progression of the prevailing traffic from the very beginning up to the present, and the implementation of published future trend studies with regard to the next ten years (until 2015), are used for fatigue analysis (variation of traffic volume and variation of the notional truck weight). Figure 4.57 shows the increase of freight traffic volume at the Europabrücke. According to *Verkehrsentwicklung in Tirol - Berichte 1984–2006* (undated), traffic volume in 2006 increased to 472% relative to 1964 and is expected to grow 2.9% per year until 2015

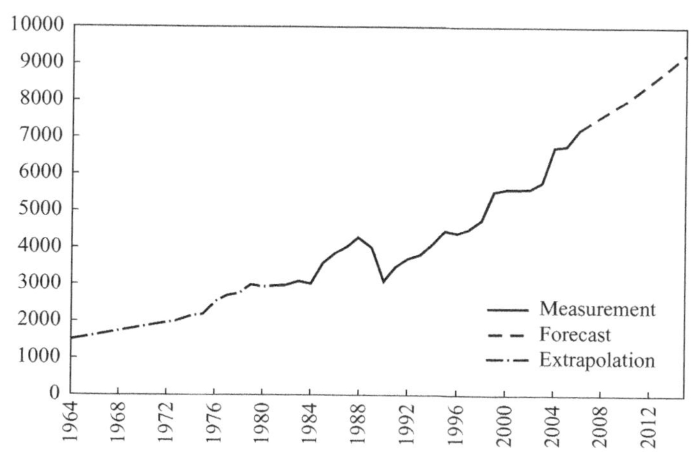

Figure 4.57 Progression of freight traffic volume at the Europabrücke from 1964 until 2015 (trucks/day)

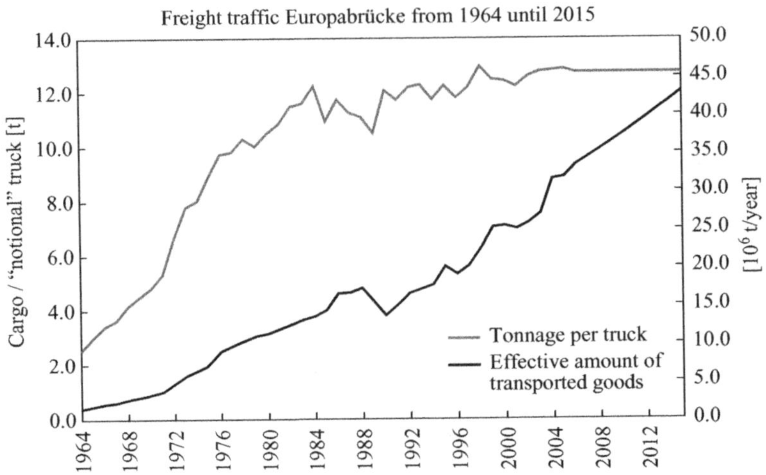

Figure 4.58 Effective amount of transported goods on the Brenner route compared to a calculated cargo per notional truck

(*Verkehrsprognose 2015 – vorläufige Ergebnisse hochrangiges Strassennetz Österreich* 2000). To derive every considered year's Damage Matrix affected by the variation of traffic volume, fatigue analysis approximately demands a uniform adaptation of the number of occurrences for all elements of the derived Rainflow Matrix before the extrapolation techniques of the measured impact for the whole lifetime are applied.

Figure 4.58 shows the increase of the effective amount of transported goods compared to a calculated cargo per notional truck. The calculations show that this truck weight in 2006 increased to 503% relative to 1964 and is assumed to have already reached a maximum. This means that a further increase of transported goods is likely to be a consequence of the still growing traffic volume. An adaptation for fatigue analysis due to the variation of the notional truck weight is realized by rescaling the Rainflow Matrix's subdivision.

The information included in both of these functions can be broadened crucially, as the DYGES-based freight traffic classification delivers exact truck weight data for the analysed bridge object, starting with the summer of 2004.

4.1.7 Conclusions and Future Work

The present contribution explicitly deals with measurement data and procedures, and how they were provided and conditioned for continuative performance analysis. The methodology provides a strictly in-situ-based loading input parameter for continuous fatigue analysis leading to a "tailor-made" performance prediction. In addition, strain gauge measurements in each of the considered levels of analysis (Level I to Level III) are done for verification purposes of the FE based stress-life approach. Thus the main goal – substitution of the standard's premises, referring to loading – has been reached in a quite innovative manner with regard to determining the consumption of the structure's overall capacity per year by means of a three-level approach.

One of the main innovations within the present methodology is the DYGES algorithm, which is necessarily based on a pattern recognition algorithm by Wenzel and Veit-Egerer (2008) in order to reproduce vertical cantilever deformations from accelerometers (a major feature of the permanent monitoring system). The development of the DYGES algorithm was nominated for the Austrian Award for Telematics, by the Federal Ministry of Transport, Innovation and Technology.

Besides the traffic-induced impact, additional monitoring campaigns revealed that there is a strong influence due to sun radiation. In addition to the described laser-supported global deflection measurements, the progression of the offset of the bridge's reference accelerometer in the lateral direction was transformed into an angle of inclination. The resulting temperature gradient function induces additional axial forces and deformations explicitly into the outer bridge spans. Approximate analysis indicates that this constraint's intensity level can occur up to the range of the traffic impact itself. The affect due to temperature and radiation impact needs to be analyzed separately before it is necessarily superimposed with the strictly traffic-induced fatigue impact.

Contrary to the general doctrine on the structural performance analysis of welded components, the authors assume that the influence of temperature leading to varying mean stresses will have to be considered. The authors are convinced that the already discussed shrinkage effects that lead to residual stresses in the welds (Section 4.1.2.1) are minimized or have already vanished in this almost 40-year-old and constantly stressed structure. Thus, varying mean stresses become relevant even for welded structural members and demand the application of appropriate correction rules.

Starting in May 2007 the in situ distribution of temperature impact as well as solar radiation itself is assessed via permanent monitoring using a profile of temperature sensors over a certain box girder's cross-section (Figure 4.59). An explicit temperature load case is to be created and again implemented into the FE calculations.

The implementation of strictly measurement-based loading impact into FE calculations strongly supports quantitative estimation of the service lifetime via separation of fatigue-relevant loading cycles from the randomly occurring overall traffic (Figure 4.60). The Palmgren-Miner-based accumulation of all calculated Damage Matrices from the very beginning of the bridge's existence up to now leads to the remaining capacity of loading cycles for the analyzed detail. As a superior conclusion in addition to a quantitative estimation of the service lifetime, another key parameter, fatigue relevance (FR), is derived:

$$FR = \frac{\sum n_i}{ADTV} \tag{13}$$

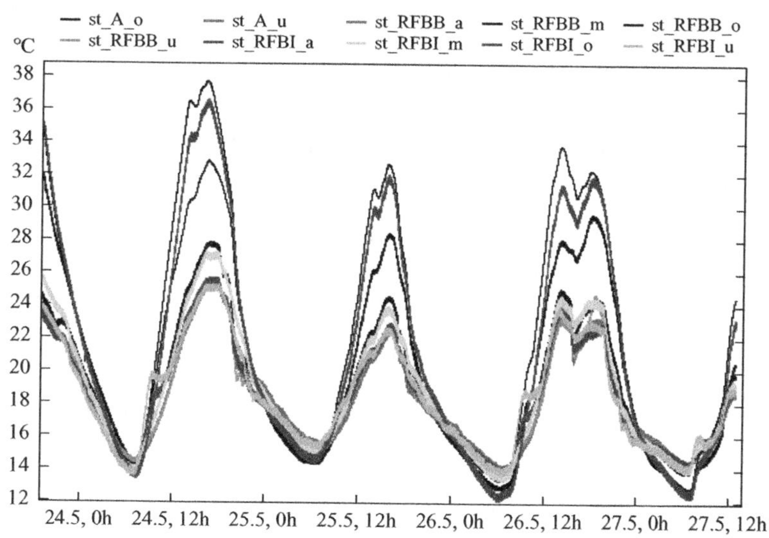

Figure 4.59 Integral assessment of environmental conditions: steel temperature and air temperature along a certain cross-section (pier V): 19:00 hours, May 23, 2007 to 13:00 hours May 27, 2007

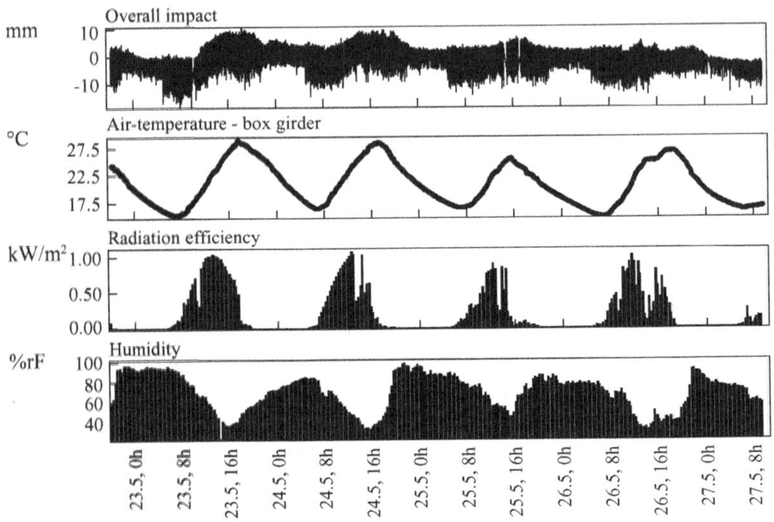

Figure 4.60 Integral assessment of traffic impact and comparison with environmental conditions: global vertical deformations versus air temperature versus radiation efficiency versus humidity: 20:00 hours May 23, 2007 to 12:00 hours May 26, 2007

It separates the remaining fatigue-relevant loading cycles n_i (registered by sensors and taken from the Damage Matrix) from the randomly occurring traffic ($ADTV$, average daily traffic volume). It is obvious that the investigation's results are going to be improved by progressive stages the longer the observation period lasts.

This whole conception assures the determination and observation of slowly progressing processes in the structure, which might lead to local damage or to deterioration of the structure's operational integrity. The results of these hot-spot analyses with regard to the fatigue resistance itself will be undertaken and discussed in the course of further publications.

4.2 Condition Compensation in Frequency Analyses

4.2.1 Prelude

Recent publications raised doubts that damage in structures can be detected by the application of frequency analyses. In fact temperature changes very often show larger reactions in spectra rather than any smaller damage. Besides temperature, other environmental influences such as solar radiation create changes in the structural systems that have to be considered.

The present contribution demonstrates the capabilities of new compensation methods in frequency analyses. When the environmental conditions are monitored together with the structural response, a proper reaction can be predicted. The described compensation process in general deals with the following sources of input:

- temperature (daily and annual cycles);
- compensation of live load (moving vehicles, etc.);
- influence of wind loads;
- bearing friction;
- restoring forces;

- change of boundary conditions;
- impact energy;
- instrumentation.

After elimination of all operational and environmental factors, stable frequencies are achieved:

$$f_{total} = f_0 + \sum_{i=factors} f_i \tag{14}$$

where f_{total}, f_0 and f_i are vectors of total, structural and influenced frequencies and i denotes the operational and environmental factors considered. Any deviation can be interpreted as damage or an extraordinary event. This procedure opens up new possibilities for structural management and lifetime prediction. The present contribution demonstrates some investigation on two sources, which are assumed to be the major ones.

4.2.2 *Compensation for Temperature*

The following analysis is based on investigations on the Europabrücke – a well-known Austrian steel bridge near Innsbruck, opened in 1963 – which is part of one of the main alpine north–south routes for urban and freight traffic. A long-term preoccupation of VCE with BRIMOS on the Europabrücke (since 1997) led to the installation of a permanent monitoring system in 2003 (Wenzel and Pichler 2005).

The bridge's reference sensor (3D forced balance accelerometer) is installed within the main span, at a distance of 0.4 times the span length from pier II. At this base point, global stiffness and its dependence on several environmental influences are assessed (sampling rate = 100 Hz, file length = 330 s).

By evaluating the results (frequency spectra) of several measurements, telescoping them together and viewing them from above (so-called trendcards), the visuals in Figure 4.61 are obtained, which show the main span's relevant vertical stiffness patterns of a particular day with a distinctive progression of temperature. For the sake of completeness, the corresponding frequency spectra themselves are shown in Figure 4.62, again over the same day. An individual procedure has been developed in IMC (2005), which contains some measurement preconditioning (offset elimination and bandpass filtering). To enable more stabilized, automatic peak-picking in different ranges of frequency, the response spectra are smoothed in the course of frequency assessment.

The complementary relation between stiffness and the air temperature itself (registered at the bridge's base point directly above pier II) is obvious and can be interpreted as a long sinusoidal wave of the main span in the vertical direction. In the course of the described procedure the described bandpass filtering could be replaced by a further optimized smoothing of the frequency spectra to stabilize the accuracy of peak-picking.

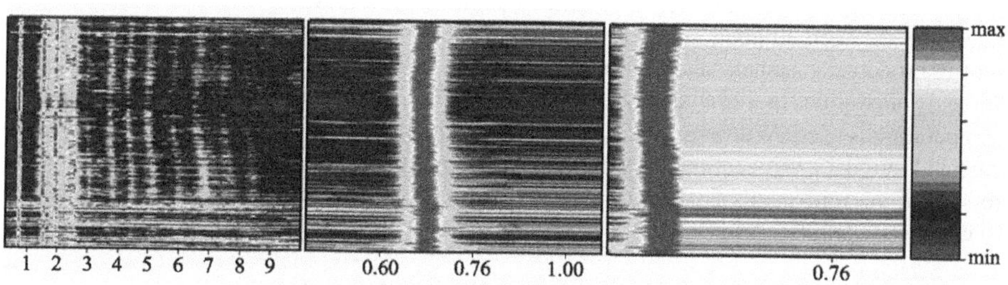

Figure 4.61 Trend of stiffness during one day: 0.30–10/0.30–1.10/0.60–0.80 Hz

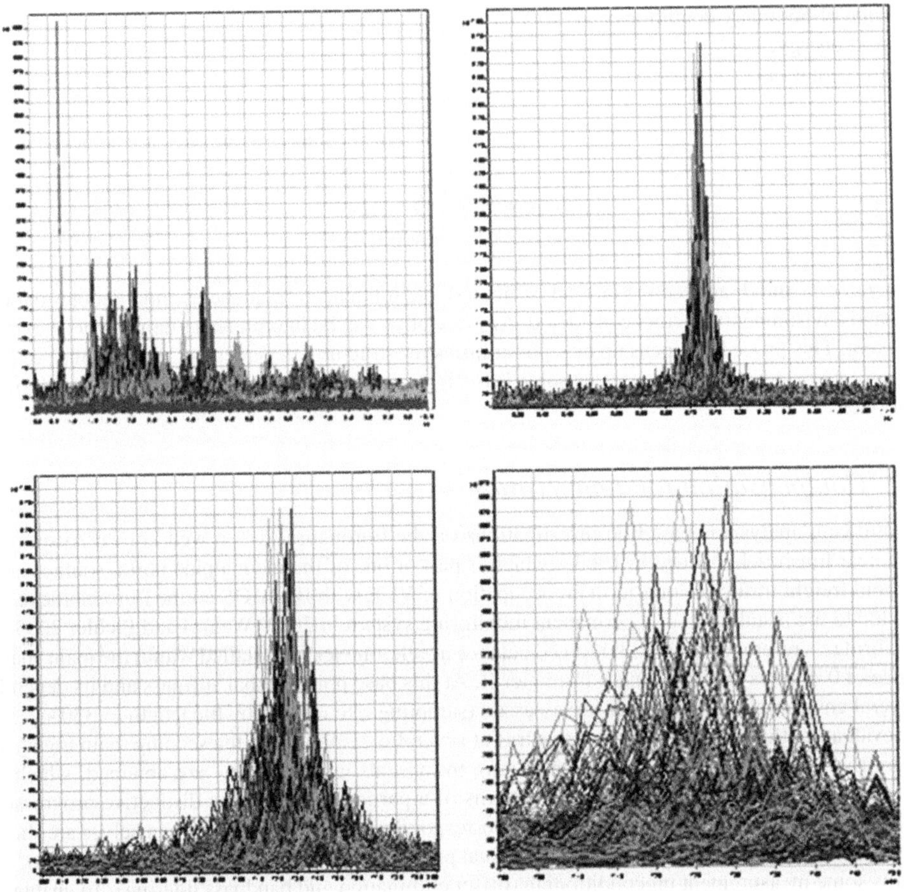

Figure 4.62 Front views of trendcards for one day: 0.30–10/0.30–1.10/0.60–0.80/0.68–0.74 Hz

The permanent monitoring system exhibits a remarkable loading impact, as the bridge is currently stressed by more than 30 000 motor vehicles per day (approximately 20% of them are freight traffic). By applying the previously described method to the reference sensor's measurement data for the whole day, a progression of stiffness that consists of 281 single peaks is obtained (Figure 4.63: left) and represents randomly occurring ambient and forced vibration conditions (i.e. scatter).

To describe the verified phenomenon mechanically, it has to be focused on the temperature dependence of the roadbed's asphalt layer, because the change of steel characteristics under varying climate conditions is negligible. In a first step, the characteristic relationship of the dynamic Young's modulus with temperature is used (Willberg 2001). Due to this relation, the temperature-sensitive asphalt layer is implemented into the cross-sections of the global structural analysis model, which leads to a distinctive progression of the mid-span's flexural rigidity (Figure 4.64: right). According to the following widely known equation:

$$f_i = \frac{\lambda_i^2}{2\pi L^2}\sqrt{\frac{EI}{m}} \tag{15}$$

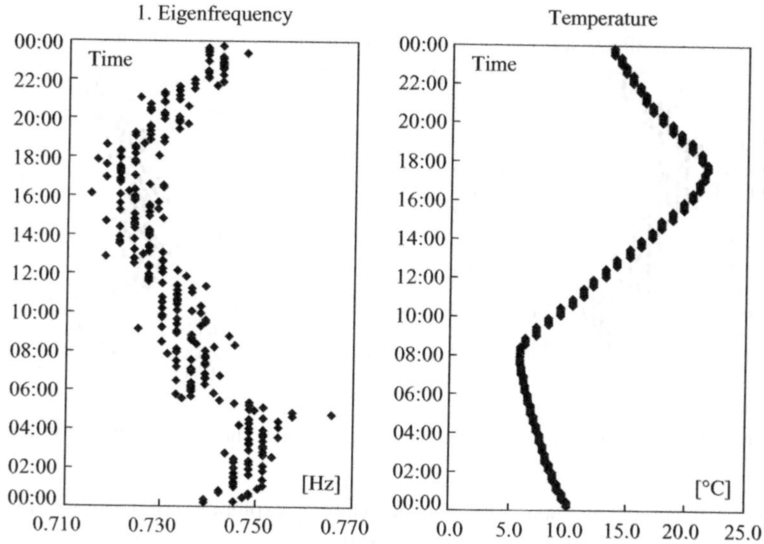

Figure 4.63 Pattern of first eigenfrequency and its obvious dependence on temperature

the frequency of vibration is proportional to the square root of the moment of inertia (Blevins 1979). For that reason, a curve in terms of frequency needs to be generated (Figure 4.65: middle) for the next step, when a temperature-based stiffness path is eliminated from the overall trend (from Figure 4.65, left, to Figure 4.65, right). The trend obtained shows very clearly the remaining impact of freight traffic itself, which strongly affects the timeframe between 5 am and 10 pm, when trucks are allowed to pass over the bridge and cause two characteristic offsets during the course of the day.

4.2.3 *Compensation for Additional (Moving) Masses*

The modified trend of the main span's stiffness already includes lots of characteristics of the prevailing freight traffic progression (Figures 4.66 and 4.67). Unfortunately traffic data from the competent

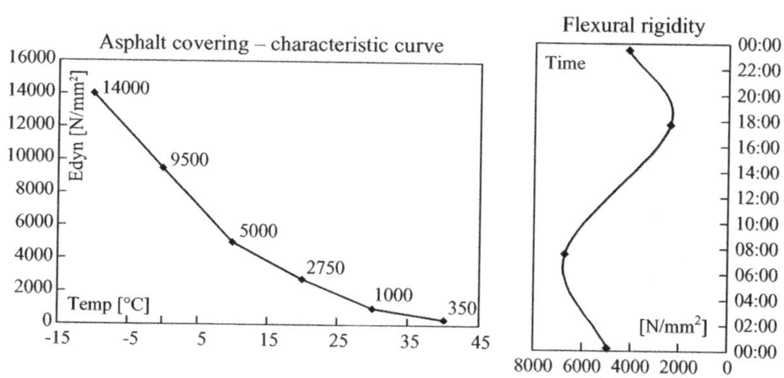

Figure 4.64 Progression of the asphalt layer's flexural rigidity with temperature

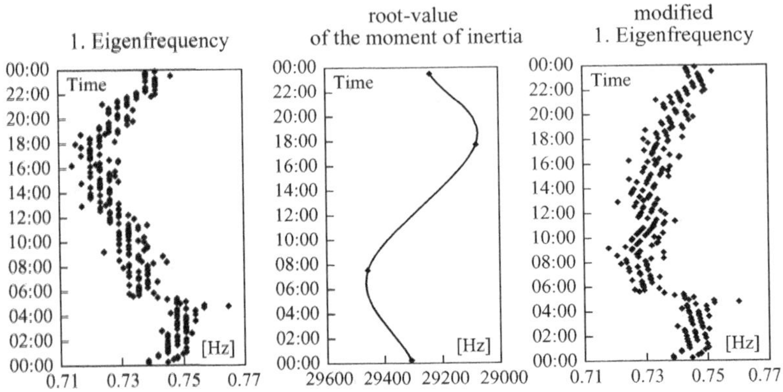

Figure 4.65 Pattern of first eigenfrequency before and after the compensation for temperature

authorities are available only per hour. For some introductory exploration of the approximate additional mass compensation, further steps need to be undertaken.

Frequency of vibration, based on Equation (15) again, is inversely proportional to the square root of the mass. This means that live loads cause an increase in effective mass, which leads to hourly

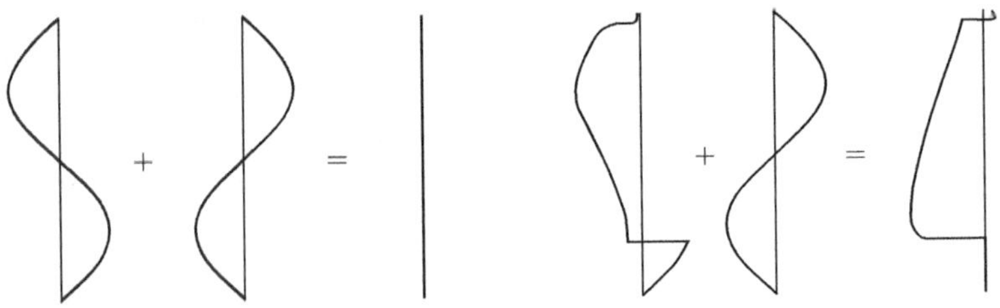

Figure 4.66 Comparison of expected (left) and actual (right) consequences of temperature-compensated natural frequency patterns

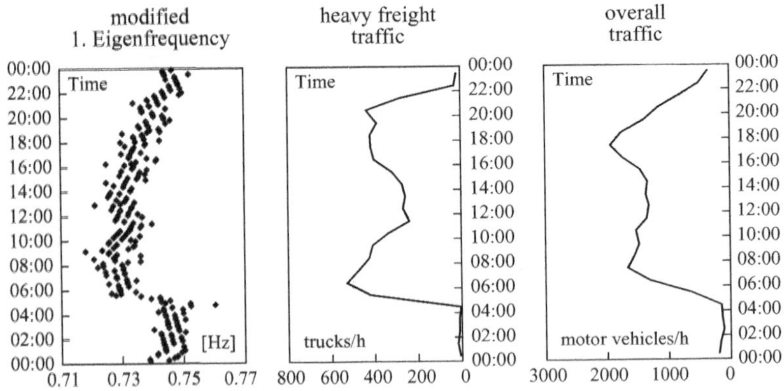

Figure 4.67 Modified pattern of stiffness strongly affected by traffic loading (additional moving masses)

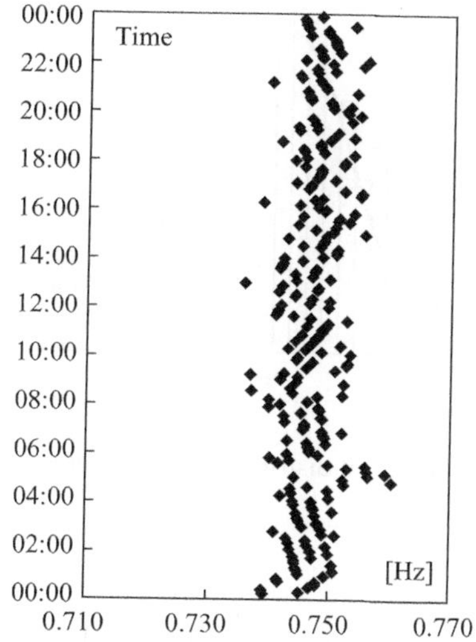

Figure 4.68 Stiffness pattern after approximate compensation for additional masses

calculated factors to modify the fluctuating frequency. Due to that relation the scattered frequency trend is straightened in accordance with the modal contribution of trucks per hour (Figure 4.68).

In fact the permanent monitoring system's present configuration provides the possibility to develop a more sophisticated and more reliable, strictly measurement-based method. Forced balance accelerometers located at a defined distance along the cantilever's outer edges – in both directions of traffic – enable the verification of recurring truck passages and their related velocity and tonnage without any disturbance of traffic. A dynamic freight traffic registration system was developed (a certain pattern recognition procedure introduced in Veit and Wenzel (2006)), which utilizes accelerometer-based, reproduced cantilever deformations. In this way the moving loads within each measurement file – passing the main span simultaneously – could be identified and would lead to a shifting of the single peak in the frequency response spectrum, which represents the registered time-history in each measurement file.

4.2.4 Conclusion

As permanent monitoring systems produce huge amounts of data, they have to be processed systematically in order to exploit the information fully. For easy handling it is proposed that statistically based threshold levels are calculated. In this way continuous monitoring systems – providing information about changed modal parameters under "normal" operational conditions – can be used to trigger warning and alarm levels with regard to damage assessment.

The acquisition and elaboration of the quantities provided by the installed instrumentation allow a structural behavior model to be set up that is considered the "regular model" (baseline model). The periodic elaboration of the acquired measurements and the comparison with the baseline model allow indicators of potential structural damage to be pointed out. The availability of periodic surveys of the cause quantities allows statistical models of the structural behavior, to be set up whereby the structural response is statistically correlated with the trend of the cause quantities. These models allow a control in time of the structural response by pointing out the "weight" of the cause quantities. The definition of alert threshold levels from analysis of the historical database (e.g. extreme-value analysis) may be illustrated as in Figures 4.69 and 4.70.

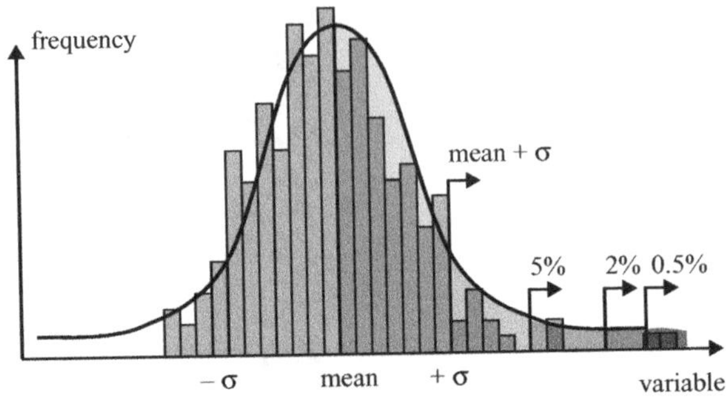

Figure 4.69 Histogram and best-fit distribution-based determination of threshold values

4.2.5 Outlook

The discussed approach, benefited by permanent monitoring, allows the conducted goals to be reached even if the experience already gained with the applied methods is still of some approximation in character. The results are very promising, although using air temperature instead of the temperature of the structural elements. The approach represents an innovation in stiffness assessment appropriate for long-term application. The goal of generating frequency progressions over time without major environmental and operational impact has come within reach. The procedure will be optimized in progressive stages as soon as the cantilever-sensor-based approach is implemented.

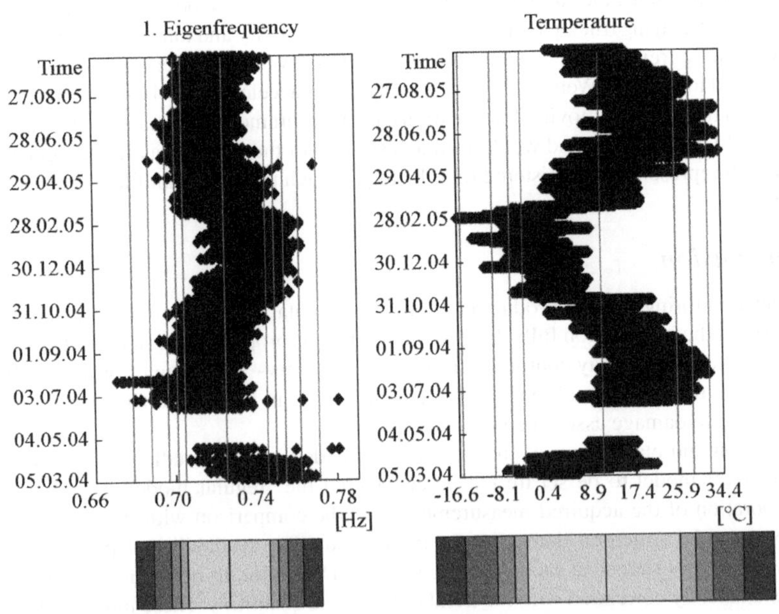

Figure 4.70 Response for 18 months using threshold levels (statistical time-history) with 5/2.75/2.5/1/0.135% probability of exceedance

4.3 Model Updating and System Identification

Considerable progress has been made in system identification and model updating. Techniques and approaches are discussed in Chapter 7. It comprises a very wide field with many valid approaches. Various associations and interest groups are extensively discussing this subject:

- IMAC, a US initiative with an annual conference;
- EVACES, a European initiative with bi-annual events;
- IOMAC, another European initiative on this subject with bi-annual events;
- ISHMII, a global initiative with bi-annual events;
- and many dedicated workshops and symposia.

The vision is a future integrated system that holds all the necessary information on our structures to allow permanent online model updating and subsequent condition assessment.

4.4 Performance Assessment (Damping, Time-Histories)

4.4.1 Introduction

The design of civil engineering structures is characterized by two main features: load-carrying capacity and serviceability. However, each structural system undergoes various environmental and loading influences during its service life, which can cause a significant damage accumulation. Consequently, the structural carrying capacity and serviceability are enormously affected. Therefore the need for reliable nondestructive evaluation techniques and detection of damage at the earliest possible stage has been pervasive throughout the civil engineering community in the last decade. The process of implementing damage detection strategies can be referred to as "structural health monitoring." The so-called vibration-based health monitoring techniques rely on the fact that damage causes changes in the local structural damping (energy dissipation) and stiffness. As a consequence, the global dynamic properties of the structure, e.g. eigenfrequencies, mode shapes, modal damping, etc., should be influenced.

Several methods of damping estimation are known today. However, their performance is heavily influenced by the quality of the recorded data: length of the time series, presence of measurement and system noise, system excitation, etc. In addition, different types of energy dissipation could be present at any given time instant. Some of them can be associated with material properties, others with the system boundary conditions. Effects of contact friction can be observed in some cases as well. Thus, the damping estimation procedure requires a very careful use of numerical procedures. Furthermore, an engineering understanding and critical considerations are important for a reliable identification of the presented damping properties.

This section deals with the relationship between the sensitivity of four common damping estimation techniques (see Section 4.4.2) and the quality of the recorded structural response. For this purpose, the routines were developed and tested on some numerical models where input, presence of noise and energy dissipation sources could be controlled well. The results obtained were promising enough to apply the procedures on vibration response data recorded on different types of real civil structures in a second step. The latter results are the main topic of the following pages.

Besides some satisfying examples, several other problems will be mentioned such as multiple closely spaced predominant frequencies. Furthermore, parameters such as filter order, overlapping, reference frequencies, etc. will be varied in order to stress the importance of a careful parameter choice to ensure reasonable results (see Section 4.4.3).

Finally, the basic conclusions are summarized in Section 4.4.4, with some advice for further interest and development.

4.4.2 Damping Estimation Techniques

Today, different methods for estimating the damping coefficient are known. In order to provide a general overview, the author has chosen five of them to be explained in the following sections.

4.4.2.1 Half-Power Bandwidth Method

This common method is very simple and quick and has been practiced already. It is based on the fact that the width of the response amplitude of a single-degree-of-freedom (SDOF) system is proportional to the system's damping ratio. Let f_r be the resonant frequency, i.e. the forcing frequency at which the largest response amplitude occurs. Furthermore, define frequencies f_1 and f_2 as the forcing frequency at which the amplitude is $1/\sqrt{2}$ times the resonant amplitude. Then, for small damping ratio ζ:

$$\zeta = \frac{f_2 - f_1}{f_r} \tag{16}$$

Further details can be found in Bendat and Piersol (1993). Put into practice, this method often does not seem to be a reliable technique for estimating the damping ratio, as reported by Peeters (2000). A possible problem is, for example, when the spectral peak is missing. In fact, there is a good chance that the exact spectral peak will not be available. If the peak is missing, the corresponding bandwidth tends to be greater than the real value and damping is likely to be overestimated.

4.4.2.2 Random Decrement Technique (RDT)

The random decrement approach is a time-domain method developed at NASA and was first introduced by Henry Cole (1973). The basic idea is to pick out time segments and average them whenever the time series fulfills a given so-called "trigger condition":

$$\hat{D}_{XX}(\tau) = \frac{1}{N}\sum_{i=1}^{N} x(t_i + \tau)|x(t_i) = a \tag{17}$$

When making use of Equation (17), the random part of the time series tends to average out. Thus, the obtained RDT signature $\hat{D}_{XX}(\tau)$ can be interpreted as free decay, which is the fundamental idea behind extracting damping ratios by means of RDT.

The method is not very effective when two or more outstanding system frequencies coexist. In this case, it is highly recommended to use a bandpass filter before processing the data. As the filter order greatly influences the quality of the results, an additional analysis was performed in which filters of 1st and 10th order were compared (see Figure 4.79).

4.4.2.3 Logarithmic Decrement

When interpreting the RDT signature as free decay, the damping ratio can be estimated by means of the logarithmic decrement δ described by the following relationship (see Chopra 2000):

$$\delta = \ln \frac{x_n}{x_{n+1}} = \frac{2\pi\zeta}{\sqrt{1-\zeta^2}} \tag{18}$$

where x_n and x_{n+1} are amplitudes of two successive free vibrations of like sign. Thus, for small damping ratios, one finally gets:

$$\zeta \approx \frac{\delta}{2\pi} \tag{19}$$

4.4.2.4 Curve Fitting

The application of curve fitting on the system decay has been recommended particularly when coexisting frequencies are closely spaced. This is done by making use of:

$$u(t) = e^{-\zeta\omega_n t}\left[u(0)\cos\omega_D t + \frac{\dot{u}(0) + \zeta\omega_n u(0)}{\omega_D}\sin\omega_D t\right] \tag{20}$$

where:

$$\omega_D = \omega_n\sqrt{1-\zeta^2} \tag{21}$$

By fitting the function $u(t)$ in a least-squares sense to the RDT signature $\hat{D}_{XX}(\tau)$, the damping ratio ζ can be obtained directly.

4.4.2.5 Stochastic Subspace Identification

This method was developed by Van Overschee about 20 years ago. The general stochastic identification problem can be described as follows. Given s measurements of $y_k \in \mathbb{R}^l$ generated by the unknown stochastic system of order n:

$$x_{k+1}^s = Ax_k^s + w_k \tag{22}$$

$$y_k = Cx_k^s + v_k \tag{23}$$

with w_k and v_k zero mean, white vector sequences with covariance matrix:

$$E\left[\begin{pmatrix} w_p \\ v_p \end{pmatrix}\begin{pmatrix} w_q^T & v_q^T \end{pmatrix}\right] = \begin{pmatrix} Q & S \\ S^T & R \end{pmatrix}\delta_{pq} \tag{24}$$

determine the order n of the unknown system as well as the system matrices $A \in \mathbb{R}^{n\times n}$ and $C \in \mathbb{R}^{l\times n}$. By using the developed technique, all of the matrices in Equations (22–24) can be determined.

The damping value can finally be estimated by making use of the eigenvalues λ_i of matrix A and the relationship:

$$\mu_i, \mu_i^* = -\zeta_i\omega_i \pm i\omega_i\sqrt{1-\zeta_i^2} \tag{25}$$

where:

$$\mu_i = \frac{1}{\Delta t}\ln(\lambda_i) \tag{26}$$

Stochastic subspace identification has become a common approach. The mathematical concepts behind it as well as additional information may be found in van Overschee and De Moor (1996), or Peeters (2000).

4.4.3 Results

When developing an RDT routine several trigger conditions were implemented, four of which can be found in Asmussen (1997). Most of the results presented in this paper were obtained by using the positive point trigger condition. Thus, a segment was picked out each time the time series was greater than a given threshold. (see Figure 4.79, which shows the level-crossing trigger condition, i.e. a segment is picked out every time the threshold is crossed).

Since there are several outstanding frequencies in both of the examples that will be presented later on, a Butterworth bandpass filter was used that filters a band of width 2% of the Nyquist frequency. Unless anything else is specified, the order was chosen to be 1.

In the present section, the vibration response of two real structures was investigated: a cable and a pre-stressed concrete bridge. For the latter example two structural states were analyzed, namely undamaged and damaged. The structural damage was artificially induced by eliminating some of the tendons. The sampling time rates were $\Delta t = 0.005$ s for the cable and $\Delta t = 0.01$ s for the bridge, respectively.

The acceleration signals were windowed with a length of 4125 sample points and an overlapping of 50%. Then, the damping for each window was estimated and plotted by squares. The damping for the whole signal was calculated and depicted by a solid line over the whole frequency amplitude range. Note that the damping was obtained for the fundamental eigenfrequency, which was selected from the frequency domain representation of the whole signal.

The choice of method and, in the case of RDT, the different parameters cause both useful and useless results. One aim of this section is to provide support in the final step of interpreting the results and eliminating any wrong ones.

4.4.3.1 Cable Structure

The first example shows the measurement data of cables of the "Berliner Brücke" in Halle/Saale, Germany. The time domain can be seen in Figure 4.71. It is a good example of the "frequency separation problem" that tries to extract the frequency of interest for the late damping estimation. Figure 4.72 shows the frequency response function. Since many outstanding frequencies are very closely spaced, many of them still pass the filtering. Thus, one often obtains inaccurate damping values, e.g. negative damping ratios as in Figure 4.73.

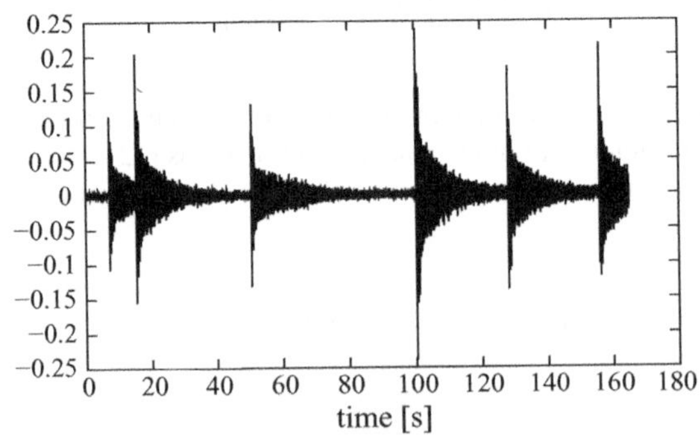

Figure 4.71 Vertical channel of the cable of "Berliner Brücke"

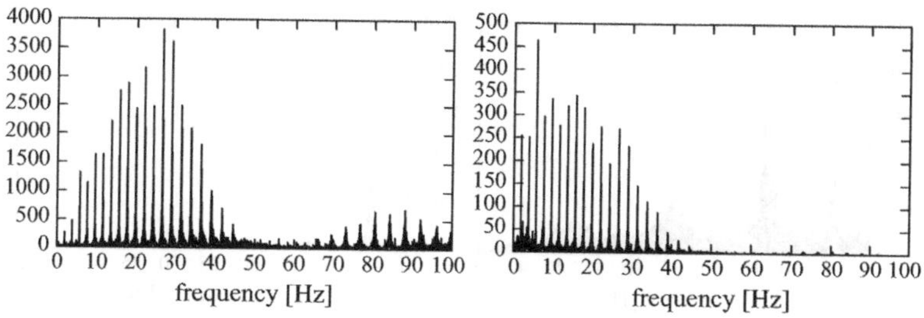

Figure 4.72 Cable structure: frequency response function unfiltered (left) and filtered according to the fundamental eigenfrequency at 2.3 Hz (right)

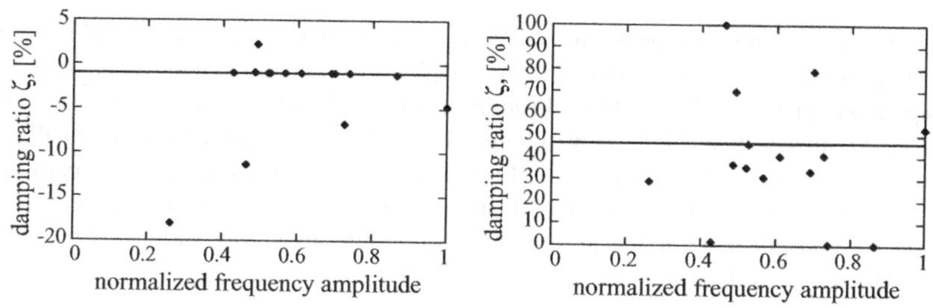

Figure 4.73 Cable structure: results from the damping estimation via RDT and positive point triggering. On the left, the values are obtained by using the logarithmic decrement (average: −1%) and on the right by using curve fitting (average: 46.35%)

4.4.3.2 Pre-stressed Concrete Bridge

The second example in question is a well-known single-span post-tensioned concrete bridge crossing a highway near Regau, Austria, which represents a typical monitoring case. The measurements were done in two steps: before and after artificially inducing structural damage by eliminating some of the tendons. Again the signals were windowed with a length of 4125 sample points and an overlap of 50%. Figure 4.74 shows the time series of the measurements. The signal was filtered between 4 and 5 Hz by a first-order Butterworth filter (see Figure 4.75).

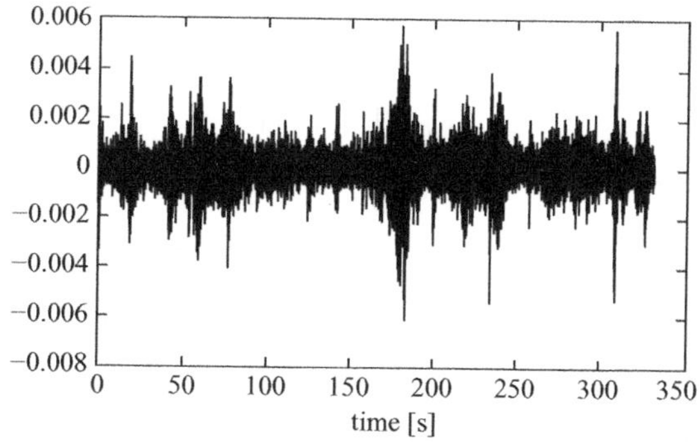

Figure 4.74 Bridge structure, undamaged

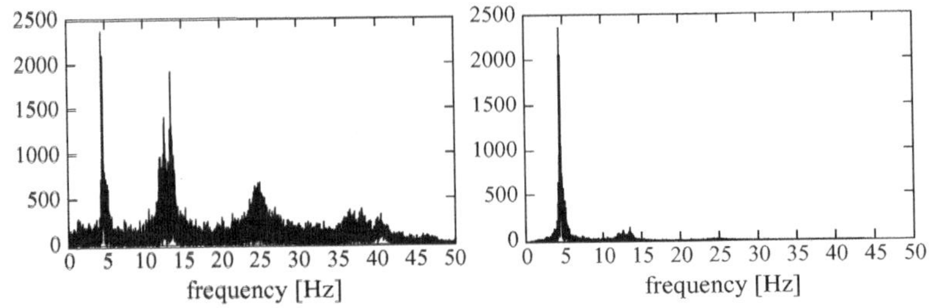

Figure 4.75 Bridge structure, undamaged: frequency response function unfiltered (left) and filtered according to the fundamental eigenfrequency at 4.5 Hz (right)

Figures 4.76 and 4.77 present the results of the four different methods mentioned in Section 4.4.2. Again, the damping values of the different windows are plotted by squares and the damping with respect to the whole signal is depicted by the solid line. In addition, the trend of the damping values is represented by a dashed curve. Figures 4.76 and 4.77 show that the obtained damping correlates well in all the methods used. However, using half-power bandwidth often leads to discrepancies, as was already reported by Peeters (2000). Nevertheless, the theory-based progression "frequency amplitude vs. damping ratio" can be clearly recognized: low amplitudes correspond to higher damping, and vice versa. This trend is depicted in all figures by a dashed curve.

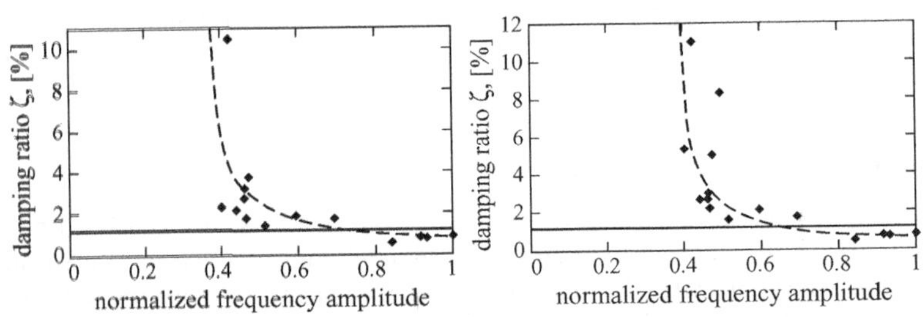

Figure 4.76 Bridge structure, undamaged: results from the damping estimation via RDT and positive point triggering. The values on the left were obtained by using the logarithmic decrement (average: 1.25%) and on the right by using curve fitting (average: 1.19%)

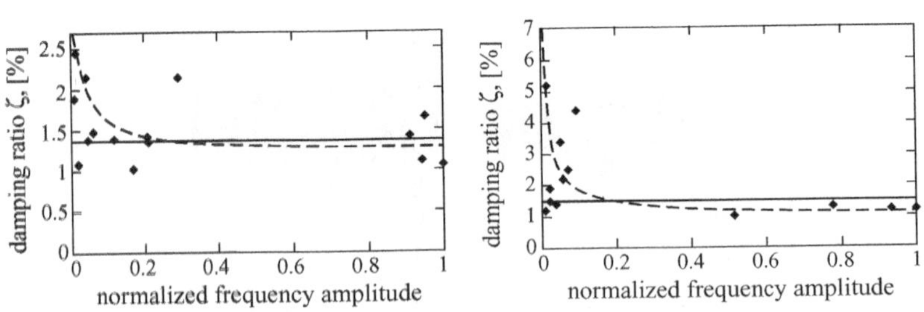

Figure 4.77 Bridge structure, undamaged: results from the damping estimation; on the left by using half-power bandwidth (average: 1.37%) and on the right by using stochastic subspace identification (average: 1.50%)

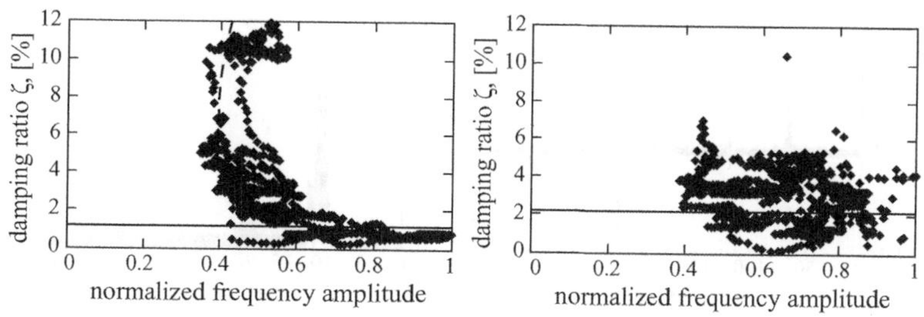

Figure 4.78 Bridge structure, 99.61% overlapping: results from the damping estimation via RDT, positive point triggering and curve fitting. On the left, data of the undamaged bridge were analyzed (average: 1.19%) and on the right measurements of the damaged bridge were investigated (average: 2.21%)

Looking at Figures 4.76 and 4.78 it seems useful to ignore windows in which normalized frequency amplitudes are smaller than 60% of the maximum, since results become more and more unstable below this threshold (especially for undamaged structures).

In a next step parameter studies were performed. Firstly the overlapping range was increased by 99.61%. The results from an RDT analysis with a positive triggering condition and curve fitting of the signature of the bridge structure, both undamaged and damaged, are presented in Figure 4.78. The damping progression for the undamaged structural state shows a very good correlation between theory and the previous analysis with 50% overlapping. For the damaged state no trend for the damping ratios can be estimated. This is probably caused by the presence of serious damage. However, the damping ratio obtained for the whole files can be used as a damage indicator: $\zeta_{undamaged} = 1.19\%$ and $\zeta_{damaged} = 2.21\%$. A loss of about 28% from undamaged to damaged state can be detected in the corresponding frequency amplitudes. These facts verify the theoretical considerations regarding the presence of damage.

As mentioned previously, the influence of the filter order for the purpose of the RDT-based damping estimation is presented in Figure 4.79. In other words, a higher filter order leads to an enormous discrepancy. Very good results are obtained for a first-order filter. This is also verified by means of theoretical signals generated by the Newmark time-integration algorithm with predefined damping.

Damping can be estimated with respect to different eigenfrequencies. The selection of the reference eigenfrequency requires some engineering know-how. Estimated damping for two different eigenfrequencies selected from the spectrogram of the damaged bridge structure (Figure 4.80) is shown in Figure 4.81.

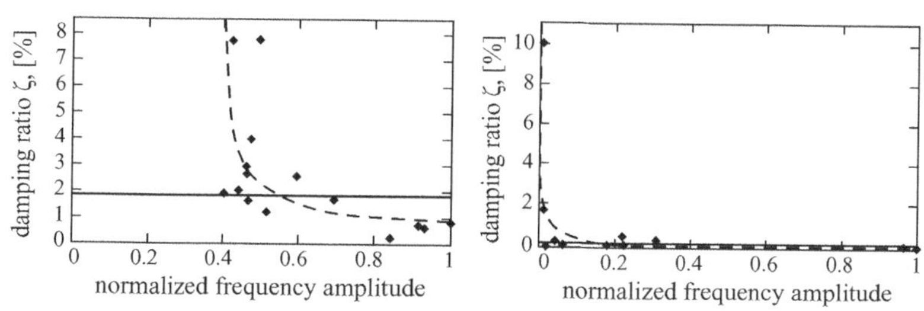

Figure 4.79 Bridge structure, undamaged, 50% overlapping: results from the damping estimation via RDT, level-crossing triggering and logarithmic decrement. On the left, a first-order Butterworth filter was used (average: 1.81%) and on the right a tenth-order Butterworth filter was used (average: 2.21%)

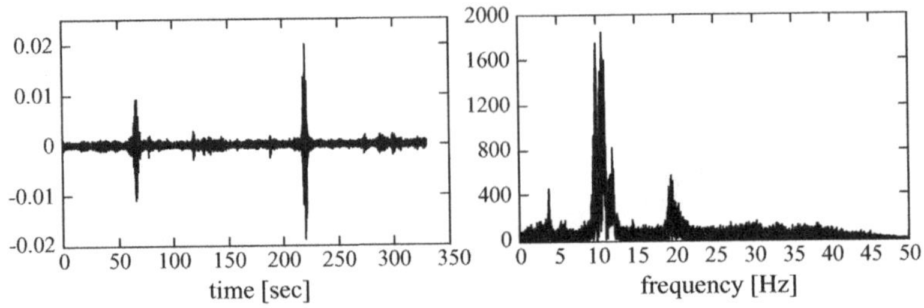

Figure 4.80 Bridge structure, damaged: the time domain is depicted on the left and the frequency response function on the right, showing the two outstanding frequencies $f_1 = 3.76\,\text{Hz}$ and $f_2 = 9.90\,\text{Hz}$ respectively

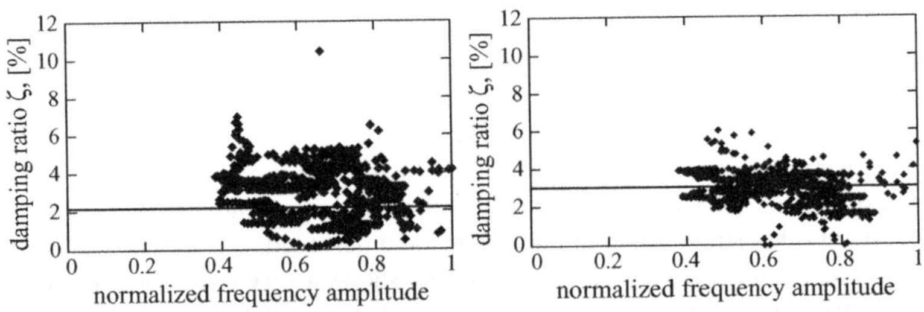

Figure 4.81 Bridge structure, damaged, 99.61% overlapping: results from the damping estimation via RDT, positive point triggering and curve fitting. The data on the left were filtered according to 3.76 Hz (average: 2.21%) and on the right according to 9.90 Hz (average: 3.14%)

Finally, Figure 4.82 shows the damping evolution over the damage time-history. The increasing damping ratio can clearly be recognized and might be used as a damage indicator. Here, no windowing was used and the reference frequency was selected by its maximum amplitude. The single extreme values shown in the graph suggest that for reasonable damage detection longer records are necessary in order to enable the elimination of unreasonable results. Furthermore, it can be said that the comparison between damping values and the characteristic of damping rather than the absolute damping value is critical. The absolute value in this method depends very much on the number of events in the record. The elimination

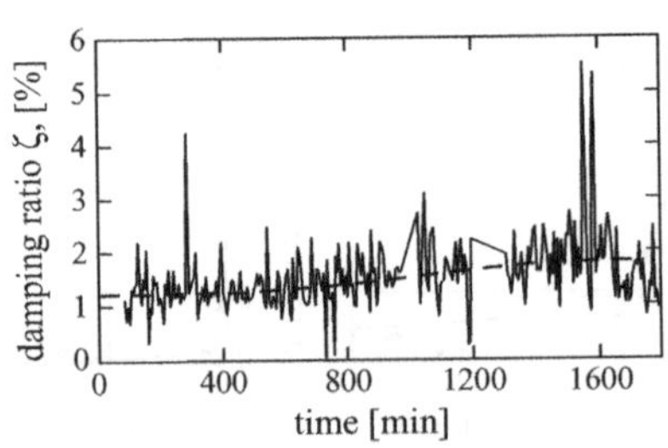

Figure 4.82 Bridge structure, damping as recorded over time from undamaged to damaged: results from RDT, positive point triggering and curve fitting. The reference frequency is nonfixed (i.e. selected by its maximum amplitude)

of all external influences leads to the loss of important indicators that subsequently leads to a critical limitation of information and thus reduced reliability of the results.

4.4.4 Conclusion

This section presented five different common methods for damping estimation and compared their results to each other. When studying several different structures, RDT and stochastic subspace identification seem to be equally good. However, when using the half-power bandwidth method one should pay attention because since it is not very stable and thus is not reliable in practice.

Several possible problems were studied and the results analyzed. In the following list, the main conclusions for damping estimation are summarized.

- The separability of different predominant frequencies plays a major role in the reliability of the results. The choice of the filter is critical.
- If the corresponding frequency amplitude is small, the method becomes instable. Therefore it is advisable to restrict the study to amplitudes of at least 60% compared to the maximum amplitude.
- Both the measurement data and the obtained results have to be analyzed from an engineering point of view, since numerics and measurement technique leave possible error sources behind that are not always obvious at first sight.
- In general, there are several reasons to believe that damping is a possible (good) damage indicator.

Moreover, cases have been observed in which no conclusive result could be achieved. This has mainly been attributed to the special conditions due to the time of measurement. However, this does not invalidate the method but has to be taken into account when applying the procedure. In general, longer measurements with a higher sampling rate are recommended in order to improve the results. Currently, a 500 Hz sampling rate and 22 min of recording length are recommended.

In the future, extension of the RDT will be attempted, which has been the main focus for damping estimation. The intention is to push the development of methods for an automatic frequency separation and a quality check of the random decrement signatures. The effects of heavy input events will also be investigated for the purpose of developing a reliable damping estimation technique. Finally, the main result of this work should be an improved damage detection methodology.

4.5 Discussion of the SHM Axioms

At the Stanford SHM Workshop in 2005, Farrar and Worden (see Farrar et al. 2005) specified axioms for structural health monitoring that are an attempt to formulate common rules and understanding to support the "fundamental truth" that has been argued by the community. These axioms do not represent operators for SHM. In order to generate methodologies it will be necessary to add a group of algorithms that carry the SHM practitioner from data to a decision. The discipline of statistical pattern recognition is proposed for this approach. The axioms formulated are:

Axiom 1. The assessment of damage requires a comparison between two-system states. In order to come to an agreement on this axiom it is necessary to define what a baseline is. Some of the civil SHM methodologies do not require a baseline coming from measurement at the undamaged system. This baseline can simply be created from the theoretical elaboration of the desired system behavior. The comparison is done between a theoretical and a practical behavior. In the usual pattern recognition approaches to SHM a training set is required. In the case of damage detection where two detection approaches can be applied, the training set is composed of samples of features that are representative of the normal condition of the system of the structure of interest (Figure 4.83). For higher levels

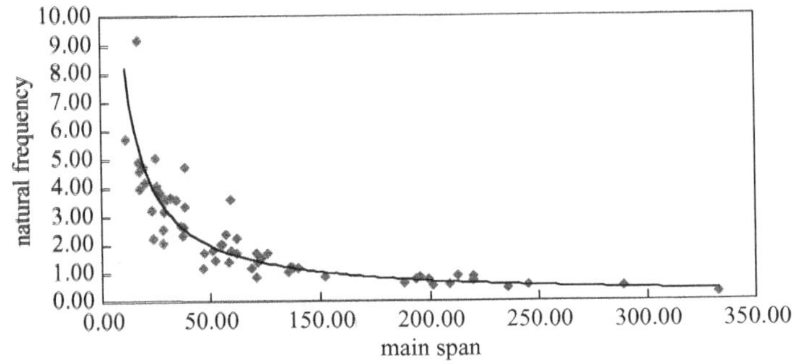

Figure 4.83 Eigenfrequency versus span width

of diagnosis requiring estimates of damage location or severity, the training data must also contain samples of normal condition data, but also must be augmented with samples of the various damage conditions. In this case there is no argument that normal condition data constitute the baseline. Within the civil engineering infrastructure an undamaged condition is as good as impossible. It has to be assumed that theoretical models will provide the baseline for this undamaged condition. Nevertheless the difficulties experienced have been in finding an appropriate model with reasonable effort. The baseline can also be found from experience with similar structures that can be administered in a database. This means a comparison to an average performance of similar structures. Furthermore the dynamic properties can be defined from historic tests and any deviation is referred to as damage. It will be useful to state the baseline in terms of its character whenever a damage detection approach is taken.

Axiom 2. Identifying the existence and location of damage can be done in an unsupervised learning mode, but identifying the type of damage present and the damage severity can only be done in a supervised learning mode. For civil engineering structures the most important question is: Is there damage? The other questions about location and severity are, under normal circumstances, not so important because the pure existence of damage will trigger a completely changed procedure. This is due to responsibility questions, which is different for mechanical aerospace or automotive engineering fields. Therefore these facts do not play such a high role in civil SHM (Figure 4.84).

The statistical model development portion of SHM is concerned with the implementation of the algorithms that operate on the extracted damage-sensitive features to quantify the damage state of the structure. The algorithms used in statistical model development usually fall into three categories. When data are available from both undamaged and damaged structures, the statistical pattern recognition algorithms fall into the general classification referred to as supervised learning. Group classification and regression analysis are categories of supervised learning algorithms and are generally associated with either discrete or continuous classification. Unsupervised learning refers to algorithms that are applied to data not containing examples from the damaged structure. Outlier or novelty detection is the primary class for algorithms applied in unsupervised learning applications. All of the algorithms analyze statistical distributions of the measured or derived features to enhance the damage detection process.

Axiom 3. Without intelligent feature extraction, the more sensitive a measurement is to damage, the more sensitive it is to changing operational and environmental conditions. The sensitivity of civil engineering structures towards operational and environmental conditions is demonstrated in detail in a separate section. The better the environmental and loading conditions, the better the damage detection will work. Ideally these conditions will be eliminated by the proposed compensation routines. Nevertheless there remains considerable space for uncertainty because of observation periods

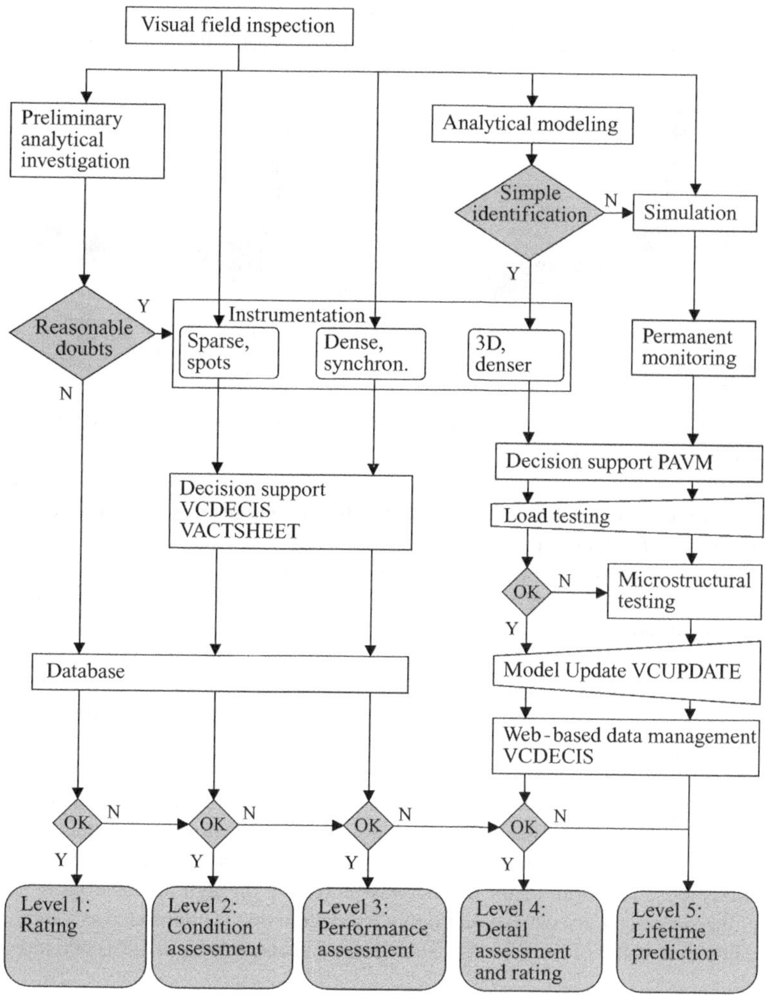

Figure 4.84 Structural health monitoring procedure diagram

that are too short and considerable exceptional conditions that cannot be foreseen. Therefore it is now understood that, for example, stiffness degradation might not be a good damage indicator because of its low sensitivity. As stiffness degradation is not the only indicator, a combination of approaches might be more successful. Particular changes in behavior might be more representative and how these changes can be documented will be discussed.

Axiom 4. There is a trade-off between the sensitivity to damage of an algorithm and its noise rejection capability. The noise to signal ratio is a main subject of interest. This ratio is expressed as a percentage of the maximum over the clean normal pattern vector. For a given noise ratio, the training set and testing set for the outlier analysis are generated as follows. The training set is composed of 1000 samples taken of the clean normal pattern. The testing set is simply compared with this average training set. This training set can be a baseline measurement of the structure in an undamaged condition considering the real condition present. The example of the post-tensioned beam is provided here where damage has been introduced in various steps (Figure 4.85).

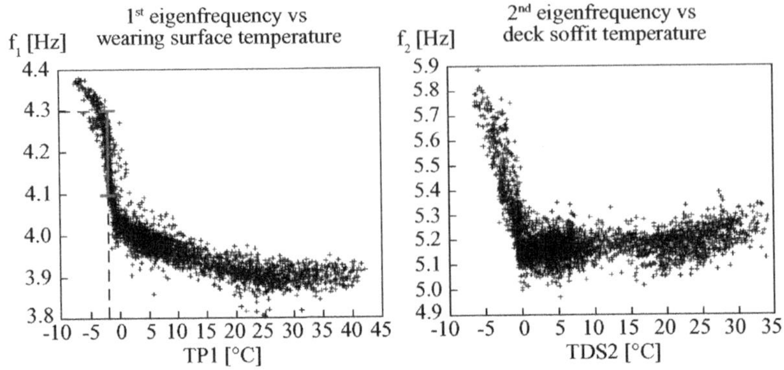

Figure 4.85 First eigenfrequency versus wearing surface temperature (left) and second eigenfrequency versus deck soffit temperature (right)

Axiom 5. The size of damage that can be detected from changes in system dynamics is inversely proportional to the frequency range of excitation. It is obvious that the size of damage is dependent on the extent of the measurement campaign. This includes the number of measurement points on a structure as well as the sampling rates obtained. Damage can only be detected if its basic features are actually measured. This means that slight damage in a steel structure, for example, will not be detected by the normal ambient vibration methodologies sampling with 100 Hz. Therefore it will be advisable to introduce stepwise damage identification starting from the global behavior via the cross-sectional behavior into the structural member behavior. Each step will require a different measurement campaign and approach. Mainly the incurred costs will limit this subject (Figure 4.86).

It has to be mentioned here that nonlinearities involved in the behavior and their changes when damage is present can lead to difficult identification procedures. The relationship between damage sensitivity and wavelength can be extended to more general types of vibration-based damage detection methods. In these applications the wavelength of the elastic wave traveling through the material is replaced by the wavelength of the standing wave pattern set up in the structure that is interpreted as a mode of vibration. The technical literature is full of evidence that such lower frequency loads are not good indicators of local

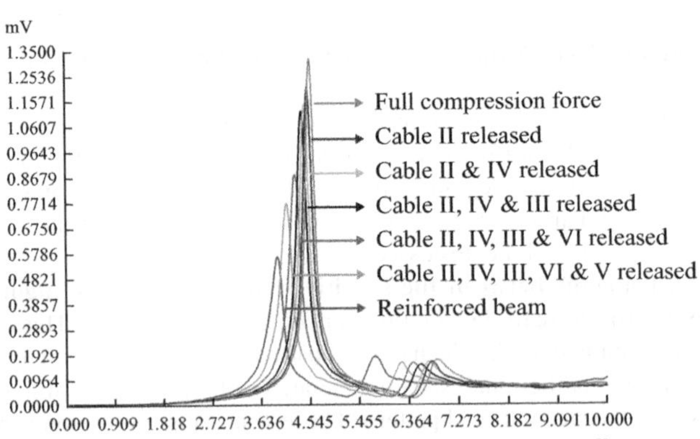

Figure 4.86 Frequency and amplitude changes with increased damage ratio

damage. The lower frequency global modes of the structure that have a long characteristic wavelength tend to be insensitive to local damage. For the case of civil engineering infrastructures such as suspension bridges these mode-shape wavelengths can be of the order of hundreds of meters and damage such as fatigue cracks, which are of the order of centimeters, will not be detected. Nevertheless the proposed observation concept going from global to local can cover this gap.

4.6 Safety Assessment

Casas (2006) reported on the current European status of safety assessment. Assessment of existing structures can be either deterministic or probabilistic. The review of existing methods mostly used, undertaken by the road and railway agencies, shows that they are mainly based on deterministic or semiprobabilistic (partial safety factors) assumptions. From the analysis of the methods used worldwide, the following conclusions can be drawn:

- Few countries (UK, Denmark, USA, Canada) are using specific guidelines or standards for structural safety assessment of highway bridges. In European countries with a huge stock of highway bridges, such as France, Germany, Italy, Poland and Spain, there are no specific methods of structural safety assessment of highway bridges and in general the basis of assessment calculation is the same for the design of new bridges. Specific codes for bridge safety assessment also are not available for Asian and South American countries.
- There is consensus that the structural safety assessment should be based on a limit state format as already assumed in the design.
- There is a consensus that the most efficient assessment process is based on the application of different and increasingly sophisticated assessment levels. A five-level model is presented in BRIDGE, Cost 345, Highways Agency 1998 and SAMARIS. These levels of assessment, numbered 1 to 5 with level 1 being the simplest and level 5 the most sophisticated, are summarized in Figure 4.87.

This step-level philosophy has also been assumed in Germany (Hille et al. 2005) in the current preparation of a guideline for the assessment of existing structures and in the future European guideline for the capacity safety assessment of existing railway bridges:

<table>
<tr><th>Assessment level</th><th>Strength & load models</th><th>Calculation models</th><th>Assessment methodology</th></tr>
<tr><td>1</td><td>Strength and load models as in design code</td><td>Simple, linear elastic calculation</td><td rowspan="3">LRFD-based analysis, load combinations and partial factors as in design code</td></tr>
<tr><td>2</td><td>Material properties based on design documentation and standards</td><td rowspan="4">Refined, load redistribution is allowed, provided that the ductility requirements are fulfilled</td></tr>
<tr><td>3</td><td rowspan="2">Material properties can be updated on the basis of in situ testing and observations using Bayesian approach</td></tr>
<tr><td>4</td><td>LRFD-based analysis, modified partial factors are allowed</td></tr>
<tr><td>5</td><td>Strength model including probability distribution for all variables</td><td>Probabilistic analysis</td></tr>
</table>

LRFD, Load and resistance factor design.

Figure 4.87 General scheme of the five-level assessment (BRIME)

Table 4.3 Target values of reliability index at member level for ultimate limit stress (ULS) and reference period of 1 year and normal consequences of failure

	Canada	USA	Eurocode	JCSS	Denmark	ISO
Design	3.75	3.75	4.7	4.2	4.2	4.7
Assessment	3.25	2.5	-	-	4.2	4.7

JCSS, Joint Committee on Structural Safety; ISO, International Organization for Standardization.

- A clear consensus also exists on the fact that the same resistance formulas for undamaged elements cannot be assumed for deteriorated ones. However the consensus fails when dealing with how the influence of deterioration in damaged bridges should be considered.
- A clear consensus on how the loading test must be used in bridge assessment does not exist. Because experience with bridge testing has often revealed bridge behavior that deviates from results expected from the conventional analytical methods, some codes reflect the load testing possibility. It is generally accepted in many countries that load tests may be used as a supplementary source of information in the theoretical assessment process, providing information on the actual structural behavior of the bridge. However, complete evaluation using only experimental load testing is normally not considered. The exception to this general rule is the LRFR evaluation code in the USA (*Manual for Condition Evaluation and Load Resistance Factor Rating (LRFR) of Highway Bridges (LRFR-1)* 2003). Road tests also carry the disadvantage that they are mainly performed in the clearly linear elastic behavior of the structure and do not reveal the nonlinear elements that are guiding the process close to failure.
- It seams clear that target reliability levels different to those assumed in the design of new structures may be considered in the safety assessment of existing bridges. As an example, in Table 4.3 the target reliability levels used in some countries and proposed by international organizations are shown. The lower reliability levels for assessment are justified in the fact that evaluation is performed for a much shorter exposure period (inspection every 2–5 years), consideration of site realities and the economic consideration of rating versus design.

The actual and future trend in safety assessment is the use of reliability-based methods. Also there have been a number of applications of reliability-based assessment for bridges in some countries (Casas 2000; Lauridsen 2004) and some countries have developed standards, guidelines and codes for the evaluation of existing bridges based on probabilistic models (SAMCO: Canada, Denmark, Slovenia, USA). The practice is not yet widely used, mainly due to the lack of information and standardization. An important step in the direction of the necessary standardization of this work developed by the Joint Committee on Structural Safety (*Probabilistic Model Code* 2001) is in the probabilistic model code. Initiatives like this and the SAMCO guideline on bridge monitoring and assessment will provide the bases needed for the increasing future application of reliability methods in bridge capacity assessment.

Another future trend where much research is needed is related to the following question: How to model in a capacity assessment calculation format the deterioration detected in the bridge during the inspection and summarized in a condition state? If the concept of condition factor is used, the question is how this factor can be derived from a condition index obtained basically from visual inspection. This condition factor multiplies the nominal resistance of the element evaluated in the real dimensions and strength. The general rule is to use the engineering judgment and there is no direct relation between the condition rating as derived from the inspection process and the condition factor. Only in Slovenia and the USA (*Manual for Condition Evaluation and Load Resistance Factor Rating (LRFR) of Highway Bridges (LRFR-1)* 2003) is there a direct link between condition rating and condition factor.

A special issue that potentiality is recognized in many codes and guidelines, but still has a lot of work to be developed, is the integration of load testing results in bridge assessment. Despite many countries already recognizing the usefulness of diagnostic loadtesting in the updating of available theoretical models used in the evaluation process, only the USA has standardized the possibility of using a load test to directly assess the bridge capacity or safety via so-called proof load testing. The most important European contribution to this item has been the efforts of the SAMCO network [www.samco.org].

Further Reading

Asmussen JC (1997) *Modal analysis based on the Random Decrement Technique – Application to Civil Engineering Structures.* Department of Building Technology and Structural Engineering, University of Aalborg, Denmark.

Bendat JS and Piersol AG (1993) *Engineering Applications of Correlation and Spectral Analysis*, 2nd edn. Wiley-Interscience.

Blevins R (1979) *Formulas for Natural Frequency and Mode Shape*. Van Regional Company, New York.

Bronstein IN, Semendjajew KA and Musiol G (2001) *Taschenbuch der Mathematik*, 5th edn. Harri Deutsch-Verlag, Frankfurt am Main.

Casas JR (2000) Permit vehicle routing using reliability-based evaluation procedures. *Transportation Research Record* **2**(1696), 150–157.

Casas JR (2006) Bridge management: Actual and future trends. In *Bridge Maintenance, Safety, Management, Life-Cycle Performance and Cost* (ed. Cruz PJ, Frangopol DM and Neves LC), pp. 21–30. Taylor and Francis, London.

Chopra AK (2000) *Dynamics of Structures: Theory and Applications to Earthquake Engineering*, 2nd edn. Prentice Hall.

Cole HA (1973) *On-line Failure Detection and Damping Measurement of Aerospace Structures by Random Decrement Signatures.* Technical Report NASA CR-2205, US National Aeronautics and Space Administration.

Des Industries Mechaniques CT (2001) *Pressure Components Fatigue Design in the Framework of Directive 97/23/EC on Pressure Equipment – Work Package 6.* Final Report, Centre Technique des Industries Mechaniques, Mulhouse.

Drešler K, Gründer B, Hack M and Köttgen VB (1996) *Extrapolation of Rainflow Matrices.* Technical Report SAE Technical Paper No. 960569, Fraunhofer-Institut für Techno-und Wirtschaftsmathematik, Kaiserslautern.

Ermittlung von Dauerschwingfestigkeitskennwerten für die Bemessung von geschweišten Al-Bauteilverbindungen auf der Grundlage örtlicher Strukturbesanspruchngen (2002) Final report, Instituf für Schweitechnik der Technischen Universitt Braunschweig.

Farrar C, Worden K, Mansin G and Park G (2005) Fundamental axioms of structural health monitoring. *Proceedings of 5th International Workshop on Structural Health Monitoring*, September, Stanford, CA.

Fatigue Strength of Steel Structures (2002) Technical Report, Comité Européan de Normalisation (European Committee for Standardization). prEN 1993-1-9: 2002.

Haibach E (2002) *Betriebsfestigkeit - Verfahren und Daten zur Bauteilberechnung*, 2nd edn. VDI-Verlag.

Hille F, Rohrmann R and Rucker W (2005) Guideline for the assessment of existing structures. In *Bridge Management*, Vol. 5, pp. 227–234. Thomas Telford.

Hobbacher A (2003) *Recommendations for Fatigue Design of Welded Joints and Components.* Technical Report XIII-1965-03/XV1127-03, International Institute of Welding (IIW), Paris.

IMC (2005) *FAMOS V. 5.0* Reference Manual, IMC Data Works, Maddison, WI.

Lauridsen J (2004) Bridge owner's benefits from probabilistic approaches – experiences and future challenges. *Bridge Maintenance, Saftey, Management and Cost - Proceedings of IABMAS 04*, pp. 3–10. Balkema, Rotterdam, Kyoto.

Leuven KU (undated) *ESDEP – European Steel Design Education Program: WG12 Fatigue.* Technical Report, Katholieke Universiteit Leuven.

Manual for Condition Evaluation and Load Resistance Factor Rating (LRFR) of Highway Bridges (LRFR-1) 2003. Technical report, American Association of State Highway and Transportation Officials.

Naubereit H and Weihart J (1999) *Einführung in die Ermüdungsfestigkeit.* Carl Hanser Verlag, München-Wien.

Niemi E (2000) *Structural Stress Approach to Fatigue Analysis of Welded Components – Designer's Guide.* Technical Report. IIW doc. XIII-1819-00/XV1090-01.

Peeters B (2000) *System identification and damage detection in civil engineering* PhD thesis Katholieke Universiteit Leuven Leuven.

Probabilistic Model Code (2001) Technical Report, Joint Committee on Structural Safety. http:www.jcss.ethz.ch/JCSSPublications/PMC/PMC.html.

Ramberger G (1998) *Stahlbau.* Manz-Verlag.

Simonsen B (2001) *Procedure for Calculating Hot Spot Stresses in Aluminium Constructions.* Technical Report, Department of Naval Architecture and Offshore Engineering, Technical University of Denmark.

Spaethe G (1992) *Die Sicherheit Tragender Baukonstruktionen*, 2nd edn. Springer-Verlag, Wien - New York.

Verkehrsentwicklung in Tirol - Berichte 1984–2006 (undated) Technical Report, Amt der Tiroler Landesregierung - Abteilung Gesamtverkehrsplanung, Innsbruck.

Verkehrsprognose 2015 – vorläufige Ergebnisse hochrangiges Strassennetz Österreich (2000) Technical Report, BMVIT-Abteilung II/A/1, Wien.

Van Overschee P and De Moor B (1996) *Subspace Identification for Linear Systems: Theory–Implementation–Applications*, 1st edn. Kluwer Academic Publishers, Dordrecht.

Veit R and Wenzel H (2006) Measurement based performance prediction of the Europabrücke against traffic loading. *Proceedings of the 16th European Conference of Fracture ECF16*, Alexandroupolis, Greece.

Veit R, Wenzel H and Fink J (2005) Measurement data based lifetime-estimation of the Europabrücke due to traffic loading–a three level approach. *Proceedings of the 58th International Conference of International Institute of Welding*, Prague.

Wenzel H and Pichler D (2005) *Ambient Vibration Monitoring.* J. Wiley & Sons Ltd., Chichester.

Wenzel H and Veit-Egerer R (2008) Measurement based traffic loading assessment of steel bridges – a basis for performance prediction. *International Journal of Structure and Infrastructure Engineering.*

Willberg U (2001) *Asphaltschichten auf hydraulisch gebundenen Tragschichten - Untersuchungen zum Tragverhalten.* PhD thesis, Technical University Munich.

5

Decision Support Systems

Our transportation systems already contain millions of sensors that produce data permanently. Only a very small fraction is actually used in the decision-making process. This is mainly due to missing resources and only partly due to the lack of knowledge. Decision support systems (DSS) can filter the data and convert them into information. The operator receives options and only has to make the final decision. Structural health monitoring with its many permanently instrumented structures produces terabytes of data daily. They are ideal cases for the application of decision support systems. The reason why they have been applied rarely is that standard solutions are hardly applicable in civil engineering. The system introduced here has been borne out of the necessity described above.

5.1 Decision Support Systems for SHM

Monitoring assessment and decision support of structures has become a major issue in our modern society. The current practice is a case to case approach by experts to specific problems. A well-organized procedure based on standards has not been brought to maturity yet. The VCDECIS system is based on VCE's experience with the monitoring and assessment of bridges (Figure 5.1). It consists of several modules and provides all the necessary tools for decision making when monitoring data are available.

5.2 Architecture

The system consists of five modules (Figure 5.2):

- the operation mode (operator's tool);
- the databases;
- the decision support module;
- the alert system;
- the output module (takes action if appropriate).

The function of each module is explained in the following sections.

5.3 The Operation Modes

The idea is to allow the operator to interact with the process. His individual assessment is important as well as the tasks to be performed. The choice of mode has an influence on the determination of the risk level, thus the operator can directly influence the procedure.

Health Monitoring of Bridges Helmut Wenzel

Figure 5.1 Start page of VCDECIS

Currently there are eight operation modes (Figure 5.3):

- Normal operation performs the predefined standard procedures (default).
- Low margin operation activates additional checks of the data and in doubt provides an assessment on a higher risk level.
- Warning operation is done at the stage when a problem has already been identified. The time steps for reassessment are shortened and the risk levels are increased.
- Emergency operation means permanent online assessment and warning. Operator attendance is expected.
- Forensic analysis means the analysis of events in order to learn from them. The data will systematically be scrutinized on indicators for failure.
- Other applications allow combining existing methodologies or adding new items. It is also the development platform.
- Scientific use allows the use of the data for any scientific purpose. No risk assessment is done and no warning is issued.
- Prognosis module allows a prognosis from data if sufficient time series are available. This module will mainly be activated in systems working over a certain period.

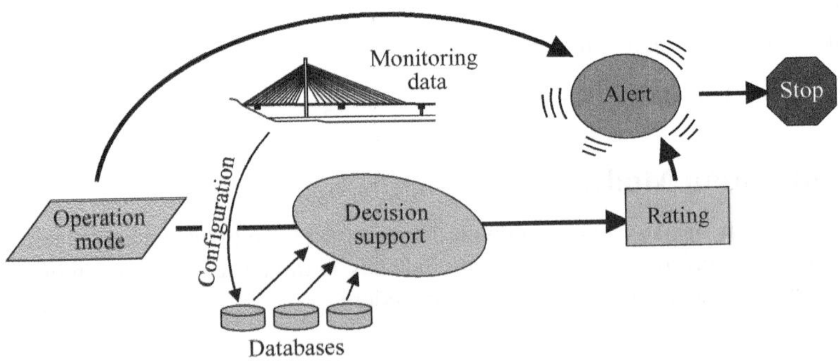

Figure 5.2 System architecture of VCDECIS

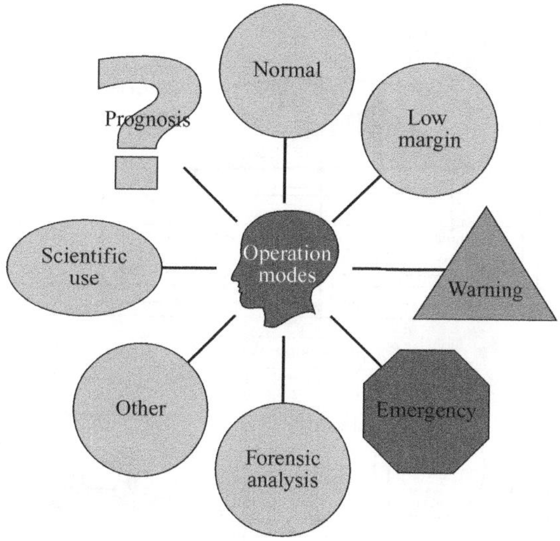

Figure 5.3 Operation modes of VCDECIS

It has been recognized that a very strict application of rules is not feasible for our structures, each of them being prototypes. The knowledge and intuition of the operators will be taken into account to produce more useful results. A major problem is to avoid false alarms, which are costly and provide problems for the reputation and acceptance of the system. During annual maintenance, adjustments are made to use the growing experience and the increased data quantity. Learning routines also grow in effectiveness with time and input from monitoring.

5.4 Monitoring System and Databases

5.4.1 Data Acquisition

The data acquisition systems should be carefully designed according to the requirements of the structure. All types of input are desired, such as (Figure 5.4):

- Remote sensing data, which are purchased from ESA, NASA or other sources, might form the basis for a very rough assessment of the structure, the exposure and the geographic settings. It is intended to generate information on environmental conditions from these data.
- Permanent installations are online systems installed in the structures according to the designer's plans. It is useful to design a runtime version of the software for every permanent installation. This allows a much quicker elaboration of the results, because the input is well known in its structure and content. Such systems will also be able to produce frequent reports automatically. Standard procedures for the correlation of data and eventual warnings (i.e. wind speed) will be designed individually. This will be the standard application of VCDECIS.
- A mobile monitoring unit will provide the necessary additional data to obtain more information about the structure or to look at specific elements of it. This application is only temporary. For example eigenmodes are recorded from sensors at many varying locations.
- Environmental data will be collected from whatever sources are available. Cross-correlations between stations in remote places and the actual conditions for the structure will be designed. A couple of

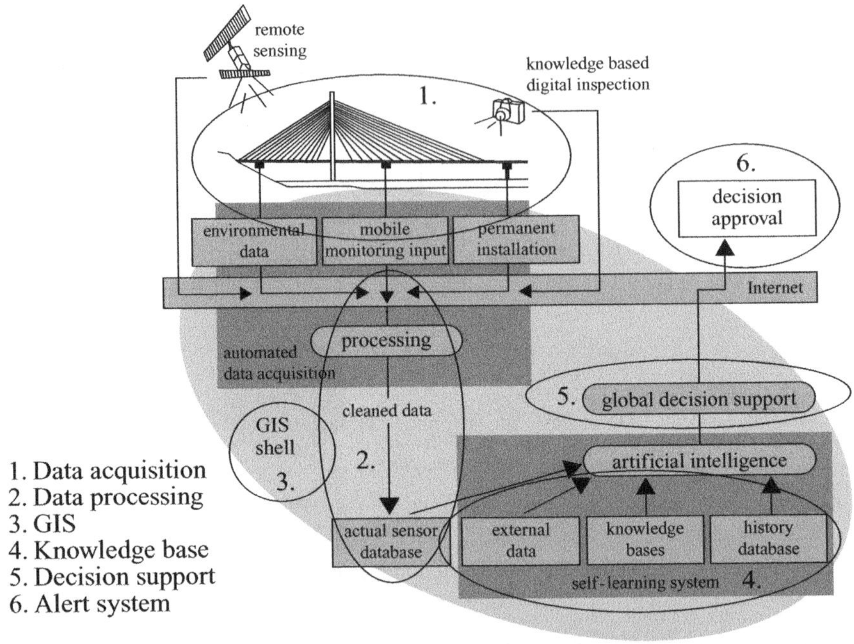

Figure 5.4 Database and assessment concept

standard transfers are offered. A final target is to use data stored in the database already for a rough first assessment of any new project. Connections to eventual data sources will be established for each case separately.

5.4.2 Data Processing

The data processing module takes over the incoming data from the internet. A plausibility check is carried out in order to identify false readings or failed sensors. These obvious wrong data will be eliminated. All other data will be taken over and stored in its raw format for future work. Original raw data will be kept permanently.

A predefined metadata protocol will be completed (Figure 5.5). All necessary information has to be provided by the responsible engineer. Data fusion also happens at this level. All data are stored in a relational SQL database.

5.4.3 Geographic Information Systems

Geographic Information Systems (Figure 5.6) provide the option to organize huge amounts of data in a clear and structured way. It is used here to bring all information together, which is necessary to conduct serious monitoring assessment and decision making. The list of information comprises the geographic information as well as all other necessary documents, which are:

- maps in various details;
- drawings of the structures;

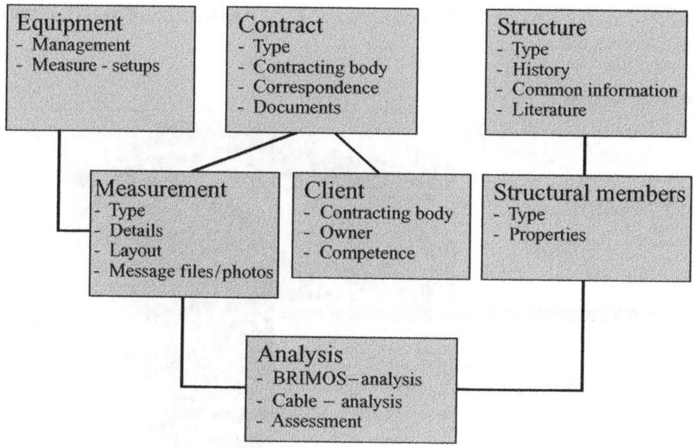

Figure 5.5 Data model in VCDECIS

- photographs, videos and other communication;
- reports and documentation;
- results of assessment and eventual summaries;
- auto-CAD drawings or other related documents;

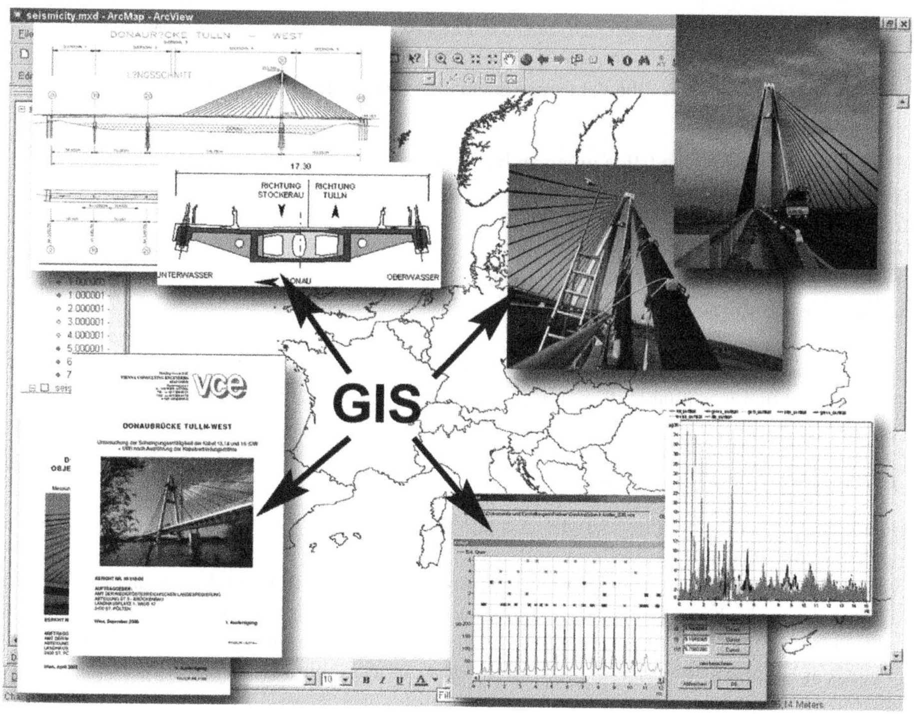

Figure 5.6 The GIS environment (drawings, pictures, reports, results)

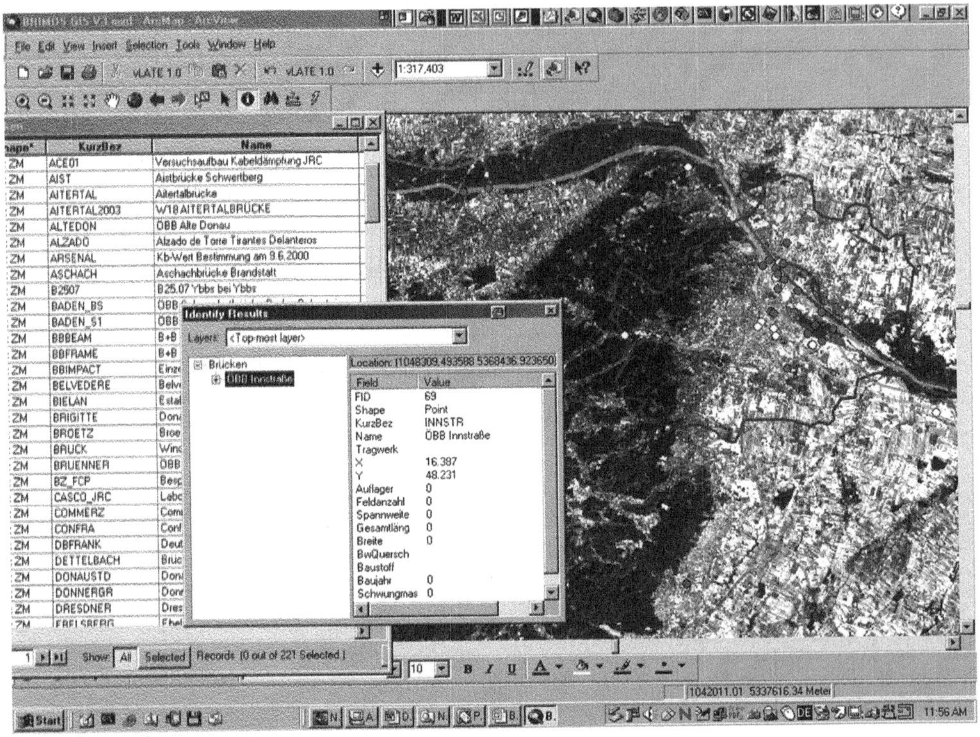

Figure 5.7 Access the database (get bridge data from clicking on the location)

- hazard maps for seismic landslide or storm hazards;
- satellite images of various kinds;
- infrastructure information;
- topographic information;
- references and relation to the literature database;
- any other useful information.

The design is made to enable intuitive handling of the information. The user should not require a handbook to find everything that is available. Examples of applications are provided in Figure 5.7.

5.4.4 Knowledge Databases

The knowledge part of the system is subdivided into three databases (Figure 5.8), namely the knowledge, external and history databases.

It is a fact that knowledge is not sustainable. Information received through the internet might be invalid after a short time. The basic information starting from physics is stored in the knowledge database to enable the user to obtain information quickly on basic facts and figures. The database can be searched by keywords as usual. Documents in rather short form are available, which are supported by references and literature subsequently.

The database for external data provides the opportunity to merge external data received from meteorological stations (Figure 5.9), other databases (Figure 5.10) or any external source with the actual

Figure 5.8 Content of the knowledge base

assessment process. The target is to fill gaps in the monitoring chain, particularly on environmental conditions. The final target would be to have a standing network an meteorological data input that provides sufficient information to cover any new structure to be assessed.

The history database includes all cases performed so far with their results and comments. All results are stored in a form so that they can be used by any new assessment procedure. A combination of previous results with new records is possible and statistics from available assessment procedures can be produced. This also allows predictions on results of incoming structures, when similar projects have already been performed.

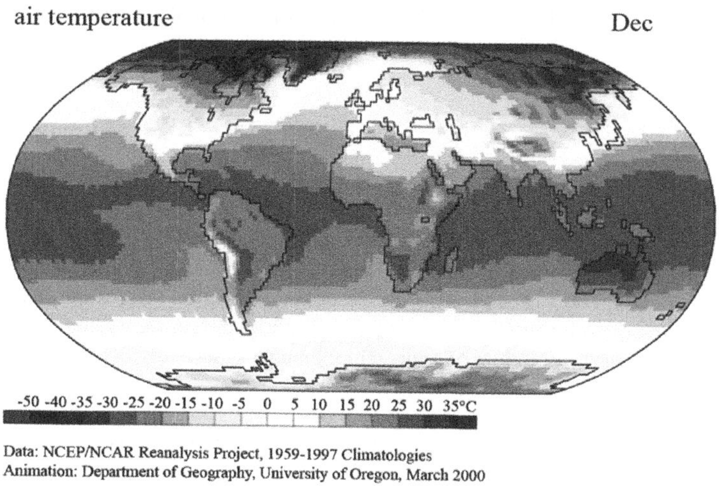

Figure 5.9 Annual global temperature cycles

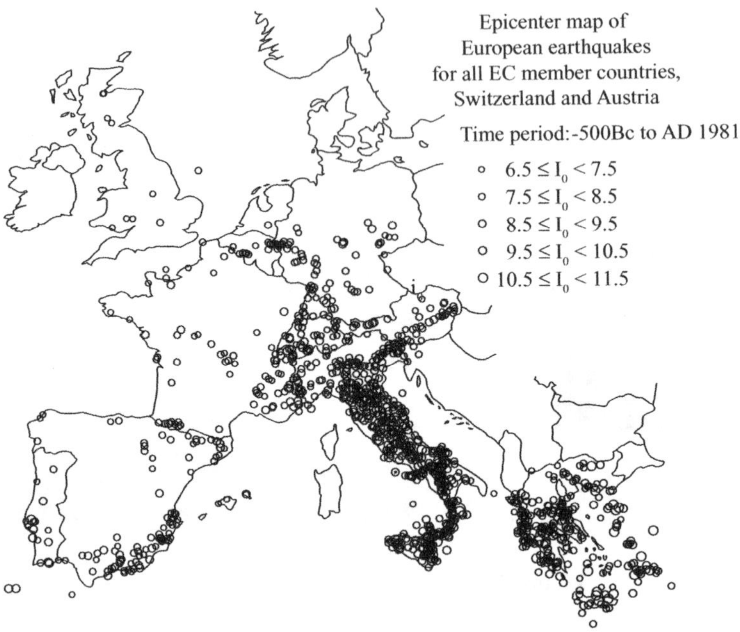

Figure 5.10 Earthquake epicenter data of Europe (sample)

5.4.5 Decision Support System

Many different applications are available to assess monitoring data (Figure 5.11). There still exists a wide array of open questions where new methodologies will be produced that can be introduced into the process. The system is defined by rules, which allow an assessment on finding facts and indicators in the available data. Some of the methodologies are complementary and it is intended to conduct parallel processing and a comparison of the results. The building of a mean also might make sense in several cases.

5.4.5.1 Set of Rules

The rules are the core of the assessment system. They consist of methodologies developed in order to get the maximum information out of the data. There are several approaches to achieve the same result, which can be applied in parallel, averaged or compared.

The values of the various quantities computed from the monitoring campaign are combined to a total rating over a set of rules. The various factors are described in Section 3.2 for the BRIMOS approach. Every identified assessment routine is documented and introduced to the system as a rule. A typical example follows, showing the principle of temperature compensation.

Short Description

It is well known in the health monitoring community that changing boundary conditions (temperature) sometimes have considerable effects on the measured modal parameters, particularly the natural frequency. By employing a monitoring concept that evaluates the structural integrity as a function of the measured frequencies, this dependency could be the crucial point in the assessment. The main task is

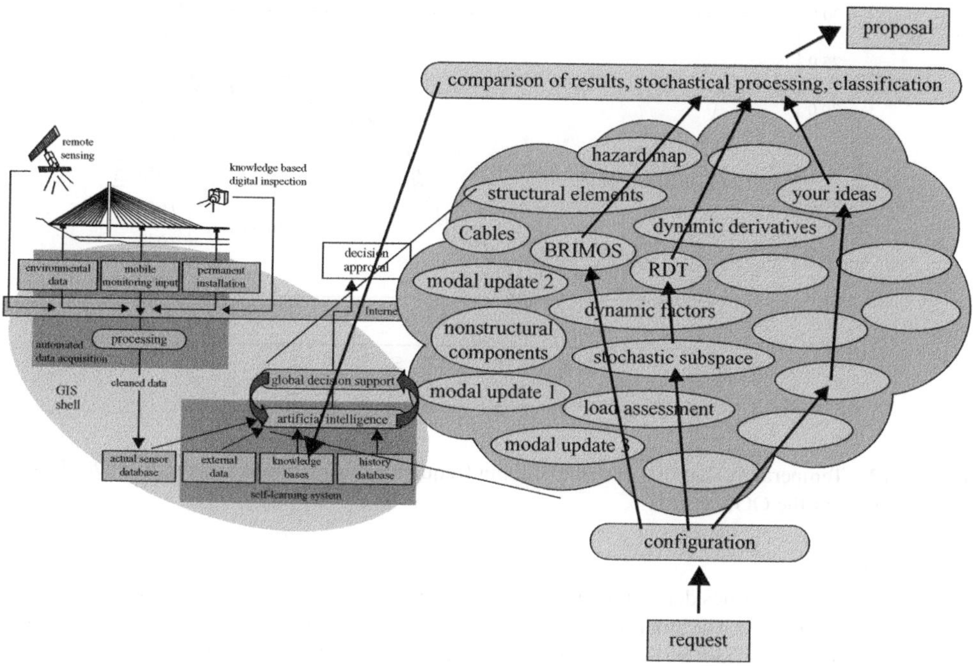

Figure 5.11 Decision support module of VCDECIS

to separate normal (environmental) changes in the dynamic response from abnormal changes caused by progressive damage of some load-bearing parts.

Scientific Base

From August 1998 till August 1999 a continuous operating monitoring system was installed on the Olympic Grand Cable-Stayed Bridge (OGB) in Korea. This monitoring system acquires the dynamic response of the deck structure and some selected cables, as well as temperature data, over time. The aim of the monitoring system was to provide both environmental and vibration data. From these measurements data lines have been derived, which are presented in Figure 5.12.

From Figure 5.12 it can be clearly identified that the temperature is varying between $-3°$ C minimum and $+25°$ C maximum temperature over the one-year monitoring period. The first vertical bending frequency shows during this period a characteristic trend to increasing frequencies for lower temperatures. The consideration of this fact is important to avoid wrong interpretation of the acquired dynamic response.

Rule

Thus, for structural assessment using ambient vibration monitoring it is a must to consider temperature effects. The data collected so far (Figure 5.13) lead to the following rules:

1. Increasing temperature leads to decreasing natural frequencies.
2. Structural temperatures dropping down below $0°$ C are critical, which results in the following expression for a reference temperature of $20°$ C:

$$f_{norm}(t_i)[\text{mHz}] \cong f(T; t_i) + 0.32(T - 20°\text{ C})$$

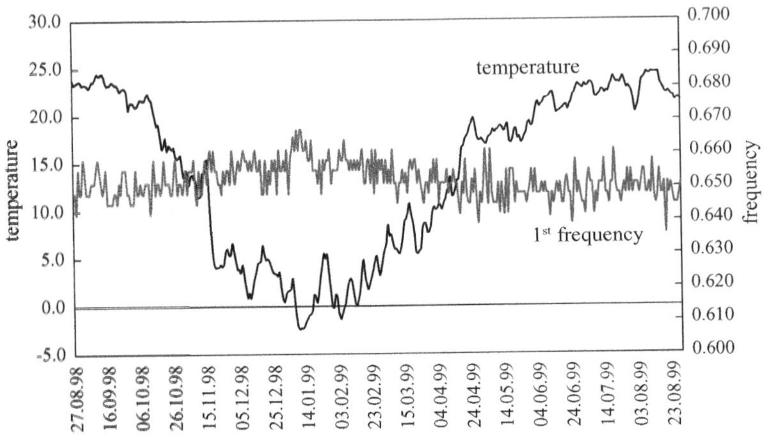

Figure 5.12 Temperature and first vertical bending frequency during one year (August 1998–1999) of measurements for the OGB, Korea

Application

The practical use of the rules defined requires measurement of the air and structural temperature during the monitoring procedure. The aim is to monitor a specific structure under nearly the same environmental conditions as during the last assessment. Therefore the same month in the following year should be selected, which is an easy but efficient method. If this was considered, and the temperature difference is still not negligible, the natural frequencies identified have to be updated, using the equation above, to the reference temperature.

Background Information

The effect of temperature variation on the modal parameters has been presented in several research papers. All investigations have shown on average, an increase of 4–5% in the natural frequencies of all modes for a decrease in temperature. In this context it must be mentioned that the influence of changing temperatures could be much stronger if the structure is under constraint (abutment, piers, expansion joint, etc.). The temperature drop mainly affects the elasticity modulus of the whole structure. In this context it is also important to catch the temperature course of the asphalt layer if there is one applied. Below 0° C, the asphalt layer seems to be responsible for frequency variations.

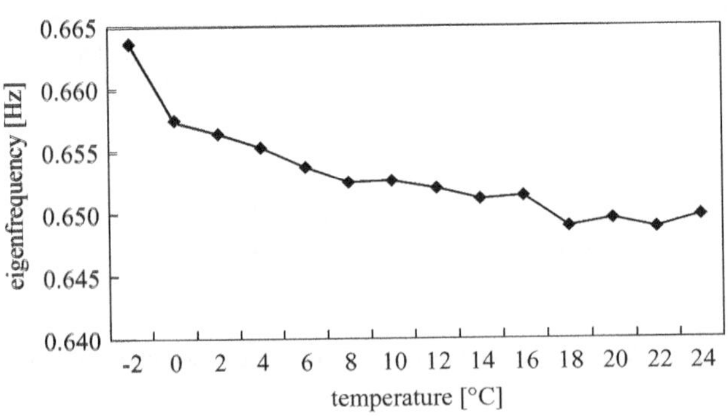

Figure 5.13 Eigenfrequency values for different temperatures

Based upon the data analysis it should be outlined that it is not possible to formulate a reliable, transferable rule for the compensation of temperature differences. Each structure behaves differently under varying boundary conditions, therefore basic knowledge concerning the structural response under changing temperatures should be available before temperature compensation models can be applied.

5.4.5.2 Analysis Modules

The set of rules is permanently updated and enlarged and for protection of knowledge the details cannot be provided here. An overview is given as follows:

- The BRIMOS approach developed by VCE over the last ten years is well documented. It represents a standard approach towards data analysis. The main result is a classification of the structure following the traffic light scheme.
- A number of methods of frequency analysis have been developed because there is plenty of information in the signals that has to be extracted, ranging from a simple comparison of responses to frequency decomposition over a certain recording period.
- Methodologies in the time domain are becoming of growing importance. With longer records, patterns can be recognized and compared.
- Mode shapes are an interesting tool for assessment when a sufficient number of points has been recorded. Particularly well represented are changes in the boundary conditions.
- Model update can take various forms. For cables, an automatic model update is available that allows direct comparison of the results with a theoretical model that generates itself. Other updating procedures require an export of results to the FE model code. Model assurance criteria are computed in order to find matching modes.
- Damping is analyzed in various ways, to extract material on system damping. For comparison it is necessary to have the same recording parameters in order to overcome numerical problems.
- Displacement is computed by double integration of the acceleration signals. In many cases verification through a laser is applied, which allows accurate displacement calculations. This provides a deep insight into the behavior of the structure.

For determination of the response spectrum, three methodologies are applied: traditional peak-picking, the stochastic subspace method and the ARMA model.

5.4.5.3 Other Applications

The data are suitable for a couple of other applications. Derivatives are frequently developed and implemented and the following are worth mentioning:

- Compensation for environmental influences, as applied to the data at entry to the decision support module, can also furnish information on the behavior of the structure.
- Dynamic factors are often a wanted side product of analysis.
- Load determination is very often wanted by the client to find out overloads on the structures.
- Fatigue and remaining lifetime assessment is a big issue for steel structures currently. They are mainly implemented in permanent monitoring systems.
- The assessment of structural elements is possible by isolating a portion of the frequency response. Stochastic methods allow an assessment of these elements.
- Nonstructural components very often show their vibrational signature in the response. If it can be isolated, there can be an assessment on these items too.
- By sequencing the record dynamic, derivatives can be determined showing the development of the structure over time. This application is particular helpful when the source of excitation is desired.
- Wavelet applications allow an analysis of the system to find damage in the structure.

- Damage detection, location and analysis is the current fields of development worldwide. Various routines are available but there is still much to do with regard to automation and reliability.

This list is consequently updated with every new development and detection of new knowledge.

The system is started through a configuration file, where the user is asked which methodologies are to be applied and which combinations are desired. A default menu is available that directs the user to a standard evaluation. Each assessment step provides results that are then stored in the knowledge database for statistical use.

There is an option to go with the process through a neural network that has been trained by the previous applications. Nevertheless this step is not yet mature due to a lack of training possibilities for the network and ongoing development work. It is anticipated that in the future support can be provided for the user by this system.

5.4.6 Alert System

Information on the risk level is provided to the user. The system has been chosen in connection with alert systems used for natural hazards. Landslide or avalanche alert systems have five distinct risk levels (Figure 5.14) and the color-coding system follows international practice. When risk level 5 (Extreme) is reached, automatic action by the system can be triggered. This could eventually be a red traffic light at the approach to a bridge being monitored. The other risk levels provide information for the operator and also ask for his input. As explained in the operation mode, the operator has the opportunity to interfere with the system here and to add his subjective impression to the process. The actions to be performed at each level will have to be defined from case to case. It has to be coordinated with the standard procedure of the bridge owners.

The example provided in Figure 5.15 is a typical periodic report for a structure that is permanently monitored. This one-page report is sent by email to the owner at the agreed schedule: weekly, monthly, quarterly or once a year. The main function is to inform the owner of the performance over a particular period (particularly if any of the thresholds have been exceeded) and to provide an updated rating of the structure. The elements of the report are the major window containing all the variables monitored with the respective thresholds in a normalized way. It is immediately visible whether the records are within allowable limits. In addition, special information requested by the owner, such as wind records, is provided. The main information on this report is the rating system and the expected lifetime.

5.4.6.1 BRIMOS Rating

The basic philosophy is to provide a rating similar to the company rating system, so that the situation can be identified at a glance. The rating can vary from A to C, with a number of intermediate steps possible. It consists of three letters that represent:

	Risk Level	*Why*	*What to do*
green	Low	Info	Regular operation
	Moderate	Info	Long term action
yellow	Considerable	Show development	Mid term action
	High	Demonstrate risk	Immediate action
red	Extreme	Show fault	Automatic alert, ACTION

Figure 5.14 Risk levels with cause and consequence

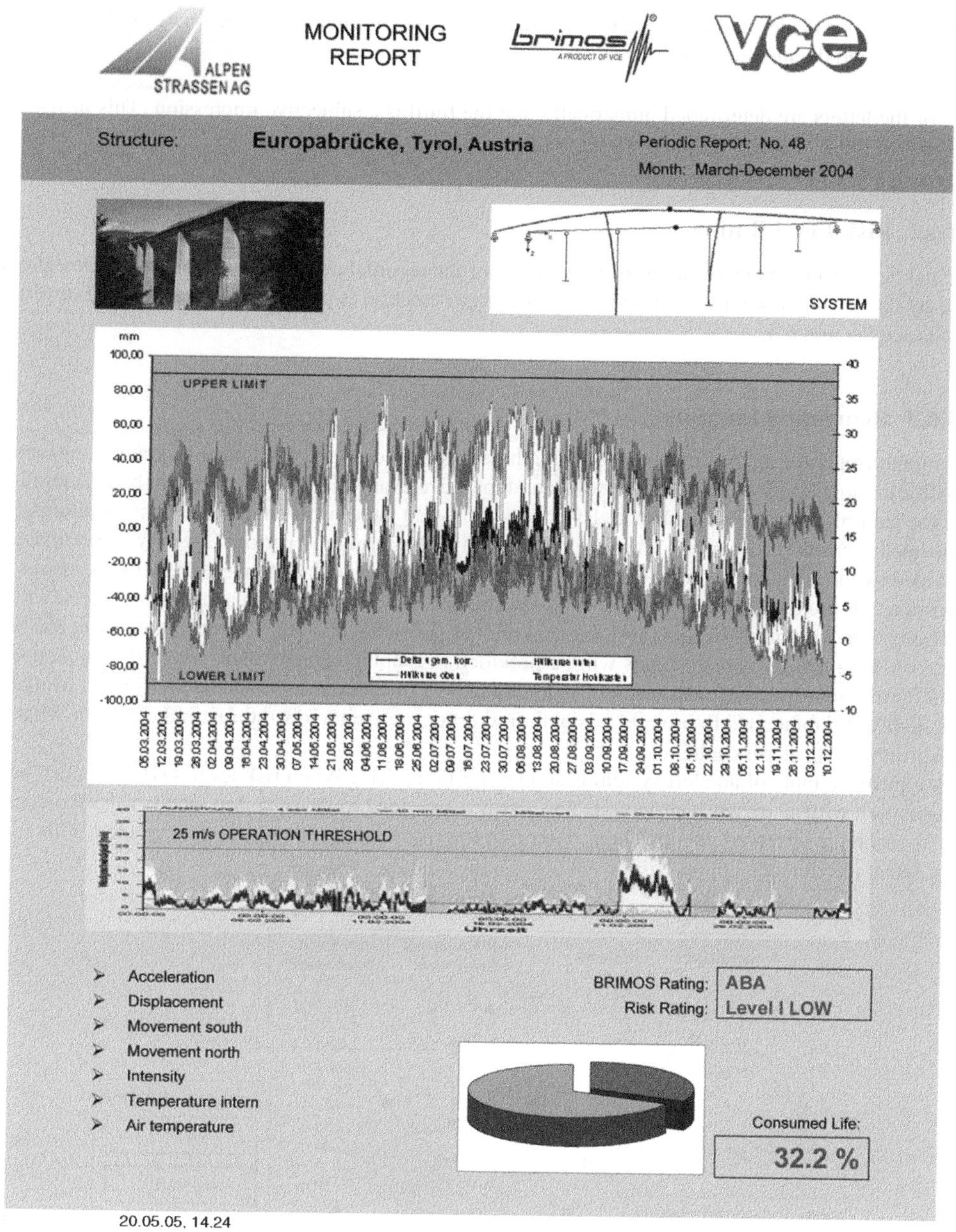

Figure 5.15 Periodic report of the Europabrücke

- Result of the numerical assessment of the records, ending in the typical traffic light assessment according to BRIMOS.
- Result of the inspection and subjective engineer's impression during the measurement or using additional information on the structure that is available.

- Result of the model update process, showing the quality of comparison between the intent of the designer and the measured performance.

Two of the letters are determined numerically and the third is a subjective impression. This provides a balance for those approaches that are not yet mature enough.

5.4.6.2 Risk Level Rating

The risk level is not only a result of the assessment of the record, but it also considers development over time by comparing new with existing records. Change detection is performed and assessed. For further details see Section 6.3.

5.4.6.3 Remaining Lifetime

For monitoring systems that are equipped with the necessary sensors, a remaining lifetime assessment is possible through Rainflow Counting. In separate steps the relevant fatigue assessment has been determined and the patterns have been established. By counting the real patterns through permanent monitoring the consumed lifetime is computed. This is done automatically during the period. Once a year an update is provided by the engineer in order to adjust the methodology to eventual changes. The term expresses the percentage of life consumed already. This allows ranking of structures and planning of budgets.

The system not only delivers information on the risk level, but also on what this assessment has been made on. Eventual outliers or trends will be provided in graphic form to show where the threat comes from. Furthermore, a direct activation of video images of the critical area will be provided if available. It is highly recommended to provide this option in order to give the operator an indication of what the background of an alarm could be.

A typical example of an alert system is shown in Figure 5.16. When risk level 4 (High) is reached a consultation with an expert is proposed. The alarm is provided to the operator, who is asked to consult an expert on the observed phenomenon. After that the risk level can be up- or downgraded. This is the only active interference option.

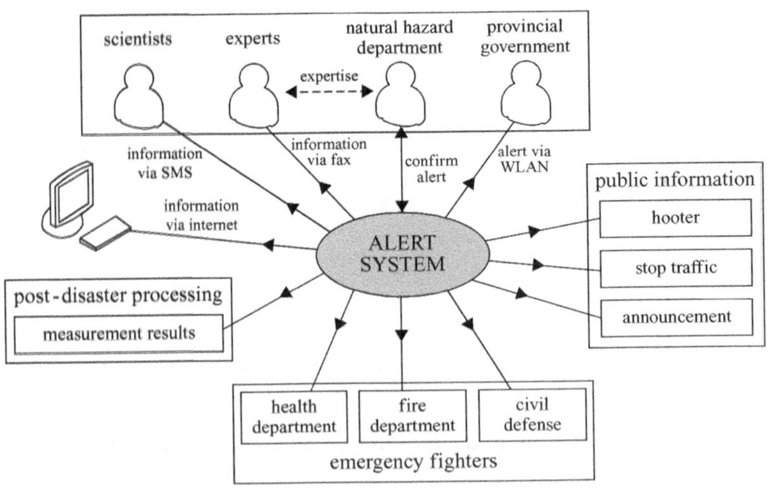

Figure 5.16 Alert system components of VCDECIS

The alert system is engaged in various activities:

- The relevant information is sent to a number of addresses predefined in the system. This includes email but also fax messages.
- New calculations are performed and the history database is consulted.
- The respective authorities, including civil protection, eventual ambulance organizations and traffic authorities, are warned.
- Direct action can be provided by traffic lights, hooters or loudspeakers, where pre-prepared messages can be distributed.

This process has to be well defined for each case separately. The worst scenario is to have a number of false alarms, which would lead to the situation where the system would not be taken seriously and action would not be triggered when needed.

5.5 Current Status of the System

The backbone comprises the databases, which are erected and then filled. The existing data of VCE are inserted step by step. The knowledge base is filled with material, but it will take considerably more time to complete. Other tasks, particularly, in the decision support system, are approached through international collaboration where colleagues are invited to introduce their methodology into the system. A wide sharing of knowledge would be beneficial for every partner.

It is obvious that the system is a living thing. With every improvement the results will be better, but the thresholds might vary. The idea is that thresholds should not be treated as totally inflexible and should be adapted to the new knowledge available. This is not a contradiction. The learning system shall also allow learning about the best applicable threshold values.

5.6 Data Treatment

Many factors influence the quality of our monitoring signals. In the data processing module of a current monitoring system a check routine can help to assess the quality of the data and to eliminate errors. For this purpose all files introduced into the database are scrutinized and the results are reported to the operator. The following process is usually performed:

- the file name is given and a link created;
- statistical pattern recognition methods check whether any of the following phenomena existing: drift, spikes, breaks, jumps, saturation;
- the number of channels contained in the record is provided;
- the energy content is computed, subdivided into vertical, horizontal and longitudinal, and percentages are provided with the option to compare only internally or with related files or to compare with the database;
- a graphic display of the signal and the spectra is given;
- pattern recognition routines assess the character of the file: ambient, road traffic, train traffic or other categories:
- the noise floor is determined and compared to normal values;
- the maximum acceleration is given with an indication of the channel and time;
- the values for vibration intensity, human comfort and damage limits are calculated and displayed graphically, and there is a color indication on the result;
- a category for the file is created, indicating whether it is an average or interesting file;
- eventually a photo of the campaign could be helpful for subsequent identification;

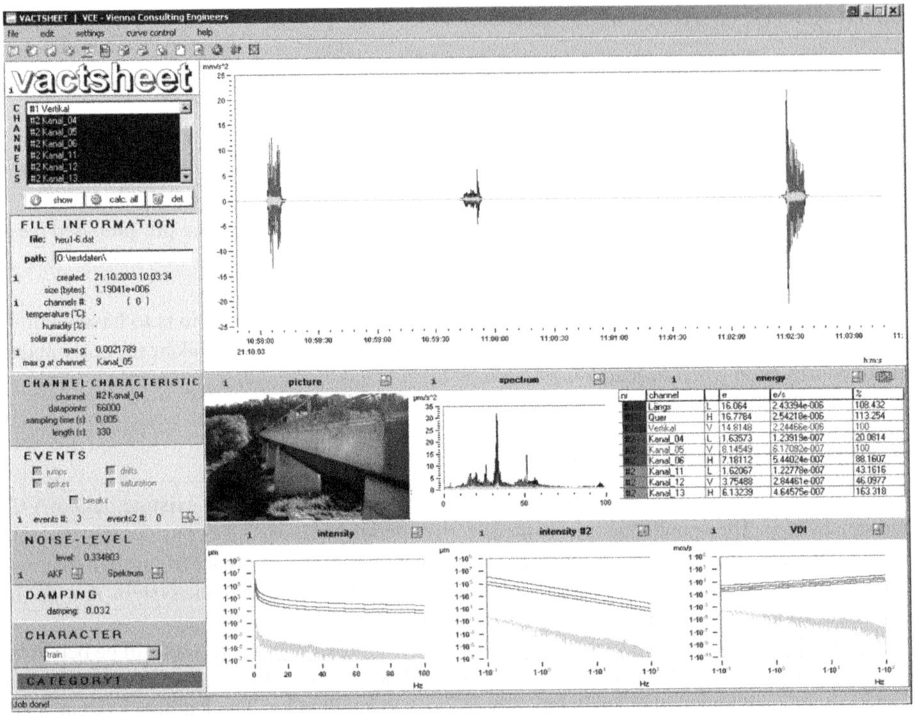

Figure 5.17 Data check record

- other types of records might require different approaches, creating different values to be checked in this routine;
- damping of the maximum frequency is determined by the half-power bandwidth method.

A typical data check record is shown in Figure 5.17. This routine can be performed for a single file or a group of files. The options allow files of the same project to be compared but also files with stored knowledge in the database. Such a check routine should be flexible in order to accommodate new ideas and approaches.

5.7 Data Storage

Customer approval and the long-lasting successful application of BRIMOS have made professional data handling necessary: First, integration of the BRIMOS technology to facility management, including GIS, requires well-defined interfaces as well as organized archival storage; and, Second, technology management needs to be aware of the risk of a slowdown in the evolution of BRIMOS expertise without a proper data and knowledge base.

The basic concept of the development of the BRIMOS database is:

1. Management of data (Figures 5.18 and 5.19): structures (general information, history, condition, maintenance planning, etc.); measurement data; measurement information (equipment, photos, videos, etc.); results and expert report; clients.

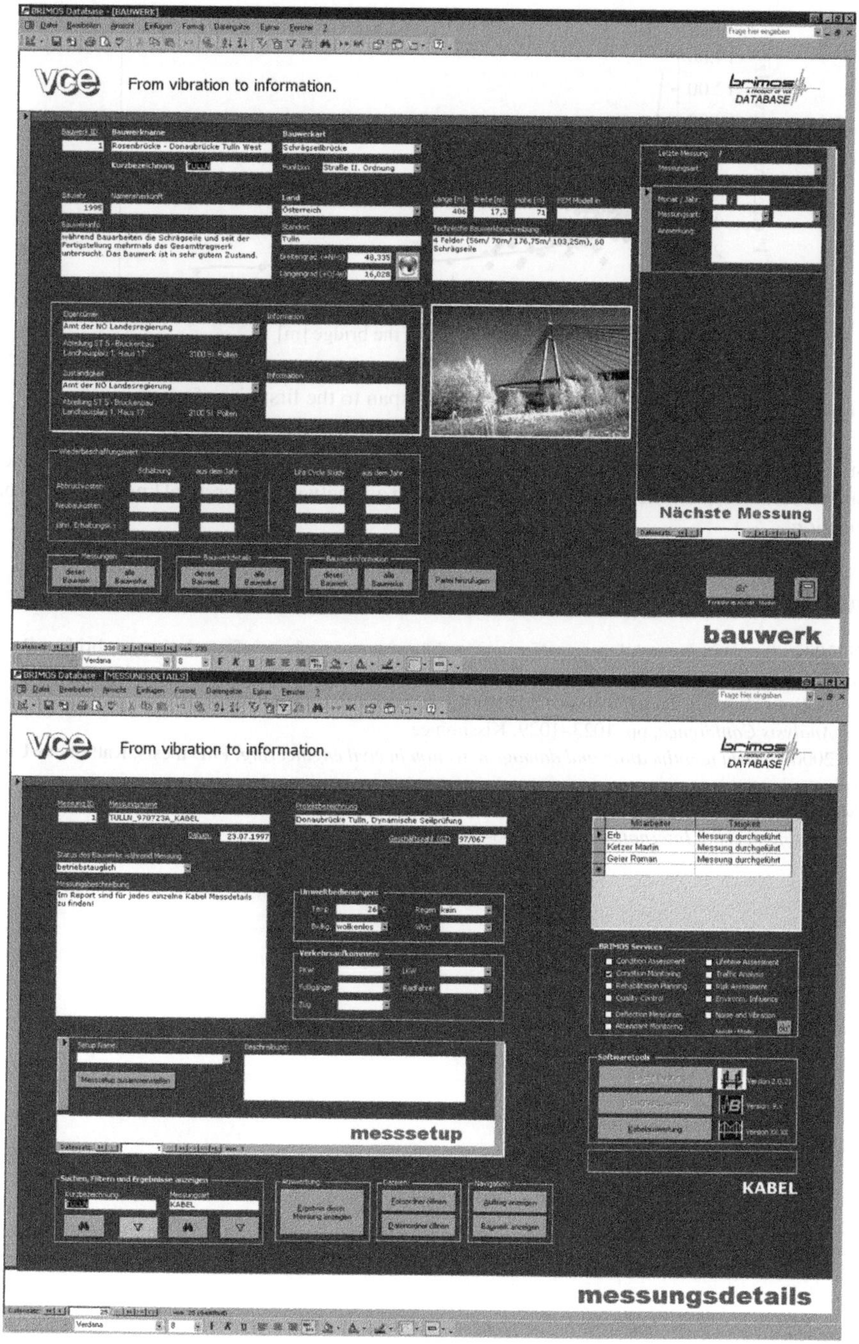

Figure 5.18 Typical bridge project template

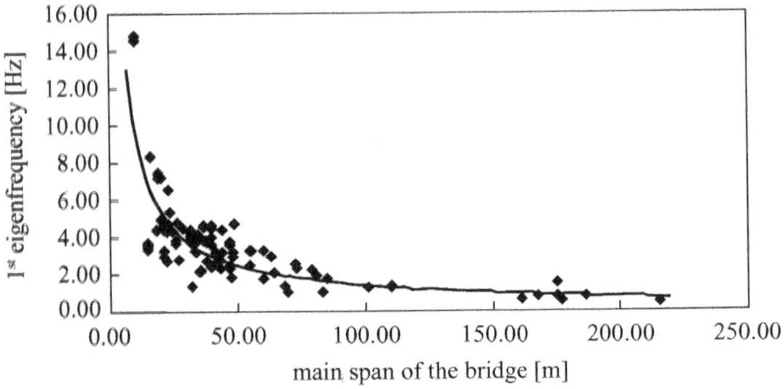

Figure 5.19 Relationship of main span to the first eigenfrequency

2. Feasibility to perform queries to gain knowledge from various statistics: the database access for researchers and scientists needs to overarch the arising gap between theoretical know-how and practical verification.

Further Reading

Feltrin G (2002) Temperature and damage effect on modal parameters of a reinforced concrete bridge. *Proceedings of the Eurodynamics Conference EURODYN 2002*, pp. 373–378, Munich.

Krämer C, De Smet C and De Roeck G (1999) Z24 Bridge damage detection tests. *Proceedings of the 17th International Modal Analysis Conference*, pp. 1023–1029, Kissimmee.

Peeters B (2000) *System identification and damage detection in civil engineering*. PhD thesis, Katholieke Universiteit Leuven.

Wahab MA and De Roeck G (1997) Effect of temperature on dynamic system parameters of a highway bridge. *Structural Engineering International* 7(4), 266–270.

6

Lifetime Assessment of Bridges

Current lifetime models for the constructed infrastructure neglect reality. Consideration of figures estimated for Germany immediately reveals that it is imperative to extend the lifetime of structures considerably, given the prevailing budgetary and feasibility constraints. The value of the constructed environment in Germany is estimated to be €20 trillion. Considering 100 years of lifetime the replacement rate per year would than be €200 billion. In fact the total German construction market is only €60 billion per year, out of which 50% goes into new construction and 50% into maintenance repair and replacement. On the other hand monitoring experience has shown that some structures may have an infinite lifetime because of good design and favorable environmental conditions. One of the priority tasks of SHM therefore has to be the identification of those structures that require intervention in order to achieve the desired lifetime. Fortunately the majority of bridges monitored in the past 10 years can be rated sustainable. If it is considered that half the bridges rated deficient are just functionally inadequate and have to be replaced anyway, an approximate 90% of bridges show infinite lifetime expectations. The opinion among the bridge engineering community that 1% preventive maintenance input per year will ensure this infinite life is technically correct, but practically impossible due to the difference in budgetary figures discussed above. Structural health monitoring has the responsibility of identifying those structures where less maintenance input will not lead to unacceptable deterioration and also to accurately assess those structures where critical stages have been or will be reached at any moment.

The usual life-cycle assessment methods are not very reliable and therefore tend to predict towards the safe side. Differences between theoretically calculated and observed life-cycles may differ an order of magnitude. Life-cycle assessment by means of parallel structural health monitoring can considerably improve the accuracy of these predictions. Relevant examples for comparison are nuclear power plants, where many of them are reaching their predicted lifetime of 30 years, but cannot be taken off the net. In a similar way, if all bridges where the predicted lifetime has ended were closed the transportation network would breakdown completely. The pressure on the SHM community to produce prediction of longer lifecycles will dramatically increase in the near future.

Actually lifetime assessment has not yet developed beyond its infancy stage, and there is a lack of lifetime cost assessment which would favor SHM approaches considerably. The large scientific community working in this field has also not responded sufficiently to this challenge. The main deficiencies are:

- Closed solutions are sought, which due to the complexity of the subject are almost impossible to identify.
- Statistical pattern recognition approaches, as practiced in other sectors, have not really entered SHM, leaving a deficiency of indicators to be included in lifetime assessment routines.

Health Monitoring of Bridges Helmut Wenzel

- The step from laboratory to the field has not been successfully managed. In many approaches environmental conditions govern the response in the field, which completely invalidates laboratory assessment results.

There is a clear need for recognition of health indicators, that work in the field and are derived from simple measurement campaigns. Ideally an intelligent system mounted on a bridge directly computes these indicators and wirelessly transmits this information to the control office.

6.1 Lifetime Assessment Procedure

A flowchart (Figure 6.1) for life-cycle assessment by means of SHM has been proposed by Peil et al. (2006). At the beginning of any monitoring campaign a careful check and assessment of the bridge condition and its eventual accumulated damage is necessary. It is assumed that this is possible using conventional inspection methods complemented by non-destructive testing performed at critical points of the structure. It should not only be focused on visible damage but also on symptoms and indicators an experienced bridge engineer can assess. Experience has shown that the input of the experienced bridge engineer in any monitoring campaign is an important quality indicator. This requirement will be necessary for a long time, until much of this knowledge can be introduced into automatic knowledge systems.

Experience has shown that the conditions on site very often differ considerably from the documentation available. As this affects the assessment dramatically it is necessary to check the actual geometry and boundary conditions as well as material properties of the structure carefully before carrying out the assessment. This requirement is a separate monitoring task beyond the scope of this book. Nevertheless the potential for laser scanning, photogramatric methods and non-destructive testing (NDT) should be assessed.

6.2 Hot-Spot Detection

Probabilistic assessment of critical hot spots requires refined monitoring methodologies. Identifying critical parts of the structure is one of the most important tasks in monitoring. Hot spots or weak spots are areas of the structure that are prone to damages or where possible damages cause nontolerable consequences. Hot spots on older structures, usually designed with very different safety levels, are normally well known and can be determined by existing structural calculations or by experience. New structures, however, show an equally distributed safety level over a large number of critical details, hiding potential hot spots.

One of the most important feedbacks of monitoring to bridge design should be that any moderate overdesign, eliminating the appearance of hot spots, will have an essential influence on the lifetime expectation of a bridge. The mistakes made in the 1970s and 1980s, when only the cheapest price counted and materials were used to their threshold limits, demonstrated the high price to be paid for adopting this strategy. Almost all of these structures had to be retrofitted before they reached 30 years of age, and some of the retrofit measures have been more expensive than the initial investment. Unfortunately these mistakes are currently being repeated in many countries where the erection of transportation infrastructure is booming. One of the generic tasks of future research on this subject should be the production of statistics that support and disseminate this knowledge widely.

Because of the high costs of monitoring every hot spot, critical details should be ascertained using probabilistic methods, leading to the optimum number and location of samples. The procedure aimed at reliability in the determination of hot spots should classify critical weak spots as those that contribute most to the overall failure probability of the structure, and it is these points that must be monitored. To

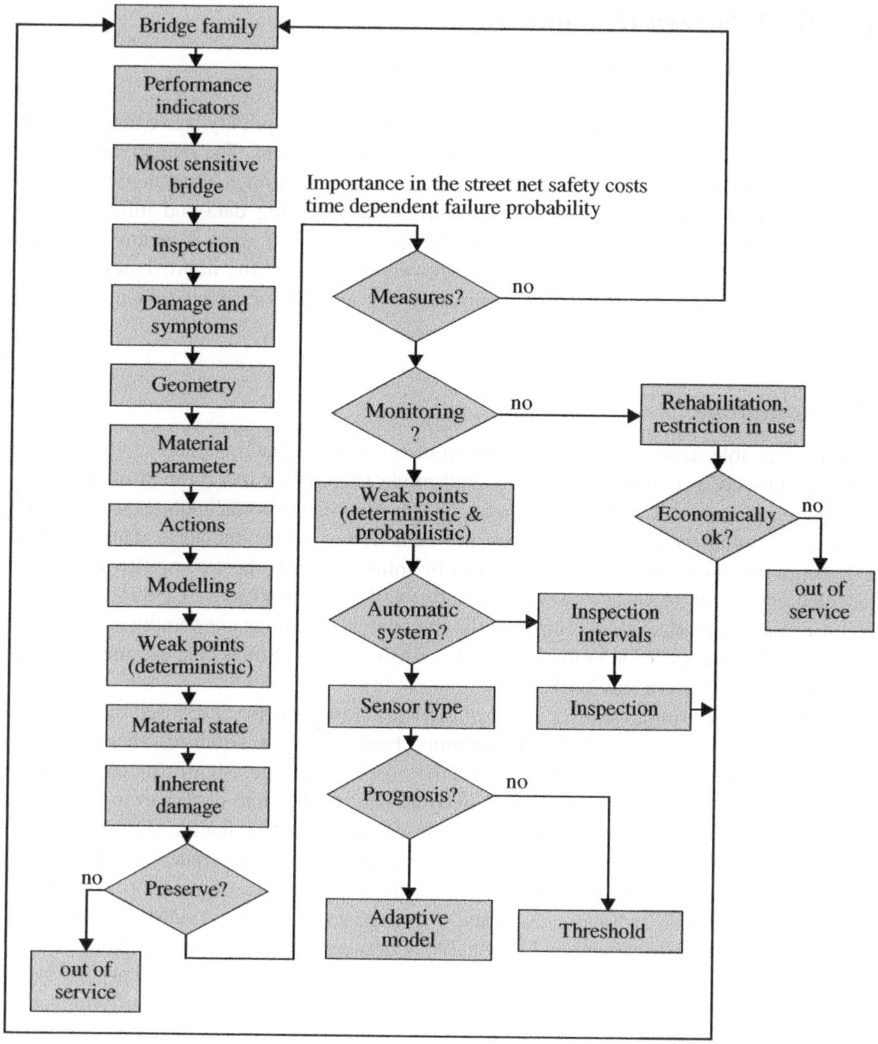

Figure 6.1 Overview of the procedure

determine the failure probability, descriptions of the limit state functions and limiting values, stochastic and mechanical modeling, and analyses of sequences of events and fault trees are required.

The most critical hot spots should be permanently monitored, which will take a considerable amount of time before reliable results are available. In order to overcome this shortcoming, Peil et al. (2006) have proposed building physical models of the detected hot spots in order to test them in the laboratory. This downscaling of time allows results to be obtained much more quickly but with the disadvantage that the environmental conditions are not sufficiently represented. Nevertheless such experiments have the potential to detect those areas where cracking may occur first, which could improve the monitoring system to be installed.

6.3 Statistical Pattern Recognition

Contributed by Professor Anne S. Kiremidjian, Stanford University

If we follow the idea of permanently monitoring hot spots we could quickly arrive at a situation where too many monitoring systems produce unmanageable amounts of data. The human capacity to assess large amounts of data is very limited, so statistical pattern recognition methodologies need to be introduced into the procedure. The procedure scrutinizes the incoming data and informs the operator of unusual events, but it will only be as good as the specification of what is meant by unusual. It is therefore necessary to keep the raw data in order to allow reassessment after improvement of the methodology. In the following sections some of the currently applied pattern recognition methodologies are described.

6.3.1 Introduction

Most currently available structural damage detection methods are global in nature, i.e. they track changes in the dynamic properties (natural frequencies and mode shapes) for the entire structure from input–output data using global structural analyses and system identification techniques (see Doebling et al. 1996). However, global damage measures are not sensitive to minor and local damage. Also, these damage detection methods require that all the data be transmitted to a host computer, which presents a significant challenge for the low-data rates of currently available sensor radios (e.g., 240 kbs for 802.15.4 wireless sensor radio modules). Most importantly, the analysis methods are computationally expensive and do not lend themselves to be embedded at the sensing nodes. Some of these methods have been discussed in earlier chapters of this book.

A pattern classification framework for SHM of civil structures was first proposed by Sohn and Farrar (2001). Such methods rely on the signatures obtained from vibration, strain or other data recorded at sensing nodes to extract features that change with the onset of damage. Significant developments have followed the initial work by Sohn and Farrar, thereby advancing the-state-of-the-art in statistical pattern recognition methods for SHM. In particular, a simple damage sensitive feature (DSF) from an autoregressive moving average (ARMA) model and a localization algorithm have been proposed by Nair and Kiremidjian (Nair et al. 2006). A Gaussian mixture model for feature discrimination was introduced by Nair and Kiremidjian (2007), who also proposed a damage extent measure (DM) that is based on the Mahalanobis distance (see Brockwell and Davis 2002) between feature vectors. Wavelet based models are discussed in Nair et al. (2006). In this section we summarize the statistical signal processing methods described in these papers to provide the reader with an understanding of the underlying fundamental mathematical formulats. The section starts with a discussion on the differences between the conventional cable-based structural health monitoring systems and presents the main reasons why decentralized computational tools have become necessary. We use the American Society of Civil Engineers (ASCE) Benchmark Structure (see Johnson et al. 2004) to illustrate the various methods, and therefore a description of the structure and the experiments conducted with it are presented next. In subsequent sections the various algorithms are described in detail, with illustrative examples from the ASCE Benchmark Structure.

6.3.1.1 Wireless Structural Health Monitoring and the Need for Decentralized Damage Diagnosis

In traditional structural health monitoring systems (HMS) sensors are distributed at key location on a structure connected via cables to a data acquisition system where data are collected and stored. Interrogation of collected data may be performed on-site or data may be transferred to an office or control center where diagnosis and prognosis is typically performed by experts. Figure 6.2 shows schematically

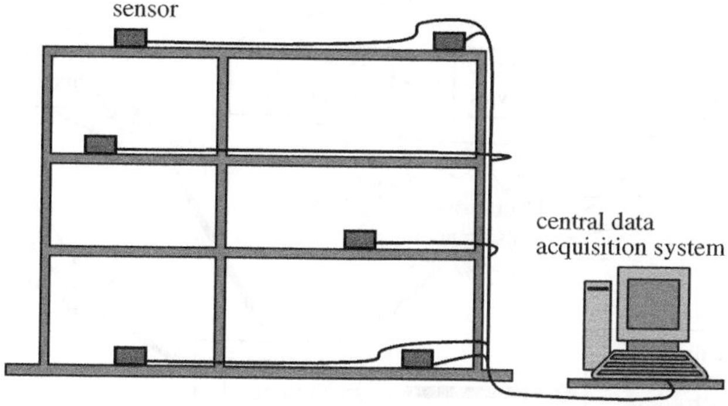

Figure 6.2 Schematics of current cable-based structural monitoring systems

the layout for such systems. Because data cables require high fidelity, they are expensive and moreover costly to install. In existing structures thick walls or floors often make it difficult for cables to be installed. As a result, sensing systems are sparse and provide data only at limited locations on a structure.

With recent developments in wireless sensing devices (Straser and Kiremidjian 1998; Lynch et al. 2004; Wang et al. 2006) dense sensor networks with computational capabilities at sensing nodes can now be realized. Figure 6.3 shows the schematics for a wireless sensor network. A key feature of such a network is its scalability to higher density, since the network does not depend on multiple cables and is not limited by the number of channels available in the data acquisition system. In contrast, the cost of wired networks increases significantly with greater sensor density and multiple data acquisition system will need to be used.

The distributed computing environment provided by wireless networks also provides an opportunity to develop analysis capabilities at the sensing nodes, enabling multi-tiered diagnostic and prognostic decision making. Thus, damage diagnosis and prognosis can be performed at sensing nodes using individual sensors, then extracted information is fused using outcomes from multiple sensors at the nodes, followed by diagnosis by combining information from several sensing nodes, and finally analysis results

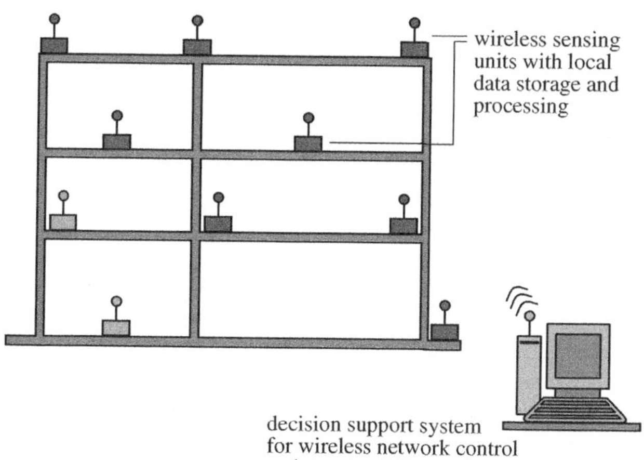

Figure 6.3 Schematics for wireless structural monitoring system

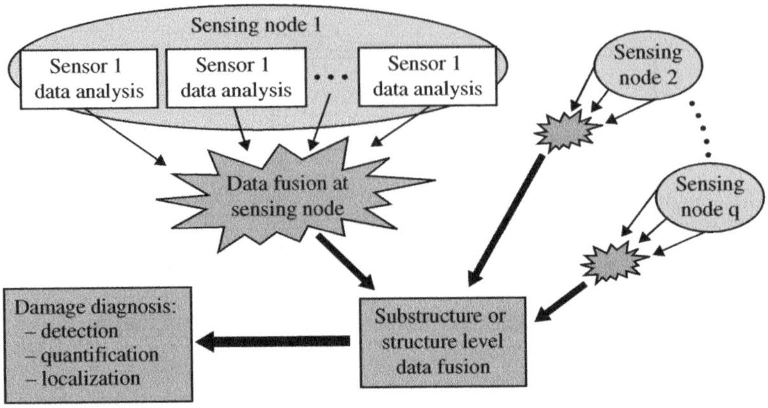

Figure 6.4 Multi-tiered, distributed diagnostic and prognostic decision making paradigm for structural monitoring systems

are assembled at the system level. Figure 6.4 shows the multi-tiered decision analysis paradigm that can be achieved through the dense sensor networks communicating through wireless sensor radios.

6.3.1.2 Decentralized Damage Detection Paradigm and Statistical Pattern Recognition Methods

With wireless sensing becoming more of a reality, the need and advantages for diagnosis and prognosis at the sensor location are multifold. Damage diagnosis performed at the sensing node requires that only the results of the assessment need to be transmitted to the central decision support system. As a consequence, there is a significant decrease in power consumption by the sensing node, which greatly reduces communication clutter across the network. Statistical signal processing and pattern classification methods are particularly suitable for analysis at sensing nodes because they rely on individual signals to perform damage diagnosis and prognosis. Statistical pattern classification methods have been developed over the past two decades for applications in engineering, biology and finance. In the past decade, developments in the engineering field have been fueled by the need for image reconstruction for medical and computer visualization applications, automated speech recognition, finger print identification, and much more (Duda et al. 2001). Classification schemes are broadly divided into supervized learning and unsupervized learning schemes (Hastie et al. 2001). In supervized learning schemes, the algorithm is trained on a dataset whose outcome variables are observed and predictions are made with respect to the training. On the other hand, unsupervized learning schemes are algorithms where no outcome variables are observed, and thus the main aim is to classify or cluster the data.

Although pattern classification techniques have been applied to identify faults in machinery or discrimination of vibrations arising from different rotating components (Farrar and Duffey 1999), there are many challenges in extending this paradigm to civil engineering structures. Civil engineering structures generally have complicated geometry; are built with different materials such as steel, reinforced concrete and composites, whose behavior is not well understood; are strongly affected by local environmental and geological conditions such as temperature, humidity and local soils; and subjected to extreme loads from earthquakes and hurricanes that push them well beyond their design limits. Diagnosis and prognosis of civil engineering structures are further complicated because these systems are composed of components or substructures, where damage to components or substructures would lead to a forced redistribution, a phenomenon not generally observed in mechanical systems. While force redistribution increases stresses in remaining members of a structure, they will not necessarily cause damage or failure in these components. Further analyses of the subsystems and the global system are needed to identify failure mechanisms that can lead to partial or complete collapse of the structure.

6.3.1.3 The ASCE Benchmark Structure

In order to illustrate the various steps in statistical pattern recognition and classification, we need data from a well organized experiment. The ASCE Benchmark Structure experiment (Johnson et al. 2004) is well documented and as such provides data that are particularly suitable for demonstrating the various steps in the damage diagnosis presented in this chapter. The ASCE Benchmark Structure is a four story, two-bay by two-bay steel braced frame, illustrated in Figure 6.5 There are 16 sensors (measuring acceleration) in the building, and their placement and direction of the measured acceleration are shown in Figure 6.6. Damage is simulated by removing braces in various combinations, resulting in a loss of stiffness. Damage patterns (DP) include:

- 0: Undamaged structure
- 1: Removal of all braces on the first floor
- 2: Removal of all braces on the first and third floors
- 3: Removal of one brace on the first floor
- 4: Removal of one brace on the first and third floors
- 5: Damage pattern 4 + loosening of bolts
- 6: Reduction of the stiffness of a brace to one-third of its original value

Damage patterns 1 and 2 are major damage patterns, whereas damage patterns 3 and 6 are minor damage patterns.

We use the results from the numerical simulation primarily because these data are obtained under well controlled conditions. Example applications with other laboratory and field data can be found in

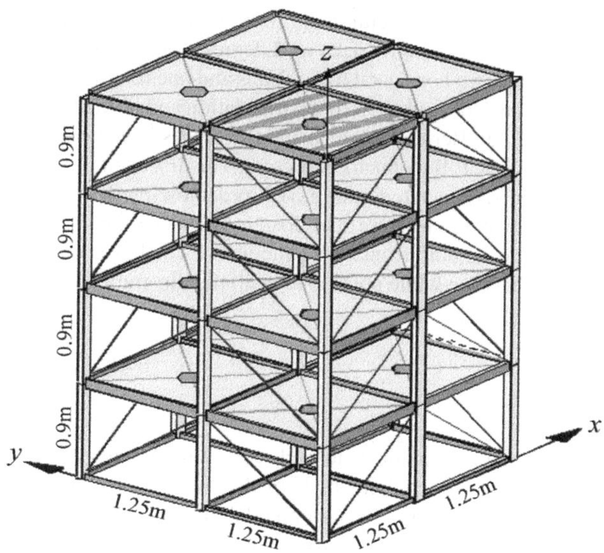

Figure 6.5 Illustration of the ASCE Benchmark Structure (Johnson et al. 2004)

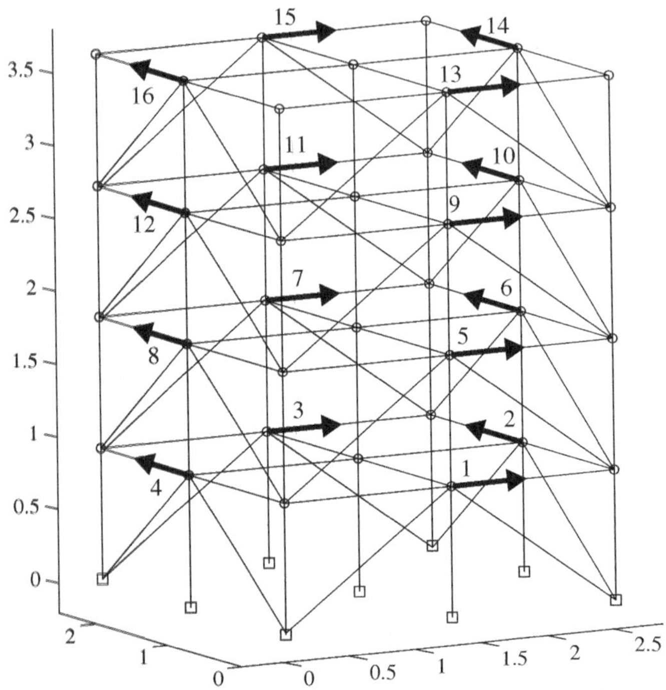

Figure 6.6 Sensor location and direction of acceleration signals in the ASCE Benchmark Structure (Johnson et al. 2004)

Nair and Kiremidjian (2007), Noh and Kiremidjian (2008) and Cheung et al. (2008). Two FE models were used to generate the simulated response data: a 12 degree of freedom (DOF) shear-building model that constrains all motion except two horizontal translations and one rotation per floor; a 120 DOF model that requires that floor nodes have the same horizontal translation and in-plane rotation. The columns and floor beams are modeled as Euler–Bernoulli beams and the braces have no flexural stiffness. There are two loading conditions on the ASCE Benchmark Structure. The first excitation is a series of independent filtered Gaussian white noise loads generated using a sixth-order low-pass Butterworth filter with a 100 Hz cutoff and applied at each story of the structure. This load is intended to model wind or ambient vibration forces. The second loading is a random excitation generated by a shaker on the roof-top of the center column. We use the filtered Gaussian white noise load in all the illustrations presented in the subsequent sections.

6.3.1.4 Outline

Section 6.3.2 presents the general steps of pattern classification as they pertain to structural damage diagnosis. In Section 6.3.3 feature extraction methods are discussed. Two types of models are considered– time domain that fall within the autoregressive (AR) model family and time–frequency domain that are described by the wavelet transform. Classifications methods used to discriminate features from signals belonging to damaged and undamaged structures are presented in Section 6.3.4. Damage quantification is the topic of Section 6.3.5.

6.3.2 *Overview of Statistical Pattern Classification Based Damage Detection Algorithms*

Physical and geometric changes in a structure caused either by normal everyday loads or by extreme events result in changes in the static and dynamic behavior of the system and its components. Statistical pattern classification methods for damage diagnosis rely on the premise that measurements of structural response, such as vibrations and strain, will reflect these changes. These methods track specific features in observed data to identify and quantify the changes. Thus, for SHM a set of baseline measurements need to be acquired from which a database with feature vectors for various environmental and loading conditions needs can be created. The same features are then extracted from measurements taken periodically during the life of the structure and are compared to the most appropriate set of baseline features to determine if changes have taken place. The basic steps in a pattern classification algorithm for SHM are as follows:

1. Signal conditioning through detrending, filtering, normalization and standardization.
2. Signal modeling and features extraction.
3. Classification through features discrimination for damage identification.
4. Damage localization.
5. Damage quantification.

Steps 1, 2 and 3 are performed for both the baseline and the subsequent signals. Feature discrimination requires statistical modeling to determine if change/damage has taken place. Selection of appropriate signals from the baseline data needs to be performed carefully to insure that the statistical discrimination is performed on signals coming from similar environmental and loading conditions. Additional models are required for damage localization. Damage quantification can be achieved through extensive laboratory and field testing to enable correlation of changes in the feature vector to specific changes in the structure. In particular, systematic tests need to be performed to identify the specific feature changes to corresponding damage state in the structure.

Signal conditioning is needed because of the inherent noise in measurement data, temperature variation during measurements, and instability of sensors. Trends in signals are removed by fitting a curve through the signal. There is a wealth of information on filtering methods for reducing and virtually eliminating the noise in such data and the reader can select from a multitude of textbooks on the subject.

Feature extraction can be performed either in the time domain or the frequency domain. The features of the models representing the signal must be such that they can be correlated to physical parameters or changes in such parameters of the structure. Examples of time domain methods include autoregressive (AR), autoregressive moving average (ARMA) and time varying autoregressive moving average (TV-ARMA) models. In the frequency domain the Fourier, Hilbert and wavelet transforms can be used. In Section 6.3.3 we explain how to apply an ARMA model and the wavelet decomposition of vibration signals to extract appropriate features. Hypothesis testing and Gaussian mixture models (GMM) are the two methods used for feature discrimination presented in Section 6.3.5.

6.3.3 *Signal Modeling and Feature Extraction Methods*

Structural damage affects the dynamic properties of a structure, resulting in a change in the statistical characteristics of the measured acceleration time-histories. Thus, damage detection can be performed by extracting useful information from the vibration signal (measured from a structure before and after damage) in either the time domain or the spectral domain. Methods for modeling measurement signals in the time and frequency domain for the purposes of extracting damage sensitive features and discriminating these features are summarized in this section.

6.3.3.1 Time Domain Models – AR Family of Models

Time domain models that have been considered in relationship to SHM include autoregressive (AR), autoregressive moving average (ARMA) and autoregressive models with exogenous inputs (ARX). A fundamental assumption in these models is that the data come from linear systems. Damage in structures, however, occurs as a result of nonlinear behavior. For example, when a member is loaded beyond its elastic limit the structure will deform permanently. This permanent deformation can be associated with damage to the member and/or of the subsystem. In order to apply linear models such as those considered herein, it is assumed that damage is localized at a joint or at specific points on a member and after damage has taken place the structure continues to behave linearly. For example, for a steel member, when the elastic limit is exceeded by overload and 10% of the cross section is now plastic, the member will still act linearly when the overload is removed. The member will continue to act linearly but most likely with a different stiffness and as long as the elastic limit is not exceeded again. Thus, measurement obtained in the vicinity of plastic hinge formation can still be represented by linear time-domain models as long as those measurements are not obtained during the overload process.

If damage results in large geometric changes in the structure, such as those caused by loosening of several bolts leading to permanent displacements, the static or dynamic behavior of individual members will still be linear. The structural system will also behave linearly provided it is still stable, but now the system would have changed from the one originally in place. Thus, using time-domain models to represent measurements from linear systems is again deemed appropriate. Sequences of component failures leading to system collapse require careful modeling and consideration prior to SHM system installation in order for the diagnostic model to identify such events and report impending collapse. Understanding of the structural behavior is critical when feature discrimination criteria are selected and applied.

In the following development the type of measurements are not specified. The data can be either strain or acceleration. These data are treated as time-varying signals that are sampled at appropriate rates to capture the properties of the structure. Autoregressive and ARMA models are used when the forcing function to the structural system is unknown. This is the case when ambient vibrations are measured. The ARX model is used when the input to the structure is known. Most frequently, we do not know the input motion, thus, our focus will be on the AR or ARMA models. The ARMA model is discussed in detail. The AR model is a subset of the ARMA model and thus is presented only briefly. For more details on time-series modeling, the reader is referred to Brockwell and Davis (2002).

A typical vibration signal obtained from a structure is shown in Figure 6.7. Environmental effects, such as temperature and humidity, and loading effects will distort and change the signal. To reduce the likelihood of reporting change due to these effects rather than actual damage, the signals need to be standardized and normalized. After standardization and normalization the features extracted from the signals from undamaged and damaged structure would have similar statistical characteristics and can be compared.

Let $x_i(t)$ be the acceleration data from sensor i. This sensor's data is then partitioned into different streams, $j = 1, 2, \ldots, N$, where j denotes the jth stream of data from the sensor I and N is the total number of data streams used in the analysis. The signal $x_{ij}(t)$ is standardized by subtracting its mean value and is normalized with respect to the standard deviation to obtain the signal $\tilde{x}_{ij}(t)$ as follows:

$$\tilde{x}_{ij}(t) = \frac{x_{ij}(t) - \mu_{ij}}{\sigma_{ij}} \tag{1}$$

where μ_{ij} and σ_{ij} are the mean and standard deviation of the jth stream for sensor i, respectively. For notational convenience, $x_{ij}(t)$ will be used hereafter instead of $\tilde{x}_{ij}(t)$.

A second assumption of the ARMA model is that the signal is stationary (Brockwell and Davis 2002). Compliance with the stationarity condition is performed by observing the autocorrelation function (ACF). Figure 6.8 shows that the autocorrelation function of the normalized data has a cyclical trend that will

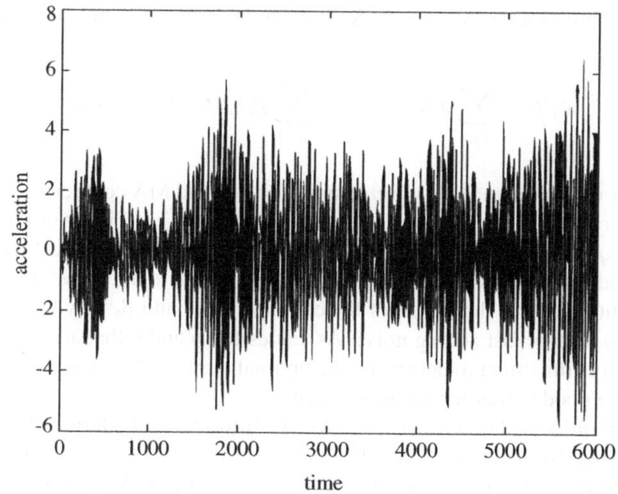

Figure 6.7 Plot of a typical acceleration time history before filtering, denoising and detrending

need to be removed. For detrending the data, three methods are used: harmonic regression, simple average window and moving average window (Brockwell and Davis 2002).

Nair and Kiremidjian (see Nair et al. 2006) found that harmonic regression could not remove the trends when they considered simulated acceleration data. They used a combination of the simple average window and the moving average window to obtain a stationary signal. In general, several different methods need to be tested to determine the window sizes and the combination of methods that will remove any nonstationary trends in the data. A review of the residuals of the signal or performing the Ljung-Box statistic test can provide a further verification test for compliance of the stationarity condition.

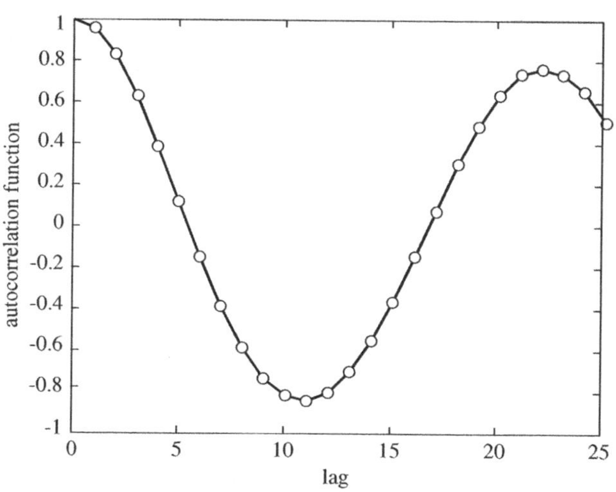

Figure 6.8 Autocorrelation function of the normalized data

The general formulation for the ARMA model is given by Brockwell and Davis (2002):

$$x_{ij}(t) = \sum_{k=1}^{p} \alpha_k x_{ij}(t-k) + \sum_{k=1}^{q} \beta_k \varepsilon_{ij}(t-k) + \varepsilon_{ij}(t) \tag{2}$$

where $x_{ij}(t)$ is the normalized signal, α_k and β_k are the kth AR and MA coefficient, respectively; p and q are the model orders of the AR and MA processes respectively and $\varepsilon_{ij}(t)$ is the residual term. It is critical that the orders p and q are optimal. If the selected values of p and q are too large, then the computation of the coefficients α_k and β_k becomes expensive without adding value to the model. In general the larger the order the better the fit, however, the values of the coefficients α_k and β_k may be very small, raising the issue of whether they are not just adding noise. As values for p and q that are too low will not capture the signal correctly, it is important to determine the optimal numbers for p and q, and to test that indeed these orders provide a good representation of the signal.

The optimal model order is obtained using the Akaike Information Criteria (AIC) (see Brockwell and Davis 2002) for detailed formulation). The AIC consists of two terms, one is a log-likelihood function and the other is a penalty function for the number of terms in the ARMA model. Figure 6.9 shows the variation of the AIC values with the AR model order. From this figure it can be observed that an AR model order of 5–8 and MA model order of 2–4 is appropriate for the analysis (the figure is only for AR). Various data were analyzed by Cheung et al. (2008) and the same orders for p and q were found to be appropriate for applications to ambient vibrations on structures. For any other applications, however, similar tests will need to be performed before settling on specific values of p and q.

A cross-validation analysis can also be carried out to check the accuracy of the modeled signal. For a particular data stream, the data set is split in two, one is used for the analysis and the other is used for forecasting. In the analysis part, the coefficients of the ARMA model are ascertained. Using these coefficients, the values of the acceleration data are predicted. The error between the predicted values and actual values are obtained to be a minimum using the above model orders.

The algorithms most frequently used for estimating the ARMA coefficients are the Innovations and Burg algorithms (Brockwell and Davis 2002). In the examples presented in this chapter, the Innovations algorithm is used. After estimation of the AR coefficients, the residuals obtained are tested to determine

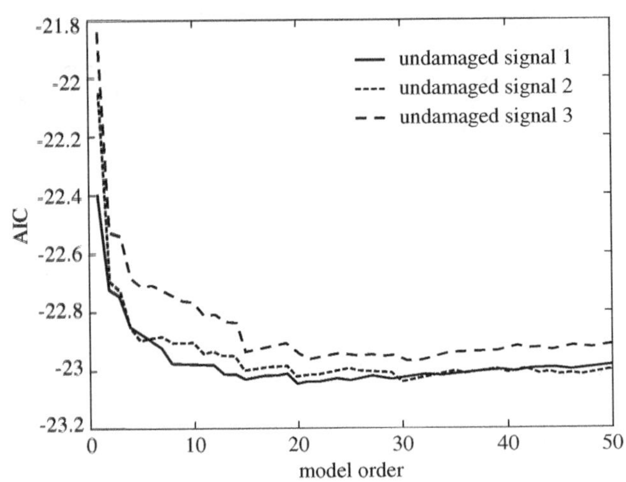

Figure 6.9 Variation of AIC with model order

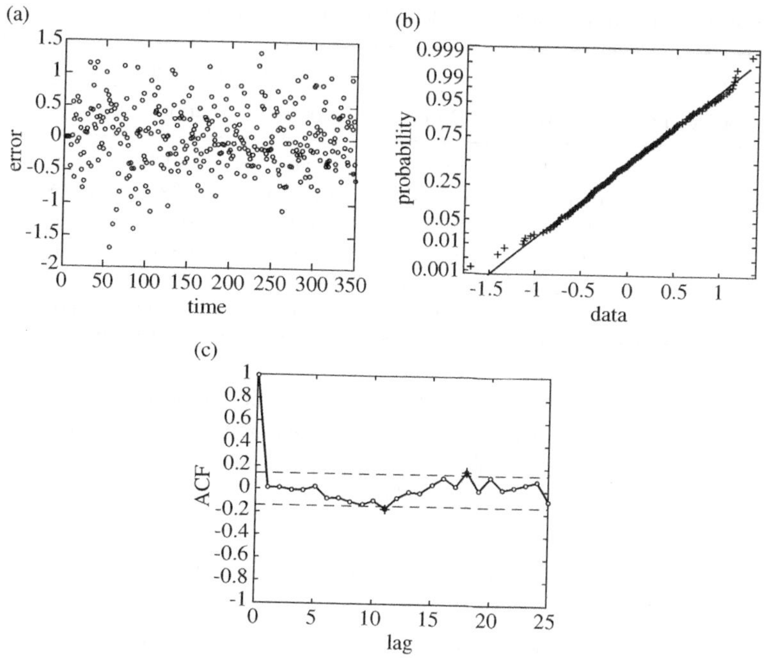

Figure 6.10 Verification of the independent and identically distributed characteristics and normality of residuals. (a) Variation of residuals with time. (b) Normal probability plot of the residuals. (c) Variation of the autocorrelation function of the residuals with lag

if they are normal, and independent and identically distributed (i.i.d.). Figure 6.10a shows the normal probability plot of the residuals. The straight line variation indicates a normal distribution of the data, which is mildly violated at the tails. Figure 6.10b shows the variation of the residuals with time. It is seen that there is no trend, therefore, indicating homoskedasticity. Figure 6.10c shows the autocorrelation function (ACF) of the residuals, from which it is observed that the values of the ACF at lags >1 are not statistically significant. The Ljung-Box statistic is also used to test the i.i.d. assumption of the ARMA residuals. The Ljung-Box statistic is defined as follows:

$$Q_{LB} = n(n+2)\sum_{j=1}^{h}\frac{\rho^2(j)}{n-j} \tag{3}$$

where n is the sample size, $\rho(j)$ is the autocorrelation function at lag j, and h is the number of lags being tested. The null hypothesis of randomness is rejected if $Q_{LB} > \chi^2_{1-\alpha,h}$, where α is the level of significance of the hypothesis test and $\chi^2_{1-\alpha,h}$ is the $(1-\alpha)$th percentile of the χ^2 distribution with h degrees of freedom. For this particular dataset, the null hypothesis is accepted. Thus, the assumptions made on the residuals are satisfied.

The total duration of the record $x_i(t)$ is 480 s. The record is divided into 80 segments, denoted by $x_{ij}(t)$, $j = 1, 2, \ldots, 80$, each having six seconds duration sampled at 1000 Hz resulting in 6000 data points per segment. The ARMA coefficients are computed for each 6-s segment of the acceleration data and the first three AR coefficients are used in the computation of the damage sensitive feature. To determine the sensitivity of the coefficients to the number of data points in the signal, analyses were performed in

Table 6.1 Sensitivity of AR coefficients to the number of data points

	Number of data points					
Value of AR coefficient	1000	2000	3000	4000	5000	6000
Mean of α_1	1.0441	1.5087	1.0566	1.0453	1.0359	1.0301
(SD α_1)	(0.1947)	(0.1369)	(0.1088)	(0.0981)	(0.0831)	(0.0788)
Mean of α_2	1.0359	1.0502	1.0517	1.0459	1.0403	1.0358
(SD α_2)	(0.1966)	(0.1373)	(0.1002)	(0.0840)	(0.0710)	(0.0582)
Mean of α_3	1.2204	1.2644	1.2772	1.2762	1.2761	1.2712
(SD α_3)	(0.1366)	(0.0861)	(0.0608)	(0.0451)	(0.0360)	(0.0338)

the range 1000–6000 points in increments of 1000. The AR coefficients were found to reach stable values at about 3000 points; however, 6000 points were used in the analysis presented in this study. The stability of the first AR coefficients with the number of data points is presented in Table 6.1. Both the mean and standard deviation of the coefficients are listed in that table.

6.3.3.2 Definition and Development of the Damage-Sensitive Feature using AR Coefficients

In this section, the ARMA time-series model is used to develop features that discriminate between the damaged and non-damaged state of a structure. Several damage-sensitive features (DSF) were considered. Of those various DSFs considered, those depending on the first three AR coefficients appeared to be most promising because these coefficients are statistically the most significant among all the coefficients of the model. After testing several different combinations with the first three coefficients, it was found that the first AR coefficient normalized by the square root of the sum of the squares of the first three AR coefficients provides the most robust damage-sensitive feature. Thus the proposed damage-sensitive feature (DSF) is defined as follows:

$$DSF = \frac{\alpha_1}{\sqrt{\alpha_1^2 + \alpha_2^2 + \alpha_3^2}} \tag{4}$$

where α_1, α_2 and α_3 are the first three AR coefficients. Variation of the DSF with the record number for different damage patterns is illustrated in Figure 6.11.

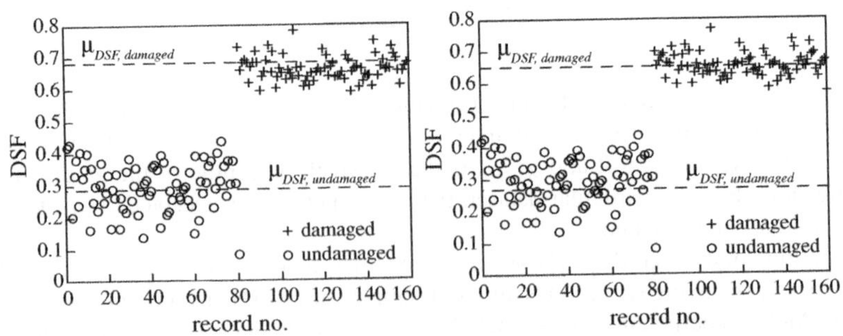

Figure 6.11 Variation of DSF with record number for different damage patterns

From these figures it can be seen that for all damage patterns there is a significant difference in the mean levels of the *DSF*s of the damaged and the undamaged states. Thus, to test the statistical difference between the means of two groups of data, the standard t-test is used Rice (1999). The relation of the AR coefficients to the physical parameters of the system is given below (Nair et al. 2006).

Applying the z-transform to both sides of Equation (2) and ignoring the effect of the error term, we obtain

$$X_{ij}(z) = \sum_{k=1}^{p} \alpha_k z^{-k} X_{ij}(z) + \sum_{k=1}^{q} \beta_k z^{-k} \Xi_{ij}(z) \tag{5}$$

where $X_{ij}(z)$ and $\Xi_{ij}(z)$ are the z-transforms of the $x_{ij}(t)$ and $\varepsilon_{ij}(t)$ respectively. Then, the transfer function $H(z)$ is derived as

$$H(z) = \frac{X_{ij}(z)}{\Xi_{ij}(z)} = \frac{\beta_1 z^{-1} + \beta_2 z^{-2} + \cdots + \beta_q z^{-q}}{1 - \alpha_1 z^{-1} - \alpha_2 z^{-2} - \cdots - \alpha_p z^{-p}} \tag{6}$$

The denominator of the transfer function $H(z)$ is a polynomial equation of order p known as the characteristic equation. The roots of the characteristic equation, known as the poles of the system, are expressed as follows:

$$z^p - \alpha_1 z^{p-1} - \alpha_2 z^{p-2} - \cdots - \alpha_p = 0 \tag{7}$$

The poles, z_{pole}, of the characteristic equation are related to the modal natural frequencies and the damping ratios as follows:

$$z_{pole} = \mathrm{e}^{-\zeta\omega_n \Delta t \pm j\sqrt{1-\zeta^2}\omega_n \Delta t} \tag{8}$$

where, ζ and ω are the damping ratio and natural frequency of the particular mode and Δt is the sampling time of the signal. Equation (8) can also be rewritten as $z_{pole} = re^{j\phi}$, where the amplitude r and phase angle ϕ are expressed as

$$r = \mathrm{e}^{-\zeta\omega_n \Delta t} \tag{9}$$

$$\phi = \sqrt{1-\zeta^2}\omega_n \Delta t \tag{10}$$

Using simple theory of polynomial roots, it can be shown that

$$\sum_i z_{pole,i} = \alpha_1 \tag{11}$$

$$\sum_{i,j} z_{pole,i} z_{pole,j} = -\alpha_2 \tag{12}$$

$$\sum_{i,j,k} z_{pole,i} z_{pole,j} z_{pole,k} = \alpha_3 \tag{13}$$

Equations (8) and (11–13) show the relationship between AR coefficients and the modal frequencies and damping. Thus, the AR coefficients are good candidates for feature vectors as an indicator of damage.

Further details of the sensitivity of the AR coefficients to stiffness coefficients are given in Nair et al. (2006).

Figure 6.11 illustrates the migration of the first three AR coefficients as damage is induced in the ASCE Benchmark Structure. The circles represent the values of the first three AR coefficients of the undamaged system and the crosses show the values after damage has been introduced to the system. The mean values for each of the signals are shown by the dashed line. It can be observed from these plots that the mean values of the damage-sensitive feature defined by Equation (4) are very different for the damaged and undamaged signals.

A second damage-sensitive feature that can be used in conjunction with the Gaussian mixture model, which will be discussed in Section 6.3.4.2, is the vector of the AR coefficients defined as $\alpha = \{\alpha_1, \alpha_2, \alpha_3\}$. These coefficients are found to migrate in three-dimensional space as damage increases. Figures 6.12–6.14 show the migration of these coefficients as damage increases in the structure. The distributions of α values appear further and further apart. The centroid of each distribution is represented by the dark cross, showing more clearly the separation between the coefficients from the damaged and undamaged signals. As we will see in Section 6.3.4.2, these distributions will be modeled as samples from Gaussian distributions and the distance between the distributions will be used to quantify damage degree.

For strain measurements, Noh and Kiremidjian (2008) found that the first AR coefficient α_1 is much larger in numerical value than the higher order coefficients. They were successful in discriminating damage from strain measurement using only that coefficient as the damage-sensitive feature. Other

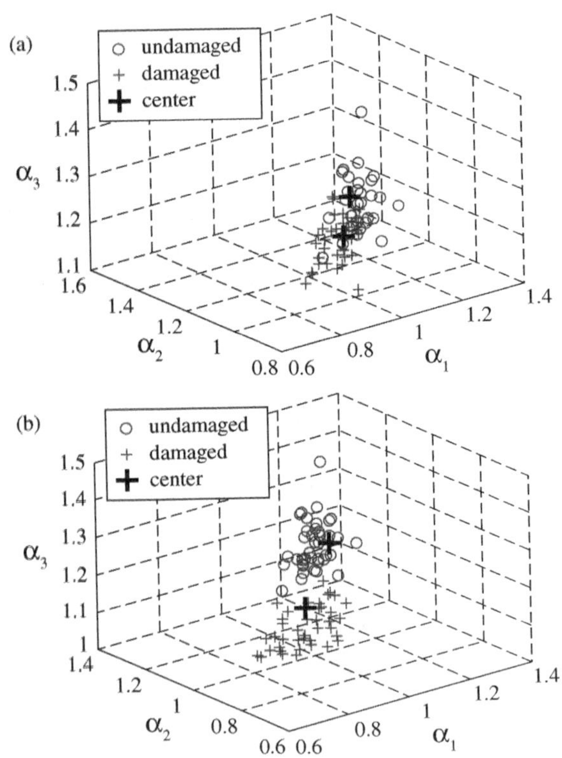

Figure 6.12 Migration of the feature vectors with damage for minor patterns: (a) Damage pattern 6 and (b) damage pattern 3

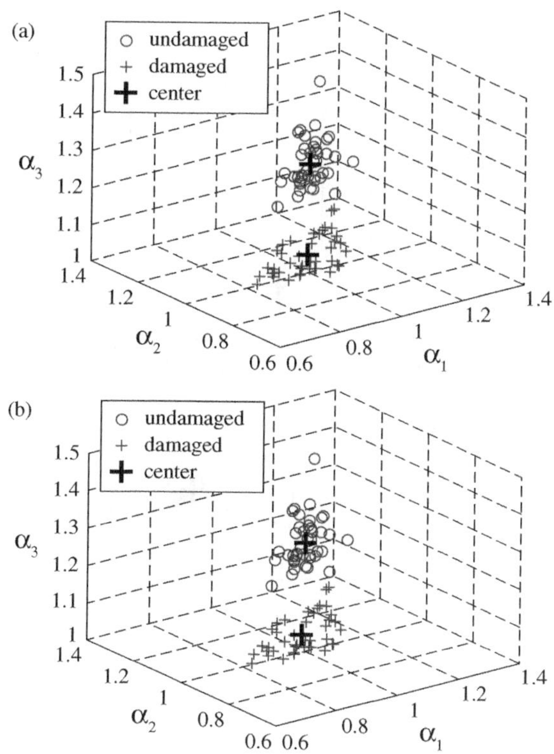

Figure 6.13 Migration of the feature vectors with damage for moderate patterns: (a) Damage pattern 4 and (b) damage pattern 5

formulations are possible and those will depend on the type of structure, the type of signals gathered and the relative value of the coefficients. Thus, to select an appropriate DSF, the user should study the signals that are collected, the relative value of the coefficients and the stability or variability of these coefficients. For some applications, more than three coefficients may be needed to capture the variation of the signal from damaged to undamaged state. Factors that may affect the stability of the coefficients are sampling rate and duration of signal. As discussed and demonstrated in the previous section, these parameters should be tested prior to sensor installation or be modified so that they produce stable AR coefficients and robust DSF.

6.3.3.3 Time- and Frequency-Domain Models – Wavelet Analysis of Vibration Signals

Another approach for modeling measurement signals is through wavelet analysis. While the AR model captured the characteristics of the measurement purely in the time domain, tracking changes in the frequency domain can also be a desirable approach. Frequency domain methods require that the signal be transformed using conventional spectral methods, e.g. Fourier. The main advantage of wavelet analysis over purely spectral methods is that the data are represented in both time and scale domains (Mallat 1999). At lower scales, the wavelet basis function has a smaller support and thus is better able to capture

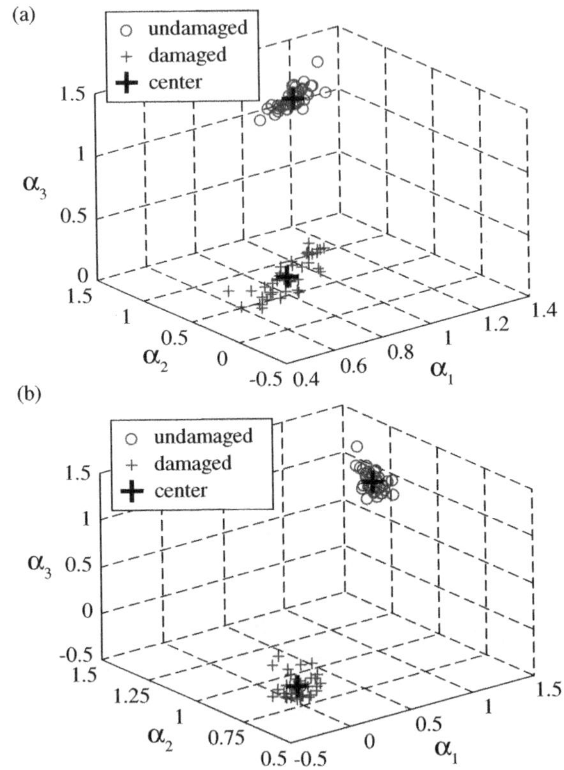

Figure 6.14 Migration of the feature vectors with damage for major patterns: (a) Damage pattern 1 and (b) damage pattern 2

transient phenomena such as discontinuities in the dataset. Similarly, at higher scales, the wavelet basis function has a wider support and is thus helpful in identifying long-range phenomena.

In the context of SHM, earlier work was carried out in wavelet-based system identification of nonlinear structures by Staszewski (2000), Ghanem and Romeo (2000) and Kijewski and Kareem (2003). From a signal processing viewpoint, initial work was done by Hou et al. (2000), where the discrete wavelet transform was used to study the transient phenomenon when the stiffness of the structure is abruptly changed. Sun and Chang (2002) used the wavelet packet transform for decomposition of the signals, where the wavelet packet component energies were used to detect damage and were then used as inputs to a neural network for damage assessment. In this section we present a brief review of the wavelet transform and then utilize the scales of the transform for SHM purposes.

A wavelet is a function $\psi(t) \in \boldsymbol{L}^2(\mathfrak{R})$, where $\boldsymbol{L}^2(\mathfrak{R})$ is the space of square integrable functions, with the following properties Mallat (1999):

$$\int_{-\infty}^{\infty} \psi(t)\, dt = 0 \text{ and } \|\psi\| = 1 \tag{14}$$

A mother wavelet function $\psi(t) \in \boldsymbol{L}^2(\Re)$ that is dilated/scaled by a and translated by b, denoted as $\psi_{a,b}(t)$, is given as

$$\psi_{a,b}(t) = \frac{1}{\sqrt{a}}\psi\left(\frac{t-b}{a}\right) \tag{15}$$

Here again, $\|\psi_{a,b}\| = 1$. The continuous wavelet transform (CWT) of a function $f(t) \in \boldsymbol{L}^2(\Re)$ (Mallat 1999) is given as:

$$Wf(a,b) = \int_{-\infty}^{\infty} f(t)\frac{1}{\sqrt{a}}\psi^*\left(\frac{t-b}{a}\right)dt \tag{16}$$

where * represents the complex conjugate.

There are several mother wavelets that have been defined (Mallat 1999). Of these, the Morlet wavelet is found to be particularly suitable to be used for feature extraction (Nair et al. 2006). The Morlet wavelet is defined as follows:

$$\psi(t) = \mathrm{e}^{j\omega_0 t - t^2/2} \tag{17}$$

The Fourier transform of the Morlet wavelet can be shown to be equal to:

$$\Psi(s) = \sqrt{2\pi}\mathrm{e}^{-(s-\omega_0)^2/2} \tag{18}$$

In general, the value of ω_0 is chosen to be five, which satisfies the admissibility condition Mallat (1999). Thus, from Equations (16) and (18), we obtain

$$Wf(a,b) = \frac{1}{\sqrt{2\pi}}\int_{-\infty}^{\infty} F(s)\sqrt{a}\mathrm{e}^{jsb-(5-as)^2/2}\,ds \tag{19}$$

The Morlet wavelet and its Fourier transform are shown in Figure 6.15.

In order to select appropriate scales for feature vectors, we need to investigate the relationship between the various scales of a wavelet to parameters of the structural system. For that purpose a single degree

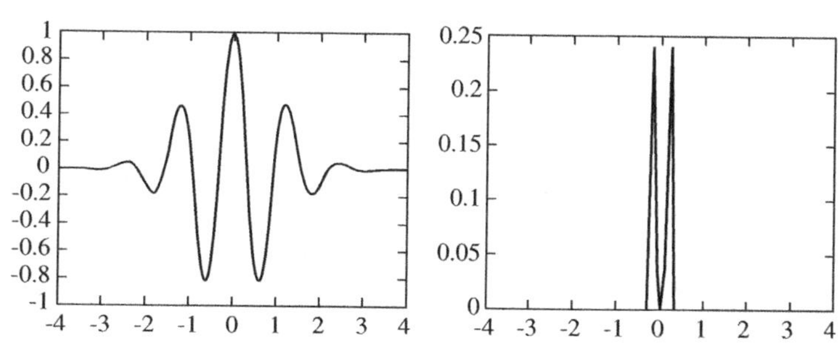

Figure 6.15 (a) Morlet wavelet and (b) its Fourier transform

of freedom (SDOF) system, with mass m, damping coefficient c and undamaged stiffness coefficient k, subject to a forcing function $g(t)$ is considered. The equation of motion for this system is given as:

$$m\ddot{x} + c\dot{x} + kx = g(t) \tag{20}$$

The Fourier transform of Equation (20) is obtained and the derivative rule

$$F\left(\frac{d^n f(t)}{dt^n}\right) = (js)^n F(s) \tag{21}$$

is applied to obtain:

$$(-s^2 m + jcs + k)X(s) = G(s) \tag{22}$$

where $X(s)$ and $G(s)$ are the Fourier transforms of the displacement and forcing functions, respectively. The system is assumed to be linear and the forcing function is stationary. The first assumption is valid, since when we compare the damage with undamaged system we are looking at an equivalent linear system with reduced stiffness. The second assumption is not always valid in practice; however we can segment the signal so that each signal is quasi-stationary.

With these assumptions, we can estimate the Fourier transform of the acceleration as:

$$\ddot{X}(s) = FT(\ddot{x}(t)) = -s^2 X(s) = \frac{-s^2 G(s)}{-s^2 m + jcs + k} \tag{23}$$

Thus, using Equations (16) and (23), we can show that

$$W\ddot{x}(a, b) = \frac{1}{2\pi}\int_{-\infty}^{\infty} \frac{-s^2 G(s)}{-s^2 m + jcs + k}\sqrt{a}e^{jsb}\Psi^*(as)\, ds \tag{24}$$

The energy of the Morlet-based wavelet coefficients of the acceleration signal at scale a, E_a^{Morlet}, can be derived as

$$E_a^{Morlet} \approx \Lambda_{Morl}(a)\frac{e^{-2\zeta\omega_n \Delta t}(1 - e^{-2K\zeta\omega_n \Delta t})}{1 - e^{-2\zeta\omega_n \Delta t}} \tag{25}$$

where, $\omega_0 = 5$ and

$$\Lambda_{Morl}(a) = e^{-\omega_0^2 + 2a\omega_0\omega_d}\frac{a}{2(1-\zeta^2)}\left[|G(p)|^2 e^{a^2\omega_n^2(2\zeta^2-1)} + |G(q)|^2 e^{a^2\omega_n^2(1-2\zeta^2)}\right] \tag{26}$$

Similar relationships have also been developed for the energies of the wavelet transform of the normal modes of a multidegree of freedom system (see Nair and Kiremidjian (2007) for more detailed treatment).

Equations (25) and (26) are used to investigate the energies of the wavelet transform to develop a damage sensitive feature. The energy E_a of the wavelet coefficients of a signal at scale a, is defined as follows:

$$E_a = \sum_{b=1}^{K} |W\ddot{x}(a, b)|^2 \tag{27}$$

where, $W\ddot{x}(a, b)$ is the wavelet coefficient of the acceleration signal at the ath scale and bth time-step, K is the number of data points in the signal and $|.|$ is the absolute of the quantity. The lower scales (e.g., the first second and third scales) are highly sensitive to the loading and environmental conditions to which the structure may be subjected. As a result these scales are not good candidates for damage detection. The higher scales have wider support and are affected by changes in the vibration characteristics. It is found that the fifth, sixth and seventh dyadic scales are particularly suitable for damage detection Nair et al. (2006). An intuitive understanding of why these scales are chosen has to do with the support of the scaled wavelet basis. We have to choose an appropriate scale to capture the desired phenomenon (both long term and short term). In cases when we choose higher scales, we increase the support of wavelet basis, thus increasing its capacity to detect long-term phenomenon. Also, at higher scales, we do not pick up transients and thus problems with noisy data could potentially be avoided. For a physical understanding of the wavelet coefficients of vibration signals, the reader is referred to Nair and Kiremidjian (2007).

Figure 6.16 illustrates the plot of the damage-sensitive feature E_5 for a damaged and undamaged case in the situation of noise-free vibration data. The distribution ellipses clouds are populated with the damage-sensitive features as defined in Equation (27). The vector v is plotted at 45° to the x-axis. Along the direction of v, the uncertainty of the operational conditions (such as loading direction and noise) is shown. Also, in the case of damage, the value of $E_{damaged}$ and $E_{undamaged}$ would be different and thus deviate away from the vector v. Thus, the variance in the direction v' and v'' would give an indication of damage. To this end, the principal directions should be used to obtain the directions of highest variances (Mardia et al. 2003). This is illustrated in Figure 6.16. Thus, the lower singular value would be a good indicator of damage and is chosen as the damage-sensitive feature.

In practice, data are collected under different operational conditions (loading and environmental conditions such as temperature and humidity). In order to compare signals under various operational conditions, we will use the energies of the wavelet coefficients at the first dyadic scales to determine whether there is any difference in the loading conditions of the signal or not. The reason why lower scales are chosen is that these will be able to take into account the transients in the loading conditions. As before, we will use E_1 to determine how close the new signal is to the signal in the database. Thus the lower the second singular value is, the closer the signal in the database is to the new signal.

Figure 6.17 shows the use of principal components with real data. In general, the angle of the first principal component will not be at 45° since the data have to be suitably normalized to account for various operating conditions. In such cases, the variance in the directions v_2 and v_4 will give an indication whether

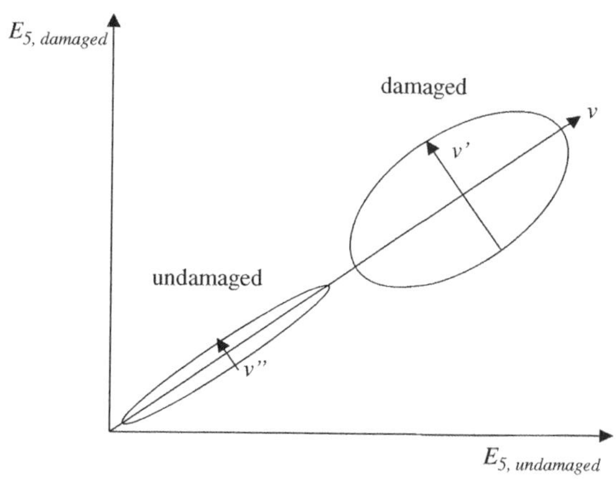

Figure 6.16 Illustration of distribution ellipses for damaged and undamaged populations for noise-free data

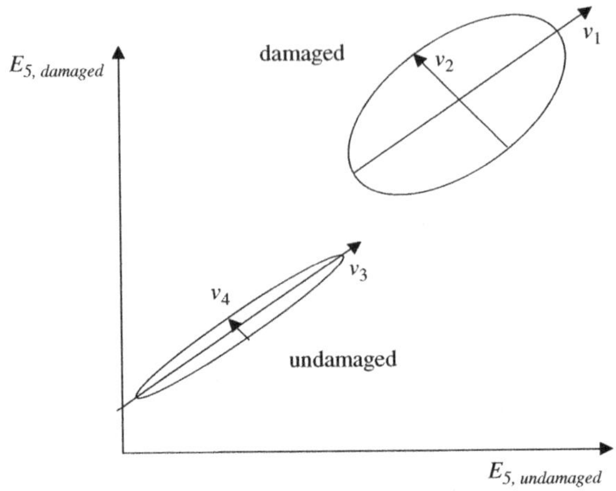

Figure 6.17 Illustration of distribution ellipses for damaged and undamaged populations using principal components analysis

there is damage in the dataset/signal. It can be observed that there is a clear difference in the clouds of the undamaged dataset and the damaged datasets. Figure 6.18 shows the migration of the wavelet energy for the seventh dyadic scale, E_7, with increasing damage. The circles in these plots correspond to E_7 obtained from the undamaged signal and the crosses to E_7 estimated from the damaged signal. The dashed lines correspond to the mean values of the energies in each set. These figures show the clear difference between the damaged and undamaged wavelet energies. Similar observations are made with the fifth and sixth scale energies.

6.3.4 Classification Schemes

Once a damage-sensitive feature is selected and the features are extracted from the signals, a statistical procedure has to be developed to discriminate between damaged and undamaged states. Statistical pattern classification methods are particularly suitable for this purpose. Frequently used classification schemes include hypothesis testing, Gaussian mixture models (GMMs), hidden Markov models (HMMs) and

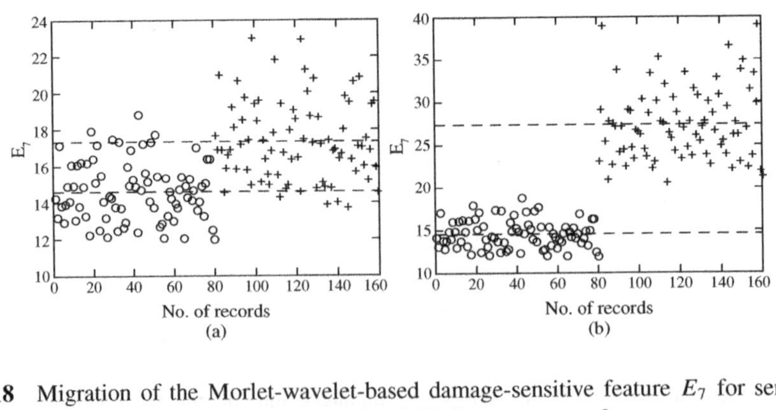

Figure 6.18 Migration of the Morlet-wavelet-based damage-sensitive feature E_7 for sensor 2 with damage for minor patterns: (a) Damage pattern 6 and (b) damage pattern 3

self-organizing maps. In this section we will formulate and use hypothesis tests and GMMs damage classification through feature discrimination and briefly discuss the other methods.

6.3.4.1 Hypothesis Testing

Figure 6.11 shows the results from the application of the proposed damage algorithm to the numerically simulated datasets of the ASCE Benchmark Structure. From Figure 6.11a and b, it can be observed that there is a significant difference between the mean values of the DSFs obtained from the damaged and undamaged cases. If $\mu_{DSF,damaged}$ and $\mu_{DSF,undamaged}$ are defined as the mean values of the DSFs obtained from the damaged and undamaged case, respectively, then a hypothesis test may be set up as follows to determine if their differences are significant:

$$H_0 : \mu_{DSF,undamaged} = \mu_{DSF,damaged} \qquad H_1 : \mu_{DSF,undamaged} \neq \mu_{DSF,damaged} \tag{28}$$

where H_0 and H_1 are the null and alternate hypothesis respectively. H_0 represents the undamaged condition and H_1 represents the damaged condition. The significance level of the test is set at 0.05. The hypothesis used in Equation (28) is called a two-sided alternative. For testing the above hypothesis, the t-statistic is used (Rice 1999). The t-statistic is defined as follows:

$$t = \frac{\mu_{DSF,undamaged} - \mu_{DSF,damaged}}{s\sqrt{1/n + 1/m}} \tag{29}$$

where m and n are the number of samples obtained from $DSF_{damaged}$ and $DSF_{undamaged}$ respectively; and s is the pooled sample variance, given as

$$s^2 = \frac{(n-1)S^2_{DSF,undamaged} + (m-1)S^2_{DSF,damaged}}{m+n-2} \tag{30}$$

where $S^2_{(.)}$ is the sample variance of (.). For H_1, the region for rejection is defined as

$$|t| > t_{n+m-2}(\alpha/2) \tag{31}$$

where $t_{n+m-2}(\alpha/2)$ is the value of the t-distribution with $n+m-2$ degrees of freedom obtained at $\alpha/2$.

An example application taken is that from the numerically simulated datasets of the ASCE Benchmark Phase I Structure. Damage pattern two is chosen for analysis. Table 6.2 shows the results of the hypothesis tests.

In Table 6.2 it is seen that the p-values are extremely low, indicating a difference in the values of $DSF_{damaged}$ and $DSF_{undamaged}$, thus concluding that there is a higher probability of damage. In other words, the p-value is the probability that the DSF does not predict damage, given in fact that there is damage in the structure. Since the p-values are all significantly much less than the significance level of 0.05, the null hypothesis H_0 is rejected.

Application of the two-sided hypothesis test to all of the sensor measurements from the ASCE Benchmark Structure showed similar results. Furthermore, positive damage identification was achieved at all but the lowest damage level, which corresponds to the bolt loosening.

In practical applications, however, it may not be sufficient to use only hypothesis testing. A combination of methods may be implemented to increase the reliability of the prediction and to reduce the likelihood of a false positive or a true negative prediction.

Table 6.2 Results of damage decision for damage pattern two

Sensor No.	Damage decision	p-value
1	H_1	≈ 0.0
2	H_1	≈ 0.0
3	H_1	≈ 0.0
4	H_1	≈ 0.0
5	H_1	≈ 0.0
6	H_1	≈ 0.0
7	H_1	≈ 0.0
8	H_1	≈ 0.0
9	H_1	≈ 0.0
10	H_1	≈ 0.0
11	H_1	≈ 0.0
12	H_1	≈ 0.0
13	H_1	2.204×10^{-6}
14	H_1	≈ 0.0
15	H_1	8.265×10^{-6}
16	H_1	≈ 0.0

6.3.4.2 Gaussian Mixture Models

Figures 6.12–6.15 illustrate the migration of feature vectors (Figures 6.12–6.14 use the first three AR coefficients, whereas Figure 6.15 uses the lower singular values of the wavelet energies at higher scales) with damage. Gaussian mixture models (GMM) are frequently used as clustering algorithms in pattern classification (Hastie et al. 2001). A Gaussian mixture model with M classes (or mixtures) has the following form:

$$f(\boldsymbol{x}_{1:N}) = \sum_{i=1}^{M} \pi_i \phi_i(\boldsymbol{X}; \theta_i), \tag{32}$$

where, $\boldsymbol{X}$ is the collection of N feature vectors, $\phi_i \sim N(\mu_i, \Sigma_i)$ is a Gaussian vector with mean vector μ_i and covariance matrix Σ_i and π_i is the non-negative mixture weight for each class. The unknown parameters of the GMM $\Theta = \{\mu_i, \Sigma_i, \pi_i, i = 1, 2, \ldots, M\}$ can be estimated using the expectation maximization (EM) algorithm. Consider a random variable $\mathbf{I}_i$ $(i = 1, \ldots, M)$ whose realization is an M dimensional indicator vector, which would show 1 in the jth component of I_i $(I_{ij} = 1)$ if $\boldsymbol{x}_i$ corresponds to the jth mixture. The EM algorithm in the context of the parameter estimation of the GMM is given in Figure 6.19. In this algorithm, d is the dimension of the feature vector space and $\log L_{(t+1)}$ is the log-likelihood function at the $(t+1)$th time-step. Steps two and three constitute the Expectation (E-step) and step four is the Maximization (M-step). In the above algorithm, the number of mixtures M has been fixed. Also, note that the inputs to the algorithm are initial values of $\mathbf{I}_i$, μ_i, Σ_i, π_i $(i = 1, 2, \ldots, M)$. The outputs of the algorithm are the converged values of the above initialized parameters.

To obtain the optimal number of mixtures, the gap statistic is used (Tibshirani et al. 2001). This involves the generation of a uniformly generated dataset over the support of the actual dataset in the feature vector space. The gap statistic measures the differences of the within-cluster distances of the actual dataset and the uniformly generated dataset. Additional information on the theory and implementation of the gap statistic can be obtained from Tibshirani et al. (2001).

Step 1: Initialize the values of $E(I_{ij})$, π_i, $\boldsymbol{\mu}_i$ and $\boldsymbol{\Sigma}_i$.
Step 2: for $i = 1 : N$ do
Step 3: for $j = 1 : M$ do

$$p\left(\boldsymbol{x}_i | I_{ij} = 1; \Theta\right) = \frac{|\Sigma_j|^{-1/2}}{(2\pi)^{d/2}} \exp\left(-\frac{1}{2}\left(\boldsymbol{x}_i - \boldsymbol{\mu}_j\right)^T \Sigma_j^{-1}\left(\boldsymbol{x}_i - \boldsymbol{\mu}_j\right)\right)$$

$$E\left(I_{ij}\right) = \frac{p\left(x_i | I_{ij}=1;\theta\right)\pi_j}{\sum_{k=1}^{M} p\left(x_i | I_{ik}=1;\theta\right)\pi_k}$$

end of Step 3
end of Step 2
Step 4: for $j = 1 : M$ do

$$\boldsymbol{\mu}_j = \frac{\sum_{k=1}^{N} E\left(I_{kj}\right)\boldsymbol{x}_k}{\sum_{k=1}^{N} E\left(I_{kj}\right)}$$

$$\boldsymbol{\Sigma}_j = \frac{\sum_{k=1}^{N} E\left(I_{kj}\right)\left(\boldsymbol{x}_k - \boldsymbol{\mu}_j\right)\left(\boldsymbol{x}_k - \boldsymbol{\mu}_j\right)^T}{\sum_{k=1}^{N} E\left(I_{kj}\right)}$$

$$\pi_j = \frac{\sum_{k=1}^{N} E\left(I_{kj}\right)}{N}$$

end of Step 4
Step 5: Repeat until convergence $\frac{|\log L_{(t+1)} - \log L_{(t)}|}{|\log L_{(t)}|} < \varepsilon$.

Figure 6.19 Expectation maximization algorithm for GMMs

Figure 6.20 shows that the gap statistic with a maximum separation is found at two, leading to the conclusion that there are two clusters. Two or more clusters indicate that the data are coming from signals that have distinct characteristics, which would then lead to a conclusion that there has been a change in the structure and it is most likely due to the onset of damage. Application of the gap statistics to the lowest damage pattern (damage pattern 3) of the ASCE Benchmark Structure shows that although there is an intersection of the distributions the data come from two mixtures. It should be noted that the probability of detecting two clusters is inversely proportional to the overlap of the two distributions (Tibshirani et al. 2001). For all other damage states, the gap statistics clearly identify that there are two clusters as well.

6.3.4.3 Other Classification Schemes

Hidden Markov models (HMM) are stochastic models used for making sequential decisions (Duda et al. 2001), and have been used extensively in speech and gesture recognition (Rabiner 1989). Any finite state HMM is characterized by the number of states, the number of distinct observation symbols per state, and the state transition probability matrix.

To estimate the parameters of a HMM for modeling normal system behavior, feature vectors obtained from the baseline/undamaged configuration are used as training data. Once the HMM has been trained, thresholds for probability measures for detecting damage have to be proposed. Yeung and Ding (2002)

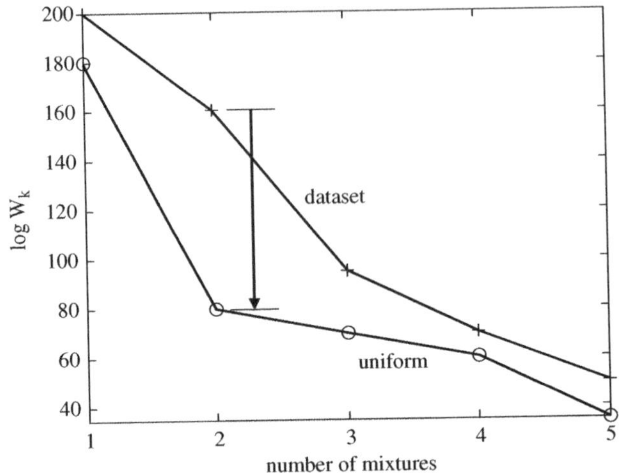

Figure 6.20 Illustration of the gap statistic for the undamaged and damage pattern 3 signals

describe the use of HMMs for novelty detection in computer security (intrusion detection) and propose the following approaches.

- Given a trained HMM, the sample likelihood of an observed sequence can be computed with the forward or backward algorithm (Rabiner 1989). A threshold on the probability can discriminate between the undamaged (baseline, corresponding to the trained HMM) and damaged state.
- The probability distribution of the undamaged state behavior observed over a period of time is modeled. The behavior of the system being monitored is modeled in the same way. An information-theoretic measure known as cross-entropy can be used to measure how dissimilar the two distributions are (Cover and Thomas 1991). A threshold can determine whether the observed behavior is deviating from the undamaged state. A validation set is used to determine the threshold. In the case of HMM, the threshold will be chosen to be the minimum likelihood among all training sequences (Yeung and Ding 2002). Also, the cross-entropy value is computed between the entire training set and each vector in the training set. The threshold is chosen to be the maximum cross-entropy. In the study conducted in Yeung and Ding (2002) the information-theoretic technique outperformed the HMM approach.

Self-organizing maps (SOM), proposed by Kohonen (2001), form an unsupervized learning scheme. In most SOM-based methods, similar to statistical clustering methods (such as the k-means algorithm), some form of cluster membership value is thresholded to determine whether a sample belongs to a cluster or not. Ypma and Duin (1998) employ a SOM to develop a novelty detection technique used for the detection of faults in a fault monitoring application. First, the SOM is trained with samples from a healthy structure. Then samples from a structure under question are used to see whether there is damage or not. A large distance indicates damage. Thresholds again have to be determined. Applications of these models to damage detection in SHM have shown some promise and are currently under consideration by various researchers.

6.3.5 Damage Quantification and Localization

Damage quantification and localization using statistical pattern recognition methods are still in their infancy. While some attempts have been made to develop methods for damage quantification with some success, damage localization has proved to be significantly more challenging.

Most current approaches correlate the extent of damage to some measure of the change in a signal gathered at the same location on the structure. The difficulty stems from the great diversity of damage characteristics and their dependence on the specific structure. Thus, the initial models focus on damage extent quantification in relative terms of the degree of damage. For example, minor damage imposed on the ASCE Benchmark Structure, such as a partial cut in one of the braces or removal of a single brace, should in principle result in a small change in the damage sensitive feature. Similarly, if more critical damage has taken place, such as the failure of all the braces on a floor, then the change in the damage sensitive feature is expected to be significantly greater than that of a single brace breaking. The key is to find analytical methods that capture and quantify these changes. In the next section a method for damage quantification is presented that uses the Mahanalobis distance.

Hypothetically, if damage occurs near a sensor, that sensor should reflect that event by exhibiting larger changes in the damage-sensitive feature than sensors that are further away from the region where damage has taken place. With this hypothesis, one would assume that localization can be realized with denser sensor placement. The reality, however, is that this phenomenon will be observed only if the sensor is able to measure structural parameters that reflect a localized behavior and not more global structural response to the loading. For example, if the sensors are strain gauges, then the measurements are strictly of the strains at the locations of the sensors and these data should reflect damage in the vicinity of the gauge. If the sensors are accelerometers, these sensors capture the more global response of the structure and localization will prove to be particularly challenging. In some cases, the prediction of damage location based on acceleration measurements can be misleading because softening of one part of the structure may cause increased vibrations in another part of the structure that is far away from the damaged region. Other sensing methods, such as acoustic emission and impact echo, have as one of their primary goals to localize the damage. These techniques, however, are beyond the scope of this section and will not be considered. Ultimately, a combination of sensors and methods will need to be used to provide complete and robust damage diagnosis that includes also localization of damage.

6.3.5.1 Damage Extent using Mahalanobis Distance

The Mahalanobis distance is a metric frequently used in multivariate analysis to determine the separation of two distributions (Mardia et al. 2003). The Mahalanobis distance between two vectors **a** and **b** with a covariance matrix $\boldsymbol{\Sigma}$ is defined as follows:

$$\Delta(\mathbf{a}, \mathbf{b}, \boldsymbol{\Sigma}) = \sqrt{(\mathbf{a} - \mathbf{b})^T \boldsymbol{\Sigma} (\mathbf{a} - \mathbf{b})} \tag{33}$$

In the present study, the Mahalanobis distance is used to define a measure of the damage extent. More specifically, the damage metric (DM) used is defined as $\Delta(\boldsymbol{\mu}_{undamaged}, \boldsymbol{\mu}_{damaged}, \boldsymbol{\Sigma}_{undamaged})$ where $\boldsymbol{\Sigma}_{undamaged}$ is the covariance matrix of the undamaged dataset, and $\boldsymbol{\mu}_{undamaged}$ and $\boldsymbol{\mu}_{damaged}$ are the means of the undamaged and damaged dataset respectively. These values are obtained after modeling the feature vectors as a GMM. Mathematically, the DM can be defined as:

$$\begin{aligned} DM &= \Delta(\boldsymbol{\mu}_{undamaged}, \boldsymbol{\mu}_{damaged}, \boldsymbol{\Sigma}_{undamaged}) \\ &= \sqrt{(\boldsymbol{\mu}_{undamaged} - \boldsymbol{\mu}_{damaged})^T \boldsymbol{\Sigma} (\boldsymbol{\mu}_{undamaged} - \boldsymbol{\mu}_{damaged})} \end{aligned} \tag{34}$$

The above formulation is valid for a two-mixture dataset. However if the number of mixtures is greater than two, the value of DM is chosen as the maximum of the values of this measure computed between the undamaged mixture and the other mixtures.

Close examination of the Mahalanobis distance would reveal that it corresponds to the Euclidean distance between the centroids of two mixtures that is weighted with respect to the covariance matrix, in this case the covariance matrix of the undamaged signal. The Euclidean distance can also be used as a damage-extent measure, but one can argue that it does not include the information on the dispersion of the coefficients. As damage increases the dispersion of the distributions is expected to increase as well, reflecting the more nonlinear behavior of the structure. Thus it is important to include the dispersion of the data in the damage-extent measure. Several other forms of the damage extent defined by Equation 34 can be defined that consider the covariance between the damaged and undamaged mixtures and the reader is referred to Noh and Kiremidjian (2008). The damage extent using Equation 34 can be used either with the AR model with the Gaussian mixture model or the wavelet model. In the following examples we show the results of the application of the Mahanalobis distance computation to the AR damage models in conjunction with the ASCE Benchmark Structure.

The results of the *DM* calculated from the AR model for sensor 2 of the ASCE Benchmark Structure are shown in the semi-log plot in Figure 6.21. From this figure it can be observed that *DM* increases for damage patterns 6, 3, 4, 5, 1 and 2, which corresponds to a progressive increase in damage. Also, the *DM* varies from 2.95 (corresponding to damage pattern 6) to 4591.56 (corresponding to damage pattern 2).

The variation of *DM* for various sensors and different damage patterns is given in Table 6.3. In this table, NA is used in cases where only one mixture is identified, indicating no damage. From the analysis of vibration signals obtained from the ASCE Benchmark Structure, the following observations are made.

- Damage patterns 1 and 2 are detected consistently at all sensor locations. These damage patterns are characterized by large values of *DM* indicating that the baseline and new mixtures are significantly separated.
- Since damage patterns 4 and 5 have braces removed on Faces 1 and 2 (Figure 6.5), it is seen that damage is consistently detected at all sensor locations. Although bolt loosening takes place near sensor 3, damage patterns 4 and 5 show no difference, thus indicating that bolt loosening has not been detected.
- Damage patterns 3 and 6 have the reduction of a brace stiffness on Face 2 (see Figure 6.6). Thus, on the first two floors of the structure, it is observed that damage has not been detected at sensors that are located on Faces 1 and 3. On the higher floors, damage has been detected at all sensors. Examination of the structure and its vibration characteristics would point to potential torsional motion that could

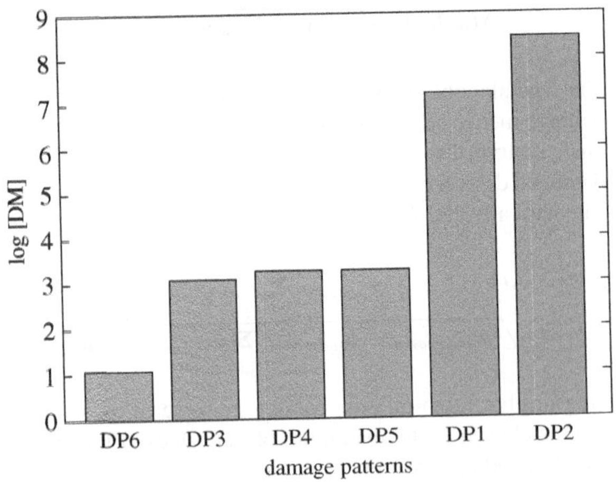

Figure 6.21 Variation of the damage metric with damage pattern for sensor 2

Table 6.3 Variation of *DM* for various sensors and different damage patterns of the ASCE Benchmark Structure

Sensor	DP1	DP2	DP3	DP4	DP5	DP6
1	431.38	1415.53	NA	50.90	50.90	NA
2	1374.21	4591.56	22.54	25.85	25.85	2.95
3	454.63	1673.77	NA	34.46	34.46	NA
4	1096.92	5993.50	32.33	48.73	48.73	2.78
5	295.61	826.69	NA	25.67	25.67	NA
6	1350.03	4306.10	21.79	24.04	24.04	2.64
7	265.79	703.07	NA	20.73	20.73	NA
8	858.93	3224.43	9.52	10.75	10.75	1.59
9	263.74	1068.42	14.25	46.43	46.43	NA
10	822.55	727.49	4.08	6.74	6.743	NA
11	340.21	1360.80	11.97	48.51	48.51	NA
12	506.16	344.10	4.79	4.73	4.73	NA
13	1444.02	4485.12	15.33	12.21	12.21	NA
14	463.91	1282.90	2.82	3.15	3.15	2.389
15	694.23	2632.63	3.69	49.29	49.29	2.09
16	756.74	2691.54	10.31	9.50	9.50	2.32

result from the removal of braces. Such torsional motion would particularly be severe at the higher stories, resulting in larger differences in vibration responses at these sensor locations.

Euclidean distance was used successfully by Noh and Kiremidjian (2008) to quantify damage from strain measurements obtained from a four-story steel-frame structure test. They also explored various other forms of the Mahalanobis distance with vibration measurements. These included consideration of the cross covariance between the damaged and undamaged signals, separate normalization of the damaged and undamaged signals by their respective covariance matrices, and use of the covariance matrix of the difference between the damaged and undamaged signals. Equation 34 appears to be the best candidate, with moderate success with the other definitions of mixture separation.

The above discussion and all the illustrations used thc AR coefficients to quantify damage. The energies of the wavelets at higher scales can also be used to quantify damage. In the previous sections, the fifth, sixth and seventh scale energies were used as the damage sensitive feature. The Mahalanobis distance using these scales was used by Nair and Kiremidjian (2007) to show that they predict the degree of damage correctly for the ASCE Benchmark Structure. Table 6.4 shows the *DM* based on Equation (30) using the Morlet wavelet energies. The table lists the results for all the sensor locations. It is recalled that progression of damage is DP6, DP3, DP4, DP5, DP1 and DP2 going from the smallest to the largest value. For all sensor locations, this progression can be directly correlated to the value of *DM*. The absolute values of *DM* appear to be quite different for the various locations of the sensors. However, these values could not be correlated to the damage location and have to do more with the global vibration characteristics of the structure.

6.3.5.2 Effect of Noise

A question that arises continuously in structural health monitoring and is of particular concern in applications of statistical pattern recognition methods is the effect of noise. In particular, one often wonders if the signal is so masked by noise that the comparison may be between one noisy signal and another

Table 6.4 Variation of *DM* for the Morlet wavelet based damage sensitive feature for various sensors and different damage patterns

	Damage metric					
Sensor	DP1	DP2	DP3	DP4	DP5	DP6
1	9.68	21.52	0.85	0.95	0.95	0.33
2	99.63	146.00	7.71	7.49	7.49	1.63
3	9.12	18.43	0.98	2.57	2.57	0.35
4	97.10	123.32	5.89	5.58	5.58	1.33
5	5.85	10.96	1.27	1.49	1.49	0.39
6	62.01	72.05	6.70	6.60	6.60	1.46
7	7.13	12.65	1.87	2.13	2.13	0.39
8	63.94	80.00	4.66	4.55	4.55	1.08
9	6.18	11.35	0.95	6.65	6.65	0.25
10	50.89	43.29	5.46	5.49	5.49	1.52
11	5.54	9.51	1.22	5.49	5.49	0.25
12	53.05	43.02	5.02	5.06	5.06	1.11
13	3.12	8.12	0.97	1.60	1.60	0.27
14	48.64	43.83	5.74	5.52	5.52	1.29
15	4.53	10.89	1.06	1.48	1.48	0.24
16	40.76	41.20	3.87	3.90	3.90	0.88

noisy signal. As discussed in the introduction of this chapter, care is needed to filter the signal from large amounts of noise. If some noise were still present, however, we may want to know how it affects the identification and quantification of damage. Thus, the effect of zero mean additive Gaussian white noise on the damage detection algorithm is briefly discussed next. The ratio of the root mean square (RMS) value of the noise to the RMS value of the signal is defined as the noise to signal ratio, and is denoted as NSR. We vary the NSR from 0.05 to 0.15 and observe the results of the *DM* predictions. We are showing only the results from the wavelet-based analyses, however, similar results were also obtained with the investigations of the AR model. The values of *DM* with these noise levels for sensor 2 are presented in Table 6.5. From this table the values of *DM* correlate well to the amount of damage. For damage patterns 3 (DP3) and 4 (DP4), there is very slight difference between these values since at sensor 2, DP4 is DP3 + removal of a brace in a direction perpendicular to the direction of acceleration measured at sensor 2. Similar conclusions can be made for DP4 and DP5. The values of *DM* are sensitive to noise and only major damage patterns 1 and 2 can be detected at high ranges of noise levels. As expected, the separation of the clouds decreases with increased noise and at the 10% noise level, the lower damage levels cannot

Table 6.5 Variation of *DM* for sensor 2 with different noise to signal ratios (NSR) for damage patterns DP1-6

	Damage metric					
NSR	DP1	DP2	DP3	DP4	DP5	DP6
0.0	103.26	151.25	7.87	7.66	7.66	1.68
0.05	57.00	78.17	2.11	1.84	1.84	0.36
0.1	20.11	35.16	0.88	0.53	0.53	0.00
0.15	10.49	17.13	0.22	0.46	0.46	0.00

be distinguised from each other. With 15% damage the value of *DM* decreases by a factor of 10 leading to the conclusion that damage detection at that level is not likely to be reliable. Thus, care needs to be taken to filter the noise from the signals prior to application of these methods.

6.4 Application Example: Steel Bridge

A good example where lifetime prediction of a steel bridge has been performed is the Europabrücke at the Brenner Highway in Austria. Supported by various research projects monitoring campaigns have been carried out since 1997 ending in permanent monitoring with the target of progressive lifetime assessment. The campaign and details are described in detail in Section 4.1.

6.5 Ongoing Research and Development Projects

The need for further development in SHM has been recognized worldwide. The initiatives are dependent on the prevailing conditions in each region and its stage in economic development. The largest and most relevant programs are established in the USA (LTBP program of FHWA), in Japan and in Korea (Seoul National University Kaist), with high profile programs being implemented recently. Europe lacks a drive towards regional standardization but instead deals with this subject at a national level, thus bridges can be assessed completely differently from one country to another, which can create unusual situations in cross-border bridges.

6.5.1 The FHWA Long-Term Bridge Performance Program

In 2005 the SAFETEE-LU initiative passed US Congress and the FHWA Office of Infrastructure Research and Development is initiating the Long-Term Bridge Performance (LTBP) program. The LTBP program is an ambitious 20 year (2005–2025) research effort that is strategic in nature and has both specific short-term and long-range goals. The funding target is US$100 million for this program. It will include detailed inspection, periodic evaluation and testing, continuous monitoring and forensic investigation of representative samples of bridges throughout the USA in order to capture and document their performance. This program will result in a high-quality quantitative database, which will have an impact on the value, success and efficiency of bridge management systems in the future. A total of 2000 bridges should be monitored and partly treated by forensic engineering after they are taken out of service.

The FHWA's national bridge inspection standards (NBIS) have facilitated the creation of one of the most comprehensives sources of bridge information in the world, the national bridge inventory (NBI). The NBI contains information on the condition of more than 595 000 bridges, tunnels and culverts located on public roads (Figure 6.22). In 2005, according to the NBI, there where approximately 156 000 structurally deficient or functionally obsolete bridges. This number is linkely to increase in coming years due, in large part, to an increasing traffic demand, continued bridge aging and deterioration, and limited funds for rehabilitation and maintenance.

A number of States in the USA have adapted and implemented the PONTIS bridge management program of AASHTO, or a similar system with advanced management decision-making capabilities. Some States are augmenting the NBI data used in these systems by also collecting element level bridge data. Even with these bridge management tools and data, however, there are still many unknowns regarding how structures and materials perform or degrade over time, and current effective maintenance, repair and rehabilitation strategies apply to a given component or a complete bridge system. In addition, with the recent move to higher performance materials and advanced structural systems, high-level long-term performance and durability are assumed, but currently are not demonstrated.

In order for the nation's bridge network to meet the increasing demands made upon it without similar increases in funding, future bridge management systems will require improved life-cycle cost and

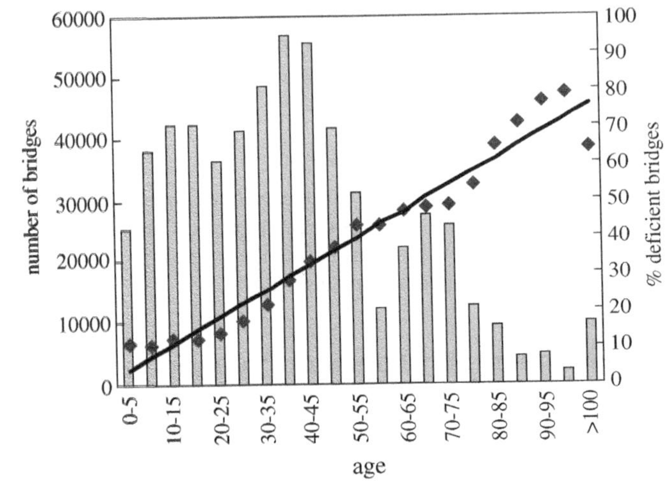

Figure 6.22 Ages versus deficiency

performance models, and information on the effectiveness of maintenance and repair strategies. However, such improvements will require high-quality quantitative data on which to base the development of new models and decision-making algorithms.

The objective of the LTBP program is to collect, document and make available high-quality quantitative performance data on a representative sample of bridges nationwide. Data will be collected through detailed inspections and evaluations, supplemented by a limited number of continuously monitored structures and forensic autopsies on decommissioned bridges. In the later years of the program, the data collected will be analyzed in order to improve knowledge regarding bridge performance and degradation, develop better design methods and performance prediction models, and advanced management decision-making tools.

Specifically, it is anticipated that the LTBP program will provide a better understanding of bridge deterioration due to numerous effects, including corrosion, fatigue, weather exposure, and loads. The program will also provide information regarding the effectiveness of current maintenance and improvement strategies, and should lead to an improvement in the operational performance of bridges, with the potential to reduce congestion, delay and accidents.

The program's schedule foresees a draft framework to be finalized in 2010. The specific issues to be addressed by the framework include:

- Program management and administration;
- Specific data to be collected;
- Types and number of bridges to be inspected and monitored;
- Data quality and collection strategies;
- Data management and archiving;
- Data mining and analysis;
- Data and information dissemination;
- Opportunities for participation and collaboration.

As noted the LTBP program will have three components of bridge monitoring and evaluation:

- Periodic inspection of bridges: detailed inspection of these bridges will include visual inspection as well as the use of advanced NDT to detect and document deterioration. These inspections will be conducted periodically with additional inspection when appropriate.

- Instrumentation of bridges: continuous monitoring of bridges will be conducted using sensing technology to measure and record their performance characteristics under routine traffic conditions as well as during and after rare or extreme events.
- Decommissioning of bridges: forensic autopsies of these bridges will be conducted to learn more about their capacity, reliability and failure modes.

The LTBP program is highly ambitious and will require considerable synergy and cooperation among and between FHWA, bridge owners, the bridge industry and academia. If successful, the program will drive efforts that result in bridges that last longer, require less maintenance, and can be modified to accommodate changes in traffic or function much more quickly and far less intrusively than current technology allows. The program is frequently updated and its status can be obtained from www.fhwa.dot.gov.

6.5.2 *The European Structural Assessment Monitoring and Control Initiative*

The Structural Assessment Monitoring and Control (SAMCO) network was formed under the European Union Fifth Framework Program for Research and Development. From 2001 till 2006 a prominent community of stakeholders involved in monitoring assessment and control of structures was assembled. The initiative produced valuable reports, which are now standard references in the sector. Details are still available for download under www.samco.org, which also provides information on the SAMCO Association established after the end of the initiative. The guideline included in this book also has been developed by this group.

6.5.3 *Other Programs – EU, Japan, Korea, China*

As the transportation infrastructure is the backbone of any economy, SHM initiatives on bridges have reached wide attention in many countries, as demonstrated by the following examples:

- Several major research projects are devoted to this subject in the various framework programs of the European Union, for example the program 'Sustainable Bridges,' which can be analyzed under www.sustainablebridges.org and is mainly dedicated to railway bridges.
- The Republic of Korea has launched a 10-year bridge program hosted at the Seoul National University. This program systematically deals with SHM issues on bridges and has major field activities including a test track with three bridges, where valuable data and methodologies are produced.
- In Japan the focus has shifted from new construction to maintenance and management. This includes huge programs, tests and initiatives to improve the management of bridges.
- In China recent collapses have triggered a national bridge program dedicated to improve the safety and performance level of the bridges nationwide.
- The best known monitoring initiative is that of Hong Kong, having been established for more than 10 years with a number of major monitoring systems in place. These systems of exceptional size produce valuable data that can be used in the improvement of codes and guidelines for the design of major structures.

Further Reading

Brockwell P and Davis R (2002) *Introduction to Time Series and Forecasting*, 2nd edn. Springer-Verlag, New York.

Cheung A, Cabrera C, Nair K, Kiremidjian A and Wenzel H (2008) Application of statistical pattern recognition methods for damage detection to field data. *International Journal of Materials and Structures*.

Cover T and Thomas J (1991) *Elements of Information Theory*. Wiley Interscience.

Doebling SW, Farrar CR, Prime MB and Shevitz DW (1996) *Damage Identification and Health Monitoring of Structural and Mechanical Systems from Changes in Their Vibration Characteristics: A Literature Review*. Technical Report LA-13070-MS, Los Alamos National Laboratory, Los Alamos, New Mexico.

Duda RO, Hart PE and Stork (2001) *Pattern Classification*, 2nd edn. John Wiley & Sons, New York.

Engelberg S (2008) *Digital Signal Processing: an Experimental Approach (Signals and Commucation Technology)*. Springer, London.

Farrar C and Duffey T (1999) Vibration based damage detection in rotating machinery and comparison to civil engineering applications. *Proceedings of Damage Assessment of Structures*, Dublin.

Ghanem R and Romeo F (2000) A wavelet based approach for identification of linear time varying dynamical systems. *Journal of Sound and Vibration* **234**(5), 555–576.

Hastie T, Tibshirani R and Freidman J (2001) *Elements of Statistical Learning: Data Mining, Inference and Prediction*. Springer Verlag, New York.

Hou Z, Noori M and Amand RS (2000) Wavelet-based approach for structural damage detection. *Journal of Engineering Mechanics* **126**(7), 677–683.

Johnson E, Lam H, Katafygiotis L and Beck J (2004) Phase IIASC–ASCE structural health monitoring benchmark problem using simulated data. *ASCE Journal of Engineering Mechanics* **130**(1), 3–15.

Kijewski T and Kareem A (2003) Wavelet transforms for system identification in civil engineering. *Journal of Computer-Aided Civil and Infrastructure Engineering* **18**, 339–355.

Kohonen T (2001) *Self Organising Maps*. Springer-Verlag.

Lee Z, Wu T and Loh C 2003 System Identification on the Seismic Behavior of Isolated Bridge. *International Journal of Earthquake Engineering and Structural Dynamics* **32**(12), 1797–1812.

Lynch J, Sundararajan A, Law K, Kiremidjian A and Carryer E (2004) Embedding damage detection algorithms in a wireless sensing unit for attainment of operational power efficiency. *Smart Materials and Structures* **13**(4), 800–810.

Mallat S (1999) *A Wavelet Tour of Signl Processing*, 2nd edn. Academic Press.

Mardia K, Kent J and Bibby J (2003) *Multivariate Analysis*. Academic Press, London.

Nair KK and Kiremidjian AS (2007) Time series based structural damage detection algorithm using Gaussian mixtures modeling. *ASME Journal of Dynamic Systems, Measurement and Control* **291**, 285–293.

Nair KK, Kiremidjian AS and Law KH (2006) Time series-based damage detection and localization algorithm with application to the ASCE benchmark structure. *Journal of Sound and Vibration* **291**, 349–368.

Noh H and Kiremidjian A (2008) Application of a time series based damage detection algorithm to the benchmark experiment at the National Center for Research on Earthquake Engineering (NCREE) in Taipei, Taiwan. *International Journal of Smart Structures and Systems* p. under review.

Peil U, Mehdianpour M, Frenz M and Weilert K (2006) Life time assessment of bridges In *Bridge Maintenance, Safety, Management, Life-Cycle Performance and Cost* (ed. Cruz PJ, Frangopol DM and Neves LC).

Rabiner LR (1989) A tutorial on hidden Markov models and selected applications inspeech recognition. *Proceedings of the IEEE* **77**(2), 257–286.

Rice J (1999) *Mathematical Statistics and Data Analysis*. Duxbury Press.

Sohn H and Farrar CR (2001) Damage diagnosis using time series analysis of vibration signals. *Smart Materials and Structures* **10**(3), 446–451.

Sohn H, Farrar C, Hunter H and Worden K (2001) *Applying the LANL Statistical Pattern Recognition Paradigm for Structural Health Monitoring to Data from a Surface-Effect Fast Patrol Boat*. Technical Report LA-13761-MS, Los Alamos National Laboratory, Los Alamos, New Mexico.

Staszewski W (2000) Identification of non-linear systems using multi-scale ridges and skeletons of the wavelet transform. *Journal of Sound and Vibration* **214**(4), 639–658.

Straser EG and Kiremidjian AS (1998) *A Modular, Wireless Damage Monitoring System for Structures*. Department of Civil and Environmental Engineering Report No. 129, The John A. Blume Earthquake Engineering Center, Stanford.

Sun Z and Chang CC (2002) Structural damage assessment based on wavelet packet transform. *Journal of Structural Engineering* **128**(10), 1354–1361.

Tibshirani R, Walther G and Hastie T (2001) Estimating the number of clusters in a dataset via the gap statistic. *Journal of the Royal Statistical Society: Series B* **63**(2), 411–423.

Wang Y, Lynch J and Law K (2006) In *Intelligent Computing in Engineering and Architecture*. Springer-Verlag. Wireless sensing, actuation and control – with application to civil structures, 116–121.

Yeung D and Ding Y (2002) Host-based intrusion detection using dynamic and static behavioral models. *Pattern Recognition* **36**, 229–243.

Ypma A and Duin R (1998) Novelty detection using self-organizing maps. *Progress in Connectionist Based Information Systems* **2**, 1322–1325.

7

Bridge SHM Methodologies

As explained in previous chapters each bridge is a prototype, and the decision on the SHM methodology to be applied therefore has to be done individually. Standardization in this field is more difficult than in other civil engineering areas, and therefore some of the axioms used in other fields might not be applicable in bridge engineering – as explained in this chapter.

7.1 Ambient Vibration Monitoring

7.1.1 Lessons Learned

The following is a collection of lessons learned from monitoring of over 400 bridge structures since 1997. The structures monitored are mainly situated in central Europe and represent the typical design for this region. Nevertheless most of these lessons can be extended to structures worldwide.

7.1.1.1 Conservative Design

The monitoring results of over 400 bridges clearly show different behavior characteristics of structures designed following different philosophies. Bridges designed conservatively are not affected by dynamic phenomena that generate concern or trigger damage. Bridges with a design focus on economy very often do not have any reserves to cover the extraordinary loadings that appear in reality. The difference in dynamics becomes obvious in the following (Figures 7.1 and 7.2). Conservative bridges show a high system damping with distinctive characteristics when monitored. Economic designs, on the other hand, very often come close to displaying resonance, which might be a very local and limited effect, but over time it will lead to structural damage.

From bridge management we know that an additional investment in 10% higher quality construction specification makes a difference of over 200% in costs over the 100-year life-cycle of a bridge. A drastic example is the bridge composed of single-span I-girders, designed to the limit, that has consumed 220% of the investment costs in retrofit over a period of 25 years. An example of the opposite is the resistance of a conservatively designed box-girder bridge that has survived the displacement of one of its piers by 110 cm after retrofitting at a reasonable cost.

7.1.1.2 External versus Internal Pre-Stressing

Damage found on grouted internal cables has triggered a dramatic change of design philosophy. Some countries, such as Germany, have specified that new bridges should be built using external cables only, for

Health Monitoring of Bridges Helmut Wenzel

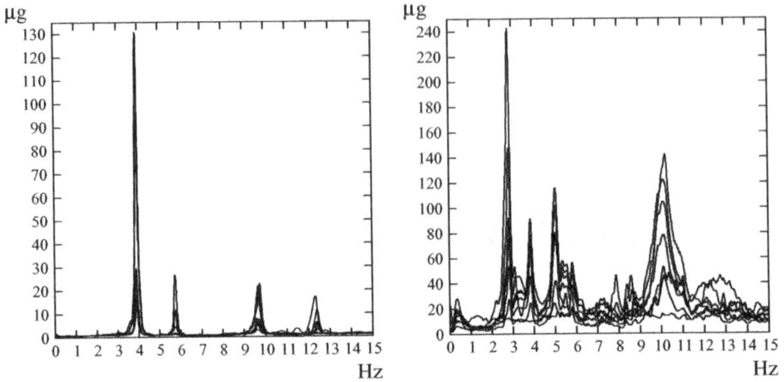

Figure 7.1 Spectrum of a sound bridge (left) and spectrum of a damaged bridge (right)

the purpose of inspectability and eventual replacement. Testing 30 bridges with grouted internal cables built in the late 1950s and early 1960s showed damage to cables in only one of the structures. In all the other bridges, damage was suspected, but no evidence of corrosion or wire breaks was detected. The bridge showing damage, however, was far from malfunctioning. On the other hand, bridges with external pre-stressing often show cracks in their anchoring parts and often display unequal stress patterns. The best results have been received on bridges with grouted tendons embedded in concrete in combination with external cables.

7.1.1.3 Influence of Temperature

The design codes for bridges provide clear instructions for considering temperature effects in bridge design. These instructions normally require a high and a low temperature threshold to be considered and eventually a temperature gradient between the bottom and top of a structure. No reference is made to the type of bridge or the material used. Monitoring provides the chance to record exactly the actual effects of temperature on structures, and the following lessons have been learned.

- Slender structures react very close to the provisions of the codes.
- Stiff structures very often deviate considerably from the expected stress distribution.

Figure 7.2 Resistant box-girder (left) and costly I-girders (right)

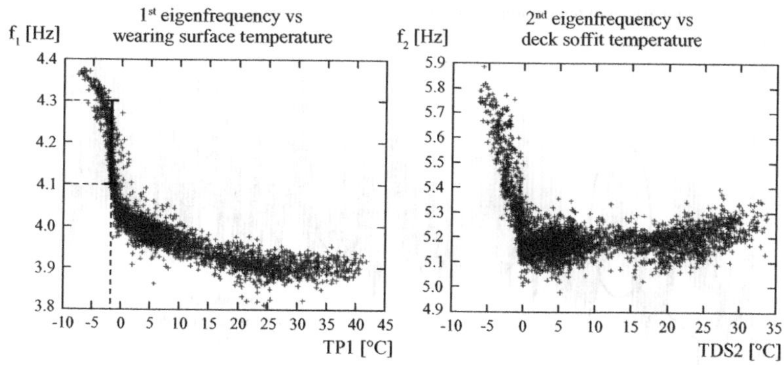

Figure 7.3 First eigenfrequency versus wearing surface temperature and second eigenfrequency versus deck soffit temperature (Peeters and DeRoeck 2000)

- The temperature gradients actually recorded on stiff structures by far exceed the values accounted for in the design.
- Temperature load cases can be the decisive load cases.
- Temperature changes do not trigger a linear behavior. Below 5°C a clear stiffening effect is recorded.

The stiffness of concrete bridges is dependent on temperature. The relation is given in Figure 7.3, which shows an almost bi-linear condition measured on a classic post-tensioned concrete box-girder bridge. This has to be considered in the interpretation of monitoring results.

Steel bridges quickly react to changing temperatures. The records of a 5 m high steel box-girder are shown in Figure 7.4. There is a difference in response between heating or cooling periods. The sensor data in Figure 7.4 represent the outside temperature and the inside temperature on the bottom slab and on the deck slab, and the patterns shown here are representative.

The annual cycle of temperature changes is rather homogeneous, as demonstrated by Figure 7.5, which shows the behavior of a concrete mass supported on bridge bearings in a railway tunnel, i.e. there is no influence from direct sunlight. Temperature conditions of the surrounding soil are rather stable, but the trains transport air from outside through the tunnel. The graph shows how homogeneous the structure behaves year by year. The maxima and minima values measured are actually higher than expected. Consideration of a concrete structure (Figure 7.5), where concrete is used in the shape of a box girder with cantilever arms, also reveals a rather homogeneous cycle. The maximum and minimum temperatures according to the design codes are never reached, and the structure reacts moderately to both warming and cooling, with no clear daily cycle being apparent.

A typical steel structure, however shows a rather violent reaction to temperature changes (Figures 7.6 and 7.7), which is rapid and produces strain in the system. This might be because of bearing friction, which is released suddenly causing a displacement of the structure. This effect can be particularly harmful to the expansion joints, and also to the bearing. Bridges that are not straight in plan might introduce exceptional forces into weak axes of the outfitting.

The consequence of these monitoring results should be an individual application of temperature loads depending on the type of structure and the conditions they have to experience. The following implications might be considered.

- Increase the loads from differential temperatures in stiff structures.
- Increase the temperature range considered in steel structures.
- Consider the effects of quick temperature changes on the global behavior of the structure.

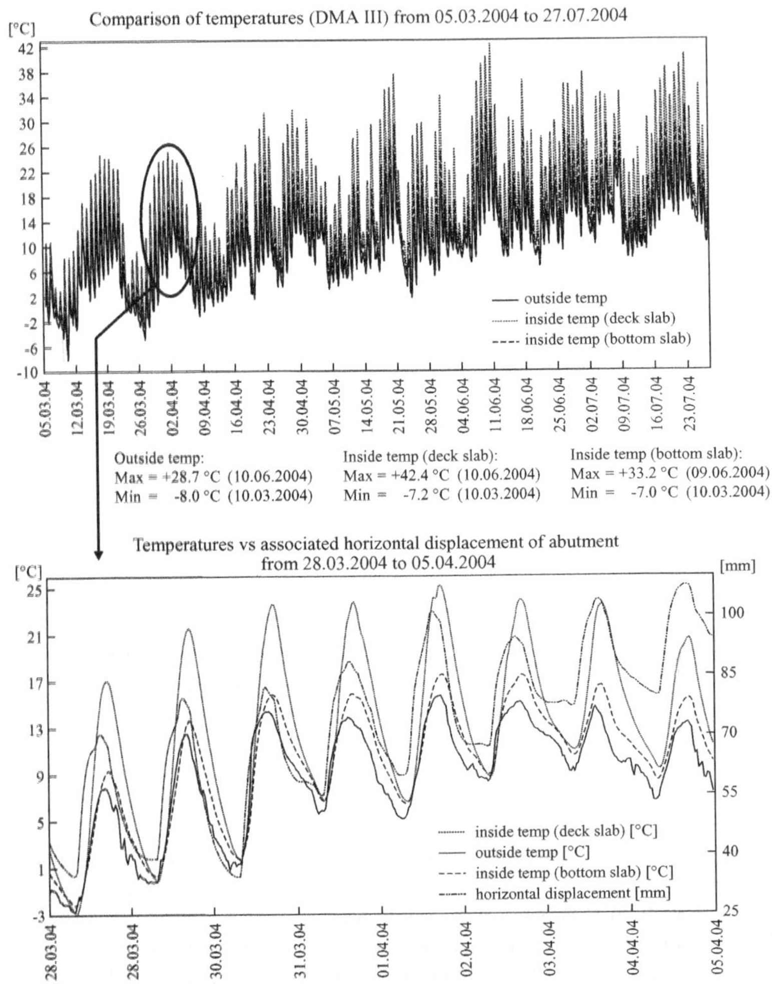

Figure 7.4 Representative temperature sensor records and longitudinal displacement of a steel bridges' abutment

7.1.1.4 Displacement

Bridges are flexible and displace under various loads, and the extent of displacement which can be calculated using structural models or FE calculations. The models used very often do not reflect reality, as exemplified by the modeled displacement of a steel bridge due to temperature changes shown in Figure 7.8, compared with displacement provided by the monitoring results shown in Figure 7.9. Figure 7.10 demonstrates especially that the displacements of a steel bridge's abutment can be approximately twice as much as those obtained from theoretical, linear elastic calculations, with the differences being mainly related to the following facts.

- The stiffness of the columns very much depends on the degree of fixation of the pier in the foundation.
- Bearings do not show a linear behavior at all times and rather tend to be stiff until a certain minimum force has been reached.

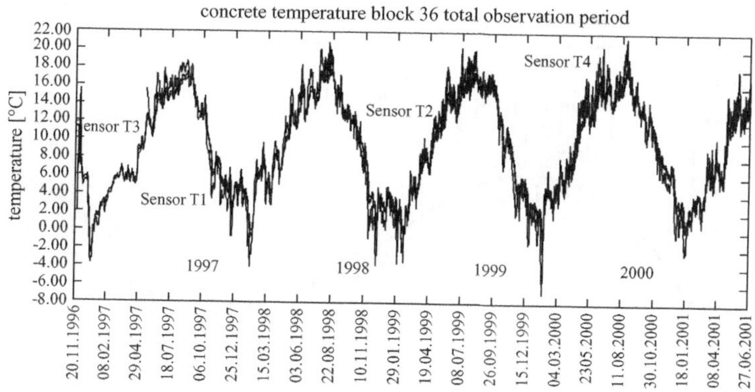

Figure 7.5 Long-term-behavior of a concrete mass supported on bridge bearings in a railway tunnel

- A certain stress limit has to be reached before restoring forces are activated, particularly when elastomeric bearings are provided.

In major bridges sudden displacements of ±50 mm have been recorded that have to be attributed to restoring forces of bearings being suddenly released (Figures 7.11–7.15). This displacement normally is within the regular limits of allowable displacement, but the sudden reaction might trigger secondary problems, such as restraints in the expansion joint. A number of failures of expansion joints could be attributed to this phenomenon. The frequency of such a phenomenon is not yet sufficiently documented but in a 6-month record of a major steel bridge, three such occasions have been detected (Figures 7.11 and 7.12).

A consequence of these examples is that realistic behavior of a structure can be recorded through monitoring, which might explain damage observed in the superstructure. It is also apparent that displacements of bearings and expansion joints calculated by theory might be not sufficient to account for extraordinary events the center of expansion of a structure can be dozens of meters away from the theoretical centre, (Figure 7.9), which needs to be taken into account in the design of bearings and expansion joints.

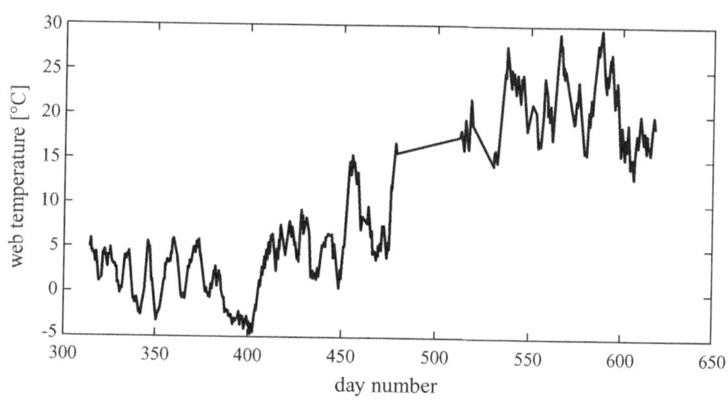

Figure 7.6 Variation of web temperature of the Z24 bridge observed over a period of one year (DeRoeck et al. 2000)

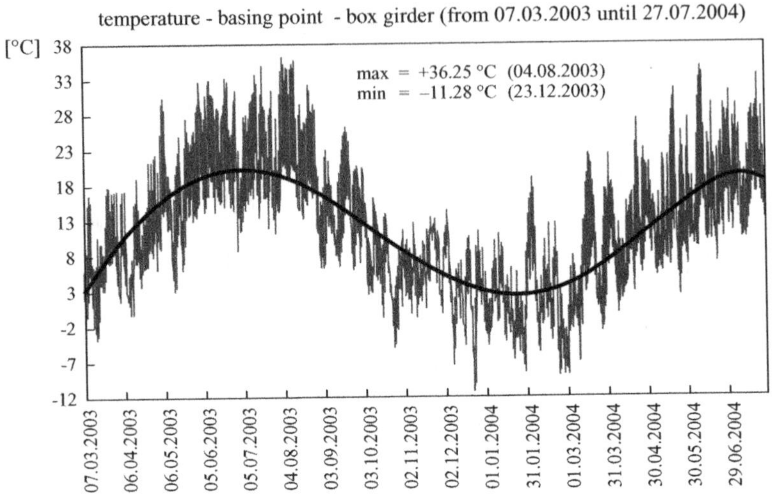

Figure 7.7 Temperature conditions inside the steel-box girder at the Europa Bridge of the Brenner Motorway

7.1.1.5 Large Bridges Versus Small Bridges

Monitoring concentrated initially on large and important bridges, and results of this monitoring led to the impression that bridges normally perform very close to the theoretical behavior determined and based on the design assumptions. Subsequent assessment of small bridges has shown that it is considerably more difficult to achieve good results the smaller the structure is, because of the different approaches taken

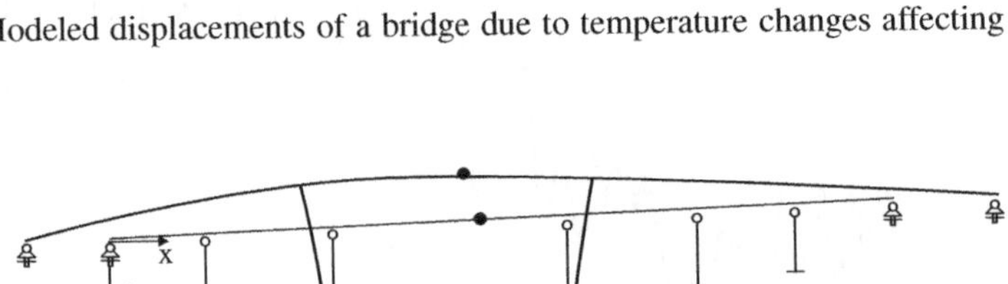

Figure 7.8 Modeled displacements of a bridge due to temperature changes affecting only the superstructure

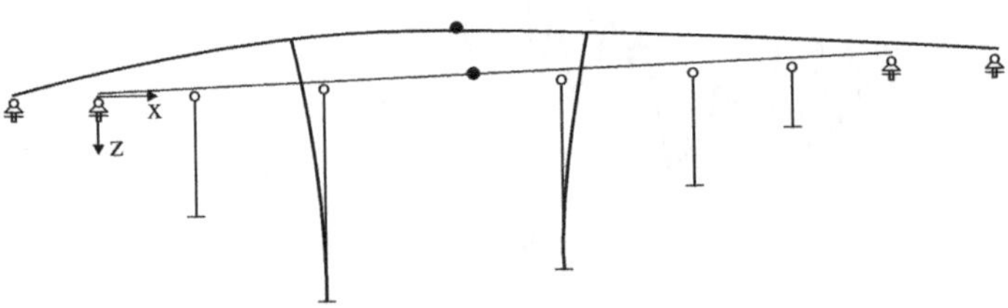

Figure 7.9 Displacements of the bridge in Figure 7.8 due to temperature changes recorded by monitoring

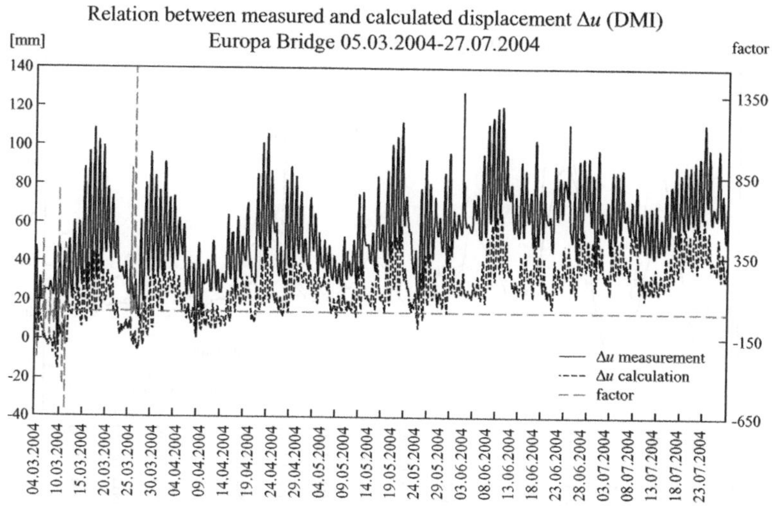

Figure 7.10 Comparison between measured and FE-based displacements of a steel bridge's abutment

towards these, not so important, structures. In addition, boundary conditions are much clearer in larger structures.

The lesson learned from monitoring is that greater attention should be paid to smaller bridges and that although a number of provisions of construction codes apply very well to large structures, they are less applicable to small structures. Here in particular the subject of temperature, as detailed in Section 7.1.1.3, has to be emphasized, especially the correct modeling of the boundary conditions.

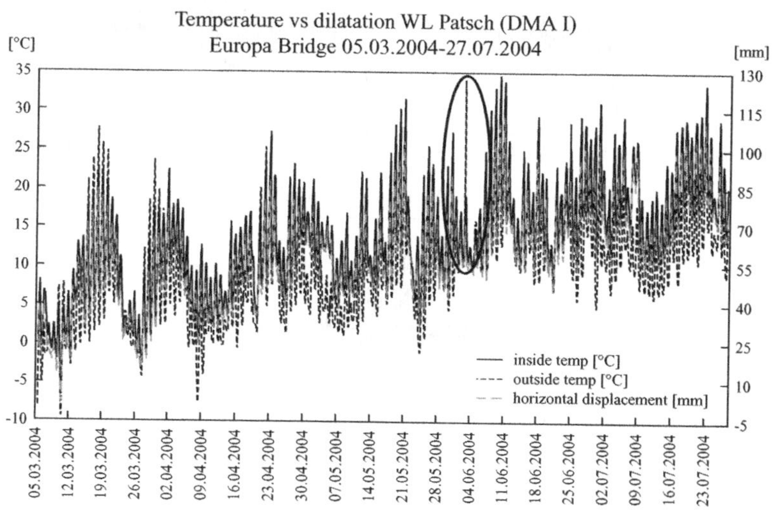

Figure 7.11 Uncommon, sudden reactions in the abutment's displacement recordings over a period of five months

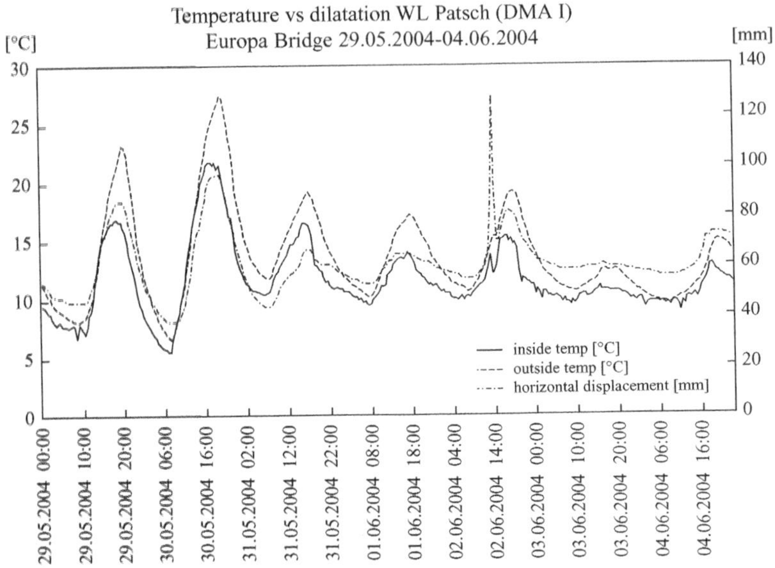

Figure 7.12 Focus on Figure 7.11 over a 1-week period, including one of the observed sudden reactions

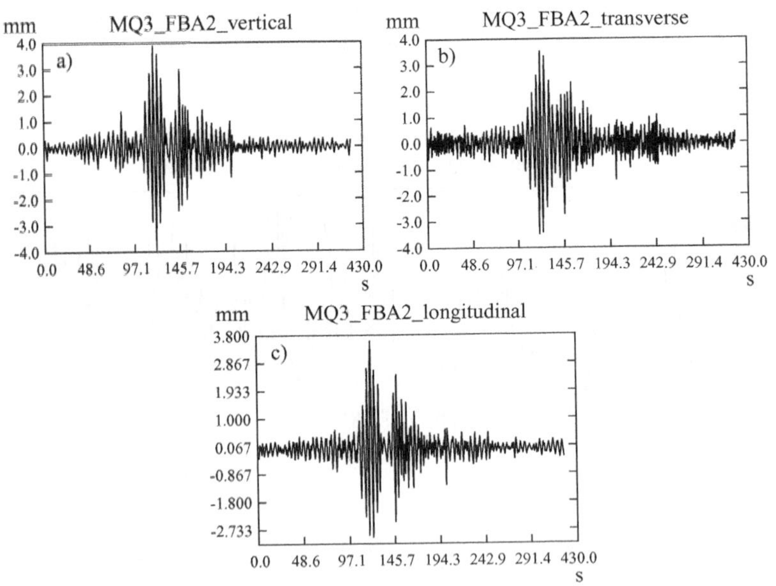

Figure 7.13 Relative displacement due to sudden occasions of restraint recorded with a 3D-acceleration transducer at the top of a 200 m high pier subdivided into the vertical (a), transverse (b) and longitudinal (c) direction

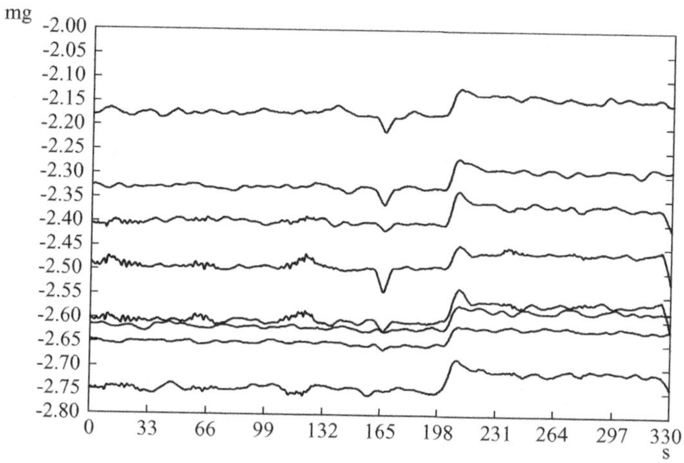

Figure 7.14 Displacement of the system's neutral axis due to bearing reset forces recorded for the St. Marx flyover (basis: acceleration sensors)

7.1.1.6 Vibration Intensities

The subject of resonance in pedestrian bridges is well known and is accounted for in design codes. Frequencies close to resonance, particularly those of structural members such as cantilever slabs, are not yet a subject of consideration. Experience has shown that evaluation of the vibration intensities measured for a structure can provide considerable information on fatigue and related problems. Assessment of vibration intensities therefore can produce indicators on the expected lifetime of a structure and on local problems to be expected on structural elements in the near future. It has been clearly demonstrated that bridges, where high vibration intensities have been recorded (Figure 7.16), most probably develop local

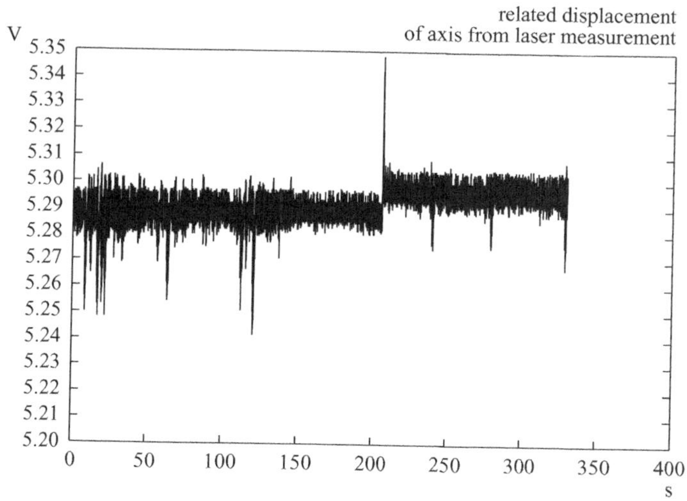

Figure 7.15 System displacement due to bearing reset forces of the St. Marx flyover (basis: longitudinal laser-displacement sensors)

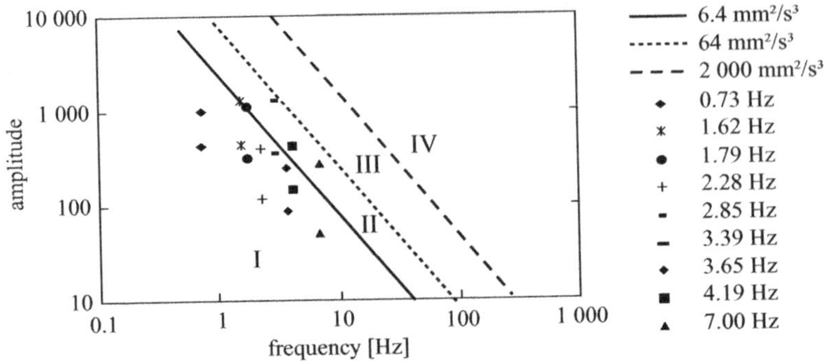

Figure 7.16 Vibration intensity chart for the Europa Bridge of the Brenner Motorway (representing high vibration intensities). I, no damage; II, possible plaster cracks; III, possible damage to load-bearing structural parts; IV, damage to load-bearing parts

problems in expansion joints, bearings, outfitting and, particularly, waterproofing. For comparison a low-vibration intensity record is shown in Figure 7.17.

7.1.1.7 Damping Values of New Composite Bridges

Measurements taken at a number of new composite bridges show that the damping values determined at the newly built structure are considerably higher than the normal values of comparable concrete bridges (Figure 7.18) or steel bridges. This might be attributed to the fact that the composite effect has to be established through a number of load cycles. After some time the damping values of these bridges stabilize to normal ranges. Firmer conclusions on this phenomenon have not been drawn yet, but it might be expected that a sharp drop in damping indicates eventual problems with bonding or the triggering of hidden local damage.

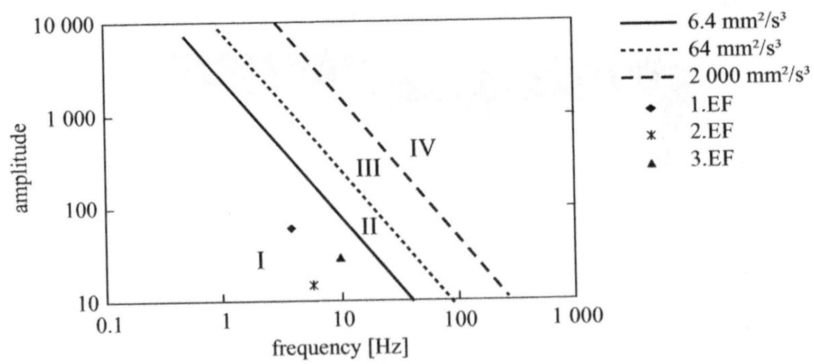

Figure 7.17 Vibration intensity chart for the S36 Bridge of the A1 Motorway (representing low vibration intensities). I, no damage; II, possible plaster cracks; III, possible damage to load-bearing structural parts; IV, damage to load-bearing parts

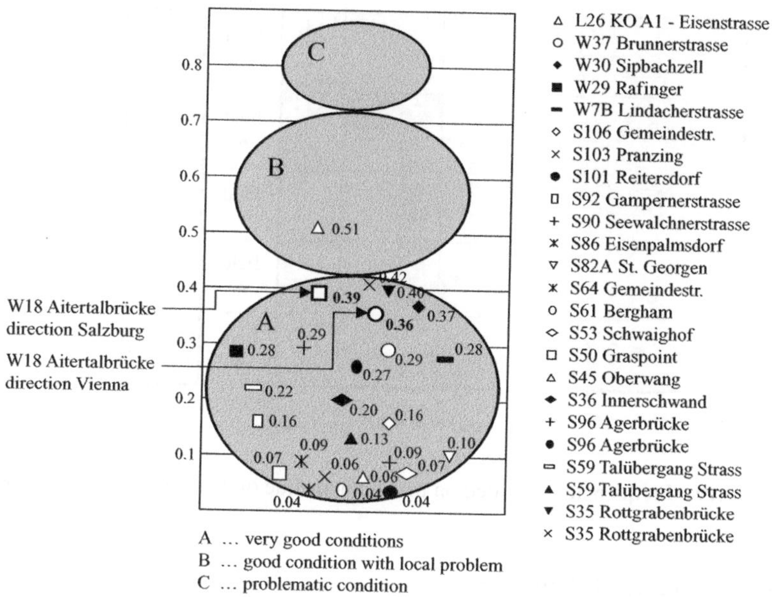

Figure 7.18 Classification of pre-stressed concrete and composite bridges according to their damping values

7.1.1.8 Value of Patterns

Certain elements of bridges are repetitive and it is to be expected that all members of a particular element show the same dynamic performance under service. A valuable approaches of monitoring is to recognize patterns and to observe the performance of comparable components. Any deviation from the pattern for individual components indicates a malfunction or an extraordinary situation that should be investigated.

As an example the case of a concrete box girder bridge with a distinct cantilever is considered (Figures 7.19–7.22). The monitored performance of the cantilever minus the action of the global system (Figure 7.21) provides information on the cantilever eigenmodes. Related symmetric modes can be determined and displayed. This should provide a distinct pattern, where every deviation indicates a problem. On the basis of colored frequency cards, so called trend cards, the relevant cantilever eigenfrequencies have been determined, which are marked in Figures 7.19 and 7.20 for separate box girders. By comparison of the response spectra of both box girders and their cantilevers the share of cantilever vibration can be displayed directly. A detailed evaluation procedure analyzing the relation between the response spectrum (Figure 7.22) and its energy content within the relevant frequency ranges leads to a certain behavior pattern of the cantilevers along the bridge. Deviations from this pattern are typically indications of irregularity.

Figures 7.23 and 7.24 show the pattern of an undamaged cantilever compared to a cantilever with minor corrosion damage of the transverse reinforcement. Although this method cannot provide detailed localization information on the problem, it does provide sufficient information on the functional quality of a structural element and by using a very quick and cheap test it can be determined whether action is required or not.

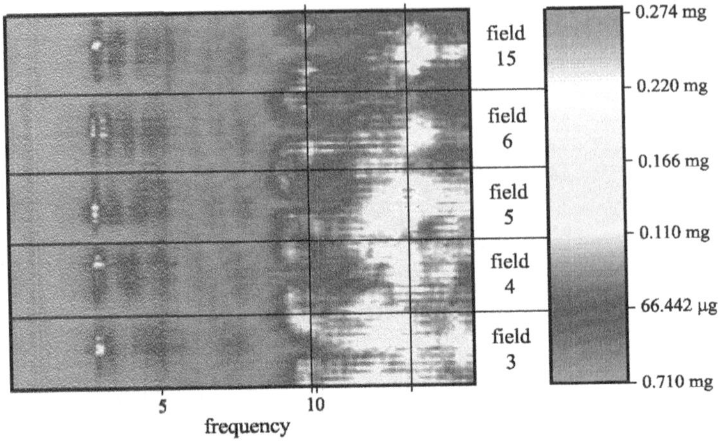

Figure 7.19 Frequencies recorded on the southern side of a concrete box-girder bridge

7.1.1.9 Understanding of Behavior

Complex bridge structures are often modeled in a rather simple way thereby neglecting behavior of the structure in three-dimensional space. Monitoring is the recording of the actual behavior of a structure, which comprises eventual drift or strain in response to temperature variation, as well as eventual responses to original construction mistakes that eventually, such as incorrect placement of bearings or nonrelease of restrainers. Figure 7.25 shows a case where a temporary fixture during construction had not been removed at the time of handover. The performance of the bridge was considerably different to what was expected, but monitoring has detected the problem and it was rectified immediately.

Another important aspect is information on the actual displacement of a structure, particularly with regard to complicated cable-supported structures, where such displacements could generate problems in traffic clearance or related interfaces.

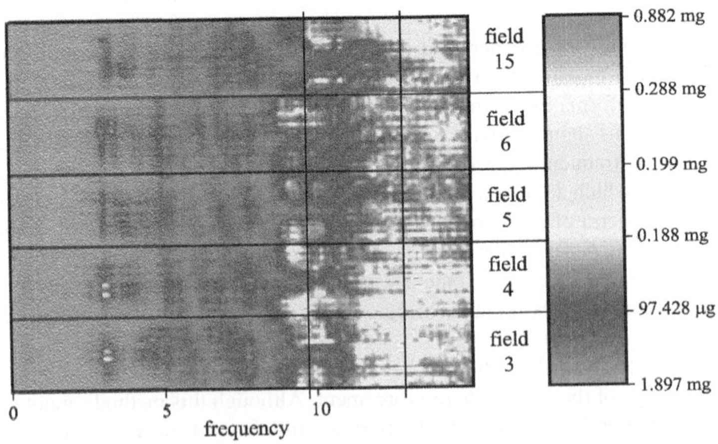

Figure 7.20 Frequencies recorded on the northern side of a concrete box-girder bridge

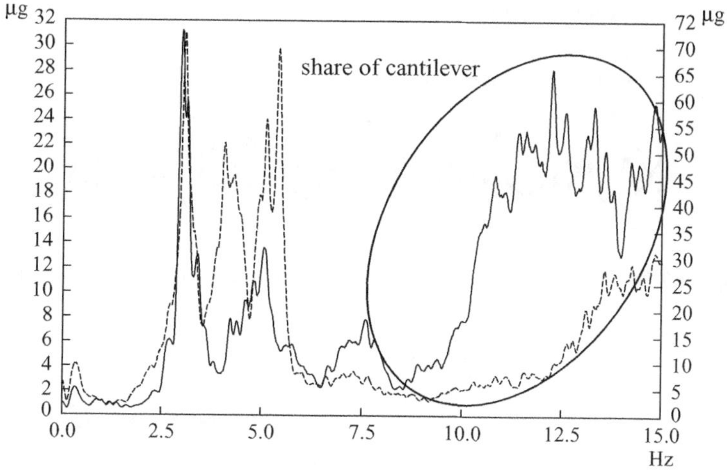

Figure 7.21 Spectrum of cantilever (solid line) and box-girder (dashed line) bridge

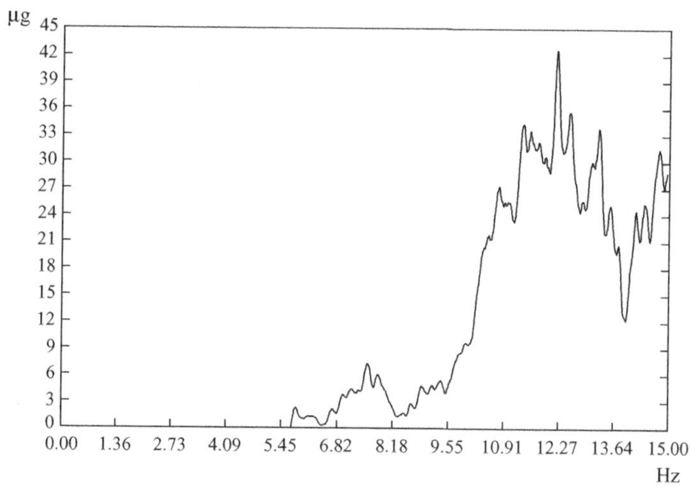

Figure 7.22 Response spectrum of cantilever vibration

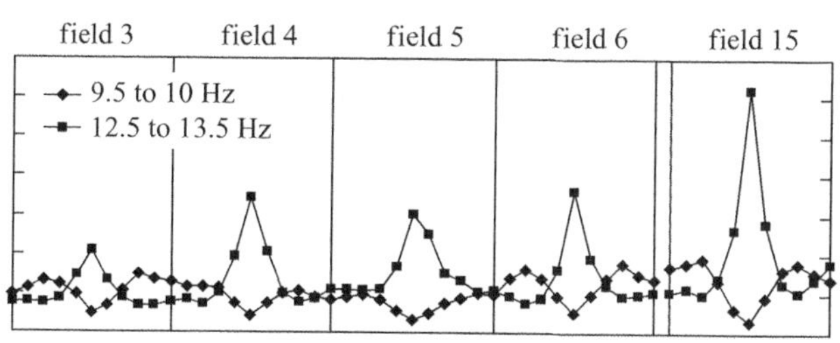

Figure 7.23 Acceptable behavior pattern of the cantilevers along the bridge

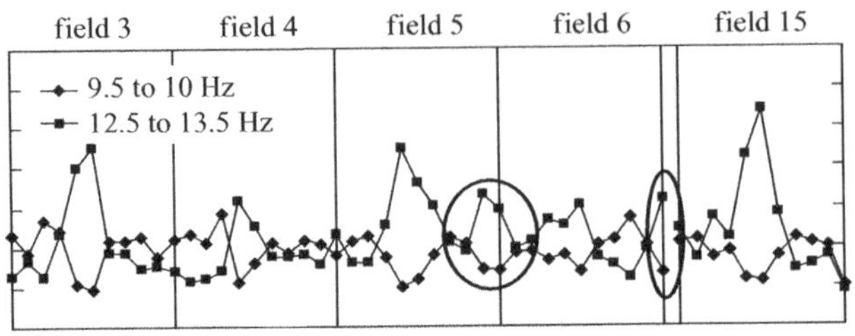

Figure 7.24 Behavior pattern of the cantilevers with indications of irregularity

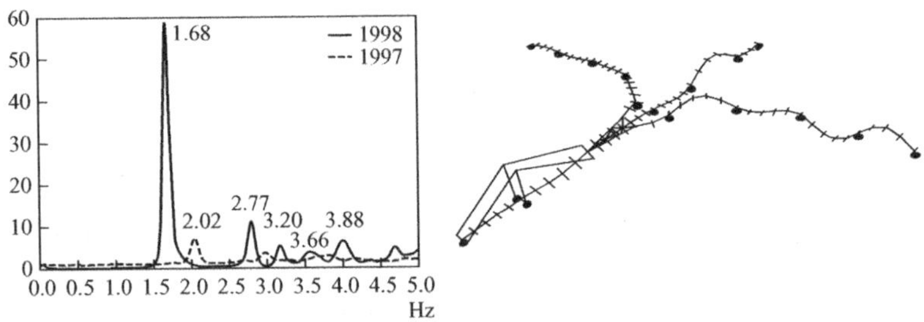

Figure 7.25 Frequency spectrum of Inn Bridge Hall West 1997 and 1998

One way of detecting problems is to compare expected against measured behavior. An example is provided by stay cables at Steyregg Bridge (Figure 7.26). The stay cables are protected by steel tubes against vandalism, but monitoring detected contact of the cable with the tube, thereby shortening the effective vibration length of the cable. Such a problem can lead to drastic damage of the cable through

Figure 7.26 Steyregg Bridge

provision of a sharp edge, which introduces unintended bending. Effective monitoring can identify these problems.

7.1.1.10 Dynamic Factors

Current bridge design codes require dynamic factors that mainly apply to the bridge span. The factor is considered to be 1.40 for components or directly effected members, but varies between 1.00 and 1.40 depending on the span of a bridge. The following lessons have been learned from monitoring.

- The dynamic factor provision within the design code according to the span length is actually conservative. All bridges have so far shown smaller dynamic factor values.
- The measured dynamic factor value for individual components sometimes exceeds the design-code value considerably. The highest dynamic factor value measured has been 2.20.
- The dynamic factor is also considerably dependent on the speed of traffic. This can eventually be controlled by speed limits.

Consequently overloaded vehicles that drive at low speed will not produce harmful stresses. The moderate increase always has to be considered in conjunction with eventual speed effects. Consequently dynamically sensitive elements should be avoided in design.

Another lesson learned is that the dynamic behavior also depends on the type of structure designed. Bridges with box girders (Figure 7.27) are considerably less vulnerable to dynamic effects than bridges of other types of design (Figure 7.28). The dynamic vulnerability of a structure is dependent on the mass acting upon it. This is clearly shown in monitoring records. Concrete bridges with a mass of $1.5\,t/m^2$ or more are very little affected by dynamic amplification, and continuous girders are less reactive to impacts. A combination of elements with major differences in stiffness, however, produce an inharmonic behavior detrimental to the structure.

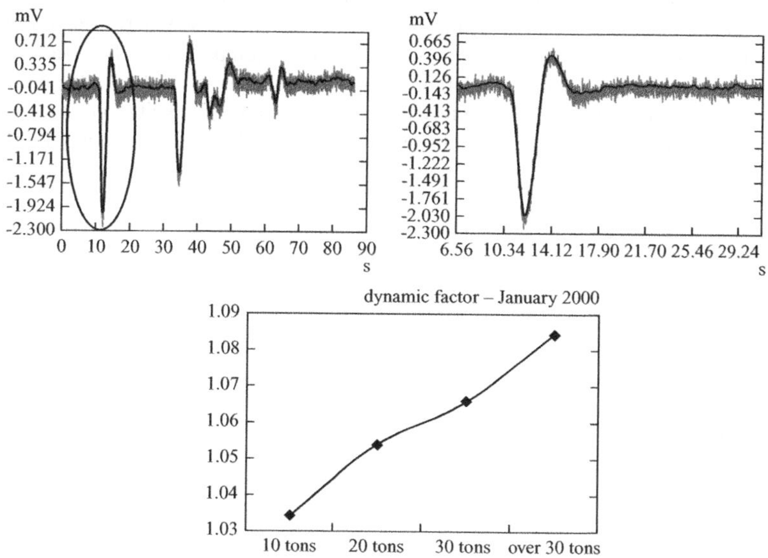

Figure 7.27 Dynamic factor of the St. Marx flyover

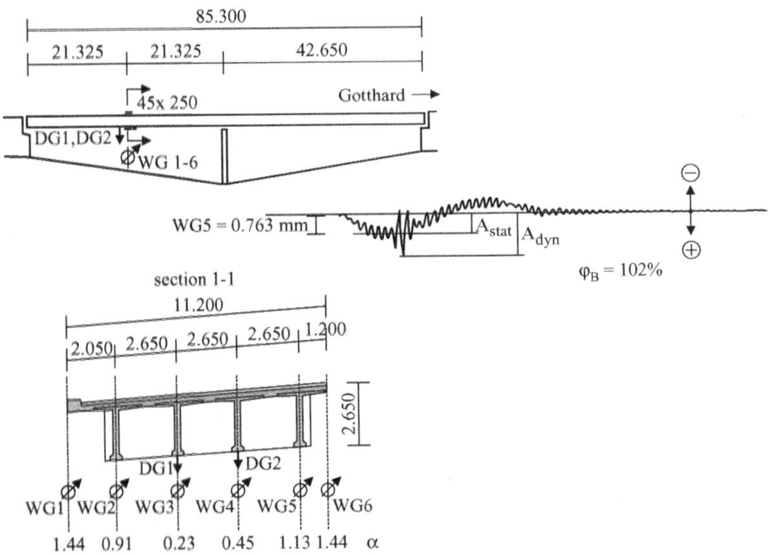

Figure 7.28 Dynamic factor of the Boeschrüti Viaduct due to induced impact loading (Cantieni 1983)

7.1.2 Monitoring of Stiffness

When spans are over 1000 m the hangers of suspension bridges also reach critical lengths. They are subject to vibration from wind and related phenomena. For example, interesting observations have been produced from the monitoring system of the Storebelt Bridge project in Denmark. The longest hangers are 177 m long, and about 35% of the hangers are affected by vibrations associated with wind and related phenomena. Amplitudes greater than 0.5 m are categorized as large, and in the case of the Storebelt Bridge, 52 incidents have been recorded since the year 2000. Amplitudes have ranged up to 2 m. Of particular interest is the correlation of large amplitudes to low temperatures. The occurrence of large amplitude vibration events has mostly been at temperatures between -5° and $+2^\circ$. Ice buildup has been observed on these occasions and there definitely has been no parametric excitation.

The threshold criteria for cables at Storebelt Belt were set as follows: vibration amplitudes should be smaller than $L/500$ or smaller than three cable diameters, but this was impossible to achieve for this type of hanger.

The problem identified by the Storebelt Bridge monitoring system is the top anchor, where through local bending fatigue caused damage to the cables. In order to measure the conditions tilt meters were placed on top of some of the cables at 25 cm distance from the anchor to measure the displacement, after which different types of dampers were installed.

- A spiral rope that was put around the hangers from spacer to spacer worked but was in adequate.
- Horizontal wind ropes were applied but as they work only with plain vibrations, which do not happen at this structure.
- A tuned liquid damper was applied which also proved not to be fully effective.
- The best results have been achieved by hydraulic dampers installed from the bridge deck, with damping increasing from the previous 0.19% to 0.63%, which is just above the required 0.6%.

The lesson learned in this six-year-long study was that there is no really efficient working damping system. The lifetime expectation of the hangers therefore becomes a critical consideration. Given the observation that these phenomena occur only for about 8–10 hours per year (estimation) a lifetime expectation of 5 – 30 years has been calculated. This presents a problem for the owner because better lifetime prediction methodologies are required. The solution could be system identification by monitoring and then the implementation of a counting system (e.g. rainflow counting as described in Section 4.1).

The mechanism of excitation is not clearly understood. It mainly occurs at wind speeds of 8–10 m/s, indicating that there is vortex shedding and some wake buffeting effects. The accompanying low temperatures indicate that ice is also a part of the problem. Considering that damping measures might be more costly than replacement of hangers every 15 years, a suitable strategy should be devised.

7.1.3 Finite Element Model Updating

7.1.3.1 Introduction

Structural Health Monitoring is nowadays a major concern due to the constant aging of infrastructures. Requests for evaluations of structural reliability and for remaining life assessment are, therefore, assuming major importance in order to reduce the maintenance costs and to increase the safety level of the structures. In particular, the use of nondestructive dynamic testing for the evaluation of the structural performances is a well-established methodology for SHM, improving the data reliability and overcoming the limitations of traditional visual inspection methods.

The structural response obtained from monitoring can be used in conjunction with several different numerical methods, which can be grouped into two main categories: model-based and parameter-based. The first relies on a reference structural model, e.g. FE model, whereas the main characteristic of the second is a general mathematical description of specified parameters or system features, e.g. analysis of time-series details by means of wavelet theory.

From the numerical point of view, it has to be considered that the FE method is nowadays a very powerful computer-aided simulation technique, allowing the user to study any kind of problem virtually without limitation in model size and complexity. On the other hand, every FE model is a numerical approximation of the real structure by several assumptions, which can lead to great differences between the model and the real structure. Through the comparison of experimental data from dynamic testing and numerical data, these differences usually can be well established.

Finite element model updating (FEMU) is a model-based numerical technique used to minimize the differences between the real structure and the FE model. The basic idea is to use the recorded structural response to update some selected structural parameters of the numerical model (such as stiffness, mass and internal forces) as well as some boundary conditions (such as translational or rotational springs), until an adequate agreement between numerical and experimental results is achieved. The resulting structure presents a better dynamic agreement with the physical reality. Moreover, since damage is an intentional or unintentional change to the material and geometric properties of a given system which affects the system performance (Inman and Farrar 2005), the parameter distributions obtained as outcomes can provide useful information about the possible structural damage. Consequently, the FEMU can be employed for the purpose of damage detection. In fact, by using experimental data from a damaged structure and applying the updating procedure, the resulting structural parameter distributions (e.g. elastic modulus) can identify a deficiency in the structural properties that indicates the occurrence of damage. In this fashion, the FEMU can be applied proficiently in the field of SHM.

The updating procedure is based on a mathematical optimization problem: the difference between numerical and experimental data should be minimized through iterative modal analysis. Practical experience has shown that the vibration response represented by accelerations provides suitable and sensitive structural information for this purpose. The recorded acceleration time series should be processed, that is, a transformation from the time-domain to the frequency-domain is required. The general problem of extracting eigenfrequencies and corresponding mode shapes regarding the effects of noise, excitation sources and number of instrumented points is widely discussed in the civil engineering community and several different approaches can be found in the technical literature.

The solution of an optimization problem by iterative procedures often requires a large number of calculation steps. In practice, engineers and owners of structures are interested in answering the basic questions: "Is there any damage and if yes, where is it located?" The damage extent and the remaining load-carrying capacity of the structure (levels 3 and 4 in the damage detection procedure; (Rytter 1993) are of course important, but currently not essential from the practical point of view. Considering these aspects, the use of simplified models and procedures is well suited within the FEMU in order to correctly describe the structural behavior and to effectively detect and locate the damage without requiring excessive computational resources.

In this section, the software solution VCUPDATE for FEMU is presented and some applications to practical problems are reported and discussed. VCUPDATE is based on the open-source environment Scilab (Consortium S Undated) and allows the use of different FE codes (OpenSees (Center Peer Undated) and ANSYS (Inc. SI Undated)) and different updating methods (penalty method, coupled local minimizers (CLM), subproblem approximation method and first-order optimization method). The implementation of the CLM method (Teughels et al. 2003) is still not completed and the theoretical part will not be discussed here. The code has been applied with good results to several different structures (Wenzel and Mordini 2006); (Mordini and Wenzel 2007a,b); (Mordini et al. 2007); (Mordini et al. 2008). VCUPDATE is developed as a numerical framework. It comprises several different modules, which can be modified separately and updated. In this way, the development of a part can be undertaken without modification of the others. Moreover, additional modules can be added in order to enlarge the software capabilities. For example, new optimization algorithms or connections with different FE codes could be easily implemented. The VCUPDATE architecture is shown in Figure 7.29. The data accepted as input are frequencies and mode shapes. The updating procedures, however, are performed by using frequencies and modal assurance criterion values, see Equation (11). In this way, the data can be easily normalized by their initial values, providing the numerical solution with an improved stability.

The use of two different FE codes provides VCUPDATE with a wider flexibility. In fact, the commercial code ANSYS has very powerful modeling capabilities and allows the graphical visualization of the model and of the results, including the outcomes from the updating procedure. On the other hand, the updating procedure using OpenSees is much faster and therefore, when possible (i.e. when the model is simple), the use of OpenSees is preferred.

In addiction, it has to be considered that many different commercial FEMU solutions are available in the market at high licensing prices but both Scilab and OpenSees are open source and free of charge

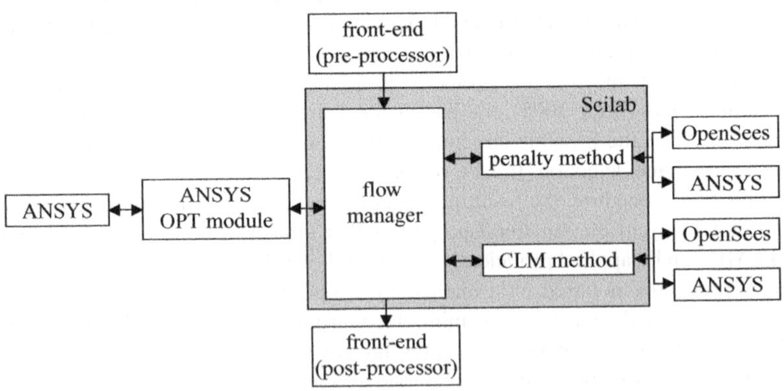

Figure 7.29 VCUPDATE architecture

software. This difference can improve the accessibility of the FEMU to a wide range of engineering companies interested in the method but not able to afford the costs of a commercial software solution (optimization + FE code).

7.1.3.2 Theoretical Basics of the Penalty Method

The Penalty Method implemented in this work is a sensitivity-based iterative method (Friswell and Mottershead 1995). A non-linear Penalty Function is minimized through subsequent linear steps. Depending on the analyzed problem, this procedure can be quite time-consuming from the computational point of view. In Figure 7.30 the main steps of the code are shown.

The theoretical formulation of the Penalty Method is rather simple. It is a gradient-based algorithm and therefore it is sensitive to the chosen starting point. In fact, this code provides very good results if the starting point can be correctly estimated and located close to the global minimum of the problem. One possibility to overcome this problem is the use of the CLM method (Teughels et al. 2003). In civil engineering applications, however, the structural analyst should be able to evaluate, following his experience, a plausible starting point according to the real structure.

Sensitivity Analysis

The sensitivity analysis is performed to select the most sensitive parameters for the FE model. Moreover, the sensitivity matrix is also used in the updating algorithm. At the jth iteration, the sensitivity matrix can be written as

$$S_j = \frac{\partial d_j}{\partial p_j} \tag{1}$$

where ∂p_j is the perturbation in the parameters and ∂d_j is the change between measured and numerical data.

The matrix S is numerically computed by using the forward difference of the function with respect to each parameter. This means that the knowledge of the system matrices is not required and the calculation can be carried out using results from multiple FE analyses only. If the problem has n parameters, $n + 1$

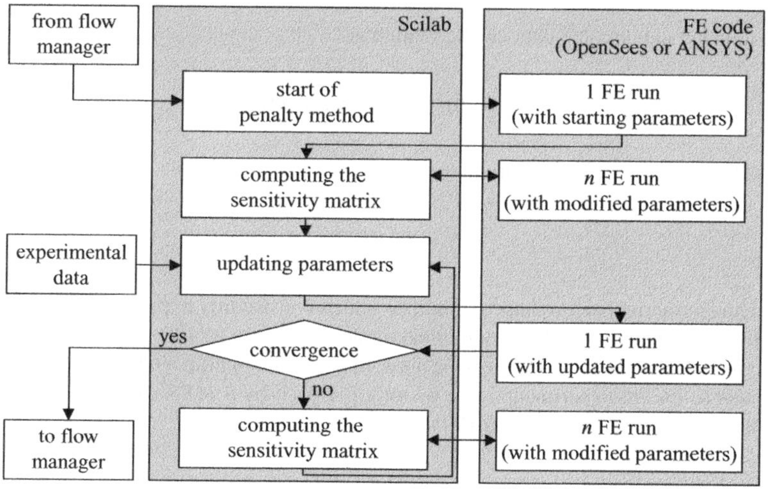

Figure 7.30 VCUPDATE Penalty Method flow chart

FE runs are required: the first with the starting values for the parameters and n different runs perturbing one parameter only in each of them.

The forward difference method has the form (Center Peer Undated)

$$S = \frac{\partial d}{\partial p} \approx \frac{\Delta d}{\Delta p} = \frac{d(p + \Delta p) - d(p)}{\Delta p} \tag{2}$$

The perturbation in the ith parameter is

$$\Delta p_i = \frac{\Delta D}{100}(\overline{p}_i - \underline{p}_i) \tag{3}$$

where ΔD is the forward difference step size and $(\overline{p}_i - \underline{p}_i)$ is the difference between the upper and lower bound for the ith parameter. Each vector $\Delta d/\Delta p$ gives one column of the sensitivity matrix.

Sensitivity Matrix and Mode Shape Scaling

Using different structural properties with very different values as parameters may lead to numerical problems in the iterative solution (Dascotte et al. 1995). This problem arises from the ill-conditioning of the sensitivity matrix and is more probable when both frequencies and mode shapes are used as data. An appropriate scaling of the sensitivity matrix can improve the stability and speed up the convergence. Within VCUPDATE, the sensitivity matrix is scaled by normalizing parameters and data by their initial values.

The scaling matrices are defined as

$$D_p = \begin{bmatrix} \frac{1}{p_1} & 0 & \cdots & 0 \\ 0 & \frac{1}{p_2} & \cdots & 0 \\ \vdots & \vdots & \ddots & \vdots \\ 0 & 0 & \cdots & \frac{1}{p_n} \end{bmatrix} \quad \text{and} \quad D_d = \begin{bmatrix} \frac{1}{d_1} & 0 & \cdots & 0 \\ 0 & \frac{1}{d_2} & \cdots & 0 \\ \vdots & \vdots & \ddots & \vdots \\ 0 & 0 & \cdots & \frac{1}{d_n} \end{bmatrix} \tag{4}$$

The scaled parameters and scaled data can be written as

$$\Delta\overline{p} = D_p \Delta p \quad \text{and} \quad \Delta\overline{d} = D_d \Delta d \tag{5}$$

respectively. Therefore the scaled sensitivity matrix is

$$\overline{S} = D_d S D_p^{-1} \tag{6}$$

Experimental and numerical mode shapes should be scaled consistently to properly perform the updating algorithm. Moreover, experimental and numerical mode shapes can be 180° out of phase. To solve these problems, the modal scale factor (MSF) can be used. In particular, within VCUPDATE, the ith numerical mode is rescaled to the ith experimental one by multiplying it by a MSF calculated as (Allemang and Brown 1982); (Friswell and Mottershead 1995):

$$MSF(\varphi_{exp,i}, \varphi_{num,i}) = \frac{\varphi_{exp,i}^T \varphi_{num,i}}{\varphi_{num,i}^T \varphi_{num,i}} \tag{7}$$

Updating Algorithm

The updating problem can be represented as the minimization of a Penalty Function $J(\Delta p)$ subjected to the constraint $\Delta d = S\Delta p$ (Consortium S Undated). For each parameter an upper and a lower bound can be specified $\underline{p}_i \leq p_i \leq \overline{p}_i$. Inserting the weighting matrices W_d and W_p for data and parameters respectively, the penalty function can be written as

$$J(\Delta p) = \varepsilon^T W_d \varepsilon + \Delta p^T W_p \Delta p \tag{8}$$

Substituting the error $\varepsilon = \Delta d - S\Delta p$ and minimizing $J(\Delta p)$ with respect to Δp, gives the updated parameter values as

$$p_{j+1} = p_j + \left[S_j^T W_d S_j + W_p\right]^{-1} S_j^T W_d \Delta d_j \tag{9}$$

where $\Delta d_j = d_0 - d_j$ is the difference between experimental and numerical data.

The sensitivity matrix should be computed at each iteration, but this can lead to an excessive computational time. VCUPDATE allows the user to choose the frequency with which the sensitivity matrix has to be updated. For some extremely time-consuming cases, the initial matrix can be used for the entire analysis.

According to experience, not all the experimental data are measured with the same accuracy. Usually lower frequency data are more reliable than higher frequency data. In order to express the user confidence in measured data, weighting matrices are used. Parameters can also be weighted separately. Using weighting matrices is very powerful, but engineering insight is required.

Five different ways to create the W_p matrix according to (Dascotte et al. 1995) are implemented in VCUPDATE. Several different tests were performed in order to verify the methods: in particular some of them should be used carefully, since they provide, in some cases, an increased convergence speed but, in other cases, the convergence is worse.

Convergence Criterion

The iterative scheme presented above is repeated until a convergence criterion is satisfied. The criterion can be based on parameters or on data. Two different convergence criteria are implemented within VCUPDATE according to (Dascotte and Vanhonacker 1989). The first is based on the frequency deviation

$$CC_{abs,j} = \frac{1}{n}\sum_{i=1}^{n} \frac{\left|f_{exp,i} - f_{j,i}\right|}{f_{exp,i}} \qquad \text{(at } j\text{th iteration)} \tag{10}$$

where $f_{exp,i}$ is the ith experimental frequency and $f_{j,i}$ is the ith numerical frequency at the jth iteration.

A different way to check the convergence is to also include the mode shape correlation. The most important indexes for this purpose are the Modal Assurance Criterion (MAC; Allemang and Brown 1982) and the Normalized Modal Difference (NMD; Maya and Silvia 1997), which, for two vectors φ_j, φ_k are defined as:

$$MAC(\varphi_j, \varphi_k) = \frac{\left|\varphi_j^T \varphi_k\right|^2}{\left(\varphi_j^T \varphi_j\right)\left(\varphi_k^T \varphi_k\right)} \tag{11}$$

and

$$NMD(\varphi_j, \varphi_k) = \sqrt{\frac{1 - MAC(\varphi_k, \varphi_k)}{MAC(\varphi_k, \varphi_k)}} \tag{12}$$

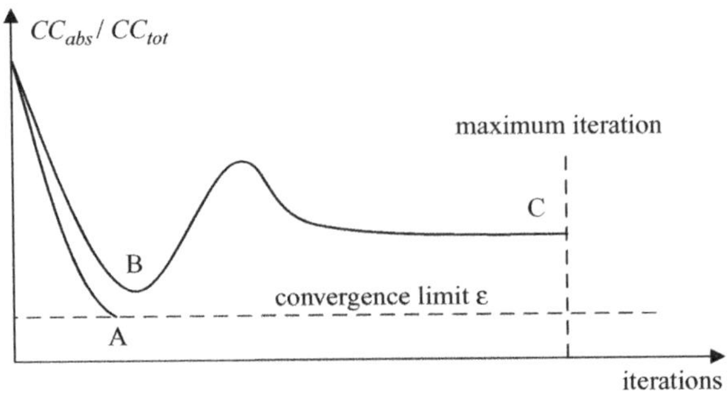

Figure 7.31 Convergence criterion

respectively. The closer the MAC is to 1, the better the correlation is. Since the NMD is much more sensitive than the MAC to the differences in similar vectors, it can be used when the mode shapes are highly correlated, while it is less useful for uncorrelated vectors. The closer the NMD is to zero, the better the correlation is.

The second convergence criterion implemented in VCUPDATE includes the MAC index:

$$CC_{tot,j} = \frac{1}{2nW_{max}} \sum_{i=1}^{n} \left(W_f \frac{|f_{exp,i} - f_{j,i}|}{f_{exp,i}} + W_{\Phi}(1 - MAC_i) \right) \quad \text{(at } j\text{th iteration)} \tag{13}$$

where W_f and W_{Φ} are the weights for the frequency and MAC deviation respectively, and $W_{max} = \max(W_f, W_{\Phi})$. By using the weights, the balance between the importance of frequencies and mode shape can be directly established.

The closer the selected CC approaches to zero, the better the agreement is between experimental and numerical data. The convergence is achieved if $CC \leq \varepsilon$ (point A in Figure 7.31). During the analysis all the information (updated data, updating parameters, CC) are stored and if the convergence is not attained, the updating procedure stops when a maximum number of iterations is reached. Then, the information corresponding to the iteration with the minimum value of CC (point B) is used as output.

A restart option is available in VCUPDATE. If the convergence is not achieved within the maximum iteration number (point C), the analysis data can be stored and used as starting point for a subsequent analysis. In particular, the parameters corresponding to the last iteration are taken as starting point for the next analysis. The model should not be modified between the two analyses.

7.1.3.3 Theoretical Basics of the ANSYS OPT Processor

In this section, the theoretical basics of the ANSYS OPT processor are described. Only a short review is reported here based on the integration in VCUPDATE. The ANSYS documentation (Inc. SI Undated) provides a more detailed description.

Objective Functions

An objective function is a measure of the deviation between the real structure and the FE model. The updating procedure is performed through the minimization of this function. Two different objective functions are implemented in VCUPDATE. The first is based on frequencies (Jaishi and Ren 2005):

$$\Pi_i = \sum_{i=1}^{n} \alpha_i \left(\frac{f_{exp,i} - f_i}{f_{exp,i}} \right)^2 \tag{14}$$

where α_i is the weighting factor for the ith frequency. The second objective function is based on frequencies and MAC index (Moller and Friberg 1998); (Jaishi and Ren 2005):

$$\Pi_2 = \Pi_1 + \sum_{i=1}^{n} \beta_i \frac{(1 - \sqrt{MAC_i})^2}{MAC_i} \tag{15}$$

where MAC_i is the diagonal term of the MAC matrix related to the ith mode shape:

$$MAC_i(\varphi_{exp,i}, \varphi_{num,i}) = \frac{\left|\varphi_{exp,i}^T \varphi_{num,i}\right|^2}{\left(\varphi_{exp,i}^T \varphi_{exp,i}\right)\left(\varphi_{num,i}^T \varphi_{num,i}\right)} \tag{16}$$

and β_i is the weighting factor related to the ith mode shape. α_i and β_i can be used to express the user confidence on the measured data.

Optimization Techniques (OPT)
The updating parameters are

$$X = \{x_1, \ldots, x_i, \ldots, x_n\} \tag{17}$$

Up to 60 parameters can be defined in OPT. A constraint is specified for each parameter:

$$\underline{x}_i \le x_i \le \overline{x}_i \tag{18}$$

where underbars and overbars represent the lower and upper bounds. The minimization problem can be written as:

$$\min \Pi(x) \tag{19}$$

subjected to

$$\begin{aligned} g_i(x) &\le \overline{g}_i & i &= 1, \ldots, m_1 \\ \underline{h}_i &\le h_i(x) & i &= 1, \ldots, m_2 \\ \underline{w}_i \le w_i(x) &\le \overline{w}_i & i &= 1, \ldots, m_3 \end{aligned} \tag{20}$$

where $g_i(x)$, $h_i(x)$ and $w_i(x)$ are the state variables that depend on the parameters and are used to constrain the problem. No state variables are used in VCUPDATE.

Two different minimization methods are implemented in OPT, the Subproblem Approximation Method and the First-Order Optimization Method.

The Subproblem Approximation Method is an advanced, zero-order method which requires only the values of the dependent variables (objective function and state variables) and not their derivatives. The dependent variables are first replaced with approximations by means of least squares fitting and the constrained minimization problem described in Equations (18)–(20) is converted to an unconstrained problem using penalty functions. Minimization is then performed every iteration on the approximated, penalized function (called the subproblem) until convergence is achieved or termination is indicated. For this method, each iteration is equivalent to one complete analysis loop. As a first step, each dependent variable is written by an approximation such as $\hat{\Pi}(x) = \Pi(x) + error$ for the Objective Function. Then, the constrained optimization problem is written as an unconstrained problem by using penalty functions:

$$F(x, p_k) = \hat{(}\Pi) + \Pi_0 p_k \left(\sum_{i=1}^{n} X(x_i) + \sum_{i=1}^{m_1} G(\hat{g}_i) + \sum_{i=1}^{m_2} H(\hat{h}_i) + \sum_{i=1}^{m_3} W(\hat{w}_i) \right) \tag{21}$$

where $F(x, p_k)$ is the unconstrained Objective Function, which varies with the parameters and p_k, while X, G, H and W are the penalty functions used to force the parameters and the state variables constraints. A sequential unconstrained minimization technique is used to solve Equation (21). The convergence is based on the Objective Function value as well as on the parameter values. In particular, the convergence is assumed at the jth iteration when one of the following conditions is satisfied:

$$|\Pi^j - \Pi^{j-1}| \leq \varepsilon \tag{22}$$

$$|\Pi^j - \Pi^b| \leq \varepsilon \tag{23}$$

$$|x_i^j - x_i^{j-1}| \leq \rho_i \tag{24}$$

$$|x_i^j - x_i^b| \leq \rho_i \tag{25}$$

where the superscript b indicates the best design set and ε and ρ_i are the convergence tolerance for the Objective Function and the ith parameter respectively. The program execution is also stopped if the maximum number of iterations is reached without convergence.

The First-Order Optimization Method is based on the computation of derivative information. The constrained problem statement expressed in Equations (18)–(20) is transformed into an unconstrained one via penalty functions. Derivatives are formed for the objective function and the state variable penalty functions, leading to a search direction in design space. Various steepest descent and conjugate direction searches are performed during each iteration until convergence is reached. Each iteration is composed of subiterations that include search direction and gradient computations. In other words, each optimization iteration will perform several analysis loops. Compared with the subproblem approximation method, this method is usually seen to be more computationally demanding and more accurate. An unconstrained form of the problem can be written as

$$Q(x, q) = \frac{\Pi}{\Pi_0} + \sum_{i=1}^{n} P_x(x_i) + q \left(\sum_{i=1}^{m_1} P_g(g_i) + \sum_{i=1}^{m_2} P_h(h_i) + \sum_{i=1}^{m_3} P_w(w_i) \right) \tag{26}$$

where $Q(x, q)$ is the dimensionless unconstrained Objective Function and P_x, P_g, P_h and P_w are the penalties applied to the parameters and the state variables. The constraint satisfaction is controlled by the surface parameter q.

At the jth optimization iteration, the subsequent new parameter values are written as

$$x^{j+1} = x^j + s_j d^j \tag{27}$$

where d^j is the search direction vector and s_j is the line search parameter. The global minimization is obtained through the sequential generation of the search directions and the internal adjustment of the response surface parameter.

The convergence is based on the Objective Function value. In particular, the convergence is assumed at the jth iteration when both the following conditions are satisfied:

$$|\Pi^j - \Pi^{j-1}| \leq \varepsilon \tag{28}$$

$$|\Pi^j - \Pi^b| \leq \varepsilon \tag{29}$$

In order to check the sensitivity of the Objective Function to the parameters, a gradient analysis is implemented. Starting from the starting parameter values, the gradient of the Objective Function

$$\nabla\Pi(x) = \left[\frac{\partial\Pi}{\partial x_1}, \frac{\partial\Pi}{\partial x_2}, \ldots, \frac{\partial\Pi}{\partial x_n}\right] \tag{30}$$

is computed by a forward difference approximation as follows:

$$\frac{\partial\Pi}{\partial x_i} = \frac{\Pi(x + \Delta x_i) - \Pi(x)}{\Delta x_i} \qquad \Delta x_i = \frac{\Delta D}{100}(\overline{x}_i - \underline{x}_i) \tag{31}$$

where ΔD is the forward difference step size in percent.

7.1.3.4 Discussion on the Computational Efficiency

The first version of VCUPDATE was developed by using OpenSees. Subsequently, the connection with ANSYS was accomplished in order to provide the system with better modeling capabilities. In addition, ANSYS allows the graphical visualization of the model and results, including the outcomes from the updating procedure.

One of the main problems highlighted in the practical use of VCUPDATE is the computational time required for the calculation.

The variable dominating the computational performances is the number of parameters, since, if the problem has n parameters, $n + 1$ FE runs are required for the computation of the sensitivity matrix. The problem arises in the data exchange between the FE code and the Scilab code. The data are in fact written by the Scilab code and read by the FE code (parameters) or vice versa (modal results) by using ASCII files. For a simple model, it was observed that more than half the total running time is spent in the write–read procedures. In order to improve the time performances, VCUPDATE provides the user with the possibility to set how often the sensitivity matrix is updated. In some extreme cases, the initial sensitivity matrix can be used along the entire analysis.

In order to overcome this problem, the connection with the ANSYS OPT processor was implemented. In fact, the computational performances are greatly improved since all the calculations are performed within ANSYS and no data exchange is necessary. As a drawback, OPT can be used only with ANSYS and it can manage a maximum of 60 parameters, which could be, in some cases, inadequate to describe the property distributions along the structures.

7.1.3.5 Applications to Beams

A Bilfinger Berger Pre-Stressed Reinforced Concrete Beam

The first presented application of VCUPDATE is the investigation of a pre-stressed reinforced concrete (PRC) beam designed and produced by Bilfinger Berger, Germany. The PRC beam is shown in Figure 7.32. The properties are: elastic modulus $E = 34\,000$ MPa, cross-sectional dimensions 1.45×1.00 m for an area $A = 1.45\,\mathrm{m}^2$ and a moment of inertia $J = 0.120833\,\mathrm{m}^4$ and mass per unit volume $M = 2500\,\mathrm{kg/m}^3$. The beam is pre-stressed by six tendons with a pre-stressing load $P = 300.468\,kN$ for each tendon. The cross-section is shown in Figure 7.33.

Figure 7.32 The investigated PRC beam

The dynamic tests were performed in seven different steps by relaxing the tendons one by one according to the following order: C1, B2, A1, A2, C2 and B1 in order to simulate the service life of the beam. After each relaxation, the beam was excited with a falling weight and the dynamical response was recorded. The first three frequencies were measured and are used in the updating procedures. In Figure 7.34 the stiffness decrement related to the tendon relaxations is clearly readable in the first frequency variation.

The FEMU procedures are performed by using OpenSees and the Penalty Method. The structural model is shown in Figure 7.35. The results obtained are compared to a previous investigation of the same problem carried out by using the software FEMTools (Solutions DD Undated). The FEMTools application has been described by Savov and Wenzel (2004). Therefore only the most important results are reported here.

According to the beam second-order theory (Equation 32), when the elastic modulus is constant, an external compressive axial force should, in general, decrease the stiffness and therefore the frequencies. In the beam investigated, we have the opposite effect: this can be explained by considering that a decrement in the pre-stressing force causes a decrement in the elastic modulus. The two combined effects lead to a stiffness (and frequency) decrement.

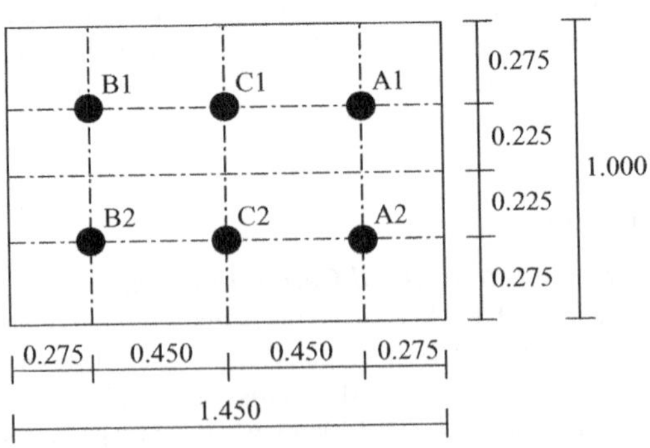

Figure 7.33 The PRC beam cross-section

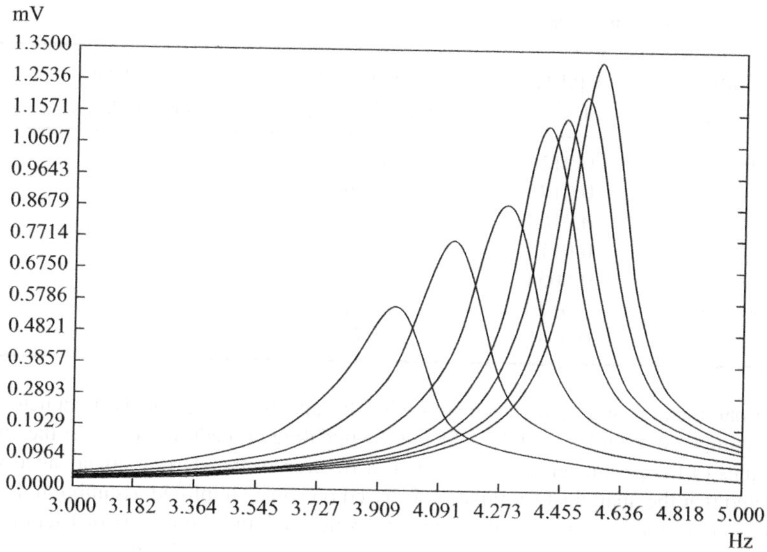

Figure 7.34 Variation of the first frequency as a function of the axial force

The beam is modeled within OpenSees by 60 beam elements taking the eccentricity of the pre-stressing axial force into account. A horizontal spring is inserted at each support to simulate friction effects. In order to evaluate their stiffness, a preliminary updating analysis is performed considering the case 0 (all tendons active) as undamaged. A value of $K = 918\,850\,\text{kN/m}$ is obtained.

The procedures are performed with the Penalty Method. Two different updating analyses are performed: in the first the same elastic modulus is taken for the whole beam (one parameter analysis), whereas in the second a different elastic modulus for each element (60-parameter analysis) is used. This means that the damage due to increased cracking after each relaxation is simulated through a reduction of the concrete elastic modulus. The results are compared in the former case with the analytical solution of the Euler beam and in the latter case with a previous numerical study (Savov and Wenzel 2004).

According to the Euler theory the global elastic modulus can be written as (Clough and Penzien 1993)

$$E = \frac{1}{J\lambda_i^2}\left[m\left(\frac{2\pi f_i}{\lambda_i}\right)^2 + N\right] \tag{32}$$

where m is the mass per unit length, f_i is the ith frequency, N is the pre-stressing force, λ_i is the ith coefficient function of the boundary conditions (tuned on case 0) and J is the moment of inertia. The results for the one-parameter analysis are reported in Table 7.1; the initial (calculated with the starting

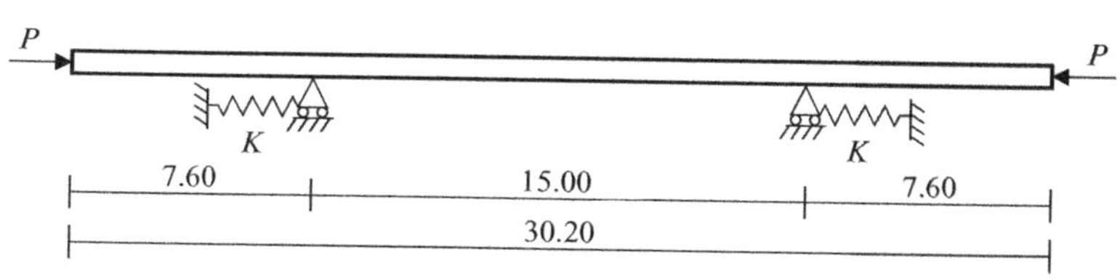

Figure 7.35 Bilfinger Berger PRC beam geometry (dimensions in meters)

Table 7.1 One-parameter analysis results

Case	Initial CC_{abs}	Updated CC_{abs}	Analytical modulus E_{anal}	Numerical modulus E_{num}	$\frac{E_{num}-E_{anal}}{E_{anal}}$ (%)
0	0.0130	0.0126	1	0.97	−2.79
1	0.0325	0.0173	0.96	0.93	−3.12
2	0.0438	0.0164	0.94	0.91	−3.32
3	0.0637	0.0115	0.91	0.87	−4.14
4	0.0853	0.0054	0.88	0.84	−4.99
5	0.1280	0.0120	0.81	0.77	−5.21
6	0.1919	0.0045	0.73	0.68	−6.62

values of the parameters) and final (calculated with the updated values of the parameters) values of CC_{abs} can be compared to understand the effect of the updating procedure. Moreover the analytical and numerical values of the parameter E are written as a fraction of the starting, undamaged value. It has to be considered that the analytical formulation is based on some simplified assumption. In fact, it does not take into account the eccentricity of the pre-stressing load and the effects of friction at supports. Therefore, it is included here only as an indication and not as an exact reference.

The 60-parameter analysis is quite time consuming as an average time of about 1500 s is required on a Celeron 2.80 GHz computer to perform a single case. Therefore the initial sensitivity matrix is used to eliminate the computational cost of recalculating it at each iteration. In this way the mean computational time is reduced to about 50 s for each case, which is 1/30 of the previous case. The difference between outcomes from the two methods (updating or not updating the sensitivity matrix) is negligible.

Since the experimental data are frequencies only, the damage distribution is symmetrical. Thus only the half beam is shown in Figure 7.36. The damage distribution can be compared with those obtained by the FE code FEMTools (Solutions DD Undated) as depicted in Figure 7.37. Moreover, the results attained on frequencies are compared with those obtained by FEMTools in the already mentioned previous study (Table 7.2). By comparing the results from the two different analyses, some indications can be obtained. The results are very good in all the cases. The experimental frequencies are captured with very high accuracy as can be immediately noted from the CC_{abs} values. A very similar damage distribution is obtained by means of FEMtools and VCUPDATE. In both cases, the damage is concentrated mainly at the supports and secondarily at the midspan.

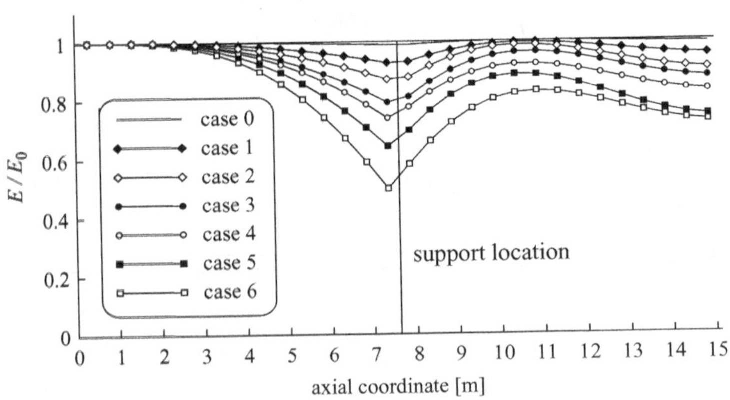

Figure 7.36 Damage distribution for PRC beam (half model)

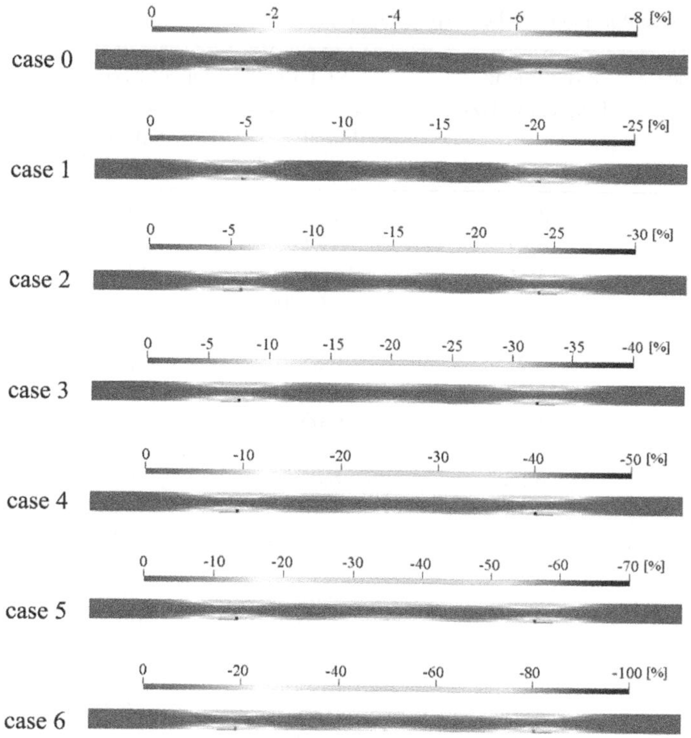

Figure 7.37 Results using FEMtools in terms of elastic modulus with respect to the initial value

A Pre-Stressed Reinforced Concrete Beam

A pre-stressed reinforced concrete beam tested in the laboratory is investigated by VCUPDATE. In general, the damage in pre-stressed elements is difficult to detect since, after the load is removed, the cracks tend to close and this leads to relatively small changes in modal data. The experimental data as well as the test description is taken from a previous study (Teughels 2003) The results obtained by VCUPDATE are compared, in terms of damage distribution, with those obtained in that study. The investigated beam is 17.60 m long and 0.8 m high. Along the member the cross-section has an I-shape, while at the ends it has a solid square section (Figure 7.38). The beam is pre-stressed by an inclined post-tensioned cable.

The beam was artificially damaged by increasing static load levels in the configuration shown in Figure 7.39a. The external load Q is applied in six different steps with an increasing amplitude of 45, 67.5, 125, 140, 150 and 154 kN respectively. The first visible crack on the beam corresponds to $Q = 67.5$ kN while the reinforcement yielding is achieved for $Q = 150$ kN. After each step, the beam was dynamically tested in a free–free configuration (Figure 7.39b). The structure is excited by the impact of a falling weight at its first edge. During the dynamic test procedures, 74 sensors were placed on the structure at both longitudinal upper sides of the beam, with a distance of 0.5 m between them. In most load steps the dynamic behavior of the beam was found to be pure flexural with small torsional components. Therefore, the values from two sensors placed at the same longitudinal distance are averaged in order to obtain 37 values only. In this way, the torsional components of the mode shapes are ignored.

From the experimental data it can be observed that before the steel yielding, the changes in frequencies and mode shapes are very small. This is probably because, after the load removal, the cracks tend to close and they produce a slight effect on the dynamic behavior. It can be observed, however, that the

Table 7.2 Result comparison for the 60-parameter analysis

Case	Mode	Experimental frequency [Hz]	VCUPDATE (OpenSees) Initial frequency [Hz]	VCUPDATE (OpenSees) Updated frequency [Hz]	FEMtools Updated frequency [Hz]
0	1	4.57	4.56	4.56	4.53
	2	6.82	6.84	6.82	6.90
	3	13.61	14.09	14.04	13.68
		CC_{abs}	0.0130	0.0115	0.0085
1	1	4.52	4.58	4.52	4.47
	2	6.72	6.85	6.72	6.76
	3	13.22	14.09	13.78	13.37
		CC_{abs}	0.0325	0.0144	0.0095
2	1	4.46	4.59	4.47	4.44
	2	6.68	6.86	6.62	6.70
	3	13.10	14.10	13.50	13.25
		CC_{abs}	0.0438	0.0138	0.0063
3	1	4.41	4.60	4.41	4.39
	2	6.49	6.87	6.46	6.56
	3	12.94	14.10	13.17	13.01
		CC_{abs}	0.0637	0.0074	0.0069
4	1	4.29	4.61	4.34	4.32
	2	6.37	6.88	6.34	6.45
	3	12.81	14.10	12.88	12.79
		CC_{abs}	0.0853	0.0073	0.0070
5	1	4.13	4.62	4.22	4.18
	2	6.19	6.89	6.12	6.21
	3	12.25	14.11	12.33	12.24
		CC_{abs}	0.1280	0.0132	0.0054
6	1	3.97	4.64	4.10	4.03
	2	5.73	6.90	5.76	5.82
	3	11.72	14.11	11.69	11.61
		CC_{abs}	0.1919	0.0131	0.0134

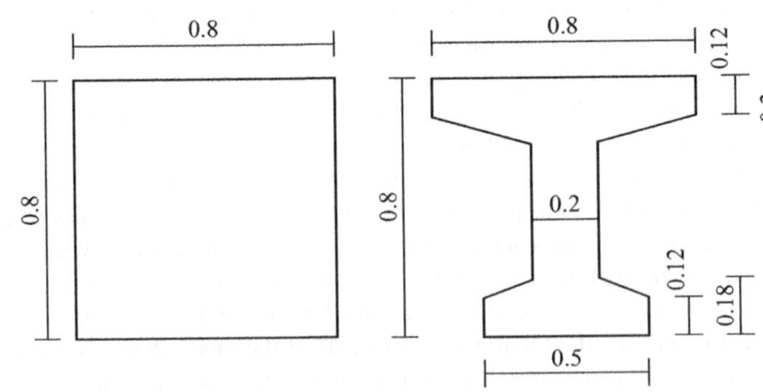

Figure 7.38 Cross-sectional sketches (dimensions in meters)

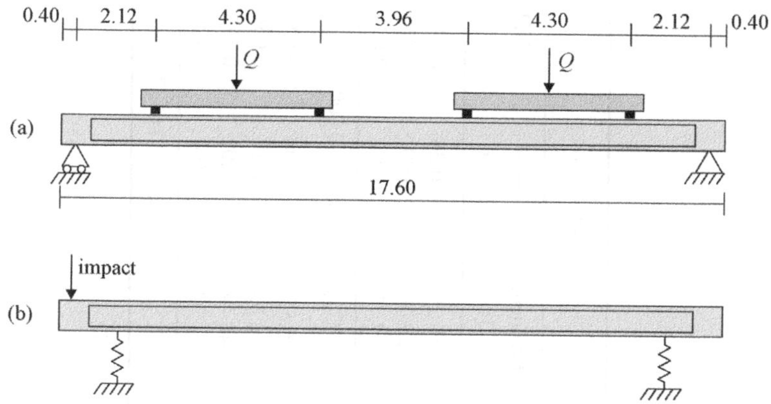

Figure 7.39 Static and dynamic test configuration (dimensions in meters)

damage detection procedures are more important when the structure is subjected to an irrecoverable excitation.

The beam is modeled in OpenSees by means of a simple Beam model. Thirty-six uniaxial beam elements with five integration points and three degrees of freedom (DOF) for each node are used. In order to simulate the free–free boundary conditions, two springs with very low stiffness are placed at the supports. The reinforced concrete is modeled as homogeneous and isotropic. The section properties are: area $A = 0.301\,\text{m}^2$ and $0.64\,\text{m}^2$, moment of inertia $J = 0.02246\,\text{m}^4$ and $0.03413\,\text{m}^4$ for the I- and the rectangular shape respectively. Adopting a mass per unit volume $M = 2500\,\text{kg/m}^3$, a mass per unit length $\text{m} = 752.5\,\text{kg/m}$ and $1600\,\text{kg/m}$ is obtained for the two sections respectively. The value of the elastic modulus for each element is used as the updating parameter with a starting value of $E_0 = 37\,500\,\text{MPa}$. Therefore, a total number of 36 parameters is used.

Two different updating procedures are performed: with the first one, the initial FE model is updated to the undamaged state. Then, this undamaged model is updated to the damaged state corresponding to the post reinforcement yielding step ($Q = 154\,\text{kN}$). The first four frequencies and flexural mode shapes are used in the second updating process while, due to the poor quality, the fourth mode shape is not used in the first updating process.

The outcomes from the numerical procedures in terms of frequencies and mode shapes correlation are reported in Table 7.3. Since the mode shapes are highly correlated, the NMD is used instead of the MAC. The effect of the updating procedures is smaller for the first step. This means that the initial model fits the undamaged one quite well, both in terms of frequencies and mode shapes. In the second step, where the beam is damaged, the effect of the updating procedures is much higher as can be seen by comparing the CC_{abs} values. The largest frequency deviation is recorded in the first mode shape, probably due to the presence of a torsional component in the experimental data.

The effect of the updating algorithm on the mode shapes also can be clearly seen in Figure 7.40 where the outcomes of the second numerical process are shown. It can be noted that, starting from the undamaged state, the mode shapes are transformed to the damaged state, which is closer to the experimental results.

The damage distribution is reported in Figure 7.41 in terms of the ratio between the elastic modulus values before and after the updating process. In the same figure, the VCUPDATE outcomes are compared with the distribution taken from the previous study (Teughels 2003). It should be noted that even if the static load is symmetrical, the corresponding damage distribution is not symmetrical. In this case, the mode shapes are fundamental for the damage location.

Table 7.3 Updating procedure results for the OpenSees analysis

Step	Mode	Experimental frequency f_0[Hz]	Before update			After update		
			Frequency f[Hz]	$\frac{f_i - f_{0i}}{f_{0i}}$ (%)	NMD	Frequency f[Hz]	$\frac{f_i - f_{0i}}{f_{0i}}$ (%)	NMD
Initial-to-undamaged	1	10.32	10.54	2.11	0.049	10.29	−0.25	0.049
	2	29.49	29.91	1.41	0.032	29.49	0.00	0.038
	3	57.77	59.84	3.59	0.039	57.77	0.01	0.025
			CC_{abs}	0.0237		CC_{abs}	0.0009	
Undamaged-to-damaged	1	6.83	10.29	50.71	0.082	6.55	−4.08	0.027
	2	22.66	29.49	30.14	0.185	22.63	−0.12	0.056
	3	48.45	57.77	19.25	0.209	48.44	−0.02	0.179
	4	70.99	100.48	41.55	0.395	71.03	0.05	0.120
			CC_{abs}	0.3541		CC_{abs}	0.0107	

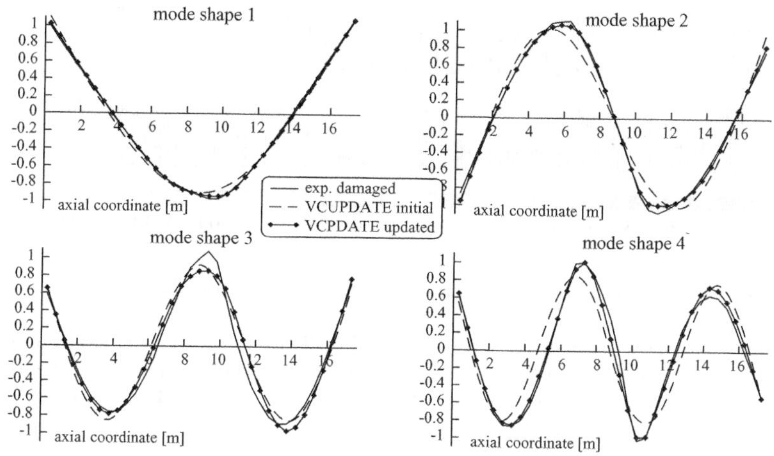

Figure 7.40 Experimental and numerical mode shapes for the undamaged to damaged step

From the undamaged to damaged step graph (Figure 7.41b) four damaged zones can be recognized: two near the beam center with a higher level of damage and two closer to the beam ends with a lower damage level. The position of these points corresponds to the loading points in the static test. The two main plastic hinges are situated close to the beam center where the reinforcements yielded. Between the loading points, no damage is revealed, probably because between the plastic hinges the steel does not yield and therefore the cracks tend again to close after the load removal.

The same structure is investigated by VCUPDATE by means of an ANSYS model. A single step analysis is considered: the starting model is directly updated to the damaged beam. This is, in fact, the standard way to assess the existing structures when no test data of the initial state are available. This version of the code provides the user with an increased modeling power. In fact, the FE mesh can be freely created and there is no necessity to place a node of the FE model in the same location of a sensor. In this way, the best FE model can be chosen without taking care of the sensor position. The code requires the position and the modal displacements for each sensor. Subsequently, the numerical mode shapes are computed in the sensor positions as a function of the mode shapes of the entire FE model.

Both a beam and a plane model are used. The first model contains the BEAM3 element: a uniaxial element with three DOF for each of its two nodes. Furthermore, the plane model is based on the PLANE42 two-dimensional element, which has four nodes with two DOF respectively.

In Figure 7.42 the analysis outcomes in terms of elastic modulus reduction for the beam model with two different meshes (0.4 and 0.2 m) are shown. Analogously, for the plane model, the results are shown

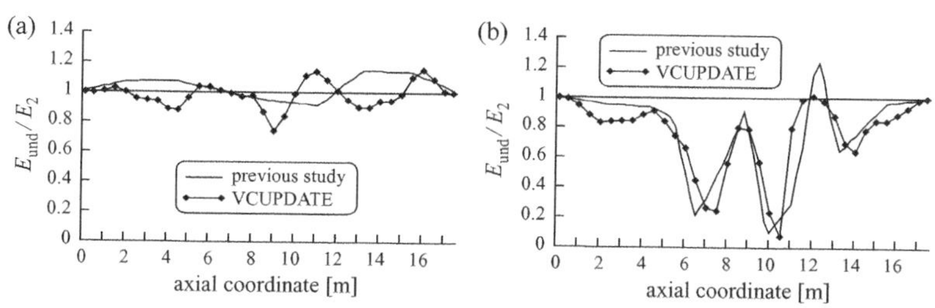

Figure 7.41 Damage distribution for (a) the initial to undamaged step and (b) the undamaged to damaged step

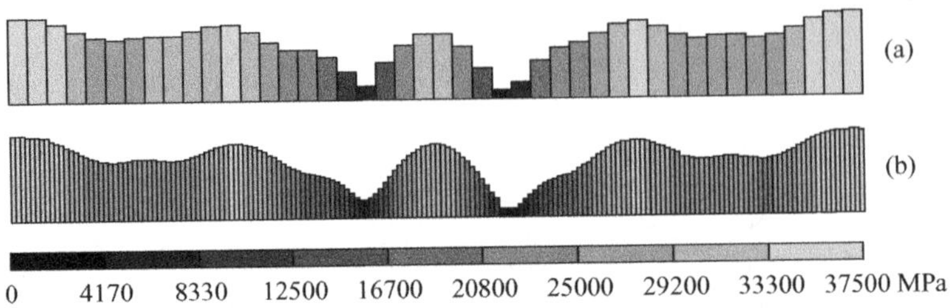

Figure 7.42 Elastic modulus distribution for the beam model with element sizes 0.4 and 0.1 m

in Figure 7.43 for the same element sizes of 0.4 and 0.2 m. The damage distribution of these figures can be compared with the one from the OpenSees analysis (Figure 7.41). It can be noted that the agreement is very close. The results in terms of frequencies, NMD values and mode shapes are very good but they cannot be extensively reported here due to lack of space.

Several different meshes are used in the calculations for both the beam and the plane analyses. In general, it can be noted that the outcomes are not dependent on the mesh used as long as the mesh can effectively describe the structural response.

This application shows the purposes of FEMU applied to SHM: by combining eigenfrequencies and mode shapes the damage detection procedure is fully achieved. In this respect, it is important to emphasize that even if the structure configuration was symmetrical, it was found that the damage distribution was not symmetrical, and therefore the use of mode shapes is fundamental to localize the damage. The experimental structural behavior could be captured with very high accuracy by using different FE codes, different model dimensionality and different meshes.

Reinforced Concrete Beam

A reinforced concrete beam tested in the laboratory is investigated. The beam is part of an extensive laboratory program carried out on five identical beams with different damage locations (Maeck 2003). The beam indicated in the experimental program was number four. The beam was artificially damaged by applying a vertical static load at $x = 4$ m in the three-point bending configuration shown in Figure 7.44a. Then, the modal test was performed in a free–free boundary condition configuration (Figure 7.44b). This means that the beam supports were very flexible allowing free modal displacements. This procedure was repeated for different load levels: in this work the experimental data related to step number five are considered. The corresponding imposed load is 25 kN.

The structural properties are: length $L = 6$ m, cross-sectional area $A = 0.05$ m^2, moment of inertia $J = 1.93 \times 10^{-4}$ m^4 (considering the steel contribution, Figure 7.45) and elastic modulus $E = 37\,500$ MPa.

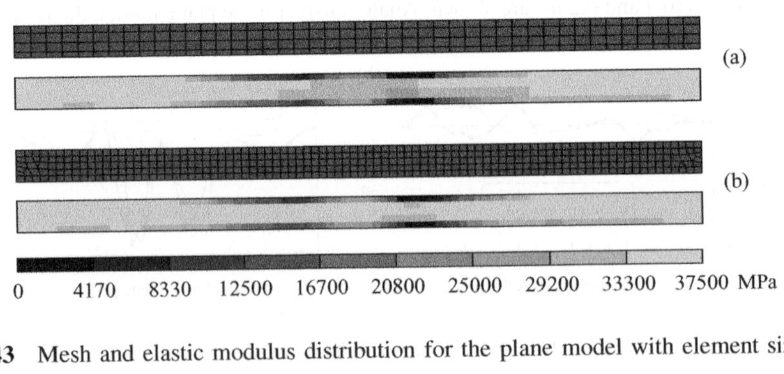

Figure 7.43 Mesh and elastic modulus distribution for the plane model with element sizes 0.4 and 0.2 m

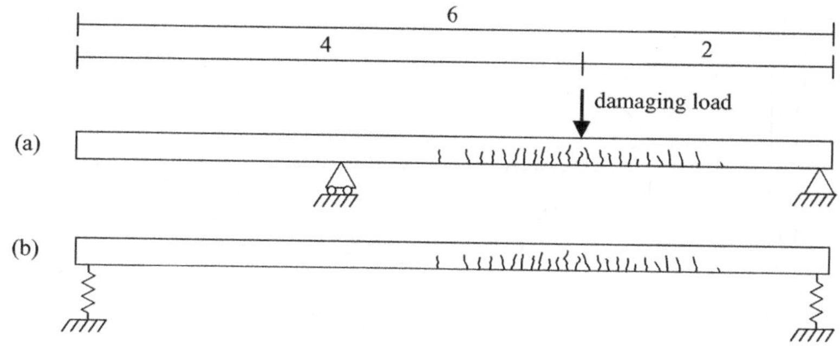

Figure 7.44 (a) Static and (b) dynamic test configuration with experimental crack pattern (dimensions in meters)

To record the dynamics outcomes, 31 accelerometers are placed at both sides of the upper surface of the beam, for a total number of 62 sensors. The values from each couple of sensors are averaged to obtain only 31 values. The first four frequencies and mode shapes are recorded. Due to the damage, a reduction in frequencies and significant changes in mode shapes are observed. In the first mode, a slight torsional component is recorded but it cannot be observed in the modes shapes as the average value from the sensors with the same axial coordinate is reported.

The beam is modeled in OpenSees using 30 two-dimensional beam elements and the elastic modulus for each element is taken as the parameter. In order to reproduce the experimental conditions, the beam is supported by very flexible springs. This leads to the first four mode shapes shown in Figure 7.46, where there are no points with zero displacement common for all the modes.

Two different updating processes are performed with the Penalty Method, namely the initial to undamaged and the undamaged to damaged steps. In the first, the initial FE model is updated to the undamaged beam in order to eliminate the initial modeling errors. In the second, this undamaged model is updated to the damaged state. A value of 0.0005 for the convergence limit is used. The convergence ratio is very fast: for the first updating process, two iterations only are necessary, whereas for the second the convergence is achieved in three iterations only.

The results in terms of frequencies and NMD are reported in Table 7.4. The improvement in frequency values is very high and the FE model can be taken as very close to the experimental model. This also can be seen from the updated CC_{abs} values, which are almost zero.

The improvement of the updating procedures on mode shapes can be seen in Figures 7.47 and Figure 7.48. The correction in the mode shapes is evident. The worst case is the first mode, probably due to the torsional component being neglected. It can be concluded, therefore, that VCUPDATE is able to bring the numerical model to be almost coincident with the real structure in terms of dynamic behavior.

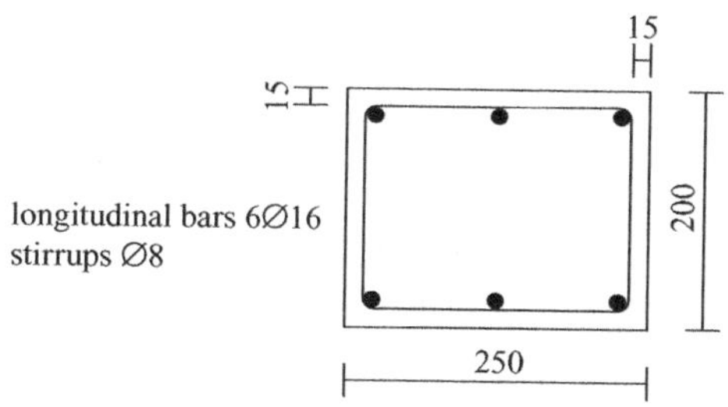

Figure 7.45 Beam cross-section (dimensions in millimeters)

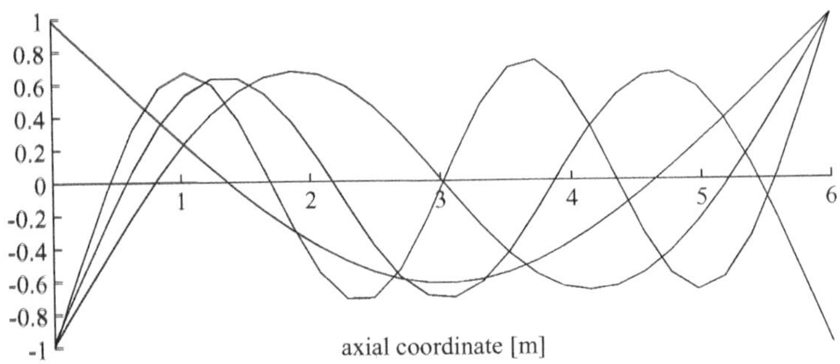

Figure 7.46 First four mode shapes

The comparison between the elastic modulus distribution obtained by VCUPDATE and a previous study (Teughels 2003) is reported in Figure 7.49. This is clearly strictly related to the damage since the loss of stiffness due to the progressive external load can be properly modeled as an elastic modulus decrement. This damage distribution also can be directly compared to the experimentally obtained crack pattern reported in Figure 7.44. The damage can be successfully located with a smooth distribution. The initial to undamaged step is characterized by a reduction of E in the central part of the beam, probably due to the crack pattern induced by the self weight. In contrast, in the undamaged to damaged step, the maximum change is located close to the applied load, as can be seen from Figure 7.44. A parameter decrement can be recognized near the first support ($x = 2$ m): this could be due to the effect of the beam self weight.

The damage distribution provided by VCUPDATE is in good agreement with the previous study where a more sophisticated algorithm was implemented (Teughels et al. 2002): the FE properties are not corrected separately, but a damage function obtained combining seven triangular shape functions is used. This procedure should avoid a nonrealistic stiffness pattern with many peaks instead of a smooth distribution and other numerical problems. Moreover, the optimization problem has only seven variables and this should lead to a saving in computational time. However, the VCUPDATE procedures are also very effective since the solution is achieved in a few iterations.

The same structure is investigated by VCUPDATE by means an ANSYS model. In this case, a single step analysis is considered: the starting model is directly updated to the damaged beam. In the practical assessment of safety levels for SHM purposes, in fact, this is the standard way to evaluate the existing structures when no previous test data are available.

The use of VCUPDATE in conjunction with ANSYS provides the user with increased modeling capabilities. The FE mesh can be freely chosen and there is no necessity to place a node of the FE model in the same location of a sensor where the eigendata are recorded. In this way, the best FE model can be chosen without taking care of the sensor position. The code requires the position of each sensor and, subsequently, the numerical mode shapes are computed in the sensor positions as a function of the mode shapes of the entire FE model.

Both a beam and a plane model are used. The first contains the two-node, two-dimensional elastic beam BEAM3 element. It is a uniaxial element with tension, compression, and bending capabilities, with three degrees of freedom (two translations and one rotation) for each node. The plane model is based on the four-node, two-dimensional structural solid PLANE42 element. It is a plane element used in plane stress case with two translational degrees of freedom for each node. For both analyses, the springs are modeled with the spring-damper COMBIN14 element. In order to test the mesh independence of VCUPDATE, several different mesh refinements are used in the calculations, for both the beam and the plane models.

In Figure 7.50, the analysis outcomes in terms of elastic modulus distribution for the beam model with two different meshes (0.1 and 0.05 m) are shown. Analogously, for the plane model, the results are shown

Table 7.4 Updating procedure results for the OpenSees analysis

Step	Mode	Experimental frequency f_0[Hz]	Before update			After update		
			Frequency f[Hz]	$\frac{f_i - f_{0i}}{f_{0i}}$ (%)	*NMD*	Frequency f[Hz]	$\frac{f_i - f_{0i}}{f_{0i}}$ (%)	*NMD*
Initial-to-undamaged	1	22.02	23.72	7.72	0.057	22.03	0.05	0.054
	2	63.44	65.20	2.78	0.035	63.43	−0.01	0.030
	3	123.27	127.51	3.44	0.060	123.26	−0.01	0.030
	4	201.92	210.25	4.13	0.084	201.91	0.00	0.048
			CC_{abs}	0.0451		CC_{abs}	0.0002	
Undamaged-to-damaged	1	19.35	22.03	13.86	0.094	19.35	0.02	0.065
	2	56.90	63.43	11.48	0.104	56.89	−0.03	0.043
	3	111.64	123.26	10.41	0.130	111.62	−0.02	0.031
	4	185.22	201.91	9.01	0.171	185.18	−0.02	0.038
			CC_{abs}	0.1119		CC_{abs}	0.0002	

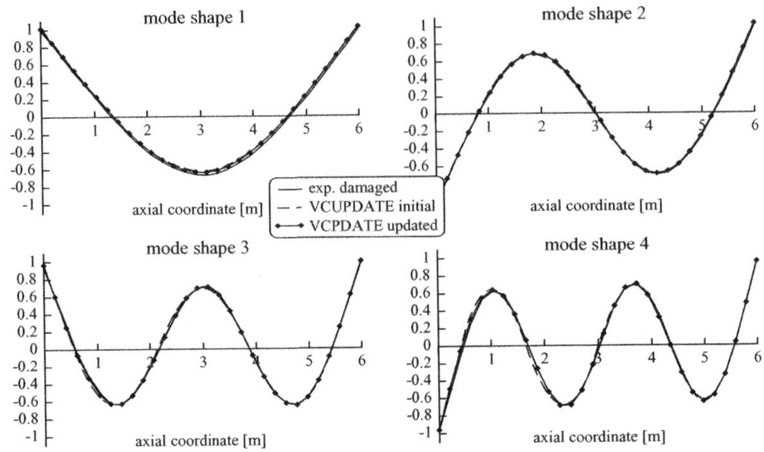

Figure 7.47 Experimental and numerical mode shapes for the initial to undamaged step

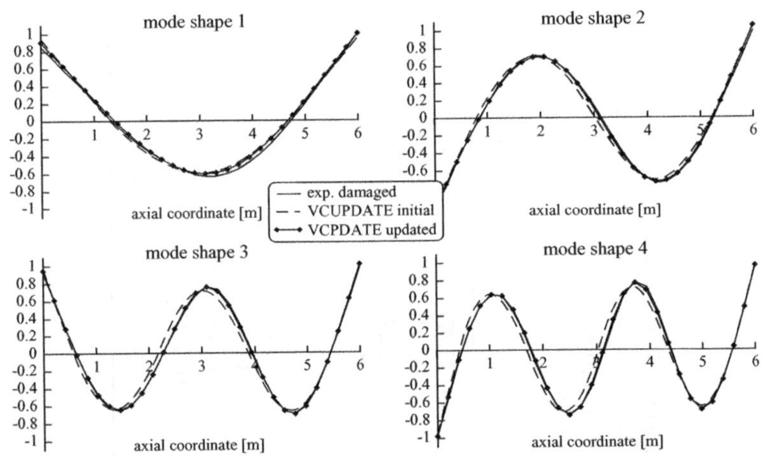

Figure 7.48 Experimental and numerical mode shapes for the undamaged to damaged step

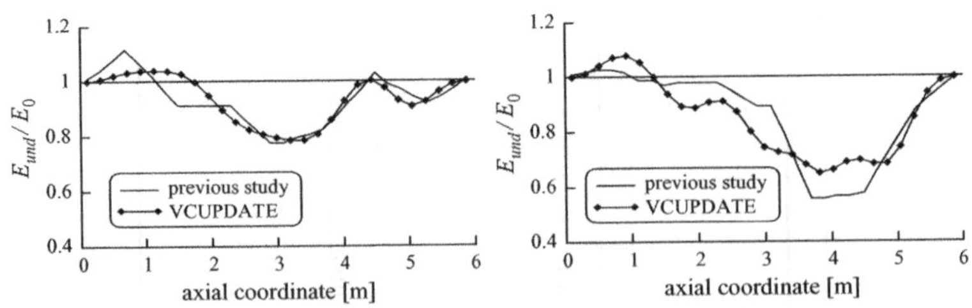

Figure 7.49 Mesh and elastic modulus distribution for the plane model with element size 0.4 and 0.2 m

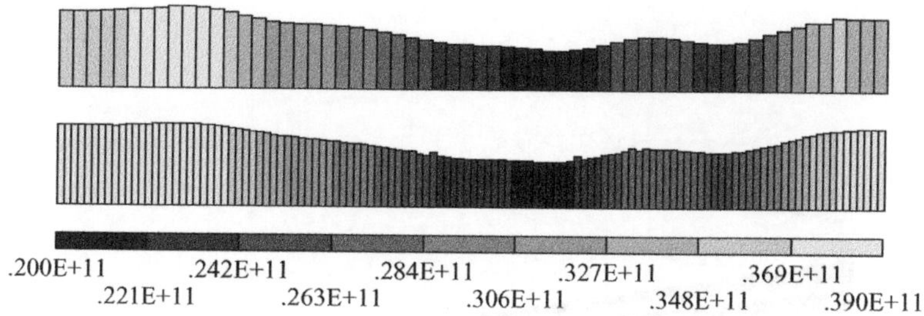

Figure 7.50 Elastic modulus distribution for the beam model with element sizes 0.1 and 0.05 m

in Figure 7.51 for element sizes of 0.1 and 0.05 m respectively. The damage distribution of these figures can be compared with that from the OpenSees analysis (Figure 7.49). It can be noted that the agreement is, in general, very satisfactory. The results in terms of eigenfrequencies, *NMD* values and mode shapes are also very good but they cannot be extensively reported here due to lack of space.

In general, both models and all the meshes capture the physical structural behavior very well. Moreover, it can be noted from the figures that mesh size has a negligible influence on the outcomes. Obviously, it is always required that the adopted mesh can effectively describe the structural response.

In Figure 7.50 it can be noted that the left part of the beam presents an elastic modulus higher than that with which it started. This is, in principle, not possible, but the same phenomenon is revealed also in the OpenSees calculation and in the previous example (Figure 7.49). Therefore, it is possible that this discrepancy is due to the experimental data or to a real nonhomogeneous distribution of the structural stiffness, owing, for example, to a locally different concrete-drying process. Furthermore, it is clear that the effect is less noticeable in the beam model with the refined mesh. This is probably due to the fact that the fine mesh can better describe the structural behavior. Moreover, the higher the number of elements, the higher the number of parameters in the updating procedure and therefore the better is the possibility to reach the minimum of the objective function. In the plane model (Figure 7.51), the same problem is even more pronounced. But, again, in the case of more refined mesh this phenomenon is limited to a very few elements.

In conclusion, it has been shown that by combining frequencies and mode shapes it is possible to accurately locate and quantify the damage by using different FE codes.

7.1.3.6 Applications to Cables

Lanaye Bridge Cables

The Lanaye Bridge is a cable stayed bridge situated in Belgium, between the cities of Liege and Maastricht (Figure 7.52). The bridge has a single pylon and a 177 m long main span, which is balanced by a

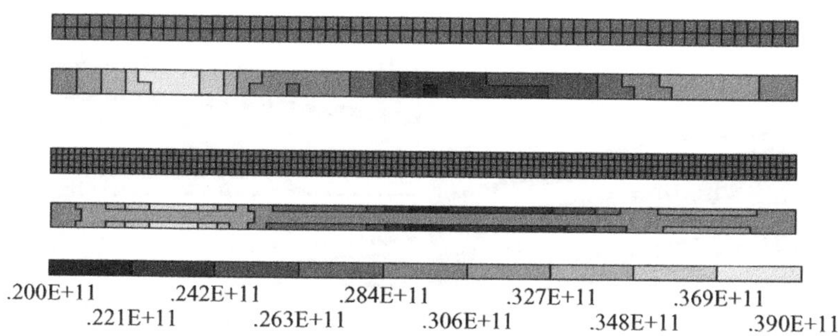

Figure 7.51 Mesh and elastic modulus distribution for the plane model with element sizes 0.01 and 0.05 m

Figure 7.52 Lanaye Bridge

counterweight abutment anchored to the ground. There are 20 fan-shaped cables connected to the main span with internal forces ranging between 2750 and 5400 kN. Ten cables are located on the counterweight side with an internal force of about 8800 kN. In the following, the two ends of the bridge are referred to by the name of the city they face. The cable pairs are numbered from 1 to 15 starting from the end furthest away from the pylon.

The bridge has been investigated several times by monitoring tests because it has some damaged cables that require frequent study and verification. For these reasons, this structure was included in the Integrated Monitoring and Assessment of Cables (IMAC) program under the Fifth European Framework Program (Consortium I Undated). Within this research program, an extensive experimental program was performed on all the bridge cables measuring their dynamical behavior in terms of eigenfrequencies. Three-dimensional sensors recording the acceleration were used. Subsequently, by employing the Peak Picking method, the eigenfrequencies of the cables were evaluated.

Three cables are investigated by VCUPDATE: cable M1 is known to be damaged by corrosion (Figure 7.53) and it was surrounded by a security structure to catch it if it collapsed. After a visual inspection of cable L1, a damaged support system was suspected. Therefore, changes in the dynamical behavior were expected, but in the test results no significant changes were observed, indicating that the cable is not damaged and can carry its full load. Cable L2 is undamaged and presents a standard behavior.

The cables were modeled with 60 beam elements, which can be adequately used to model stay cables (Mordini et al. Undated) (Figure 7.54). To eliminate the effects of sag and of self weight, the out-of-

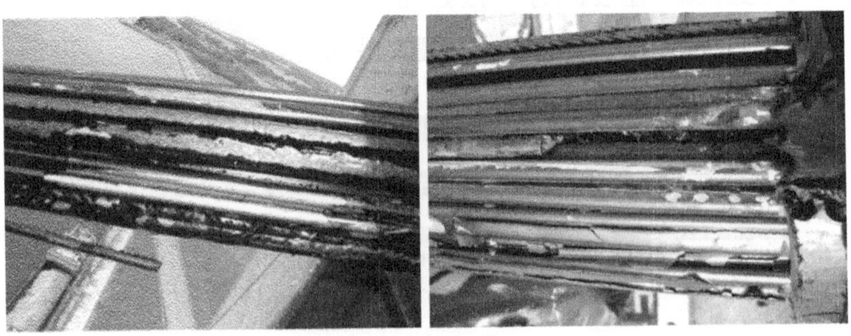

Figure 7.53 Damage on cable M1

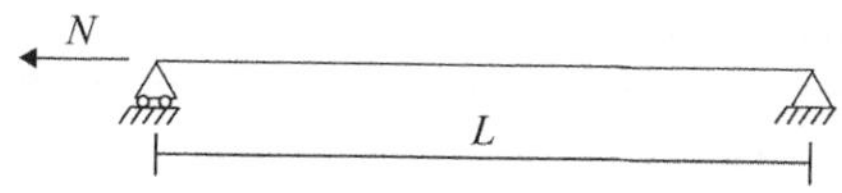

Figure 7.54 Cable model with hinged ends

Table 7.5 Cable properties and analysis results

Cable	Length [m]	Cable properties			Analysis results	
		Mass per unit length m [kg/m]	Moment of inertia J [mm^4]	Design force N_D [kN]	$N_{VCUPDATE}$ [kN]	$\frac{N_{VCUPDATE}-N_D}{N_D}$ (%)
M1	165.07	43.80	3 927 262	2668	1944	−27.15
L1	165.07	43.80	2 000 850	2668	2826	5.93
L2	150.44	78.20	7 125 824	5042	5079	0.74

plane frequencies were used. Different analyses were carried out using the first, the first five and the first ten experimental frequencies. The axial force (N) was used as the updating parameter with a value of 5×10^{-4} for the convergence limit.

The computation of the moment of inertia (J) of a cable is not a straightforward issue. In this investigation, the bending stiffness (elastic modulus (E) and J) was obtained from the software VCE_Kabel based on the measured frequencies. The code provides, as a result, the axial force and the bending stiffness. Because cable M1 is damaged, a different value of J is obtained than for cable L1, as reported in Table 7.5. This is an unphysical assumption since the two cables should be equal. In order to evaluate the effect of this imprecision, a preliminary sensitivity analysis was performed and, in the investigated cases, very little relevance was found for the bending stiffness parameters (Figure 7.55). This means that the assumption has little influence on the results.

The damage in cable M1 is clearly detected by a strong axial force (N) reduction (Table 7.5). Cable L1 presents a slight increment in N that could be due to the lower N in cable M1 and to the consequent stress

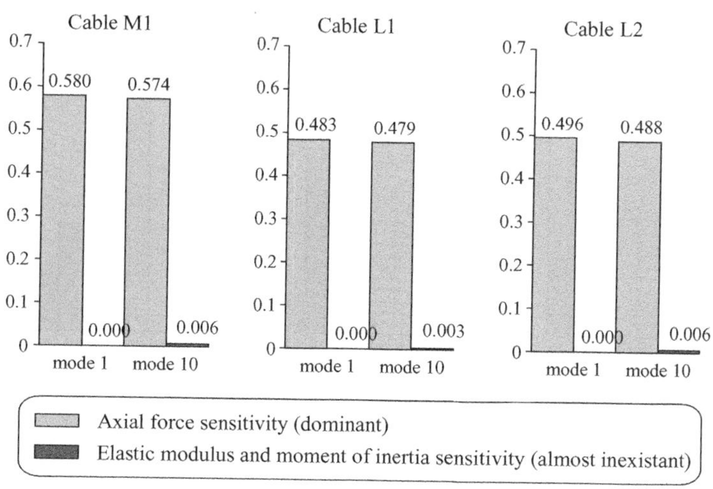

Figure 7.55 Lanaye Bridge cable structural sensitivity

Table 7.6 Analysis result comparison

Cable	Design force N_D [kN]	N_{VCE_Kabel} [kN]	$\frac{N_{VCE\_Kabel}-N_D}{N_D}$ (%)	$N_{VCUPDATE}$ [kN]	$\frac{N_{VCUPDATE}-N_D}{N_D}$ (%)
M1	2668	1895	−28.98	1944	−27.15
L1	2668	2795	3.62	2826	5.93
L2	5042	4948	−1.86	5079	0.74

redistribution, since M1 is the corresponding cable to L1 on the other side of the bridge. The increment in N is a signal of the cable integrity. Cable L2 is also undamaged as its N value is almost equal to the design force.

The results obtained from VCUPDATE are compared, in terms of N, with those from the software VCE_Kabel based on the standard method of estimating cable force from frequencies (Table 7.6). The agreement between the two codes is very close, leading to a maximum discrepancy of 2.6%.

In addition, from Figure 7.56, it is evident that the results are not influenced by the number of frequencies used, i.e. using the first, the first five or the first ten frequencies. The percentages indicating the difference to the design force provide illustrate good measurement quality in the field of higher frequencies. Figure 7.56 also shows the number of iterations necessary to obtain the convergence and, as a general trend, the updating procedure is faster using more experimental data.

By using eigenfrequencies only, three stay cables from Lanaye Bridge in Belgium have been investigated. One of the cables presented showed damage, which was detected by a considerable reduction in the value of N. Moreover, in order to evaluate the influence of the unknown structural properties, a sensitivity analysis was performed that indicated for the cases investigated the effect of bending stiffness was negligible (Figure 7.55). This outcome is very important in practical engineering applications since cable properties are in general not well known. By investigating two equal cables, but one with damage and the other undamaged, it was demonstrated that even with a low confidence in the knowledge of the structural parameters it is possible to detect the damage through a reduction in N.

Rosenbrücke Cable

In order to better understand the effect of bending stiffness, a shorter cable from Rosenbrücke over the Danube River in Tulln, Austria, is examined. The bridge (Figure 7.57) is a typical representative of

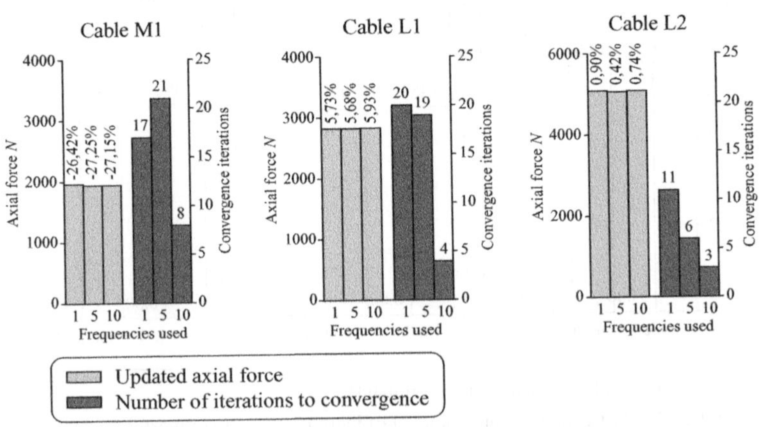

Figure 7.56 Lanaye Bridge cable analysis results as a function of the number of frequencies used

Figure 7.57 Rosenbrücke crossing the Danube at Tulln

modern cable-stayed bridges, with 60 single-strand cables applied. This structure, completed in 1995, is interesting because of the several cable tests performed over a period of five years. Furthermore, detailed studies regarding the vibration sensitivity of individual cables were performed (Consortium I Undated).

The investigated cable has the following properties: length $L = 44.27\,\text{m}$, elastic modulus $E = 205\,000\,\text{MPa}$, cross-sectional area $A = 6.88 \times 10^{-3}\,\text{m}^2$, moment of inertia $J = 3.77 \times 10^{-6}\,\text{m}^4$, linear mass $m_L = 59.89\,\text{kg/m}$ and design axial force $N = 1957\,\text{kN}$. In order to evaluate the effect of bending stiffness and of boundary conditions, four different analyses are performed varying the number of parameters and the support conditions. Two different types of boundary conditions are taken into account, namely hinged and clamped ends (Figures 7.54 and 7.58 respectively). The first eight eigenfrequencies are used as experimental data.

The updating procedure results are reported in Table 7.7. By updating E and J it is possible to reach a more accurate solution, as can be seen from the CC_{abs} values. Obviously, in the case of clamped ends the cable has greater stiffness because the changes in parameters are smaller.

In Figure 7.59 the structural sensitivity to the parameters for cases 2 and 4 is reported. These values can be compared with those obtained for the (longer) Lanaye Bridge cables (Figure 7.55): the sensitivity matrices clearly demonstrate important differences in the bending stiffness of long and short cables.

As an effect of this difference in sensitivity, it is observed that the convergence ratio for E and J is much slower than for N (see Figure 7.60). Hence, in cases 2 and 4, according to Table 7.7, the maximum number of iterations is increased to allow E and J to reach a satisfactory precision.

In Table 7.8, the numerical results for the axial force (N) obtained by both VCUPDATE and VCE_Kabel are compared for all cases. Evaluation of these results enables the best boundary conditions to be chosen:

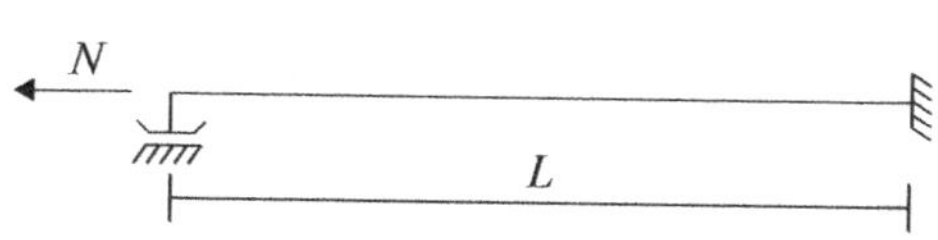

Figure 7.58 Cable model with clamped ends

Table 7.7 Cable analysis results for Rosenbrücke

Case	Parameters	Cable ends	Initial CC_{abs}	Updated CC_{abs}	Normalized updated N	Normalized updated E, J
1	N	Hinged	0.0555	0.0037	1.125	–
2	N, E, J	Hinged	0.0555	0.0009	1.117	1.100
3	N	Clamped	0.0293	0.0028	1.065	–
4	N, E, J	Clamped	0.0293	0.0009	1.055	1.067

both codes indicate that the model with clamped ends reproduces the connection better between the cable and the anchorage. However, it has to be noted that the model with hinged ends also is capable of reproducing the modal response, but with larger (and maybe unphysical) changes in the parameters for the VCUPDATE code.

From the analyses performed it can be stated that the best structural system for cable investigation is the one with clamped ends. In fact, these boundary conditions can better describe the behavior of non-slender cables, whereas slender cables are less influenced by the boundary conditions.

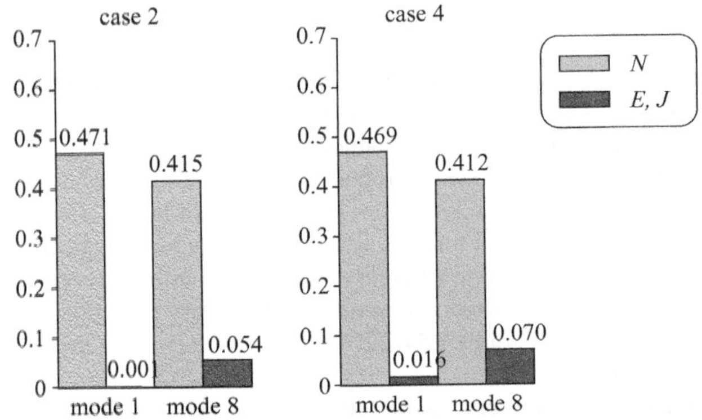

Figure 7.59 Rosenbrücke cable structural sensitivity

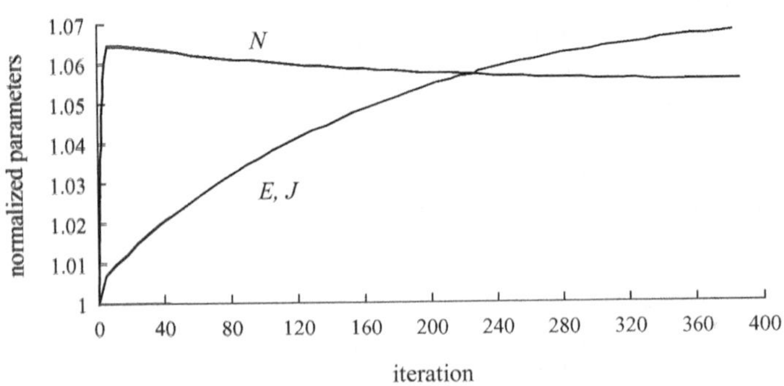

Figure 7.60 Parameter convergence for Rosenbrücke cable, case 4

Table 7.8 Analysis result comparison for axial force (N)

Case	Design force N_D [kN]	N_{VCE_Kabel} [kN]	$\frac{N_{VCE\_Kabel}-N_D}{N_D}$ (%)	$N_{VCUPDATE}$ [kN]	$\frac{N_{VCUPDATE}-N_D}{N_D}$ (%)
1	1957	2071	5.83	2202	12.52
2	1957	2071	5.83	2185	11.66
3	1957	2071	5.83	2083	6.46
4	1957	2071	5.83	2064	5.47

Moreover, the difference in the updated values is very small, regardless of whether the updating applied to only N, or N, E and J together. Hence, if the mode shapes are unknown, in order to save computational time the best solution is to update only N. However, if mode shape information is available, E and J have to be updated in order to locate the damage on the structure.

7.1.3.7 Application to Tendons

Rümmecke and Berbke Bridge External Tendons

The external pre-stressing tendons of two German pre-stressed reinforced concrete bridges were measured in April 2007 by Vienna Consulting Engineers (VCE 2007). Both bridges are located on the A46 highway, in the North Rhine region of Westphalia: the Rümmecke Bridge near the city of Meschede and the Berbke Bridge near the city of Arnsberg (Figure 7.61). The tendon axial forces had previously been measured by a lift-up test and values of 2900 kN and 2448 kN were found for the Rümmecke Bridge and Berbke Bridge respectively. The measurements were performed using the BRIMOS technology (see Section 3.2). The BRIMOS Recorder is transportable and can work as a standalone unit or connected to a laptop (Figure 7.62a). The recorder supports one 3D external sensor as well as three 1D external sensors. The sensors comprise the 3D Kinemetrics EpiSensor ES/T and the 1D Kinemetrics EpiSensor ES/U (Figure 7.62b). Both types of sensor are force balanced accelerometers with full-scale recording ranges of ± 0.25 to $\pm 4\,g$ (user selectable) providing on-scale recording of structural motion for a wide range of structure types. For the tendon measurements, the adopted sample rate is 500 Hz and the measuring time for each sensor layout is 330 s for a total of 165 000 points.

From the operative point of view, the BRIMOS measurements can be used proficiently for a quick but reliable structural investigation. The two examples reported in this section were measured in just half a day. This is of great benefit from the economic point of view of the bridge owner. After the data acquisition phase, the data are analysed by the BRIMOS software and through signal processing the

Figure 7.61 (a) Rümmecke Bridge and (b) Berbke Bridge

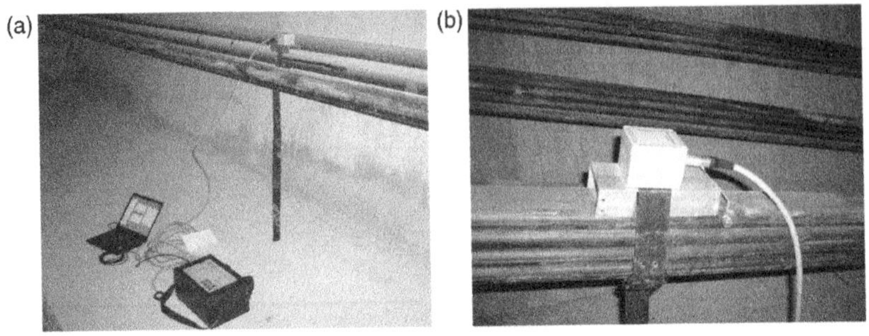

Figure 7.62 (a) The BRIMOS Recorder at work and (b) the external 1D sensor during the Rümmecke Bridge measurements

eigendata, such as frequencies, mode shapes and damping, are extracted. Finally, the modal information obtained (in this case the frequencies) is used within the FEMU procedures outlined earlier in Section 7.1.3. When necessary, the first estimation of axial force (N) and bending stiffness (E, J) used in the updating procedures are obtained using the software VCE_Kabel (Forstner and Wenzel 2004).

The cable investigated on the Rümmecke Bridge is a Vorspann-Technik type VT-CMM $4 \times 04 - 150$ D with a low relaxation steel ST 1570/1770. In the numerical analyses, a structural scheme of a simply supported beam with clamped ends is adopted with a length between the anchorages of 6.04 m, corresponding to the geometrical length measured in situ. Six frequencies are used as experimental data in the updating procedures. The updating procedures analyses are performed with OpenSees (through the Penalty Method) as well as with ANSYS (through the First-Order Optimization Method), and the results obtained for the frequencies used are almost coincident.

A preliminary sensitivity analysis was performed in order to evaluate the most important parameters with respect to the structural behaviour. At first, the axial force, the length and the bending stiffness are considered as parameters (Figure 7.63a). The relative sensitivity results indicate that the effect of a bending stiffness variation on the structural behaviour is much less important than a change in length and

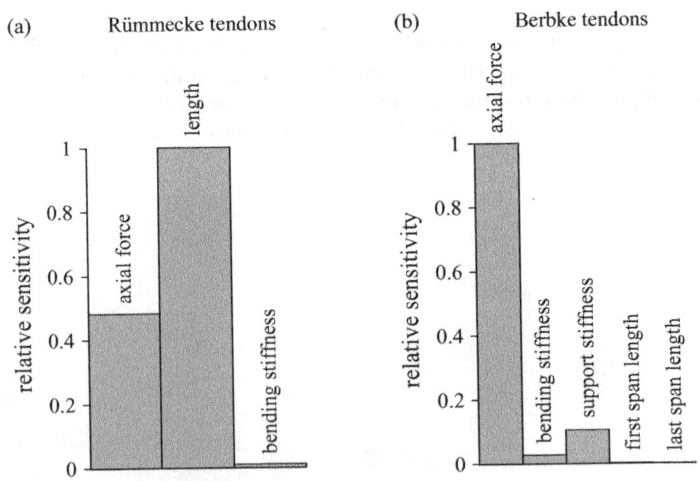

Figure 7.63 Sensitivity analysis results for the (a) Rümmecke and (b) Berbke tendons

Table 7.9 Finite element model updating results in terms of frequencies

Bridge	Mode	Experimental frequency f_0 [Hz]	Before update		After update	
			Frequency f[Hz]	$\frac{f_i - f_{0i}}{f_{0i}}$ (%)	Frequency $f[Hz]$	$\frac{f_i - f_{0i}}{f_{0i}}$ (%)
Rümmecke	1	27.85	28.73	3.16	27.86	0.03
	2	55.90	57.50	2.87	55.76	−0.25
	3	83.30	86.37	3.68	83.74	0.53
	4	111.80	115.36	3.19	111.85	0.05
	5	140.90	144.54	2.58	140.13	−0.55
	6	168.30	173.94	3.35	168.61	0.19
			CC_{abs}	0.0314	CC_{abs}	0.0027
Berbke	1	17.77	16.81	−5.40	17.60	−0.95
	2	35.60	33.63	−5.53	35.15	−1.26
	3	52.00	50.10	−3.66	52.21	0.41
	4	68.40	65.52	−4.21	68.52	0.18
			CC_{abs}	0.0470	CC_{abs}	0.0070

axial force. Therefore, in the subsequent analysis, the bending stiffness is not updated. In this respect, the sensitivity analysis is a powerful tool for selecting the updating parameters.

The analysis results in terms of frequencies are reported in Table 7.9. The effect of the updating procedure is clearly detected by the CC_{abs} change. The updated parameter values are reported in Table 7.10. The parameter values obtained seem to be reliable: the axial force is not far from that previously measured in the lift-up test and the length variation is compatible with the cable arrangement.

The tendons investigated on the Berbke Bridge is BBRV type SUSPAVI with steel ST 1470/1670. Some images of the tendon measurement instruments are shown in Figure 7.64. The corresponding adopted structural scheme is reported in Figure 7.65. The analyses are performed in ANSYS with the Penalty Method as well as with the First-Order Optimization Method. In general, outcomes are almost coincident. The Penalty Method, due to the different software implementation, requires a much longer computation time since good convergence is obtained by using a high number of iterations.

As a first step, the cable supports were considered fixed in the vertical direction. The axial force resulting from the updating procedures is 1676 kN, which seems to be too small compared to the value from the lift-up test and incompatible with the bridge design. Therefore, a different structural scheme was adopted where the supports are modeled by vertical springs. A sensitivity analysis is performed (Figure 7.63b) and the influence of bending stiffness and lateral span length is found to be negligible: for this reason, axial force and support stiffness are chosen as updating parameters. The results of the

Table 7.10 Finite element model updating results in terms of updated parameters

Bridge	Parameter	Before update	After update
Rümmecke	Axial force [kN]	2907.00	2864.83
	Length [m]	6.04	6.19
Berbke	Axial force [kN]	2009.00	2246.80
	Support stiffness [kN/m]	2000.00	1694.39

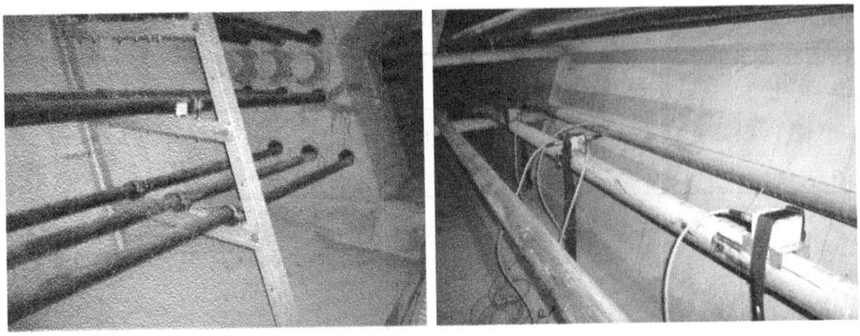

Figure 7.64 Berbke Bridge measurements

updating procedure are presented in Tables 7.9 and 7.10. The dynamic behaviour of the tendons was well captured by the updated model, as confirmed by the low frequency deviation obtained. Moreover, the axial force is now closer to the measured value. In this case, the FEMU procedure provided valid support to identifying the real structural behavior.

In summary, the approach adopted for these examples, starting from the BRIMOS measurements, through the BRIMOS data evaluation, ending with the VCUPDATE updating procedures, provides a powerful, flexible as well as quick tool for structural investigations. In the same way, the use of experimental testing in conjunction with advanced numerical techniques can assure a reliable investigation of many different structures. After an experimental campaign consisting of data acquisition through the BRIMOS Recorder and modal information extraction through the BRIMOS software, the structural indicators obtained are used as input for FEMU procedures. The updating procedures are performed by using VCUPDATE, software developed by Vienna Consulting Engineers, which provide the user with a large range of possibilities, such as three different updating algorithms and two different FE codes, ANSYS and OpenSees.

In both cases, a sensitivity analysis is performed as a first step, in order to evaluate the most important structural parameters. This step is very important, assuring that the chosen parameters to have a physical meaning and saving computational time by eliminating the negligible parameters.

In the case of Rümmecke Bridge, axial force and length are used as parameters. The parameter values obtained are in good agreement with previous measurements and cable geometrical arrangement. For the Berbke Bridge a first analysis revealed that the adopted structural scheme did not correspond to the physical reality. Then, the cable supports were modeled as vertical springs and axial force and support stiffness were used as parameters. In this way, an improved model capable of better representing the real physical behaviour was obtained. Moreover, the new updating procedure provided a more reliable value of the axial force. In this respect, one of the most important applications of FEMU is shown: supporting the structural analyst in creating the best possible structural model.

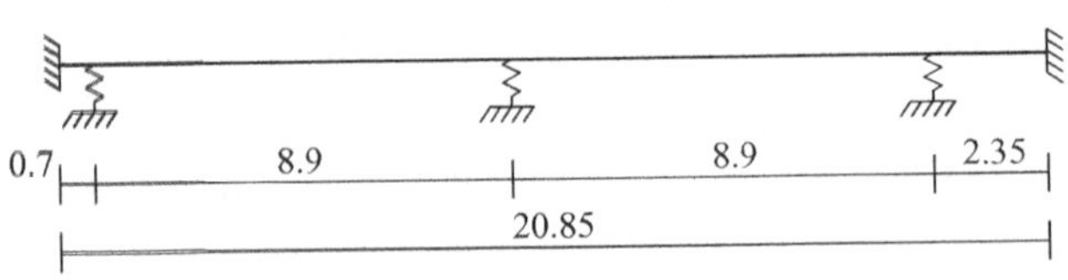

Figure 7.65 Structural scheme of the Berbke Bridge tendons (dimensions in meters)

7.1.3.8 Concluding Remarks

In this section, the application of FEMU to the SHM of civil engineering structures is presented. The updating procedures were implemented in the software VCUPDATE, a Scilab code interfaced with the FE codes OpenSees and ANSYS. Different applications have been presented and the main aims of the FEMU have been illustrated: structural evaluation, damage detection and helping the structural engineer in finding the most reliable numerical model, thereby qualifying the method as a valuable tool for SHM.

7.1.4 Ambient Vibration Monitoring

7.1.4.1 System Identification

Calculation models for the determination of stresses, and consequently their application to structures, represent only an approximation of reality and therefore have to be calibrated. For determining conformity between model results and the actual load-bearing behavior, frequent stress tests (for example at railway bridges) have been carried out and the measured deformation data (flexures) have been compared with the calculated reference values. From this comparison conclusions can be drawn on the load-bearing safety and performance capability of the structure. A simpler method for the determination of these parameters is based on the determination of the dynamic characteristics of structures by so-called ambient vibration measurements (AVM). From these measurements the vibration behavior of a structure is recorded, evaluated and interpreted under ambient influences, i.e. without artificial excitation, by means of highly sensitive acceleration sensors.

The methodology to make conclusions on the load-bearing capacity of a structure by measuring its dynamic behaviour and to check mathematical model assumptions is well established. In (Eibl et al. 1988) there is a report on stress tests between 1922 and 1945 in Switzerland, where tests by free oscillations at the aerial Beromünster in 1941 are described. The results were used for checking the calculation assumptions, deviations between measured and calculated results were interpreted and statements for similar future towers were produced. The checking of structures by means of dynamic measuring methods has a long tradition in Switzerland, and until the beginning of the 1990s was carried out in the form of tests by free oscillations by means of initial strains or intermittent stresses and by means or excitation with unbalance exciters or hydraulic shakers (Wenzel and Pichler, 2005). Such tests were also carried out in Austria and Germany for scientific purposes but at a smaller scale, and they were not extensively applied for system identification or to check and calibrate results of calculation models. In (Cantieni 1996) it is, however, suggested to further develop dynamic procedures for the assessment of the maintenance condition of structures.

The rapid development of measuring technology on the one hand and computer technology as well as software on the other enables us to carry out dynamic measurements of ambient structure vibrations and their evaluation very quickly and with relatively little expenditure. Vibrations influencing the structure due to natural excitation sources such as microseismic phenomena, wind, waves, etc., are regarded as ambient causes. The measuring and evaluation system BRIMOS® (Bridge Monitoring System) takes advantage of this progress and opens a wide field of application to technology in general.

Measurement of the dynamic characteristics of a structure at internals enables more than a single check against calculation models to be made, allowing in addition statements on the chronological development of the load-bearing capacity and therefore estimates of the remaining service life. Individual measurements supply snapshots of the integrity of a structure and can be used in combination with parallel mathematical analyses to determine possible damage to the structure.

Decisive dynamic parameters to be determined for system identification are described below. During monitoring all analyses of system identification are applied. In addition to the procedures of structure mechanics and dynamics, statistical methods have to be used that determine trends from large data sets. The use of so-called trend cards, which clearly present an eventual change of individual parameters

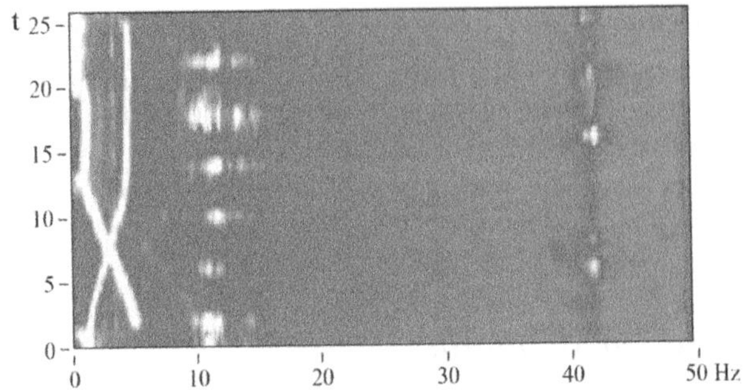

Figure 7.66 Transposition of the system according to damage – moment of failure

by means of the time–frequency diagram, has proved successful. This trend card shows the shift of eigenfrequencies over time due to the damage of individual pre-stressed cables in a test at the motorway flyover Regau (Object S123a; Figure 7.66).

Eigenfrequencies and Mode Shapes

The eigenfrequencies are an essential parameter for the description of the vibration behaviour of a structure in the linear elastic field. A mode shape, i.e. a vibration form in which the structure oscillates with the respective eigenfrequency, is provided for every eigenfrequency. The actual oscillation of a real structure is composed of the respective shares of the individual mode shapes.

Mathematical modal analysis supplies both the eigenfrequencies and the mode shapes of a structure – in experimental modal analysis the eigenfrequencies are obtained as well and the mode shapes can be determined point by point (at the measuring points). Both types of analyses have to be carried out for system identification. The actual static system is obtained by comparing the measuring results with the calculated values and adaptation of the calculation model to the measurements. In order to get a correct image on the actual load-bearing system, it is required to consider not only the first eigenfrequency and the respective modal form but also higher frequencies and their respective forms (Figure 7.67).

Nomenclature of the Dynamic-Response Characteristics

The eigenfrequencies – recurring harmonic responses in terms of vibration – that are extracted from the measurement signal represent the effective dynamic stiffness of the structure (Figure 7.68). To assure a consistent understanding of the occurring eigenfrequencies a specific nomenclature was developed for BRIMOS. The naming convention is described on the basis of the dynamic measurement at the Aschach Danube Bridge (Figure 7.69). That bridge structure is 325 m long and consists of three spans with lengths of 96.30 m, 132.40 m and 96.30 m.

Table 7.11 includes all eigenfrequencies that were considered for diagnostic evaluations and demonstrates how the specific names are created. Additionally Figure 7.70 explains how to understand the three-character BRIMOS® nomenclature stated in the last column of Table 7.11. The first character of the three-character code indicates the number of half cycles of the described mode shape; the second character describes the type of occurring mechanical motion; and the third character assigns the associating rotation axis or translation axis belonging to the reference coordinate system (i.e. nature of dynamic response). The introduced nomenclature is clarified by means of Figures 7.71–7.76 showing some representative measured and computed eigenmodes and frequencies of the Aschach Danube Bridge.

Damping

All real structures have a damping (Figure 7.77) which results in a continuous decay of vibrations after excitation until a static equilibrium is reached. The damping properties are dependent on frequencies

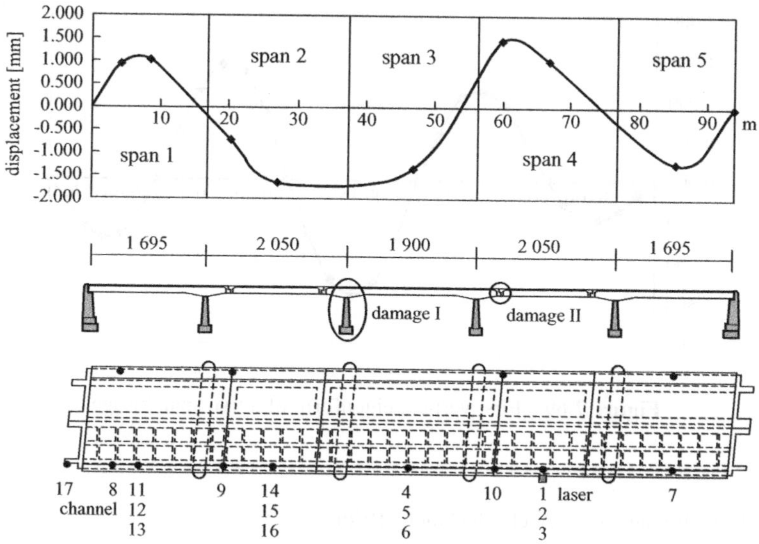

Figure 7.67 Measured first mode shape of a defect structure (settlement of supports due to heavy traffic)

and represent a significant value for system identification. In particular they are an indicator for the current degree of utilization of the load-bearing capacity of a structure, as with increasing utilization of the maximum load-bearing capacity, i.e. at the transition from the elastic into the elastoplastic range, the damping coefficients rise considerably (Eibl et al. 1988). In addition damping conditions have an influence on the eigenfrequencies themselves, which is negligible in the usual damping values of structures in use, but which gain in importance with increasing utilization of the load-bearing capacity. In the course of a dynamic test of a structure the determination of damping properties is therefore necessary to obtain a complete picture of load-bearing behavior.

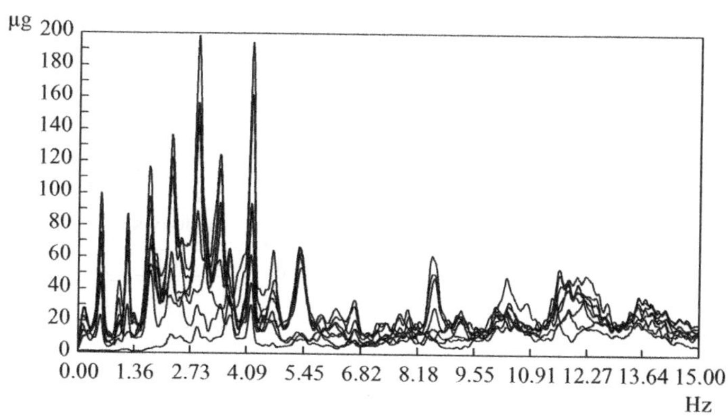

Figure 7.68 Frequency spectrum vertical 0–15 Hz

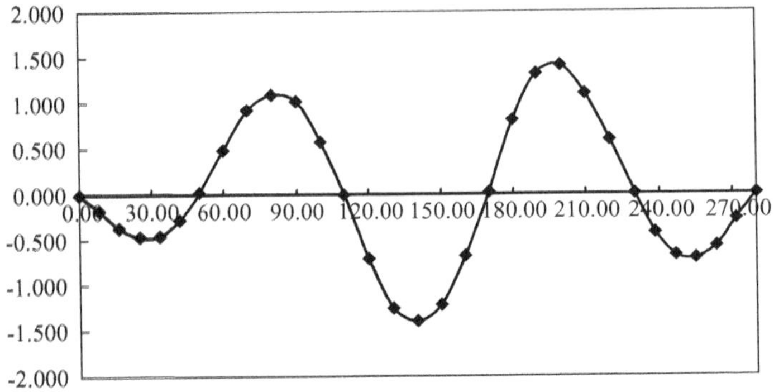

Figure 7.69 First vertical eigenform of a five span bridge

Table 7.11 Eigenfrequencies – Aschach Danube Bridge

Eigenfrequency [Hz]	Typification	Number of alternations	Initiated in span number	Brief description	Sequential mode nomenclature	Specific mode nomenclature
0.85	*BT*	1	2	Main span	(1)*BT*	1*BT* main span
1.24	*BT*	1	1; 3	Side-span	(2)*BT*	1*BT* side-span
1.52	*BT*	1		Uniformly	(3)*BT*	1*BT* uniformly
2.09	*TL*	1	2	Main span	(4)*TL*	1*TL* main span
2.90	*BT*	2	2	Main span	(5)*BT*	2*BT* main span
3.81	*BT*	2	1; 3	Side-span	(6)*BT*	2*BT* side-span
4.02	*BT*	2		Uniformly	(7)*BT*	1*BT* uniformly
6.40	*TL*	3	2	Main span	(8)*TL*	2*TL* main span
7.24	*BT*	3	1; 3	Side-span	(9)*BT*	3*BT* side-span
8.69	*BT*	4	2	Main span	(10)*BT*	3*BT* main span

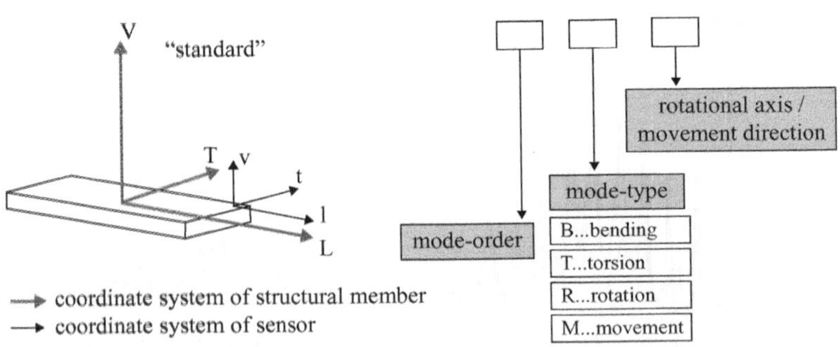

Figure 7.70 Nomenclature of the dynamic-response characteristics

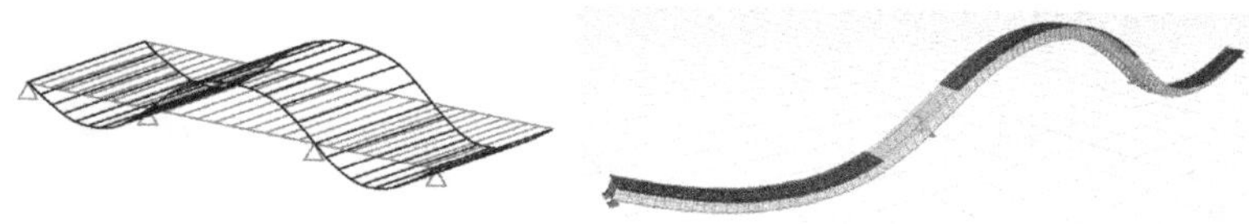

Figure 7.71 First mode shape 0.85 Hz (measurement) and 0.81 Hz (computation) – 1*BT* symmetric, main span

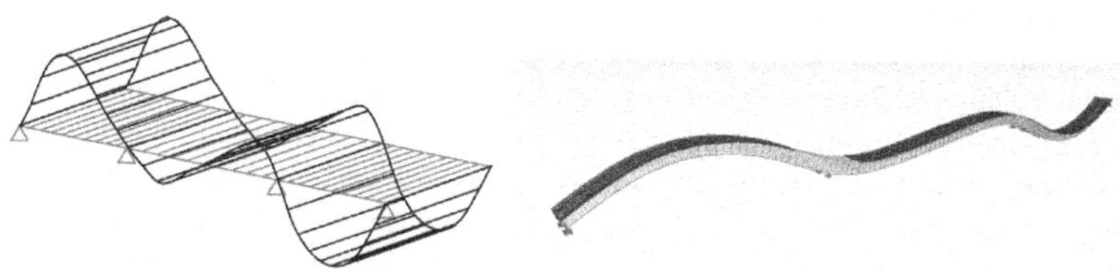

Figure 7.72 Second mode shape 1.24 Hz (measurement) and 1.22 Hz (computation) – 1*BT* side-span

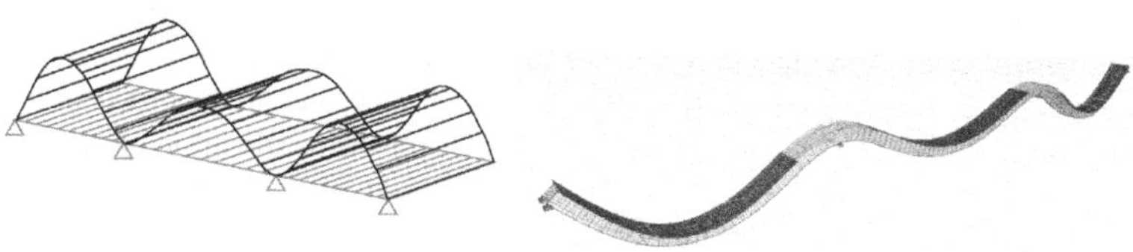

Figure 7.73 Third mode shape 1.52 Hz (measurement) and 1.53 Hz (computation) – 1*BT* uniformly

Figure 7.74 Fourth mode shape 2.09 Hz (measurement) and 2.29 Hz (computation) – 1*TL* main span

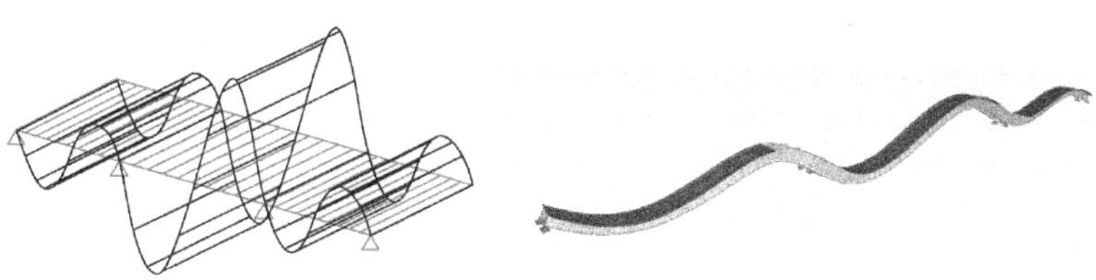

Figure 7.75 Fifth mode shape 2.90 Hz (measurement) and 2.66 Hz (computation) – 2*BT* main span

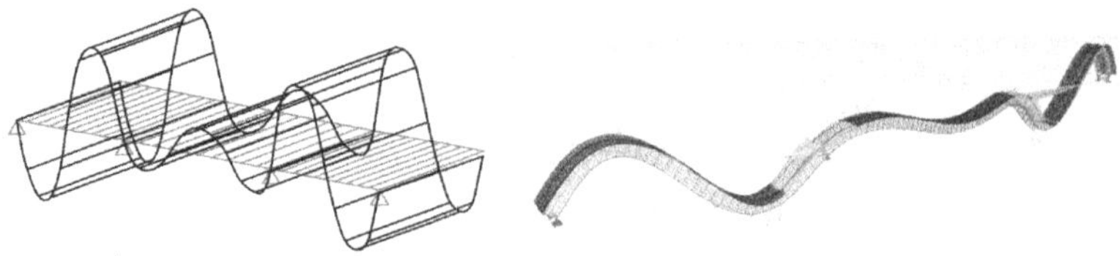

Figure 7.76 Sixth mode shape 3.81 Hz (measurement) and 3.93 Hz (computation) – $2BT$ side-span

Deformations and Displacements

Traditional measuring methods mainly measured deformation under defined loads and the results were later compared to calculated forecast values (e.g. test loads on railway bridges). The information content of this method is essentially limited to the flexural stiffness of the tested structure, which is definitely an important assessment criterion. Dynamic measurements also include information on deformation of the structure during the measuring period. By registering the three-dimensional vibration behavior it is even possible to extract the deformation information from the measurements in all three directions in space. Therefore not only vertical flexures but also horizontal (transversal and longitudinal) displacements are finally available for the assessment of structural integrity (Figures 7.78–7.80).

Vibration Intensity

The vibration intensity is a very good indicator of the stress a structure experiences by dynamic loads. High vibration intensities for individual structures or structural members are very susceptible with regard to fatigue relevant damage mechanisms. In BRIMOS the measured values are recorded and classified in an intensity chart (Figure 7.78). This is appropriate in particular for the localization of detailed inspections at especially sensitive parts of the building. In case of the Voest Bridge in Linz the decisive sections were localized on a small part of the total structure. In addition it is possible to identify and monitor individual contributions to the vibration intensity more accurately by observing them (e.g. through a traffic monitoring system).

Trend Cards

Trend cards represent a signal in the frequency–time domain by means of area mapping. In order to be able to distinguish the individual frequency peaks, colouring of the card is required so that the energy content of the vibration and therefore the respective intensity can be determined.

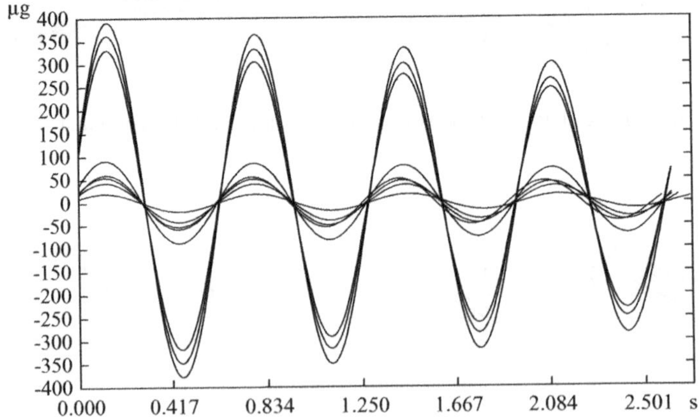

Figure 7.77 Damping window, first vertical eigenfrequency

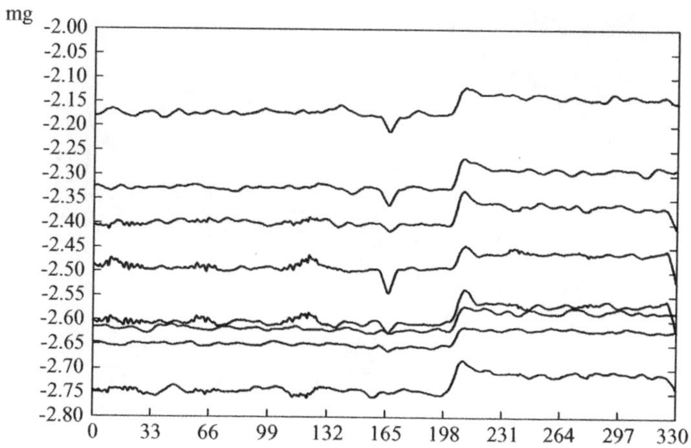

Figure 7.78 System displacement due to bearing reset forces of the St. Marx flyover

7.1.4.2 Stress Test

The knowledge of the current stress condition of a structure and its individual load-bearing elements is of particular interest. Examination is required on the one hand to determine the existing current load-bearing safety and possibly to be able to introduce necessary immediate measures. On the other hand it is an essential basis for the forecast of future maintenance costs. An important assessment criterion in this context is the evaluation and interpretation of the vibration intensity of the respective structure, application of which can lead to classification of the extent to which a structure is endangered with regard to damage.

Determination of Static and Dynamic Stresses

Global stress condition has to be determined statically and dynamically. From knowledge of these two stress types the dynamic capacity ratio for traffic loads can be determined. This is not only an indication

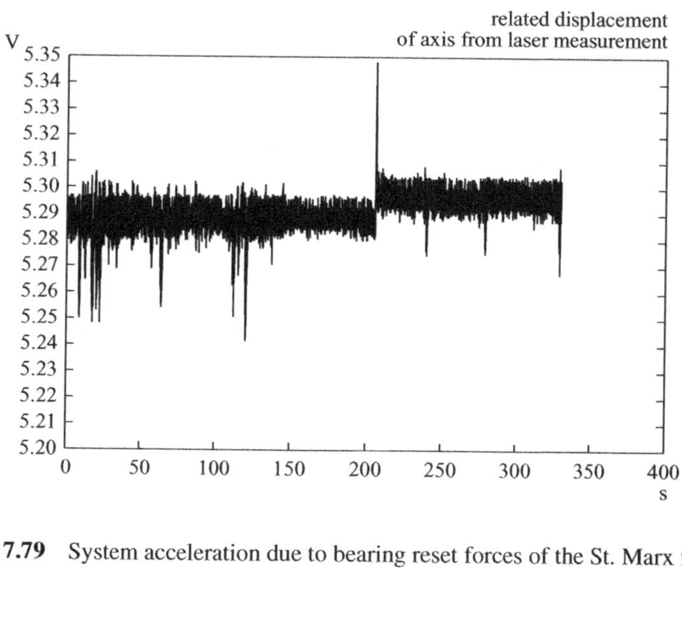

Figure 7.79 System acceleration due to bearing reset forces of the St. Marx flyover

Figure 7.80 St. Marx flyover section of the A23 motorway

of the dynamic sensitivity of the structure, it also allows the establishment of specific measures for the reduction of the stress level (e.g. by speed limits).

Determination of the Vibration Elements

Individual elements of a structure frequently show particularly strongly developed vibration behavior (e.g. rim beams of deck cantilevers, stay cables of guy-wired constructions). Such structural members have a decisive influence not only on the convenience of use of a structure, but also because they are particularly susceptible to fatigue fractures caused by continuous changes in stress in their load-bearing elements. Recognition of such damaged structural members by modifying the vibration behavior within the scope of a dynamic measuring program enables timely rehabilitation, thereby preventing greater damage. A simple dynamic analysis supplies information on such elements in a structure and thus enables specific measurement (Figures 7.82 and 7.83).

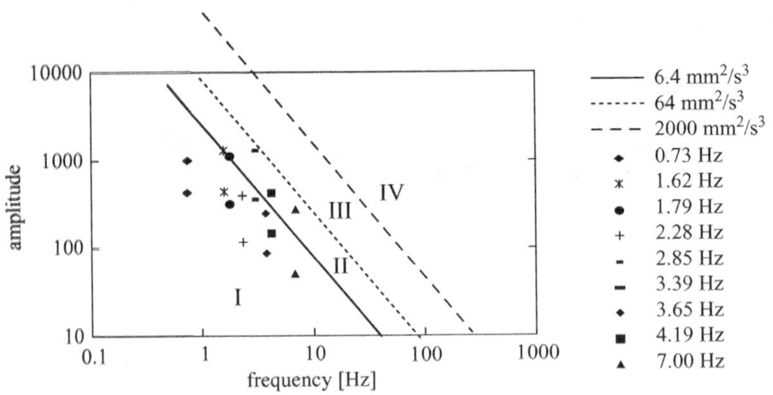

Figure 7.81 Vibration intensity chart for the Europa Bridge of the Brenner Motorway. I, no damage; II, possible plaster cracks; III, possible damage to load-bearing structural parts; IV, damage to load-bearing parts

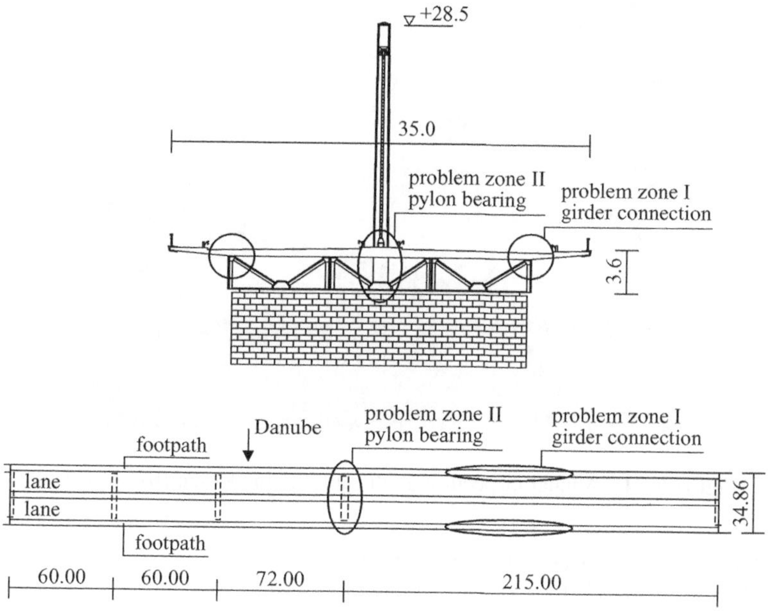

Figure 7.82 Vibration-intensive areas at the Voest Bridge across the Danube

Stress of Individual Structural Members

Apart from testing the global stress situation and that on individual structural members in general, particularly susceptible to vibrations, a specific stress determination for individual structural members is required as part of an extensive test of the structure. For all load-bearing elements with a direct relation between eigenfrequency and stress level (stay cables, tendons, tension members under compression), determination of the stress by means of an evaluation of the vibration measurement is possible. If the relations are more complicated, it is more advisable to determine the stresses by means of the optimized calculation model included in the scope of system identification outlined in Section 7.1.4.1.

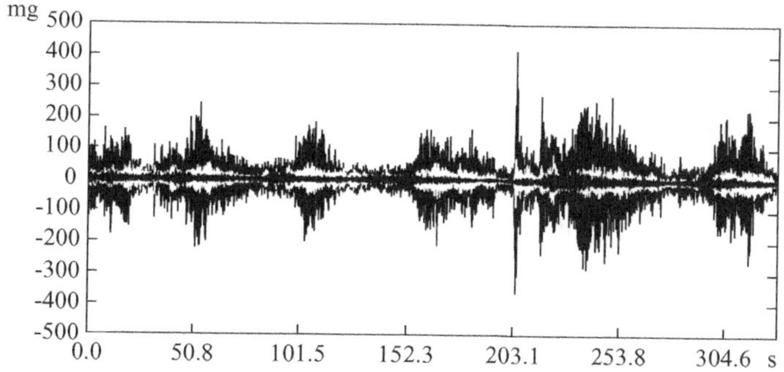

Figure 7.83 Acceleration signal of a bridge cross-section with high values for the cantilever vibration

Figure 7.84 Annual inspection of all cables at the Rosen Bridge in Tulln

Determination of Forces in Tendons and Cables

Knowledge of the actual tensile forces in the tendons of guy-wired structures (e.g. pylons, pyramid type roofs) or in the cables of cable-stayed bridges, in the suspenders of arched bridges or in external tendons is required for the assessment of individual elements and for examination of the global stress of the structure (Figure 7.84). The determination of these forces by lift-off tests is very costly and carries with it a risk of damages. The works at the anchorages can unfavourably influence the durability of these critical elements. Therefore fast and non-destructive methods for the determination of stress are required, such as measurement of vibration characteristics because there is a simple, quasi-linear relation between the eigenfrequencies of a cable and the inherent force.

7.1.4.3 Assessment of Stresses

An essential task is the assessment of stresses of the total system as well as of individual structural members. This assessment must consider both the actual structural condition and the predicted future development of the structural condition. Ambient Vibration Monitoring, offers the possibility of assessment on the basis of objective parameters. If the measurement results are combined with calculation models very good predictions can be made by applying probabilistic approaches. However, statements with varying accuracy are possible, and the scope of the executed test determines their representiveness.

Structural Safety

Dynamic characteristics contain information on the global structural condition as well as on localized conditions. During assessment of the structural safety the whole system is viewed and its behavior analyzed. System identification concentrates on the lower frequencies in the spectrum. The scope of tasks is the determination of the frequencies and the respective mode shapes of the whole system and to calculate damping values that describe the system behavior. The total structure can be assessed by means of these results.

For consideration of the safety of a structure there are numerous excellent models in the literature, but those by Frangopol (USA) and by Das (England) are particularly apt. They analyze the service life of a bridge starting from construction up to the slow decline of resistance by use and finally to the so-called intervention points where activities have to be taken to keep the safety level above a critical limit. Such a chart is shown in Figure 7.85. A curve for the theoretical service life, however, cannot

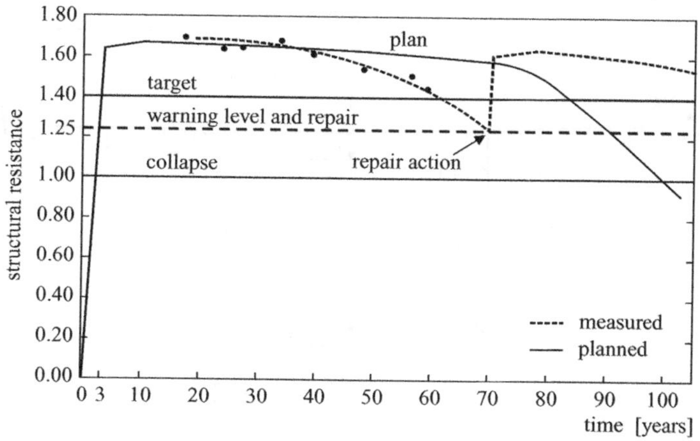

Figure 7.85 Theoretical service-life chart of a structure

be transferred to reality. A simple approach is the assessment of the existing load-bearing safety by the system identification of the eigenfrequencies. It results in a value – in relation to the planned value – representative of the moment of measuring. After consideration of the environmental influences and using periodic measurements as the basis, values can be determined that lead to construction of a curve. Future models will consider further values from parameter classification apart from the assessment of the modal parameter eigenfrequency and therefore enable even better predictions.

The aim of these considerations is to find those structures requiring rehabilitation most urgently from a large random sample of structures. The current development of the BRIMOS® recorder, which can register and assess a large number of structures at low cost, complies with this aspect.

Structural Member Safety

The bases of whole-structure safety can be analogously applied also to individual structural members. Every member has its own characteristic, which can be isolated from the total-structure measurement. So, for example, the vibrations of decks are very distinct. During a sufficiently accurate measurement, i.e. check points at every member, their characteristics can be unequivocally determined. This isolation makes it possible to produce statements on the quality of individual structural members. An isolated local damping value is calculated, which gives information on the quality of the respective structural member.

Measurements at several cross-sections of the structure were carried out at the Europa Bridge with the individual elements instrumented according to the sketch in Figure 7.86. Thus both the load displacement and individual stress can be derived.

Maintenance Requirements and Intervals

The increasingly aging engineering structures in the transport infrastructure entails growing significance of maintenance problems. Bridges are checked by periodic monitoring measures with the aim to minimize the safety risk on the one hand and to keep the costs for the maintenance as low as possible on the other hand – by carrying out rehabilitation measures at the right moment. In Austria the bridge tests for the federal road network are carried out on the basis of a formal guideline for the monitoring and checking of road bridges at periodic intervals. In order to fix an optimum duration of checking intervals, it is advisable to combine conventional bridge tests and dynamic monitoring.

Although parameters are better and more objectively quantified by dynamic tests, when combined with a qualitative visual inspection it is possible to extend the inspection intervals for bridges with a good maintenance condition. A further attractive possibility is the application of simple verification procedures

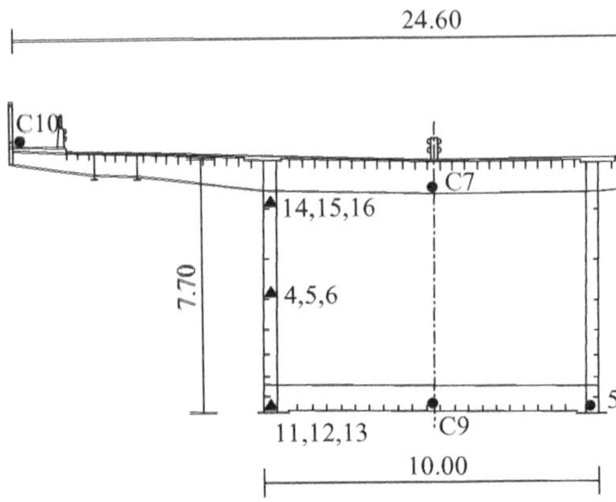

Figure 7.86 Cross-section of the Europa Bridge with sensors

that supply information on changes in the system by means of comparison to the basic measurement. In this way the requirement of a main inspection can be determined at a relatively low cost, thus essential maintenance costs could be reduced without having to accept a lower safety level. Special cases in this connection are the cable-stayed bridges where information on the structural condition can be gained by simply monitoring the cables at shorter intervals.

Remaining Operational Lifetime
For the maintainer and the user of a structure the question of the expected remaining operational lifetime is of special significance. As described Section 2.3.1, the remaining operational lifetime can be extracted from measurements at periodic intervals. Safe forecasts are possible for triple the measurement period, i.e. in case of measurements over three years, an extrapolation of nine years. From this point of view it seems important to carry out first measurements as soon as possible even if there is no immediate reason (problem) for them. This type of application of the system will be significant only after several years of its use.

7.1.4.4 Load Observation (Determination of External Influences)

The objective of the determination of external influences (also called load observation) is the complete registering of traffic loads (Figure 7.87) or other influences acting on the structure. In this connection the induced loads are not registered by means of a special balance but by the dynamic reaction (response) of the structure. This requires knowledge of the dynamic system behavior of the structure acquired by model calculations and/or experimentally (measurement). The dynamic reactions and the determined traffic loads can be used for other tasks in the assessment procedure.

Collective Load
This is the determination of the collective load by recording the acting stresses according to location, type, size, duration and frequency. The influences of wind and temperature also can be considered. Sophisticated load models allow more precise statements on the service limit state and remaining lifetime of damaged structural members.

Stress Characteristic
Load models, which can be estimated only with difficultly in the design phase (especially for structures such as cable guy-wired structures, suspenders or lateral braces of bridges), can be improved dramatically

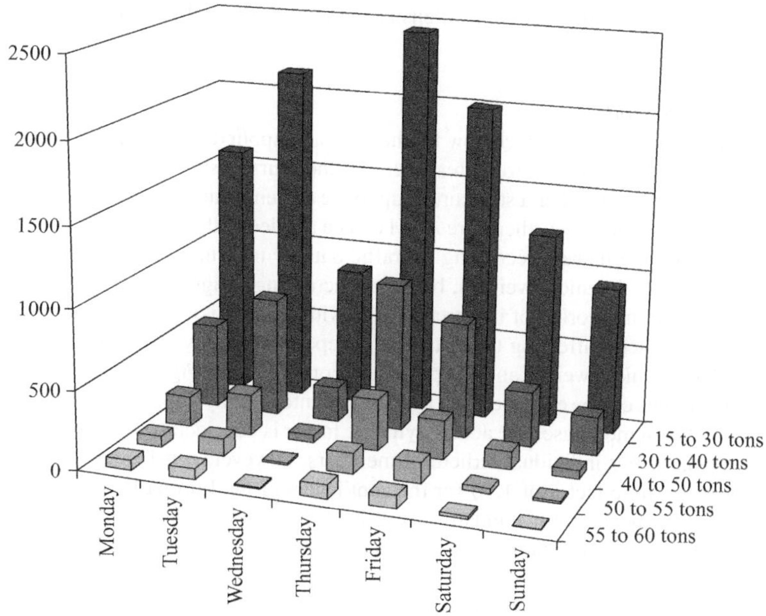

Figure 7.87 Passage events during a week (January 4–10, 1999)

with the incorporation of measurement data, as knowledge of the stress characteristic enables improved and clearly more efficient designs.

An example is provided by a railway bridge (Figure 7.88) where load displacement was followed by measurements. The reason for the displacement was the particularly high noise radiation of the structure because of the absence of ballast at the structure. The contribution of the individual elements to the total sound spectrum needed to be determined with the aim of applying damping measures at

Figure 7.88 Detailed examination at the Rohrbach Bridge

the right location. Therefore the vibration intensity experienced by all structural members had to be measured.

Verification of Load Models

A further application of AVM is to check how far theoretically applied load models correspond to reality. This is especially important for older structures where the then current standards required design levels for lower traffic loads. Proof of whether a structure is up to the current actual loads can be determined on the basis of dynamic measurements. Furthermore AVM can be applied to the assessment of plans with regard to widening a structure. Permanent recording of traffic data by measuring for example, traffic intensity and density as well as total vehicle weights, becomes increasingly significant for strategic planning. A further possibility is the monitoring of important roads with regard to overloaded vehicles.

The actual dynamic factor affecting the structure is dependent on several parameters, but essentially the driving speed, the vehicle weight and the pavement condition (see Figure 7.27). From measurement data the respective values can be extracted and be included into a safety check. One of the most important findings of tests is that in most cases the actual dynamic factor is considerably lower than that prescribed by the design standard. For individual structural members, however, considerably higher factors have been found (e.g. a dynamic factor of 1.90 for the cantilevers of the Europa Bridge). For old structures in particular where the then current standards prescribed relatively low stresses, such an assessment is often essential for being able to establish the load-bearing safety.

Determination of Environmental Influences

Determination of the influences from the environment, such as aerodynamic vibrations, running water, shocks from neighboring industrial plants or other excitations is possible through dynamic measurements. An external influence often leads to unusual behavior of a structure. In areas prone to earthquakes it is possible to determine dominant site frequencies, which are to be avoided in the design of the structure.

Influences from the environment are determined by means of an external sensor being included in the scope of the measurement campaign. This sensor is located outside the structure system and measures the vibrations transferred via the ground. This has proved particularly successful in the vicinity of a quarry (Mur Bridge Kraubath). Registering of the ground spectrum is regarded as one of the most essential development steps of the system, considering above all the safety of buildings in compliance with Eurocode 8. As part of a large project in West Java (Indonesia) the adoption of this approach for earthquake mapping based on dynamic measurements has proved to be successful.

Determination of Specific Measures

In case of problems or changes in design requirements (e.g. extensions), AVM can be applied for the deduction of specific measures for stress reduction. This can be achieved by a reduction in the source of the stress or a change in the construction that enables utilization of reserves built into the design specification. An example is the structure of the F9 Donnergraben Bridge on the Tauern Motorway (Figure 7.89) where a damaged concrete deck led to the structure experiencing increased vibration intensity. If the value recorded for the vibration intensity is outside the safety limit, measures have to be taken, for example the rehabilitation of defective spots. The scope of the damage can be localized very specifically by AVM.

Check on the Success of Rehabilitation Measures

By conducting measurements before and after rehabilitation, information on the quality of the work and the success of the rehabilitation can be gained. This is possible by a simple comparison of the dynamic characteristics. An example is the Inn Bridge Hall West (Figure 7.90) where a clear deviation of the system was noticed during a basic measurement program. This was attributed to the unintentional fastening of the system in an area of the abutment. After the removal of this fastening the calculated and expected frequency value was measured for verification, which established that rehabilitation was successful.

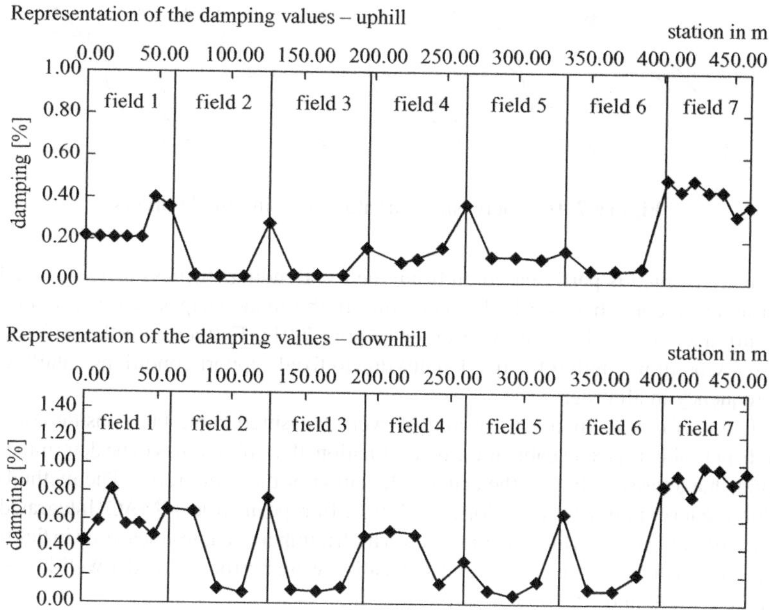

Figure 7.89 Damping progress of the F9 Donnergraben Bridge

Dynamic Effects on Cables and Tendons

The dynamic behavior of a cable is clearly defined by its free vibration length, cable mass and the inherent force according to cable theory if the cable is devoid of bending stiffness. Since a typical cable which is used in civil engineering does have a considerable stiffness a method was developed for AVM cable

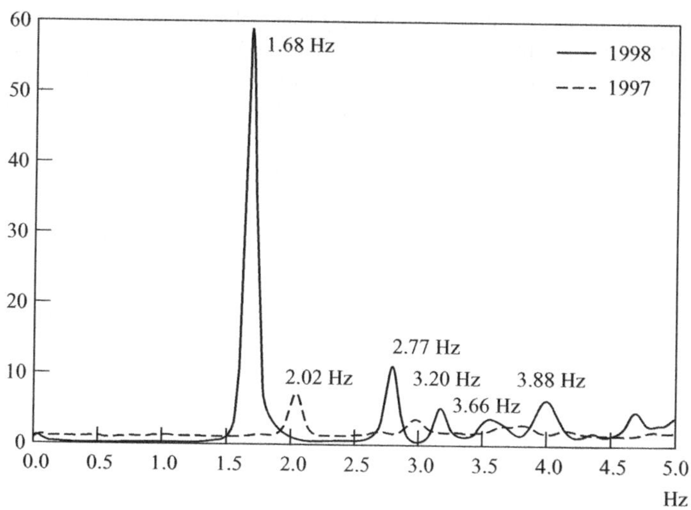

Figure 7.90 Frequency spectrum of Inn Bridge Hall West 1997–1998

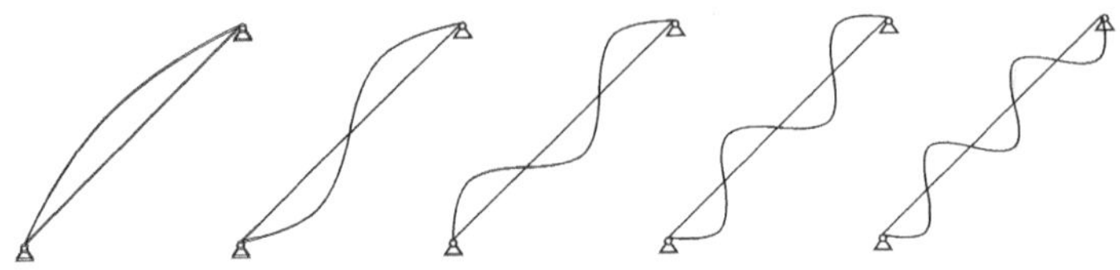

Figure 7.91 Increasing curvature in rising mode orders

force determination which is partly based on beam theory. Bending stiffness increases the frequencies, especially in higher modes, because higher curvature in the mode shapes causes a nonlinear relation between eigenfrequency and its respective order (Figure 7.91). Furthermore the stiffness of a cable shortens its free vibration length (effective length) due to fixed support conditions entailing an increase of the eigenfrequency in all orders.

Since bending stiffness is unknown for almost every measured cable but consideration of bending stiffness is indispensable for exact cable force determination, the stiffness has to be determined in analyses of the dynamic characteristics before the actual calculation of the cable force. The method used for the AVM cable force determination was developed within the European project IMAC (Integrated Monitoring and Assessment of Cables), for which not only were a high number of cables measured in field tests but also investigations and verification on tendons and stay cables were performed under well-known laboratory conditions (Figures 7.92 and 7.93).

Determination of cable forces enables accurate, quick and cheap quality control under construction, after pre-stressing procedures and for periodical supervision for safety and maintenance reasons.

Parametric Excitation

Problems with the vibration behavior of individual structural members such as cables, struts or delicately proportioned elements can often be attributed to parametric excitation. In this process the eigenfrequency of a structural member is stimulated by a frequency of the total structure. Consequently an uncontrolled vibration of cables or a fracture of individual stays in steel construction occurs. These phenomena can be quantified by measurements so that the assessment basis for rehabilitation can be derived.

Parametric excitation has particularly been observed in the higher eigenfrequencies of cables. Therefore it is not sufficient to harmonize the first eigenfrequencies (basic frequencies) of the cables with the deck frequencies. The fact that cables permanently oscillate (as for example in case of the Erasmus Bridge in Rotterdam) is attributed to the constant supply of energy via parametric excitation at a higher eigenfrequency. The cable vibration itself mostly takes place in the second eigenfrequency, with superimposed

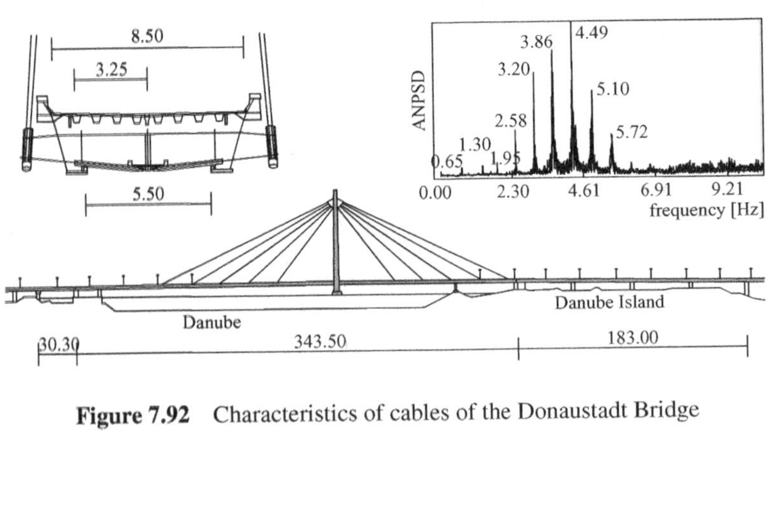

Figure 7.92 Characteristics of cables of the Donaustadt Bridge

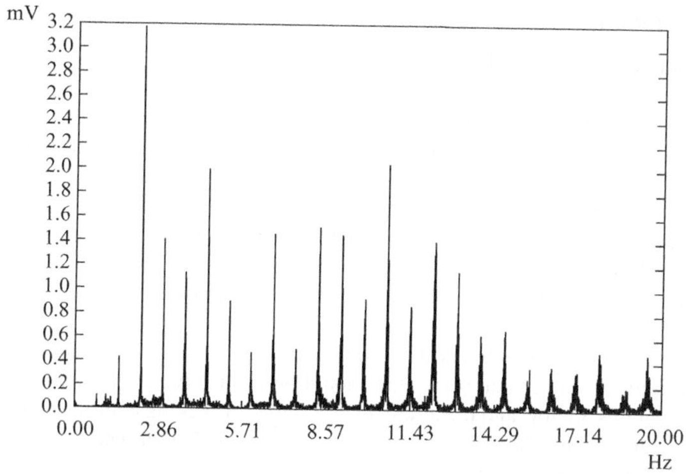

Figure 7.93 Spectrum of a cable of Donaustadt Bridge

higher frequencies. It is, however, inadmissible to draw conclusions on the energy transfer from the visually recognizable vibration form.

The possibilities and examinations mentioned in the foregoing do not represent the whole potential of the dynamic vibration method for cables and tendons. In the future, further phenomena probably will be identified and covered. An extension, particularly on the behavior of road joints, has already occurred (Technical University Innsbruck, Professor Tschermanegg), which covers structural members that are frequently damaged. Therefore a continuously growing extension of the application field for AVM is to be expected.

7.2 Deflection and Displacement Monitoring

Deflection and displacement are valuable indicators of the performance of a bridge. In contrast to the opinion that deflection is a strict function of bending, monitoring has shown that unequal temperatures or breaking forces contribute considerably and practically change the structural systems. It is therefore imperative to know the actual values for deflection and displacement. The calculation of deflections through double integration of the accelerograms is only applicable in simple arrangements and in most of the cases requires adjustments. Therefore a good measurement campaign also contains elements of deflection and displacement recording.

7.2.1 Deflection

Knowledge of the specific deflection of a bridge under a specific load is part of successful system identification. The relative displacement can be monitored easily using a laser or similar instrument. It is advisable to integrate this instrument into the monitoring campaign to establish the relationship with other elements. When lasers are used it has been proven more successful to have the target at the bridge and the laser on firm ground. This eliminates eventual errors from displacement of the laser due to the bridge activity. A successful example is discussed in Chapter 4 at the Europa Bridge. It also should be considered that deflection very often is not only in one direction. An indication of the other components often provides additional information on the bridge performance.

7.2.2 Displacement

The displacement of a structure can have many reasons. There are:

- Elongation and shortening due to temperature change.
- Displacement due to unequal temperature in a structure. This very often leads to torsional strain and horizontal displacement.
- Displacement due to breaking loads.
- Displacement due to damages. Weak sections lead to a diversion of the forces and this often is expressed by a specific displacement pattern.
- Displacement associated with external loads such as moving slopes or adjacent structures.

The character of these activities implies that changes happen slowly. To detect them properly needs monitoring campaigns of longer duration. This should be annual cycles at a minimum. For structures where the safety margins are low and the assessment is aimed at seeking additional capacity, monitoring campaigns with a duration less than one year are difficult to be applied properly. The pattern of displacement resulting from daily and seasonal cycles has to be properly established in order to find patterns of deviation indicating damage or abnormal behavior.

The traditional instruments for these campaigns are displacement transducers of various kinds. In large bridges the displacement due to temperature can be several decimeters, which requires instruments such as wire transducers for measurement. They work reliably but practise has shown a high probability of destruction by maintenance personnel or wildlife. The typical force balance accelerometer can provide information on rotations from which the displacements could be derived. This also has been applied successfully.

7.2.3 Drifting

Drifting is a phenomenon that can appear in long cycles. It can be recorded in a typical measurement campaign and is often represented in the spectrum as a very low frequency. This can only be an indication that there is something changing the system slowly. A quantitative interpretation is difficult. This subject should follow the standard approach, namely making a quick and cheap measurement first and deciding after reasonable doubts have been established.

It is found important to identify eventual drift phenomena in detail because they very often impose a great strain on structures. A typical example was the gradual displacement of a bridge supported on elastomeric bearings until the restoring force triggered a sudden release and set back. This repeated activity has led to the damage of expansion joints, which is a very costly to repair.

Another important incident of drift is susceptibility to earth movements and the subsequent displacement of piers. Deck structures show an expression of drift in the direction of the movement, which can be identified by monitoring.

7.2.4 Tilting

Tilting is a special case of deflection or displacement and needs to be identified, particularly in soft structures. The extent as well as direction of tilting can be used for damage identification. Nevertheless tilt meters are rarely seen in typical bridge monitoring campaigns.

7.3 Fatigue Assessment by Monitoring

Fatigue assessment has gained importance with increasing age of steel structures. A very good example is elaborated in detail in Chapter 4 where the case of the Europa Bridge is discussed, and where the superior performance of monitoring versus visual inspection is demonstrated. System identification techniques to find local fatigue phenomena have proven successful in various Japanese applications (see Fujino 2005). It is feasible to find the actual cause of fatigue when the exact behavior of structural components is known. The application of laser vibrometers is useful in this procedure. Knowledge of the mechanisms leading to fatigue aids in finding the suitable retrofit for each case. Fatigue assessment can be performed successfully by monitoring but fatigue prediction requires more sophisticated approaches, that currently are not feasible due to their cost.

7.4 Corrosion, Carbonization, Chlorite Content

The monitoring of corrosion, carbonization and the chlorite content has developed differently to other monitoring approaches. Even though these are permanent processes, permanent monitoring does not easily produce satisfactory results. This is mainly due to the fact that the lifetime of a structure and the lifetime of a monitoring system differ by a factor of ten at least. Very attractive corrosion monitoring systems have been designed and applied to huge structures with foundations in seawater, e.g. the Greatbelt Link and the large bridge projects in Hong Kong. Nevertheless where well-designed projects consider corrosion properly, the monitoring system only demonstrates that this consideration is correct. Major problems with corrosion have been in encountered older bridges, where concrete cover and concrete quality do not fulfill today's requirement. There are procedures for the detection of corrosion in existing structures, but they are rather expensive, and require access close to the structural element, which is very costly and leads to eventual disturbance of traffic. It is therefore recommended that monitoring concentrates on structural performance, without taking into consideration the cause of a particular defect, i.e. it is not important to know whether a steel bar is missing or damaged due to corrosion or to fatigue cracking, but it is important to know what needs repair or replacement and how immediately that should be done.

The same applies for carbonization and the chlorite content. It is advisable to use currently practiced conventional methods at those structures not performing adequately. Monitoring a large number of structures will be better financially and any technical assessment can concentrate on the combination of problematic cases revealed by the monitoring program.

It is therefore not expected that any of the monitoring systems for corrosion, carbonization or chlorite content will be applied widely on permanent basis.

7.5 Load Transfers

In case of insufficient bearing capacities loads are often transferred to other systems. This particularly applies to those bridges composed of a number of independent prefabricated elements. The intended effect of load transfers by connections is not always sufficiently provided for, and an intended transfer might overload an adjacent element. When the flow of loads within a structure is known assessment is able to proceed better.

Attention is drawn here to the nonlinear behavior of such multi-element structures. Under ambient vibrations no load transfer takes place. When a certain energy level is exceeded the elements decouple and might behave unfavorably, but this is difficult to detect by an ordinary AVM campaign. Engineering judgment is necessary in order to modify the procedure of the AVM accordingly. If the bridge is sufficiently loaded with heavy traffic this problem should not appear. In quiet situations the passage of a truck or a train will help to solve this problem.

The pre-stressed concrete road bridge at Komořany is a typical example of such a structure (Figure 7.94). The system considerably changes its behavior with the passage of a train. To monitor this behavior needs considerable engineering judgment as standard routines will fail.

Figure 7.94 Komořany Bridge

7.5.1 Assessed Bridge Structure

The pre-stressed concrete road bridge Komořany is a two-lane crossing of a multi-track connection of the Czech Federal Railways (Figure 7.94). The main structure consists of two spans with lengths of 16.72 m and 46.43 m (total length approximately 63 m and a total width of the bridge deck of approximately 13 m). In the main span a suspended span made of nine prefabricated, pre-stressed concrete elements (I-beams) is integrated into the bridge structure via two Gerber joints. Both the suspended beam and the monolithic parts of the bridge are reinforced with internal pre-stressing in the longitudinal direction. The bridge was completed in 1961 and represents one of the first practical applications of pre-stressed concrete in the former Czechoslovakia.

7.5.1.1 Scope of Work

The one-day measurement campaign with BRIMOS® was performed with regard to the global maintenance condition of the structure. Parallel to conventional bridge inspections the results from following, representing the vibration behaviour of the structure, were used to determine and localize problematic zones.

- The relevant eigenfrequencies and corresponding mode-shapes of the bridge structure extracted from sensor layouts distributed in longitudinal and transverse direction (Figures 7.95 and 7.96).
- Pattern of damping values in the longitudinal and transverse direction of the structure.

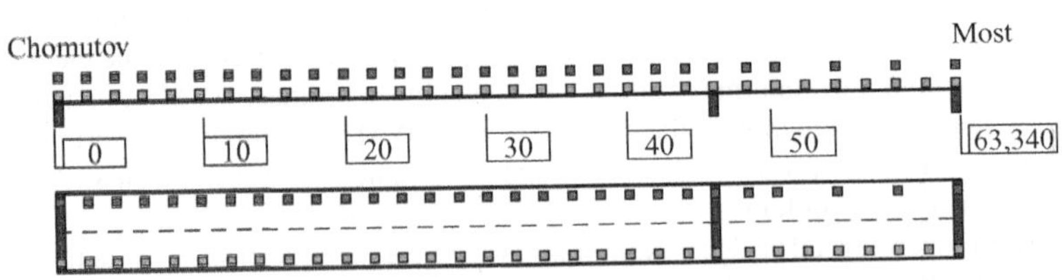

Figure 7.95 Sensor layout of the measurement in longitudinal direction

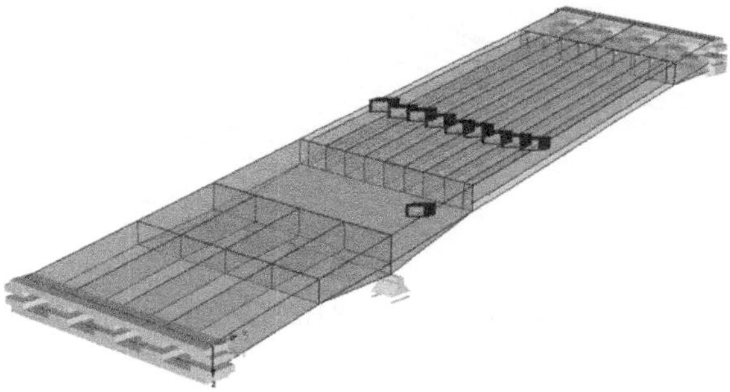

Figure 7.96 Sensor layout of the measurement in transverse direction

- Vibration intensities at the entire bridge deck.
- Total vertical displacements at the main span due to traffic loading.

The load-bearing capacity of the bridge structure had to be defined by means of the analysis performed and recommendations for the further maintenance provisions had to be given. Special attention was required regarding the suspended beam and its integration into the bridge structure (Gerber joints) as well as the condition of the transversal pre-stressing of the prefabricated concrete elements.

7.5.2 Visual Inspection

Each BRIMOS® measurement campaign was also combined with an accompanying rapid visual inspection of the structure in order to document obvious damage or irregularities. This rapid assessment showed that the Komořany road bridge structure was in a bad condition, in particular the condition of the transversal pre-stressing designed for the assembly of prefabricated suspended beams (see Figure 7.97).

Figure 7.97 Suspended span formed by prefabricated concrete beams pre-stressed in transverse direction

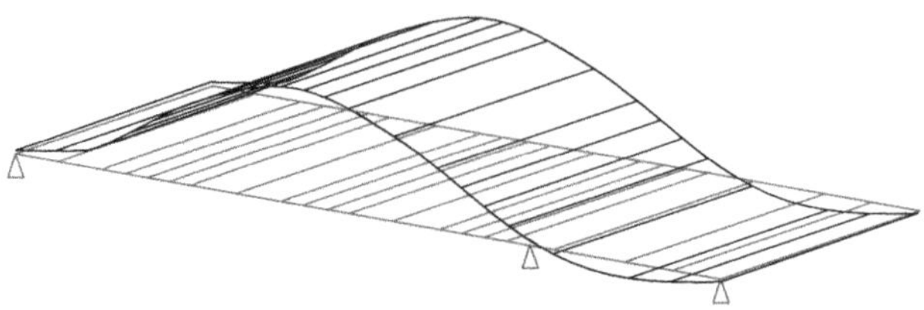

Figure 7.98 First mode shape – 2.8 Hz (1*BT* symmetric, main span)

7.5.3 Dynamic Analysis Mode Shapes

The determined characteristic mode shapes in general correspond to the expected modes of vibration. While the global mode shapes confirm the functionality of the bridge bearing capacity (Figures 7.98, 7.100, 7.102 and 7.104) the lateral mode shapes indicate the deficiency of the transversal pre-stressing (Figures 7.99, 7.101, 7.103 and 7.105). Several areas of the transverse assembly – represented by the prefabricated concrete beams – no longer contribute to the loading transfer. This is especially evident – where the bridge deck's torsional resistance is activated – in the course of comparing the analyzed measured mode shapes with those expected in the transverse direction (Figures 7.100, 7.101, 7.104 and 7.105).

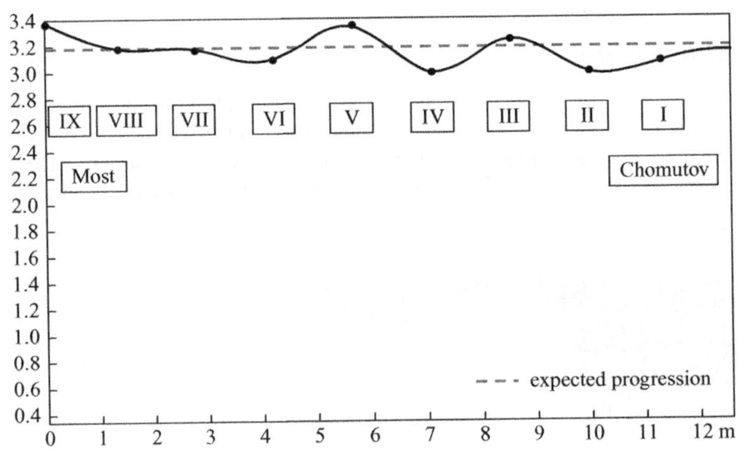

Figure 7.99 First mode shape – transversal direction

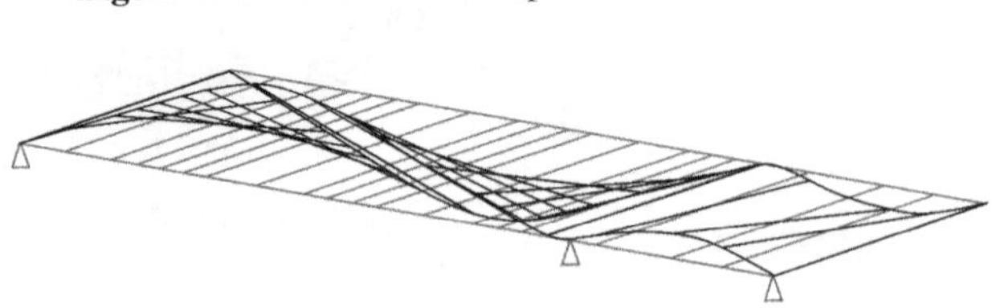

Figure 7.100 Second mode shape – 5.6 Hz (1*TL*, main span)

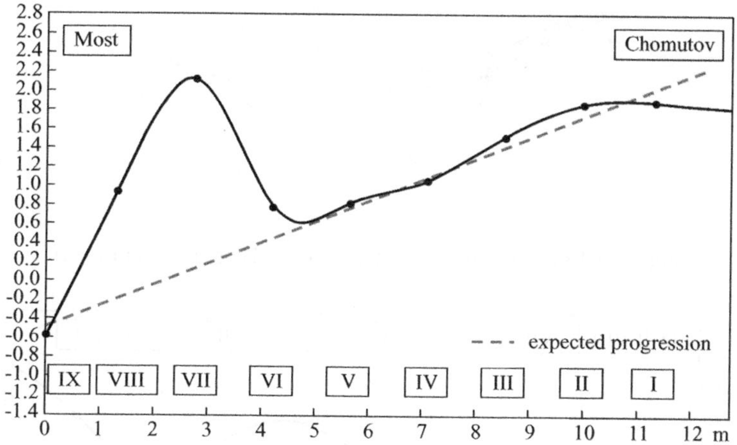

Figure 7.101 Second mode shape – transversal direction

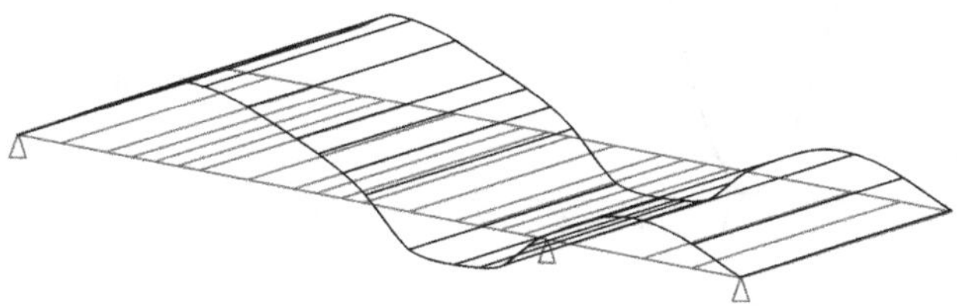

Figure 7.102 Third mode shape – 6.24 Hz (2*BT*, main span)

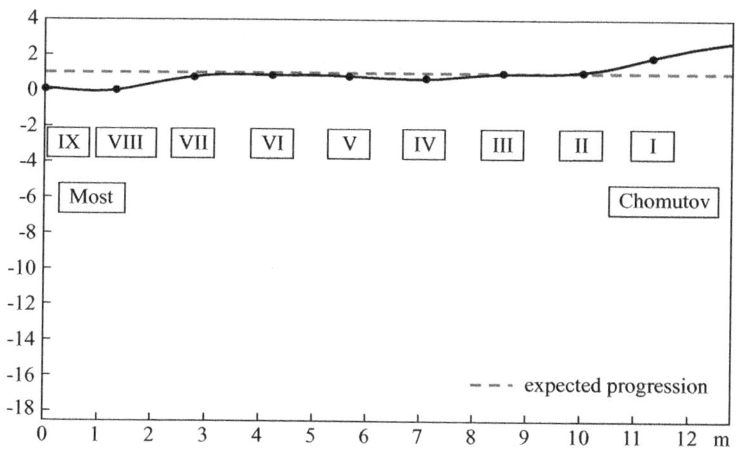

Figure 7.103 Third mode shape – transversal direction

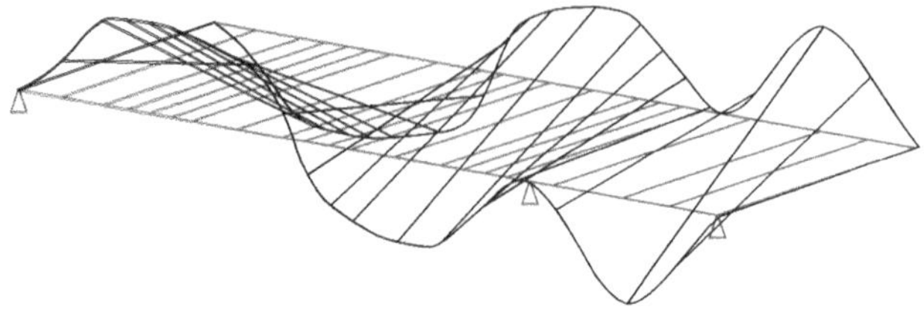

Figure 7.104 Fourth mode Shape – 10.22 Hz (2*TL*, main span)

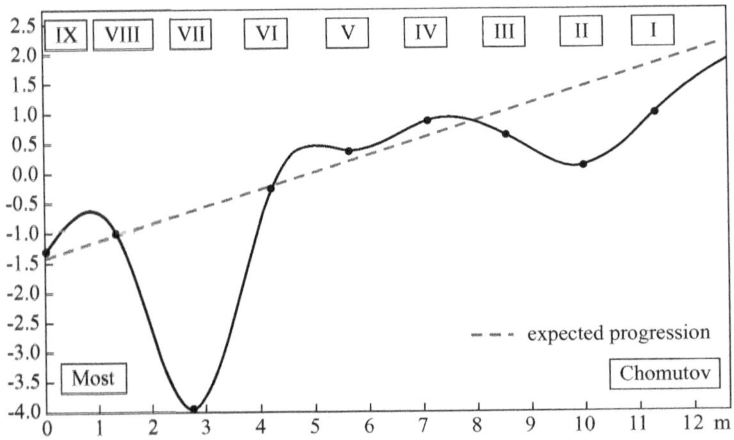

Figure 7.105 Fourth mode shape – transversal direction

7.6 Material Properties

On several occasions it has been found that material properties do not match with expectations. This is partly attributable to incorrect information in the available documents but also partly on stability changes in material over time. Particular concrete structures might experience considerable changes under adverse environmental conditions. For testing purposes it is advisable to take samples by simple means (i.e. Schmidt-Hammer).

Steel also has shown sometimes to be of different quality. This is related to the typical practice in construction to look for the cheapest solution. Unexpected low quality often has led to considerable problems. In particular the problem of the early high strength steel production should be borne in mind. In the 1950s and early 1960s the new methods to produce high strength steel produced materials that tended to brittle failure after some time. If this concerns the post-tensioning cable it presents a considerable assessment problem. In general it is recommended that a material property database be established, which is frequently updated with any new finding.

Further Reading

Allemang RJ and Brown DL (1982) A correlation coefficient for modal vector analysis. *Proceedings of the First International Modal Analysis Conference (IMAC)*, Orlando, Florida.

Cantieni R (1983) *Dynamic Load Tests on Highway Bridges in Switzerland. 60 Years Experience of EMPA*. Technical Report EMPA-Report Nr.211, EMPA - Materials Science and Technology.

Cantieni R (1996) *Untersuchung des Schwingungsverhaltens groer Bauwerke Technische Akademie Esslingen*.

Center PEER (Undated) *OpenSees, Open System for Earthquake Engineering Simulation*. http://opensees.berkely.edu.

Clough RW and Penzien J (1993) *Dynamics of Structures*, 2nd edn. Mcgraw-Hill College.

Consortium I (Undated) *IMAC – Integrated Monitoring and Assessment of Cables*. http://www.vce.at/imac/imac.htm.

Consortium S (Undated) *Scilab, a free scientific software package*. http://www.scilab.org.

Dascotte E and Vanhonacker P (1989) Development of an automatic mathematical model updating program. *Proceedings of the 7th International Modal Analysis Conference (IMAC)*, Las Vegas, Nevada.

Dascotte E, Strobbe J and Hua H (1995) Sensitivity-based model updating using multiple types of simultaneous state variables. *Proceedings of the 13th International Modal Analysis Conference (IMAC)*, Nashville, Tennessee.

DeRoeck G, Peeters B and Maeck J (2000) Dynamic monitoring of civil engineering structures. *Computational Methods for Shell and Spatial Structures IASS-IACM 2000*, Greece.

Eibl J, Henseleit O and Schlter FH (1988) Baudynamik. *Betonkalender 1988* **Band II**, 665–774.

Esveld C (2001) *Modern Railway Track*, 2nd edn. MRT Productions.

Forstner E and Wenzel H (2004) *IMAC – Integrated Monitoring and Assessment of Cables*. Final Technical Report, IMAC Project, VCE Holding GmbH, Vienna, Austria.

Forstner E, Wenzel H and Furtner P (2004) *IMAC – Integrated Monitoring and Assessment of Cables*. Deliverables d23 - part 1, IMAC Project, VCE Holding GmbH, Vienna, Austria.

Friswell MI and Mottershead JE (1995) *Finite Element Model Updating in Structural Dynamics*, 1st edn. Kluwer Academic Publishers, Dordrecht, The Netherlands.

Fujino Y (2005) Monitoring of bridges and transportation infrastructures. *Proceedings of the SAMCO Summer Academy*, Zell am See, Austria.

Inc. SI (Undated) *Ansys 10.0*. http://www.ansys.com/.

Inman DJ and Farrar CR (eds) (2005) *Damage Prognosis*. John Wiley & Sons Ltd, Chichester, England.

Jaishi B and Ren WX (2005) Structural finite element model updating using ambient vibration test results. *Journal of Structural Engineering* **131**(4), 617–628.

Maeck J (2003) *Damage assessment of civil engineering structures by vibration monitoring*. PhD thesis, Katholieke Universiteit Leuven.

Maya NMM and Silvia JMM (eds), (1997) *Theoretical and Experimental Modal Analysis*. Research Studies Press Ltd, Taunton, England.

Moller P and Friberg O (1998) Updating large finite element model in structural dynamics. *AIAA Journal* **36**(10), 1861–1868.

Mordini A and Wenzel H (2007a) Bringing experimental tests and numerical modeling together for a reliable investigation of cables. *Proceedings of the Seventh International Symposium on Cable Dynamics*. Vienna, Austria.

Mordini A and Wenzel H 2007b VCUPDATE a numerical tool for Finite Element Model Updating *Proceedings of the second International Operational Modal Analysis Conference IOMAC*, Copenhagen, Denmark.

Mordini A, Savov K and Wenzel H (2007) The finite element model updating: a powerful tool for structural health monitoring. *Structural Engineering International* **17**(4), 352–358.

Mordini A, Savov K and Wenzel H (2008) Damage detection on stay cables using an open source-based framework for Finite Element Model Updating. *Structural Health Monitoring* **7**(2), 91–102.

Peeters B and DeRoeck G (2000) One year monitoring of the Z24-bridge: environmental influences versus damage events. *Proceedings of the IMAC 18, the International Modal Analysis Conference*, San Antonio, Texas, USA.

Pichler D (1998) Concrete based floating track slab systems – modelling and reality. *Proceedings of the EURO-C 1998 Conference on Computational Modelling of Concrete Structures*, pp. 665–671, Badgastein, Austria.

Pichler D (2003) Vibration attenuating measures for railway lines – floating track slab systems. *Proceedings of the Rail-Tech Europe Conference*, Utrecht.

Pichler D and Huber P (1997) *Reduction Measures for Tunnel Lines*. Report for renvib ii phase 1 to erri, Vienna Consulting Engineers and Rutishauser Ingenieurbro.

Pichler D and Zindler R (1999) Development of artifical elastomers and application to vibration attenuating measures for modern railway superstructures. *Proceedings of the First European Conference on Constitutive Models for Rubber*, Vienna, Austria.

Pichler D, Mechtler R and Plank R (1997) Entwicklung eines neuartigen Masse-Feder-Systems zur Vibrationsverminderung bei Eisenbahntunnels. *Bauingenieur* **72**, 515–521.

Riessberger K (2002) Festere Fahrbahn auf Schotter. *Eisenbahntechnische Rundschau* **4**(2), 183–192.

Rutishauser G and Pichler D (2001) Masse-Feder-Systeme: Erfahrungen und Stand der Technik. *Proceedings of the Getzner-Congress*, Brand, Austria.

Rutishauser G, Steinhauser P, Honeger C, Flesch R, Kalivoda MT, Hasslinger HL, Schilder R and Pichler D (2003) LEO – low noise and low vibration track. *Proceedings of the LEO Seminar*, Vienna, Austria.

Rytter A (1993) *Vibration based inspection of civil engineering structures*. PhD thesis, Department of Building Technology and Structural Engineering, Aalborg University.

Savov K and Wenzel H (2004) Damage detection in a prestressed concrete test beam by means of FE-model updating. *Proceedings of the 11th International Workshop of the European Group of Intelligence Computing in Engineering*, Weimar, Germany.

Schilder R (2004) USP – under sleeper pads. *Proceedings of the VG Tagung Salzburg*, Salzburg, Austria.

Solutions DD (Undated) *FEMtools*. http://www.femtools.com.

Steinhauser P (1996a) *Rmerbergtunnel – Ergebnisse der VibroScan © Untersuchung zur immissionsmigen Abstimmung des Oberbaus*. Report to hl-ag, Vienna, Austria.

Steinhauser P (1996b) *Zammer Tunnel – Ergebnisse der VibroScan © Untersuchungen auf der Betonsohle*. Report to bb, Vienna, Austria.

Steinhauser P (1997a) *Rmerbergtunnel – Ergebnisse der Erschtterungsimmisionsmessungen des Bahnverkehrs auf dem Masse-Feder-System*. Report to hl-ag, Vienna, Austria.

Steinhauser P (1997b) *Rmerbergtunnel – Ergebnisse der VibroScan © Untersuchung auf dem Masse-Feder-System*. Report to hl-ag, Vienna, Austria.

Teughels A (2003) *Inverse modeling of civil engineering structures based on operational modal data*. PhD thesis Katholieke Universiteit Leuven. http://www.kuleuven.be/bwm/pub/phdt.htm.

Teughels A, Maeck J and DeRoeck G (2002) Damage assessment by FE model updating using damage functions. *Computers and Structures* **80**(25), 1869–1879.

Teughels A, DeRoeck G and Suykens J (2003) Global optimization by Coupled Local Minimizers and its application to FE model updating. *Computers and Structures* **81**(24-25), 2337–2351.

VCE 2007 *Testmessungen an externen Spanngliedern, Talbrücken Berbke & Rümmecke*. Short Report 06/2303-02, Vienna Consulting Engineers (in German).

Wenzel H and Mordini A (2006) Automatic FE update after monitoring of eigenmodes of cables. *Proceedings of the Symposium on Mechanics of Slender Structures MoSS*, University of Northampton.

Wenzel H and Pichler D (2005) *Ambient Vibration Monitoring*. Wiley, Chichester.

Wenzel H, Geier R and Eichinger EM (2001) Untersuchungen anlsslich des Abbruchs ausgewhlter Tragwerke. Endbericht in kooperation von vce und tu wien, VCE Holding GmbH, Wien, Austria.

Wenzel H, Pichler D and Rutishauser R (1997) Reduktion von Lärm und Vibrationen durch Masse-Feder-Systeme für Hochleistungseisenbahnen. Oral presentation at the *D-A-CH-Meeting in Zürich, SIA-Dokumentation D 0145*, pp. 123–132.

Wenzel H, Pichler D and Schedler R (1999) Ambiente Schwingungsmessungen zur System- und Schadenserkennung an Tragwerken. *Bauingenieur* **74**(3), 115–123.

8

The Business Case for SHM of Bridges

New technologies are supported during their development phases by research grants. Only a few of them manage the transformation into business (less than one out of 100). The reason why this is particularly difficult in a construction environment is the conservative environment and the competitive situation in this sector. Even the best idea might not make it to the market under these conditions.

Structural health monitoring has undergone a long development period and many useful results have been produced. Nevertheless by 2008 the transformation of these results into a business application has rarely been managed, for three main reasons.

- Structural health monitoring of bridges is a very complex issue. The principal developments concentrate on issues that the ordinary bridge owner is not interested in. A common language between technology developers and bridge owners has not been found and method statements that appeal to bridge owners are lacking.
- The gap between the expectations of bridge owners and the actual services that can be provided for the budget available is considerable. The development community has not been able to explain that the new methods do not eliminate the problem of aging or damaged bridges but are only better at being able to identify problems. Very often the monitoring campaigns that are performed, however, are so expensive they are of only scientific interest.
- The hardware involved is still very expensive and not robust, with a life expectancy of 3 years, which when compared with the life expectancy of a typical bridge of 100 years is in most situations unacceptable.

In addition there are other aspects related to differences in culture and in national practices of bridge management.

8.1 Incentives for SHM of Bridges

Three main incentives have been identified to encourage application of a reasonable SHM campaign.

- Responsibility, i.e. the existence of a standard or recommendation that obliges the owners to monitor their bridges.

Health Monitoring of Bridges Helmut Wenzel

- Economics, i.e. if SHM of bridges can prove that ultimately it saves money.
- Curiosity, i.e. owners of bridges are willing to spend money on learning more about their bridge stock, especially if in the past there has been doubt the structural condition.

The SHM community has managed to issue some guidelines and recommendations, which could form the basis for eventual application. Nevertheless they do not provide an obligation to apply SHM, and the current economic situation does not favour voluntary adoption. It therefore remains for the SHM community to show its benefits and where favourable economic returns may be made.

A good example of SHM of bridges is one that has been promoted and implemented in Austria, where regulations for bridge management allow both visual inspection and a monitoring campaign. In cases where monitoring achieves better quantified results the inspection period can be increased up to 100%. This saves money by reducing the need for inspection, which in turn can be invested in monitoring, resulting in a better service for the same costs.

8.2 The Costs of SHM of Bridges

Costs will depend on the depth of investigation. Initially it is necessary to offer a variety from simple quick investigations till permanent online structural health monitoring can be made available for cost comparison it helps to know that an in-depth visual inspection currently costs approximately € 10 000 per 100 m of bridge. This includes the bridge inspection equipment and has the added disadvantage that a lane on the bridge has to be closed. A team of two inspectors can investigate 200 m of bridge per day. They require the support of the driver of the equipment, safety personnel and very often a representative of the client. These costs are not contained in above figure. The results are a visual inspection report with an appraisal including photographic documentation and results of any tests deemed necessary (e.g. carbonatization, chloride content).

If it can be demonstrated that an SHM campaign can match this price and at the same time produce better results, two factors may be attractive to the client.

- There is no closure of a lane necessary and therefore no disruption of the traffic.
- The campaign will include a simplified visual inspection to satisfy compliance with standards.

The latter is a key point of the business case, because monitoring cannot be seen to be isolated from current practice activities, i.e. a SHM compaign will be acceptable only if it is performed by bridge engineers that are able to provide the standard of services required.

A good example where SHM shows considerable advantages against visual inspection are cable-stayed bridges. A SHM campaign provides measurements of the actual cable forces, which when compared in subsequent campaigns can provide a baseline for assessing the bridge condition. Here the alternative for the client is a lift off test, which cannot be classified as nondestructive.

In order to be able to offer these services an enterprise will require sufficient capital to invest in the expensive monitoring equipment necessary. The cost for a 32-channel SHM system is in the region of € 100 000 with a life expectation of 3 years there have to be at least 100 major contracts to justify this investment. Under current conditions this is not feasible. Therefore the business case for SHM can be justified only if it is supplied in conjunction with other services. In case of VCE Holding GmbH (Vienna), for example, SHM and related services comprise less than 10% of the total turnover. This provides the opportunity to use the SHM resource in various ways, particularly as monitoring itself would be too seasonal to become commercially successful. It should also be observed that companies offering SHM hardware cannot survive on these sales alone, which comprise only a minor part of the total business.

8.3 The Future of the SHM Business

It is difficult to say how long it will take, but a bright future can be anticipated. The direction this subject is taking is towards fully digital database solutions with geo-information systems on the surface. For every structure an online model would be available to allow automatic updating. Any record of any monitoring inspection or other campaign whatsoever would be available through the internet. Routines would be widely automated and the operator kept informed about the structural condition. Developing problems would be indicated and solutions proposed as part of an integrated SHM system. At this stage it will no longer be necessary to argue the business case for SHM of bridges. By 2020 the first such systems may be implemented and running.

8.4 Typical SHM Service Catalogue

A hierarchical approach is envisaged with four main categories, in order to satisfy existing standards and client requirements.

- Spot observation: a simple campaign using mobile equipment with one or two isolated sensors. The duration is only a couple of minutes and the costs are mainly born by traveling to and from the site. Assessment is mainly done by statistical pattern recognition methods.
- Synchronous measurements: a number of sensors are applied and monitored synchronously. This provides additional information on mode shapes and performance. Two engineers are able to perform this campaign on a 200 m long bridge within one day. The assessment is done by comparison of the measured value against a theoretical model.
- Permanent monitoring: where in-depth information is required it will be necessary to record the performance under the prevailing environmental conditions. This ideally would require a campaign of one year to cover a closed annual cycle. The assessment is done as before but the results will be improved by the consideration of environmental influences.
- Online monitoring: where there reasonable doubt about structural integrity the SHM system has to be applied online in order to deliver permanent assessment from site. This is only possible after the previous step has been performed and performance indicators have been defined. Such systems perform the assessment on line permanently and deliver results through the internet to a control station. An SMS warning is available in case of emergency.

The costs of SHM application for scientific purposes, e.g. system development, are not considered here. There are several installations, particularly in China, where systems worth several million Euros have been installed. Their primary use is to verify the design considerations and they can help to save money in future designs of major structures – they should be considered in the sense of the business case as discussed here.

As an example the service catalogue of BRIMOS® is presented here as it is one of the leading systems worldwide. It should be noted that the services are frequently updated following developments detected in the market.

8.4.1 Introduction to BRIMOS®

8.4.1.1 Ambient Vibration Monitoring

Ambient vibration monitoring (AVM) can be used for system identification and damage detection in bridge structures, due to the dynamic response of a structure to ambient excitation such as wind, traffic and microseismic activity. The BRIMOS® technology is based on AVM and has been used for many years in the field of SHM.

Structural Health Monitoring through AVM comprises the recording of the dynamic behavior by the use of measuring instruments as well as the evaluation and analysis of the measured signals. The fundamental tools of SHM are system identification (SI), damage determination and localization, as well as safety assessment and maintenance management of the infrastructure. The analysis provides the determination of the modal parameters, namely the structure's natural frequencies, its mode shapes and its damping coefficients. These parameters, which are gained from the measurements, represent the real condition of a structure and are used to update mathematical models of a structure or are simply compared to reference data from earlier measurements. The measurements are, however, so precise that they can offer reference data with a high qualitative value for every future evaluation method.

8.4.1.2 The BRIMOS® 3by3 outline

The 'BRIMOS® 3by3 outline' outlined below exemplifies the simplicity and convenience of the BRIMOS® structural assessment method.

1. **Three Structural Parameters** determine the dynamic behaviour:
   - Geometry (size, shape, moment of inertia, ...)
   - Material properties (specific weight, damping coefficients, ...)
   - Boundary conditions (support conditions, load, ...)
2. **Three Modal Parameters** describe the inherent dynamic properties:
   - Natural frequencies
   - Mode shapes
   - Damping coefficients
3. **Three Main Advantages** of AVM:
   - Testing is less time consuming and more economic
   - Testing does not interrupt the operation of the structure
   - The measured response represents the real operating conditions

8.4.1.3 Classification

BRIMOS® offers a well-defined rating system (Figure 8.1) for investigated structures. This classification allows a fast identification of the structure's integrity as well as the corresponding risk level based on modal parameters, visual inspection, FE model update and reference data. The questions to be answered by SHM are:

Detection: Is there any damage?
Localization: Where is the damage located?
Quantification: Is further use of the structure possible?
Prediction: Safety level for user.

Although there are more political approaches for facility management than there are countries, every competent authority and owner of structures is interested in:

- Economical inspection
- Reliable structural assessment
- Expert reports to assign priorities
- Savings due to well-timed and focused rehabilitation work
- Integration of RAMS (reliability, availability, maintainability and safety)

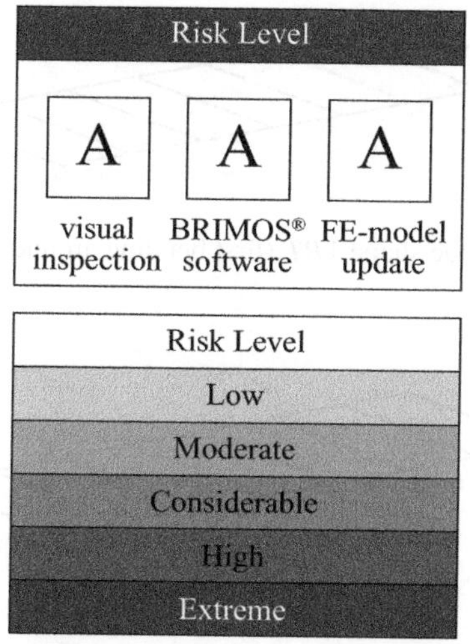

Figure 8.1 The BRIMOS rating system

8.4.1.4 Theory on Vibration Based SHM

Each structure has its own typical dynamic behavior, known as its 'vibrational signature'. Changes in a structure, such as all kinds of damage leading to a decrease in the load-carrying capacity, have effects on the dynamic response. This suggests that the measurement and monitoring of the dynamic response characteristics can be used for evaluation of structural integrity.

Different types of bridge vibration tests exist: (i) excite the bridge with a heavy shaker or dropped weight (forced vibration testing); or ambient excitation (AVM, wind, traffic, etc.). This AVM method as used in the BRIMOS® technology has the advantage that no expensive equipment is needed to excite the bridge and that the traffic has not to be interrupted.

The system identification consists of extracting the dynamic characteristics of bridges and other civil engineering structures from the vibration data. These dynamic characteristics serve as input to damage identification and model updating. The following decisive dynamic parameters are determined for system identification.

Frequencies and Mode Shapes

Natural frequencies of the structure are uniquely decided by the structural parameters and are termed the eigenfrequencies. These structural parameters are geometry (dimension, shape, moment of inertia, etc.), material properties (specific weight, damping coefficients, etc.) and the boundary conditions (support conditions, load, etc.). However, their magnitude also depends upon the way the structure vibrates, which is called the mode of vibration. For example, a bridge can vibrate during vertical bending, horizontal bending or in torsion. In vertical bending mode alone the bridge can vibrate with its maximum dynamic deflection at the span centre, or with no deflection at the mid-span but significant movement at the quarter span points. Each vibration mode has its own corresponding natural frequency (Figures 8.2–8.4).

Natural frequencies are usually evaluated under the standard design conditions, in which the structure is free of live loads and extreme temperature effects. Since the instantaneous mass and stiffness of the structure in reality could be different from the design assumptions, the natural frequencies of the bridge in service can be different from the values calculated earlier.

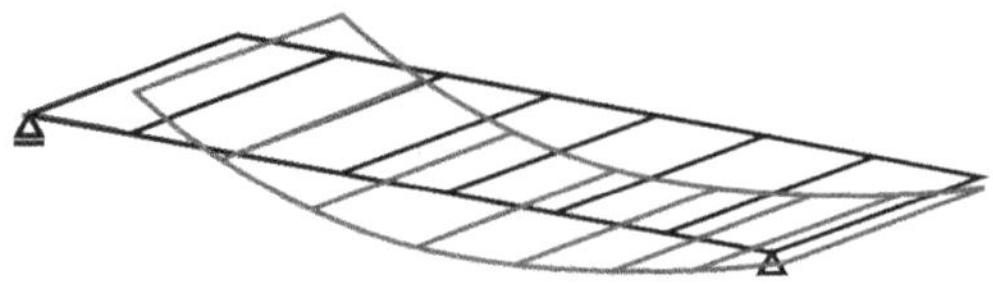

Figure 8.2 Mode shape $1BT$ (first bending around transverse axis)

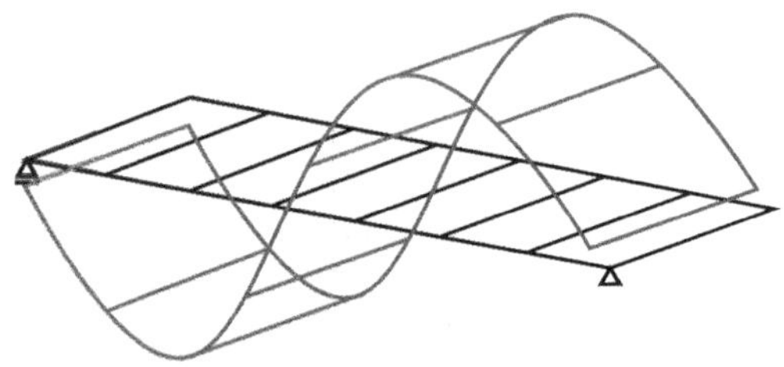

Figure 8.3 Mode shape $2BT$ (second bending around transverse axis)

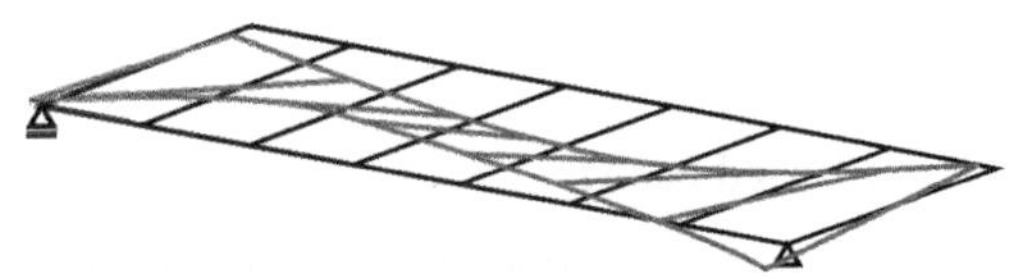

Figure 8.4 Mode shape $1TL$ (first torsional around longitudinal axis)

Damping

Damping is the capacity of structures to dissipate energy imparted by external forces (Figure 8.5). The dissipation of dynamic energy during vibration results from many different sources, such as the imperfect elasticity and internal friction of structural materials, friction of structural members at their joints and

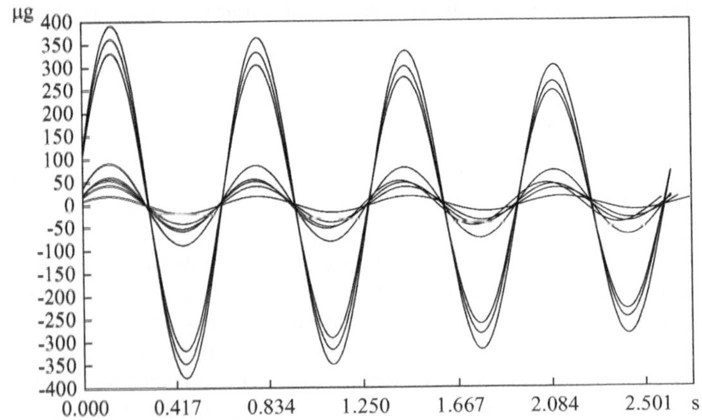

Figure 8.5 Damping pattern taken from bridge measurements after RDT processing

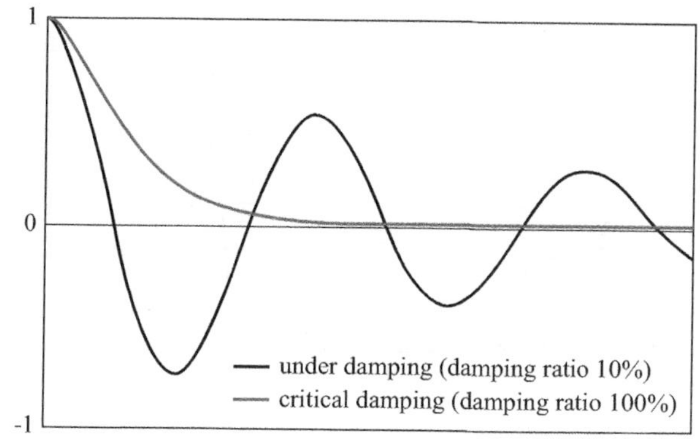

Figure 8.6 Theoretical damping patterns

support mechanisms, aerodynamic and hydrodynamic damping due to surrounding environment, the nonlinear structural characteristics, energy dissipation through foundation and substructures, damage sustained, and so on. Although the mechanism of damping is quite diverse, their overall effects on vibration is usually characterized by considering an equivalent viscous damping, crystallized in a single number of damping ratio (ζ) as a fraction of critical (Figure 8.6). If the overall damping of the system is 1% of critical, for example, the free vibration amplitude will be reduced to a half after 11 cycles, whereas the 10% damping will reduce the amplitude to a half at each cycle. When damping is at or beyond critical, there is no vibration.

Vibration Intensity

The vibration to which a structure may be subjected is usually considered with respect to its effect on the structure itself. The consideration of vibration limits is therefore becoming increasingly important for the maintenance of structural integrity for purpose. It is important to appreciate that even when the level of structural vibration is considered intolerable by the user, the risk of structural damage from sustained vibration is usually very small. Structural vibration limits for particular damage risks can be classified according to the level of vibration intensity (Figure 8.7).

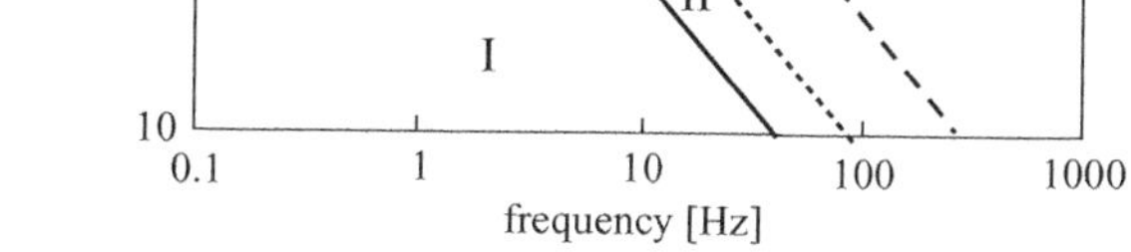

I ... no damage
II ... possible plaster cracks
III ... probable damage to load-bearing structural parts
IV ... damage to load-bearing parts

Figure 8.7 Classification of vibration intensity

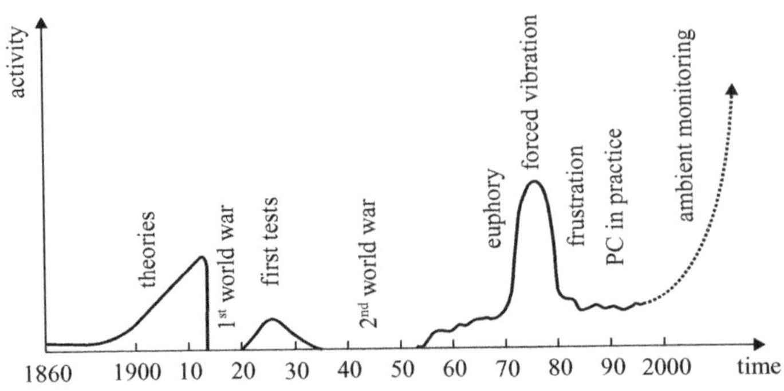

Figure 8.8 History of theoretical and practical structural dynamics

8.4.1.5 History of AVM

The possibility to draw conclusions of the condition of the structure by eigenfrequences has been known for a long time (Figure 8.8). Practical experiments on steel pylons took place in the 1920s. The first dynamic experiments were performed on radio antennas in 1941. Practical realization failed because of insufficient measurement technology at the time. In the late 1970s EMPA (Eidgenössische Materialprüfung - und Forschungsanstalt) produced a servo-hydraulic exciter with data-recording equipment, and the first series of experiments was conducted on the Deibuel Bridge in Switzerland.

Chronology of AVM

19th century: Development of the relevant structure dynamics
1920 – 1945: Execution of simple tests at clearly defined structures
1965 – 1975: Development of the linear FE method
1970 – 1980: Development of the 'forced vibration method'
1975 – 1990: Promotion of computer technology, PCs
1990 – 2000: Integration of nonlinear FE analysis
1992 – 1995: Introduction of the AVM
1993 – 1996: Introduction of computer measuring technology for data recording
Since 1994: Application of the AVM by EMPA in Switzerland, by the province of Quebec in Canada and by EDI in Vancouver
Since 1995: Further development of the method by VCE and KUL
Since 1996: Commercial utilization by VCE (BRIMOS®)
2000: More than 120 structures measured and assessed
2002: BRIMOS Recorder®
2004: More than 400 structures assessed
2005: BRIMOS® Database

History of Brimos®

BRIMOS® developed from long-standing monitoring activities of VCE. The most important development steps are listed below chronologically.

BRIMOS® 1.0: Tension measurements in tunnel shells and concrete structures (underground railway construction at Vienna , Olympic Grand Bridge in Korea, 1989).

BRIMOS® 2.0: Static monitoring of cable forces at the cable-stayed bridge in Tulln, monitoring of cracks as a result of underground railway construction (1993).

BRIMOS® 3.0: Development of frequency analysis based on FAMOS (fast analysis and monitoring of signals), FFF (Forschungsfoerderungsfond) project VCM (vibration characteristic method), switch to a multichannel system, development of ANPSDs (1996), EMPA training.

BRIMOS® 4.0: Introduction of RDT for the determination of damping, calculation of mode shapes, automatic evaluation of data records (1998).

BRIMOS® 5.0: Laser calibration, generalization of input data (channel allocation), animation of results, *MAC* (modal assurance criteria) assessment, trend analyses, intensity analysis (1999).

BRIMOS® 6.0: Change to programming language C++, development of the BRIMOS® Recorder (2001), development of classification according to BRIMOS®.

BRIMOS® 7.0: Automatic data import from measurement layout, automatic sensor calibration before every measurement, determination and animation of the mode shapes, trend investigations of the eigenfrequencies, improvement of RDT, graphical result illustration system was equipped with new sensors (2004).

BRIMOS® 8.0: Integration of a camera for videos synchronized to measurement data, Connection to the database to withdraw information (structure details, sensor layouts, channel assignment) and archive results.

BRIMOS® 9.0: New channel assignment module for the increase of channel number and high variety of sensors, extension of the BRIMOS® Database.

BRIMOS® 10.0: Introduction of a 3D coordinate system for measured structures, measurement layout and mode shape animation; integration of a meteorological station (temperature, radiation, humidity).

8.4.2 BRIMOS® Measurements

For the implementation of all BRIMOS® services VCE distinguishes three types of dynamic measurements: detailed measurement, periodic measurement, and permanent monitoring. All these types of measurements are performed by VCE. The company also develops most of its measurement equipment.

8.4.2.1 Detailed Measurement

Detailed measurements are performed by means of a compact sensor layout (Figure 8.9). These measurements allow the identification of the global system (natural frequencies) as well as local behavior (mode shapes, damping) for damage detection and localization. The measurement of 200 m of structure

Figure 8.9 Measurement grid of accelerometers along the bridge structure

Figure 8.10 BRIMOS® Recorder applied for analyzing dynamic structural behaviour of a bridge deck

per day has resulted as an average value for the more detailed basic measurement. A detailed measurement is carried out to establish a comprehensive reference data set of a structure. The modal parameters of detailed measurements can be used for exact system identification and model updating of the structure. In general detailed measurements are applied for: condition assessment, rehabilitation planning, risk assessment and quality control.

8.4.2.2 Periodic Measurement

Monitoring and verification – the most efficient employment of the system – is the execution of measurements over a time interval with the BRIMOS® Recorder. Modal parameters from periodic measurements can be used for comparison with results from detailed measurements, design values, analytical or numerical calculations as well as values from the BRIMOS® knowledge base. Dynamic examinations (Figures 8.10 and 8.11) at several demolished structures clearly showed that one sensor alone, mounted at a favorable spot on the bridge, supplies very high information content with regard to possible damage.

Periodic measurement can be applied for the monitoring of structures and single structural elements such as stay cables (Figure 8.12), tendons, etc. In general periodic measurements are applied for: quality control, condition monitoring, maintenance scheduling and attendant monitoring.

8.4.2.3 Permanent Monitoring

For special cases permanent monitoring should be taken into consideration, if there are reasons to doubt the load-bearing capacity or if high impact from traffic, wind or seismic activity threatens the structure.

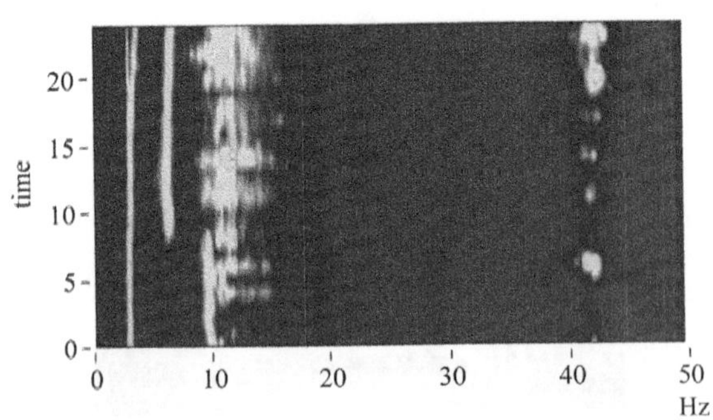

Figure 8.11 Sudden change in structural response

Figure 8.12 BRIMOS® Recorder applied for analyzing dynamic behavior of a stayed cable

For such tasks effective, tailored multichannel measurement systems are designed and installed that provide information about changed structural parameters. They encourage an advanced decision support for bridge owners in the course of maintaining their infrastructure and are based on structure-specific, objective online information by representing the bridge's hot spots. A good example is Europabrücke, see Figures 8.13 and 8.14.

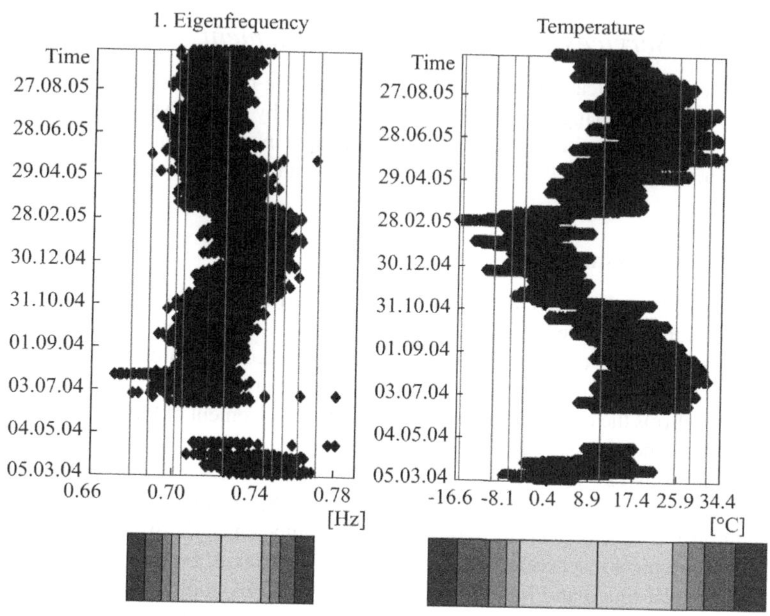

Figure 8.13 Comparison of structural stiffness (left) with the corresponding temperature loading (right)

Figure 8.14 Europabrücke, Tyrol (Austria)

Permanent monitoring benefits the preparation of statistics on traffic or environmental influences and the investigation of the resulting effects to the structures in order to trigger warning or establish alarm levels – cases registering an alert level are immediately reported to the appropriate authority via telemetry. Knowledge received from permanent monitoring data contributes essentially to the further development of SHM. In general permanent measurements are applied for: condition monitoring, lifetime assessment, traffic analysis and environmental influences.

8.4.3 BRIMOS® Services for Life-Cycle Management

There is a broad field of applications for AVM in SHM and life-cycle management of infrastructures. This section outlines the following services enabled and offered by the BRIMOS® technology: condition assessment (CA), condition monitoring (CM), rehabilitation planning (RP), quality control (QC), lifetime assessment (LA), traffic analysis (TA), risk assessment (RA) and environmental influences (EI).

8.4.3.1 Condition Assessment

Life-cycle management can be understood as an extension of the typical life-cycle cost (LCC) planning process over a structure's lifetime. Usually the LCC planning is completed within the design phase of structures without considering the actual structural conditions. Recent revaluation of LCC from existing civil structures reveals an extensive between planned maintenance actions and those that are actually necessary, which confirms the opinion that proper life-cycle management of long-term buildings requires an ongoing evaluation and assessment process (Figure 8.16).

BRIMOS® condition assessments comprise a detailed measurement, system identification for detection and localization of damage, and an expert report (Figure 8.15) explaining the current condition of a structure. Condition assessment meets the requirements of competent authorities and owners of structures by providing an economical inspection, reliable structural assessment, expert reports to assign priorities and Savings due to well-timed and focused rehabilitation work. Examples of condition assessment for Voest Bridge, an industrial smokestack at Skoenergo and a Bronze Statue (Archduke Karl memorial) are detailed below.

Figure 8.15 BRIMOS® report

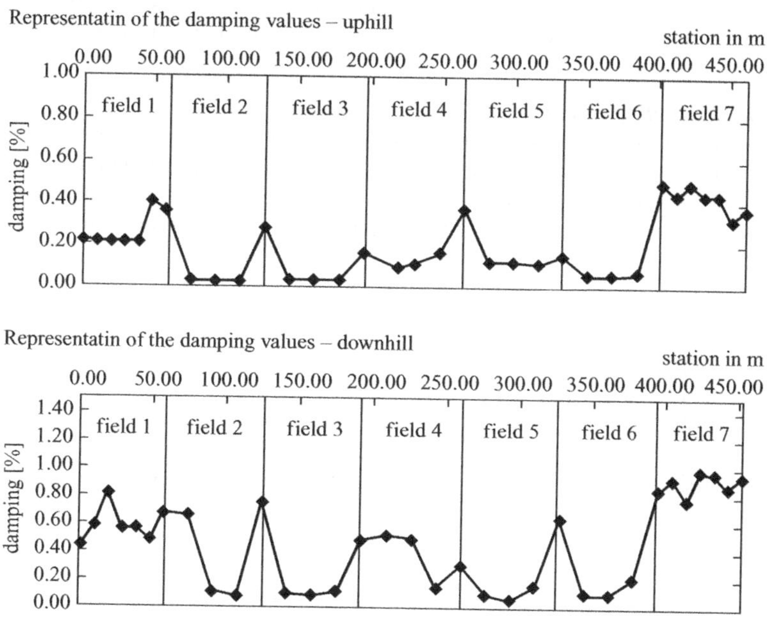

Figure 8.16 Damping pattern along a bridge structure in the longitudinal direction

Figure 8.17 Voest Bridge in Linz (Austria)

Voest Bridge

The Voest Bridge across the Danube in Linz is a six-tracked road bridge for the A7 Mühlkreis motorway with pavements on both sides (Figures 8.17–8.19). The basic system is a central beam cable stayed bridge with one pylon constrained in the dividing of the girder and three parallel cables conducted over the pylon. The guying is asymmetrically arranged. The girder is a continuous steel girder with an orthotropic deck. The stiffening girder consists of four main girders. The pylon has total height of 65.0 m. The effective span lengths amount to 2 × 60 + 72 + 215 m, the total width of the structure is 34.9 m. The overall condition assessment of the structure included the bridge substructure and superstructure as well as pylon and stay cables.

Industrial Smokestack at Skoenergo

The industrial smokestack (Figures 8.20 and 8.21) with a height of 200 m was subject to a BRIMOS® condition assessment. Detailed measurement of the structure was conducted in order to use its modal parameter for a FE model update (Section 7.1.3). The FEMU technique uses a numerical model defined by structural parameters, i.e. mass, stiffness and damping. The matrices for mass, stiffness and damping

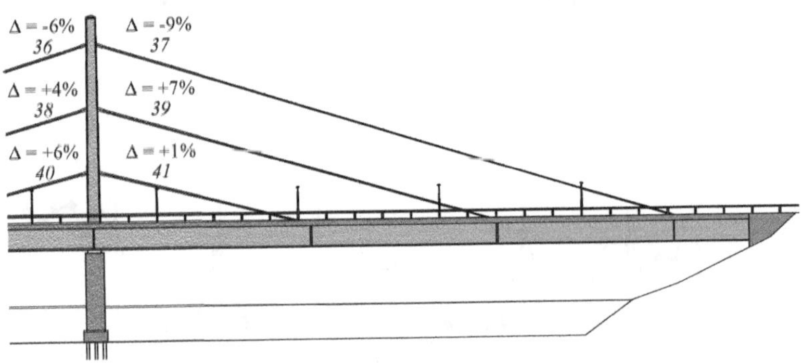

Figure 8.18 Actual cable forces versus design values

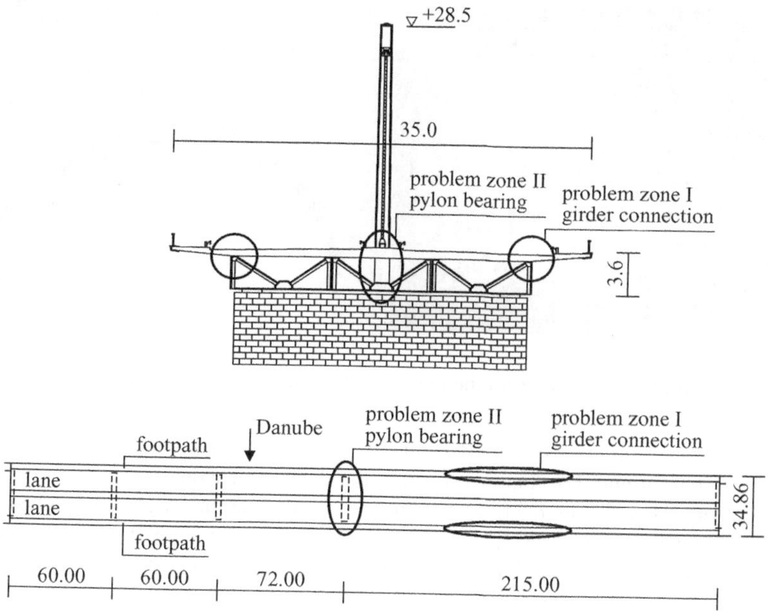

Figure 8.19 Identification of problematic zones

in the model are formulated in such a way that the model response will be almost similar to the measured dynamic response of the structure. The matrices are updated with new dynamic measurements and the updated stiffness as well as damping matrix can be compared to the original stiffness and damping matrix, respectively, in order to detect the location and intensity of damage in a structure (Figure 8.22).

Archduke Karl Memorial Bronze Statue

The approximately 12-meter-high bronze statue of Archduke Karl is placed on the Heldenplatz in Vienna (Figure 8.23). It has a weight of many tons and is one of very few worldwide monuments of a horse and rider that rests only on the hind legs of the horse. It is made of bronze with a thickness of 30 mm, whereby the most important bending stressed parts are based on forged steel. Functional efficiency and preservation are very difficult to judge just by a visual inspection.

Figure 8.20 Smokestack at Skoenergo (Czech Republic)

Figure 8.21 View from the platform (left) and from the ground (right)

With the dynamic measurement system BRIMOS® it is possible to make an assessment on the condition in respect to material fatigue. For this investigation the damping characteristics from the acceleration signal have been analysed (Figure 8.24). Furthermore a valid vibrational signature has been established and documented to generate reference parameters for future measurements (Figures 8.25 and 8.26). In comparison with an analytical model it can be concluded that there are no substantial damages in the nonvisible kernel of the statue which might provoke a dissipation of registered vibration.

8.4.3.2 Condition Monitoring

Condition monitoring is the comparison of the current dynamics parameters from periodic (Figure 8.27) or permanent measurements with reference data from earlier ambient vibration measurements. In many cases condition monitoring is applied after the detailed measurement included in the BRIMOS® condition assessment. Periodic measurements for condition monitoring can also be performed by competent authorities or operators using the BRIMOS® Recorder developed for that purpose, which is based of the following criteria.

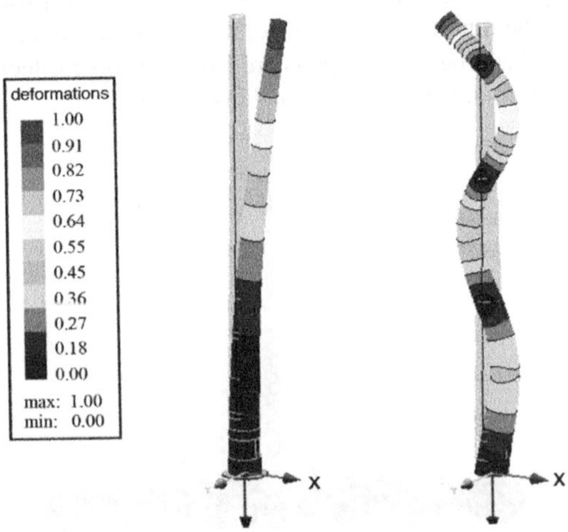

Figure 8.22 Application of FEMU to an industrial smokestack at Skoenergo

Figure 8.23 Bronze statue 'Archduke Karl' in Vienna

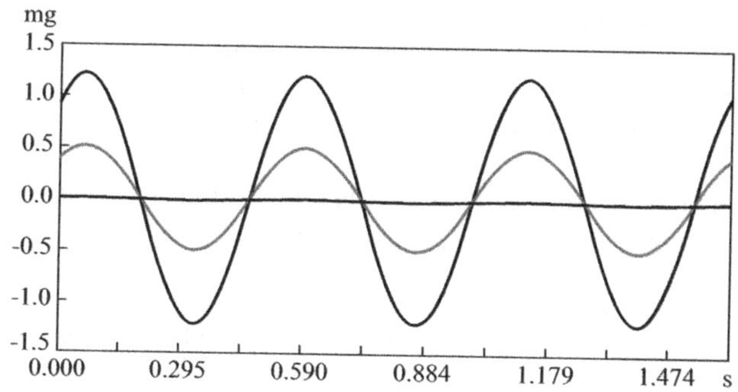

Figure 8.24 Decay curve for damping determination

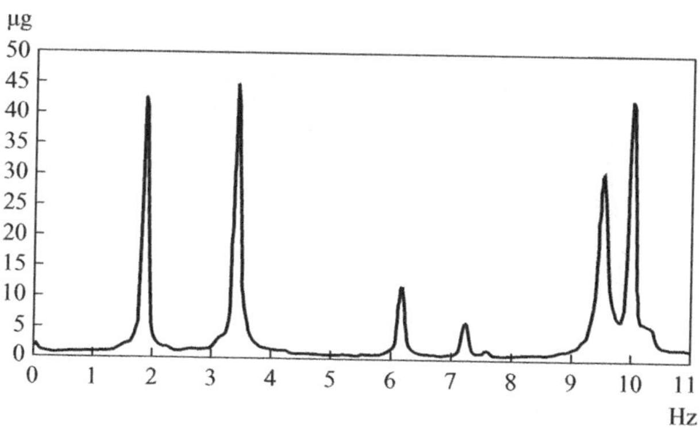

Figure 8.25 Frequency spectra

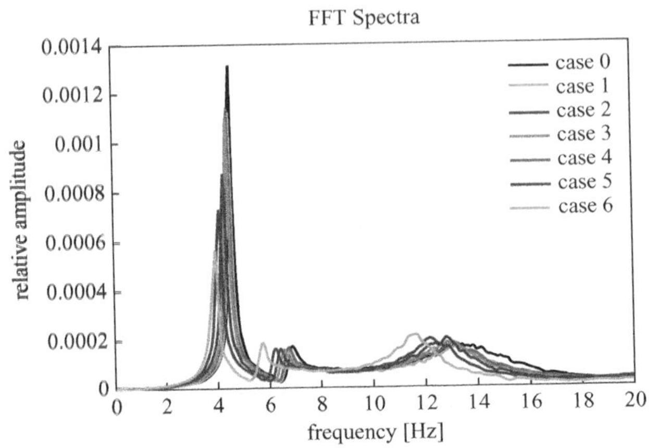

Figure 8.26 Loss of stiffness in a beam (experimental)

- A sensor at the correct location registers damages early and reliably.
- An easily operable and robust measurement unit.
- Sufficient information in measurement data from a few minutes.
- The files are read by a laptop or PC and sent to VCE by E-mail.
- The interpretation and representation is carried out by VCE.

Examples of condition monitoring for the stay-cable Rosen Bridge (periodic), stay-cable Taichung Bridge (permanent) and multispan Melk Bridge (periodic) are detailed below.

Stay-cable Rosenbrücke, Tulln

Rosenbrücke (Figure 8.28) is a new cable-stayed bridge across the Danube near Tulln and was finished in 1995. The maximum span is 177 m and the bridge consists of a 70 m high A-shaped pylon. During construction cable measurements were applied for the purpose of quality control. Since 1998 the condition of the bridge has been monitored at regular periods. Condition monitoring comprises:

Figure 8.27 Periodic measurement

Figure 8.28 Stay-cable Rosenbrücke, Tulln

- Stay-cables are checked every year to determine cable force (Figure 8.29 and Table 8.1).
- The bridge deck is measured every 3 years and compared to modal parameters derived from earlier detailed measurement.
- The pylon is also investigated every 3 years.

A BRIMOS® report about the current structural condition and statements of changes relative to all earlier measurements are provided to the client.

Stay-cable Taichung Bridge

The cable-stayed Taichung Bridge (Figure 8.30) opened in 2003 for urban traffic. Owing to the requirements to monitor the cable forces, the global condition of the structure and the dynamic behaviour of the pylon, a permanent monitoring system was installed at completion of construction. Taichung Bridge has 44 cables and a total length of 189 m. The superstructure is represented by steel girders and an orthotropic deck.

The monitoring system consists of:

- determination of the cable force of eight selected cables;
- measurement of temperature, wind speed and wind direction;
- dynamic measurement of the main girders and the pylon top;
- three-dimensional acceleration measurement of the pylon base (seimic activity).

The software supplies the cable forces in a way that the client can easily check the status of the tension force in the form of a signal light (Figure 8.31).

Table 8.1 Determination of the cable forces for Rosenbrücke

Cable number	$f1$ [Hz]	Length [m] (dampers)	Sheath [mm]	g total [kg/m]	Total cable force [kN]
1 OW	2.119	44.27	180.00	59.89	2108
2 OW	1.735	50.23	180.00	59.89	1819
3 OW	1.607	56.29	180.00	59.89	1960
4 OW	1.362	62.68	200.00	75.70	2207
5 OW	1.409	69.40	180.00	61.42	2349

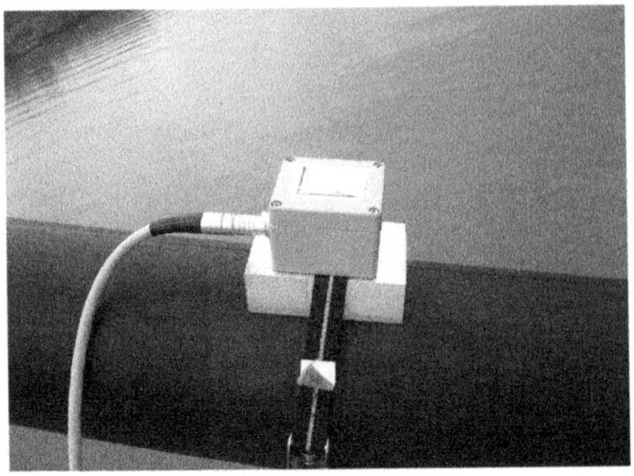

Figure 8.29 Cable measurement at Rosenbrücke using a 3D-accelerometer

Figure 8.30 Stay cable bridge Taichung

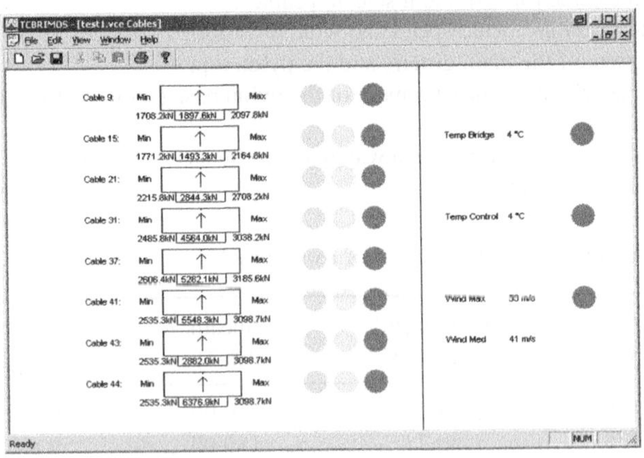

Figure 8.31 Automatic alarm system

Figure 8.32 Multispan Melk Bridge during dynamic measurement under representative conditions

Multispan Melk Bridge

Steel with an internal tendons with a strength of 145/160[KP/mm^2] called "Sigma Oval" and "Neptun *N*40" was used for in a post-tensioned bridges up to 1965. Subsequent material testing determined that certain batches were susceptible to stress cracking and corrosion. The structures concerned, such as the Melk 3Ba Bridge (Figure 8.32) are to be checked continuously by BRIMOS®. A detailed measurement of Melk Bridge for condition and risk assessment was conducted by VCE in the year 2000, and will be measured every 2 years to assess the structural condition and to assure safety.

8.4.3.3 Rehabilitation Planning

Rehabilitation planning is an additional service offered in combination to the BRIMOS® condition assessment and enables well timed and focused maintenance work (Figure 8.33). Therefore the preceding detailed measurement necessary for the condition assessment is also used for the establishment of a detailed rehabilitation plan.

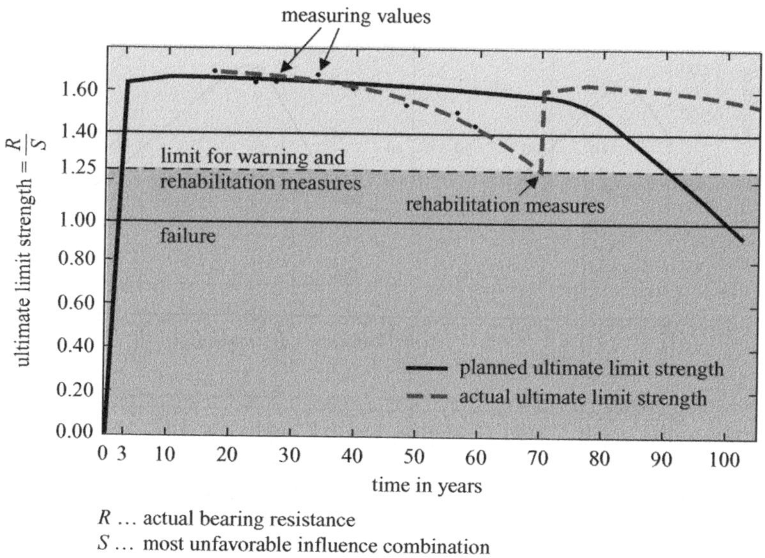

Figure 8.33 Ultimate limit strength versus service lifetime

Figure 8.34 Multispan bridge across Gurk River (left) with temporary support (right)

The preceding AVM based condition assessment avoids the following badly timed, and therefore, costly cases.

- Damage on the surface can initiate expensive rehabilitation measures without being necessary because there is actually no bad structural condition at all.
- The need of repair is recognized at a very advanced stage of deterioration, which increases the renovation expenses.

Rehabilitation measures recommended in BRIMOS® are classified into immediate, short-term, medium-term and long-term actions. An example of rehabilitation planning for a multispan bridge cross the Gurk River is detailed below.

Multispan Bridge across Gurk River

The bridge crossing the Gurk River near Rain is a multispan slap beam structure (Figure 8.34). During a BRIMOS® condition assessment a settlement of the support due to heavy traffic was identified in the analysis of the first mode shape (Figure 8.35). Furthermore deficiencies in a joint have been detected. The results from the condition assessment led to various actions:

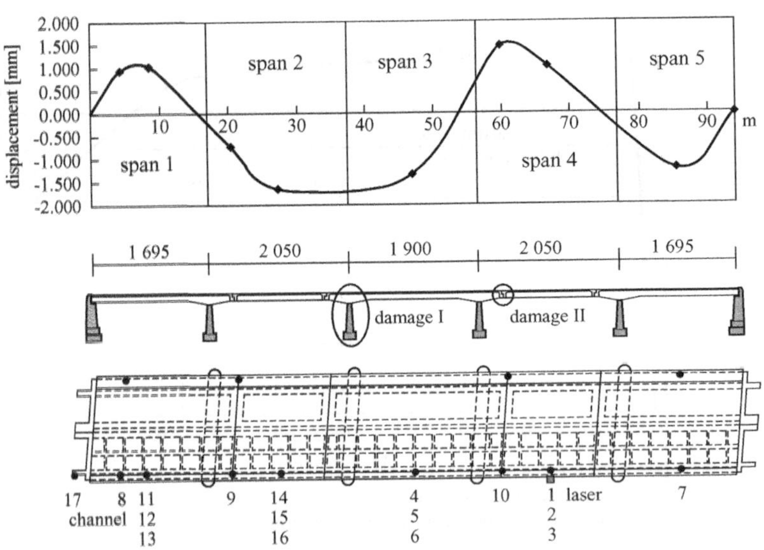

Figure 8.35 Identified deficiencies: settlement of support due to heavy traffic

Figure 8.36 Kao Ping Hsi Bridge, Taiwan (left) and Gersbach viaduct, Germany (right)

Short term	Long term
installation of a temporary support	replacement of the superstructure
speed limit for heavy vehicles	rehabilitation of the pier foundation.
annual periodic measurement until replacement	

8.4.3.4 Quality Control

BRIMOS® quality control is based on AVM to check the actual dynamic behaviour of a structure. In other words the measured modal parameters of a physical model (structure) are compared to the determined values from design (analytical or numerical model). Quality control is applied during or immediately following construction or rehabilitation in order to verify the works and services. The cantilever method in bridge engineering, where every construction phase is characterized by new load cases, is especially suitable for quality control to be done at every phase.

Another application of quality control is after construction is finished. This dynamic analysis ensures that an erected structure satisfies the design and that every element, especially support systems, work properly. Many owner and operators already a request a dynamic analysis by contract before the hand over of the structure. Examples (Figure 8.36) of quality control applications are detailed below.

Stay-cable Kao Ping Hsi Bridge

The Kao Ping Hsi Bridge in Taiwan is a cable-stayed bridge with a record-breaking cantilever of 330 m (Figure 8.37). The cross-section in the main span is a closed steel box girder and was designed to be able

Figure 8.37 Kao Ping Hsi Bridge, Taiwan

Figure 8.38 Kao Ping Hsi Bridge, construction phase

to resist extreme dynamic loads. The stay cables are located in the centre of the deck. The design of the bridge follows old Chinese principles of harmony utilizing modern methods and materials. In the design phase environmental aspects were considered in detail.

The BRIMOS® technology was applied for the observation of the changing states due to different construction levels (Figure 8.38). Stay-cable bridges built in the cantilever method especially require ongoing monitoring of the redistribution of cable forces due to changing load situations. Knowledge of the actual tensile forces is required for the assessment of every element and also for the examination of the overall construction quality of the structure.

Determination of these forces by lift-up tests with hydraulic jacks entails considerable expenses and the risk of damage. The mounting works at the anchorages can have an unfavorable influence on the durability of these critical elements. Cable force determination using BRIMOS®, however, is performed by the measurement of the dynamic parameters of the stay-cables (Figure 8.39). Considering the geometry,

Figure 8.39 Observation of cable forces

Figure 8.40 Donnergraben viaduct, Austria

mass and the bending stiffness of a cable its tension can be calculated with an accuracy of 1%. Besides the accuracy and the fact that vibration-based cable force measurement is a nondestructive method, the economic application makes it to an indispensable tool in the construction of stay-cable bridges.

Multispan Donnergraben Bridge

In 2002, for the Donnergraben Bridge (Figure 8.40) a seven span pre-stressed concrete structure with a total length of 425.15 m, rehabilitation measures through additional external pre-stressing tendons were advised based on detailed measurement. The installation of the external tendons has been monitored by VCE for quality control.

The monitoring of the pre-stressing activity was aimed at:

- Investigation of the force loss due the tendon anchoring and friction on a saddle.
- Comparison of the tension values indicated on the hydraulic jack (pre-stress protocol) and the tension values determined by dynamic parameters.

Figure 8.41 is an illustration of the inherent stress condition on tendon E2/2 in section 5/2–6, which is the first section after the hydraulic jack, and section 6/1–6/2, which is the third section after the hydraulic jack. The darker gray columns represent the tendon stress immediately after the pre-stressing procedure, which was applied according to the hydraulic jack pressure. The dynamic measurements confirm the application of the design force (99.8%) in the first section but also reveal an actual tension of 95.8% of the design load after only two deviation saddles. The lighter gray columns describe the tension when the tendon was fixed by wedges.

The pre-stressing procedure for multiple tendons gradually applies load on one tendon after another. Consequently there is possibility of force loss due to the structure shortening in the previous pre-stressed tendon even this is considered in the pre-stress sequence and design load. Since the estimated loss of the tension is based on structure stiffness and cannot be controlled after the pre-stress process is completed (hydraulic jack removed) there is a high uncertainty of the actual loss of tendon force.

Figure 8.42 presents the difference in the tension of 500 kN or about 15%. The arrangement of the results in the graph corresponds to the chronological pre-stress sequence (first pre-stressed tendon on the left, last pre-stressed tendon on the right). The eigenfrequencies which were used for the calculation of the tendon force were determined with the cable evaluation software. This automatic tool provides

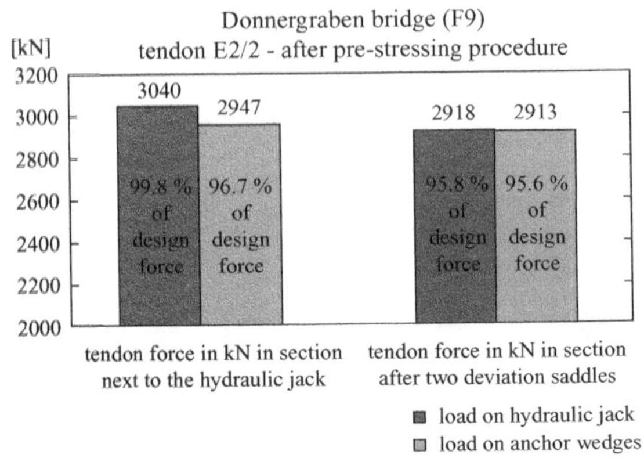

Figure 8.41 Tension force on jack and anchor wedges, Donnergraben viaduct

eigenfrequency values which are adjusted to the influence of the support conditions as well as the bending stiffness of the cable.

8.4.3.5 Lifetime Assessment

The lifetime of a structure depends on the number of consuming events experienced, such as dynamic loads, and the resistance available against this usage (Figure 8.43). Neither parameter is well defined during the design phase and they represent underlying changes during the whole lifetime of a structure. Long-term monitoring (periodically or permanent) of an infrastructure in combination with traffic analysis enables lifetime assessment based on AVM. Reasons for the divergence of the lifetime defined in the design phase and the actual lifetime are for example:

- new rehabilitation methods, better materials
- different load conditions (e.g. traffic growth, higher dynamic loads)

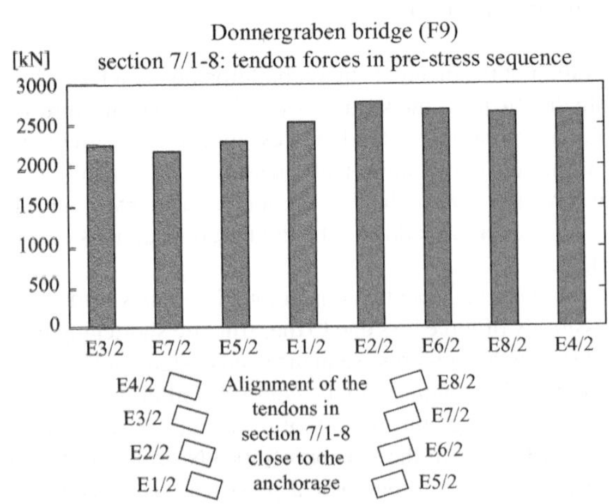

Figure 8.42 Tension forces in one section, Donnergraben viaduct

Figure 8.43 Bridges as consumable (left) and consumed lifetime (right)

- quality of building construction
- unsteadiness of long-term changes in material properties (e.g. fatigue, relaxation)
- short-term changes due local damages
- changes in environmental influences.

An example of lifetime assessment is detailed below.

Large Span Steel Box Girder Europabrücke

The Europa Bridge near Innsbruck Austria, which opened in 1963, is one of the main alpine north–south routes for urban and freight traffic (Figure 8.44). Currently the bridge stress load is contributed by more than 30 000 motor vehicles per day. The combination of measuring and analytical calculation over the past years has led to detailed system identification. Due to the requirement to assess the prevailing vibration intensities with regard to fatigue problems and possible damage, a permanent measuring system was installed in 2003. Assessment of the remaining lifetime considering the actual traffic development is of major interest to the operator. The assessment is done by combination of measurement data and analytical calculation. Analysis of the measured data with the help of the rainflow algorithm provides information about recurring response cycles in different categories of intensity and occurrence (Figure 8.45). The major goal is to determine the relation between the randomly induced traffic loads (vehicles per day) and the fatigue-relevant, dynamic response of the structure. As lifetime predictions in modern standards

Figure 8.44 Europabrücke, Innsbruck, Austria

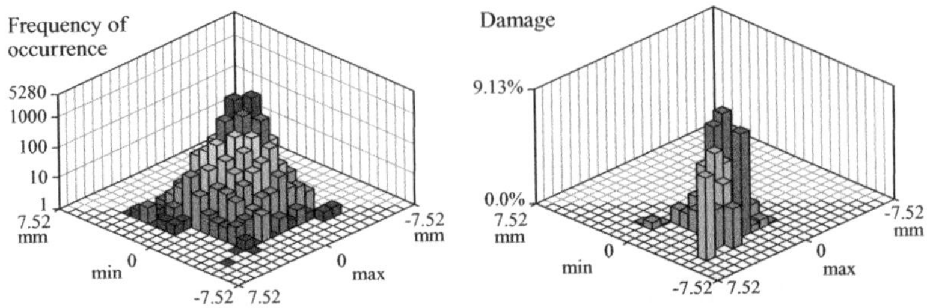

Figure 8.45 Rainflow matrix (counting, left) and damage matrix (assessment, right)

depend on many assumptions, the emphasis is to replace all estimates by measurements, with focus on the following contexts.

- Global behaviour in response to all relevant loading cases.
- Cross-sectional behaviour with special consideration of the cantilever regions.
- Local systems analyzing the interaction between tires and the beam-slab connections.

At each of these levels of analysis the consumption of the structure's overall capacity per year is to be determined.

On the basis of statistically varying Wöhler curves the number of occurring stress ranges at a certain level is compared to the relating number of allowable cycles. In this way the structure's gradually developing material-fatigue is represented by remaining capacities of loading cycles for certain analyzed details. Additionally the fatigue-relevant traffic is extracted from the known, randomly induced traffic volume.

8.4.3.6 Traffic Analysis

An ongoing increase of traffic volume and higher cruising speeds result in a redefinition of dynamic loads to bridges and affects the service life of structures. For this reason traffic analyses are often implemented in combination with lifetime assessment. BRIMOS® traffic analyses systems measure the dynamic load introduced to the superstructure (Figure 8.46). As bridges have a distinctive dynamic response and

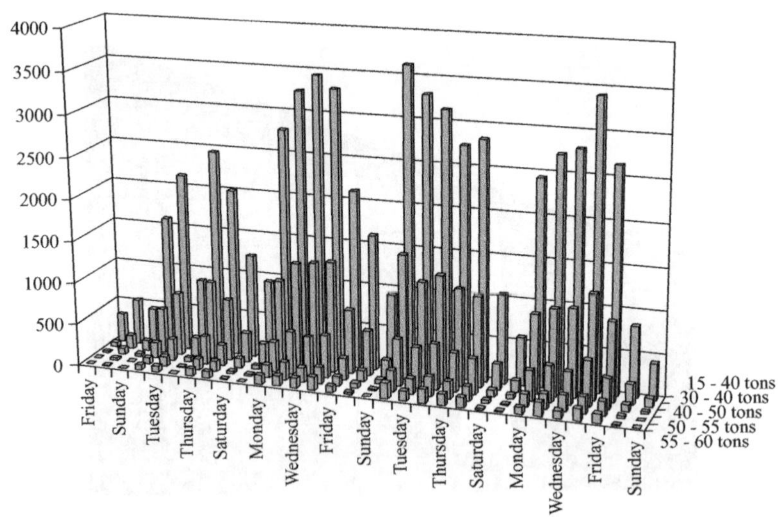

Figure 8.46 Load-dependent event counting

Figure 8.47 Video monitoring at St. Marx

constitute bottlenecks in the infrastructure they provide an ideal location for traffic analysis. This system does not require any construction work and traffic is not affected at any time. Currently VCE is involved in national research projects regarding traffic analysis and infrastructure telematics. For research purposes traffic analysis systems are often equipped with video monitoring (Figure 8.47). Examples of traffic analysis are detailed below.

Large Span Steel Box Girder Europabrücke

The Europa Bridge near Innsbruck, opened in 1963, is one of the main alpine north-south routes for urban and freight traffic (Figure 8.48). Currently the bridge stress load is contributed by more than 30 000 motor vehicles per day. The combination of measuring and analytical calculation over the past years has led

Figure 8.48 Europabrücke, Tyrol, Austria

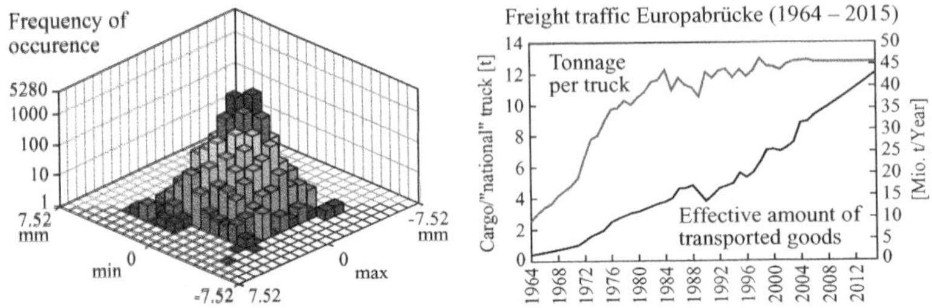

Figure 8.49 Rainflow matrix (counting, left) and freight traffic from 1965 to 2015 (right)

to detailed system identification. Due to the requirement to assess the prevailing vibration intensities with regard to fatigue problems and possible damage, a permanent measuring system was installed in 2003.

The major goal is to determine the relation between the randomly induced traffic loads (vehicles per day) and the fatigue-relevant, dynamic response of the structure. As lifetime predictions in modern standards depend on many assumptions, the emphasis is to replace all these estimates by traffic analyses. Data evaluation with the help of the rainflow algorithm provides information about recurring response cycles in different categories of intensity and occurrence (Figure 8.49). The fatigue-relevant traffic is extracted from the known, randomly induced traffic volume. A synergetic usage of the method with prevailing, past and future traffic data leads to a prediction of the remaining operational lifetime, which is considerably more precise and with less scatter than is likely by conventional methods.

Multispan St. Marx flyover

The St. Marx flyover in Vienna, Austria, built from 1973 to 1978, is located between the Danube Canal and the traffic node Landstrasse (Figure 8.50). The bridge is considered one of the most busiest sections of the A23 South-East Highway. The total traffic volume averages about 240 000 motor vehicles per day, but an increase of the ratio of heavy loads has been detected. Thus, in order to detect the passing heavy loads, which cause damage, a traffic analysis system in combination with a video control system were installed in 1998. On the basis of a permanent analysis of the dynamic structural behavior possible issues to be considered are as follows:

- Determination of passing heavy loads causing structural damage.
- Verification and update of the existing load models.

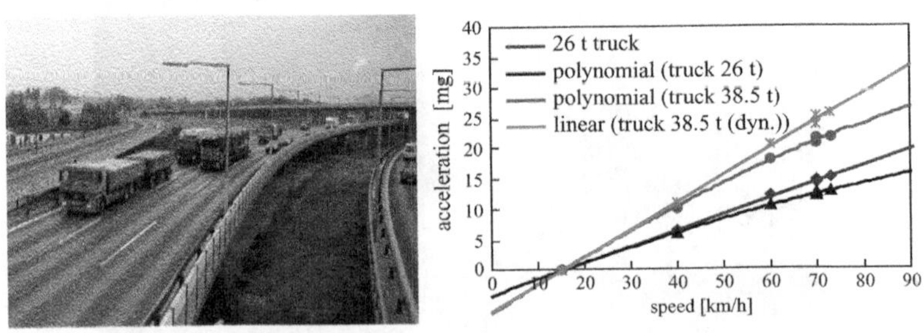

Figure 8.50 Süd-Ost Tangente near St. Marx, Vienna (left) and system calibration with test runs (right)

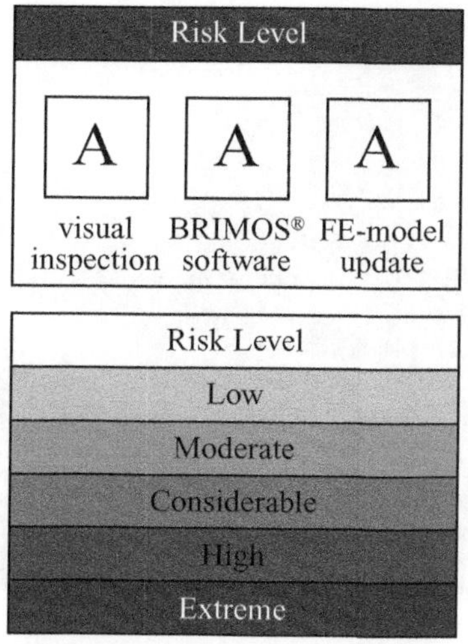

Figure 8.51 BRIMOS® classification

- Determination of the overall load configurations and vibration coefficients, with wind and temperature effects as optional considerations.
- Monitoring of the structural loading capacity and serviceability by means of structural identification.

8.4.3.7 Risk Assessment

BRIMOS® services dedicated to the risk assessment of a structure investigate not only the condition of a structure in respect to maintenance but determine the overall actual safety level to the users (Figure 8.51). Depending on the kind of a structure as well as its current safety level different BRIMOS® measurements (detailed, periodic and permanent) are applied for risk assessment. If a safety level is classified to be critical, short-term action has to be taken. Such immediate action can be constructive measures, or restrictions for utilization, or a ban on operation. Under these circumstances ongoing monitoring (permanently or periodically) is recommended for early identification of possible impairment until the risk level is classified to be safe. Examples of risk assessment are detailed below.

Concrete Schlegeis Dam

Earthquakes and rock slides are the most dangerous dynamic loads a dam can be exposed to. To analyze the resistance to such forces the Schlegeis dam in Tyrol (Figure 8.52) was subject to risk assessment based on the BRIMOS® technology (Figure 8.53). The modal parameters obtained by BRIMOS® have been used for a FEMU (see Section 7.1.3). The model matrices for mass, stiffness and damping are formulated in such a way that the dynamic parameters of the numerical dam model are similar to the measured dynamic response of the structure. This procedure allowed determination of the structural condition as well as the definition of unknown boundary conditions (e.g. dam–subsoil and dam–water interfaces). The obtained FEMU (Figure 8.54) is suitable for simulation of different natural hazard scenarios. An ongoing monitoring (permanently or periodically) is recommended for early identification of possible impairment until the risk level is classified to be safe.

Figure 8.52 Schlegeis dam, Tyrol

Rinterzelt Waste Treatment Plant

Safety is an important aspect for technical systems such as buildings and has to be considered in the design phase as well as during operation. The roof of the waste treatment plant at Rinterzelt (Figure 8.55) is supported by a wooden structure, load-bearing capacity of which is influenced by intense loads. Dynamic loads resulting from intense winds are weakening the junctions of the wooden structure. In 1997 a permanent system was installed to monitor the safety level of the building. The data acquired (Figure 8.56) are used to study the influence of higher wind loads on the structure as well as assess risk assessment and update the structure's safety level.

Damaged Railway bridge, Krems

In December 2005 a Danube push tow crashed into pier 6 of the railway bridge crossing the Danube near Krems. The pier was heavily damaged (Figures 8.57 and 8.58) and the bridge was closed for operation

Figure 8.53 Sensors at the dam crest

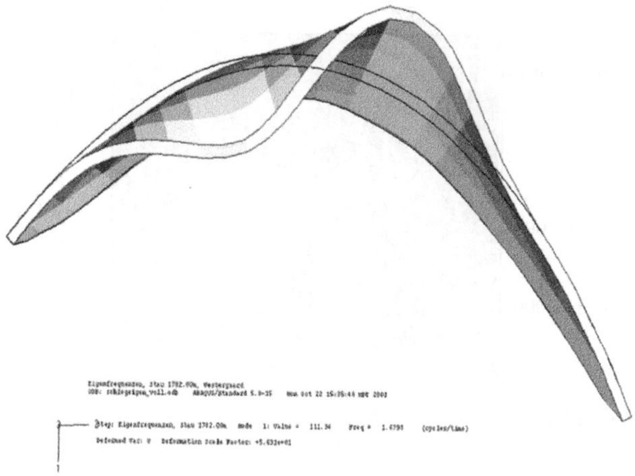

Figure 8.54 Finite element model update, Schlegeis dam

immediately. The pier sustained a shear fracture shifting the upper part 2 m upstream. On the surface of the lower part many cracks have been identified by scuba divers.

Two scenarios for the collapse of the bridge have been considered.

- Failure of the substructure, which could be identified early by tilting.
- Fracture of the upper part of the pier due to the load of the structure.

BRIMOS® was applied for a more detailed assessment of the structural condition of the bridge. Constraints which had been introduced to the bridge structure due to the displacement of the support system were determined. After the initial condition assessment, a system for continuous risk assessment was installed for safety reasons in respect of workers doing the preparation for the removal of the superstructures.

Figure 8.55 Rinterzelt Waste treatment plant

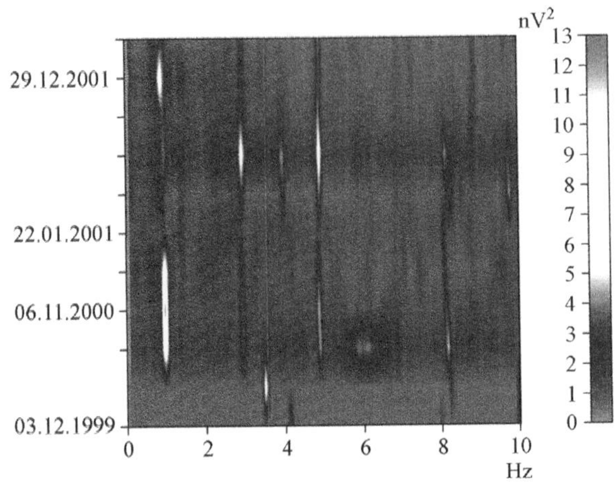

Figure 8.56 Trend card acquired at Rinterzelt waste treatment plant

8.4.3.8 Environmental Influence

Environmental influences such as temperature (Figure 8.59) or humidity mean considerable stress for a structure as heavy additional loads can be induced to the structure. If measurement of the dynamic behavior is called for in the assessment of the condition of the structure and if SHM is the aim, it is necessary to distinguish between normal changes of the dynamic behavior and extraordinary changes (damage). Normal changes in the dynamic behavior of a structure are caused by variation of environmental influences such as humidity, wind, traffic, solar radiation and, as a decisive factor, temperature. Temperature exerts a decisive influence on the boundary conditions of the structure (Figure 8.60) as for example frozen ground or the elastic modulus of the building material.

It is obvious that normal variations in dynamic behavior should not lead to a false interpretation of the actual condition of the structure and the degree of damage. Normal variations exert a harmless influence

Figure 8.57 Damaged pier, Krems

Figure 8.58 Shifted rail track (left) and damaged support system (right), Krems

on the load-bearing behavior, whereas extraordinary changes can lead to a critical condition for safety. Such extraordinary changes of the dynamic behavior are often caused by a loss of stiffness through damages (e.g. crack formation) or by changed bearing conditions. Examples of environmental influences are detailed below.

Olympic Grand Bridge, Korea

The Olympic Grand Bridge in Korea (Figure 8.61) is a symmetric cable-stayed bridge with one pylon. The stay cables are arranged fan-like in the center of the structure and were stressed from the 88 m high pylon. The form of the pylon head symbolizes the Korean imperial crown. Environmental conditions, above all temperature, influence the results of the dynamic investigation. Therefore an accurate knowledge of these influences is required. For the establishment of a characteristic relationship between temperature and frequency a permanent monitoring system was installed on the Olympic Grand Bridge for long-term investigations. Results from measurements on various structures show that the change in frequencies due to temperature essentially depends on the total stiffness of the structure. For more flexible structures, such as the Olympic Grand Bridge (OGB) this influence amounts to 0.2% per 10°C temperature change

Figure 8.59 Rosenbrücke at Tulln in winter

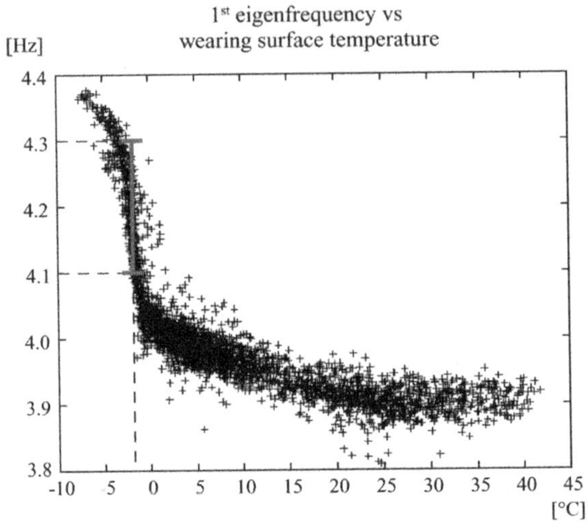

Figure 8.60 Influence of temperature on frequency; record from Bridge Z24, see Section 9.15

(Figure 8.62). What has to be particularly considered is, however, the phenomenon of stiffening in the case of negative temperatures. A separate study is to be dedicated to this phenomenon and its effects.

High Rise Buildings, Dresdner Bank

Changing environmental influences refers to the variation in the Earth's global climate or regional climates over time. It describes changes in the variability or average state of the atmosphere – or average weather – over timescales ranging from decades to millions of years. These changes may come from internal processes, be driven by external forces or, most recently, be caused by human activities. For this reason of climatic change worldwide, but especially in major cities, high rise buildings must resist higher wind loads. Furthermore mutual dynamic effects with alternating wind velocities playing an important part in their dynamic stresses if high rise buildings stand close together.

Figure 8.61 Olympic Grand Bridge, Korea

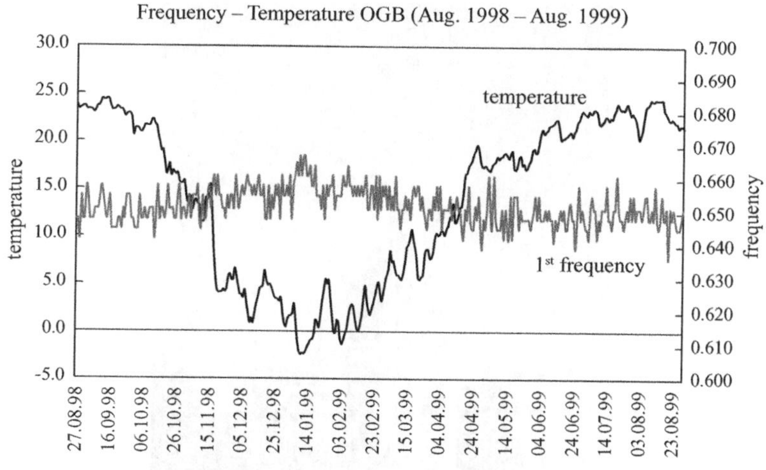

Figure 8.62 Relationship between temperature and frequency, Olympic Grand Bridge

For a research study within the European project SAMCO (Structural Assessment Monitoring and Control) detailed analyses to investigate the impact of wind loads on high rise buildings and towers have been performed. Monitoring systems have been installed in the Dresdner Bank and the Commerzbank (Figures 8.63 and 8.64) in Frankfurt for permanent recording of dynamic parameters as well as wind velocity and wind direction in order to obtain information in this context.

Figure 8.63 Dresdner Bank, Germany

Figure 8.64 Commerzbank, Germany

8.4.4 Special Measurements

8.4.4.1 Attendant Monitoring

Attendant monitoring using the BRIMOS® technology is usually applied when a planned extraordinary load is introduced to a structure. This higher loading can result from either a high dynamic or static load. Heavy construction machines working on the removal of pavement (Figure 8.65) for rehabilitation reasons may damage a structure due high dynamic loads. If a superstructure is to have a renewed pavement, the contractors are informed of the maximum dynamic stress allowable. With the aid of attendant monitoring during demolition works using the BRIMOS® Recorder, immediate reaction is possible when accelerations overstep the determined safetey limit (Figure 8.66), thereby avoiding damage.

Figure 8.65 Monitoring of demolition work

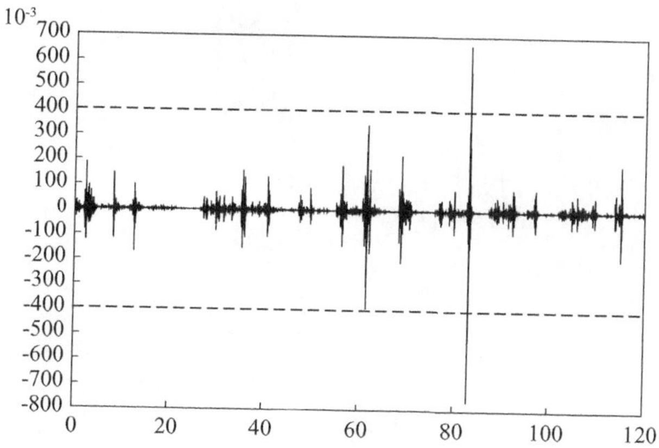

Figure 8.66 Assessment of effective acceleration with regard to predetermined threshold levels

Another application of Attendant Monitoring is the monitoring of high static loads such as heavy load crossings on bridges. In such an application the BRIMOS® technology can be used to verify whether a structure is harmed during passage of a heavy load (Figure 8.67). Therefore dynamic measurement before and after the passage is carried out. In this example the results of the subsequent measurement are compared with the initial records in order to identify possible changes to the structure due to the high static load.

8.4.4.2 Noise and Vibration

Noise and vibration is a sensitive subject between any railway operator and property owners alongside the track (Figure 8.68). This subject concerns not only the construction period but also the operation. Therefore measurements of prevailing noise and vibration are of great interest for both parties. In recent years the superstructure of rail track systems is implemented as a mass-spring-system (Figure 8.69),

Figure 8.67 Monitoring the passage of a heavy load

Figure 8.68 Measurement of railway-induced vibration, Vienna

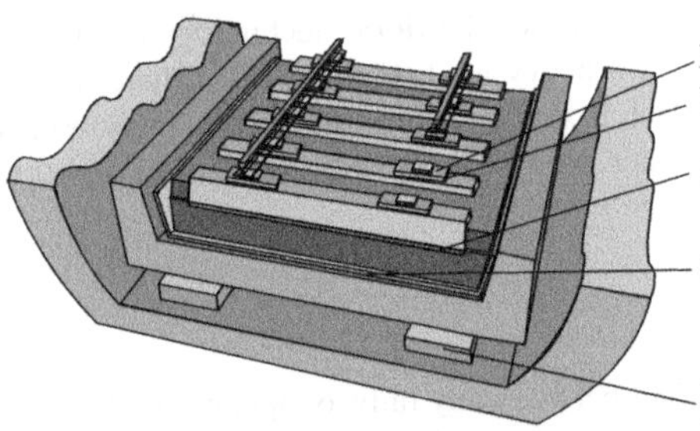

Figure 8.69 Mass-spring system

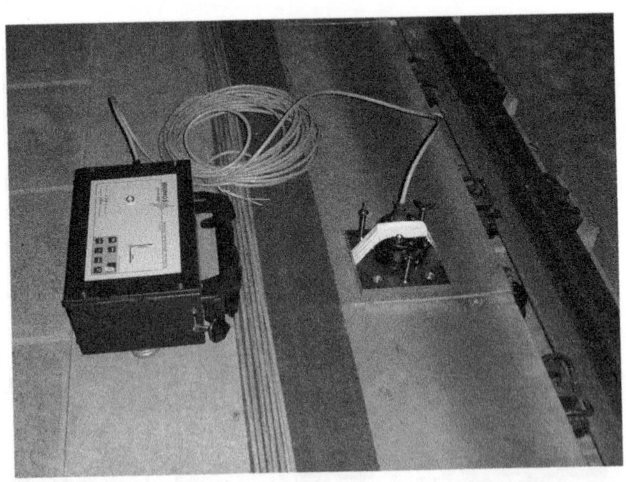

Figure 8.70 Dynamic measurement of a mass-spring system

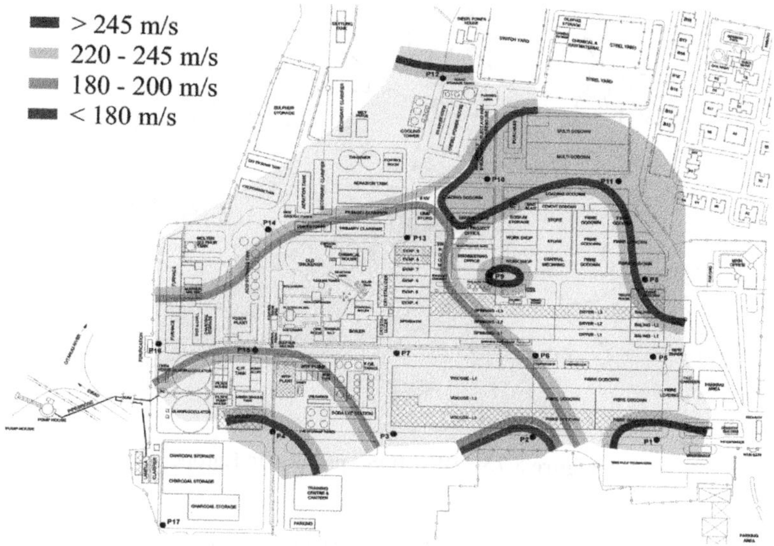

Figure 8.71 Map of wave propagation (by VCE)

Figure 8.72 Measurement of soil acceleration

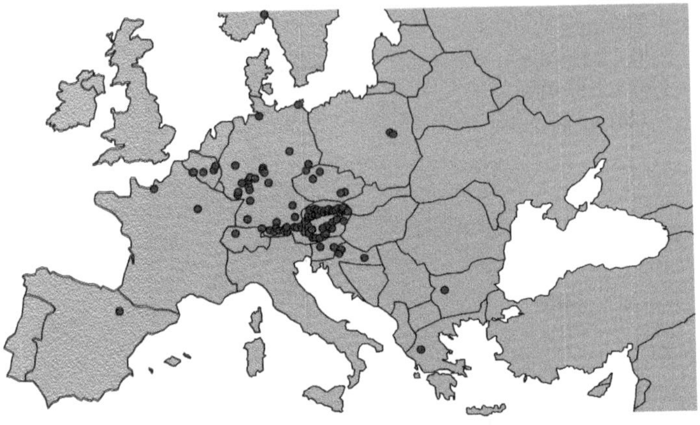

Figure 8.73 BRIMOS® projects in Europe

especially in urban areas. Railway-induced vibrations are generated by the vehicle-track interaction. The whole vehicle–track–soil system is excited by irregularities. There are very different irregularities of the vehicle, the track and the soil. At low frequencies, there are the track irregularities due to imperfect track alignment. At high frequencies, there is the unevenness of the rail (for very high frequencies, it is called the rail roughness). A specific excitation due to the track is the excitation with the sleeper-passage frequency.

Measurements of noise and vibration using the BRIMOS® equipment are applied to:

- verify the proper function of mass-spring-systems (Figure 8.70);
- supporting the research and development dedicated to the improvement of mass-spring-systems.

Figure 8.74 BRIMOS® projects worldwide

The results can also be used for the determination of the effective vibration level, with the results (k_B values) compared with the limit values of valid standards (DIN 4150, OENORM 5100).

8.4.4.3 Seismics

Since earthquakes place extraordinary stress on buildings, assessment regarding earthquakes using Nakamura's Method is another application of dynamic analysis (Figure 8.71). Since the natural soil frequencies overlap the relevant frequencies of a building, collapse during seismic activity can be the consequence. Knowledge about the dynamic characteristics of a building as well as the ground it stands on provides engineers with the necessary information to adapt the structural design to the prevailing ground condition.

Nakamura hypothesized that site response could be estimated by dividing the horizontal component by the vertical component of noise spectra. This ambient approach is attractive given the relatively simple procedure of data collection and the fact that it can be applied in areas of low or even no seismic activity. In addition, impact tests can be performed in order to more accurately identify ground response. Several studies have shown that Nakamura's procedure can be successful in identifying the fundamental resonance frequency of sedimentary deposits. This hypothesis has been tested by results of theoretical modeling. The peak observed in the H/V ratio could be related to the horizontal polarization of the fundamental mode of Rayleigh waves at the resonance frequency of the stratigraphy. The BRIMOS® measurement equipment also can be used for the investigation of the dynamic parameters of the subsoil (Figure 8.72).

8.4.5 Distribution of BRIMOS® Projects

VCE Holding GmbH has successfully finished more than 400 BRIMOS® projects worldwide (Figures 8.73 and 8.74).

9

Applications

9.1 Melk Bridge M6 Austria

Contributed by Arsenal Research GmbH

Project Description

In the framework of an Austrian research project focused on noise emission and vibration transmission, the project was related to the global and local vibration behavior of railway bridges to study the sound emission radiated by steel structures. A detailed monitoring campaign was implemented consisting of ambient and forced vibration testing of the global structure, testing of the steel webs and measurement of the related noise emission. Using these data, an extensive comparison between the noise radiation of steel and concrete bridges was performed.

Brief Facts

Name and location: Melk Bridge M6, Melk, Lower Austria
Owner: Hochleistungsstrecken (HL) AG, Austria
Structure category: large span composite bridge
Spans: 5, 53.0 + 53.0 + 79.0 + 53.0 + 36.0 m; total length 274.00 m
Structural system: composite bridge, steel main girder, concrete slab deck
Start of SHM: July 2001
Number of measurment points: 40
Instrumentation design by: Arsenal Research GmbH, Business Area Transport Route Engineering, Vienna, Austria

Description of Structure

The bridge (Figure 9.1), erected in the year 2000, is an integral part of the track upgrade of Austrian Federal Railways. The design and realization of the project was by the Austrian HL AG. The bridge consists of a continuous composite beam with five spans. The total length of the structure is 274.0 m. The deck width is sufficient to bear parallel tracks for the high speed railway traffic. The bridge plan view is slightly curved; therefore the bridge must bear high transversal loads. The high design speed of the track requires dynamic investigations for the structure.

Purpose of Inspection

Main purpose of the investigation performed in this case was the check and further development of the technology to assess the noise emission. Therefore the noise emission was measured during a train passage. This measurement was compared to the frequency response of the girder webs. From the investigation it was shown, that the vibration of the web is mainly responsible for the noise emission (sound

Health Monitoring of Bridges Helmut Wenzel

Figure 9.1 Melk Bridge M6, Austria

speaker effect). Moreover an initial dynamic investigation was performed to assess the global response of the structure. This measurement could serve as base for future implemented monitoring concepts, modal updating for damage assessment and localization and for upgrading the noise emission from the structure.

Measurement Equipment and Data Management (Table 9.1)

Type of system: mobile measurement system based upon ambient and forced vibration technology.
Data management:

- data pre-analysis (natural frequencies) on site in order to check the obtained data quality
- main analysis, graphical presentation and documentation in office
- data transfer by data cables to main station
- use of Reaction Mass Exciter VICTORIA in order to excite all relevant modes of vibration.

Data Analysis Procedures

Type of analysis: identification of modal parameters by system identification, additional forced vibration testing.
Software: stochastic subspace identification (MACEC).
Additional features: application of the Reaction Mass Exciter VICTORIA and additional measurement of sound emission during train passage.

Table 9.1 Sensor details for Melk Bridge M6

Type of sensors	Number	Location
Acceleration Sensors	8	Aligned on bottom flange and girder web of the structure
Reaction Mass Exciter	1	Close to each pile of the structure
Load Cell	1	Integral part of VICTORIA
PT100 (Temperature)	1	Surface temperature

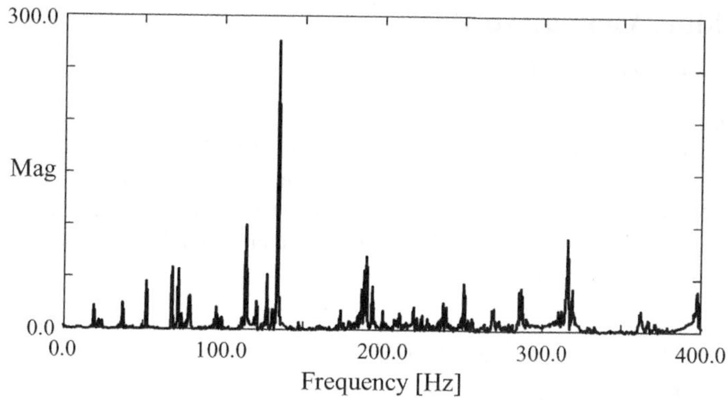

Figure 9.2 Response of a web element, Melk Bridge M6

Examples of Outcomes

The comparison between ambient and forced vibration testing has shown very good correspondence. The frequency of the first vertical mode is $f = 1.86\,\text{Hz}$. The second vertical mode was identified at $f = 3.31\,\text{Hz}$ representing typical vibration shapes. Due to the curved shape of the structure, all modes have a major transversal and torsion component. The analysis of the web elements has shown frequencies in the range of 19.27 Hz, 21.91 Hz and 24.22 Hz (Figure 9.2). Main energy is represented by the first mode of the web (Figure 9.3). The modes of the web elements have been identified by a detailed FE model.

Benefits of Using SHM Technologies in the Project

By performing these global and local vibration tests, an initial measurement of the structure was established which could be used for further monitoring activities. Moreover a FE model was created which was updated to the experimental results. Based upon both data sources (measurement and simulation) a reliable monitoring concept could be applied for the structure. Moreover it turned out that the sound emission of steel structures mainly comes from the local vibration behavior of the girder webs.

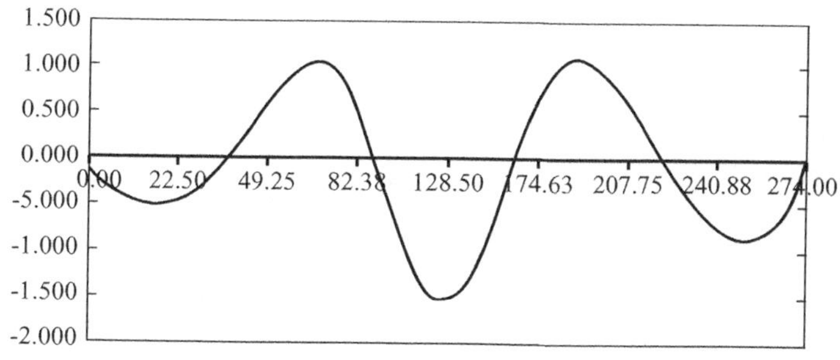

Figure 9.3 First vertical bending mode of Melk Bridge M6

9.2 Porr Bridge, Vienna, Austria

Contributed by Vienna Consulting Engineers

Project Description

In 1975 a segment-construction bridge (Figure 9.4) was erected by the A. Porr AG for research purposes. The bridge crossing a major road in Vienna consists of single segments which are glued and stressed together. The single supported box girder with a span of 44.0 m consists of 18 prefabricated parts. In the framework of a research project it was possible to introduce artificial damage and study the changes to the dynamic behavior of the structure.

Brief Facts

Name and location: Porr Bridge, Vienna, Austria
Owner: Porr AG, Austria
Structure category: short span post-tensioned box girder
Spans: one span: 44.0 m
Structural system: post-tensioned box girder, segment construction
Start of SHM: October 2002
Number of measurment points: 36
Instrumentation design by: Arsenal Research, Business Area Transport Route Engineering, Vienna, Austria and VCE Holding GmbH and Technical University Wien - Institut für Massivbau

Description of Structure

The structure was designed as a precast segmental bridge with glued joints (Figure 9.5). The single supported structure (approximately 380 t weight) with a span of 44.0 m and a free length of 43.5 m offers a total width of 6.2 m. The width of the carriageway is 4.8 m. The monocellular box girder was designed with a constant web width of 40 cm and a construction height of 2.10 m. The width of the box girder is 3.80 m, thus the cantilever arm has a length of 1.20 m on both sides.

Purpose of Inspection

The main target of the investigation was to evaluate and determine the changes to the dynamic properties of the structure caused by artificial damage. Damage was applied to the concrete itself as well as to selected pre-stressing tendons. The main interest was how the dynamic properties (frequencies, modes

Figure 9.4 Porr Bridge, Vienna, Austria

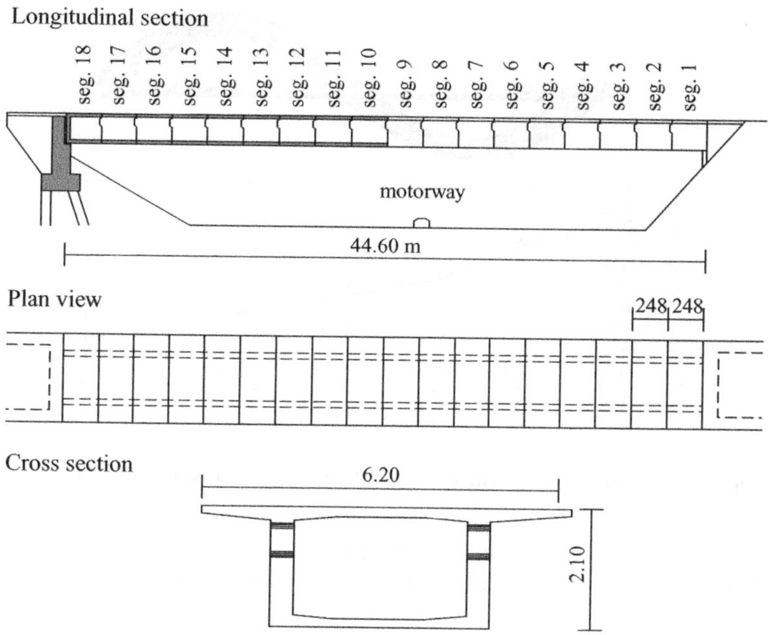

Figure 9.5 Geometry of Porr Bridge

and damping coefficients) change due to loss in pre-stressing forces. Another question was to assess if vibration testing technologies can be used as early diagnosis tools for failure of single tendons.

Measurement Equipment and Data Management (Table 9.2)
Type of system: mobile measurement system based upon Ambient and Forced Vibration Technology.
Data management:

- data pre-analysis (natural frequencies) on site in order to check the obtained data quality
- main analysis, graphical presentation and documentation in office
- data transfer by data cables to main station
- 24 h measurement.

Data Analysis Procedures
Type of analysis: identification of modal parameters by System Identification, Correlation to damage state.

Table 9.2 Sensor details for Porr Bridge

Type of sensors	Number	Location
Acceleration sensors	6	Aligned on both sides of the carriageway
Displacement sensors	2	Mid-span
Displacement sensors	8	Measuring track widths
Strain gauges	8	Attached to the tendons
PT100 (Temperature)	1	Inside box girder

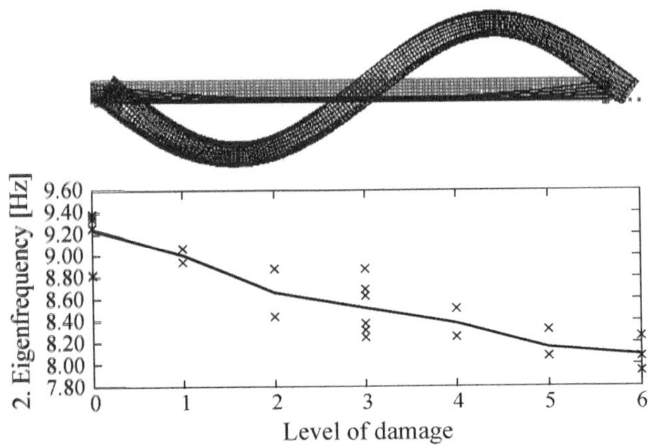

Figure 9.6 Correlation of damage stage (bottom) to second bending mode (top)

Software: stochastic subspace identification (MACEC).
Additional features: correlation of the results to the current damage stage and the applied load.

Examples of Outcomes

The first two modes were identified as reliable by AVM. Both frequencies were reduced with increasing damage stage in the loaded condition (Figure 9.6). In the unloaded stage the frequency of the first mode is below that of the loaded stage, this effect is mainly caused by the stiffness produced from stressing the construction. An important damage indicator is the mode shape, which in particular is sensitive to smaller scale damage. In addition the higher order modes are relevant if local damage is to reliably detected, as focusing only on the basic modes is not sufficient.

Benefits of Using SHM Technologies in the Project

It was shown within this project that structural health monitoring and modal updating could be a powerful tool for assessing the condition of a structure. Moreover an important outcome of the investigation was that focusing on basic modes is not sufficient for damage detection and localization. The difference in the dynamic response between the loaded and unloaded stage for post-tensioned structures is important to consider.

9.3 Warth Bridge, Austria

Contributed by Arsenal Research GmbH

Project Description

To convince the engineering community that vibration monitoring is a valuable technique for structural assessment, proof of its applicability is essential. Therefore it is of paramount importance that existing functional bridges are tested. In the framework of the European research project SIMCES covering several bridges, one bridge was extensively instrumented to set up a long-term test for quantifying the degree of variance due to environmental influences and also due to differences induced by the parameter choice of the selected system identification methods. In addition some initial measurements were carried out in order to test the capabilities of the system identification techniques. Moreover the intention was to create a initial measurement of the structure, which might be used for future monitoring and maintenance applications.

Figure 9.7 Warth Bridge, Lower Austria

Brief Facts

Name and location: Warth Bridge, A2, 63 km south of Vienna
Owner: Municipal Authorities of Lower Austria
Structure category: multispan post-tensioned structure
Spans: 7, 62.0 m + 5 × 67.0 m + 62.0 m = 459.0 m
Structural system: Post-tensioned box girder
Start of SHM: April 1999
Instrumentation design by: Arsenal Research, Business Area Transport Route Engineering, Vienna, Austria in the Framework of the European SIMCES project

Description of Structure

"Talübergang Warth" is located on motorway A2, 63 km south of Vienna. One of the noncoupled twin bridges (direction towards Graz) was tested. Warth Bridge has seven spans and a total length of 459 m (Figures 9.7 and 9.8). The continuous cross section of the bridge has a carriageway width of 14.0 m. The height of the regular cross section is 5.0 m. The box girder bottom width is 6.2 m, which is extended to 8.0 m under the carriageway.

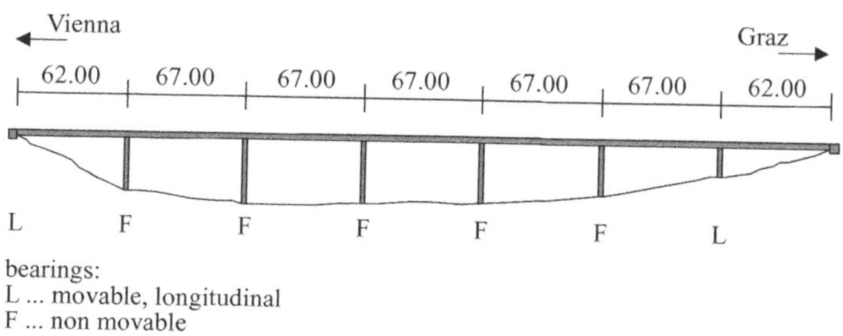

Figure 9.8 Geometry of Warth Bridge

Table 9.3 Sensor details for Warth Bridge

Type of sensors	Number	Location
Velocity transducer Hottinger SMU 30A	8	Multiple setups along the bridge axis to identify mode shapes
Acceleration sensors	4	Expansion joints
Displacement sensors	2	Expansion joints
Temperature sensors	1	Air temperature
Load cell	1	Reaction mass exciter

Purpose of Inspection

Service loads, environmental and accidental actions may cause damage to constructions. Regular inspection and condition assessment of engineering structures are necessary so that early detection and localization allows maintenance and repair works to be properly programed, thereby minimizing costs. Vibration monitoring of civil engineering structures (e.g. bridges, buildings, dams) has gained a lot of interest over the past few years, due to the relative ease of instrumentation and the development of new powerful system identification techniques. The first goal was to demonstrate the high capability of measurements using swept sinusoidal force excitation. The experiments have been carried out under operational conditions. The second goal for the measurement campaign was to compare the results obtained by forced vibration and by ambient vibration testing.

Measurement Equipment and Data Management (Table 9.3)

Type of system: mobile measurement system based upon ambient and forced vibration technology

Data management:

- data pre-analysis (natural frequencies) on site in order to check the obtained data quality
- main analysis, graphical presentation and documentation in office
- data transfer by data cables to main station
- reaction mass exciter.

Data Analysis Procedures

Type of analysis: identification of modal parameters by system identification.

Software: simple peak picking approach to identify the modal parameters. MACEC for ambient testing.

Additional features:

- additional forced vibration testing of the structure;
- comparison of data quality from forced and ambient vibration tests.

Examples of Outcomes

Examples of outcomes are provided in Figures 9.9 and 9.10. The modal properties can be identified by forced vibration under traffic disturbance quite well. Due to the large number of sweeps, the traffic

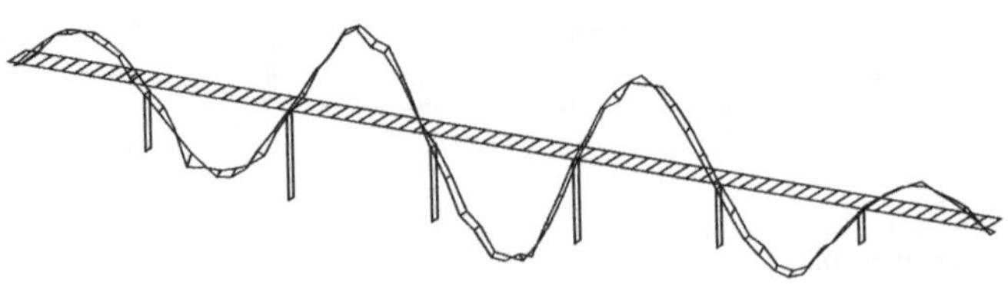

Figure 9.9 Frequency differences between shell model and measurement, Warth Bridge

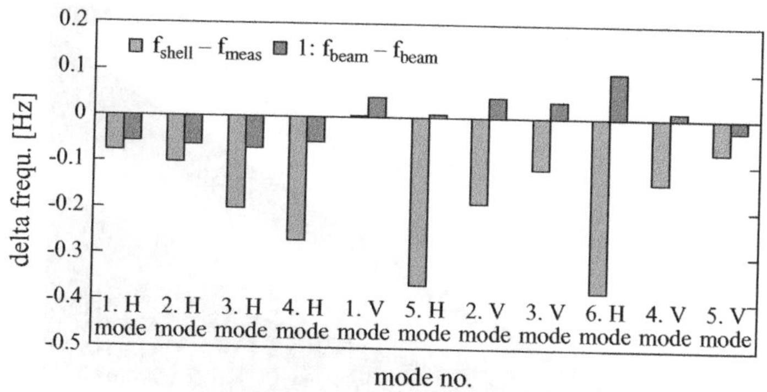

Figure 9.10 Frequency differences between beam model and shell model, Warth Bridge

disturbances can be filtered out at least to a great extent. The higher modes (in this case torsional + mixed modes), which could be very important for model updating, are sometimes not excited by traffic in the case of ambient testing. This is plausible because trucks mainly using the rightmost lane (the closed lane is only for emergency stops) which is almost above the center of gravity. In general comparable results were obtained by both ambient and forced vibration testing.

Benefits of Using SHM Technologies in the Project

Within the project modal updating was performed in order to adapt the numerical results to the real vibration behavior obtained. Thus, a realistic simulation exists that can be used for future monitoring applications. Moreover an initial measurement was established which could be very important for damage detection if required. The results indicate that ambient vibration testing employing an advanced system identification technique is a powerful tool in civil engineering practice. The use of an additional reaction mass exciter is particularly important if higher order modes are to be determined by vibration testing.

9.4 Putlitz Bridge, Berlin, Germany

Contributed by Federal Institute for Materials Research and Testing (BAM), Division of Buildings and Structures

Project Description

The Putlitz Bridge in Berlin (Figure 9.11), Germany, opened in 1977, and is part of a main route for urban traffic which is also used for heavy loads. Currently the bridge is stressed by the transport of heavy gas turbines with maximum loads of 500 t (Figure 9.12), which is much more than the design loads. To ensure the bearing capacity was not being exceeded, experimental investigations including SHM were required.

Brief Facts

Name and location: Putlitz Bridge, Berlin, Germany
Owner: City of Berlin, Germany
Structure category: medium span bridge
Spans: 9 spans, 25.1 + 34.7 + 31.5 + 30.2 + 32.6 + 30.1 + 30.2 + 30.4 + 29.2 m
Structural system: steel box girder with orthotropic deck and steel columns
Start of SHM: September 2001
Number of sensors installed: 21
Instrumentation design by: BAM, Division of Buildings and Structures, Berlin, Germany

Figure 9.11 Putlitz Bridge, Berlin, Germany

Description of Structure

The superstructure comprises of two steel box girders and an orthotropic deck plate. The bridge, with a total length of 270 m, consists of two parts, separated by a lengthways joint. There are two lanes of traffic in each direction. Every bridge part is 14 m wide. For heavy load transports the western part of the bridge is used.

Purpose of Inspection

Static calculations show that the ultimate limit state of the bridge is reached under these heavy loads. These results have to be verified by experimental investigations measuring continuously maximum strains due to heavy loads and temperature (Table 9.4). Additional fatigue stresses at endangered points of the bridge are of interest. Furthermore it is important to know the real static and dynamic loads which can be calculated from the results of the available strain measurements.

Examples of Outcomes

It was affirmed that the dynamic loads, as assumed for the static calculations, can be neglected. Within the time of observation no exceedance of the limit state was noted. The global condition state of the bridge is not yet affected (Figure 9.13).

Benefits of Using SHM Technologies in the Project

All data from stress measurements due to normal and heavy traffic loads as well as temperature loads are available so that exceeding limit states and the occurrence of damage can be detected immediately.

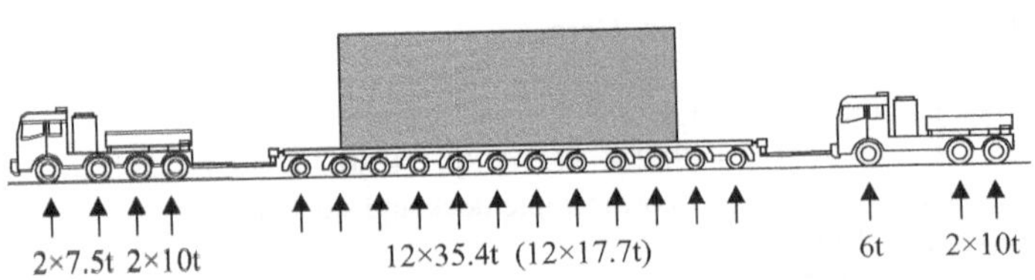

Figure 9.12 Heavy load transport of 490 t

Table 9.4 Sensor details for Putlitz Bridge

Type of sensors	Number	Location
Strain gauges	11	At the main girders and the orthotropic bridge deck
Velocity transducers	4	Fixed at the main girders
Position sensitive detectors	2	Fixed at the main girders
LVDT	2	Fixed at the main girders
PT100	2	Fixed at the main girders

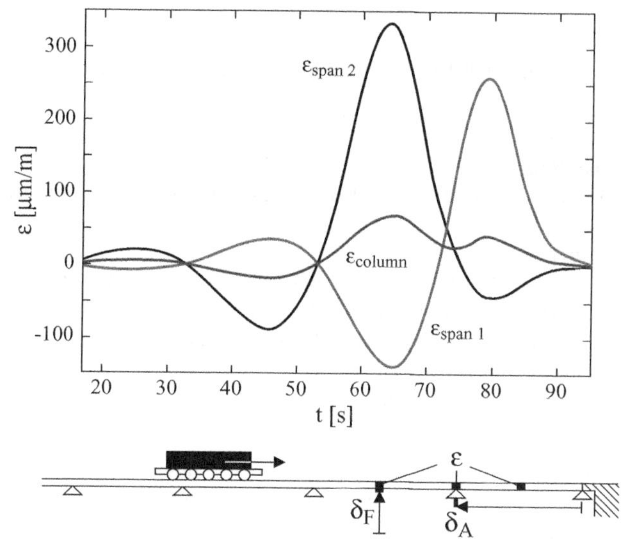

Figure 9.13 Strain distribution at a main girder of Putlitz Bridge during the crossing of a heavy load vehicle measured by SHM

9.5 Westend Bridge, Berlin, Germany

Contributed by Federal Institute for Materials Research and Testing (BAM), Division of Buildings and Structures

Project Description

The Westend Bridge (Figure 9.14) is part of the Berlin city highway that connects downtown Berlin with the airport in the north. The 38-year-old structure is a pre-stressed concrete bridge. In the past the bridge had to be strengthened multiple times due to cracks and open connecting joints within the floor slab. After the reunification of Berlin the commercial traffic into the city increased considerably and, hence, doubts arose whether the bridge would sustain this new loading.

Brief Facts

Name and location: Westend Bridge, Berlin, Germany
Owner: City of Berlin, Germany
Structure category: medium span bridge
Spans: 8, 25.0 + 36.3 + 37.5 + 31.1 + 38.1 + 38.0 + 31.6 + 5.0 m
Structural system: pre-stressed concrete box girder and reinforced concrete columns

Figure 9.14 Westend Bridge, Berlin, Germany

Start of SHM: January 1994
Number of sensors installed: 36
Instrumentation design by: BAM, Division of Buildings and Structures, Germany

Description of Structure

The 243 m long superstructure is continuous and consists of a three-cell box girder with a maximum width of 14 m. It is supported by seven reinforced concrete columns with a hollow cylindrical cross section. The column in the middle of the bridge which is fixed at both ends takes the horizontal loads. All the remaining columns are pin-ended. The whole bridge is built on foundation slabs. The two bridge abutments are formed by reinforced concrete walls.

Purpose of Inspection

The purpose of the inspection of the Westend Bridge was to assess its condition with respect to the presence of multiple cracks within the girder. The inspection was performed by using a monitoring system that records permanently the current traffic loads, stresses and the structure's health (Table 9.5 and Figure 9.15). This system has worked since 1994, with an increasing number of measurement channels, and has continuously supplied monitoring data on the structural health of the bridge.

Examples of Outcomes

It was found that, in general, the dynamic loads acting on the bridge are dependent on the weight of vehicles. For Westend Bridge it was shown that the increasing dynamic loads are correlated to the quality of the road surface.

Analyzing the traffic data obtained by load monitoring, changes in the load spectra could be observed and the increasing number of heavy load vehicles and their weight could be quantified.

Table 9.5 Sensor details for Westend Bridge

Type of sensors	Number	Location
Strain gauges	4	Fixed within the box girder of span 2 and 3
Velocity transducers	20	Fixed within the box girder of span 2 and 3
Accelerometers	3	Fixed within the box girder of span 2 and 3
Crack sensors	1	Fixed within the box girder of span 2 and 3
Inclinometers	2	Fixed within the box girder of span 2 and 3
Position sensitive detectors	1	Fixed within the box girder of span 2 and 3
PT100	5	Attached at the web of one the girders

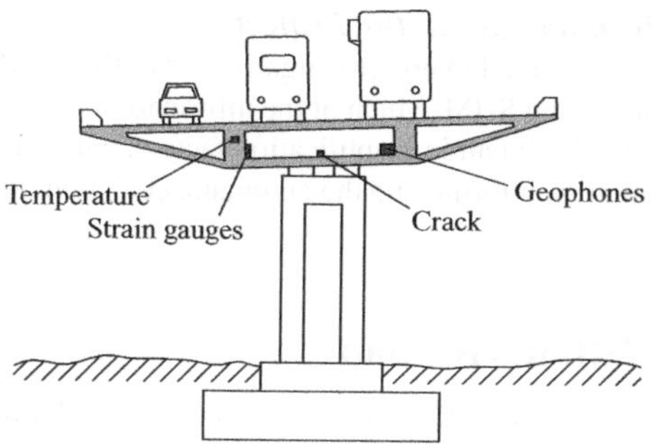

Figure 9.15 Cross section and sensor distribution within the Westend Bridge superstructure

Performing global condition monitoring it was found that the natural frequencies of the bridge are varying with respect to changes of the structural temperature (see Figure 9.16), which means that changes of the bearing capacity can be assumed.

Local condition monitoring proves a strong temperature dependence but reversible behavior of the cracks in the slab of the girder.

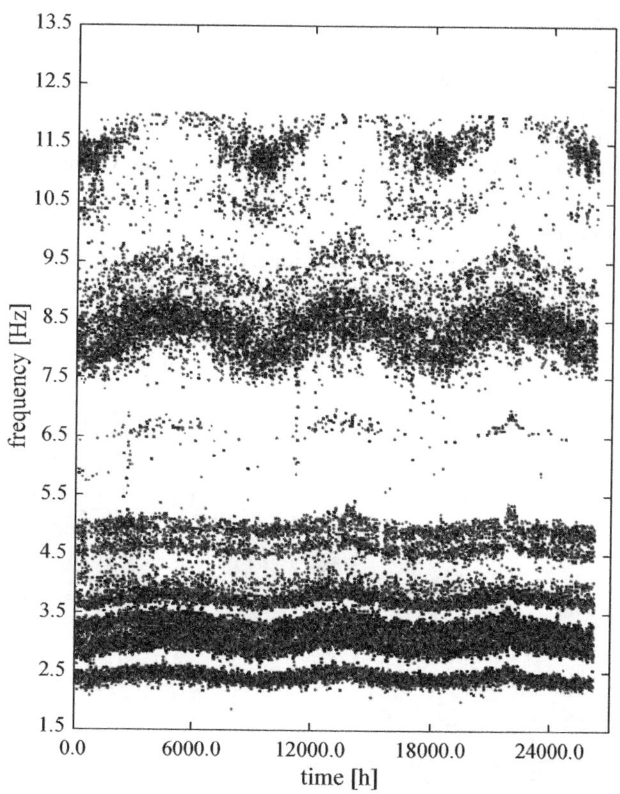

Figure 9.16 Observed natural frequencies of Westend Bridge for a duration of three years

Benefits of Using SHM Technologies in the Project

At Westend Bridge BAM has performed many investigations over time to develop dynamic approaches for inspection of the bridge, including SHM. Much of the information required by bridge owners, such as the actual acting static traffic loads, dynamic amplification factors, combined loadings due to traffic and temperature, structural condition monitoring and the automatic detection of damage, cannot be provided without SMH.

9.6 Neisse Viaduct, Zittau, Germany

Contributed by Federal Institute for Materials Research and Testing (BAM), Division of Buildings and Structures

Project Description

The Neisse railway viaduct (Figure 9.17) was built in 1859. It crosses the Polish–German border near the town of Zittau. A dramatic lowering of groundwater level has led to pier foundation settling that caused wide cracks to appear in the superstructure above the piers.

Brief Facts

Name and location: Neisse viaduct, Zittau, Germany
Owner: German Rail
Structure category: Masonry arch bridge
Spans: 34 arches, each between 17 and 23 m
Structural system: arch bridge on masonry piers
Start of SHM: November 2000
Number of sensors installed: 12
Instrumentation design by: BAM, Division Buildings and Structures, Germany

Description of Structure

The Neisse viaduct is a natural stone masonry arch bridge. It is 750 m long and between 3 and 25 m high. The width of the superstructure is 8 m. The bridge is used for public railway transport. The structure was strengthened in the past by a concrete construction.

Figure 9.17 Neisse railway viaduct at Zittau, Germany

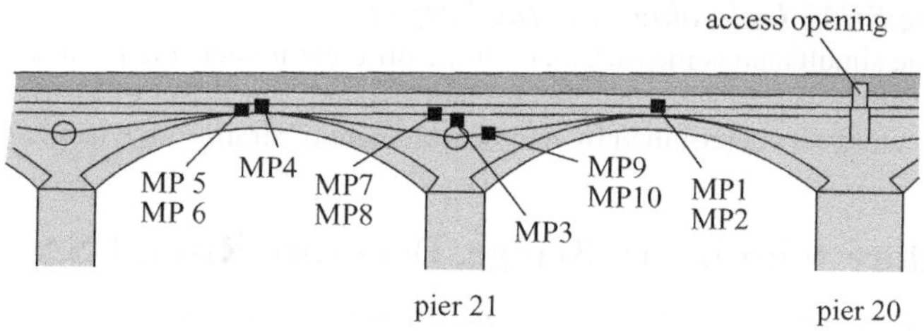

Figure 9.18 Distribution of sensors within the Neisse Viaduct superstructure

Purpose of Inspection

Visual inspection revealed that existing cracks were continuing to open. Through SHM (Table 9.6) the cause for the continuing crack movement should be identified, and it should be possible to assess, the degree to which the bearing capacity of the structure was affected.

Examples of Outcomes

The long-term monitoring of the crack width together with other monitored parameters showed that the whole cross section of the superstructure was cracked. The crack widths change with the temperature course of the structure (Figure 9.19). The progressive foundation settling as well as the traffic have no obvious irreversible influence on crack evolution.

Table 9.6 Sensor details for Neisse Viaduct

Type of sensors	Number	Location
Velocity transducers	4	All sensors are fixed at the superstructure of the bridge (Figure 9.18)
Crack sensors	2	
Strain gauges	2	
PT100	2	

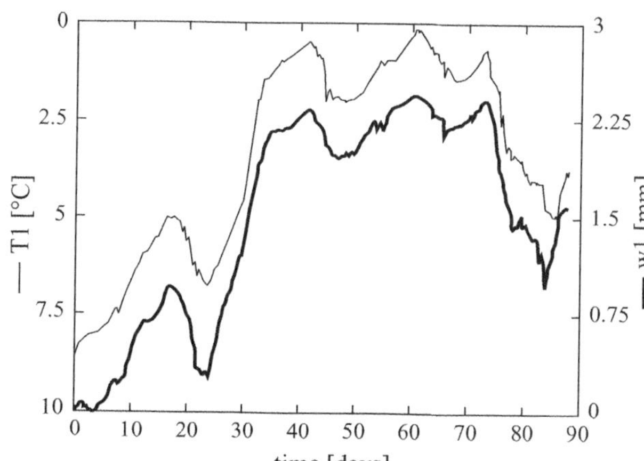

Figure 9.19 Crack width versus structural temperature at Neisse viaduct measured by SHM

Benefits of Using SHM Technologies in the Project

Only because of the simultaneous measurement of the different influencing parameters on crack movement over a long period was a cause and effect relation found. The effect of the cracks on the bearing capacity of the structure was determined by the measurement of strain and dynamic parameters.

9.7 Commodore John Barry Bridge, Delaware River, USA

Contributed by Drexel Intelligent Infrastructure and Transportation Safety Institute

Project Description

The Commodore John Barry Bridge (CBB) spans the Delaware River between Chester, Pennsylvania and Bridgeport, New Jersey (Figure 9.20). The bridge has five traffic lanes and currently serves more than six million vehicles annually, a significant percentage of which is heavy truck traffic. It was opened to traffic in 1974.

Brief Facts

Name and location: Commodore John Barry Bridge (CBB) spans the Delaware River between Chester, Pennsylvania and Bridgeport, New Jersey, USA
Owner: Delaware River Port Authority (DRPA) of Pennsylvania and New Jersey, USA
Structure category: long-span steel truss bridge
Spans: 3, 822 ft + 1644 ft + 822 ft
Structural system: steel truss bridge
Start of SHM: 1998
Number of sensors installed: 97
Instrumentation design by: Drexel Intelligent Infrastructure Institute

Description of Structure

The total length of the bridge is 13 912 ft. The study focuses on the long span through-truss section of the bridge, which is 3288 ft long. The substructures of the through-truss, comprised of four reinforced concrete piers, were constructed on pile foundations in the river bed. The main truss has 73 panel points spaced at 45.7 ft intervals. The two principal trusses of the through-truss are spaced 72.5 ft apart. The floor system of the bridge is an 8-inch-thick lightweight reinforced concrete deck that is composite with nine steel beams laterally spaced at 6.9 ft.

Figure 9.20 Commodore John Barry Bridge through-truss structure

Table 9.7 Sensor details for Commodore John Barry Bridge

Type of sensors	Number	Location
Ultrasonic wind sensors	4	On tower and at midspan
Strain gauges, tilt meters, and crack meters	231	Close to each pile of the structure
Accelerometers	16	For the vibration dampers

Purpose of Instrumentation

The purpose of monitoring (Table 9.7) the truss bridge is to evaluate the following:

- The actual stresses of the critical elements that govern the structural safety performance such as the hangers, stringers and the truss members that were constructed with an electro-slag welding process.
- Ambient environmental conditions at the bridge.
- Performance of the primary movement systems.
- Performance of the truss hangers and the auxiliary support system that was added as a retrofit.
- The effectiveness and condition of approximately 1000 vibration dampers.

Examples of Outcomes

The layout of the information system which was designed in conjunction with the monitors for this project provides a user-friendly, intuitive and secure interface (Figure 9.21). The basic building blocks of the health and performance monitoring system can be summarized as follows:

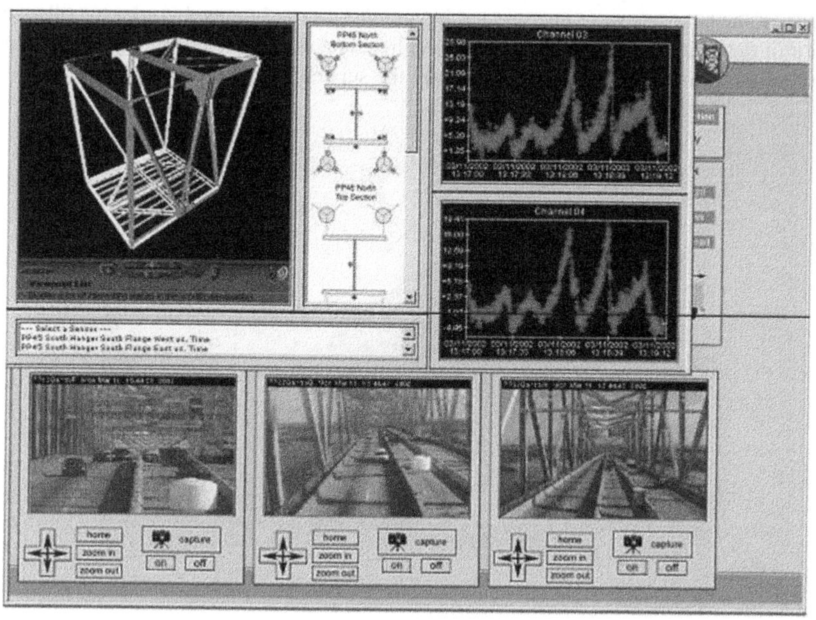

Figure 9.21 Real-time synchronized hanger strain data and live load imaging

- Sensing, data acquisition and control.
- Data processing and information management.
- Human and organizational interfacing for adoption as a management tool.

The data and information processing challenges are influenced by the necessity of providing proper training for the human operators, which is, in fact, the key to organizational acceptance and adoption. The latter is naturally the true measure of success for any technological innovation.

Benefits of Using SHM Technologies in the Project

Monitoring the Commodore John Barry Bridge shows the following:

- The damper should have a useful life of at least 50 years if controlled by the durability of neoprene.
- No changes were observed in the conditions of any of the defects that were identified a decade ago.
- A closer scrutiny of the measured strain and temperature histograms indicated that the hanger intrinsic strains were affected by the complex movement and force-release systems at and in the vicinity of these members. A distinctly unsymmetrical behavior at the long-term strains of the two-instrumented hangers was attributed to a difference in the behavior of the movement systems at their respective boundaries on the north and the south trusses. In addition, an out of plane behavior was noted in the hangers that were expected to be concentrically loaded due to radiation and temperature changes

9.8 Bridge BE 109/21, Bützberg, Switzerland

Contributed by Swiss Federal Laboratories for Materials Testing and Research (EMPA)

Project Description

Bridge BE 109/21 (Figure 9.22) in Bützberg, Switzerland, was built in 1970 as a part of the railway route between Bern and Zürich. Recently the bridge was demolished because of changes in the railway routing by the Swiss Federal Railways (SBB). As a part of the SBB initiated research program ZEBRA the load capacity of the bearings were investigated in laboratory experiments. In order to collect data about the behavior of the bearings in response to temperature changes, two monitoring periods measuring the deformation of the bearings and the temperature of the superstructure during summer and winter were performed.

Figure 9.22 Bridge BE 109/21, Bützberg, Switzerland

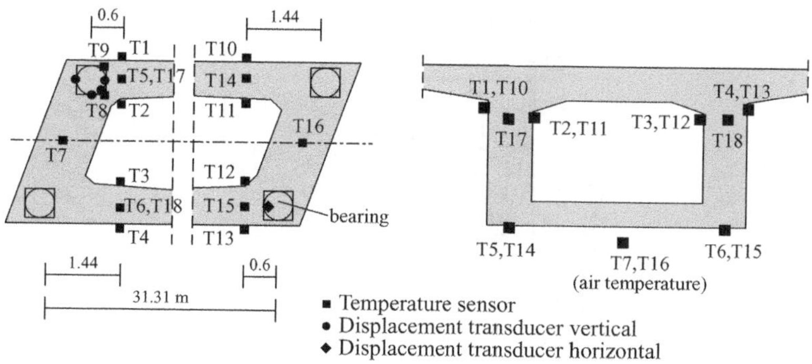

Figure 9.23 Schematic diagram of temperature sensors and displacement transducers on bridge BE 109/21

Brief Facts

Name and location: Bridge BE 109/21, Bützberg/BE, Switzerland
Owner: Swiss Federal Railways (SBB), Switzerland
Structure category: medium span bridge
Spans: 1, 31.10 m
Structural system: post-tensioned concrete box girder bridge
Two Periods of SHM: August 29 to September 4, 2002 and January 13 to January 15, 2003
Number of sensors installed: 21
Instrumentation design by: EMPA, Structural Engineering Research Laboratory, Switzerland

Description of Structure

The railway Bridge BE 109 comprises two single bridges: BE 109/21 and BE 109/22. Each of them was built as a concrete box girder and an orthotropic deck plate of a total length of 31.10 m. Each bridge was constructed for one line of railway.

Purpose of Inspection

A special type of bearing has been used for many railway bridges. The bridge owner (SBB) is especially interested on life expectation information for this kind of bearing. After demolishing Bridge BE 109 the bearings were dismantled and tested in laboratory experiments (Figure 9.23) by fatigue loads and in a later state up to failure loads. The influence of temperature conditions on deformation caused by traffic were observed over two monitoring periods.

Measurement Equipment and Data Management (Table 9.8)

Type of system: PC-based measurement system.
Data management:

- main analysis, graphical presentation and documentation in laboratory
- long-term database.

Table 9.8 Sensor details for bridge BE109/21

Type of sensors	Number	Location
Temperature sensors	16	Within and outside the box girder
Displacement transducers	5	All sensors fixed at two bridge bearings

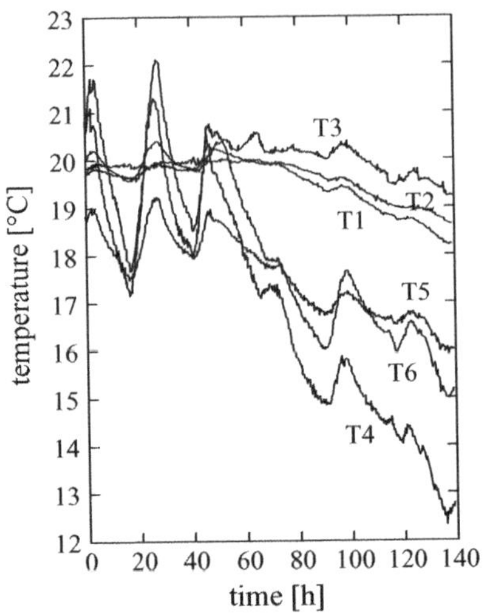

Figure 9.24 Measured temperature on T1–T6 of bridge BE 109/21 during the summer monitoring period

Data Analysis Procedures

Type of analysis: analysis of time series.

Software: self-made software; Cadman; MATLAB.

Additional features: adjustable trigger levels for different alert phases.

Examples of Outcomes

Deformation of the bearings according to temperature influence and traffic loads was observed (Figures 9.24 and 9.25). No particular deformation scenario was found. The bearings were working well and were in good general condition.

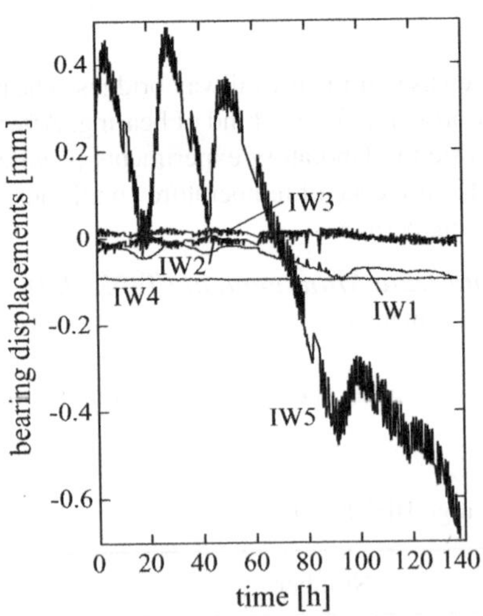

Figure 9.25 Bearing deformation at location IW1–IW5 of bridge BE 109/21 during the summer monitoring period

Benefits of Using SHM Technologies in the Project

Measurement of temperature conditions and bearing deformation caused by traffic loads and temperature effects give an idea of the long-term behavior of bearings. Together with a well-known load history (time tables of SBB) during the bridge life, the load-deformation history of the bearings life can be reconstructed. Temperature effects can also be taken into account. An enforced load history within laboratory experiments up to failure loads show the life expectation of this particular kind of bearing.

9.9 RAMA IX Bridge, Bangkok, Thailand

Contributed by Swiss Federal Institute for Materials Testing and Research (EMPA), Structural Engineering Research Laboratory

Project Description

The RAMA IX Bridge (Figure 9.26) in Bangkok, Thailand was opened in 1987. The bridge is part of Bangkok's Metropolitan Expressway. The bridge represents a crucial connection between Bangkok and Thonburi, across the Chao Phraya River. As part of a regular 15-year inspection, all 68 stay cables of the bridge were subjected to nondestructive evaluation in order to assess their condition.

Brief Facts

Name and location: RAMA IX Bridge, Bangkok, Thailand
Owner: Expressway and Rapid Transit Authority of Thailand (ETA), Bangkok, Thailand
Structure category: medium span (cable-stayed bridge)
Spans: main span, 450 m; back spans, 61.20 + 57.60 + 46.80 m
Structural system: steel box girder with orthotropic deck and steel pylons, supported on concrete piers and pile foundations.
Start of SHM: July 2001
Number of sensors installed: N/A
Instrumentation design by: EMPA, Structural Engineering Research Laboratory, Switzerland

Description of Structure

The Rama IX is a single plane cable-stayed bridge with steel box girder and orthotropic deck and steel pylons. It is comprised of a main span (450 m) and two back spans (61.2 + 57.6 + 46.8 m). The bridge

Figure 9.26 RAMA IX Bridge, Bangkok, Thailand (Photograph by Katrin Janberg, www.structurae.net)

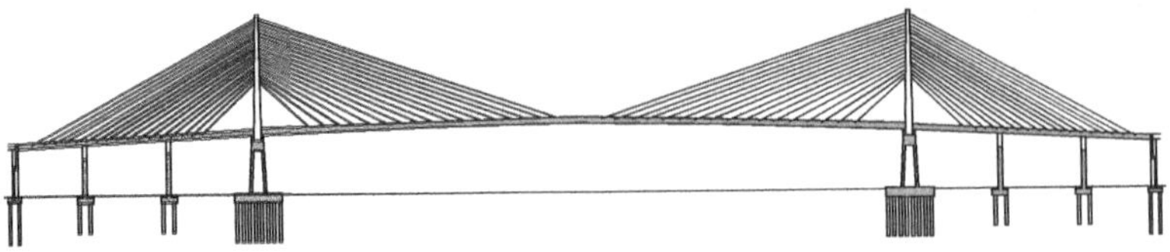

Figure 9.27 Longitudinal section of the RAMA IX Bridge

carries six lanes of traffic. A total of 68 locked coil cables (121 mm = Ø = 168 mm) are divided in four groups of 17 (one group on each back span and two groups on the main span of the bridge; Figure 9.27).

Purpose of Inspection

The purpose of the inspection of the RAMA IX stay cables was to assess their condition with respect to the presence of fractured wires and possibility of corrosion in the cross-section of the free length of the cables. The inspection was performed using the magnetic flux leakage (MFL) method. The MFL method is a nondestructive inspection method. Due to the importance of the link for the city's traffic system, the inspection had to be carried out without interrupting the service of the bridge. This condition, together with practical considerations, ruled out the use of radiation-based methods for the inspection.

Measurement Equipment and Data Management (Table 9.9)

Type of system: PC-based measurement system.
Data management:

- data preview (overview of unfiltered signal) on site
- main analysis, graphical presentation and documentation in office
- data backed up on CDs on site.

Data Analysis Procedures

Type of analysis: data post-processing (filtering of speed related effects) flaw recognition and localization.
Software: software developed in-house.
Additional features: new developments after RAMA IX: 3D localization of the flaws for multistrand cables; ANN-based recognition of flaws.

Examples of Outcomes

All of the free length of the cables was inspected. The output from the global MFL sensor yielded overall information about the presence of fractures within the cable cross-section as a function of the distance along the cable axis, as shown in Figure 9.28. In the case of locked coil cables, such as those installed on RAMA IX Bridge, and parallel wire bundles, the information obtained from the circumference resolved MFL sensors is not used to localize the position of flaws within the cross-section of the cable but as a second set of sensors for backup measurement. For multistrand systems, the mapping of the magnetic flux leakage maps on the surface of the cable can be used to localize the position of flaws within the

Table 9.9 Sensor details for RAMA IX Bridge

Type of sensors	Number	Location
Global MFL sensor	1	In the MFL device, on the circumference of the cable
Circumference-resolved MFL sensors		≤ 26

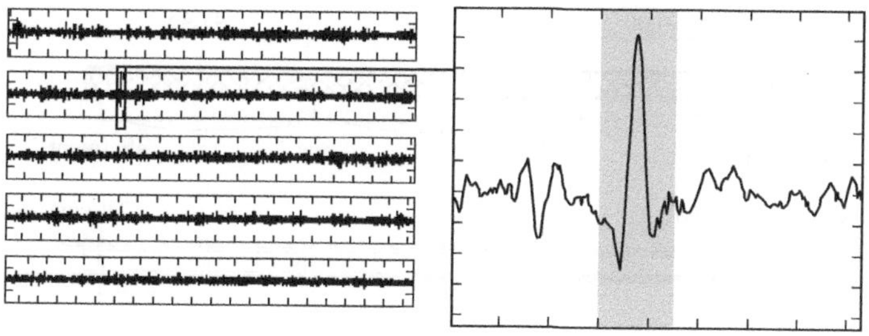

Figure 9.28 Measured temperature on T1–T5, of RAMA IX Bridge

cross-section of a cable. The identification of signals indicating the presence of flaws has now been automated to a large extent.

Benefits of Using SHM Technologies in the Project

The application of the MFL technique was at the time of its deployment the only commercially viable solution for the inspection of the cables installed on the bridge for a total length of approximately 8.5 km. At present there is no alternative to the application of magnetic methods for the inspection of large diameter steel cables. In addition to making use of a nondestructive method, the inspection procedure used on RAMA IX did not require any interruption of the traffic on the bridge

9.10 Titulcia Steel Bridge, Madrid, Spain

Contributed by GEOCISA

Project Description

The Titulcia Steel Bridge (Figure 9.29) dates from the nineteenth century, specifically from the year 1894. It was designed by Enrique Cardenal.

Figure 9.29 Titulcia steel truss bridge

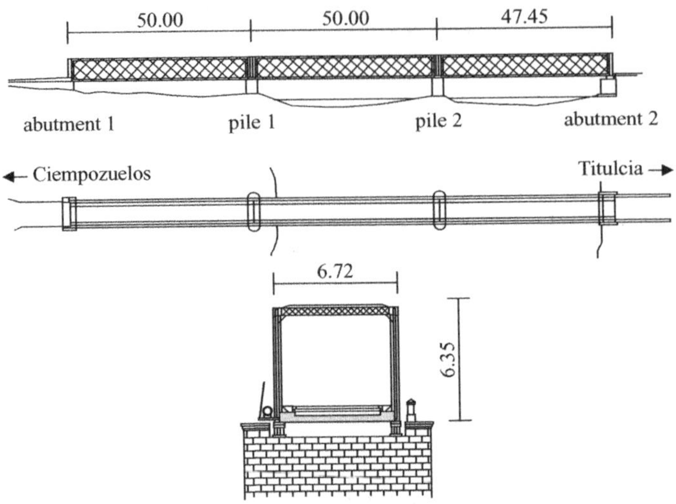

Figure 9.30 Titulcia Bridge geometry

Brief Facts

Name and location: Titulcia Steel Bridge, Madrid, Spain
Owner: Madrid Community
Structure category: Bridge
Spans: 3, 50 + 50 + 47.5m
Structural system: steel truss
Start of SHM: July 31, 2003
Number of sensors installed: 16 (topographic references)
Instrumentation design by: GEOCISA, Madrid, Spain

Description of Structure

The structure is an isostatic three-span truss bridge (Figure 9.30): two spans are 50 m long, of equal length but the third is 2.55 m shorter because it was rebuilt after being blown up during the Spanish Civil War. Abutments and piles are made of masonry and they are founded on the river bed. The total length of the bridge is 147.45 m, its width is 6.72 m and its height is 6.35 m. It supports road traffic and passes over the Jarama River.

Purpose of Inspection

This bridge is integrated in a bridge management system implemented by the Madrid Community and it is periodically inspected. In 1998 a scour problem was detected in a periodic inspection. In 1999 a subaquatic inspection was carried out, confirming the scour problem. From 2000 to 2003 periodic inspections have been carried out. In 2003 it was decided to implement a topographic control on the structure, to monitor the scour movements of pile 2. In 2004 repair work was decided due to the evolution of measured settlements. The topographic measurements were continued during the repair process to control its effectiveness.

Measurement Equipment and Data Management (Table 9.10)

Type of system: Periodic topographic measurements (manual).
Data management:

- analysis of every reading
- predefined control parameters related to the evolution of the movements.

Table 9.10 Sensor details for Titulcia Bridge

Object of monitoring	Type of sensors	Number	Location	Frequency of measurements
Vertical movements	Topographic reference	16	Both sides of the bridge: abutments, piles and midspan	Initially every 3 months, finally twice a week

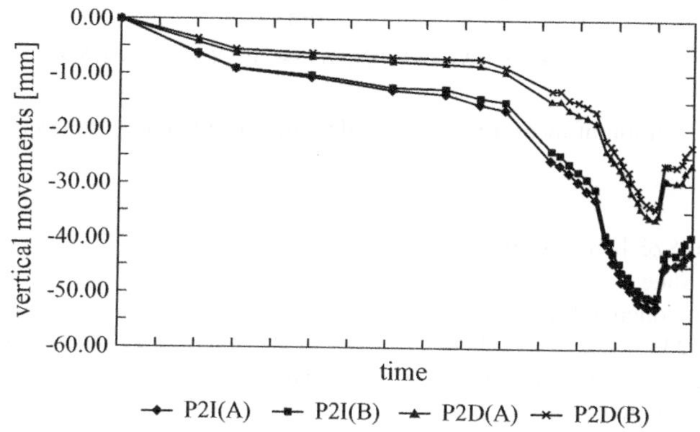

Figure 9.31 Evolution of measurements for pile 2 (four topographic references) of Titulcia Bridge

Data Analysis Procedures
Type of analysis: direct analysis of the data.
Software: no software needed.
Additional features: periodicity of the measurements and need of action were dependent on the evolution of detected movement.

Examples of Outcomes (Figures 9.31 and 9.32)
Benefits of Using SHM Technologies in the Project
These simple measurements performed over a long period of time have permitted:

- Control of the evolution of the scour process at the structure.
- Decision making for repair
- Control of the effectiveness of repair works during and afterwards.

9.11 Széchenyi Bridge, Gyor, Hungary

Contributed by Geodetic and Geophysical Research Institute of the Hungarian Academy of Sciences

Project Description
The Széchenyi Bridge (Figure 9.33) in Gyør, Hungary, opened in 1976, is a part of a main route crossing the rivers Small and Danube between the two towns Győr and Vámosszabadi. It therefore is used for heavy loads.

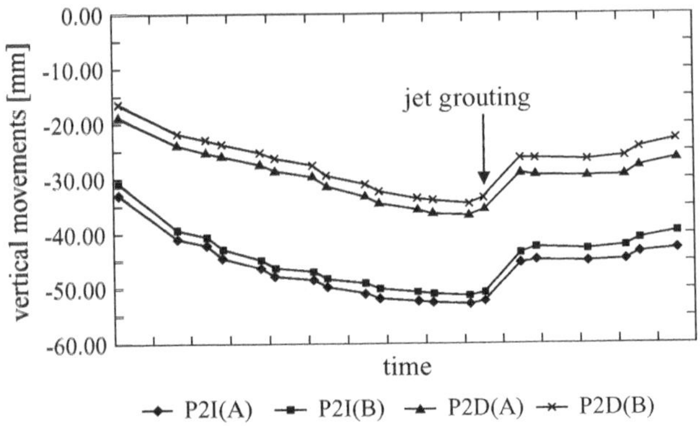

Figure 9.32 Evolution of measurements for pile 2 monitoring repair works of Titulcia Bridge

Brief Facts
Name and location: Széchenyi Bridge, Győr, Hungary
Owner: Hungarian State
Structure category: concrete bridge
Spans: 18, 23.00 + 23.00 + 23.00 + 23.00 + 23.00 + 23.00 + 23.00 + 23.00 + 23.00 + 22.70 + 45.00 + 90.00 + 45.00 + 22.70 + 23.00 + 23.00 + 23.00 + 23.00 m
Structural system: concrete
Start of SHM: October 9, 2003
Number of sensors installed: one three component velocity sensor in seven places
Instrumentation design by: Kinemetrics

Description of Structure
The bridge has a total length of 527.2 m (Figure 9.34) and consists of a road with two lanes in both directions for vehicles. The road is 14 m wide. On each side of the bridge there is a pavement 2.10 m wide.

Figure 9.33 Széchenyi Bridge, Győr, Hungary

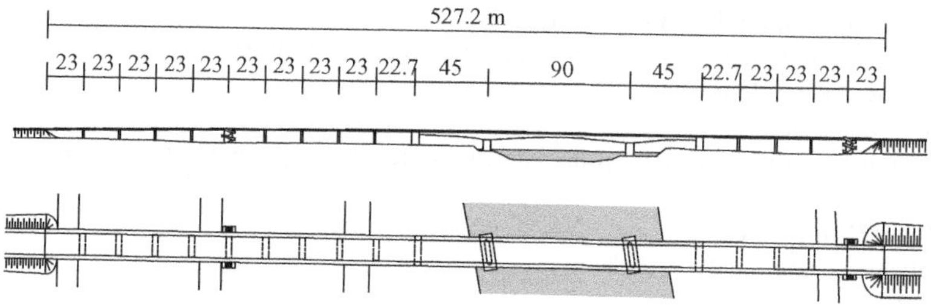

Figure 9.34 Széchenyi Bridge, geometry

Purpose of Inspection
Replacement of the visual inspection and load test by a new, reliable method. Measurements were carried on out October 9, 2003 by Tibor Czifra and Gyula Mentes.

Measurement Equipment and Data Management (Tables 9.11 and 9.12)
Type of system: PC-based measurement system.
Data management:

Table 9.11 Sensor details for Széchenyi Bridge

Instruments	Number	Sensitivity
Recorder: Kinematrics SSR1 seismometer	1	150 $V/m/s$ if gain is 100
Sensor: SS1	3 (one vertical, two horizontal)	

Table 9.12 Description of the data files for Széchenyi Bridge

File name	Sensor placement	Gain of sensor components			Duration [min]
		X	*Y*	*Z*	
Győr1	On the middle of the bridge above of the bed of the river	100	100	100	20
Győr2	On the middle of the bridge above of the bed of the river	10	10	10	10
Győr3	On the middle of the bridge above of the bed of the river	10	10	1	10
Győr4	Directly on the middle of pillar 13	10	10	1	10
Győr5	In the middle between pillars 13 and 14 in direction Revfalu	10	10	10	
Győr6	In the middle between pillars 14 and 15 in direction Revfalu	10	10	10	
Győr7	In the middle between pillars 10 and 11 in direction Győr	10	10	10	

- data control (vibration diagram) on site
- main analysis, graphical presentation and documentation in office
- data transfer via modem
- long term database.
- CMS (Civil Monitoring System) processes the static data acquired by the system

Data Analysis Procedures

Type of analysis: Fourier analyses.
Software: ORIGIN 6.0.

Examples of Outcomes

Only one measurement was carried out and therefore it was not a possible to compare different vibration records made at different times.

Benefits of Using Vibration Measurement Techniques in the Project

All data series from vibration measurements are available so that the occurrence of damage can be detected immediately.

9.12 ESK 551 Bridge, Bad Bevensen, Germany

Contributed by Infokom GmbH

Project Description

Bridge ESK 551 on (Figure 9.35) country road 232 from Bad Bevensen to Altenmedingen spans the Elbe–Seiten Canal and was built at the beginning of the 1970s. It is a threebay pre-stressed concrete bridge with bearing distances of 32.05–66.40–32.05 m; as a result of manufacturing defects it has relatively severe flexures. Considering only the loading condition of a dead load, the structure already shows bending cracks in the midspan.

Brief Facts

Name and location: ESK 551 Bridge, Bad Bevensen, Germany
Owner: Bundesamt für Wasserbau, Germany
Structure category: pre-stressed concrete bridge
Spans: 3, 32.05 + 66.40 + 32.05 m

Figure 9.35 ESK 551 Bridge, Bad Bevensen, Germany

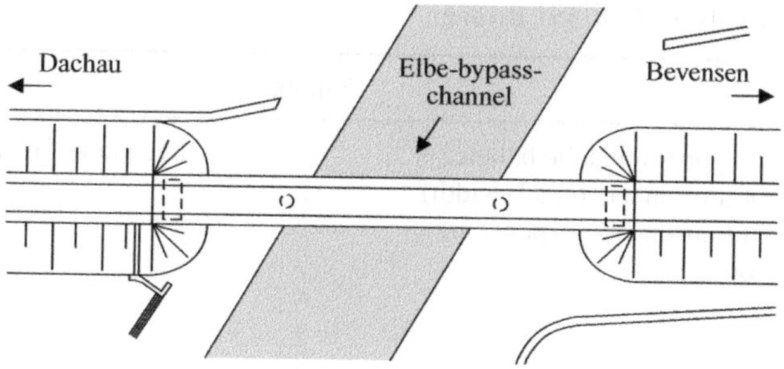

Figure 9.36 Site drawing of ESK 551 Bridge

Structural system: steel box girder with orthotropic deck and steel pylons, supported on concrete piers and pile foundations
Start of SHM: 1996 and 1999
Number of sensors installed: 21
Instrumentation design by: BAW Karlsruhe; Bauhaus University Weimar, Infokom GmbH Neubrandenburg, Germany

Description of Structure

The structure is described in Figure 9.36.

Purpose of Inspection

Monitoring (Figure 9.37) of the structural deformation as a result of general climatic variations as well as the individual deformation during vehicle crossings.

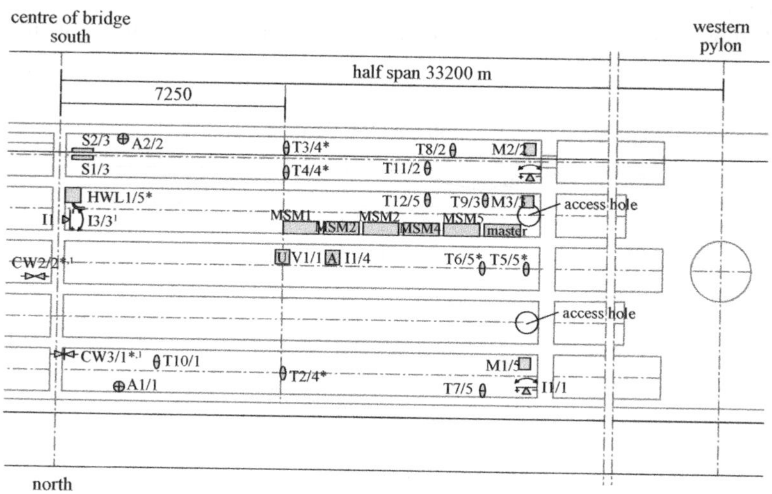

Key
1. Position - sensor type (A=acceleration, T=temperature, HWL=hose water level, M=moisture, S=strain, I=inclination, CW=crack width)
2. Position - characteristic number of sensor
3. Position - associated modul
*) - Signal from existing system
1) - inactive since 06 May 2003

Figure 9.37 Plan of measurement locations on ESK 551 Bridge

Table 9.13 Sensor details for ESK 551 Bridge

Type of sensors	Number	Location
Electronically scanned hydrostatic tube balance	1	See plan of measurement locations
Inductive displacement transducers (crack width)	2	
Inductive displacement transducers (expansion)	2	
Humidity pickup sensors	2	
PT100	6	
Acceleration sensors	2	
Inclination sensors	2	

Measurement Equipment and Data Management (Table 9.13)

Type of system: microcontroller-based stand-alone measurement system.

Data management:

- main analysis, graphical presentation and documentation in office
- data transfer via modem
- long term data base

Example of Outcomes

An example is shown in Figure 9.38.

Data Analysis Procedures

Type of analysis: statistics, general monitoring of the reformation after an incident

Software: self-made software.

Additional features: adjustable trigger levels for different alert phases; no expert system.

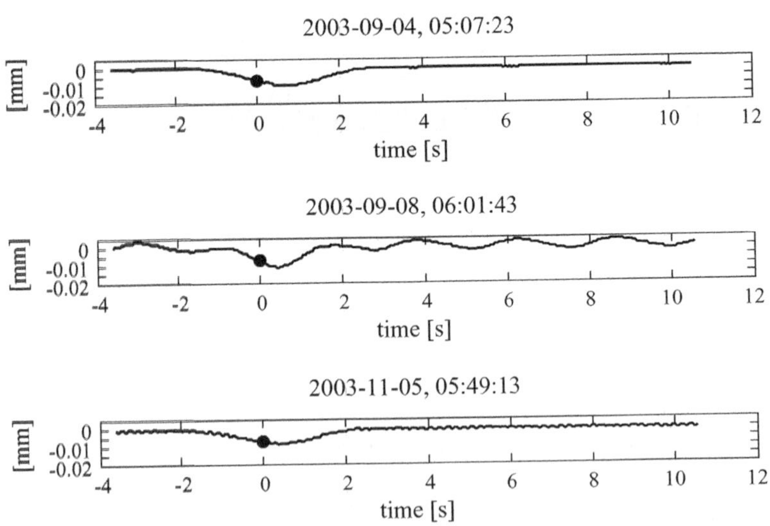

Figure 9.38 Dynamical incidents at the same displacement transducers on separate occasions

Benefits of Using SHM Technologies in the Project

- Solar power supply continuously showed to be problematic during the winter months. Measurement was just occasionally possible. Data were partly corrupted by means of spontaneous shutoffs of the solar collector calibrations and measurement.
- The system is extensively stable. Failures were noticed only at some components.
- For several years reproducible as well as comparable deformation of the load-bearing structure was located (Figure 9.38), which was repaired during temporary closure between the end of 2003 and 2005.

9.13 The New Årsta Railway Bridge, Stockholm Sweden

Contributed by Royal Institute of Technology (KTH), Department of Civil and Architectural Engineering, Division of Structural Design and Bridges

Project Description

The new Årsta railway bridge (Figure 9.39) in Stockholm, Sweden, was opened for traffic in 2006. The bridge is part of the overall upgrading from two to four tracks, between Stockholm South and a new station called Årstaberg. The purpose of the extension is to increase track capacity. The structure is a very slender and complex pre-stressed concrete bridge without any ballast. Therefore, the Swedish National Railway Administration (Banverket) initiated a measuring program to follow up and evaluate/verify stresses and deformation during construction and operation. Static and dynamic measurements/analyses are being conducted.

Brief Facts

Name and location: The new Årsta railway bridge, Stockholm, Sweden
Owner: The Swedish National Railway Administration (Banverket)
Structure category: long span bridge
Spans: 11 spans: 48 + 78 + 78 + 78 + 78 + 78 + 78 + 78 + 78 + 78 + 65 m
Structural system: continuous pre-stressed concrete troughed-beam bridge
Start of SHM: January 2003
Number of sensors installed: 86

Figure 9.39 Digital image of the new Årsta Railway Bridge

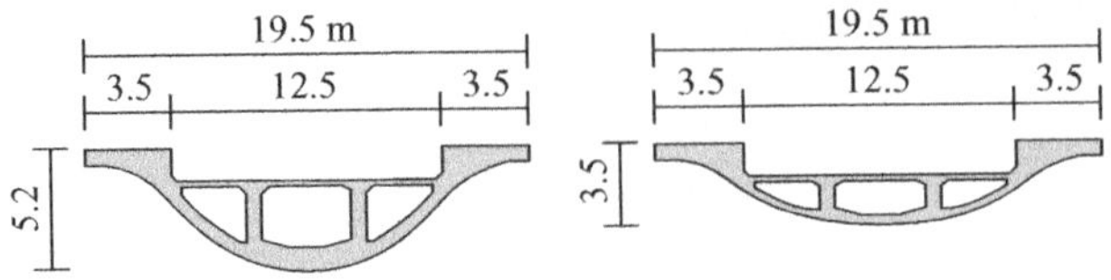

Figure 9.40 The slender design of the superstructure is thickest above the piers (left figure) and tapers of towards the center of the span (right figure)

Instrumentation design by: strain gauges – Royal Institute of Technology (KTH), Department of Civil and Architectural Engineering, Division of Structural Design and Bridges; fiber optic sensors (SOFO system) – SMARTEC.
Installations: most of the installation work was carried out by The Royal Institute of Technology (KTH), Department of Civil and Architectural Engineering, Division of Structural Design and Bridges.

Description of Structure

The bridge is 833 m long, 19.5 m wide (Figure 9.40) and has ten piers with an elliptical cross section measuring 7 × 2.5 m. The pier height varies from 9 to 25 m. The rail carriageway is embedded in a trough with 1.2 m high parapets. Running along the left-hand side of the bridge is a pedestrian and cycle way, and on the eastern side there is a road for service and rescue vehicles. In order to reduce the weight of the bridge structure and to distribute forces the beam height is decreased by eliminating the ballast and embedding the rail fasteners directly in the concrete structure. The superstructure is built in two different ways. On the north side, the curved section of the bridge, use is made of conventional fixed scaffolding, whereas from the south, the straight section of the bridge, launching formwork is used and gradually advanced as each span is completed. The design was first produced by Foster/Aarup, but reworked twice. The final design was done by the Danish consultants COWI A/S.

Purpose of Inspection

The aim is to verify uncertainties in the structure during construction and after ten years of service, leading to knowledge and perhaps updated codes, especially concerning dynamic effects. This will, in turn, give economical and safe solutions concerning similar structures in the future.

Static measuring will verify that maximum strains and stresses are kept within permissible limits; check that no cracking occurs in critical sections, according to design; study changes in strain, both during construction and in service; and compare results from fiber optic sensors with results from strain gauges. Installed monitoring equipment will: evaluate fundamental frequencies, mode shapes and damping ratios; evaluate dynamic effects of trains crossing the bridge, especially train–bridge interaction and effects of track irregularities; evaluate long-term changes in the bridge's dynamic properties.

Measurement Equipment and Data Management (Table 9.14)

Type of system: PC-based measurement system.
Data management:

- data pre-analysis (statistics) on site
- main analysis, graphical presentation and documentation in office
- data transfer via broadband
- long-term data base

Examples of Outcomes

Only some early results are presented here (Figures 9.42 and 9.43) although much more will be available shortly; more data acquisition and analysis work in the office has to be carried out before any further conclusions can be drawn.

Table 9.14 Sensor details for the new Årsta railway bridge

Type of sensors	Number	Location
Strain gauges	24	Figure 9.41 describes the locations
Accelerometers	6	
LVDT	1	
Fiber optic sensors	46	
Thermocouples	9	

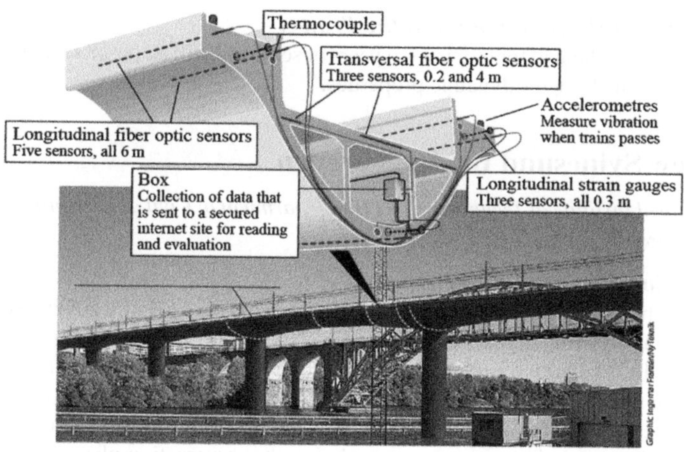

Figure 9.41 Illustration of one specific cross-section with sensors measuring strain, temperature and acceleration

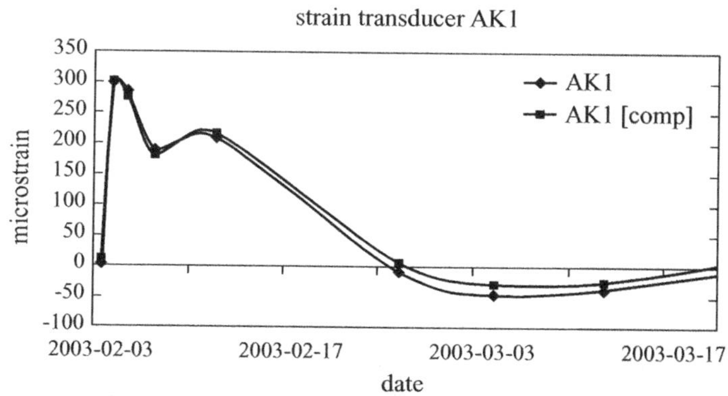

Figure 9.42 Typical results from strain transducers during construction (casting). One of the curves is temperature compensated

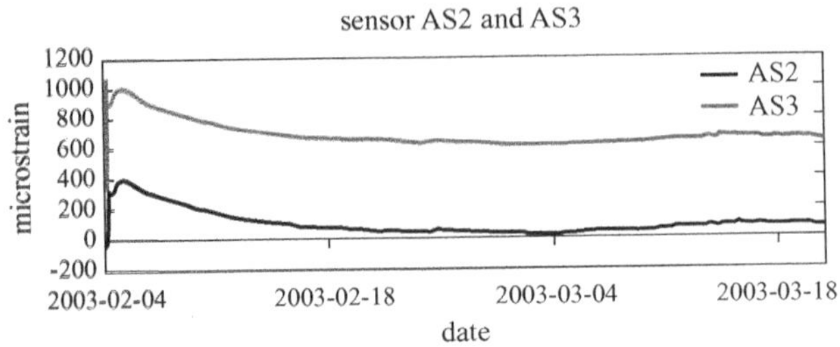

Figure 9.43 Results from two different fiber optic sensors in a very early stage

Benefits of Using SHM Technologies in the Project
Since measurements are planned for the first ten years of service it will be possible to, for example, detect long-term changes in the bridge's dynamic properties.

9.14 The New Svinesund Bridge, Sweden

Contributed by Royal Institute of Technology (KTH), Department of Civil and Architectural Engineering, Division of Structural Design and Bridges

Project Description
The world's largest bridge with a single arch is being built across the Ide Fjord at Svinesund (Figure 9.44). The bridge will form part of the European highway, E6, which is the main route for all road traffic between Gothenburg and Oslo. The bridge is elegant but structurally complicated as it combines a very slender construction with a special structural form. Due to the uniqueness of design and the importance of the bridge a monitoring project was initiated by the Swedish National Road Administration (Vägverket). The monitoring project, including measurements during the construction phase, the testing phase, and the first five years of operation, is coordinated by The Royal Institute of Technology (KTH). For more information, see the monitoring project homepage at http://www.byv.kth.se/svinesund/

Brief Facts
Name and location: The new Svinesund Bridge, joining Sweden and Norway
Owner: The Swedish National Road Administration (Vägverket)
Structure category: arch bridge
Spans: main arch span of 247 m
Structural system: two steel box girders suspended from a single concrete arch
Start of SHM: June, 2003

Figure 9.44 The new Svinesund Bridge, Sweden

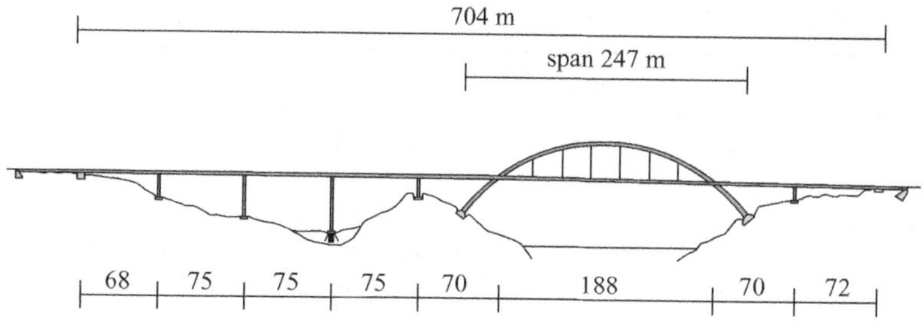

Figure 9.45 Sketch of the new Svinesund Bridge in its entirety, showing grid-line numbering and approximate dimensions

Number of sensors installed: 58 (68 when the bridge is completed)
Instrumentation design by: KTH, Division of Structural Design and Bridges, Stockholm, Sweden

Description of Structure

The new Svinesund Bridge is a highway bridge across the Ide fjord joining Sweden and Norway. The total length of the bridge is 704 m (Figure 9.45) and consists of a substructure in ordinary reinforced concrete together with a steel box-girder superstructure. The main span of the bridge between abutments is approximately 247 m and consists of a single ordinary reinforced concrete arch which carries two steel box-girder bridge decks, one on either side of the arch. The level of the top of the arch and the bridge deck are 91.7 m and 61 m, respectively. Over the part of the bridge where the arch rises above the level of the bridge decking, the two bridge decks are joined by traverse beams positioned at 25.5 m centers. The traverse beams are in turn supported by hangers to the concrete arch.

Purpose of Inspection

The primary objective of the monitoring programme is to check that the bridge is built as designed and to learn more about the as-built structure. This will be achieved by comparing the measured structural behaviour of the bridge with that predicted by theory.

Measurement Equipment and Data Management

The data acquisition system consists of two separate data subcontrol units built up of basic MGC Digital front-end modules from HBM (Hottinger Baldwin Messtechnik). The units are located at the base of the arch on respectively the Norwegian and Swedish side. The subcontrol system on the Swedish side contains the central rack-mounted industrial computer and is connected with ISDN telephone link for data transmittal to the computer facilities at KTH for further analysis and presentation of data. The logged data on the Norwegian side is transmitted to the central computer on the Swedish side via a radio ethernet link. The selected logging procedure provides sampling of all sensors continuously at 50 Hz with the exception of the temperature sensors (Table 9.15) which have a sampling of once per 20 s or 1/20 Hz. At the end of each 10 min sampling period, statistical data such as mean, maximum, minimum and standard deviation are calculated for each sensor and stored in a statistical data file having a file name that identifies the date and time period when the data were recorded. Raw data, taken during a 10 min period, is stored in a buffer if the ‘trigger’ value for calculated standard deviation for acceleration is exceeded.

Examples of Outcomes

Figure 9.46 show the strains measured at the roof of a segment close to the arch base on the Swedish side. The casting of each subsequent segment causes an elongation of the reinforcement bars. This is to be expected as the arch behaves as a cantilever and the extra weight at the end of the structure caused by the newly cast arch segments will cause tension in the top of the section at the base of the arch. In a similar manner, tensioning the support cables, represented by the dot-dashed lines, causes a contraction of the same reinforcement bars.

Table 9.15 Sensor details for the new Svinesund Bridge

Type of sensors	Number	Location
Vibrating-wire strain gauges	16	Four at arch base and four just below the bridge deck, Norwegian and Swedish side
Resistance strain gauges	8	Two at arch base, two in a segment just below bridge deck, and four at the crown
Linear servo accelerometers on concrete arch	4	Installed pair-wise and are moved to new arch segments as construction of the arch progresses. When the arch is completed, two accelerometers will be moved to the arch midpoint and two to the arch's Swedish quarter point
Linear servo accelerometers on bridge deck	6	Three at midpoint and three at quarter point
Temperature gauges	28	At the same sections as the strain gauges
Outside air temperature gauge	1	At arch base on Swedish side
Three-directional ultrasonic anemometer	1	For measuring wind speed and direction at deck level close to the first support on the Swedish side
Load cells	2	Monitor the forces in the first hanger pairs on the Swedish side
LVDT	2	Monitor transverse movement of the bridge deck at the first bridge pier supports on both sides of the arch

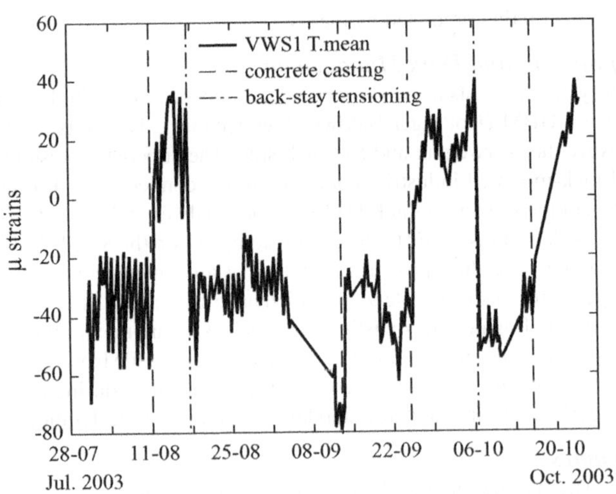

Figure 9.46 How the work on site is mirrored by the measured strains. The casting dates for segments are represented by dashed lines. The dates when tensioning of support cables occurred are shown by the dot-dashed lines

9.15 Bridge Z24, Koppigen–Utzenstorf, Switzerland

Contributed by Katholieke Universiteit Leuven, Department of Civil Engineering, Division of Structural Mechanics

Project Description

Bridge Z24, built between 1961 and 1963, spanned the A1 Bern–Zurich motorway and linked Koppigen with Utzenstorf. The three-span structure with spans of approximately 14, 30 and 14 m crossed the A1 at a slightly oblique angle (Figures 9.47 and 9.48). The superstructure consisted of a two-cell closed box girder with tendons in the three webs. The condition of the bridge was relatively good but the bridge had to be demolished to allow the construction of new railway tracks.

Within the SIMCES project the influence of the environment on the dynamic characteristics (natural frequencies and mode shapes) of the bridge was investigated, as well as the changes in dynamic characteristics due to progressive damage tests (PDT). The aim was to provide experimental proof of the feasibility of the SHM method. Vibration measurements prior to and after a damage scenario would allow conclusions to be drawn about the possibility of identifying damage from changes recorded in the dynamic characteristics.

Brief Facts

Name and location: Bridge Z24, Koppigen–Utzenstorf, Switzerland
Owner: Road Department of the Canton of Bern
Structure category: medium span bridge
Spans: 3, 14 + 30 + 14 m
Structural system: pre-stressed concrete, with two-cell closed box girder, two concrete diaphragms as main piers and concrete columns at abutments
Start of SHM: August and September 1998

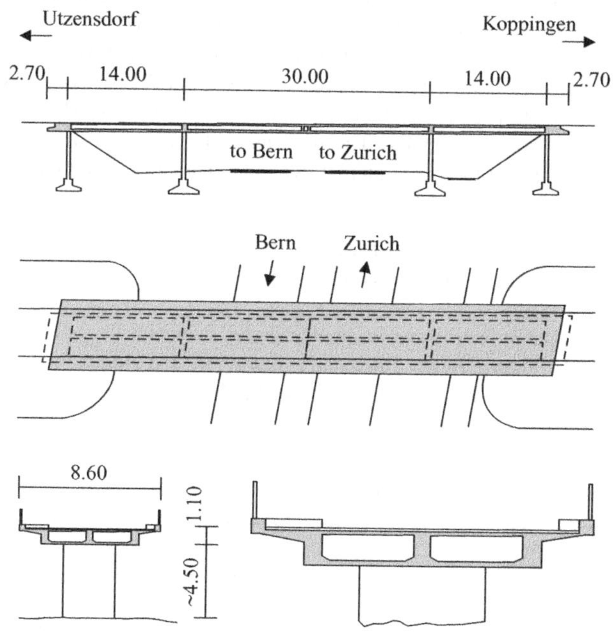

Figure 9.47 Bridge Z24, geometry, Koppigen–Utzenstorf, Switzerland

Figure 9.48 Bridge Z24, Koppigen–Utzenstorf, Switzerland

Number of sensors installed: (15 + 2) × 9 setups + 3 reference channels
Instrumentation design by: Swiss Federal Laboratories for Materials Testing and Research (EMPA)

Description of Structure

The three-span structure had a total length of 58 m, subdivided in three spans of 14, 30 and 14 m, respectively. It crossed the A1 at a slightly oblique angle. The superstructure consisted of a two-cell closed box girder with tendons in the three webs. More tendons were allocated over the main piers than in the middle of the spans. Both main piers were built as concrete diaphragms, fully connected with the superstructure. The three abutment columns were pinned at both ends. To protect the anchor heads, both ends of the bridge deck were extended.

Purpose of Inspection

The purpose of the SIMCES project was to demonstrate the feasibility of assessing the integrity of civil structures by means of evaluating their vibration (dynamic characteristics). Several damage scenarios were applied and the resulting changes in dynamic characteristics were recorded and used to detect and identify the corresponding structural damage. Full-scale ambient (AVT) as well as forced vibrations (FVT) were carried out. More information is available at http://www.kuleuven.ac.be/bwm/SIMCES.htm

One way of solving the inverse problem is by FEmu (Section 7.1.3). As an example, the damage scenario of lowering the main pier at the right-hand side by over 95 mm, which induced cracks in the main girder above this pier, was investigated. The aim was to identify the stiffness reduction of the girder by updating the FE model using the experimental dynamic characteristics. Only the AVT results were used in the updating process (Table 9.16).

Measurement Equipment and Data Management

Type of system: PC-based measurement system.
Data management:

- data pre-analysis (data reduction and frequency analysis) on site
- raw data – available on http://www.kuleuven.ac.be/bwm/IMAC/index.html
- main analysis – performed by different partners of SIMCES project (KUL results available from their office)

Table 9.16 Sensor details (AVT tests) for Bridge Z24

Type of sensors	Number	Location
Accelerometers	15	On the bridge deck
Accelerometers	2	On a pier (9 setups)
Accelerometers	3	Reference channels (9 setups)

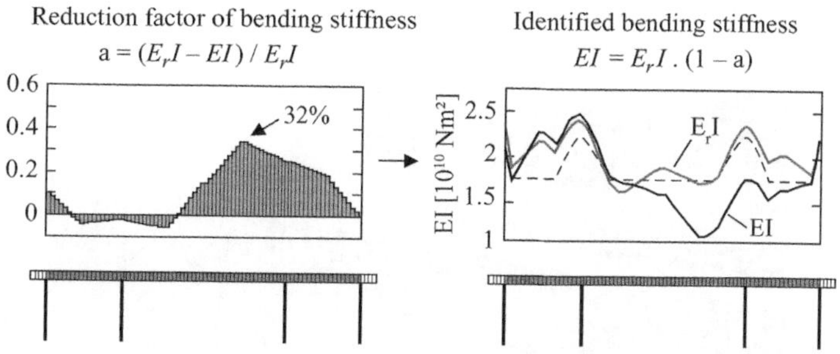

Figure 9.49 Identified relative reduction and resulting bending stiffness of main girder

- graphical presentation and documentation on http://www.kuleuven.ac.be/bwm/SIMCES.htm and http://www.kuleuven.ac.be/bwm/Z24/index.html.

Data Analysis Procedures

Type of analysis: dynamics, experimental modal analysis, damage indicators.

Software: system identification – Stochastic Subspace Identification technique; FE model updating – own developed software (in MATLAB).

Additional features: output-only system identification in time domain; natural frequencies and mode shapes are used.

Examples of Outcomes

The box bridge is modeled by a beam model with equivalent stiffness properties. The damage is represented by a reduction in bending and torsional stiffness of the constituting beam elements (Figure 9.49). A globally realistic damage pattern is identified for the bending and the torsional stiffness with the FE model updating method. A good correlation between the experimental and the updated numerical modal data is obtained (Figure 9.50).

Benefits of Using SHM Technologies in the Project

Based on the experimental modal characteristics of the bridge, the applied damage pattern could be identified through an inverse analysis. The realistic result demonstrates the viability of the nondestructive vibration-based damage identification method.

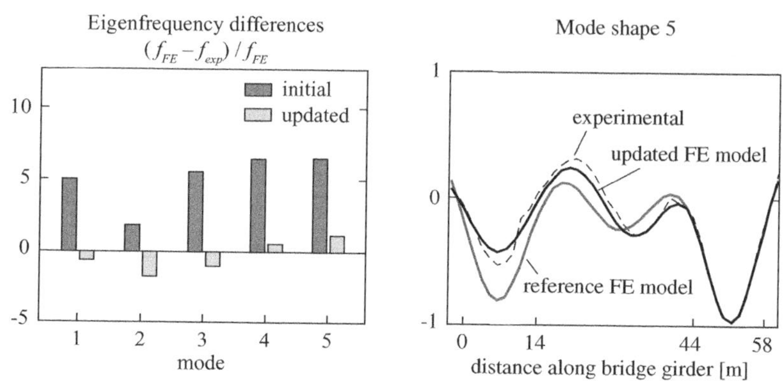

Figure 9.50 Discrepancies between numerical and experimental modal data

Figure 9.51 Roberval Bridge, Senlis, France

9.16 Roberval Bridge, Senlis, France

Contributed by Laboratoire Central des Ponts et Chaussées (LCPC), Division for Structures Behaviour and Durability, Division for Metrology and Instrumentation

Project Description

Roberval Bridge (Figure 9.51) is located on the Lille–Paris motorway (A1). This bridge is submitted to heavy traffic loading which is of interest for bridge design codes.

Brief Facts

Name and location: Roberval Bridge on the motorway A1 (Lille–Paris) near Senlis, France
Owner: SANEF (Société des Autoroutes du Nord et de l'Est de la France) District de Senlis
Structure category: medium span bridge
Spans: 16, each 33 m
Structural system: girder bridge with independent spans, post-tensioned concrete girders cross braced, concrete top slab (Figure 9.52)
Start of SHM: September 2001
Number of sensors installed: 33 channels

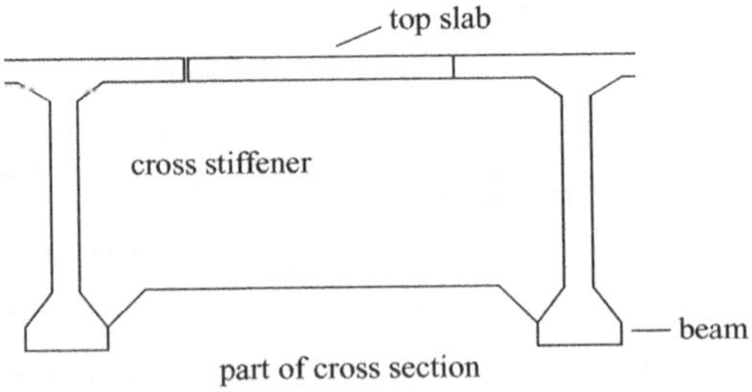

Figure 9.52 Part of the cross section of Roberval Bridge

Table 9.17 Sensor details for Roberval Bridge

Type of sensors	Number	Location
Strain gauges	31	At the pre-stressed main beams and at the top slab
WIM system	Two lanes	
Temperature sensors	2	
Complementary instrumentation: accelerometers for modal analysis in ambient conditions	6	At the pre-stressed main beams

Instrumentation design by: LCPC, Division for Structures Behaviour and Durability, Division for Metrology and Instrumentation

Description of Structure

The instrumented span is part of a one-way, three-traffic-lane section of motorway (Lille–Paris), 34 m long and 13.9 m wide with five braced I-shaped pre-stressed beams.

Purpose of Inspection

The aim of the instrumentation is to record the loads and the effects of the heavy traffic and build a database (WIM data – peak strain values) for calibration of bridge loading codes.

Measurement Equipment and Data Management (Table 9.17)

Type of system: PC-based measurement system.

Data management:

- data pre-analysis (peak analysis on site)
- main analysis, graphical presentation and documentation in office
- data transfer via modem
- long-term data base.

Data Analysis Procedures

Type of analysis: statistics, rainflow analysis, peak analysis, modal analysis from ambient vibrations

Software: self-made software (REVE) and Catman 4.0 (HBM software)

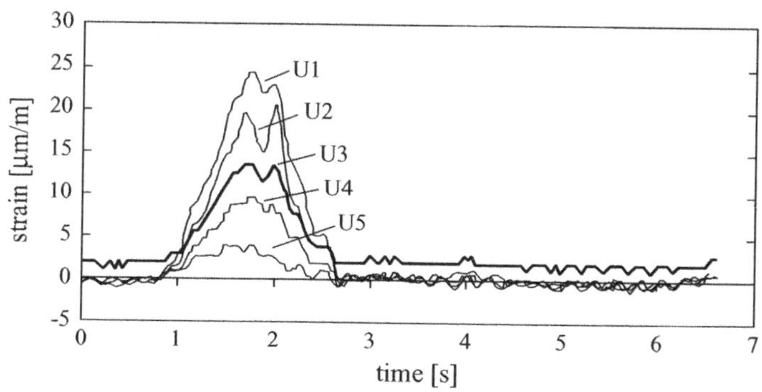

Figure 9.53 Bending strain distribution during the crossing of one heavy load vehicle on lane 1

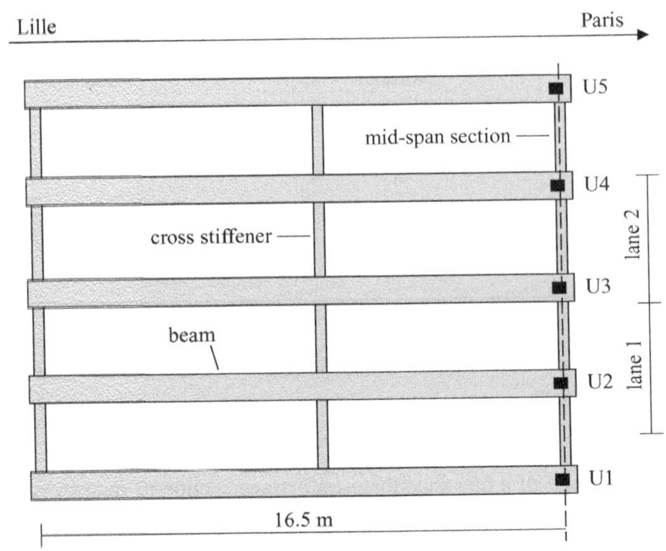

Figure 9.54 Bending strain gauges location at midspan of Roberval Bridge

Example of Outcomes

See Figure 9.53 for example of strain-gauge output and Figure 9.54 for location of the relevant strain gauges.

Benefits of Using SHM Technologies in the Project

Evaluation of the effects of traffic loading (extreme values – dynamic amplification factors).

9.17 Saint-Jean Bridge, Bordeaux, France

Contributed by CETE du Sud Ouest, Laboratoire Régional des Ponts et Chaussées & Laboratoire Central des Ponts et Chaussées

Project Description

The Saint-Jean Bridge (Figure 9.55) is located in Bordeaux, France. Opened in 1965, it was built to relieve Pierre Bridge. Like most pre-stressed concrete bridges built at the time, when thermo-mechanical behaviors were not taken into account, it is not sufficiently pre-stressed. To determine, whether pre-stress reinforcement works have to be carried out or not and to validate calculations, Bordeaux Urban Community ordered experimental investigations including SHM.

Brief Facts

Name and location: Saint-Jean Bridge, Bordeaux, France
Owner: City of Bordeaux, France
Structure category: medium span bridge
Spans: 8, 15.4 + 67.76 + 4 × 77.00 + 67.76 + 15.4 m
Structural system: triple box-girder prestressed concrete bridge
Start of SHM: 2000 – SHM described here from November 2003
Number of sensors installed: 26
Instrumentation design by: Public Works Laboratory in Bordeaux (Laboratoire Régional des Ponts et Chaussée, Bordeaux)

Figure 9.55 Saint-Jean Bridge, Bordeaux, France

Description of Structure

The Saint-Jean Bridge is located in the city of Bordeaux. It is a triple-girder bridge with a total length of 474 m. It consists of 8 spans, four are 77 m long between two 67.76 m long and two 15.4 m long. As the first and last spans are much shorter than their neighbors (0.23 times smaller), it required special attention to prevent rising of the span. Bridge girder is 3.3 m deep.

Purpose of Inspection

The purpose of this inspection was twofold. First, was to monitor the thermo-mechanical behavior of the bridge, specifically, two box-girder joints are monitored. The most-open joint (noticed during inspection and expected so by calculations) was located in the third span, whereas the barely open joint difficult to identify by inspection was located in the fourth span. Measurements were done with standard sensors, namely LVDT, and strain gauges (Table 9.18).

The second purpose was to evaluate the influence of sensor length over its accuracy and the relevance of sensor length. Sensors of lengths ranging from 10 to 400 cm therefore were installed at the two deformed mentioned above. Two types of very-long-length sensors are compared to traditional sensors: optical

Table 9.18 Sensor details (AVT tests) for Saint-Jean Bridge

Type of sensors	Number	Location
Thermal Gauges	6	First 77 m span,
LVDT	2(10 cm long)	on an open-joint
Optical Fiber Sensor (OFS)	2(60 cm, 200 cm)	Second 77 m span, on a
LVDT	1	thin joint
Vibrating Wire Sensor (VWS)	1(50 cm)	Second 77 m-span, on an
OFS	2(200 cm, 50 cm)	open-joint
LVDT	2	Plain concrete
VWS	4(2 × 50 cm, 2 × 250 cm)	
OFS	2(50 cm, 400 cm)	
Strain Gauges	1	
LVDT	1(10 cm)	
VWS	1(50 cm)	
OFS	1(25 cm)	

fiber sensors (OFS) and vibrating wire sensors (VWS). Those sensors measure strains due temperature, on a day–night cycles and also on a winter–summer cycle: every season (starting in November 2003), measurements were carried out over 3 weeks.

Measurement Equipment and Data Management

Type of system: PC based measurement system

Data management:

- data pre-analysis on site
- main analysis, graphical presentation and documentation in office

Benefits of Using SHM Technologies in the Project

After visual inspections revealed problems, calculations were made on the basis of original documents, which sometimes were missing or incomplete. Structural health monitoring technologies allow validatation of those calculations. It is economically important as prestressed reinforcement works will depend on those calculations. We also benefited from using standard instrumentation (strain gauges and LVDT) to study different types of sensors, and to compare their relevance and accuracy as a function of sensor length.

9.18 Øresund Bridge, Denmark – Sweden

Contributed by LMS International

Project Description

The Øresund Bridge (Figure 9.56) opened in July 2000. It is the most striking part of the fixed link across the Øresund connecting Copenhagen (Denmark) and Malmø (Sweden), which further includes a tunnel

Figure 9.56 Øresund Bridge, Denmark – Sweden

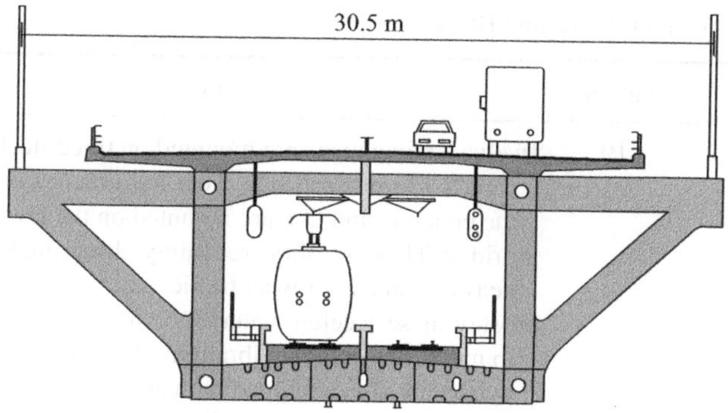

Figure 9.57 Cross-section of the Øresund Bridge

and an artificial island. The bridge owner was concerned about the stay cable oscillations under heavy wind conditions, as well as the deformation of the bridge when trains or heavy trucks are passing.

Brief Facts

Name and location: Øresund Bridge, Denmark–Sweden
Owner: Øresundsbro Konsortiet (www.oeresundsbron.com)
Structure category: cable-stayed bridge
Spans: 49 approach spans (7 of 120 m, 42 of 140 m) and a cable-stayed component with two side spans at each side (160 m and 141 m) and a main span of 490 m over the navigational channel
Structural system: the bridge has a quite unique two-level design, with a four-lane motorway placed above a two-track railway. Ten pairs of cables at each side connect the pylons of the two H-shaped towers with the bridge deck
Start of SHM: July 2000
Number of sensors installed: 55
Instrumentation design by: GeoSIG, Zürich, Switzerland

Description of Structure

Cross section shown in Figure 9.57.

Purpose of Inspection

The bridge owner was concerned about the stay cable oscillations under heavy wind conditions, as well as deformation of the bridge when trains or heavy trucks are passing. On a daily basis, the monitoring system is mainly used to record events and archive the data. As research cooperation in the framework of the SAMCO network, the dynamic cable data were used to identify the cable tension forces and the deck/tower vibrations were used to identify the structural modal parameters of the bridge.

Measurement Equipment and Data Management (Table 9.19)

Type of system: PC-based measurement system.
Data management:

- SEISLOG controls the CR-4 system and acquires data from the DSP boards
- CENTRAL provides the interface for remote access to the CR-4 systems
- CMS (Civil Monitoring System) processes the static data acquired by the system.

Data Analysis Procedures

Type of analysis: current status, statistics, alarm if out of range, operational modal analysis
Software: GeoSIG software and LMS International software

Table 9.19 Sensor details for Øresund Bridge

Type of sensors	Number	Location
Strain gauges LV3400V*S*0	19	Twelve strain gauges are mounted on three steel outriggers of the cables, one on each side. Two are mounted on the rail level in the concrete and five are mounted on the lower side of the bridge. These sensors are mainly observing torsions due to heavy wind and railway traffic
Triaxial force balance accelerometers AC-53	22	Most of these accelerometers (16) are mounted on the stay cables to measure the cable vibrations. The two tops of the east pylons are also equipped with accelerometers, as well as four locations along the deck. These sensors allow monitoring of the cable vibrations under heavy wind load and the bridge response to railway and road traffic
Thermometers PT100	12	Mounted at different locations, but mostly on the pylons. These sensors are measuring temperatures, which are correlated with the strain gauge measurements
Weather stations	2	Measuring wind speed, wind direction (1172T), air humidity and air temperature (RHA1). One is mounted on the top of a pylon, the other one at road level. The wind measurements serve as a reference for the stay-cable vibrations. The air humidity and temperature complete the meteorological information

Examples of Outcomes
See Figures 9.58 and 9.59.

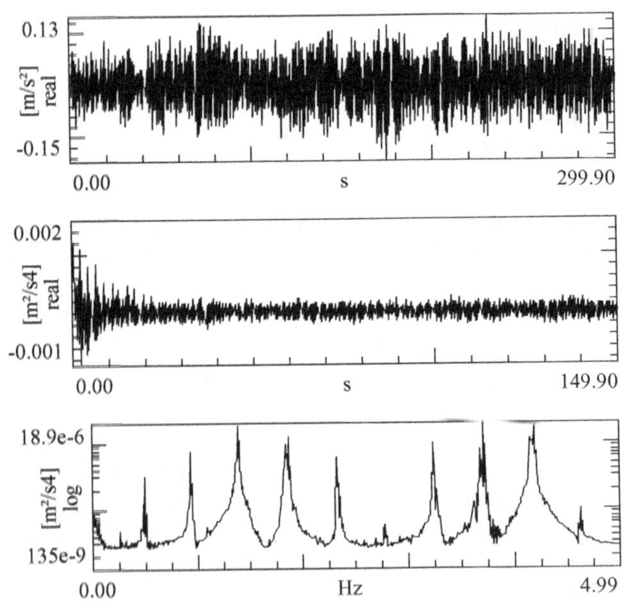

Figure 9.58 Dynamic data analysis of cable vibrations: time-history, autocorrelations, autospectrum

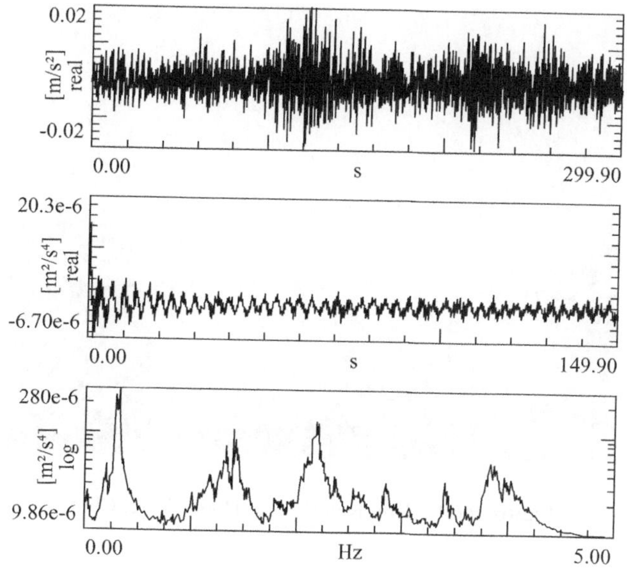

Figure 9.59 Dynamic data analysis of tower vibrations: time-history, autocorrelations, autospectrum

9.19 Ting Kau Bridge, Hong Kong, China

Contributed by The Hong Kong Polytechnic University

Project Description

The 1177 m long Ting Kau Bridge (Figure 9.60) is a three-tower cable-stayed bridge carrying a dual three-lane carriageway over Rambler Channel in Hong Kong. It provides important access connecting Hong Kong Island, Kowloon, and the New Territories and the mainland of China to the new Chek Lap Kok Airport. After 44 months in design and construction, the bridge opened to public traffic in May 1998.

Brief Facts

Name and location: Ting Kau Bridge, Hong Kong, China
Owner: Highway Department, the Hong Kong SAR Government
Structure category: cable-stayed bridge
Spans: 4, with main spans of 475 m and 448 m, and side spans 127 m each
Structural system: three single-leg towers strengthened by stabilizing cables, 384 stay cables in four planes support the two separate decks
Start of SHM: November 1998
Number of sensors installed: 236 with seven different types
Instrumentation design by: Fugro Geotechnical Service (Hong Kong) Ltd.

Description of Structure

The bridge is one of the limited instances of using multispan cable-stayed bridges in practice. It comprises two main spans of 448 m and 475 m respectively, and two side spans of 127 m each. A unique feature of this bridge is the use of slender single-leg towers that are strengthened by transverse and longitudinal

Figure 9.60 Ting Kau Bridge, Hong Kong, China

stabilizing cables. The two carriageways, with a central air gap of 5.2 m, are linked at 13.5 m intervals by I-shape main cross girders. Each carriageway grillage consists of two longitudinal steel girders along the deck edges with steel cross girders at 4.5 m intervals, and a composite deck panel on top. The deck is supported by 384 stay cables in four cable planes.

Development of Damage Detection Methodology

Since the majority of vibration-based damage detection methods need a refined or validated analytical model as a baseline reference, a three-dimensional FE model comprising over 5500 elements has been developed for the Ting Kau Bridge. The model provides a base in simulating any member-level damage, detecting damage-sensitive features and checking the feasibility of damage detection methods.

A neural network based hierarchical identification strategy has been proposed for structural damage detection in accordance with the instrumented sensors on the bridge. This multistage diagnosis strategy aims at successive detection of the occurrence, location, type and extent of the structural damage. Figure 9.61 illustrates the performance of a proposed novelty index to detect damage occurrence. By using global eigenfrequencies and local modal components, multinovelty indices are able to locate the damage region. The feasibility study indicates that using only measured natural frequencies, the novelty detector is able to signal the damage occurrence even when the damage-caused frequency change level is less than the corrupted noise level.

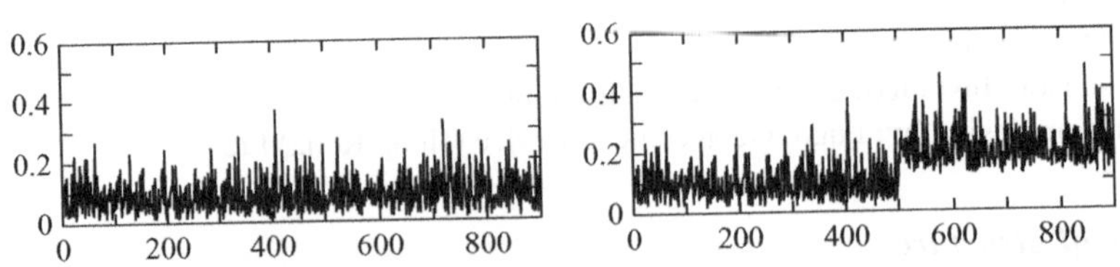

Figure 9.61 Performance of novelty index for the intact structure (on the left) and the damaged structure (on the right)

Table 9.20 Sensor details for Ting Kau Bridge

Type of sensors	Number
Anemometer	7
Temperature sensor	83
Accelerometer	45
Strain gauge	88
Displacement transducer	2
Weight-in-motion sensor	6
Global positioning system	5

Purpose of Inspection

A sophisticated long-term monitoring system, the wind and structural health monitoring system (WASHMS) has been devised by the Highways Department of the Hong Kong SAR Government to monitor the structural health and performance under in-service conditions of three long-span cable-supported bridges in Hong Kong, namely the Tsing Ma suspension bridge, the Kap Shui Mun cable-stayed bridge and the Ting Kau cable-stayed bridge. This online monitoring system consists of about 800 permanently installed sensors of various types (Table 9.20). The main objectives of this system are:

- to monitor the structural health (safety) conditions of the three bridges;
- to provide information for facilitating the planning of inspection and maintenance activities;
- to verify design assumptions and parameters for future construction of cable-supported bridges.

Examples of Outcomes

The online system accomplishes continuous 24 h monitoring per day for all sensors. The acquired raw data are accumulated at a rate about 56 MB per hour in binary format for the Ting Kau Bridge. Measured data are archived and backed-up in the form of a 1 h record and each item containing data from one channel. Based on 1-year data, a data management system has been developed. Examples of measured data are shown in the time-plot of Figure 9.62.

The influence of operational and environmental factors on modal characteristics of the bridge has been investigated. For this purpose, a total of 770 h data covering measurements in February, March, June, July, August and December of 1999 were selected. Figure 9.63 plots the average temperatures in 1-h duration

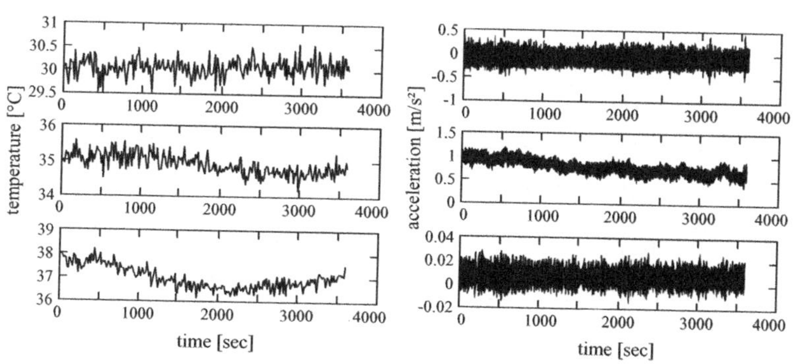

Figure 9.62 Examples of measurements for temperature (left) and acceleration (right)

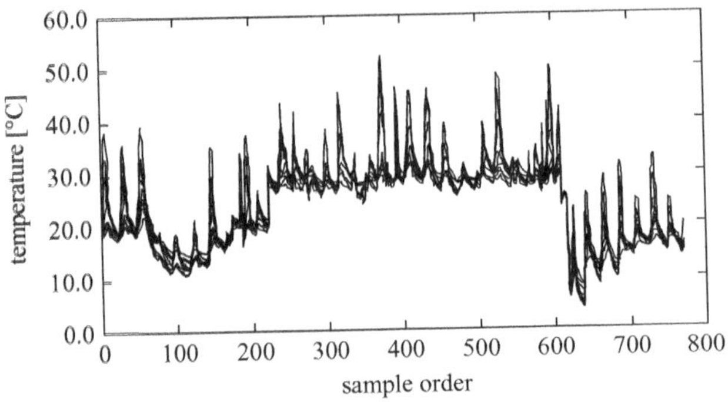

Figure 9.63 Variation in measured temperatures, Ting Kau Bridge

for 20 selected sensors. Figure 9.64 shows the identified modal frequencies. The statistical computation indicates that the standard deviation of measured frequencies can reach 4.3×10^{-3}. The variation of frequencies is attributed to the varying operational and environmental conditions. Such a variation level may mask the changes caused by actual structural damage. Therefore, for the reliable performance of damage detection methods, it is of paramount importance to discriminate abnormal changes in dynamic features due to structural damage from normal changes due to the natural variability.

Benefits of Using SHM Technologies in the Project

Using SHM technologies on the Ting Kau Bridge provides the following benefits.

- The ability to collect information of real loading effects and bridge responses, which are valuable in evaluating design parameters and assumptions.
- The ability to provide data useful in validating and updating damage-oriented structural model and in identifying damage sensitive features.
- The opportunity to provide data in verifying the feasibility and reliability of damage detection methods.
- The ability to help in maintenance and rehabilitation planning, and to give the prediction of the deterioration when combining with the analytical model.

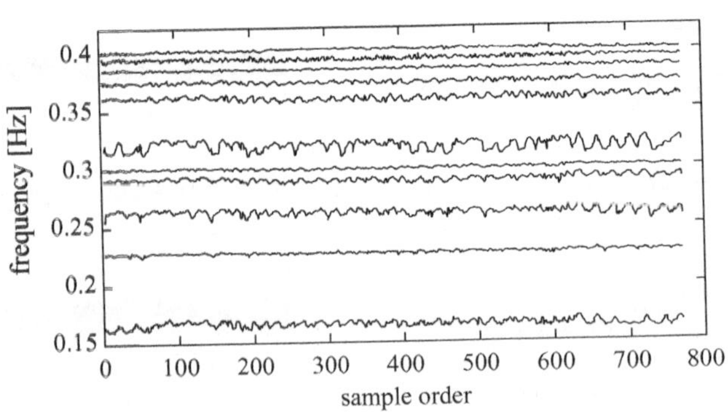

Figure 9.64 Variation in identified frequencies, Ting Kau Bridge

9.20 Skovdiget Bridge Columns, Denmark

Contributed by RAMBØLL – Bridge Maintenance and Material Technology

Project Description

The Skovdiget Bridge (Figure 9.65) north of Copenhagen, Denmark opened in 1965 and is part of a main route for urban traffic, which is also used for heavy loads. The bridge carries the busiest highway in Denmark with approximately 60 000 daily passengers over the S-train line with approximately 6000 daily passengers. Currently the columns under the western bridge are much deteriorated, just as the foundations used are of the same type that have previously failed at Fiskebæk Bridge. The columns are therefore under surveillance in order to secure their performance in the future.

Brief Facts

Name and location: Skovdiget Bridge, Copenhagen, Denmark
Owner: Danish Road Directorate
Structure category: medium span bridge
Spans: 11, 9.4 + 17.2 + 20.2 + 6 × 20.1 + 24.3 + 14.5 m
Structural system: pre-stessed concrete bridge, with hollow core girders and cross-beams, supported on concrete columns
Start of SHM: 1975 and summer 2000
Number of sensors installed: 47 in columns
Instrumentation design by: RAMBØLL, Denmark

Description of Structure

The superstructure comprises two pre-stressed concrete girders with a hollow core, with a number of closed cells. The bridge deck in each of the two parallel bridges is supported by two main girders and by 111 pre-stressed cross-beams. The superstructures are supported by a number of columns, placed under each of the main girders. The bridge with a total length of 220 m consists of two separate, parallel bridges, of which the eastern received a major renovation in 1975, whereas the western bridge received only minor repairs and renovation of the water protection. The western bridge has therefore been under surveillance since 1975.

Purpose of Inspection

It is necessary to verify that the foundations have no unexpected settlements, that the columns remain in position and that the supports work properly. It is also necessary to check the deterioration of the columns, especially so that the corrosion will not lead to an unacceptable decrease of the load-carrying capacity due to corrosion of the reinforcement.

Figure 9.65 Skovdiget Bridge, Copenhagen, Denmark

Table 9.21 Sensor details for Skovdiget Bridge columns

Type of sensors	Number	Location
Positions for inclinometers	40	Two on each column
Corrosion risk sensors (ERS)	5	
Corrosion risk sensors (CorroRisk)	20	Sensors are installed in the columns in heights from 0.1 to 2.0 m above the ground
Humidity sensors (HUM)	3	
Humidity sensors (MRE)	3	
Temperature sensors (PT100)	3	The additional sensors in the superstructure are not included in this list
Temperature sensors (others)	3	
Chloride sensors (CHL)	10	

Measurement Equipment and Data Management (Table 9.21)

Type of system: datalogger based measurement system. Manual system for reading inclinometers.

Data management:

- data pre-analysis (noise filters) on site
- main analysis, graphical presentation and documentation in office
- data transfer via modem
- long term data base in SMART light
- CMS (Civil Monitoring System) processes the static data acquired by the system.

Data Analysis Procedures

Type of analysis: averaging and additional noise filtering.
Software: SMART Light and Excel.
Additional features: no expert system.

Examples of Outcomes

It has been shown that the columns and foundations have performed well so far, as very little variations in the positions of the supports on top of the columns (beyond those variations due to temperature variations) have been observed. The inclinometer measurements over 30 years have also shown that the columns remain vertical. Continuous monitoring has been combined with the NDT mapping of resistance and corrosion potential and verifies that the NDT inspections during the autumn period tend to provide a conservative measurement of the corrosion potentials (and thus corrosion risk). Nondestructive techniques and monitoring corresponds well (Figure 9.66) and provide in combination both an overall mapping of the variations and a recording of the variation in time. The NDT mapping of corrosion potential compared with logging of corrosion potentials measured with CorroRisk sensors. The NDT mapping of corrosion potentials is correlated with the reinforcement's actual condition, and thus provides a good assessment of the reinforcement's condition and the risk of future corrosion. The evaluation shows that there is a high corrosion risk in some areas of the columns and that corrosion rates must be checked. The checking of corrosion rates is carried out annually with NDT equipment during the autumn (Figure 9.67) on selected columns in the most exposed positions and reveals a large variation over time. The mapped corrosion rates verify, however, that the corrosion rates observed will not lead to serious loss of reinforcement at the current corrosion rate levels.

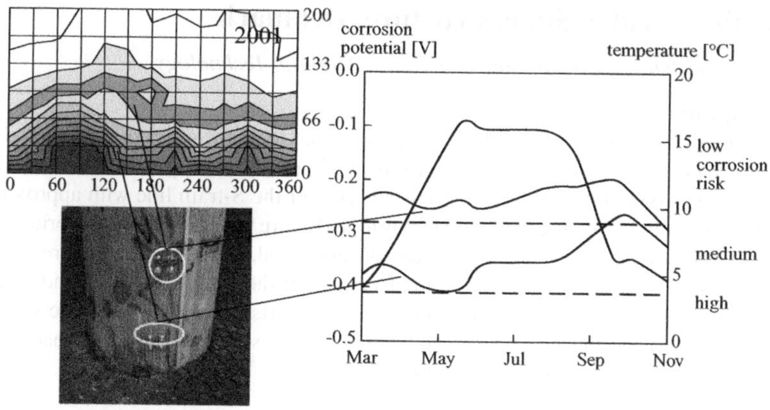

Figure 9.66 Examples of measurements for Skovdiget Bridge columns temperature

Benefits of Using SHM Technologies in the Project

The performance of the columns and foundation is checked and this verifies that the bridge has no problems with the supports and foundations. The combination of NDT mapping and continuous monitoring of corrosion risk identifies the critical positions, the correct inspection periods and the variations with time of the corrosion risk. The combination leads also to a conservative assessment of the corrosion rate, which still verifies that the loss of reinforcement area does not immediately impair the safety of the columns. Over a 5 year period the selected sensors have been found to work well in existing concrete structures and provide a good recording of the conditions in the existing concrete and of the corrosion risk of the reinforcement.

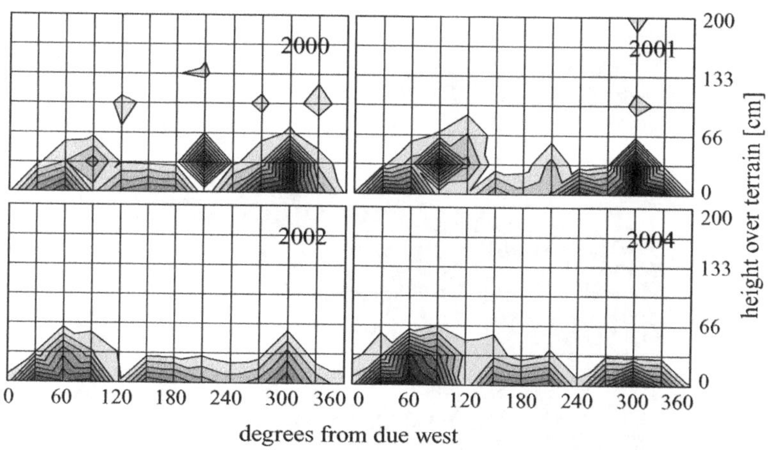

Figure 9.67 Mapping of corrosion rates using NDT in a column during the autumn, 2000 to 2004

9.21 Skovdiget Bridge Superstructure, Denmark

Contributed by RAMBØLL – Bridge Maintenance and Material Technology

Project Description

The Skovdiget Bridge (Figure 9.66), north of Copenhagen, Denmark opened in 1965 and is part of a main route for urban traffic, which is also used for heavy loads. The bridge carries the busiest highway in Denmark with approximately 60 000 daily passengers over the S-train line with approximately 6000 daily passengers. The superstructure in the western bridge (Figure 9.68) is severely deteriorated in critical positions, while at the same time facing an increased traffic load. The main girders are therefore under surveillance in order to follow the effect of the replacement of the water protection and drainage, while at the same time following the corrosion rates in the critical parts of the structure. The variations of the strains are at the same time logged, in order to generate a realistic statistic of the load variations and frequencies as well as provide a control of the FEM modeling.

Brief Facts

Name and location: Skovdiget Bridge, Copenhagen, Denmark
Owner: Danish Road Directorate
Structure category: medium span bridge
Spans: 11, 9.4 + 17.2 + 20.2 + 6 × 20.1 + 24.3 + 14.5 m
Structural system: pre-stessed concrete bridge, with hollow core girders and cross-beams, supported on concrete columns
Start of SHM: 2000, updated in 2003
Number of sensors installed: 63 in superstructure
Instrumentation design by: RAMBØLL, Denmark

Description of Structure

The superstructure comprises two pre-stressed concrete girders with a hollow core, with a number of closed cells. The bridge deck in each of the two parallel bridges is supported by two main girders (Figure 9.69) and by 111 pre-stressed cross-beams. The superstructures are supported by a number of columns, placed under each of the main girders. The bridge with a total length of 220 m consists of two separate, parallel bridges, of which the eastern received a major renovation in 1975, whereas the

Figure 9.68 Skovdiget Bridge superstructure, Copenhagen, Denmark

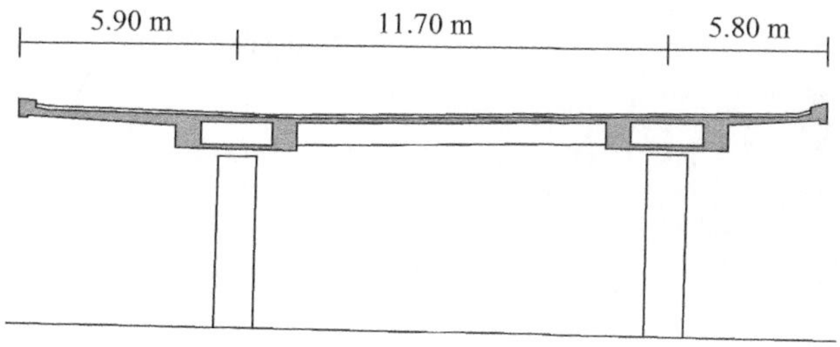

Figure 9.69 Cross-section of the Skovdiget Bridge

western bridge received only minor repairs and renovation of the water protection. The western bridge has therefore been under surveillance since 1975.

Purpose of Inspection

Initial inspection has shown severe damages in parts of the structure and the ingress of chloride and variations of humidity and corrosion potentials have been followed since 2000. It has been found necessary to determine the condition of the reinforcement in the most deteriorated parts of the main girder as well as to determine the corrosion rate. This leads to the conclusion that the traffic loads must be logged in order to generate an overview of the actual variations of the traffic loadings in the bridge.

Measurement Equipment and Data Management (Table 9.22)

Type of system: PC- based measurement system for strains. Manual system for reading CorroEye. Datalogger(s) for additional sensors.

Data management:

- data pre-analysis (evaluating, averaging and identification of extreme load, leading to storage of data) on site
- main analysis, graphical presentation and documentation in office
- data transfer via modem
- long-term data base in SMART light
- CMS (Civil Monitoring System) processes the static data acquired by the system.

Table 9.22 Sensor details for Skovdiget Bridge superstructure

Type of sensors	Number	Location
Strain-sensors (based on fibre optics)	7	3 + 2 on the two main girders plus 2 on two cross-beams
Corrosion rate sensors (CorroEye)	10	The cell over the railway and another, equally deteriorated cell over the parking area
Humidity sensors (HUM)	10	
Corrosion risk sensors (ERS)	1	
Corrosion risk sensors (CorroRisk)	12	
Humidity sensors (HUM)	7	Edge beams, main girders and some of the cross-beams
Humidity sensors (MRE)	7	
Temperature sensors (PT100)	4	
Chloride sensors (CHL)	5	

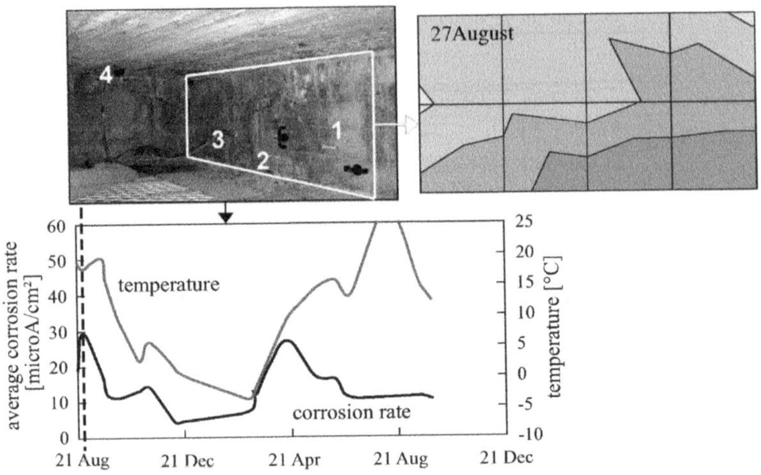

Figure 9.70 Mapping of corrosion rates using NDT in a cell on August 27, 2003 and corresponding logging of average corrosion rates by CorroEye sensors

Data Analysis Procedures

Type of analysis: transformation of strains into loading, speed and position on bridge. Transformation of corrosion current into corrosion rates.
Software: SMART Light and Excel.
Additional features SMART Light has a number of traditional Bridge Management facilities built-in.

Examples of Outcomes

The corrosion rates have been determined several times by means of NDT, using the Galvapulse equipment. This verified that the apparent corrosion rate would limit the service-life of the structure, despite the recent renovation of the waterproofing.

The NDT mapping (Figure 9.70) requires, however, regulation of the traffic on the electrically powered railway line and was therefore combined with installation of corrosion rate sensors, which enable monitoring of corrosion rates without traffic disruption. The inspection, the FE modeling calculations and the monitored corrosion rates indicate that the structure, within 10 years, will have an insufficient load-carrying capacity and will require strengthening, replacement or a more detailed essessment of the traffic loadings.

Since 2000 the deformation has been monitored continuously with sensors based on fiber optics, which allow measurements to be carried out approximately 25 times per second. These measurements are averaged over each hour. Passage of a heavy vehicle will be detected by the system and the variations of the strains from approximately 10 s before to 10 s after the passage will be kept in the records (Figure 9.71). This allows essentially a logging of the number of heavy vehicles, their loads, position on the road and their speed.

Benefits of Using SHM Technologies in the Project

The monitoring of deterioration parameters (chloride, humidity, temperature and corrosion potentials) as well as the visual inspection have identified the need for monitoring. The use of corrosion rate monitoring provides a record of the actual corrosion rates in the critical structural parts without traffic disruption. The combination of NDT mapping and monitoring with sensors provides both a general mapping of large areas of the structures and a recording of the variation in time. The logging of the passages of the heavy

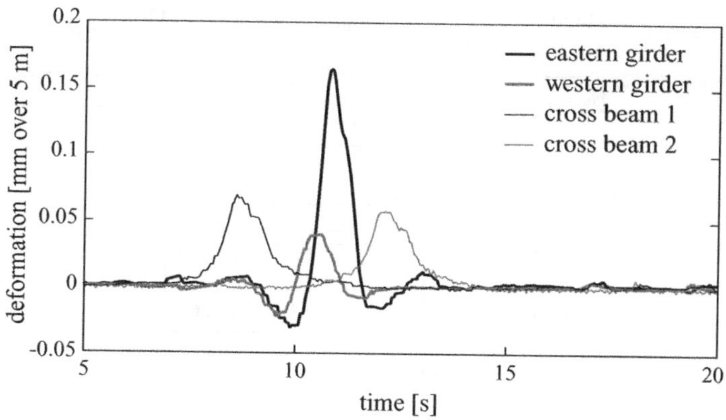

Figure 9.71 Logging of strain variations during passage of a heavy truck

trucks generates an improved load model, which will be useful in the next probalistic assessment of the structure's safety and will add some years to the structure's service life.

9.22 Bolshoj Moskvoretsky Bridge, Moscow, Russia

Contributed by Smartec SA

Project Description

Bolshoj Moskvoretsky Bridge (Figure 9.72) over the Moskva River was built in 1936–37 by the famous Soviet architect A. Shjusev. It is situated in the historical centre of Moscow, next to the Kremlin, and leads one of the main traffic lines of the city to the Red Square. The bridge has a status of 'functioning architectural heritage' protected by the state.

Brief Facts

Name and location: Bolshoj Moskvoretsky Bridge, Moscow, Russia
Owner: The Government of Moscow
Structure category: reinforced concrete arched box girder bridge with a decorative external facing of natural stone. Total length of the bridge is about 250 m
Spans: 3, 43 + 92 + 43 m
Structural system: the bridge consists of three parallel arches. The cross-section of each arch contains three boxes separated by partitions 350–450 mm thick (along the axis of the bridge) and diaphragms

Figure 9.72 Bolshoj Moskvoretsky Bridge, Moscow, Russia

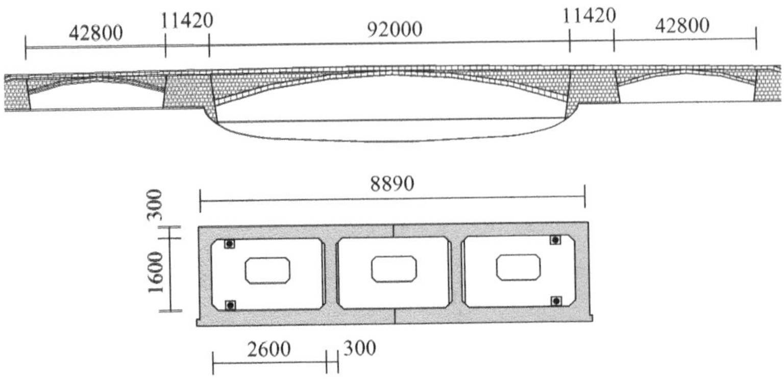

Figure 9.73 Geometry of Bolshoj Moskvoretsky Bridge

with openings for maintenance purposes (transverse to the bridge axis). The superstructure consists of the bridge deck supported by columns that rest on the above-mentioned separating partitions
Start of SHM: July 2003
Number of sensors installed: 22
Instrumentation design by: SMARTEC SA, Switzerland; ZAO Triada-Holding, Russia

Description of Structure

The bridge has eight vehicle lanes and two sidewalks separated from them by curbs (Figure 9.73). Two types of degradation are noticed on the bridge: settlement of an abutment producing cracking of the stone lining as well as the structural elements, and chloride penetration into the structures leading to reinforcement corrosion. These signs of degradation on the bridge after nearly 70 years of operation and its functional and historical importance led the authorities to decide to monitor the structural behavior of the bridge on a continuous basis.

Purpose of Instrumentation

The aim of monitoring is to increase the knowledge concerning structural behavior. Standard SOFO sensors (Table 9.23) have been installed to continuously monitor average strain along the arch curvature in both horizontal and vertical direction. Thermocouples have been installed to distinguish thermal influences.

Examples of Outcomes

The installation of the SOFO System was finished in July 2003. The long-term monitoring started. At the moment data are being accumulated for further interpretation and analysis. Examples of defects are shown in Figure 9.74 and an example of measurement is shown in Figure 9.75.

Benefits of Using SHM Technologies in the Project

- Data acquisition for better knowledge concerning the structural behavior to increase safety and to reduce maintenance costs.

Table 9.23 Sensor details for Bolshoj Moskvoretsky Bridge

Type of sensors	Number	Location
Standard SOFO Sensors	16	In the central arch
Thermocouples	6	

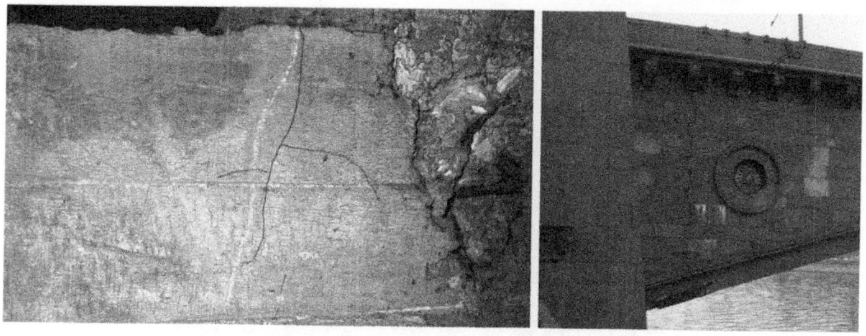

Figure 9.74 Examples of defects: cracks on structural elements (left) and external facing (right)

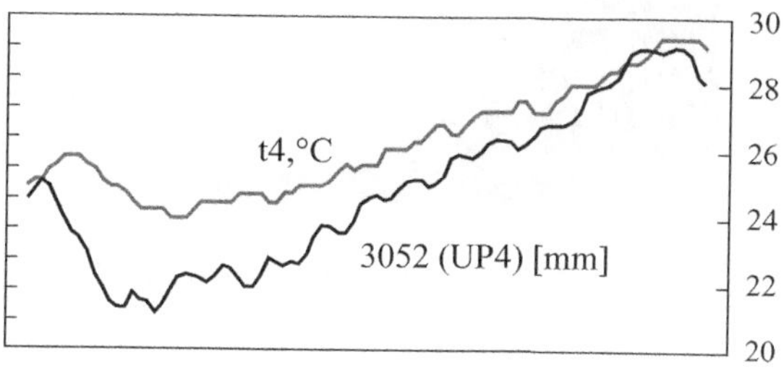

Figure 9.75 Temperature versus displacement, Bolshoj Moskvoretsky Bridge

- Continuous remote monitoring.
- Innovative system (part of which is a sophisticated software for data interpretation) allows compitation of comprehensive information on bridge behavior.

9.23 Versoix Bridge, Geneva, Switzerland

Contributed by SMARTEC SA

Project Description

The Versoix Bridge (Figure 9.76) is located near the city of Versoix, crossing the river with the same name. It lies on the A1 highway between Lausanne and Geneva, Switzerland making an important connection between two major Swiss cities (60 000 vehicles per day). From 1996 to 1998 the bridge was refurbished and enlarged to accommodate an additional security lane in both directions (Figure 9.77). Since an important amount of new concrete was added asymmetrically to an existing structure, issues of differential shrinkage could occur and decrease the structural performance.

Brief Facts

Name and Location: Versoix Bridge, Versoix, GE, Switzerland
Owner: Canton of Geneva
Structure category: short and medium span bridges (36 and 56 m)
Spans: two border spans of 36 m and four central spans of 56 m
Structural system: twin bridges, two pre-stressed continuous girders stiffened by diaphragm supports along with steel-beam cantilevered deck
Start of SHM: 1996

Figure 9.76 Versoix Bridge Switzerland

Number of sensors installed: 120
Instrumentation design by: IMAC-EPFL, Swiss Federal Institute of Technology, Lausanne

Description of Structure

The old Versoix Bridge consisted of twin concrete bridges with six spans. The deck of each bridge was laid on two continuous girders stiffened by diaphragms. Both bridges have been refurbished and widened. The girders are reinforced with new pre-stressed concrete, while the decks are cantilevered at the external side and supported by inclined steel beams.

Purpose of Instrumentation

The aim of instrumentation of the Versoix Bridge is to monitor long-term performance with particular emphasis on the consequences of interaction between the pre-existing and new parts of the structure. Thus, the following parameters were monitored: average strain in concrete including early and very early age, old–new concrete interaction, and average curvature analysis in both horizontal and vertical planes, detection of torsion and distribution of both horizontal and vertical displacements. Fiber optic sensors of type SOFO were used (Table 9.24) and long-term automatic and remote monitoring was performed. Thermocouples of type 'K' were used in order to distinguish thermal strain and load cells at abutments to control the force.

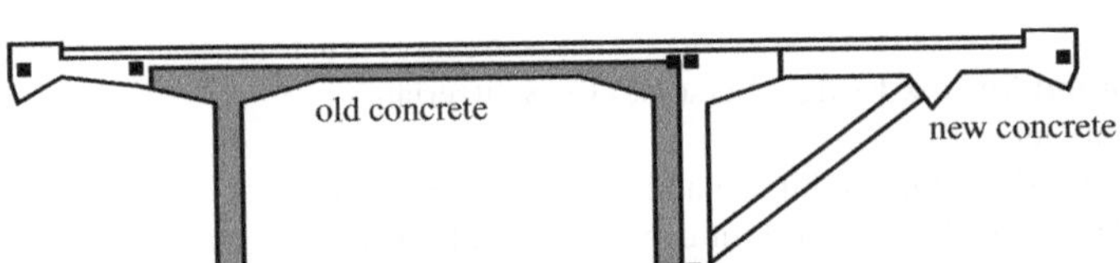

Figure 9.77 Cross-section of the Versoix Bridge and position of the SOFO sensors

Table 9.24 Sensor details for Versoix Bridge

Type of sensors	Number	Location
SOFO fiber optic sensors	104	Span 1–2
'K' thermocouples	12	Span 1–2
Load cells	4	Abutments

Examples of Outcomes

The early age measurements allowed the prediction of cracking long before it became visible and the optimization of the concrete mix for successive pours. The sensor pairs at the interface confirmed the excellent adherence between the old and new concrete. The vertical displacement was measured during the load test by double integration of the curvatures (Figure 9.78). Horizontal bending of the bridge due to asymmetrically added new concrete was successfully observed and measured. The single sensors were used to follow the long-term shrinkage of concrete and the seasonal deformations due to temperature changes (Figure 9.79).

Benefits of Using SHM Technologies in the Project

The SHM technology applied to Versoix Bridge provided the following benefits.

- Long-term data collecting concerning the structural behavior of the bridge.
- Improvement of concrete mix composition after the first pouring was done.
- Possibility to verify the interaction between the existing and new concrete.

After seven years of monitoring the following conclusions are drawn.

- The structure is perfectly monolithic and no delaminating is detected.
- The evolution of concrete performances is in the range of model prediction.

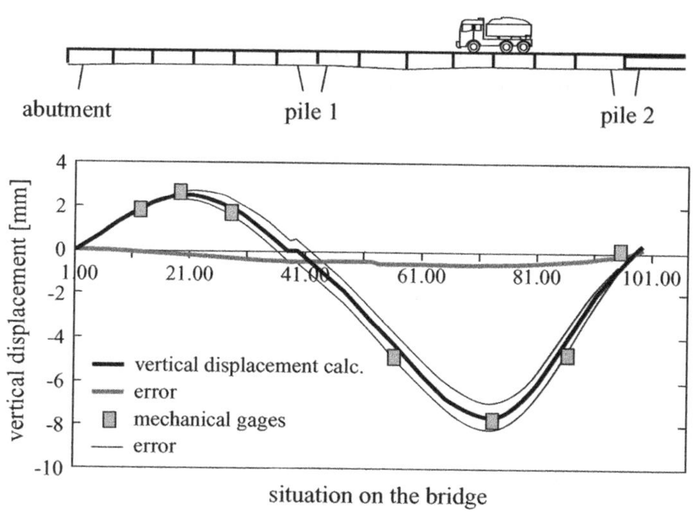

Figure 9.78 Vertical displacement during the load test, Versoix Bridge

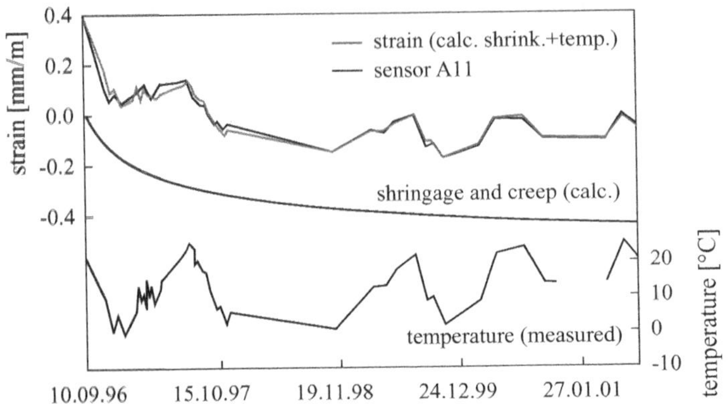

Figure 9.79 Five years measurement of single sensor and evaluation of rheologic strain

9.24 Tsing Ma Bridge, Hong Kong, China

Contributed by TNO TPD, Division Optical Instrumentation

Project Description

The Tsing Ma Bridge (Figure 9.80) is the longest suspension bridge (2.2 km) in the world for carrying both vehicle and railway traffic. Tsing Ma Bridge has a double deck: the upper deck has two three-lane highways for vehicle traffic; the sheltered lower deck includes two railway tracks and two single-lane emergency roadways for maintenance and ensuring uninterrupted traffic from/to the Hong Kong International Airport during typhoons when wind speed is still within an acceptable level. Besides the conventional sensors, Fiber Bragg Grating (EBG) sensors were installed by the Photonics Research Centre of the Hong Kong Polytechnic University to measure vibration, strain distribution and suspension cable tension.

Brief Facts

Name and location: Tsing Ma Bridge, Hong Kong, China
Owner: Suspension bridge (two main suspension cables)
Main Span: 1377 m (two high-strength concrete towers)
Overall length: 2.2 km
Main cable: 1.1 m in diameter
Shipping clearance: 62 m
Number of sensors: >350

Figure 9.80 Tsing Ma Bridge, Hong Kong, China

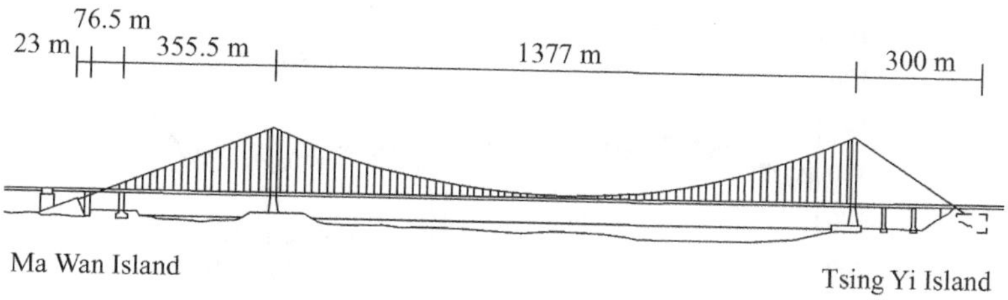

Figure 9.81 Tsing Ma Bridge geometry

Description of Structure

The Tsing Ma Bridge is a double deck suspension bridge having a fully suspended main span supported by two portal-braced, reinforced-concrete towers (Figure 9.81). The bridge deck is suspended from two main cables passing over the main towers and secured into massive concrete anchorages at each end. The bridge deck section, 41 m wide and 7.5 m high, is a hybrid arrangement combining both longitudinal trusses and cross-frames. The main-span deck and the Ma Wan side-span deck are suspended at 18 m intervals by hangers to the main cables, while the Tsing Yi side-span deck is supported by three concrete piers spaced at 72 m.

Purpose of Inspection

The sensors are the early warning system for the Tsing Ma Bridge and provide the essential information that help the Hong Kong Highways Department to accurately monitor the general health conditions of the bridge, in terms of structural durability, reliability and integrity. The sensors include strain gauges, GPS position sensors, accelerometers, level sensors, temperature sensors and weight-in-motion sensors (Table 9.25). This project is focused on the application of Fiber Bragg Grating (FBG) for strain measurement and the comparison with conventional strain gauges.

Measurement Equipment and Data Management

Type of system: TNO high-speed dense-channel demultiplexing/interrogation system for FBG sensor array.

Data management:

- Data logging
- Main analysis (statistic, frequency analysis), graphical presentation and documentation in office.

Data Analysis Procedures

Type of analysis: statistics, frequency analysis.
Software: self-made software.

Table 9.25 Sensor details for Tsing Ma Bridge

Type of sensors	Number	Location
FBG strain sensor	10	Metal structure of section 23
FBG temperature sensor	1	Metal structure of section 23
FBG strain sensor	9	Rocker bearing on tower
FBG temperature sensor	1	Rocker bearing on tower
FBG strain sensor	1	Suspension cable

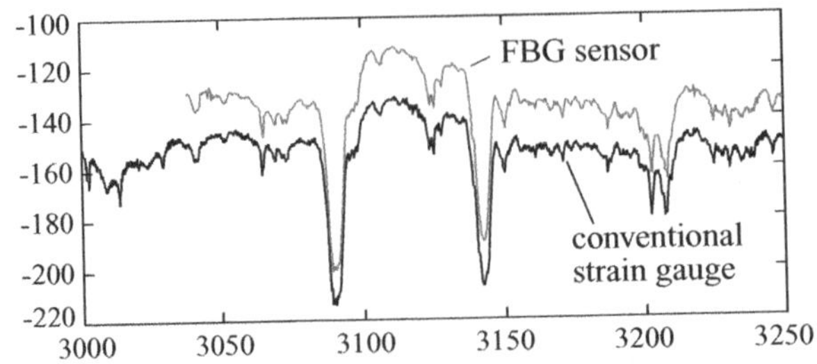

Figure 9.82 Comparison of strain measurement between FBG sensor and conventional strain gauge. An artificial off-set is applied to the FBG sensor signal

Example of Outcomes

The results of the FBG sensor are compared with that of the conventional strain gauge (Figure 9.82). Although the sensors are not located at exactly the same location, great resemblance in the results is observed. Train passages and heavy traffics can clearly be measured.

Benefits of Using SHM Technologies in the Project

- Providing information to determine distribution of strains/stresses in critical bridge components.
- Documenting abnormal loading incidents such as typhoons, earthquakes, traffic overloads and ship collisions with bridge piers.
- Detecting damage or accumulated damage in critical bridge components.
- Providing information for a cost-effective maintenance program.

9.25 A14 Huntingdon Railway Viaduct, England

Contributed by TRL Limited

Project Description

The A14 Huntingdon Railway Viaduct (Figure 9.83) is part of the Cambridge to Kettering section of the A14 dual carriageway. The structure has been the subject of a Special Inspection that indicated the presence of voids, water and chlorides in the tendon ducts, but no significant corrosion of the strands. A

Figure 9.83 Huntingdon Railway Viaduct, England

SoundPrint® acoustic monitoring system, designed by Pure Technologies Ltd of Canada, was installed to monitor tendon wire break activity in one of the cantilevers.

Brief Facts

Name and location: Huntingdon Railway Viaduct, Cambridgeshire, England
Owner: Highways Agency, England
Structure category: medium span bridge
Spans: 6, 32.3 + 32.3 + 32.3 + 64.3 + 32.3 + 32.3 m
Structural system: six-span structure of which span 4 consists of two 16.15 m cantilever sections and a 32.0 m suspended span
Start of SHM: mid-1998
Number of sensors installed: 36
Instrumentation design by: Pure Technologies Ltd, Calgary, Canada

Description of Structure

The structure has six spans; the main span consists of a 32 m long suspended span sat on half joints formed at the end of two 16 m long cantilevers extending from the adjacent piers. The viaduct spans the B1514, the East Coast Main Line and part of Huntingdon railway station (Figure 9.84).

Purpose of Inspection

Special Inspection indicated the presence of voids, water and chlorides in the tendon ducts. The SoundPrint® acoustic monitoring system has been installed to monitor tendon wire break activity in one of the cantilevers. The structure possessed further features that made it a good candidate for acoustic monitoring.

- Additional structural investigations were in progress. This would be enhanced by a clear indication of the presence or absence of actively fracturing wires.
- The structure contained features that lent themselves to monitoring, such as difficult to inspect half joints.
- The structure occupies a strategic position on the network and carried a high volume of HGV traffic.

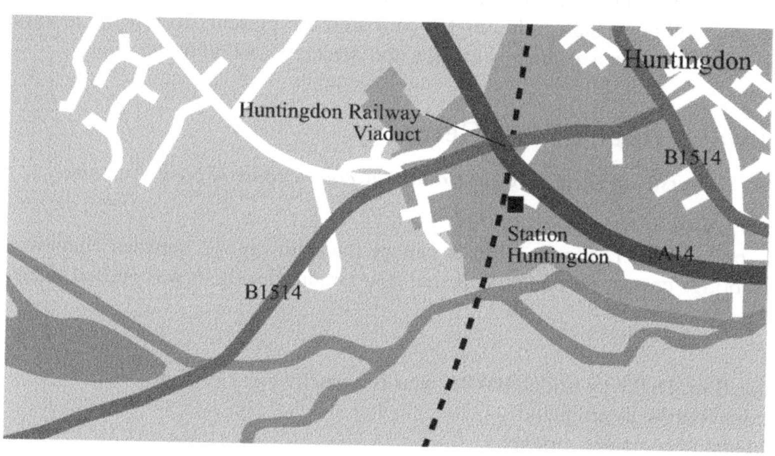

Figure 9.84 Location of Huntingdon Railway Viaduct

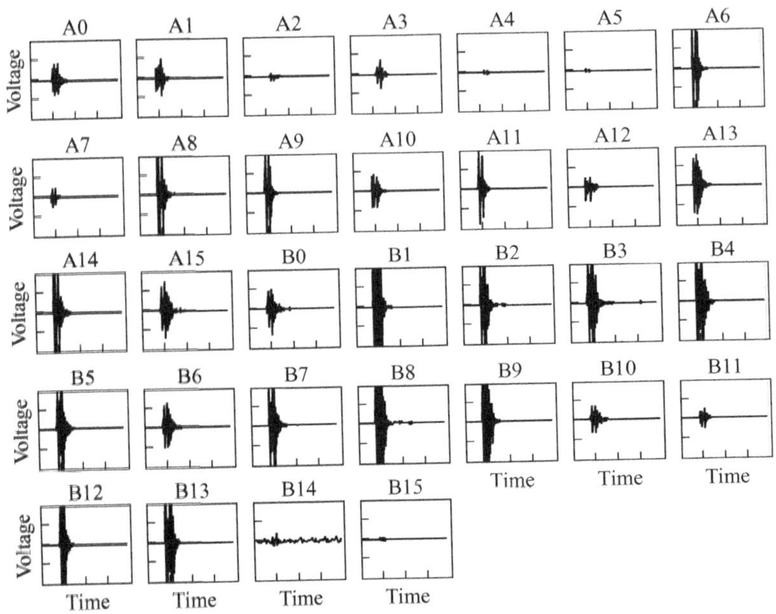

Figure 9.85 Typical acoustic response from externally mounted wire break device

Example of Outcomes
The probability of a tendon wire break occurring in the structure is very low so an external wire break device (Figure 9.85) was installed on the structure to check the operation of the monitoring system.

Benefits of Using SHM Technologies in the Project
The SoundPrint® acoustic monitoring system at Huntingdon has been in operation since mid-1998. It has provided an excellent opportunity for confronting challenges in detecting and locating post-tensioned wire breaks in noisy environments, and establishing the protocols needed to ensure the success of long-term, continuous, unattended monitoring. The viaduct has not experienced any naturally occurring wire breaks during monitoring, although the conditions for corrosion are present. To test the monitoring system, external wire breaks have been artificially created and detected in blind trials.

9.26 Highway Bridge BW91, Germany

Contributed by University of Technology at Braunschweig, Institute of Steel Structures

Project Description
The highway bridge BW91 (Figure 9.86) is part of the highway A2 between Hannover and Berlin, Germany. The bridge crosses the Mittellandkanal near Braunschweig. It was opened in 2003 as a three-lane-bridge.

Brief Facts
Name and location: Highway-bridge BW91, near Braunschweig, Germany
Owner: Bundesrepublik Deutschland
Structure category: composite bridge
Spans: 1, 56.26 m
Structural system: steel box girder with deck as a composite construction

Figure 9.86 Highway bridge BW91 near Braunschweig, Germany

Start of SHM: January 2003
Number of sensors installed: 15
Instrumentation design by: University of Technology Carolo Wilhelmina at Braunschweig, Institute of Steel Structures, Braunschweig, Germany

Description of Structure

The superstructure comprises of two steel box girders and a deck as a composite construction. The intermediate beams of the composite construction have a spacing of 3.60 m, the width of the bridge is 20 m. There are two units of this bridge, one for each direction.

Purpose of Inspection

Due to the central position of bridge BW91 the measured weights of the vehicles and their distribution in the flow of traffic are valid for a large number of other bridges of Highway A2. In addition, the measurements are carried out to obtain the strains at critical locations. Measurements are carried out within the collaborative research program SF B477 'Life Cycle Assessment of Structures via Innovative Monitoring' (www.sfb477.tu-braunschweig.de).

Measurement Equipment and Data Management (Table 9.26)

Type of system: PC-based measurement system.
Data management:

- data pre-analysis (statistics) on site
- main analysis, graphical presentation and documentation in office
- long-term database due to permanent monitoring.

Table 9.26 Sensor details for highway bridge BW91

Type of sensors	Number	Location
Strain gauges	15	At three intermediate beams (spacing: 18 m) underneath each lane including the hard shoulder (Figure 9.87)

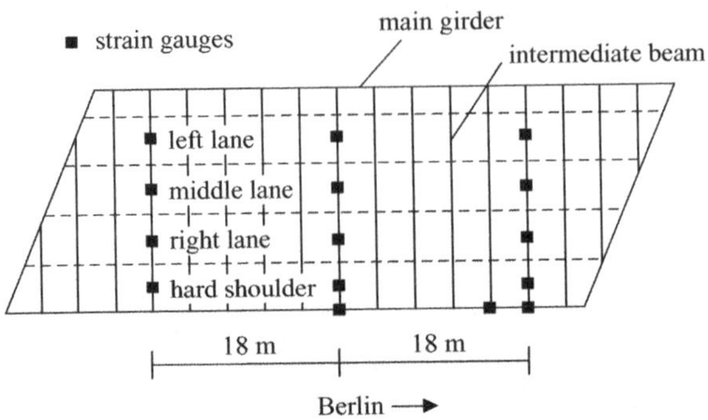

Figure 9.87 Sensor positions on highway bridge BW91

Data Analysis Procedures

Type of analysis: WIM, statistics, rainflow analysis, changes in traffic density.
Software: Self-made software.

Example of Outcomes

The calibration of the sensors was carried out by use of a 30 truck. According to the specification of the COST 323 Project the measuring system has an accuracy class of D + (20).

Benefits of Using SHM Technologies in the Project

Due to permanent monitoring long-term changes in the flow of traffic can be observed.

9.27 Herrenbrücke, Lübeck, Germany

Contributed by University of Technology at Braunschweig, Institute for Building Materials, Structural Concrete and Fire Protection

Project Description

The Herrenbrücke crossing the river Trave (Figure 9.88) was built between 1962 and 1964. A new crossing, a tunnel, was planned to be finished in 2006. The Herrenbrücke would remain in service up to

Figure 9.88 Herrenbrücke, Lübeck, Germany

Table 9.27 Sensor details for Herrenbrücke

Type of sensors	Number	Location
LDVT	29	
Fiber optic sensors	5	Foreland bridge
PT100	15	

the time the new crossing opened. The Highways Department of Lübeck contracted a consultant engineer to provide a statement assessing the life-cycle of the bridge up to the year 2006.

Brief Facts

Name and location: Herrenbrücke, Lübeck, Germany
Owner: City of Lübeck, Germany
Structure category: bascule bridge with two medium span foreland bridges
Spans: 18, 19.4 m
Structural system: pre-stressed concrete orthotropic deck, reinforced columns, steel bascule
Start of SHM: October, 2000
Number of sensors installed: 34
Instrumentation design by: Technical University Braunschweig, Institute for Building Materials, Structural Concrete and Fire Protection, Germany

Description of Structure

It consists of two side spans made from pre-stressed concrete (approximately 153 m and approximately 311 m long) as well as a balance bridge made from steel (approximately 86 m long). The bridge shows signs of corrosion damage, which is attributed to poor working quality in grouting the tendons. Tendon failure is assumed to account for up to 45%. Repair measures were executed thereupon.

Purpose of Inspection

From the point of view of the consultant engineer, a life-cycle assessment of this bridge was not possible on the basis of knowledge regarding its state of condition only. Therefore SHM systems had to be installed to monitor displacements and deformation of the bridge (Table 9.27 and Figure 9.89).

Measurement Equipment and Data Management

Type of system: PC-based measurement system.

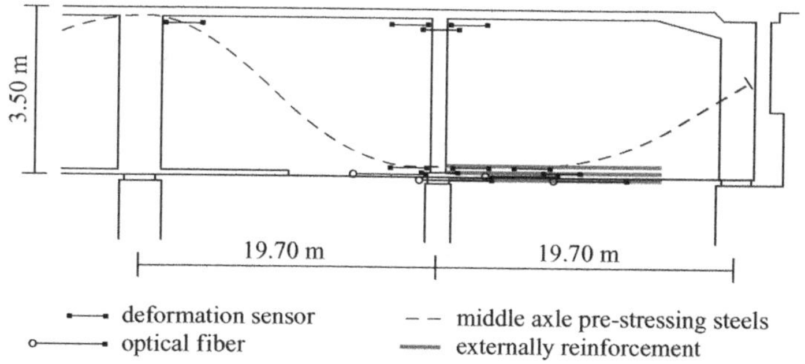

Figure 9.89 Instrumentation plan for Herrenbrücke

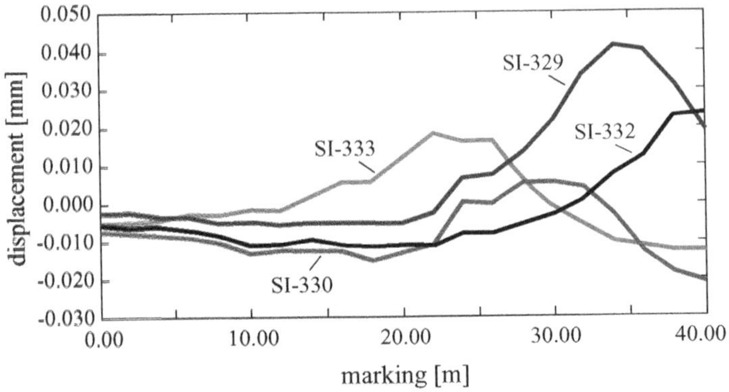

Figure 9.90 Static measurement, Herrenbrücke

Data management:

- data pre-analysis (statistics, frequency analysis) on site
- main analysis, graphical presentation and documentation in office
- data transfer via modem
- long-term database.

Data Analysis Procedures

Type of analysis: statistics, ambient analysis.
Software: self-made software and MATLAB.

Examples of Outcomes

Static as well as dynamic measurements are possible – see Figures 9.90 and 9.91.

Benefits of Using SHM Technologies in the Project

The available measurement data allow the definition of threshold values regarding maximum traffic within a defined temperature range in order to assess the crack risk due to tendon failure.

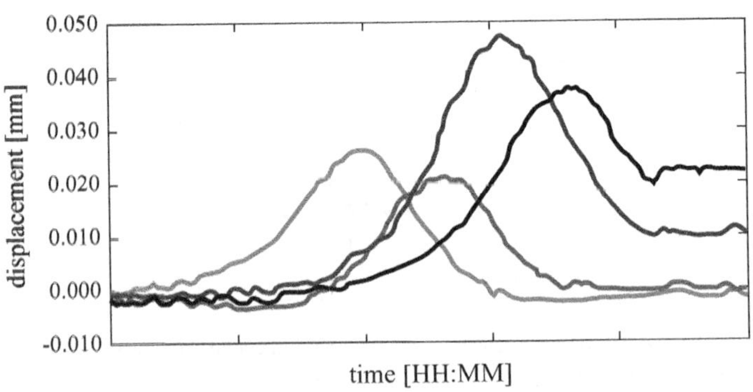

Figure 9.91 Dynamic measurement, Herrenbrücke

9.28 Pasir Panjang Semi-Expressway, Singapore

Contributed by University of Plymouth, School of Engineering

Project Description

Pasir Panjang Semi-Expressway, during construction (Figure 9.92)

Brief Facts

Name and location: Pasir Panjang Semi-Expressway, Singapore
Owner: Land Transport Authority, Singapore
Structure category: multiple short span dual carriageway viaduct
Spans: multiple, varying length, typical span 38 m
Structural system: pre-cast segmental concrete box, internal post-tensioning
Start of SHM: 2003
Number of sensors installed: 60
Instrumentation design by: SysEng (S) Pte Ltd, Nanyang Technical University, Singapore

Description of Structure

Segments of the same shape but different web dimensions are cast in West Singapore and added using the balanced cantilever method, four segments at a time. Closure strips are added at a much later date. The spans are arranged in 'bridges' of about five spans between expansion joints. The expressway was designed to carry heavy goods vehicles, principally containers in transit between two container terminals.

Purpose of Monitoring

As well as serving to validate the original design through performance evaluation, the aim of the monitoring was also to evaluate the design of a comprehensive bridge SHM system. By instrumenting span segments in one span of each bridge, the aim was to capture the performance and interaction of a large

Figure 9.92 Pasir Panjang Semi-Expressway, Singapore

Table 9.28 Sensor details for Pasir Panjang Semi-Expressway

Type of sensors	Number	Location
VWG strain gauges	40	In segment near pier of five spans as well as in piers
VWG stress cells	4	In two segments of one span
Fibre Bragg Grating	22	On deck soffit in two spans

stretch of the viaduct. Modal testing has been used to validate FE modeling of substructures during construction for extrapolating to a complete bridge. Via a validated FE modeling the aim is to interpret the static response data in terms of structural events that can be simulated and characterized by FE modeling.

A further aim was to evaluate the performance of Fiber Bragg Grating (FBG) systems for SHM of such bridges (Table 9.28).

Measurement Equipment and Data Management

Type of system: local static loggers with wireless modem and dynamic data acquisition

Data management:

- offline correlation and statistical analysis
- offline modal analysis and FEMu (see Section 7.1.3).

Data Analysis Procedures

Type of analysis: for static data statistics, correlations; for modal test – data – modal analysis.

Software: Excel/MATLAB for static data and MATLAB GUI: MODAL for modal analysis.

Additional features: ambient vibration test using human shakers for modal tests.

Examples of Outcomes

So far it has been possible to create and validate a FE model (Figure 9.93) that will be used to simulate load and structural anomalies as training for diagnosis via performance monitoring. As the bridge is constructed and spans joined, the strain/stress increments will be used for similar purposes, i.e. SHM system training.

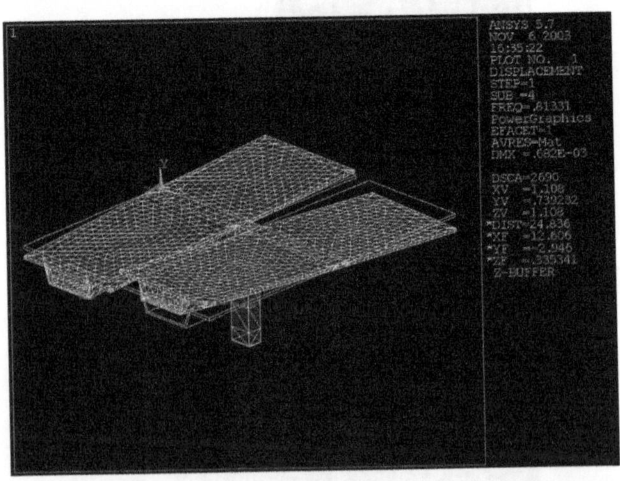

Figure 9.93 Strain distribution at a main girder during the crossing of a heavy load vehicle measured by SHM

The FBG arrays have been installed on deck soffits and will be used to monitor performance during proof testing. The exercise so far has shown the utility of remote data collection via wireless modem, and from these data traditional temperature-induced trends and stress–strain correlations have been observed.

Benefits of Using SHM Technologies in the Project

There is no other way to assess the as-built structure and to monitor performance under the possibly excessive loading of HGVs. Hence the aim is to feed back to Land Transport Authority not only the effectiveness of the design but also the assessment of loads. Further, the bridge is being used to test relatively advanced SHM systems via automated wireless data downloads and fiber optic sensors.

9.29 Pioneer Bridge, Singapore

Contributed by University of Plymouth, School of Engineering

Project Description

Pioneer Bridge, Singapore (Figure 9.94), was completed in 1970 and upgraded by Land Transport Authority in 2000–2001. It carries heavy industrial traffic to and from Jurong Port and is rated to carry vehicles of 44 t. Modal testing and short-term strain/vibration monitoring were conducted before and after the upgrade to test the need for and effect of the upgrade.

Brief Facts

Name and location: Pioneer Bridge, Singapore
Owner: Land Transport Authority (LTA), Singapore
Structure category: short span bridge
Spans: 1, 18 m
Structural system: steel box girder with orthotropic deck and steel columns
Start of SHM: October 2000
Number of sensors installed: 8 for monitoring, up to 16 for vibration test
Instrumentation design by: Infratech, Australia (monitoring system), modal test system designed by J. Brownjohn (at Nanyang Technical University Singapore)

Description of Structure

Thirty-seven pre-cast pre-tensioned inverted T-beams tied together by 25 lateral tensioning cables set in a cast in situ transverse diaphragm. The T-beams carry a deck slab having a thickness that varies from

Figure 9.94 Pioneer Bridge, Singapore

Figure 9.95 Heavy load (bridge segment), Pioneer Bridge

152 mm to 305 mm. Bearings formerly simply supported, now built in. The bridge was upgraded in line with LTA island-wide program, and carries occasional extra heavy loads (Figure 9.95).

Purpose of Monitoring

The program of two modal tests and two short-term live response monitoring exercises was aimed at validating analytical models (via modal testing and FEMu; see Section 7.1.3) then using the validated models together with live strain statistical properties to identify the load capacity before and after the upgrade. The exercise was also aimed at evaluating the procedures for doing this.

Measurement Equipment and Data Management (Table 9.29)

Type of system: HMX data logger and modal test system

Data management:

- Waveform logging
- Peak value recording
- Acceleration recording.

Data Analysis Procedures

Type of analysis: statistics, Gumbel plot, independent storm modal analysis and FEMu

Software: Self-made software, SDTools, purpose-developed MATLAB GUI: MODAL

Additional features: Forced and ambient vibration software test analysis

Table 9.29 Sensor details for Pioneer Bridge

Type of sensors	Number	Location
Strain gauges	4	At midspan on soffit
Accelerometers	4	Near strain gauges
Modal test accelerometers	16	According to test grid
Shaker	1	1/3rd span, footpath

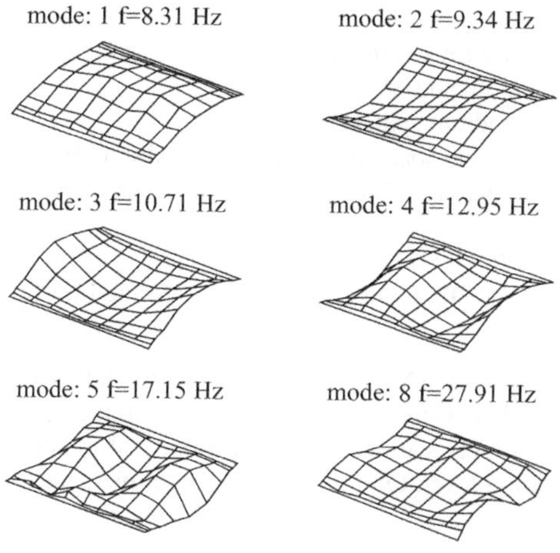

Figure 9.96 Mode shapes for the bridge after upgrade: effect of bearing rigidity is visible in mode shapes and increased frequencies

Examples of Outcomes

The mode shapes (Figure 9.96), identified from post-upgrade forced vibration testing, prove the effectiveness of the upgrade through FEMu, showing clearly the rotational stiffness imposed at the bearings. The FEMu has been used to estimate load carrying capacity.

Benefits of Using SHM Technologies in the Project

It was possible to show that the bridge was already in good shape and that n-year return period strains were a smaller percentage of capacity after upgrade.

9.30 Tuas Second Link, Singapore–Malaysia

Contributed by University of Plymouth, School of Engineering

Project Description

The Tuas Second Link, completed in 1997, is the second road access between Singapore and Peninsular Malaysia (Figure 9.97).

Figure 9.97 Tuas Second Link, Singapore–Malaysia

Brief Facts

Name and location: Tuas Second Link, Singapore
Owner: Governments of Singapore and Malaysia
Structure category: multiple medium span bridge
Spans: 27 spans totaling 1.9 km, two spans in Singapore, maximum 92 m
Structural system: Cast in situ dual reinforced concrete box with expansion joints either end of bridge
Start of SHM: March 1997
Number of sensors installed: 75
Instrumentation design by: Gage Technique UK, SysEng (S) Pte Ltd, Infratech Australia

Description of Structure

All but 170 m of the 1.9 km bridge are in Malaysian territory. Although visually similar, the Singapore side uses internal post-tensioning, whereas the Malaysia side uses external post-tensioning. On the Singapore side, spans were balanced-cantilever segmental construction, cast in situ, for the parallel carriageways that are joined only at foundation level. Of note are the slender piers and the use of only two expansion joints, one at each end.

Purpose of Monitoring

An opportunity was made available by a consultant requirement for "instrumentation" being directed to Nanyang Technical University and converted to a monitoring installation based on systems implemented in the 1980s. Arrays of thermocouples and stress cells were installed (Table 9.30), and the aim was to track performance and learn about stress behaviour, particularly associated with thermal effects.

Measurement Equipment and Data Management

Type of system: local data loggers operated by local PC.
Data management:

- logging to logger memory cards (VWGs)
- logging to RMX logger (accelerations)
- logger control by PC
- data download via modem using PCAnywhere
- data analysis offline.

Data Analysis Procedures

Type of analysis: statistics, algorithms, correlations, intervention analysis, anomaly detection ARX.
Software: self-made software, Octave 2.1.50, RSTAB 5.13.042, ANSYS 5.3.
Additional features: learning platform for SHM.

Examples of Outcomes

Monitoring during construction (Figure 9.98) provided training for "pattern recognition" systems for SHM, as well as providing basic data on structural performance and some calibration of thermal loading (design) codes.

Table 9.30 Sensor details for Tuas Second Link

Type of sensors	Number	Location
VWG Strain gauges	12	At corners of box
VWG stress cells	12	In line with strain gauges
Thermistors	48	With strain stress/strain gauges, through depth of webs and in tarmac
Triaxial accelerometer	1×3	Midspan

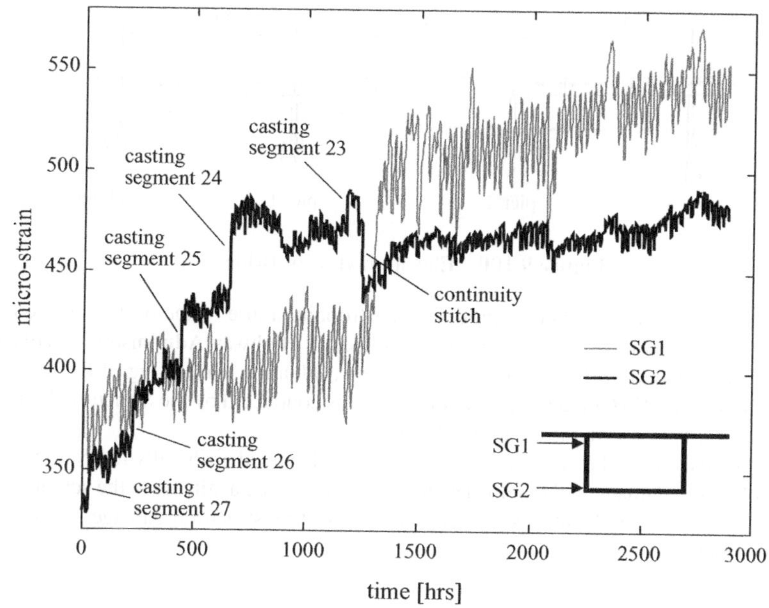

Figure 9.98 Strain variation in segment 31 (close to pier) during construction of Tuas Second Link

Benefits of Using SHM Technologies in the Project

The project has provided a great opportunity to develop strategies for recovering information from response data. In particular the experience during construction has led to insights in recognizing signatures of performance during service.

9.31 Bridge I40, New Mexico, USA

Contributed by University of Siegen, Institute of Mechanics and Automation Control-Mechatronics

Project Description

The Bridge I40 (Figure 9.99) over the Rio Grande is part of Interstate 40 in New Mexico. In the 1960s and 1970s over 2500 bridges were built in the USA with a design similar to this on Interstate 40. These bridges were built without structural redundancy and typically had only two plate girders carrying the loads.

Figure 9.99 Bridge I40, New Mexico, USA

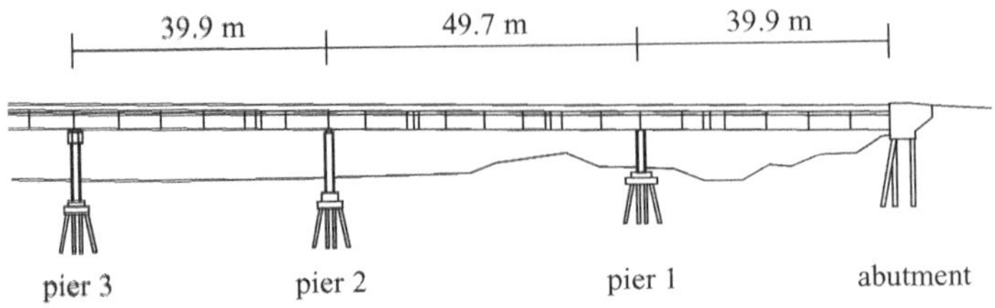

Figure 9.100 Elevation view of Bridge I40

Failure of either girder was assumed to produce catastrophic failure of the bridge; hence these bridges were referred to as fracture-critical bridges. The US Federal Highway Administration (FHWA) and the National Science Foundation (NSF) provided funds for evaluation and testing of the existing fracture-critical bridges over the Rio Grande. The investigation was conducted by the Structural Dynamics Group of Dr C.R. Farrar at the Los Alamos National Laboratories.

After a modal analysis of the undamaged bridge, it was damaged artificially to a variety states chosen to reproduce observed damage in the field. The test data were made available to the scientific community so that the bridge tests could be used as benchmarks for testing structural damage assessment methods at a full scale structure.

Brief Facts

Name and location: Bridge I40, Rio Grande, New Mexico, USA
Owner: State of New Mexico
Structure category: large span bridge
Spans: 9, 39.9 + 49.7 + 39.9 + 39.9 + 49.7 + 39.9 + 39.9 + 49.7 + 39.9 m
Structural system: steel box girder with concrete deck and concrete columns
Start of SHM: September 2001
Number of sensors installed: 26
Instrumentation design by: Los Alamos National Institute, USA

Description of Structure

The Bridge I40 consists of two separate spans for each traffic direction divided in three identical, structurally nearly independent sections in each direction. Each section is made up of three spans (Figure 9.100). The bridge comprises a concrete deck supported by two plate girders and three steel stringers. The loads from the stringers are transmitted to the plate girders by floor beams (Figure 9.101).

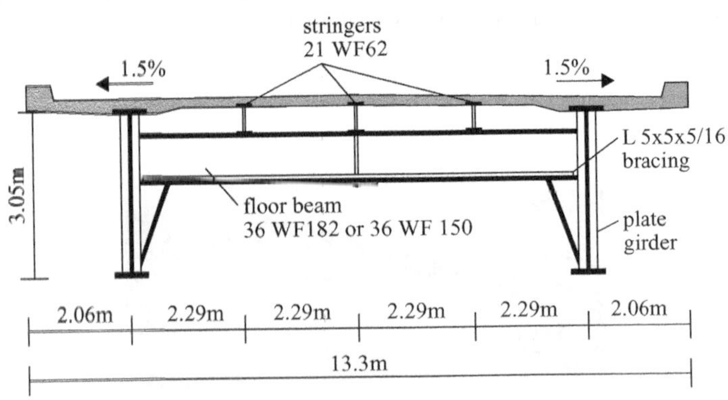

Figure 9.101 Cross section of Bridge I40

Table 9.31 Sensor details for Bridge I40

Type of sensors	Number	Location (Figure 9.102)
Acceleration	26	N1–S13
Force	12	Shaker

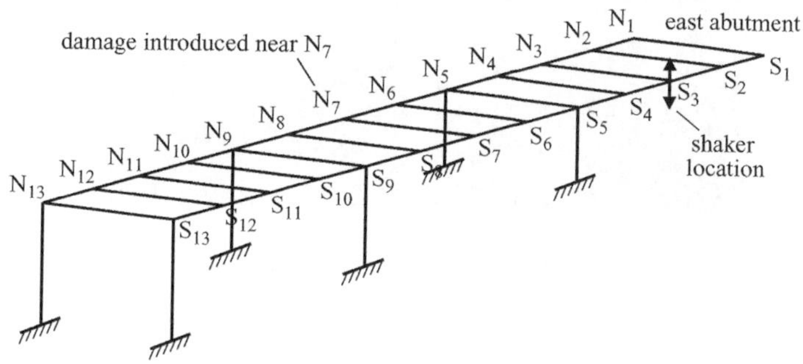

Figure 9.102 Mesh grid along the bridge section

Purpose of Inspection

The purpose of the measurement was to detect the applied damage based on measured modal data and model-based damage detection algorithms.

Data Analysis Procedures

Type of analysis: modal analysis by means of frequency response functions, excitation by shaker (Table 9.31).

Example of Outcomes

The damage could be detected, localized and quantified by means of an inverse eigensensitivity approach and frequency response function approach (alternatively) combined with parameter selection and regularization techniques.

Benefits of Using SHM Technologies in the Project

The occurrence of damage can be detected immediately. Many of the lessons learned have led to optimized model updating and damage detection algorithms.

9.32 Källösund Bridge, Goth Sweden

Contributed by Swedish National Road Administration, Section of Bridge and Tunnel Technology

Project Description

Källösund Bridge (Figure 9.103) is a part of a road link to a group of islands north of Gothenburg on the Swedish west coast. The bridge was built in 1960 as a free cantilever box beam. When an assessment of the bearing capacity was made in the late 1990s it was found that the capacity for sagging moment in the superstructure close to the abutments was too low. Special inspections showed that casting joints in those parts of the bridge were cracked but the cracks were closed. To maintain the bridge open for full traffic loads while awaiting strengthening to be designed and executed the bridge was monitored. Strengthening of the bridge was planned for 2004.

Figure 9.103 Källösund Bridge, Sweden

Brief Facts

Name and location: Källösundsbron, 50 km north of Gothenburg
Owner: Swedish National Road Administration
Structure category: post-tensioned concrete bridge built by the free cantilever method
Spans: four spans: 50 + 107 + 107 + 50 m
Structural system: concrete box girder and concrete box columns
Start of SHM: December 2000
Number of sensors installed: 72
Instrumentation design by: NGI, P.O. Box 3930, Ullevaal Stadion, N–0806 Oslo, Norway

Description of Structure

The superstructure comprises a concrete box beam that was built by the free cantilever method and post-tensioned in stages. The substructure comprises abutments in a rock slope at the ends and box section concrete columns as intermediate supports.

Purpose of Inspection

Static calculations show that the ultimate limit state of the bridge is reached under heavy loads. The bridge was not designed for a sagging moment as creep was not foreseen. When the bottom flange cracks under sagging moment there is not enough reinforcement to resist the moment but the bending moment can be redistributed to the support section of the beam. The instrumentation (Table 9.32) was used both for calibration of calculations and for monitoring of those sections.

Measurement Equipment and Data Management

Type of system: PC-based measurement system

Table 9.32 Sensor details for Källösund Bridge

Type of sensors	Number	Location
Strain gauges	72	On the bottom flange and on the webs of the main girder
Temperature gauges	2	For calibration of strain gauges

Data management:

- calibration of calculations was done with a load test
- monitoring of the strains in the crack. Alarm limit was set to give an automatic alarm when crack width reaches 0.1 mm. The alarm was placed at the 'Traffic Information Center' of SNRA in Gothenburg. This center is manned around the clock.

Benefits of Using SHM Technologies in the Project

The monitoring verified the assessment calculations. The 'full time inspection' given by monitoring with set alarm limits made it possible to keep the bridge open for heavy traffic while waiting for strengthening.

9.33 Europabrücke, Innsbruck, Austria

Contributed by Vienna Consulting Engineers (VCE)

Project Description

Bridges are aging and traffic is growing, which creates a demand for accurate fatigue life assessment. The Europabrücke – a well known Austrian steel bridge near Innsbruck, opened in 1963 – is one of the main Alpine north–south routes for urban and freight traffic (Figure 9.104). A long-term preoccupation of VCE with BRIMOS (BRIdge MOnitoring System) on the Europabrücke (since 1997) and the assessed prevailing vibration intensities with regard to fatigue problems and possible damage led to the installation of a permanent measuring system in 2003. Today's monitoring abilities enable us to measure performance precisely. High-precision sensor data of accelerations, velocities, displacements recorded by separately registered wind and temperature data and their implementation into analytical calculation provide the possibility to realize structure lifetime considerations, which are of eminent importance for bridge operators.

Description of Structure and the monitoring system

Previous measurements at the Europabrücke matched very well with the comparative analytical calculations, but they also exhibited a remarkable loading impact. Currently the bridge is stressed by more than 30 000 motor vehicles per day (approximately 20% freight traffic). The superstructure is represented by a steel box girder (width, 10 m; variable height along the bridge-length, 4.70–7.70 m) and an orthotropic deck and bottom plate. This motorway bridge with six spans differing in their length (longest span 198 m, supported by piers with an elevation of 190 m) and a total length of 657 m comprises six lanes, three for each direction distributed on a width of almost 25 m.

Brief Facts

Name and location: Europabrücke, Innsbruck, Austria
Owner: ASAG (Alpenstrassen AG), Austria
Structure category: large span bridge

Figure 9.104 Europabrücke, Innsbruck, Austria

Table 9.33 Sensor details for Europabrücke

Type of sensors	Number	Location
Displacement sensors	2	At both abutments
1D acceleration transducers	3	At the cross-section's cantilevers
3D acceleration transducers	3	Orthotropic bottom plate
Wind sensors	1	5 m from above the road surface
Temperature sensors	7	Inside and outside the box girder

Spans: 6, 81 + 108 + 198 + 108 + 81 + 81 m
Structural system: steel box girder with orthotropic deck and concrete columns
Start of SHM: May 1998
Number of sensors installed: 24
Instrumentation design by: Vienna Consulting Engineers, Austria

Purpose of Inspection

The main aim is to determine the relation between the randomly induced traffic loads (vehicles per day) and the fatigue-relevant, dynamic response of the structure. As lifetime predictions in modern standards depend on many assumptions, the emphasis is to replace estimates by measurements. In that context SMH (Table 9.33 and Figure 9.105) is focused on three goals.

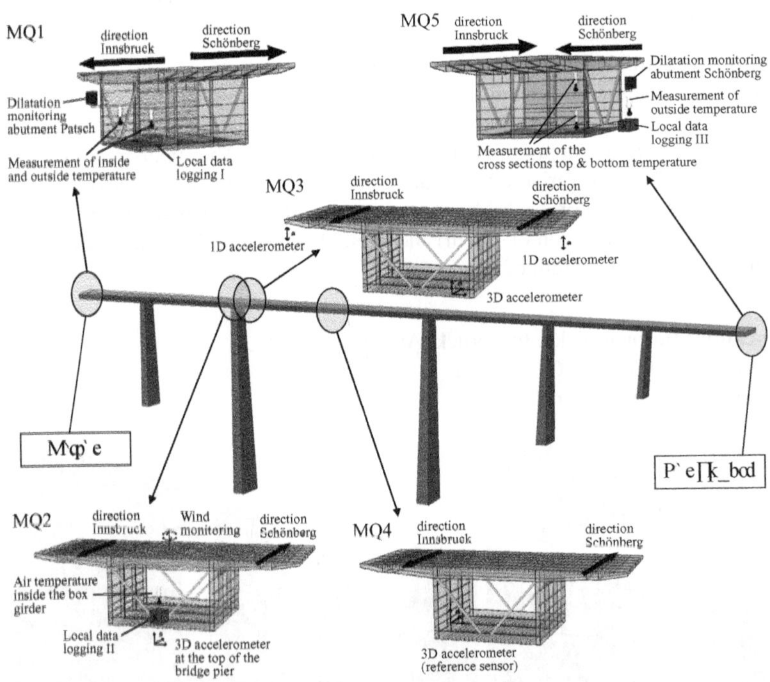

Figure 9.105 The monitoring system comprises 24 measuring channels (sampling rate 100 Hz) representing the main span's, the pier's and the cantilever's accelerations, the abutment's dilatation, wind speed and direction, and temperatures at several locations

- Global behavior in response to all relevant loading imput.
- Cross-sectional behavior with special consideration to the cantilever regions.
- Local systems analyzing the interaction between tires and the beam–slab connections.

At each of these levels of analysis the proportion used of the structure's overall capacity per year is to be determined. Additional to fatigue assessment, new compensation methods for assessment and compensation of environmental conditions (temperature, additional mass loading, etc.) using frequency analysis are developed, which provide new possibilities for structural management and lifetime prediction as well as for more accurate damage detection.

Measurement Equipment and Data Management

Type of system: PC-based and remote-access-based measuring systems
Data management:

- automatic report generating at the end of each week
- storage in a long-term database on site
- more detailed analysis (statistics, frequency analysis, etc.) and graphical presentation and documentation in office
- notification via modem about the successful operation of the measuring system.

Data Analysis Procedures

Type of analysis: ambient vibration monitoring, fatigue assessment based on rainflow analysis and statistics, assessment and compensation of environmental conditions, damage detection and lifetime calculations (Figure 9.106)
Software: Self-made software, RFEM.
Additional features: no expert system.

Example of Outcomes

At each of the previously described levels, fatigue analyses were performed by means of rainflow counting, damage accumulation, global and local FE analysis, and statistical consideration. The detailed knowledge about the progression of the prevailing traffic from the outset to the present and the implementation of published future trend studies for the next 10 years can be used for an extrapolation of the measured impact for the whole lifetime.

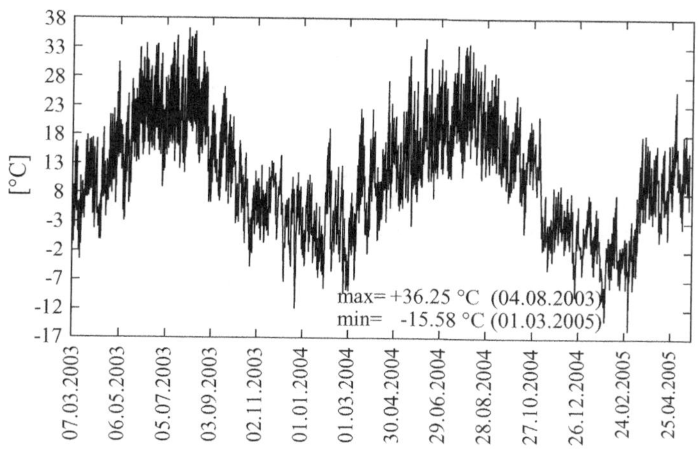

Figure 9.106 Pattern of temperature at the base of Europabrücke (assessment period 2.5 years)

Figure 9.107 St. Marx Bridge, Vienna, Austria

9.34 St. Marx Bridge, Vienna, Austria

Contributed by Vienna Consulting Engineers (VCE)

Project Description

The St. Marx Bridge (Figure 9.107) in Vienna, Austria, built from 1973 to 1978, is located between the Danube Canal and the Traffic Node Landstrasse. The bridge is counted among one of the most frequently used sections of the A23 South-East Highway. The total traffic volume averages about 240 000 motor vehicles per day, but an increase in the ratio of heavy loads has been detected. This leads to increased loading of the bridge structure. As a consequence thereof the serviceability of the expansion joints and the bridge bearings is affected. Thus, in order to detect the passing heavy loads that cause damage, a SHM system in combination with a video control system were installed in 1998. Furthermore, the remaining structural service life can be predicted.

Brief Facts

Name and location: St. Marx Bridge, Vienna, Austria
Operator: MA 29 (Bridge Maintenance by Magistrate of City Vienna)
Structure category: multiple span bridge; total length 2.70 km
Spans: 54 column sets and 24 expansion joints between single bridge girders
Spans: 2, 89.5 + 89.5 m
Structural system: pre-stressed and reinforced concrete box girder
Start of SHM: November 1998
Number of sensors installed: four accelerometer, one temperature sensor
Instrumentation design by: Vienna Consulting Engineers, Austria

Description of Structure and the monitoring system

The substructures under consideration, namely TW4 and TW5 respectively, with a total length of 205.10 m represent a seven-span continuous beam each with 29.30 m span length. The cross section, however, comprises three lanes, each with a width of 3.25 m, but the total width is 12.88 m. The construction type is pre-stressed concrete box girder with dimensions 1.96 × 4.50 m.

Purpose of Inspection

On the basis of a permanent analysis of the dynamic structural behavior possible issues to be considered are as follows.

- Determination of passing heavy loads causing structural damage
- Verification respectively update of the existing numerical load models

Table 9.34 Sensor details for St. Marx Bridge

Type of sensors	Number	Location
Accelerometers	2 sensors (4 channels) per substructure	At the box girders of spans 1 and 2 (at $0.6 \times L_{span}$ from the span beginning)
PT100	1 at substructure TW5	At the box girder of the first span (at the beginning)

- Determination of the overall load configurations and vibration coefficients, whereas wind and temperature effects are considered optionally
- Consideration of long-term trends with respect to the life loads by means of statistics
- Monitoring Table of the structural loading capacity and serviceability by means of structural identification.

Measurement Equipment and Data Management

Type of system: PC based measurement system
Data management:

- data pre-analysis (statistics, system identification)
- main analysis, graphical presentation and documentation in office
- data transfer via modem
- long-term database.
- CMS (Civil Monitoring System) processes the static data acquired by the system.

Data Analysis Procedures

Type of analysis: time domain and frequency domain SI, statistics, damage detection, FEMu, lifetime prediction
Software: self-made software, Octave 2.1.50, RSTAB 5.13.042, ANSYS 5.3
Additional features: no expert system.

Examples of Outcomes

Using the obtained accelerations time-history data (Figures 9.108, 9.109 and 9.114), system identification is carried out by time domain as well as frequency domain methods. In general the observed bridge structure is characterized by a distinct dynamic behavior (Figures 9.110 and 9.111). Therefore long-term

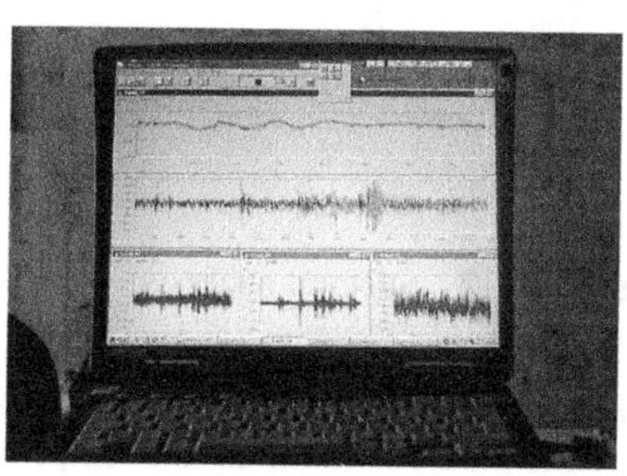

Figure 9.108 Deformation and acceleration signals during test phase

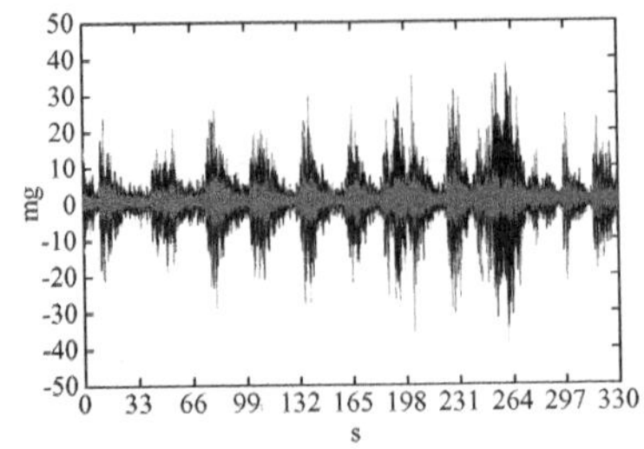

Figure 9.109 Vertical acceleration signal: all sensors during test phase

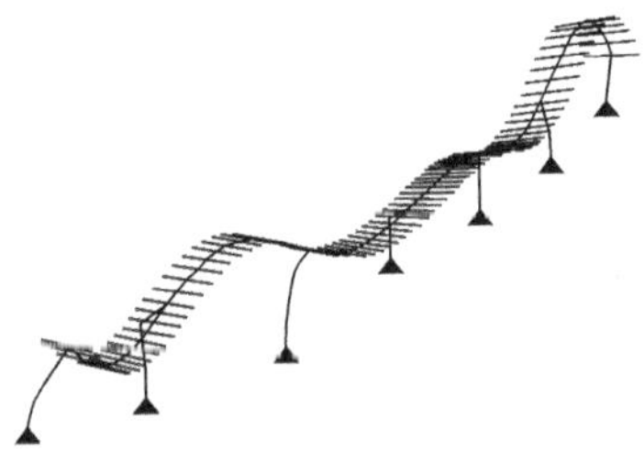

Figure 9.110 First vertical mode shape of TW4

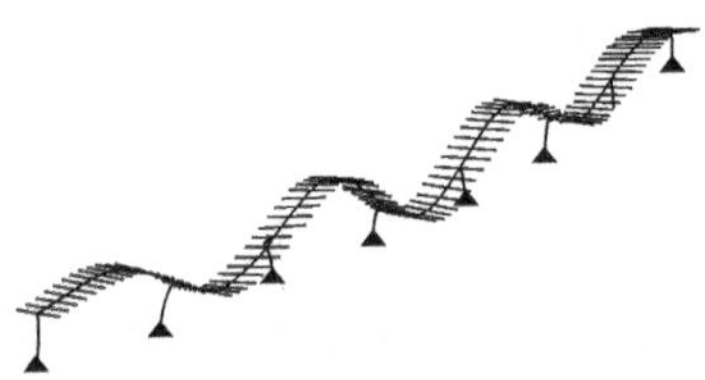

Figure 9.111 First vertical mode shape of TW5

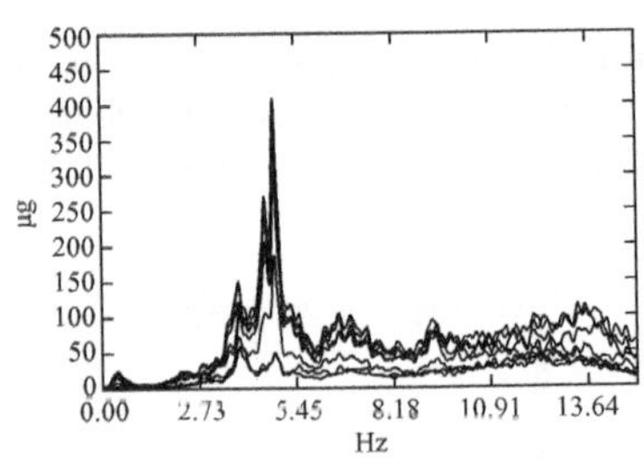

Figure 9.112 Spectral analysis: all sensors

SHM is very applicable. The implemented statistic analysis showed the relevant influence of the heavy loads. Environmental effects, e.g. wind induced vibrations and temperature influence (Figure 9.115), are recognized as well. Additionally, in order to simulate the structure numerically and to detect damage FEMu (see Section 7.1.3) is applied (Figures 9.112 and 9.113).

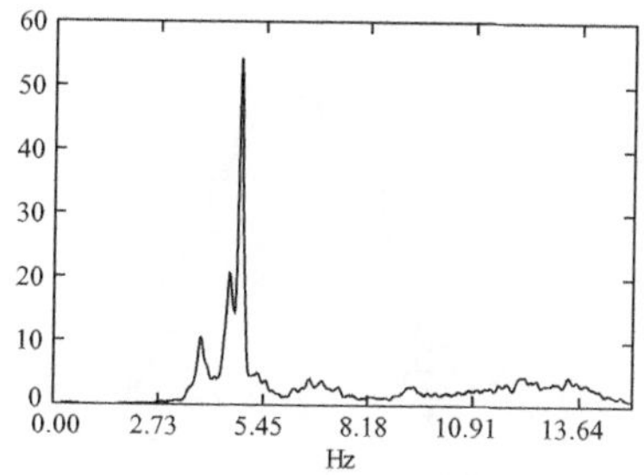

Figure 9.113 Averaged normalized power, spectra density

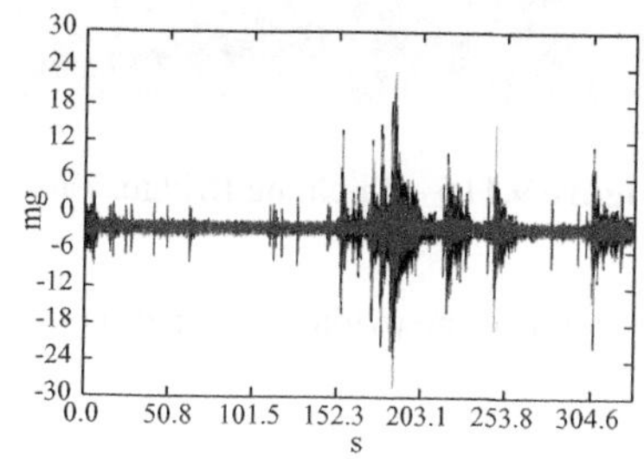

Figure 9.114 Vertical acceleration due to crossing at time instant 150 s

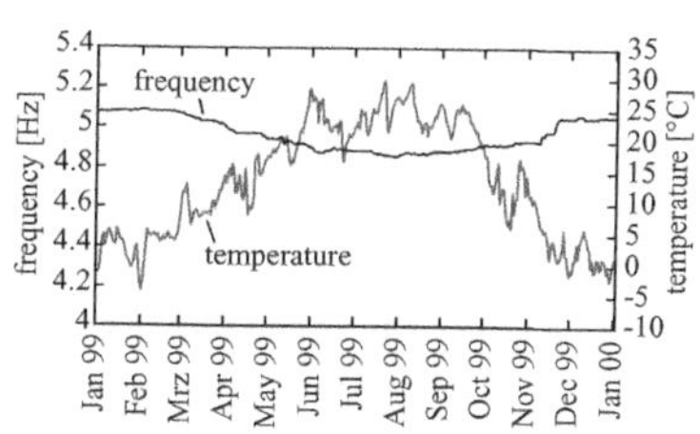

Figure 9.115 Temperature–frequency relationship over 1999

Benefits of Using SHM Technologies in the Project

The SHM allows real-time observation and the lifetime prediction of civil engineering structures. Therefore, using SHM is very cost-effective for the structural maintenance.

9.35 Taichung Bridge, Taiwan

Contributed by Vienna Consulting Engineers (VCE)

Project Description

The Taichung Bridge (Figure 9.116), opened in 2003, is a Cable-stayed bridge for urban traffic located in Taichung in the middle of Taiwan. Due to the requirement to assess the cable forces, the global state of

Figure 9.116 Taichung Bridge, Taiwan

the structure and the dynamic behavior of the pylon base a permanent monitoring system was installed in 2003.

Brief Facts

Name and location: Taichung Bridge, Taichung, Taiwan
Owner: BPI Taiwan
Structure category: cable-stayed bridge
Cables: 44
Spans: 2, 89.5 + 89.5 m
Height of pylon: 80 m
Structural system: steel girder with orthotropic deck
Start of SHM: November 2003
Number of sensors installed: 15
Instrumentation design by: Vienna Consulting Engineers, Austria

Description of structure and the monitoring system

The Taichung Bridge is a stay-cable bridge with 44 cables and a total length of 189 m which comprises four lanes and two small lanes for pedestrians and bicycles. The superstructure is represented by steel girders and an orthotropic deck.

Purpose of Inspection

The permanent monitoring system (Table 9.35) gives an overview of the global behavior of the bridge structure and supplies the actual cable forces. The monitoring system consists of following parts.

- Dynamic determination of the cable forces of eight selected cables.
- Measuring of temperature, wind speed and wind direction (Figure 9.117).
- Dynamic measurement of the main girders and the pylon top.
- Three-dimensional measurement of the pylon base.

Measurement Equipment and Data Management

Type of system: PC and stand-alone-based measuring system.

Table 9.35 Sensor details for Taichung Bridge

Type of sensors	Number	Location
Acceleration transducers	8	At 1 cable each
Velocity transducers	3	At the main girders
3D acceleration transducer	1	At pylon base
Wind sensor	1	5 m above the road surface
Temperature sensors	2	Inside and outside the box girder

Data management:

- storage in a long-term database on site
- analysis (statistics, frequency analysis) and graphical presentation and documentation in office
- control of the successful operation of the measuring system via modem.

Data Analysis Procedures

Type of analysis: ambient analysis, calculation of cable forces and lifetime calculations.
Software: self-made software.
Additional features: no expert system.

Examples of Outcomes

The permanent monitoring system at Taichung Bridge measures vibration, temperature and wind. The self-made software supplies the cable forces of eight selected cables in a way that the client can easily check the status of the cable forces in the form of a light. The green light shows immediately that all cable forces are alright (Figures 9.118 and 9.119).

Figure 9.117 Wind sensor at Taichung Bridge

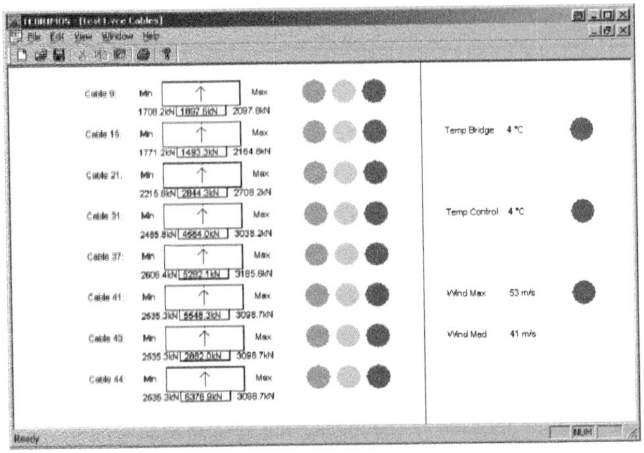

Figure 9.118 Theoretical output of the monitoring system, Taichung Bridge from left to right: green, yellow, red

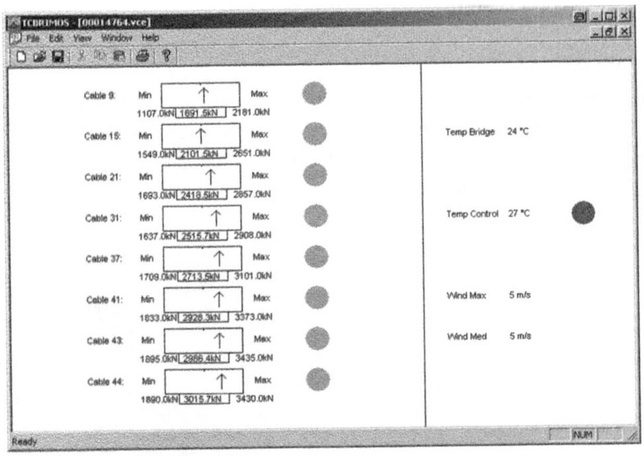

Figure 9.119 Real output of the monitoring system, Taichung Bridge, when operational temperature exceeded

Benefits of Using SHM Technologies in the Project

The ability to merge high-precision sensor data of accelerations and velocities independly of separately registered wind and temperature data provides the possibility to realize lifetime considerations, which are of greatest importance for bridge operators.

Further Reading

Aktan EA, Catbas NF, Grimmelsman K and Pervizpour M (2002) *Development of a Model Health Monitoring Guide for Major Bridges*. Technical report submitted to Federal Highway Administration, Research and Development, by Drexel Intelligent Infrastructure and Transportation Safety Institute, Philadelphia.

Benko V, Geier R and Ralbovsky M (2003) Dynamische Untersuchungen einer Segmentbrücke. *Proceedings of the D-A-CH Tagung 2003*, pp. 21–26, Zürich.

Bergamini A (2002) Nondestructive testing of stay cables – field experience in South East Asia. *Proceedings of the 3rd World Conference on Structural Control*, Vol. II, pp. 1057–1064, Como.

Bergamini A, Christen R and Motavalli M (2003) A simple approach to the localization of flaws in large diameter steel cables. *Proceedings of SPIE's 8th Annual International Symposium on Nondestructive Evaluation for Health Monitoring and Diagnostics*.

Bjerrum J, Jensen FJ and Enevoldsen I (2002) The owner's perspective in probability-based bridge management. *Proceedings of the First International Conference on Bridge Maintenance, Safety and Management IABMAS 2002*, Barcelona.

Brownjohn JMW (2003) Sensor and data management technology for structural health monitoring of civil structures. *Proceedings of the 1st International Conference on Structural Health Monitoring and Intelligent Infrastructure*, Tokyo.

Brownjohn JMW and Moyo P (2000) Monitoring of Malaysia–Singapore Second Link during construction. *Proceedings of the 2nd International Conference on Experimental Mechanics*, pp. 528–533, Singapore.

Carmen A (1989) *Historical Bridges in the Madrid Community*, Madrid. (In Spanish.)

Carracilli J (2000) Coefficients de majoration dynamique des charges routires sur les ouvrages dart, calcul et extrapolation, application au pont de Bruneseau. *Bulletin des laboratoires des ponts et chausses*. (In French.)

Catbas NF, Ciloglu KS, Grimmelsman K, Pan Q, Pervizpour M and Aktan EA (2003) Limitations in the structural identification of long-span bridges. *Proceedings of the International Workshop on Structural Health Monitoring of Bridges: Colloquium on Bridge Vibration*.

Cheng L-K (2004) Dynamic load monitoring of the Tsing-Ma Bridge using a high-spped FBG sensor system. *Proceedings of the 2nd European Workshop on Structural Health Monitoring*, Munich.

Cullington DW, MacNeil D, Paulson P and Elliot J (1999) Continuous acoustic monitoring of grouted post-tensioned concrete bridges. *Proceedings of the 8th International Structural Faults and Repair Conference*, London.

Elsener B, Klinghoffer O, Frolund T, Rislund E, Schiegg Y and Böhni H (1997) Assessment of reinforcement corrosion by means of galvanostatic pulse technique. *Proceedings of the International Conference "Repair of Concrete Structures"*, Svolvær.

Enckell-El Jemli M, Karoumi R and Larano F (2003) Monitoring of the New Årsta railway bridge using traditional and fibre optic sensors. *Proceedings of SPIEs Symposium on Smart Structures and Materials, NDE for Health Monitoring and Diagnostics*, San Diego.

Enckell-El Jemli M, Karoumi R and Wiberg J (2003) Structural health monitoring for an optimised pre-stressed concrete bridge. *Proceedings of the 1st International Conference on Structural Health Monitoring and Intelligent Infrastructure*, Tokyo.

Farrar C, Baker WE, Bell TM, et al. (1994) *Dynamic Characterization and Damage Detection in the I40-Bridge Over the Rio Grande*. Technical Report LA-12767-MS, Los Alamos National Laboratory.

Flesch R and Geier R (2004) Simulation of bridge vibrations induced by high speed train passages. *IMAC XXII, Conference and Exposition on Structural Dynamics 'Linking Test to Design'*, Dearborn.

Flesch R and Partners of the Consortium (undated) Final Report, *Work-package 2c of the National Research Project Lärm und Erschütterungsarmer Oberbau*.

Flesch R, Stebernjak B and Freytag B (1998) *Dynamic in situ testing and FE modelling of bridge Warth, Austria. International Conference ISMA 23 on Noise and Vibration Engineering*, September 16–18, Leuven.

Fritzen CP and Bohle K (1999) Identification of damage in large scale structures by means of measured FRFs – procedure and application to the I40-Highway-Bridge. *Key Engineering Materials* **167–168**, 310–319.

Fritzen CP and Bohle K (1999) Model-based health monitoring of structures. Application to the I40-Highway-Bridge. *Proceedings of the 2nd International Conference on Identification in Engineering Systems IES99*, pp. 492–505, Swansea.

Frolund T and Klinghoffer O (2004) Comparison of half-cell potentials and corrosion rate measurements- A field experience with evaluation of reinforcement corrosion. *Proceedings of the EUROCORR*, Nice.

Glisic B and Inaudi D (2003) Structural monitoring of concrete bridges during whole lifespan. *Proceediings of the 82nd Annual Meeting of the Transportation Research Board (TRB)*, CD Number 03-3012, Washington, DC.

Goltermann P (2002) *SMART STRUCTURES – Integrated Monitoring Systems for Durability Assessment of Concrete Structures*. Project Report available for downloading at http://smart.ramboll.dk/smart_eu/index.htm.

Goltermann P (2003) Managing large bridge structures in Scandinavia. *Proceedings of the SAMCO Summer School*, Cambridge.

Hariri K and Budelmann H (2004) Monitoring of the bridge Herrenbrücke in Lübeck: motivation, procedures, results and data evaluation. *Proceedings of the 2nd European Workshop on Structural Health Monitoring*, pp. 261–268, Amazeum Conference Centre, Munich.

Huth O (2003) *Topflageruntersuchungen am BE 109 in Bützberg*. EMPA-Report 202'902/1, Dübendorf.

Inaudi D, Casanova N, Vurpillot S, Glisic B, Kronenberg P and Lloret S. (2000) Deformation monitoring during bridge refurbishment under traffic. *Proceedings of the 16th Congress of IABSE* (CD), Luzern.

James G and Karoumi R (2003) *Monitoring of the New Svinesund Bridge, Report 1: Instrumentation of the Arch and Preliminary Results from the Construction Phase*. Technical Report, Royal Institute of Technology (KTH), Stockholm.

Jensen F, Knudsen A and Enevoldsen I (2000) Probalistic-based bridge management implemented at Skovdiget West Bridge. *Proceedings of the 4th International Conference on Bridge Management*, University of Surrey.

Klinghoffer O, Goltermann P and Bässler R (2002) Smart structures: embeddable sensors for use in the integrated monitoring systems of concrete structures. *Proceedings of the IABMAS 02*, Barcelona.

Ko JM, Wang JY, Ni YQ and Chak KK (2003) Observation on environmental variability of modal properties of a cable-stayed bridge from one-year monitoring data. In *Proceedings of the Structural Health Monitoring: Diagnostics and Prognostics to Structural Health Management* (ed. Chang FK), pp. 467–474. DEStech Publications, Pennsylvania.

Ko JM, Ni YQ, Zhou XT and Wang J (undated) Structural damage alarming in Ting Kau Bridge using auto-associative neural networks. In *Advances in Structural Dynamics* (ed. Ko JM and Xu YL), pp. 1021–1028. Elsevier Science, Oxford.

Link M, Rohrmann RG and Pietrzko S (1996) Experience with the automated procedure for adjusting the finite element model of a complex highway bridge to experimental model data. *Proceedings of the 14th IMAC*, Dearborn.

Luping T (2002) *Calibration of the Electrochemical Methods for the Corrosion Rate Measurement of Steel in Concrete*. NORDTEST Project No. 1531-01, SP REPORT 25, Nordisk Innovations Center, Oslo.

Mondrup AJ, Frederiksen JO and Christensen HH (1989) Load testing as an assessment method. *Proceedings of the IABSE Symposium "Durability of Structures"*.

Moyo P and Brownjohn JMW (2002) Application of Box–Jenkins models for assessing the impact of unusual events recorded by structural health monitoring systems. *International Journal of Structural Health Monitoring* **1**(2), 149–160.

Moyo P, Brownjohn JMW and Omenzetter P (2003) Bridge live load assessment and load carrying capacity estimation using health monitoring system and dynamic testing. *Proceedings of the 3rd International Conference on Current and Future Trends in Bridge Design, Construction and Maintenance*, Shanghai.

Myrvoll F and Sparrevik P (2002) *Resultater av databehandling och tolkning för hela perioden Februari 2000 – September 2002*. Technical Report, National Geological Institute, Stockholm. (In Swedish.)

Data Management of Three Cable-supported Bridges in Hong Kong Including One-year Monitoring Data. Technical Report, The Hong Kong Polytechnic University.

Peeters B and De Roeck G (1999) Reference-based stochastic subspace identification for output-only modal analysis. *Mechanical Systems and Signal Processing*. **13**(6), 855–878.

Peeters B, Couveur G, Razinkov O, Kündig C, Van Der Auweraer H and De Roeck G (2003) Continuous monitoring of the Øresund Bridge: system and data analysis. *Proceedings of IMAC XXI, A Conference and Exposition on Structural Dynamics*, Kissimmee, Florida.

Peil U and Frenz M (2003) *Lebensdauervorhersage von Ermüdungsbeanspruchten Bauwerken durch Monitoring und begleitende Versuche*. Arbeitsbericht 2000-2003 des sfb 477, beitrag tp b3, Schriftenreihe des SFB 477, S. 37-60. (In German.)

Rohrmann RG, Baessler M, Said S, Schmid W and Rücker WF (2000) Structural causes of temperature affected modal data of civil structures obtained by long time monitoring. *Proceeding of the 18th IMAC*, San Antonio.

Rohrmann RG, Said S and Schmid W (2003) Load and condition monitoring of the Putlitz Bridge in Berlin–Moabit. *Proceedings of the Symposium Topics from Civil and Bridge Engineering*, Berlin.

Rohrmann RG, Said S, Schmid W and Rücker WF (1996–1998) *Results of the Automatic Monitoring of the Westend Bridge in Berlin*. Research Reports A, B and C, Bundesamt für Materialprüfung.

Rücker W, Rohrmann RG, Said S and Schmid W (2003) Dynamic approaches used for the safety observation of bridges. *Proceedings of the Symposium of Actual Problems in the Dynamic Behaviour of Bridges*, Zürich.

Rücker WF, Said S, Rohrmann RG and Schmid W (1995) Load and condition monitoring of a highway bridge in a continuous manner. *Proceedings of the IABSE Symposium Extending the Lifespan of Structures*, San Francisco.

Stanley RK (1995) Simple explanation of the theory of the total magnetic- flux method for the measurement of ferromagnetic cross-sections. *Materials Evaluation* **53**(1), 72–75.
Swedish National Railway Administrator (Banverket) ETR (undated) *The New Årsta Bridge: a New Railway Bridge in Stockholm*. Technical Report, Swedish National Railway Administrator (Banverket), Eastern Track Region.
Teknik N (2003) *Ingjutna sensorer hÅller koll pÅ ny järnvägsbro*. Technical Report.
Teughels A and De Roeck G (2004) Structural damage identification of the highway bridge Z24 by FE model updating. *Journal of Sound and Vibration* **278**(3), 589–610.
Measurement Results of the Railway Bridge Heugasse undated. Report Submitted to the Austrian Federal Railways ÖBB.
Undated Supplier's information on the CorroEye sensor available at http://www.germann.org.
Veit R, Wenzel H and Fink J (2005) Measurement data based lifetime-estimation of the Europabrücke due to traffic loading – a three level approach. *Proceedings of the International Conference of the International Institute of Welding*, Prague.
Vurpillot S, Casanova N, Inaudi D and Kronenberg P (1997) Bridge spatial deformation monitoring with 100 fiber optic deformation sensors. *Proceedings of the SPIE 5th Annual Meeting on Smart Structures and Materials*, Vol. 3043, pp. 51 – 57, San Diego.
Wang JY, Ni YQ, Ko J and Chan THT (2002) Damage detection of long-span cable supported bridges. In *Proceedings of the Structural Health Monitoring Workshop SHM ISIS* (ed. Mufti A), pp. 299–308, Winnipeg.
Wenzel H and Pichler D (2005) *Ambient Vibration Monitoring*. Wiley, Chichester.
Wenzel H, Veit R and Tanaka H (2005) Damage detection after condition compensation in frequency analyses. *Proceedings of the 5th International Workshop on Structural Health Monitoring*, Stanford.
Wong KY, Lau CK and Flint AR (2000) Planning and implementation of the structural health monitoring system for cable supported bridges in Hong Kong. In *Proceedings of SPIE – Nondestructive Evaluation of Highways, Utilities, and Pipelines IV* (ed. Aktan EA and Gosselin S), Vol. 3995, pp. 266–275.
Work Package E (1999) *Brite-Euram Project SIMCES. A1 and A2: Long-term Monitoring and Bridge Tests*. Technical Report, Eidgenössische Materialprüfung sanstalt, Dübendorf.

10

Feedback from Monitoring to Design

Improving standards in consideration of permanent changes in the loading environment of a structure is an important ongoing task. Monitoring of can be used to determine actual loads, and the increased number of permanent monitoring stations now allows a much better measurement of the actual conditions and the identification of load phenomena, which in turn provides useful feedback for structural design.

10.1 Realistic Loads

Traffic loads have steadily increased over time. Furthermore it has been noticed that overloading is more frequent than expected. For the realistic determination of lifetime expectations it is essential to know the number of load cycles very accurately. As only the very few exceptional high load cycles are most relevant for the fatigue history it is necessary to locate them within the large number of cycles present in the data. This has been demonstrated at the Europabrücke in detail (Chapter 4). In addition to knowledge of the actual loads it also is of interest know how load transfer takes place within the structural system and to quantify the actual loads on elements of the structure, such as bearings or expansion joints. Monitoring campaigns can help considerably in providing the data required.

10.2 Environmental Conditions

As explained in detail in Chapter 4 the environmental loads sometimes exceed the limits set by the design standards. For example, temperature differences in structures might be twice what is expected. This can lead to load combinations that have not been foreseen in the structural design. As a result, structures may experience distortion, thereby adding stresses to the global stress level. Many other combinations of influences are possible. Other influences also have to be considered from displacements imposed by environmentally triggered actions such as landslides.

10.3 Conservative Design

Monitoring has clearly revealed that conservative design has considerable advantages when it comes to lifetime expectations. A comparatively small conservativism applied (say 10% overdesign) covers any of those extraordinary loads or combinations that might occur on a structure. The additional effort and cost is reasonable compared with the effect.

Health Monitoring of Bridges Helmut Wenzel

10.4 Designed-in Monitoring

Ideally monitoring systems should be considered during the design phase. The designer knows where the critical sections of a structure are and could advise the monitoring crew where to best place sensors. Furthermore a specification of information to be obtained could be provided, which would considerably improve the monitoring process. Another advantage would be that all the necessary infrastructure for monitoring, for example power supply and cables for communication, could be designed-in and optimally placed at reasonable costs. A good specification for that purpose would reduce costs and improve the results considerably. It further would help to enhance the understanding of what is required from good monitoring.

11

Guideline and Recommendations for SHM

11.1 Introduction

To guarantee the safety and reliability of civil engineering structures, permanent assessment of the structural condition is essential during the complete lifespan, in combination with a maintenance program. For assessment, both the actual loading and the structural condition must be taken into consideration. The most important premise for an assessment is the availability of relevant data.

In addition to methods of visual inspections that have been introduced in the past two decades, experimental procedures have been developed that deliver extensive essential information. These procedures, which support computational analysis by specific measurements regarding actual loading and lifetime expectation, have proved their practical suitability in many applications. Advances in sensor technology together with applications of information technology and data analysis have contributed to this development. Thus, complex instruments for providing extensive information throughout the structure lifetime are available to the structural engineer. This lifetime begins with construction, continues with operation, enhanced by specific application of maintenance action, and ends with the demolition of the structure.

11.2 Objectives and Outline of the Guideline

The purpose of the guideline presented in this chapter is to introduce existing procedures and technologies and to make recommendations for their application. These are presented systematically corresponding to the necessity for the information required for structural assessment. The focus is on the description of a systematic approach for building diagnosis, starting from available design documents up to the application of measurement technology. Thereby, a variety of proved methods for structure condition analysis and monitoring are presented. The procedures are exemplified and experience with them is described.

Another important requirement for a structural assessment is a comprehensive analysis of behavioral responses. Only with knowledge of the type, extent and duration of loading is assessment of structural strain as a basis for structural diagnosis possible. Preceded by a classification of structural responses, the potential for load monitoring to distinguish them by type, extent and duration is discussed.

The guideline also covers structural damage analysis. Knowledge of damage and its development in terms of cause, dimension and complexity allows for assessment focused on future maintenance. In addition to a description of damage, procedures for damage identification and damage assessment

Health Monitoring of Bridges Helmut Wenzel

are introduced. The potentials and the difficulties arising from the utilization of measurement data in mathematical procedures for damage analysis are also considered.

Within a coherent framework, new but previously applied procedures for obtaining information about existing structures are introduced. With knowledge of these procedures, engineers responsible for SHM can understand and make decisions about their application. The methods available and their future improvement will make an important contribution to economic and safety-oriented maintenance programs.

11.3 Analysis of Structural Responses

11.3.1 Classification of Structural Responses

11.3.1.1 Types of Structural Responses

Structures respond to loading and deformation imparted by both the structure itself and the function, it fulfills, as well as the displacements that result from their interaction with their environment. Loading and displacement cause strain in materials and hence deformation of components. Maximum structural responses are accounted for in the design of a structure, wherein the occurrence of extreme responses is analyzed by risk assessment and included in the design process. Assumptions about dimension, direction and duration of responses during the design are based on estimates and experience. Therefore, they may not necessarily correspond to the actual response of the structure.

Structural responses can be classified as mechanical, thermal and physico-chemical, and can result from external loads or develop internally, e.g. corrosion. Responses to external loads are divided with respect to their cause in terms of interaction of a structure with its surrounds, its technical environment and its utilization. Generally, structural responses are temporally and spatially variable and hence only statistically predictable and describable.

Structural responses themselves can cause static and dynamic load effects. Static effects cause no negligible mass forces. Dynamic effects originate not only from rapid load changes, but also from sudden structural changes (damage). The most important effects causing structural strain are presented below according to their cause.

Static Loads

Structure-dependent loads. Weight of the structure and its components and installations, support forces, pre-stress, abutment changes, shrinkage, creeping, construction loads, constraints.

Utilization-dependent loads. Traffic and transportation loads, construction material, silo loads, crane loads, loads from service pipes.

Natural environmental loads. Earth and rock pressure, static fluid pressure, flow pressure, groundwater pressure, pore-water pressure, snow and ice loads, wind loads, foundation and soil settlements, thermal loads, humidity, corrosion, carbonation.

Dynamic Loads

Utilization-dependent loads. Traffic loads, machine loads, brake and centrifugal forces, human excitation loads.

Loads from the natural surroundings. Wind, waves, earthquake loads, avalanches, water.

Loads from the technical surrounding. Vibrations, collision loads (vehicles, airplanes, ships), explosion loads.

11.3.1.2 Characteristics of Structural Responses

Regarding data recording, the computational treatment of the data and the modeling of loads, responses are characterized as dead loads and live loads.

Dead Loads

Dead-load responses are stationary and slow-changing in respect to their average value (e.g. self-weight, column settlement, pre-stress, earth pressure, corrosion).

Live Loads

Responses that are not constant and for which temporal and spatial changes are essential and frequent belong in this category. This includes many effects caused by the usage of structures and also by wind, temperature, snow, and other external influences.

11.3.1.3 Loads and Load Effects

Effects are stochastic quantities concerning their temporal and spatial distribution. Hence it is necessary to describe the character and magnitude of the loads by suitable statistical models. In many cases, not the load L but the load effect S is of immediate interest for structural components. The relationship between S and L is defined with the surface of influence I:

$$S_j(A, t) = \int_A I_j L \, \mathrm{d}A \tag{1}$$

whereas A is the contact area between load and surface. Often, the load L is of interest only if the local load effect matters (examples of the load L are bicycle loads on directly used components such as orthotropic plates and wind pressure on small facade components). Dynamic load components of vehicles on a bridge are an example of S.

11.3.2 Objectives and Approach to Structural Response Analysis

Knowledge about acting loads is the basis for the realistic evaluation of the structural load-bearing capacity. Further, their exact determination allows the derivation of realistic load models which can then be used for realistic statements about the fatigue strength and residual lifetime of endangered structural components. Finally, administrative arrangements for live load constraints can be derived.

The objective of the determination of external influences (load observation) is the consistent acquisition of loads acting on the structure. This requires knowledge about the behavior of the system, which can be obtained by examination of structural models or with experimental methods. The reactions or the computed loads can then be used for following tasks.

- Measurement-based permanent observation of traffic loads (e.g. traffic density and vehicle weight); these observations can also be used for verification of existing load models or for development of alternative load models.
- Statistics about the long-term trends with increasing or decreasing traffic loads.
- Determination of collective loads and dynamic factors by acquisition of acting loads according to type, location, amplitude, duration and frequency. Thereby the influences of wind and temperature can be accounted for. Such detailed load models also allow the determination of the fatigue strength and residual lifetime of vulnerable structural components.
- Improvement of load models, which within the design process could only be estimated (e.g. dynamic wind loads at cable restraints, at hangers or at wind bracings of bridges).
- Conclusions about environmental loadings such as aerodynamic excitation, temperature influences and dynamic loading.
- Derivation of specific action for reduction of load effects by change of loading or structural resistance.

Loads usually are determined only indirectly with structural and load effect models, based on information provided by the measuring parameters described in Section 11.3.3. Often it is not the absolute quantification of structural responses that is of interest, but their temporal and spatial changes to state or strain as recorded by monitoring. This applies particularly to damage analysis. Cause–effect relationships of damage based on monitoring data can often be determined only with additional knowledge of simultaneous actions.

11.3.3 Determination of Structural Responses Based on Dimension, Duration and Local Effect

11.3.3.1 Measurands for Characterization of Structural Responses

Measurement values for the determination of load effects are adjusted according to the physical conditions of the respective effects and to the underlying load model that has to be adapted for structural design or assessment. Characterizing parameters for dead/static structural responses are, for instance, the mass distributions of the weights of the structures themselves. The volume and spatial distribution of structural and nonstructural components and dead loads has to be guaranteed, possibly within the scope of the structural analysis according to Section 11.4.2. The spatial distribution of the specific weight of construction material can be determined by material tests. Changes in the specific weight are ascertained by monitoring parameters of influence (e.g. moisture penetration) or characterizing quantities (ground compaction and friction angle of earth materials). Well-known earth pressure models are applied to determine loads on horizontal surfaces and side walls (silo loads, earth pressure, rock pressure, etc.). The pressure is used as the central measuring size.

Pre-stressing loads of externally pre-stressed structures, guyed masts and towers of suspension bridges can be determined only indirectly by measurement of cable forces, possibly with the help of dynamic properties (natural frequency). The premise for determining thermal effects is knowledge of the measured temperature distributions in a structure. With the cross-sectional temperature distribution and the associated statically model, load effects can be derived. For statically undetermined structures, a mechanical model is necessary to predict the qualitative and quantitative structural behavior.

Constraint forces as a result of enforced displacements caused by settlements or abutment changes are usually the result of simultaneously constant and variable loads. For the evaluation of constraint forces, a metrological investigation of the respective displacement is necessary.

If structures are subject to air flows or water resistance, structural reaction forces result. The dimension of these loads is proportional to the kinetic energy of the flow media and the streaming surface. To identify these loads due to actions such as wind and waves, which can cause static and dynamic effects, determination of the flow velocity distribution by measurement is necessary.

Variable loads can cause static and dynamic load effects. Usually, traffic loads have static and dynamic components and their effects on the structure can be measured with an installed balance. Generally, accompanying measurement values are deformations (strains, displacements) of the structure as a scale or comparable value. In addition, information about the traffic flow, represented by the driving speed and distance travelled of vehicles, is of importance.

Vibrations, collision loads, explosion loads and disaster loads cause dynamic load effects which correspond in magnitude and dynamic properties with loads and with the structure. Measuring dimensions for such processes are vibration velocities and accelerations. These describe mass forces and/or strains. This applies to all other dynamic loads in an identical way.

11.3.3.2 Determination of Actions

Monitoring Pattern

Actions should be determined by measurement according to their dimension and frequency, their temporal and spatial distribution and character. The necessary measurement equipment for load monitoring is chosen according to the task. Monitoring is performed in a continuous, cyclic, event-dependent and load-dependent manner. Extensive information is gathered with continuous monitoring. All effects with their temporal properties are registered. If only recording of excessive load is required, inactive monitoring can be activated with trigger signals based on threshold values. For monitoring slowly variable quantities such as static loads, brief monitoring at regular intervals is often sufficient. Also, event-dependent monitoring is required where the inactive monitoring is controlled by load-independent values. To determine maximum strains, such approaches can be used for load combinations (Section 11.3.3.3).

Wind Loads

An important practical application area is the monitoring of structures that are easily excited by wind loads due to their system characteristics, such as pylons, towers, chimneys, cranes and long-span bridges. The mechanisms leading to excitation of the structure can be of different types and are listed below.

- Very slim structures with low natural frequencies can, with low damping, be excited into vibrations with large amplitudes by the gust structure of the wind.
- For circular cross sections, the cyclic flaking of turbulences (Karman's turbulences) can excite the structure to large amplitudes.
- Aerodynamically unstable cross sections, such as rectangular sections, can cause galloping vibrations. However, circular cross sections, which in general are not able to gallop, can also become aerodynamically unstable under external influences such as one-sided icing. Recent research has shown that rain water, running down bridge hangers, can cause extreme galloping vibrations, which can lead to fatigue in structural connections.
- Bridges with cross sections whose natural frequencies of torsion and transverse reaction be close to each other can be excited to flutter vibrations under certain circumstances.

The complexity of excitation mechanisms and the fact that some effects occur only under certain circumstances make it difficult to predict the loading safely in the design state. With data analysis of long-term measurements, wind load models can be calibrated for particular locations. These models can then be used as a basis for a refined estimation of the expected wind loading. The simultaneous monitoring of the weather situation is also necessary.

Wave Loads and Swell Loads

Loads from waves and swell have effects on maritime buildings such as harbors and offshore structures. Such loads usually cannot be determined by direct measurement; they can be determined only indirectly with load models. In addition to geometric data for the inflowed structural component, these models need knowledge about the kinematics of the moving water particles as arithmetic values. Input values for such calculations are the location-dependent parameters wave height and wave length. Linear and nonlinear theories can be applied depending on the application, dimension and type of waves plus water depth. Concerning load effects on structures, static and dynamic components always have to be considered.

Traffic Loads

Traffic loads on structures possess static and dynamic components. They arise, for example, with the crossing of vehicles on bridges or from loads moved on crane rails. Traffic loads have local and global effects on the strain of structural components (Section 11.3.1.3). They cannot be measured directly, but must be determined computationally with validated load models. There are a various purposes of determining traffic loads, such as the knowledge of traffic flow effects in a statistical sense, the validation of realistic load models and the determination of extreme loads.

Traffic loads acting on bridges are computed from permanently measured strain, together with calibration functions, describing the structural performance (e.g. influence line). The influence line can be determined preliminarily with proof loading or using numerical methods. The determined global load values are then classified according to the weight of the passing vehicles in each lane and the frequency is subsequently calculated for each loading class. The objective is to obtain a representative overview of the traffic occurring over a long period.

Identically dynamic traffic loads are determined. It should be noted that these dimensions have a structure-specific component, because the measured dynamic strains are the result of the interaction between the structure and each vehicle. On the other hand, measured static strains are specific for the vehicle and correspond to the weight.

It must be taken into consideration that strain, measured under traffic, represents the complete load of a structure, for example the distributed static load of all vehicles on a bridge. Hence for the automatic load recognition of single vehicles, it is important to calibrate the measurement with different traffic situations, i.e. combinations of load positions and influence lines, or to simulate these loads computationally. Then pattern recognition can be applied.

Axle loads and axle configurations of moving vehicles can be investigated with weight-in-motion (WIM) systems. It should be noted that axle load measurement results obtained using such methods in actual traffic are overlaid with dynamic components due to the vibrations of the vehicles. Such results can lead to falsification of statistics of traffic loads. With dense, slow-moving traffic up to traffic jam situations, load measurement with WIM systems is more precise, since then the determination of single vehicle weights is not necessary with strain measurements.

Loading by Displacements

Elevations and settlements of supports as a result of changes in soil reaction, construction stage, etc., cause load effects which are always proportional to the respective deformations. The determination of these loads is done computationally based on measured displacements.

Weight Loads

Mass-dependent loads from dead weight, construction material, ice and snow are determined by volume and specific weight measurements. Changes in these loads on buildings are determined by measurements of volumetric changes (e.g. height measurements with snow) and/or the changes in specific weights (e.g. by moisture absorption). Changes in load effects can also be determined by measurement of respective deformations (deflections, strains, etc.).

Impact and Collision-Loads Vibrations

During collisions, a change of kinetic energy to deformation energy occurs. In exceptional situations, the load effect is mostly dynamic and nonlinear. Hence in general, the load cannot be separated from the structural response and requires suitable models for determination. A starting point is models of elastic and plastic action effect with measured dynamic values.

Temperature Loads

Temperature loads can cause much higher strains than traffic loads, depending on structural design. If distributed nonlinearly, they cause constraint forces and residual stresses. They cause deformation and can lead to irreversible damage (e.g. cracks). Generally, they occur together with other effects (Section 11.3.3.3) as combination loads. In connection with dynamic loads, temperature loads on concrete structures lead to higher fatigue strains.

Strains caused by thermal expansion can be determined computationally only if the temperature distribution is known. Hence the prior measuring task is the determination of temperature fields and their temporal development with distributed sensors. Within the scope of inspections, temperature fields on surfaces of components can be determined with thermographic procedures (Section 11.4.2.4).

Stationary and transient temperature fields inside buildings can be determined using computational procedures if thermal material parameters and dimensions of thermal boundaries are known (outside temperature, radiation and convection conditions). The initial conditions for temporal progression must be determinable by measurement.

Effects Caused by Physico-Chemical Processes

In addition to mechanical quantities, a variety of other physical and chemical processes occur. Corrosion of reinforcement bars leads to early deterioration of concrete structures and reduces the service life. It causes a reduction in the reinforcement cross section, the cracking of concrete covers by expansion of corrosion products and the loss of bonding between steel and concrete. The main reasons for corrosion in concrete structures are chloride contamination and carbonation. The corrosion velocity within concrete structures depends substantially on the exposure and in particular on concrete humidity, electric conductivity, temperature and oxygen concentration. Spalling of concrete as a result of carbonation processes accelerates the corrosion process. Sensors are introduced for measurement of these quantities. The interpretation of the results requires great experience and expert knowledge.

11.3.3.3 Load Combinations

For the design of structures, the effect of load combination is usually assumed. In practice, the overlying load components result from constant and variable loads, exceptional loads and pre-stress. Loads in situ are usually determined by measured reactions of the structure (deformations, deflections, vibrational amplitudes) which are the result of load effects caused by load combinations. To be able to separate single load components from these results, the character of the single load component with respect to direction, propagation, duration, temporal progression and further typical properties must be known. With this knowledge, load components can be separated by data analysis procedures.

Example

For a combination of loads from traffic, temperature and settlement, the measured strains can be split into five components: a static and a dynamic component from traffic load, components from the change in the average temperature and in the temperature gradient and a component from settlement. All these components have a characteristic duration which is used for the component separation. The duration of partial load in the time domain corresponds in the frequency domain to a typical frequency. If this frequency is known, the measured total deflection can be separated into load-specific components by application of filter functions and can then be processed in accordance with Sections 2.3.2.4 and 2.3.2.5.

11.3.3.4 Use and Analysis of Measurement Data

Measurement data for slowly variable processes are recorded cyclically or continuously. They are described by statistical dimensions such as maximum values, gliding averages and variances and are stored together with the associated measuring time. Rapidly variable processes, for example effects from traffic loads, wind, waves and shocks, are continuously measured, so that all possible load processes can be recorded. Then data must be processed by procedures adapted to the information they contain. If load combinations are processed first, data separation with respect to loads must be performed (Section 11.3.3.3). These data are additionally analyzed regarding their dynamic characteristics, apart from the description by characteristic statistical values. Here the information about the frequency contents of the signals is of special interest. For these procedures, Fourier analysis and wavelet transformations are used, among others.

Traffic loads, wind loads and wave loads are separated into their static and dynamic components and are used for further statistical analysis . Subsequent to the calculation of maximum values, averages and variances, these data are used to determine frequency distributions. For the determination of wind loads, specific values of wind speed based on mean values for specified periods (e.g. 10 min) are processed. The turbulence intensities are calculated from variances of these averages and presented as power density spectra. For the description of dynamic traffic loads, maximum dynamic components of strain are applied to the simultaneously measured static traffic load. For load statistics, frequency distributions, classified in load classes, are determined. They are based on measured vehicle weights.

11.3.3.5 Load Models

Load models do not necessarily represent the physical reality of load effects, but contain a filter effect in a more or less distinctive shape, as described in Section 11.3.1.3. Load models describe loads in such a way that the effect on the structure corresponds to that of the real load. Generally, the maximum value of this equivalent action is chosen with a fixed probability and a fixed return period.

In codes for loading assumptions (see 1990 (E Undated); 1055-100 (Undated)), the elected probability level follows the return period which is associated with the average actual lifespan of the structure. If load models are updated by constant observation of effects (load monitoring), the safety level can be adjusted to typical values of the respective effects. Further, it should be noted that, considering Section 11.3.1.3, the filter effect I_j of the structure determines the actual loads and thus the load effects. Here, compliant load models have to be developed, depending on the design and construction method, and also with regard to further use of the data (e.g. assessment of the load-carrying capacity or the residual lifespan).

Calibration of Load Models

The calibration of load models has to be carried out with load monitoring results as described in Section 11.3.3.2. It refers to the statistical basis of the data (e.g. return periods) and to the calculation of loads from measured load effects (e.g. strains) with the help of load-structure models.

11.4 Diagnostics of Structures

11.4.1 Preamble

The extraction of comprehensive information about structures is the central theme of this section, which deals with structural condition analysis, monitoring of structures and numerical analysis.

Stemming from gathering and assessing information by means of structural condition analysis, methods for metrological investigations are introduced that complement and verify the available information. This concerns processes which are used predominantly locally, such as the NDT process. Information which describes the structures as a whole is obtained by application of field tests. Load-bearing capacity and global structural qualities are superficial. Nevertheless, for this information it is characteristic that it applies only to the time when it was obtained.

Time-variant values can be obtained by application of monitoring methods with repeated investigations. Monitoring methods are applied at irregular or regular time intervals or with continuous operation. They allow changes to be ascertained and to limited prognostic statements to be made.

In addition to metrological investigations, computational results are required. Nowadays, with the help of arithmetic models, comprehensive and very precise statements about the structural behavior under the influence of any effects can be given. The premise for such simulations is that the model parameters are adapted to reality as exactly as possible. The information for this process can be obtained from results of metrological investigations of structural condition and monitoring of structures.

11.4.2 Structural Condition Analysis

11.4.2.1 Description of Design and Construction of the Structure

The aim of the survey is the dimensional and constructional ascertainment of the structure and its condition in preparation for advanced structural investigations. These results serve as a basis for, among others, the assessment of the load-bearing capacity under the present loading, utility changes with increased load level, maintenance repair and structural upgrading and the assessment of existing damage under static and dynamic conditions. The purpose of the dimensional ascertainment is the actual and exact record of all geometric values, necessary for the description of the structure and its structural environment. It starts with a review of existing design documents and will be supplemented by an as realistic as possible geometric measurement of the existing stock of structures. For those parts of the structure that can be accessed only with difficulty, tools such as a laser or tacheometer should be used. The realistic dimensions of the structural members, including their cross sections, should be determined, also with consideration of the loss of material by corrosion. For thickness measurements, radiographic methods such as X-ray and ultrasound are used in addition to conventional methods. The constructional ascertainment aids the static and dynamic analysis of the structure with consideration of all members and the structural environment taking part in the load-bearing behavior. The description of a system and its condition involves the following aspects:

- identification of the load-bearing and stabilizing structural members;
- identification of coupling and connecting elements and their mechanical properties;
- specification of construction details (type of pre-stress, coupling joints, anchors, etc.);
- determination of material properties (strength, mass allocation, damping properties, humidity, chemical values, etc.);
- specification of the structure and configuration of the structural environment (ground, restraining systems, fill masses, etc.);
- evaluation of the functioning of bearings and joints;
- specification of existing loads and their spatio-temporal effect;
- recording of possible problematic spots and deficiencies;
- specification of existing damage and damage-causing circumstances.

For information acquisition, the following methods, in order, are available: visual inspection, nondestructive testing (NDT), testing with minimum destruction and destructive testing (DT).

11.4.2.2 Determination of Threshold Values for Position Stability, Serviceability and Load-Bearing Capacity

Threshold values for parameters describing the condition of a structure can be gathered from applied codes and guidelines or they can be determined taking local circumstances into consideration. Further, the accuracy of the assumptions needs to be ensured and, if necessary, be reviewed after specified periods.

Position stability

In general, position stability has to be ensured within the ultimate limits. Further, global changes of positions through settlement and tilting of foundations of structures have to be limited within the serviceability limit state. For specific types of structures such as railroad bridges, threshold values are defined by codes. Generally, threshold values for position stability within the ultimate limiting state can be determined on structural models. Threshold values for position stability within the serviceability limits have to be determined considering all boundary conditions, influencing the unconfined use of the structure.

Serviceability

The limitation of deformation values and vibration under service loading is often regulated in codes and guidelines. This is especially true for structures where deformations caused by loads restrict the serviceability. Otherwise, threshold values for deformation and dynamic behavior can be defined with discretion. To ensure the durability and serviceability of the structure, deformations, stress values and damage such as cracks need to be limited. The threshold values are often specified in design codes, but can also be self-determined.

Load-Bearing Capacity

Within the ultimate limits, threshold values of measurable condition values are only rarely specified in codes and guidelines. In fact, limits for condition values are defined, but within the ultimate limits they cannot be used as thresholds for monitoring of the serviceability state. Condition values, which indirectly indicate the limits of the state, can be developed using structural models. Fatigue stress is an exception, since here the ultimate limit state can be reached by service loading. In these cases, the amplitudes of the cyclic stress are limited depending of the number of cycles.

11.4.2.3 Structural Identification

Structural identification provides the most reliable system for characterizing a structure for analysis and decision-making. Therefore, the structural identification principle gives guidance to civil engineers for the determination of an optimized measurement system. Thereby, a structure can be characterized accurately and completely in order to establish reliably its health at serviceability and ultimate limit states. To conduct a structural identification application systematically, the following steps are specified.

Collecting Information and A Priori Modelling

If necessary for the determination of the static system, all design documents have to be consulted to check the geometric values and characteristic values of the building material. Missing details have to be completed by measurement and by nondestructive testing on the structure and on samples. A model which represents the initial knowledge about the structure is often incomplete and coarse and therefore has to be refined.

Evaluation of the Actual Condition

To assess the actual condition of the structure, the special structural features, existing documentation, known damage, results of visual inspection and of nondestructive and/or destructive tests have to be considered.

Assessment of the Existing Bearing Capacity

For the present use or an intended change of use of a structure, the bearing capacity has to be proved based on all existing information and the assumption of a safety concept for loads and structural resistances.

Preparations for Experimental Analysis

In preparation of performing full-scale tests, a sensitivity analysis by the a priori model is required in order to determine optimal excitations and responses for dynamic tests and to select acceptable ranges of measurements. Based on analytical and preliminary experimental studies, the configuration of loads for static tests and the kind, number and locations of sensors should be optimized.

Full-Scale Tests

Static load tests should be performed on account of insufficient knowledge about the structural model, the interaction of components, the effect of known damage and the effectiveness of remedial actions. Dynamic tests can be performed to verify the global system behavior and the critical mechanisms that affect the global modes of vibration. For fatigue tests, the dynamic behavior of the structure needs to be known.

Processing of Experimental Data

The processing and conditioning of measurement data from full-scale tests are an important step to achieving a higher confidence level of information about the structure.

Model Calibration
Mechanical properties, boundary conditions and continuity conditions of the model are adjusted. Model configurations agree with the physical insight observed during the experiment and obtained from the processed experimental data.

Utilization of Calibrated Models
The field-calibrated analytical model serves as the best measure of the actual condition of the structure. This may be used for load-bearing capacity rating, load assessment and evaluating internal forces, stresses and deformations under operational conditions.

11.4.2.4 Application of NDT Techniques

In structural analysis and damage analysis, nondestructive testing methods can be successfully applied. With combined application of the described processes, using methods of data fusion, various verification problems can be solved economically. By application of surface scanners, nondestructive measurement can be increasingly automated. In addition, trend towards image-producing analysis, using tomographic methods, is clearly recognizable.

Steel Structures
For locating flat separations (e.g. cracks) and for thickness determination of structural steel sections, high-frequency ultrasonic technology (usual frequency: 2 and 4 MHz) is applied successfully. The same applies to radiography, which is employed for locating volume defects (e.g. pores, inclusions, blowholes) and for locating cracks in junction plates of old trusses. To trace the temporal and spatial crack development in steel structures, sound emission analysis is used. Not only material conditions but also material-specific values can be determined nondestructively. Thus, for instance, spark-induced emission spectral analysis permits the determination of steel composition without sampling.

Reinforced and Pre-Stressed Structures
For reinforced and pre-stressed structures, numerous NDT methods are often applied in combination. Especially for locating near-surface reinforcement, static magnetic field and alternating field processes are employed up to a maximum component depth of approximately 12 cm. Locating reinforcement or tendons at greater component depth will mainly be done with radar methods. For the localization of tendon ruptures, the remanence process must be applied additionally after a successful nondestructive discovery. This is based on magnetization of the tendons and the measurement of the magnetic field. The indicator of a tendon rupture is a change in polarity. Checking of the diagnosis is possible, for instance, with a minimally invasive intervention. With a drill, an artificial opening can be created for a specific endoscopic investigation. Alternatively, the tendon area of interest can be investigated radiographically. For extensive corrosion condition investigations of the reinforcement, the electrochemical potential field method can be applied.

Various acoustic methods are also applied to solid structures. Because of the dispersing effect of the rock granulation, low-frequency ultrasound (usually 50 – 100 kHz) is used. By this means, the coating thickness and the defects (gravel nests, hollow cavities and others), also with unilaterally accessible components, can be detected. Furthermore, ultrasonic technology is employed for checking the concrete compression resistance. Additionally, the impact echo process is used for various applications. This process is employed for checking pile integrity and for coating thickness determination of tunnel shells.

Masonry Structures
An important NDT method for wide-ranging damage in masonry is radar. The objectives are the determination of the composition of the masonry (thickness, alignment, hollow cavities, metal ties and others) and the determination of moisture conditions. For a wide range of moisture investigations, IR thermography is often employed. For selective structural investigations, endoscopy, ultrasound and microseismics are

used and the compressive strength and homogeneity of natural stones and clay bricks can be evaluated. For selective moisture investigations, various electrical methods (resistor methods, capacitive methods, microwave methods, etc.), radiometric methods (nuclear magnetic resonance, neutron back-scattering and radiographic methods) are employed.

11.4.2.5 Field Tests

Field testing as part of structural identification is used both as an inspection approach as and in part in monitoring for cyclic or intermittent observations. The purpose of static field testing is predominantly to check the load-bearing capacity of a structure. During dynamic examinations, the determination of the dynamic properties of structures and the interaction between the dynamic loads and the behavior of structures is in the focus of attention.

Static Tests

Static loads are considered to be those loads that are brought on to or placed on the structure very slowly, so as not to induce dynamic effects in the structure. Static field tests can be subdivided into behavior tests, diagnostic tests and proof tests. Behavior tests are carried out either to study the mechanics of structural behavior or to verify certain methods of analysis. The objective in the latter case is to verify that analytical methods can be used with confidence for the design and evaluation of structures. A behavior test provides information regarding how the load is distributed among various components of a structure. Results from these tests can be used to calibrate analytical methods.

A diagnostic test is a test that is carried out to diagnose the effects of component interaction. For example, the diagnostic test may be conducted to establish the rotational restraint conditions at the end of a bridge column. Through a large number of tests, it has been confirmed that diagnostic testing can be used with advantage to locate the sources of problems that might exist in a structure due to inadvertent component interaction and to determine the positive effects of interaction. Diagnostic testing has the benefit of explaining why the structure is performing differently than assumed.

A proof test is carried out to establish the safe load-carrying capacity of a structure. During this test, the structure is subjected to exceptionally high static loads that cause larger responses in the structure than the responses that are induced by statically applied maximum service loads. Because of the very high loads applied to the structure in proof testing, there is always the possibility that the structure may be permanently damaged by the test. A well-planned proof test is carried out with gradually increasing loads, ensuring that the loads are not allowed to go beyond the limit of linear elastic behavior.

Dynamic Tests

Dynamic testing of structures can be subdivided into the following distinct categories: stress history tests, dynamic load tests and modal tests. Concerning their characteristics and their spatial distribution, dynamic loads are often complex and computationally cannot be described adequately. Then stress history tests are performed in order to determine the stresses experimentally in dynamically highly stressed types of structures (e.g. joint connections), which are substantial regarding fatigue loading. After preliminary numerical investigations, for the determination of "hot spots" in structures a larg number of sensors are attached and the stresses are measured under operating conditions. From the results of these investigations, optimal sensor configurations for continuous fatigue monitoring can be determined. Stress history tests are performed whenever the dynamic actions in combination with the examined structure are too complex for sufficiently exact results to be obtained by numerical simulations.

Dynamic load tests serve to determine the dynamic increment from traffic loads. Realistic information is needed in order to control design acceptance after completion. Likewise, with same traffic volume, structural changes can lead to changed dynamic stresses in parts of the structure. Also, a change of use due to planned passages of vehicles with changed dynamic characteristics requires the measurement of the new dynamic loads in advance. If during design the dynamic load effects are considered as an increase in the static stresses, dynamic load tests need to be performed by the measurement of strains in those structural parts which are of importance for the design.

Modal tests are used for the determination of the modal properties of structures. The knowledge of the modal characteristics is used for damage identification, for quality control of structures after completion, for planning and assessment of repair work for the assessment of structural safety after extreme loading and for the calibration of structural models.

The procedures for the determination of the natural frequencies, the mode shapes and the modal damping are differentiated regarding the excitation of the structures as ambient vibration tests and forced vibration tests.

In the first case, the tests are accomplished under operating conditions. The excitation energy comes from the dynamic operating load of the structures (wind, weather, traffic, ground vibration). Therefore, ambient vibration tests can also be performed with large structures, also under loads, which lead to changes in the dynamic characteristics. It is assumed that this kind of the excitation has a stochastic character with a broadband spectrum. If this is not the case, a complete identification cannot be accomplished, since only those frequencies which are present within the exciter spectrum become excited. Ambient vibration tests can be performed comparatively fast and inexpensively. Since a system's responses due to natural excitation are often small, highly sensitive sensors must be used.

The usual kinds of excitation with forced vibration tests are impulse (impulse hammer, drop weight, etc.) and Heaviside function and also regulated excitations (harmonic, periodic and stochastic) using electrodynamic and electrohydraulic exciter systems. The selection of the type of exciter depends on the dynamic characteristics of the structures and on the existing site conditions. Using the Heaviside function, the input energy is concentrated within the low-frequency range. Impulse excitations are unsuitable for large buildings. During regulated excitation, arbitrarily long measurement times are possible, with which higher frequency resolution can be achieved. It is a disadvantage that the equipment for and operation of such exciter systems are substantially more expensive and require the exclusion of the normal operating conditions (traffic). The advantage is the almost complete identification of the modal characteristics of the structures.

11.4.3 Monitoring of Structures

11.4.3.1 Objectives

The aim of automatic and permanent monitoring in the context of this guideline is the improved knowledge of the current state and long-term behavior of structures or structural component and also of the causative influences and loads. By this means, the results of previous procedures for structural monitoring are supposed to be improved and completed. Permanent monitoring is generally indicated by continuously recorded measurands, i.e. without time interruption, with permanently applied sensors. Additionally, results are compared with previously established reference data for the loading and the structural properties.

The procedures described here can be used for structure-related damage analyses. For instance, in cases of overload, of exceeding the operational lifetime or of pre-damage, which preclude normal inspection intervals, permanent monitoring procedures can allow extended use of the structure. Then, it must be guaranteed that for such safety-relevant monitoring the measuring and data processing work reliable using the available technology. Also, it is necessary to evaluate the results of the permanent monitoring constantly or at sufficient short intervals.

11.4.3.2 Specification of Monitoring Task

Monitoring of Load Effects

Permanent observation and assessment of the current load effect become highly important with strongly fluctuating and external loadings that cannot be determined sufficiently exactly (e.g. wind, traffic) and with complex structural behavior that cannot be modeled or modeled only with great effort (e.g. spatial effects).

Condition Monitoring

During condition monitoring of structures, global and local structural properties are evaluated based on continuously measured values. The objectives are to evaluate the current condition and to predict the future development of the structural condition with adequate accuracy. Another objective is to identify and record gross changes in the structural behavior.

Local structural properties are monitored if there is local pre-damage or a structural component that is exposed to special loading conditions. If pre-set threshold values, for example from codes, from experience or from an arithmetic analysis, are exceeded, further specialized investigations are usually necessary.

Generally, structural changes in single components such as the fracture of a single pre-stress tendon cause only local effects. Therefore, the success of monitoring depends on the local basis of measurement points.

Definition of Performance Parameters and Threshold Values for Monitoring

The identification of extreme load and resistance parameters is of great importance after traffic loads (e.g. heavy traffic, accidental impact) or environmental actions (e.g. storm, floods, earthquakes), since those single events can result in considerable damage to the structure. To record such events, threshold values need to be defined based on experience, measurements or calculations. Recording of events is carried out by storing time, maximum and other defined parameters and, if reasonable, the entire response of all sensors applied. The storage of the complete time response can be advantageous in these cases since possible damage can often be identified by the kind of response.

The determination of the dynamic loads can, in addition to the determination of the static loads, be of great importance for realistic evaluation of the load-bearing capacity and the remaining lifetime of a structure and also for the definition of maintenance intervals. For that reason, the dynamic factor needs to be determined. It is calculated depending on predefined load classes from the ratio of the maximum value of a strain signal to the static part of the signal (determined using a low-pass filter).

For risk assessment of material fatigue, the local stress is a significant parameter in addition to material parameters. It is allocated in stress collectives, which are established by a permanent analysis of the stress (strain) signal with a rainflow algorithm. This corresponds to classical storage of hystereses which comprise stress signals. The amplitude and shape of a stress collective indicate the fatigue potential of the stress. The mean stress amplitude should be included in the assessment for nonwelded structures compared with welded structures.

For examining current load effects, the following tasks can be of practical importance:

- monitoring of allowable static and dynamic load effects;
- determination of dynamic load effects (dynamic factors);
- classification of load effects (e.g. permanent stress analysis);
- determination of extreme stresses and the frequency of its appearance.

Based on permanent observation of load effects, following tasks can be processed:

- evaluation of maintaining lifetime;
- evaluation of the actual structural safety, e.g. according to EC 1.1 based on the reliability index;
- determination of maintenance intervals depending on the loading and the actual condition of the structure.

For structural condition monitoring in terms of this guideline, the actual structural state is evaluated by inter alia:

- measurement of strain, deflection, curvature, inclination on selected sites; this may result in conclusions about e.g. foundation settlement, global changes in stiffness, loss of continuous beam effect, etc;
- observation of selected resonance frequencies (conclusions about changes in global stiffness);
- selective monitoring of changes in dominating vibration modes.

Examples of monitoring of local structural parameters are:

- evaluation of the length and width of known single cracks;
- observation of structural parts with increased danger of cracking;
- strains on points with increased stress concentration;
- static deflection- and vibration-caused displacement of structural components (e.g. restraints of cranes and pylons);
- foundation settlements (e.g. on bridges);
- strains of pre-stress tendons.

11.4.3.3 Experimental Design of the Monitoring Task

Preliminary Procedures

The monitoring of technical systems is based on knowledge about the structural characteristics of the system under observation. Only if the system behavior is sufficiently well known can sensors be applied at the right positions and an evaluation of the measurement data concerning relevant parameters for system identification be successful. For localization of the most stressed structural region, computational and metrological methods can be suitable.

Knowledge about the structural system, obtained by methods of experimental and theoretical modal analysis, or analysis of the structural dynamics, is the best premise for configuration of the measurement equipment of a monitoring system. It is intended to demonstrate whether the application of this method is possible and economic for the structure to be monitored.

Data Acquisition and Signal Analysis

All values are measured continuously as analog signals. The necessary sampling rate for the analog–digital conversion needs to be carried out at the maximum frequency range of interest. If the data analysis takes place in the frequency domain, the sampling rate needs to be at least twice the maximum frequency. Generally, this upper limit is ensured before digitizing using a low-pass filter. For data analysis in the time domain, a higher sampling rate should be used. In practice, a sampling rate around five times the highest frequency of interest has proved to be reasonable.

For all recorded measurement signals, characteristic values or parameter functions have to be provided continuously. For interpretation, it is important that the collected data be cleansed up or intelligently processed. The following methods can be used to perform that task.

- *High- and low-pass filter*: Filtering of quasi-static and high-frequency signal parts from e.g. temperature variations, cable movements or measurement noise.
- *Band-pass filter*: Used to filter out other than determined frequency contents of the signal to which the structure is sensitive.
- *Integration*: Integrators are used to convert acceleration signals into velocity and displacement. Integrators can be realized by analog networks in digital setups.
- *Parameter generation*: Calculation of mean, standard deviation and peak value from the (probably pre-processed) signal.
- *Frequency analysis*: Determination of the spectral content of the time response signal. The frequency resolution f_0 can be determined from the length of the time range T by $f_0 = 1/T$.

- *Statistical analysis*: Determination of the probability density function of a variable or determination of the maximum of the signal. Results are presented here as histograms and density functions.

Continuous data acquisition needs direct processing on-site with the aim of reducing the amount of data, although with that approach the flexibility of subsequent data analysis is reduced.

In cases of slow-moving changes of the observed performance parameters, it can be reasonable to record only their mean values. It is also possible that a combination of data acquisition algorithms may be required so that only peak values are recorded as a general operation mode, and continuous data are recorded for discrete periods, if a threshold is exceeded. Selection of the most appropriate data acquisition algorithm is an important component of SHM and will affect both the amount of stored data and the type of diagnostic information that can be obtained.

Processing of data is also important when multiple sensory systems are used in the same SHM project. Many of these sensors may have separate signal conditioning and demodulation systems for acquiring the raw data from the respective sensors. It is important that the system is able to process the data from all inputs and relate them to a common reference such as a time stamp.

Measurement and Service Conditions

During monitoring, the measurements are taken under service conditions, where the structure is stressed by traffic loads, vibration emission, wind, temperature and microseismic influences. Although the acquisition of loads for load and load effect monitoring corresponds exactly to the monitoring requirements for condition monitoring limitations on the required measured values arise. This needs to be allowed for in the selection of the sensors and their application to the structure. Further, this can have consequences for the methods of data analysis.

The temperature range to which parts of the installed measuring and data analysis facilities might be exposed extends approximately from −40 to 50°C. It has to be estimated when temperature compensation needs to be provided within a measuring chain. Other important climatic influences are humidity and moisture. The electrical protection class of the measuring equipment (sensors, plugs, cables, control devices and computer hardware) needs to be customized to the service conditions. Generally, an electrical shield is necessary for sensors and data lines need to be protected against electromagnetic fields and currents, especially adequate protection against lightning. Cables for the transfer of analog signals should be as short as possible. For the application of sensors, cables and electronic hardware, it has to be considered that in general civil structures are not protected against vandalism. Finally, to operate the measuring system durably and economically, all components should be easily replaceable.

Sensors and Sensor Characteristics

Essential for the monitoring of structures are sensors which are robust and operate stably and reliably. It has to be assured that the characteristic qualities are not modified by environmental influences, such as temperature, humidity, mechanical influences and electrical and magnetic fields. If possible the sensors have to be protected against these influences or the effect on the measurands has to be compensated.

Sensors can be subdivided into those which concentrate on the monitoring of local properties such as the material and in those which observe structures from a global point of view. Some are embedded within the structure, others are just placed on the surface of the structure. According to the measurand's geometry and dimension, deformation, strain, force, weight, dynamic parameters, temperature and durability parameters the following instruments are the most important currently used within structural health monitoring.

Strain Gauges

It needs to be considered that the right strain gauges in terms of type and length are applied depending on the structural material involved (concrete, steel, etc.). Advice on the correct application, protection against environmental influences and the right choice of cables can be found in fib (2003). Depending on the measuring amplifier, a frequency range from 0 Hz to some kHz can be covered. Constrained to the measuring chain, a strain with a resolution up to 0.1 μm/m can be measured.

Fiber Bragg Grating

These sensors are suited for strain measurements up to 10 000 μm/m and for a temperature range from −50 to 200°C within a frequency range from DC to the MHz range. The sensors can be applied either in the structural material or on its surface. The length of the sensors can be adjusted to the measuring task. They have very good linearity and small hysteresis and they are insensitive to electromagnetic perturbation. Because of their thermal sensitivity, temperature compensation needs to be applied during strain measurements.

Piezofilm Sensors for Strain Measuring

Piezofilm sensors have, unlike strain gauges, high pass characteristics, i.e. they measure dynamic strain beyond a limiting frequency. This threshold frequency can be established by adaptation of the charge amplifier directly at the sensor. With copper-coated synthetic film, a threshold frequency of 0.2 Hz can be provided. The size of the sensor is variable, generally with dimensions of approximately 12 × 90 mm including the charge amplifier and the voltage amplifier. Piezofilm sensors including integrated amplifiers are applied with epoxy resin.

Displacement sensors for deflections

The main types of sensors which measure the relative displacement between two points based on the inductive principle are the different types of LVDT with measuring ranges between ±1 and ±50 mm and a quasi-infinite resolution within a temperature range between −20 and +120°C. For other comparable sensors basing on similar physical principles but with different technical parameters, see fib (2003).

Displacement Sensors

Based on GPS the determination of deflection in general requires a stable accessible reference location for each measurement. In cases where this is not practical, satellite-based sensors are available which measure the movement of structures. Sensor nodes mounted on the structure at sites of interest are able to observe the settlement of foundation and also long-term movements of bridges and high-rise buildings. Each sensor node consists of a GPS receiver, microcontroller and data radio. Precisions of less than 10 mm are achievable with the evaluation of the phase information of the satellite signals and use of differential GPS.

Hydrostatic Leveling Systems (HLS)

This sensor system, applicable to displacement measurements, is based on the classical physical law of connected vessels. It consists of two ore more interconnected fluid cells mounted on a structure at selected locations in which one cell is designated the datum reference. Hydrostatic leveling systems can be used only for static or quasi-static events. Within a measuring range, a resolution of 0.02 mm can be reached.

Displacement Sensors for Relative Vibration Measuring

Displacement sensors are usually applied to measure crack widths. According to the measuring principle, one can distinguish between conductive, inductive and capacitive sensors. Depending on those principles displacements, from 0.1 to 10 000 μm in a frequency range from 0 Hz to some kHz can be measured.

Vibrating Wire Strain Gauges

This kind of sensor utilizes the physical law that the square of the natural frequency of a wire is proportional to the strain. These sensors, encased in sealed steel tubes, can be used for measuring strain, strength, pressure and temperature, either fixed on the surface of a structure or produced for embedment in concrete for static and dynamic measurements. The long-term-stability is very good. The resolution is about 0.025% of the measuring range.

Vibration Velocity Sensors

For measurement of absolute motion values, these sensors are applied directly on the vibrating object. They do not need a reference point. They can be used in a limited frequency range of approximately 2–1000 Hz, although in the lower frequency range deviances in amplitude and phase could occur. Depending on the integrated mechanic–electric conversion element, the sensors can be classified into absolute displacement sensors (inductive, capacitive or strain gauge converter) and vibration velocity sensors (electrodynamic converter).

Vibration Acceleration Sensors

These are applied directly on the vibrating object. Depending on the integrated mechanic–electric conversion element, they can be used from 0 Hz (strain gauge or inductive converter and also servo-acceleration sensors) or above a low threshold frequency (piezoelectric sensors). The upper threshold frequency in both cases is at some kHz.

Laser Detector for Vibration Measurements

With these sensors, non-contact measurements over relatively wide distances can be carried out. Usually, so-called position-sensitive detectors (PSD sensors) are applied. The source is mostly a semiconducting laser with low intensity. The measurable frequency range reaches from 0 Hz to 300 kHz, depending on the measuring system. The resolution is up to 10 μm, depending on the size of the PSD, the light intensity and the measuring system.

Inclinometers for Angular Displacement Measurements

For frequencies up to 5 Hz, sensors using the capacitive principle can be applied. For higher frequency applications, servo-acceleration sensors are more suitable. The resolution of these sensors usually lies within the range 0.1–0.001°.

Fiber Optic Sensors

Depending on the applied measuring equipment, these sensors are suitable for crack width measurement, crack detection and localization. Crack width measurement and crack detection use the physical law that with weakening of the fiber cross section the incoupled light is more absorbed. The weakened light measured at the end of an optical fiber is then a measure of the existence of a crack or of the changed width of the crack. It should be noted that on reaching a certain crack width, the sensor can be destroyed. To localize cracks, one needs to measure the reflection runtime within the optical fiber. This, however, results in a large procedural investment with respect to equipment and data analysis technique.

Temperature, Humidity and Corrosion Sensors

For permanent acquisition of these measurands, commercial sensors are suitable. If these sensors are to be used in long-term measurements, they need to be applied and encapsulated with the greatest care.

The basic criteria for the selection of sensors are minimal changes of the measurand (resolution, linearity, accuracy), measuring range, type of measurement (static, dynamic, etc.), test duration (long-term stability), test environment, installation environment and financial resources.

Measurement Equipment

In general, long-term monitoring measurement equipment contains of the following components:

- signal amplifier (voltage amplifier, charge amplifier, carrier frequency-measuring amplifier, bridge amplifier);
- analog anti-aliasing filter (tuned to the necessary cut-off-frequency);
- measurement data acquisition system with analog–digital conversion (16–24 bit conversion depth);
- data analysis computer for managing, processing, data reduction and storage;
- data storage (semiconductor, flashcard, disk, floppy, streamer tape, etc.);

- uninterruptible power supply;
- unit for remote data transmission with telecommunication devices (telephone line for data or fax machine, mobile phone or transceiver for satellite communication) or data channels for traffic management systems.

11.4.3.4 Treatment and Organization of Monitoring Data

Static Measurands

Measured displacement values such as deflection, inclination, settlement, crack width and crack length and also environmental measurands (temperature, humidity, corrosion, etc.) are predominantly quasi-static since they vary only slowly with time. It has proved possible to analyze these values in the form of hourly mean values with the associated standard deviations. Additionally, it is recommended to record sudden strong amplitude changes with time using predefined threshold values.

The appearance of cracks in (fully) pre-stressed concrete structures and in major structural elements of steel structures or the crossing of a limiting crack width is often a signal for a critical structural state. The permanent monitoring of changes in width of known cracks or the determination of cracks at known weak points of a structure or at points with high stress concentrations is therefore an important element of early damage detection. For characterization purposes, determining and storing hourly maximum values have been practically proven. To correlate the crack state with acting loads (live load, temperature), strain and temperature measurements need to be monitored in the same time pattern.

Dynamic Measurands

Changes in the load-bearing behavior of structures are always associated with changes in the vibration characteristics. Changes in the static system and associated parameters of the examined structure affect the natural frequencies and also the mode shapes. They have to be determined by experimental modal analysis.

During automated long-term monitoring, the estimation of the above-mentioned values can also be carried out without artificial excitation. Therefore, the time responses of each measurement point have to be analyzed by Fourier transformation. Peaks of the power density spectrum describe approximately the natural frequencies. Further, the associated operational mode shapes need to be determined by simultaneous or successive measurements for various measurement points. The operational mode shapes are similar to the natural modes of the structure.

In practice, results of multiple successive measurements should be averaged, to eliminate interferences, outside influences and other parts of the signal which are not dependent on the structural state. Averaged normalized power spectra density (ANPSD) plots can be regarded as the signature of the structure. A change in the stiffness or mass of the structure should be indicated by changes in the pattern of ANPSD. By comparison with a reference state, it is possible to obtain qualitative information about the location and extent of structural changes.

A quantitative assessment requires the structural models under consideration to be updated. In many cases, it is recommended that additional knowledge from expert systems be used. The location of the measurement points and the considered natural frequencies and operational mode shapes should be selected in such a way that expected structural changes are reflected as efficiently as possible. This can be achieved by experimental pre-examinations or numerical simulations.

Selection, Management and Presentation of Measurement Results

A permanent monitoring system needs to be applied intelligently on-site. Everything within a defined range can be recorded as specific parameters or values. These values become stored in defined intervals. Outsider results will be stored separately with a time correlation. The stored data will be transferred continuously or at fixed time intervals via a permanent data transfer line to the monitoring computer, where all actual information is stored about the measuring setup, the measurement points and the sensors and also the channel names, the settings and the calibration factors. Subsequently, the monitoring computer performs conditioning of the results and the management of all data, and creates monitoring reports and graphical result presentations and necessary trend analysis. By linking to a database containing all residual structural information about construction and inspections, possibilities for global use of the measurement data arise, for instance for assembling a technical information system or for the development of expert systems. Furthermore, these data can be a basis for quality management.

11.4.3.5 Review of Monitoring Parameters and Consequences

Status Report

The measurement data must be examined at regular time intervals regarding functionality and optimal behavior of the hardware. Additionally, the monitoring results must be checked regarding their physical reality for correctness and compatibility. This happens on the one hand on the basis of measured time series of selected results (typical, maximum, event-oriented, etc.). On the other hand, the results are reviewed on the basis of nominal statistical values (maximum of minimum values, RMS values, mean values, sliding RMS) in the considered periods. During monitoring of dynamic values, frequency spectra are compared in addition to determined natural frequencies, and frequency changes are shown and examined for plausibility. Results of load determination are presented over the monitoring period, distributed in load classes. Furthermore, the load distributions within significant periods are computed and compared. Fatigue-effective stresses are allocated in rainflow matrices and compared with the respective stresses in analogous periods.

Analysis of Effect Coherences

The determination of effect coherences in the form of correlations is important for the analysis of monitoring results, in addition to their temporal development. By this means, important information about the dependence of the monitoring values from assessable process variables or actions can be obtained, which could be controlled during monitoring. Errors or faults in sensor technology or in the measuring technique that develop during the monitoring become rapidly visible from the presentation of correlations for different physically connected measured variables.

Alerting

In addition to periodic data transfer from the data acquisition computer on-site to a central computer, an additional event-orientated data transfer can be useful. This is the case when in consequence of special events such as threshold values being exceeded, significant changes in the structural system and deterioration of parts of the monitoring system, immediate information needs to be sent to the central computer or a fax machine.

11.4.4 Numerical Analysis

A numerical structural analysis requires an adequate model, which contains sufficiently exact representations of structural stiffness, mass distribution and support conditions. For structural evaluation, the most accepted technique for modeling and comparing measured and calculated data is the FE approach. Once the FE model has achieved a certain level of completeness and validity, it provides the best basis for analytical prediction and simulation.

Substantial tasks of the numerical analysis are the identification of structural characteristics and the simulation of the structural behavior, with the following options:

- determination of modal parameter;
- verification of measurement results;

- simulation of structural properties with action, which would be difficult and expensive to realize experimentally;
- realization of parameter studies;
- modeling of damage.

Sources of inaccuracies of models are assumptions about the structure of the model (net sizing, type of the elements, boundary conditions, linear or nonlinear behavior) and the model parameters (material parameters, continuous and discrete stiffness, masses and moments of inertia). The correct specification of the loads and other actions is a substantial assumption for close-to-reality simulation of the structural behavior.

The following procedure is recommended for examining the correctness of simulation results:

1. Review of the program code/input macros.
2. Review of analysis results, e.g. by:
   - modeling of experiments with definite boundary conditions;
   - extreme value analysis, estimation of effects;
   - statistical analysis of values with variable parameters;
   - comparison with alternative systems;
   - quantitative estimation of results.

11.4.4.1 Calibration of Structural Models

The integration of analytical modeling followed by experiments for the calibration and verification of the analytical model for reliable simulation is termed structural identification. Structural identification serves as the starting point and core of health monitoring. Calibration is conducted by progressively adjusting numerical values of groups of parameters that define the material, geometry, boundary and continuity conditions until the discrepancies between measured data and simulated behavior of the analytical model are minimized with respect to an objective function. The calibrated model has to be checked with measured data that are not applied for calibration.

Although most parameter identification processes that are based on linearity, idealized boundaries, supports and release conditions cannot fulfil the real structural conditions, a field-calibrated linear model serves as the best possible starting point for simulations by nonlinear FE analysis in order to predict possible failure modes.

Model Calibration

Mechanical properties, boundary conditions and continuity conditions of the model are adjusted. Model configurations agree with the physical insight gained during the experiment and from the processed experimental data.

Sensitivity Analysis

Sensitivity can be defined as the local gradient information obtained from an analytical formulation, differentiation or numerical computation via finite difference schemes. Sensitivity information is very valuable whenever optimization problems are to be solved. For model calibration, the input parameters should be screened to analyze which ones produce the greatest change in several response features over a range of possible values.

Utilization of Calibrated Models

The field-calibrated analytical model serves as the best measure of the actual conditions of the structure. This may be used for load-capacity rating, permit loading and evaluating internal forces, stresses and deformations under operational conditions.

11.5 Damage Identification

11.5.1 Objectives and Procedures of Damage Identification

The purpose of damage identification is to acquire as early as possible comprehensive knowledge about the damage of structures. Beyond that, the local and global effects of the damage with consideration of the damage process have to be estimated. A further goal of damage identification to find the causes of the occurrence of damage.

On the basis of a definition of damage, the principle of the procedure for damage identification is described. A classification of substantial damage to civil structures and its causes is useful. The methods that can be used differ depending on level of knowledge of the existing or expected damage in local and global procedures. For global procedures, established methods are described, which are based on results of static and dynamic measurements.

11.5.2 Definition of Damage

Damage is defined as changes introduced into a system that adversely affect current or future performance. This definition is limited to changes to the material and/or geometric properties of structures, including changes to the boundary conditions and system compatibility. Changes to the system in the sense of damage develop either directly (time invariantly) or due to time-variant processes. Time-variant damage can accumulate incrementally over long periods such as that associated with fatigue or corrosion damage. Discrete events such as earthquakes, live loads and others can lead, according to scheduled or unscheduled events, to direct, i.e. time-invariant, damage. In the sense of spatial expansion, damage has a local effect or can be recognized as distributed. The severity of damage can be represented either via a geometric description (crack geometry, etc.), by its effect on the load-carrying capacity of structures (e.g. loss of stiffness or mass) or by changes in the energy dissipation properties of a system.

11.5.3 Classification of Damage and Damage Mechanisms

Damage can be defined as partial or complete destruction of the material structure and therefore a weakening of the resistance of the affected structural component or the whole structure. Damage can be caused by several influences. Predominantly, damage is the effect of deterioration processes, mainly corrosion and fatigue. Furthermore, damage is caused by an excess material strength through unplanned high loadings.

11.5.3.1 General Causes of Damage

General causes for damage are:

1. **Overstressing (loading) with time-invariant resistance.** Possible cases:
   - accidental and seismic loads (impact, earthquake, explosion);
   - exceptionally high variable loads (excessive live load, extreme wind, wave and snow).
2. **Regular loading with reduced resistance.** The decrease in the member resistance, as considered in the design, is generally a time-dependent process, which can initiate damage by:
   - deterioration through chemical loading
     - surface corrosion of steel
     - pitting corrosion of steel
     - alkali silica reaction in concrete;

- deterioration through mechanical loading
  - submicroscopic crack formation and crack growth until excess of the lower cross-sectional limit by alternating loading of structural steel and reinforcement in concrete (fatigue)
  - stress corrosion cracking of pre-stress tendons
  - fretting corrosion of steel members
  - microscopic crack formation in concrete leads to a reduction in the transverse tensile strength and for this reason to a decrease in the compressive strength (fatigue) of concrete;
- deterioration through physical loading
  - damage to polymers by UV radiation
  - damage to concrete by frost
  - damage to materials by heat/fire;
- creeping, shrinkage, relaxation
  - reduced shear strength through loss of pre-stress by creeping.

3. **Combination of both.**

11.5.3.2 Specific Causes of Damage

Parts of the damage mechanisms, described in Section 11.5.3.1, have specific causes such as

1. **Corrosion of steel:** caused by
   - damaged anti-corrosion coating;
   - cracks in concrete and other mechanical damage of structural concrete (plus moist environment), caused by
     - accidental loads
     - poor configuration of reinforcement
     - constraints through changes in bearing conditions (foundation settlement and rotation, reduction of degrees of freedom)
     - loss of pre-stress;
   - carbonation of concrete;
   - chloride ions (from thaw salts, from sea water).
2. **Crack formation and crack growth in structural steel and microscopic crack formation in concrete:** caused by alternating loading above the fatigue strength through
   - high cyclic loading;
   - cyclic loading plus reduced cross section (cracks).

11.5.4 Concepts of Damage Identification

The identification of damage is generally described via a four-level process.

Level 1: detection. Determination that damage is present in the structure.
Level 2: localization. Determination of the geometric location of the damage.
Level 3: quantification. Determination of the severity of the damage.
Level 4: prognosis. Prediction of the remaining service life of the structure.

11.5.5 Variables and Indicators for Damage Identification

Experimental investigations have the purpose of obtaining information for tasks mentioned in Section 11.4.1, via either a single investigation in the sense of field tests mentioned in Section 11.4.2.5 or by periodic or continuous measurements with automatic monitoring systems. The processes used are distinguished by whether knowledge about damage is expected or available, leading to local damage monitoring, or whether on the basis of global structural behavior information has to be gathered. The measuring data are based on both static and dynamic investigations. A number of different analytical techniques have been developed for the identification of damage within the structural health monitoring process.

11.5.5.1 Local Procedures

The application of local procedures to weak point analysis is desirable in the following cases.

- With the knowledge of type and location of existing damage, the dimensions of the damage have to be determined and the development has to be monitored (Level 3). Then the local situation and global consequences are assessed. This applies to the present condition and also in the sense of prediction for future conditions (Level 4, Section 11.5.4).
- Likewise, this procedure applies to the situation where on the basis of preliminary existing knowledge, damage at a certain location of the structure is expected, e.g. due to overloading or fatigue (Level 1). This task can be fulfilled using load monitoring (Section 11.2).
- An additional task for local monitoring of damage is to find the causes of the damage. Furthermore, in addition to damage parameters, damage-affected values are monitored. Information about the character and causes of damage and can be obtained through correlations in a temporal context and also between the magnitudes of the measurement results. An example is the surveillance of crack widths, which can be affected by temperature, vibrations, settlement or other causes.
- If local damage has global consequences (e.g. the settlement of a bridge pillar), observations at places other than the damaged area of the structure be sensible (Levels 1–3).

With local procedures, measurement values are used that are appropriate for the type of damage or for damage effects by suitable indicators (e.g. crack widths, strains, inclinations). Here, continuously working systems are favored.

11.5.5.2 Global Procedures

If neither the existence of damage nor the possible damage position is known, the sensitive global system parameters always have to be controlled and determined. This can be done either via continuous and periodic monitoring or with uniquely accomplished field tests.

Dynamic Procedures

Vibration characteristics are global properties of the structure and, although they are affected by local damage, they may not be very sensitive to such damage. As a result, the change in global properties may be difficult to identify unless the damage is very severe or the measurements are very accurate.

The identification of a possible damage site and the severity of damage on the basis of a change in global properties derived from measurements at a limited number of sensor locations is a problem which should not be underestimated. Sophisticated and complex mathematical techniques, including nonlinear programming, need to be employed to obtain the most probable solution. Global vibration characteristics are often affected by phenomena other than damage, including environmental effects such as a change in mass and thermal effects caused by temperature variations. Additionally, boundary conditions in a structure may lead to a change in vibration if these boundary conditions are prone to changes with age.

Information about the condition of a structure is provided from measured changes in vibration properties. The more commonly used techniques are based on:

- natural frequency methods;
- mode shape and operational deflection shape methods;
- modal strain energy methods;

- residual force vector method;
- model updating methods;
- frequency response functions;
- statistical methods.

There is no optimum method for using measured vibration data for damage detection, localization and quantification. No algorithm has yet been proposed that can be applied universally to identify any type of damage in any type of structure. Additionally, no algorithm is yet available that can predict the exact service life of a structure. In addition to the application of dynamic procedures, extensive experience is required in this area.

Natural Frequency Methods

Changes in stiffness and mass and also the dynamically effective support and transition conditions of structures will usually lead to measurable changes in the natural frequencies. Usually, the associated measuring expenditure is small. Often only a few reasonably placed sensors are sufficient for acquiring the desired information. For simple structures, the measured differences in the natural frequencies form a sufficient sample for the localization of damage. Therefore, this method is suitable for online monitoring of structures for Levels 1 and 2. The largest frequency changes can be expected with those natural frequencies where the locations of the associated maximum curvatures of the mode shape match the damaged ranges of the structures. The assumption of a linear relationship between frequency shifts and damage is no longer appropriate when the severity and the number of locations of damage increase. Due to the strong environmental influence on frequency shifts, such ambient effects have to be filtered out.

Mode Shape and Operational Deflection Shape Methods (ODS)

Reasonable results of damage localization tasks (Level 2) are to be expected only if the number of measurement points is adequate in relation to the dimensions of the structure. Usually, this is only achievable by field tests (Section 11.4.2.5). Damage indicators, computed through measuring mode shapes, are based on the comparison of the amplitude changes in relation to a starting point. Two commonly used methods for comparing two sets of mode shapes are Modal Assurance Criteria (MAC) and the Coordinate Modal Assurance Criterion ($COMAC$). Depending on the location of structural changes, the direct comparison of mode shapes can lead to better information for the tasks of damage detection and localization than frequency shift measurements. Furthermore, the sensitivity due to environmental influences is lower. Good experience was obtained with the comparison of structural tests before and after repair. During the investigation of large structures, the application of artificial excitation is often not possible. Here, measured operational deflection shapes (ODS) are used for the computation of damage indicators. ODS must also be measured in cases where damage is recognizable only when operating loads are applied.

For damage identification in the presence of localized damage, the use of mode shape curvatures is more meaningful, as theoretically damage occurs here highly localized. Curvatures cannot be measured directly; they have to be calculated from measured displacement modes or can be derived from measured strains. In both cases, the requirements on accuracy and number of measuring points for global investigations of buildings are difficult to realize. The damage indicator cumulative distribution function (CDF) based on modal curvatures is computed in a similar way to $COMAC$.

Modal Strain Energy Methods (MSE)

The extension of the modal curvature method by the formulation of a modal strain energy leads to a damage indicator defined as the ratio of the modal strain energy of elements of a structure before and after the damage. The indicator employs modal curvatures computed from measured mode shapes via the second-order derivative. Taking noisy data into account, it is necessary to use interpolation polynomials for the calculation of the derivatives in combination with smoothing procedures by means of regularization methods to avoid rough curvatures. The MSE has been tested on beam-like structures and plates successfully. There have also been some enhancements of the original method by variations the damage indicator for multiple damage scenarios.

Residual Force Vector Method (RFV)

Utilizing measured natural frequencies and mode shapes for the damaged system and using the initial baseline model for the structure, an overdetermined system of equations is formulated. This allows the determination of unknown correction parameters for the initial model describing the structural damage. Expanding the measured mode shapes successfully, the RFV method seems to be a robust method for damage localization and sizing.

Frequency Response and Transmissibility Function Methods (FRF)

Modal parameters are extracted either from input–output measurements or output-only results. In order to exclude errors within damage identification procedures which emerge from modal identification, indicators can be defined directly by using FRF measurements or measured transmissibilities. The differences between the initial and damaged states are recognized by a shift of resonances and anti-resonances in the spectrum and also by different large amplitudes. Indicators for damage detection are formulated by differences from FRF amplitudes in the initial and damage states. For localization of damage, ratios of FRF and transmissibilities are used. Measurement results for FRF can be obtained from field tests whereas transmissibilities can be measured by online monitoring.

Model Updating Methods (MUM)

Many methods of model updating are used for damage identification procedures. Theoretically, MUM are able to provide solutions for the damage identification Levels 1–3. The determination of damage requires true changes in the physical properties. Problems arise with the non-uniqueness of the resultant model in matching the measured data. When damage affects both mass and stiffness properties, the parameter cannot be uniquely identified only by using modal measurements. In general, it is not recommended to use only natural frequencies for updating procedures. An advantage is to use FRF measurements directly for damage identification. Modal parameters do not have to be identified as additionally the FRF data provide much more information in the desired frequency range. Ill-conditioned matrices from adjacent points in the FRF data have to be avoided. To overcome these problems, it is often recommended to make assumptions about the location and type of damage.

Statistical Methods (SM)

The basic idea of these methods lies in the acceptance that all substantial information about the condition of a system, which affects the dynamic behavior, are contained in the measurable responses of the system. With this assumption, damage identification is a problem of statistical pattern recognition that can be solved with nonmodel-based recognition methods. Any information can be concentrated in appropriate features representing the actual state of damage. These features derived from output-only measurements made under normal operating conditions have a distribution with an associated mean and variance. A change in the distribution characteristics of the features will indicate damage, whereby it is presupposed that effects change the system performance in a non-characteristic manner. These methods are applicable to Level 1 and 2 identification.

Static Procedures

Not all kinds of damage are detectable with sufficient reliability by dynamic methods. In cases of small modifications of the initial structure and when damage emerges in the proximity of bearings, the sensitivity of dynamic damage indicators is low. With the overlaid influence of temperature changes on the measurable dynamic parameters, damage identification is very difficult.

Static measurements always require a load situation where the damage identification can be achieved using measured global parameters. Deformation measurements in most cases refer to a reference position. A proven method for damage identification of beam-like structures is proposed using differences in measured inclinations (Level 1 and 2 identification). Inclination sensors do not need any reference position. The location of only a few sensors is robust regarding the quality of the measurement results. The load is applied by a slowly moving truck crossing the structure. The damage indicator is determined from the difference in the influence lines of the inclinations in a damaged condition in relation to an initial condition. Influence lines can be measured regarding the local resolution arbitrarily exactly. For the data analysis, noise reduction of the measured signals is very important. The method is suitable also for the localization of multiple damage.

11.5.6 Determination of Damage Indicators by Measurement

For the determination of damage indicators, a redundant approach is recommended regarding, for example, for damage detection (Level 1) with dynamic processes additional damage-focused sensors are installed to enhance the informative capability of the results. Additionally it must be considered that the recorded damage indicators usually contain environmental effects, which must be 'cleaned up' computationally. A high spatial resolution for measurement of mode shapes is obtained either by the concurrent utilization of many sensors or a few, but moving sensors depending on a stationary exciter. Here measurement technology based on scanning processes is used.

11.5.7 Damage Assessment in the Sense of Condition Specification

The description of the condition of a structure first has the purpose of the qualitative evaluation of observed, i.e. already existing, or newly discovered damage concerning the magnitude, the consequences, the development and the cause of the damage. Additionally, their effects for technical systems (e.g. change of the static model) and the resulting load capacity, functionality, stability and safety have to be considered. On the basis of knowledge about the type and magnitude of essential defects and damage and the respective damage mechanisms, evaluations regarding the topical load capacity, possible rehabilitation actions and the utilization and safety planning need to be done.

11.5.8 Damage Assessment Using Threshold Values

A simple and approved method for assessing measured condition values is comparison with preliminarily defined threshold values. These values can be constituted in codes and guidelines or determined on the base of investigations and experience.

11.5.8.1 Threshold Values by Codes and Guidelines

Numerous thresholds to be met during the erection and use of a structure are specified in design codes for structures. If assessment values are measurable, they can be used directly for the assessment of monitoring results. This applies in particular to the limitation of deformation in the serviceability limit state. Further values for providing serviceability and safety against fatigue are limited in codes.

11.5.8.2 Determination of Threshold Values

If condition threshold values are not defined in codes or guidelines, they can be determined on the basis of preliminary examinations. These are primarily examinations and analysis of structural models and material testing. The determination of threshold values requires well-founded knowledge about the structural coherences and the underlying safety level.

Table 11.1 Sensor classification according to different measured values

Measured value	Sensor type	Measurement range (MR)	Resolution	Linearity	Supply	frequency range	influences
Displacement	LVDT	1–50 mm	1 ‰ MR	0.1% MR	AC/DC	0–100 Hz	cross forces
	LVDT caliper	1–50 mm	1 ‰ MR	0.1% mr	AC/DC	0–50 Hz	measurement frequency
	Triangulation sensor	2–200 mm	1 ‰ MR	0.3% MR	DC	0–10 kHz	reflecting surfaces, soil
	Cable extension transducer	50–40000 mm		0.05% MR	DC	0–500 Hz	cross fores
Inclination	Bubble level	±10 grade	1 ‰ MR	0.1 grade	DC	0.5Hz	vibrations
	Pendulum	±1 grade	1 ‰ MR	0.05% MR	DC	0.5Hz	
Settlement	Hydrostatic leveling system	0–60 mm	0.01 mm		DC	static	atmosph. pressure, temp. gradients
	PSD	0–90 mm	< 0.05 mm	< 1% MR	DC	0–500 Hz	moisture
Strain	Strain gauges	1–10000 μm/m	1 μm/m	< 1%	DC/AC	0–100 kHz	ΔT, leackage
	Fiber Bragg gratings	±10000 μm/m	1 μm/m		Laser light	0–100 kHz	ΔT, transverse stress
	Fabry–Perot fiber sensor	±5000 μm/m	1 μm/m		Broadband white light	0–1 kHz	ΔT
	SOFO system	0.5% sensor length	2 μm/m	< 1%	Laser light	static	transverse stress
	Optical string	0.5% sensor length			Laser light	0–100 Hz	
Acceleration	Piezoelectric sensors	±100 g	10 μg	< 1%	DC	0.1–2000 Hz	fixing
	MEMS-A640	±1 g	5 μg	< 1%	DC	0–250 Hz	
	B12 (differential choke)	±20 g		1%	AC	0–100 Hz	
Vibrating velocity	Geophone	100 mm/s	5 μm/s	< 1%		4–1000 Hz	fixing
	Laser vibrometer	10 m/s	1 μm/s	< 1%	DC/AC	1–20000 Hz	
Temperature	Thermocouples	−185 to 300°C	50 μV/K	1%			leads
	Pt100	−200 to 600°C	400 μV/K	< 1%	DC		temperature gradients

Table 11.2 Sensor application and comments

Measured Value	Sensor type	Comments
Displacement	LVDT	Simple installation. Avoid cross forces
	LVDT caliper	Simple installation. Small operating frequencies
	Triangulation sensor	Simple installation. Avoid reflecting surfaces
	Cable extension transducer	Simple installation. Avoid impacts
Inclination	Bubble level	Simple installation. Small operating frequencies
	Pendulum	Simple installation
Settlement	Hydrostatic leveling system	Complex installation
	PSD	Complex installation
Stress	Strain gauges	Complex installation. For short-term measurements of strain
	Fiber bragg gratings	Complex installation
	Fabry–Perot fiber sensors	Complex installation
	SOFO system	Complex installation. For long-term measurements
	Optical string	Simple installation
Acceleration	Piezoelectric sensors	Simple installation. Rugged pick-up
	MEMS-A640	Simple installation
	B12 (differential choke)	Simple installation
Vibrating velocity	Geophone	Simple installation. Reliable and rugged pick-up
	Laser vibrometer	Simple installation. Robuster Aufnehmer, inapplicable for monitoring systems
Temperature	Thermocouples	Complex installation
	Pt100	Simple installation

11.6 Qualifications of Test Personnel

The procedures described in this guideline can lead to usable results only if, before the measuring program, the equipment, the analysis hardware and especially the analysis software and the subsequent interpretation of the results and the assessment are carried out by experienced staff. This requires an expert knowledge of structural engineering and experience with structural proving and testing and with the measurement techniques. To guarantee reliability of the measurement results, the application of sensors, the realization of measurements and the maintenance of the technique needs to be carried out only by skilled personnel under the supervision of experts, who are responsible for the measurements. The measurement personnel must be permanent available. The state of the measurement technique has to be recorded.

11.7 Sensor Classification, Application and Experience

11.8 Traffic Load Identification on Bridges

Task

At a seven-span pre-stressed concrete bridge (Figures 11.1 and 11.2), the static and dynamic live loads are to be determined and classified depending upon lanes.

Procedure

The fundamental idea for the represented method is to use the bridge as a balance. On the basis of measured strains on suitable locations at the main girders of the bridge, the loads of vehicles are acquired.

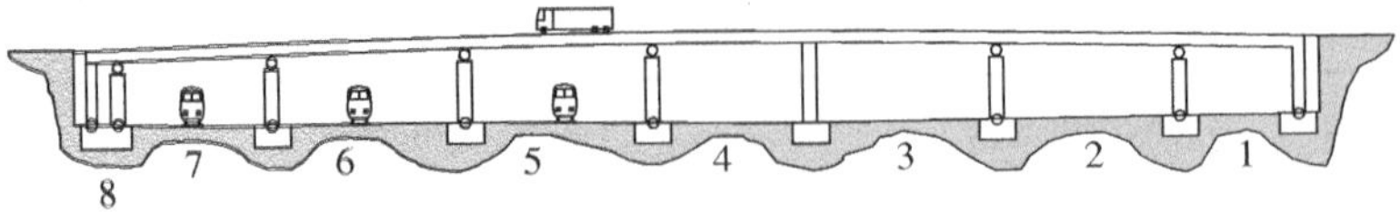

Figure 11.1 Westend Bridge, Berlin, Germany

In a first step, a validated load model has to be developed. For this purpose, the influence lines of the strains are measured during the passage of a vehicle of well-known geometry and axle loads. In a second step, those influence lines are measured and stored which contain the most important possible combinations of vehicles when passing the bridge. Possibilities of these combinations are: one vehicle behind the other, one next to each other, one behind the other plus next to each other, etc. This information forms the basic patterns for a later automatic pattern recognition.

In operation, strains caused by the weight of the passing vehicles are measured and the combinations of vehicles are identified using pattern recognition. Subsequently, the static and the dynamic portions of the traffic loads are identified by separation of the measured signals using Fourier analysis. Performing low-pass filtering of the signals with a frequency below the first natural frequency of the bridge, the static values of the measured strains are obtained. Appropriate procedures are applied to the dynamic portion (Figure 11.3). The dynamic factor is defined as the relationship of the maximum total amplitude with respect to the associated maximum value of the static amplitude.

Results

One of the results of the load identification is a representative static load for each lane. Figure 11.4 shows exemplarily the frequency of the ascertained traffic loads separated into load classes between 1994 and 2004. With such continuously recorded data, load models updated for each bridge can specifically be determined.

In the same way, the dynamic factor Φ is determined (Figure 11.5). Figure 11.6 shows the temporal change of Φ over period of 10 years. Φ is determined as a function of the magnitude of the weight of the vehicles. This information about traffic loads can be used for calculating the entire stresses in bridges that actually develop over time by means of FE models. From these results, the residual lifetime of structures can be derived. Figure 11.5 shows that the size of the dynamic factor Φ can be described with respect to the frequency of its occurrence by a statistical distribution. This distribution varies over time and gives information about the size of the fatigue-conditioned stresses. Figure 11.6 shows that the dynamic factor can reach considerably larger absolute values than this main value although with reduced frequency.

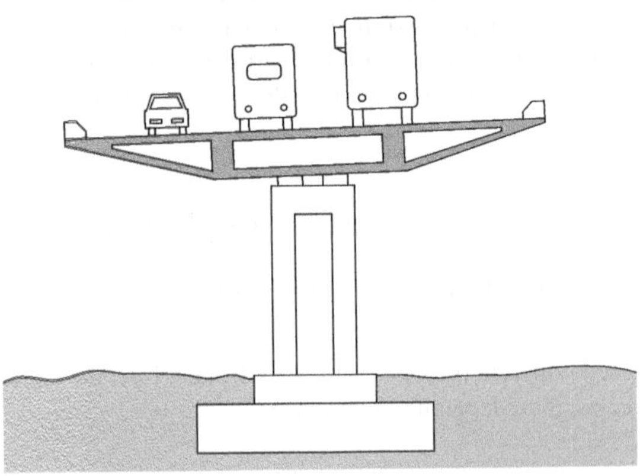

Figure 11.2 Cross section of the tested bridge

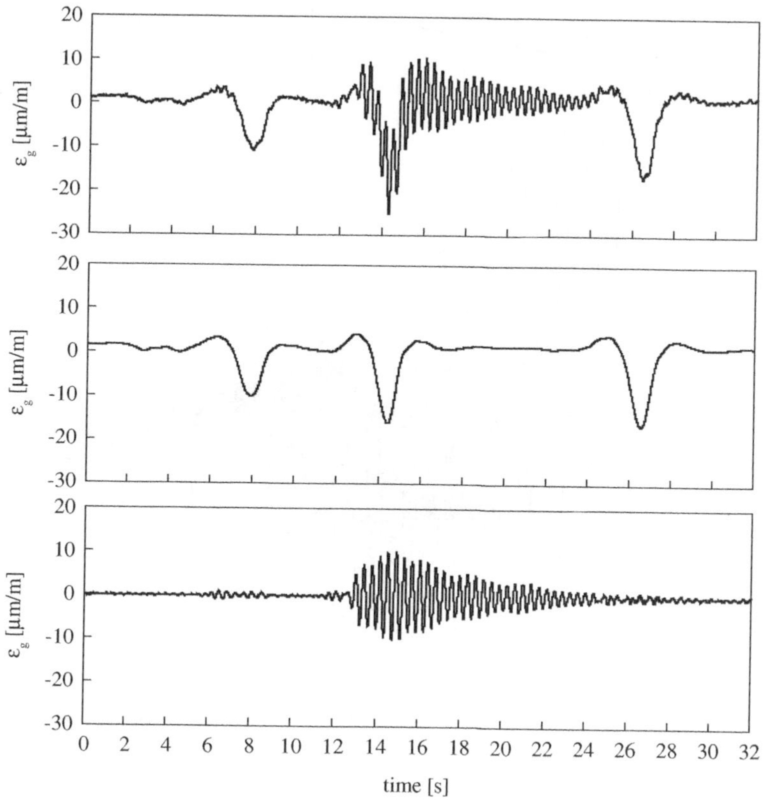

Figure 11.3 Separation of the combination load (top) into its static (center) and dynamic (bottom) portions

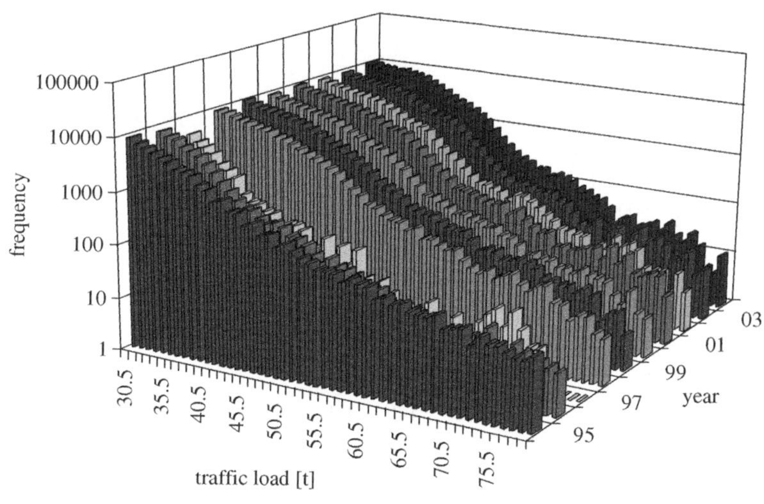

Figure 11.4 Frequency of identified traffic loads within the range 30–80 t between 1994 and 2004

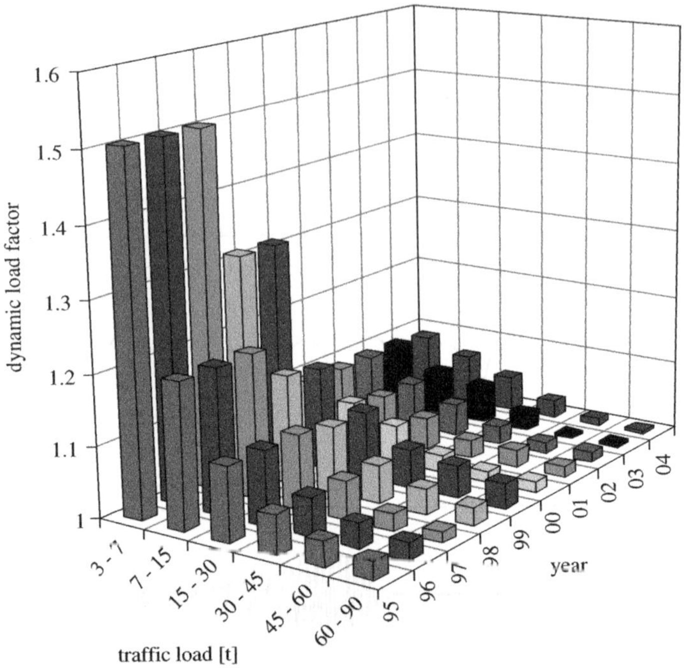

Figure 11.5 Development of the dynamic load factor Φ dependent on measured traffic loads between 1995 and 2004

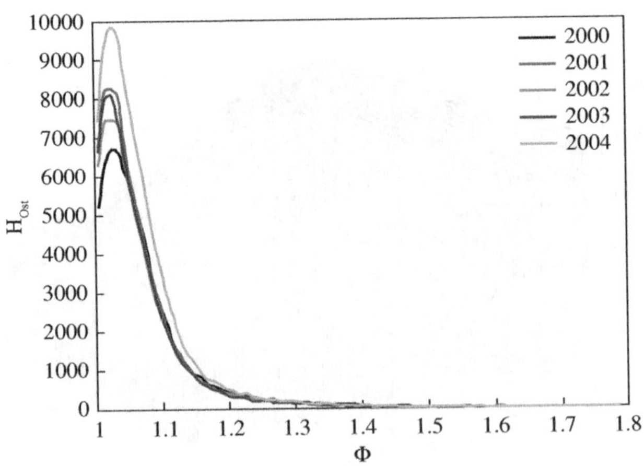

Figure 11.6 Frequency distribution of the load factor and its variations between 2000 and 2004

Figure 11.7 Construction work close to the Brandenburg Gate in Berlin, Germany

11.9 Condition Monitoring of Heritage Buildings

Task

Beneath the Brandenburg Gate construction work (tunneling) for a new underground line is carried out (Figures 11.7 and 11.8). During this time, permanent condition monitoring of the monument is to be accomplished with the goal of discovering changes and damage to the structure and the foundations.

Procedure

Tunneling in the soil in general causes vibrations at the monument and settlements of the foundations. This can cause damage to the carrying structure or to the historical foundations. Therefore, additional to the local monitoring of already existing cracks, predominantly dynamic sensors for the global surveillance are used. With the results of the vibration monitoring, damage identification on the basis of observed natural frequencies is accomplished. In addition, before beginning and after ending the construction work, experimental modal analysis is carried out to obtain information about structural damage from variations of the mode shapes in comparison with the reference state. The vibration amplitudes of the building's response excited by normal traffic before beginning and after ending the construction work are used as further indicators for changes in the vicinity of the foundations.

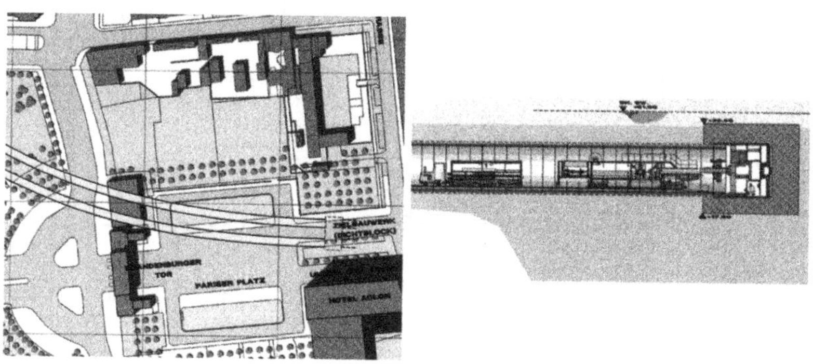

Figure 11.8 Tunneling for a new metro line underneath the monument

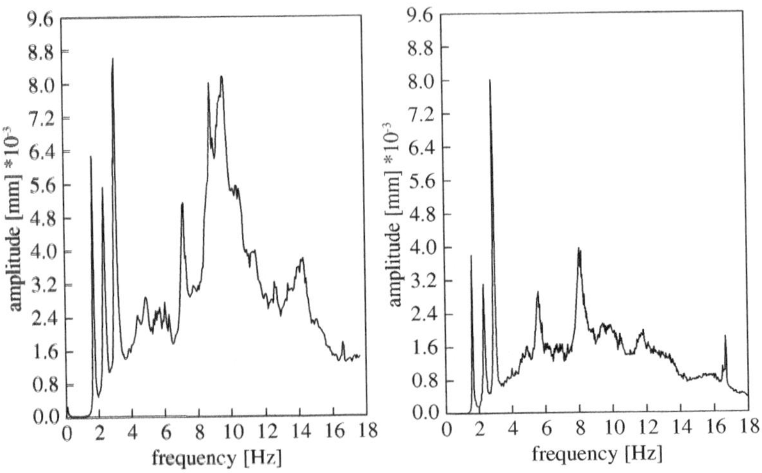

Figure 11.9 Comparison of power spectra and natural frequencies of the monument before (left) and after (right) the construction work

Results

The monitoring results show changes of the dynamic characteristics of the Brandenburg gate up to approximately the time when the tunneling in the soil had reached the foundations (Figure 11.10). From the power spectra (Figure 11.9) a decrease in the natural frequencies is recognizable. Table 11.3 shows that this change amounts to about 10% for the first natural frequency and up to 25% for the fourth natural frequency. The presented results of measurement were obtained in the context of the respective experimental modal analysis. From Figure 11.11, it can be seen that mode shape No. 1 exhibits changes regarding the directions of the movement. Before the beginning of the work, mode shape No. 1 had exclusively a horizontal component. After ending the construction work, this mode shape contained an additional lateral component and thus an additional modal degree of freedom (Figure 11.11). This trend is

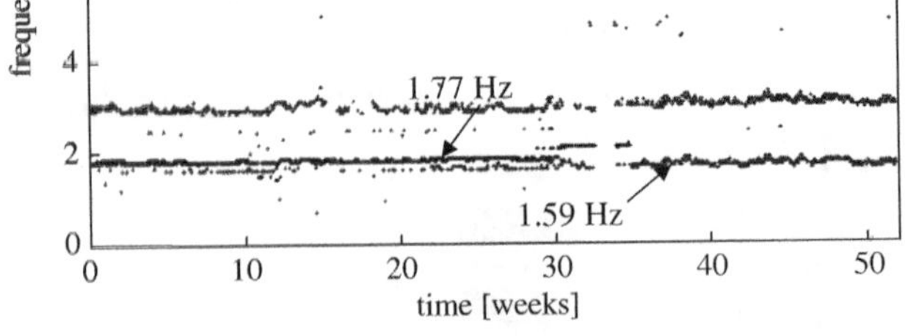

Figure 11.10 Results from continuous monitoring show changes in the first natural frequency after an observation time of approximately 30 weeks

Table 11.3 Results from the natural frequency method indicating changes in the dynamic behavior

Sensor number	Natural frequencies [Hz]	
	Before	After
1	1.77	1.59
2	2.44	2.29
3	3.17	2.90
4	7.26	5.80
5	8.91	8.2–8.4

also confirmed by the results of vibration measurements at the foundations of the monument. Figure 11.12 shows the correlation of the vibration amplitudes FR (horizontal) and FL (lateral). These results contain all frequencies of the excitation spectrum. They show also the trend of increasing amplitudes in the lateral direction after completion of the construction work.

It is assumed that tunneling changed the condition of the foundation stiffness. Such changes can be detected particularly sensitively in the global dynamic behavior. Detailed information necessary for the

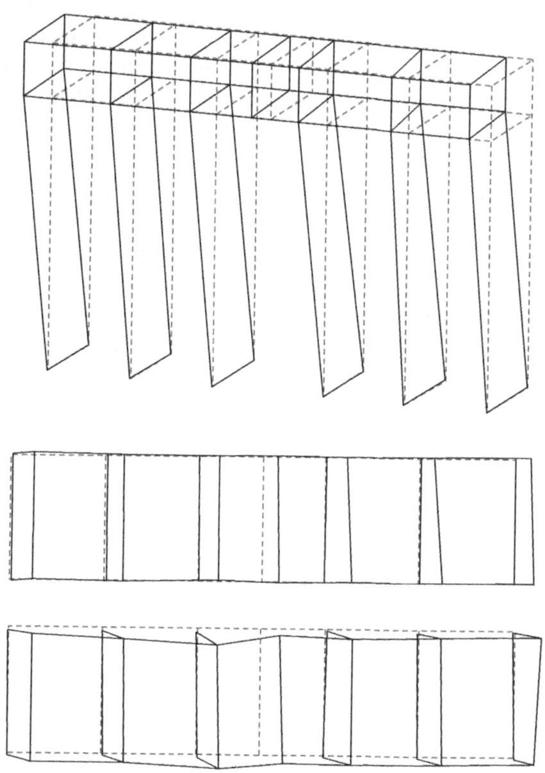

Figure 11.11 Comparison of measured first mode shape before and after construction work

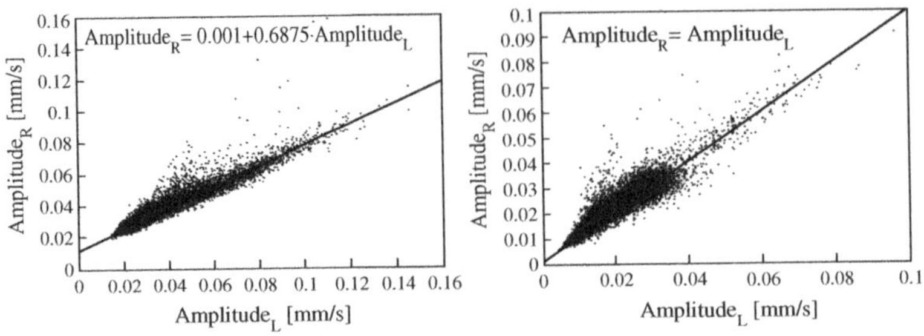

Figure 11.12 Ratio of vibration amplitudes FR/FL measured at the foundation indicates structural changes

assessment of these changes can be obtained by additional inspections of the foundations and by finite element simulation.

11.10 Identification of Local Damage and the Effect on Structures

Task

By visual inspections, cracks in the superstructure of a masonry railway bridge were discovered. With the help of a monitoring system, the changes in the crack widths under the influence of live loads and temperature are to be observed. In addition, trends for the damage development should be determined. Further, the static stresses should be measured due to the dead loads and also possible changes in the load-carrying capacity as a function of structural changes (cracks and settlements) and the influence of temperature.

Procedure

The viaduct has an overall length of 750 m and consists of 34 arches (Figure 11.13). The sensors and the measurement equipment for continuous monitoring are located within two of these arches (Figure 11.14). The measured variables used for monitoring were changes in crack widths, stresses, vibrations and temperatures. The settlements of the piers were determined by geodetic measurements in certain time intervals. Data acquisition was carried out continuously and without any gap. A first evaluation of the data for the purpose of data reduction took place in situ. Predominantly characteristic statistical values were determined. Using Fourier analysis, the results of measurements could be separated depending

Figure 11.13 Railway bridge: Neiše Viaduct, Germany

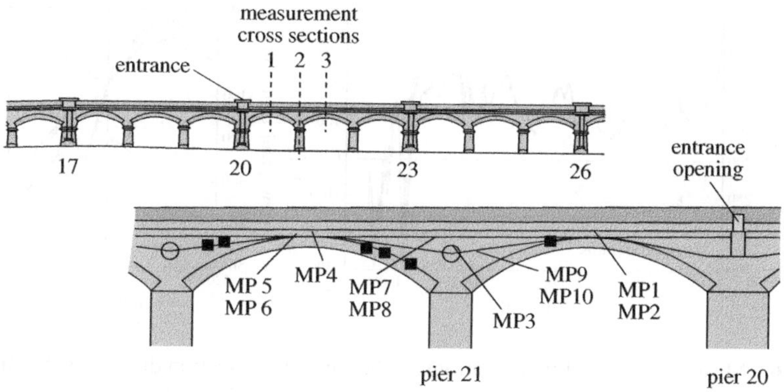

Figure 11.14 Cross sections (top) and locations for the attachment of sensors (bottom)

upon effect (setting, temperature, traffic). The causes of the damage could be analyzed using correlation functions.

Results

Generally, the monitored changes in crack widths are predominantly temperature dependent and reversible. Changes in crack widths with the passage of trains were also measured, but these are comparatively small and do not contribute to changes in structural characteristics. The maximum changes in crack width with temperature amount to 9 mm in the yearly cycle whereas those under live loads were measured as 0.13 mm. It was stated that the whole cross section is cracked and is moving like a rigid body (Figure 11.15), where w_1 and w_2 are crack widths at the two opposite sides of the superstructure. Repair work does not have any effect regarding the cracks. Changes in crack width can be described to a good approximation by a linear correlation with the building temperature: $w \cong -0.3T$. An effect of the measured settlements on the changes in crack width was not determined.

Representative strain measurements on the masonry were difficult to perform due to local temperature-dependent changes in the distributions of stress (Figure 11.16). Sensors with a sufficient measuring length compared with the dimensions of the masonry are important for good monitoring results. Comparing the

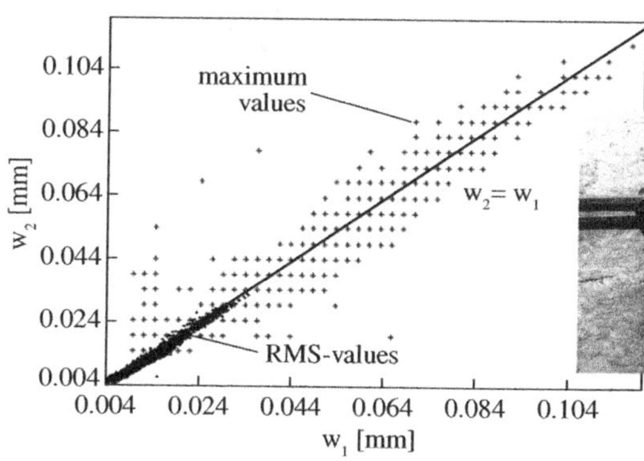

Figure 11.15 Comparison of crack widths w_1 and w_2 (left) and the sensor utilized for measuring large crack widths (right)

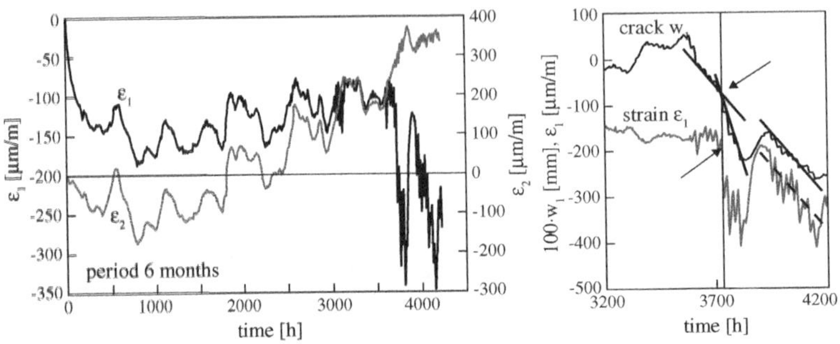

Figure 11.16 Measured strain at two locations of the same cross section due to temperature (left) and the observed rapid change in strain caused by rapid changes in crack width (right)

measured strains due to live loads and temperature, it can be seen that their ratio is 1:100. Hence it becomes clear that temperature is the crucial load case for the assessment of the viaduct concerning stress (Figure 11.17). Dynamic measurements allow control of the train traffic concerning the schedule and models of trains operating (Figure 11.18). This knowledge is necessary for comparing results induced by traffic loads. Traffic-induced vibrations are able to excite the natural frequencies of the viaduct (Figure 11.18). Modal data measured by condition monitoring provide basic information for damage-dependent structural assessment.

11.11 Damage Identification of a Steel Bridge by Dynamic Parameters

Task

At a three-span steel bridge, artificially increasing damage was generated in order to simulate the development of fatigue cracks at welding seams (Figures 11.19 and 11.20). From the results of dynamic measurements in the undamaged and damaged states, the location and severity of the damage were to be determined (Farrar et al. 1994).

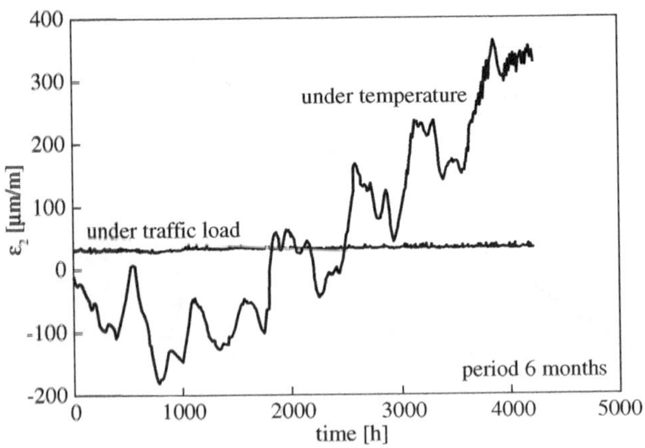

Figure 11.17 Measured strains due to the impact of traffic loads and temperature

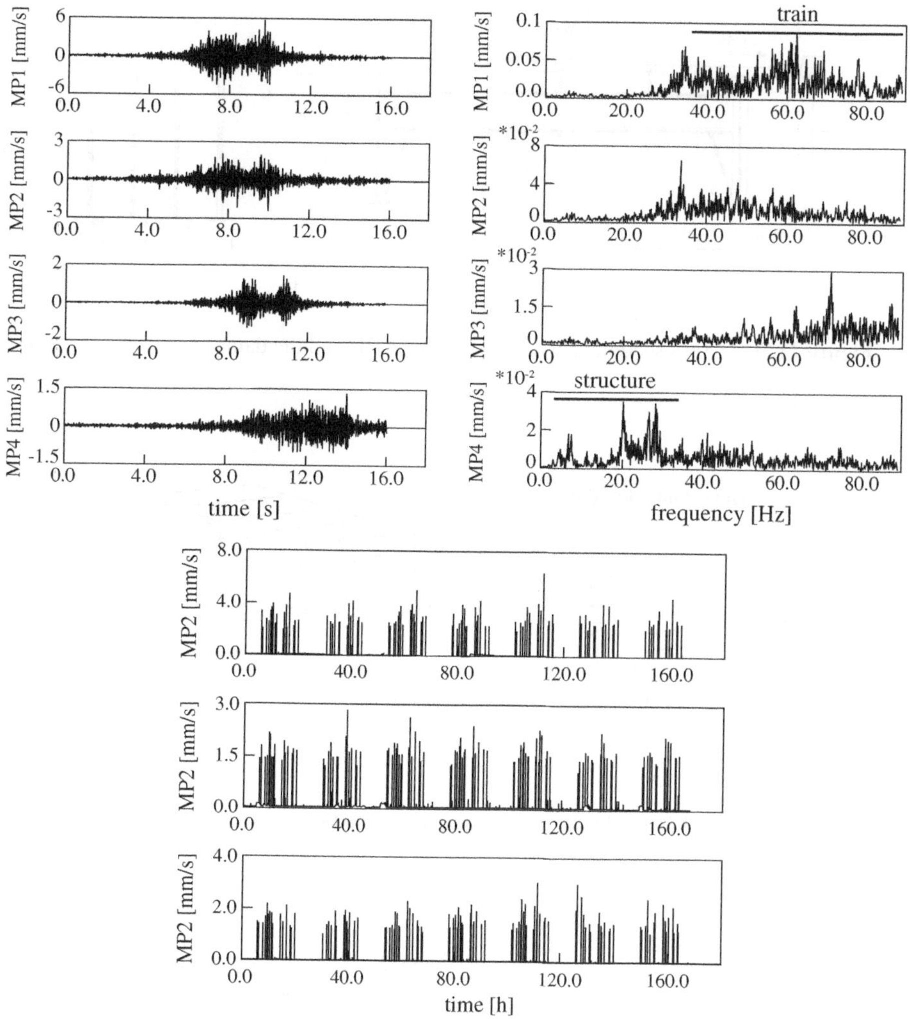

Figure 11.18 Passage of a train and the associated frequencies excited (top) and the controlled passages of trains during a period of 1 week (bottom)

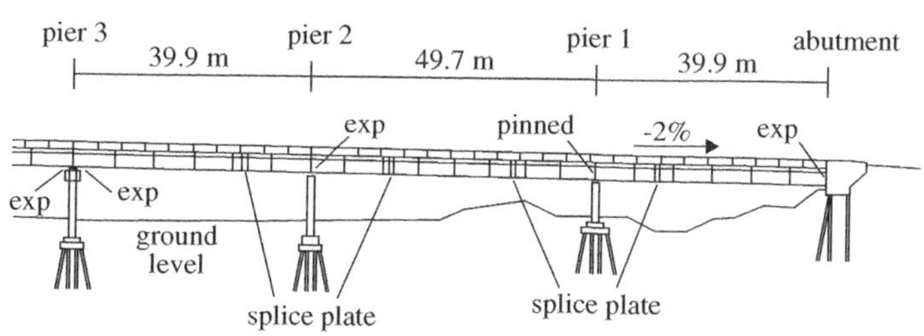

Figure 11.19 The I-40 highway bridge, New Mexico, USA

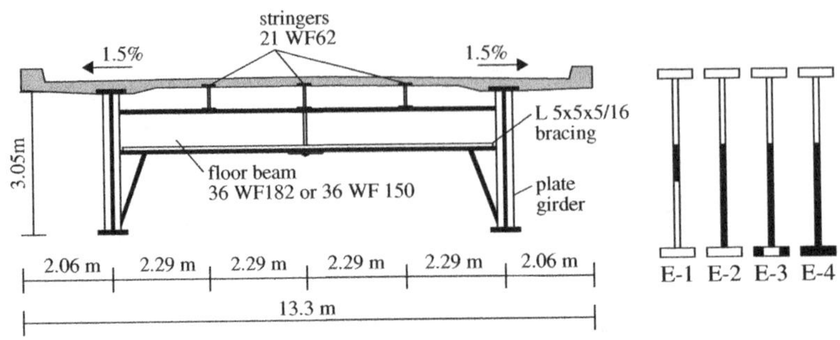

Figure 11.20 Cross section of the tested bridge (left) and the damage scenarios (right)

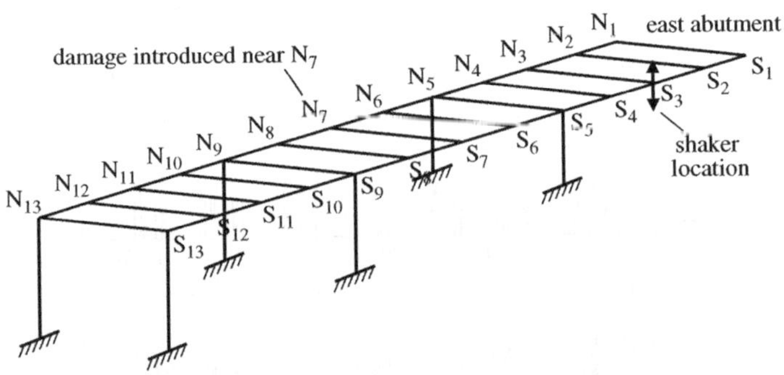

Figure 11.21 Mesh grid along the bridge section

Procedure

The bridge was excited by a shaker within the range 2–12 Hz with maximum force amplitude of approximately 8900 N. At 26 measuring points, accelerations in the vertical direction were measured (Figures 11.21 and 11.22). From these results by means of experimental modal analysis the natural frequencies and mode shapes were determined. Modal parameters were used to validate finite element models for the undamaged condition. The frequency response (Figure 11.22) functions show the degree of agreement of the dynamic characteristics of the model with reality. For damage analysis, a residue vector with six natural frequencies and the first two modes of shapes was used. The number of free parameters for the damage localization amounted to 1080 in a first step. For reduction of the number of free parameters and thus for the localization of damage, a two-stage decomposition procedure was used (Fritzen and Bohle 2000).

Results

In relation to the FE model, Figure 11.23 shows the results of the damage localization based on the parameter reduction process. By means of a consecutively updating process, the FE model in the damaged condition is enhanced. With knowledge of the sensitive parameters, the extent of the damage is quantified. The results of calculation reproduce the technical reality of the damage to the bridge very well.

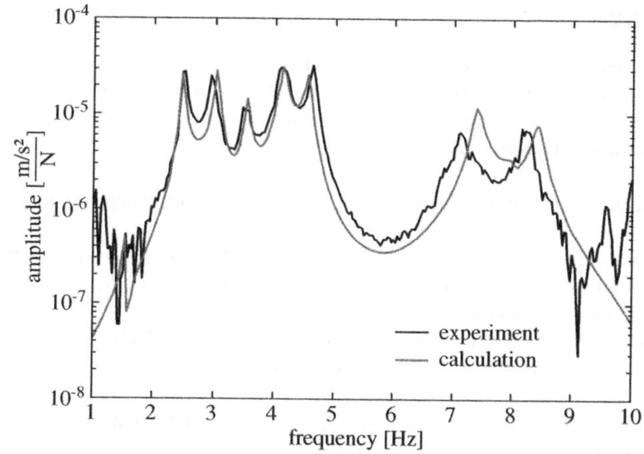

Figure 11.22 Frequency response function at measurement point N7

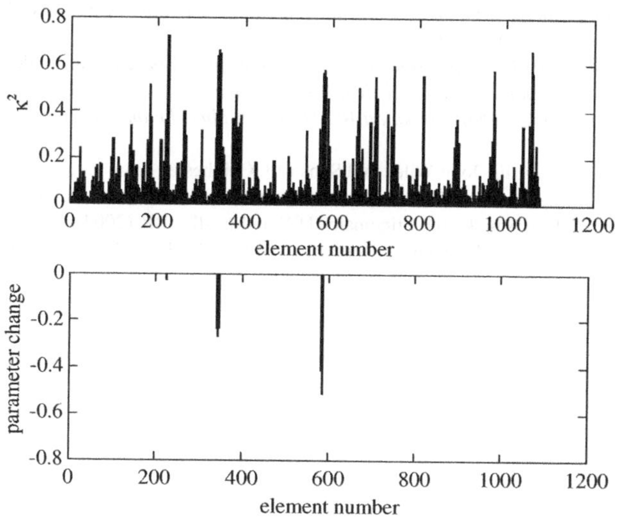

Figure 11.23 Damage indicator in correlation with model parameters (top) and damage localization results after parameter reduction (bottom)

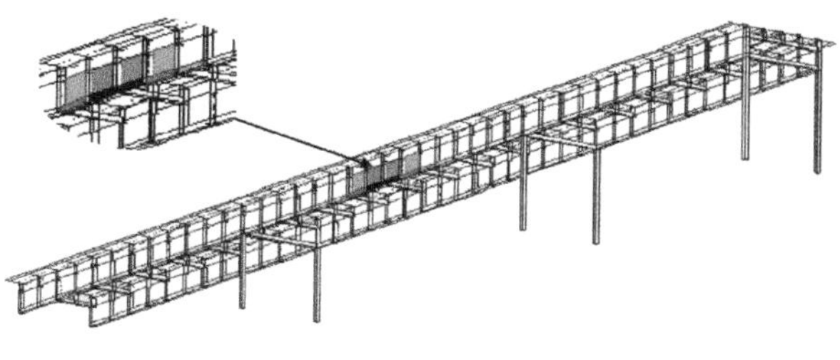

Figure 11.24 Detected damage and calculated changes of stiffness obtained by FEMU

Further Reading

1055-100 D (Undated) *Einwirkungen auf Tragwerke*. (In German.)

1076 D (Undated) *Ingenieurbauwerke im Zuge von Straen und Wegen – Überwachung und Prüfung*. (In German.)

13822 I (Undated) *Bases for Design of Structures – Assessment of Existing Structures*.

14963 I (Undated) *Mechanical Vibrations and Shock – Guidelines for Dynamic Tests and Investigations on Bridges and Viaducts*.

16587 I (Undated) *Mechanical Vibrations and Shock – Performance Parameters for Condition Monitoring of Structures*.

18430 I (Undated) *Condition Assessment of Structural Systems from Dynamic Response Measurement*.

18649 I (Undated) *Mechanical Vibrations – Evaluation of Results from Dynamic Tests and Investigations of Bridges*.

1990 E (Undated) *Basis of Structural Design*.

Balageas D, Fritzen CP and Gemes A (2006) *Structural Health Monitoring*. ISTE.

Carden EP and Fanning P (2004) Vibration based condition monitoring: a review. *Journal of Structual Health Monitoring* **3**(4), 355–377.

DAfStb-Richtlinie (Undated) *Belastungsversuche an Betonbauwerken*. (In German.)

Farrar C, Sohn H, Hemez F, et al. (2004) *Damage Prognosis: Current Status and Future Needs*. Technical Report LA-14051-MS, Los Alamos National Laboratory.

Farrar CR, Baker WE, Bell TM, et al. (1994) Dynamic Characterization and Damage Detection in the I-40 Bridge over the Rio Grande. Technical Report LA-12767-MS, Los Alamos National Laboratory.

fib (2003) *Monitoring and Safety Evaluation of Existing Concrete Structures*. State-of-Art Report, fib-bulletin 22, International Federation for Structural Concrete.

Fritzen CP and Bohle K (2000) Parameter selection strategies in model-based damage detection. *Proceedings of Second International Workshop on Structural Health Monitoring*.

Für zerstörungsfreie Prüfung DG (2000) *Merkblatt B9, Merkblatt über die automatisierte Dauerüberwachung*. (In German.)

Inman DJ (2005) *Damage Prognosis*. John Wiley and Sons Ltd, Chichester.

Mufti A (2001) *Guidelines for Structural Health Monitoring. Design Manual 2*. ISIS Canada Research Network.

Sohn H, Farrar CR, Hemez FM, Shunk DD, Stinemates DW and Nadler BR (2003) *A Review of Structural Health Monitoring Literature: 1996–2001*. Technical Report LA-13976-MS, Los Alamos National Laboratory.

12

Glossary and Derivation Criteria for SHM of Bridges

12.1 Glossary of Terms Frequently Used

Terms prefixed by ∗ are definitions according to ISO 2394 and ISO 13822.

Acceleration. The rate of change of velocity with respect to time. It is often the quantity most easily detected in vibration measurement. If the motion is ideally simple harmonic, the magnitude of acceleration is given by the amplitude of vibration multiplied by the circular frequency squared. Note that the mean of acceleration is supposed to be zero. The estimation of vibration amplitude from measured acceleration often involves significant errors.

Accelerometer. An instrument used for measuring the acceleration.
In particular, low-frequency, low-amplitude accelerometers are suitable for application in bridge dynamics. The common types of accelerometers for this application are piezo-resistive, capacitive, and force balance accelerometers. The accelerometer, in principle, is usually a high-frequency spring-mass system, in which the elastic spring is often made of a cantilevered beam of metal or ceramic material that bends under a given acceleration. The displacement is measured by strain-sensitive gauges placed on the beam, or detected by the change of electric capacitance. The gauges are usually connected in a Wheatstone bridge.

Accounted Truth. What is claimed to be true with some theoretical and/or experimental explanations that are objectively acceptable.
J.L. Austin, a British philosopher of language, says, ‘To assert that a proposition p is true is to maintain that p corresponds to the facts’. In the field of natural science, although they can be still subjected to interpretations, these ‘facts’ need to be scientific facts, which are expected to be explained by rigorous and robust theory or repeatable and replaceable experiments. Some other facts, for example from the religious point of view, may not be thus categorized.

Acoustic Emission (AE). The propagation of an elastic wave as the rapid release of energy associated with plastic deformation or development of defects within a stressed material.
Acoustic emission analysis is a useful method for the investigation of local damage in materials. It has been successfully applied to detecting and locating faults in pressure vessels or leakage in storage tanks or pipeline systems, monitoring welding applications, corrosion processes, partial discharges from components subjected to high voltage and the removal of protective coatings. Research and development

Health Monitoring of Bridges Helmut Wenzel

of AE applications currently includes monitoring of civil engineering structures, such as bridges, pipelines and offshore structures, and crack development in massive concrete structures and rocks.

Advanced Composite Materials (ACM). Materials which consist of a polymer matrix reinforced with high-strength fibers and, compared to other traditional materials, possess distinctly advantageous characteristics such as light weight and high strength.
Every composite has at least two components: reinforcements, which are high-strength, high-stiffness fibers and their surrounding matrix, which is usually a high-performance resin system that combines the reinforcement material together at a microscopic level. Three basic types of fiber reinforcement materials in use are carbon/graphite, glass fibers and aramid. Their major advantages in comparison to conventional materials include high strength and stiffness, light weight, fatigue strength, impact resistance and corrosion resistance. Major users of ACM were traditionally the aerospace industry but the market has been gradually expanding to include sporting goods and civil engineering applications. Carbon fiber reinforced polymer (CFRP) is now extensively applied to bridges for strengthening, reinforcement and repairs.

Aerodynamic Admittance Function. A transfer function to express how effectively the frequency characteristics of velocity fluctuation are picked up by the aerodynamic force components.
It is expected that the magnitude of this function is close to unity in the low-frequency range and quickly tapers off in higher frequencies. A classic example is the **Sears function**, which reflects the frequency characteristics of aerodynamic lift force in relation to a sinusoidal fluctuation of the vertical velocity component. In general, the aerodynamic admittance is not decided analytically and needs to be estimated experimentally.

Aerodynamic Instability. Dynamic instability of structures caused by aerodynamic forces.
A dynamic failure of aircraft wings caused by aeroelastic phenomena, called **flutter**, was a serious engineering concern from the early days of flight. Although the excitation mechanism was not exactly identical, the collapse of the old Tacoma Narrows Bridge in 1940 was often attributed to aerofoil flutter. Galloping instability of ice-covered power transmission lines is another example of aerodynamic instability. The term flutter is, strictly speaking, restricted to the classic flutter which is a coupled motion in bending and torsion of streamlined bodies, but it is also used rather loosely without a clear definition. It sometimes means the catastrophic structural vibration caused by fluid dynamic forces, which are coupled with the body motion.

Allowable Stress Design. A method to design structures such that allowable stresses are not exceeded when the structure is subjected to specified working loads.
Basically, an elastically computed stress from the combined nominal loads must be less than the material yield stress or the buckling stress divided by the safety factor.

Ambient Vibration Survey (AVS). A method to determine the dynamic characteristics of a structure by measurement of small vibrations, mostly microtremors, caused by existing disturbances such as earthquakes, wind and traffic, while the structure is in service.
In terms of data reliability, the forced vibration tests using shakers is probably the best method for the evaluation of dynamic characteristics of bridges. However, it usually requires a large effort, which is naturally costly, and could also mean an interruption of services. The ambient vibration survey, without any control on the input, is consequently an attractive alternative. This method is based on a few basic assumptions as follows:

(a) the input excitation is a broadband stochastic process which is adequately modelled by white-noise;
(b) the system characteristics are therefore well represented by the power spectral density function of dynamic response;
(c) the technique for measuring the dynamic response is sufficiently reliable;
(d) the data acquisition and analysis are also sufficiently reliable.

Hence, the reliability of this method is largely decided by these factors.

ARIMA Model. Autoregressive integrated moving average model.
It is one of the statistical forecasting techniques systemized by Box and Jenkins (1976). The ARIMA time series analysis uses lags and shifts in the historical data to uncover patterns and predict the future.

Artificial Intelligence (AI). Intelligence exhibited by any manufactured system. The term is often applied to general purpose computers which are expected to work on intelligent tasks that resemble to human activities.
Artificial Intelligence methods are often employed in cognitive science research, which explicitly tries to model subsystems of human cognition, whereas AI research seeks to build more useful machines. **Expert systems** and **neural networks** are two of the most common techniques used for applied artificial intelligence.

***Assessment. A set of activities performed to verify the reliability of an existing structure for future use.**

Averaged Normal Power Spectral Density (ANPSD). A method to identify all the possible natural frequencies participating in the vibration at a time by taking the average of all the normalized power spectral density functions obtained from the multipoint records.
The method was developed by Felber (1993) as a fast and effective way to identify many structural vibration modes participating in the measured ambient vibration. It is a convenient way to display the most significant frequencies at a single spot in a series of motions in a certain direction. However, it should be noted that not all the peaks identified in this method necessarily correspond to the natural frequencies.

Bayesian Statistics. A statistical method that handles all uncertainties by probability. It provides a different paradigm for both statistical inference and decision making from the conventional statistics.
The name is after Thomas Bayes (1702–61) but it may not be following particularly his idea. Bayes theorem states that the probability of A given B times the probability of B is equal to the joint probability of A and B, or

$$P(A \mid B) = \frac{P(A \cap B)}{P(B)} \tag{1}$$

The major difference between Bayesian statistics and other statistical methods is that the traditional statistics examine the probability of the data given a model or hypothesis, while Bayesian statistics examine the probability of a model given the data. This significantly enhances the power of statistical analysis. In particular, Bayesian methods make it possible to incorporate scientific hypothesis in the analysis by means of the prior distribution. It can be then applied to problems whose structure may be too complex for conventional methods to handle. The Bayesian paradigm is based on an interpretation of probability as a rational, conditional measure of uncertainty, which closely matches the sense of the word 'probability' in ordinary language.

There are three particularly important terms used in Bayesian statistics: the prior and posterior probabilities and likelihood. The **prior** is the observer's belief expressed by the probability $P(X)$ before any data D are observed. The **posterior** refers to the probability $P(X \mid D)$ after observed data have been taken into account. **Likelihood** is the probability with which D is expected to take place, that is the conditional probability $P(D \mid X)$ of the data given a particular model. Since the probability of D depends on the value of X, X is called the **parameter** of $P(D \mid X)$. Based on the Bayes theorem, the posterior probability is calculated as a product of the prior probability and likelihood divided by $P(D)$, which is called **evidence**. It also means that any events which were not observed are not involved in the computation. This is a key principle of Bayesian statistics: only what is actually observed is relevant in determining the probability that any particular model is true.

There has been a resurgence of Bayesian approaches in recent years and the methodology now plays a central role in many fields, from **expert systems, machine learning**, and **pattern recognition**, to applications in finance and law. The special situation, often met in scientific reporting and public decision making, where the only acceptable information is deduced from available documented data, is addressed by objective Bayesian methods, as a particular case. Thus it tells us, for example, the true likelihood of a person having an HIV infection if he tests positive, or that of person X being the murderer if his fingerprints turn up on the weapon.

Beating. A phenomenon where the magnitude of vibration varies with the differential frequency of the two component vibrations.
When two vibration components are present at the same time and same place and if their frequencies are very close to each other, their combined signal will vary in magnitude regularly with a rate equal to the difference of frequencies of these two components. This differential frequency is called the beat frequency.

Beaufort Scale. An empirical measure for the intensity of wind, based mainly on conditions of open sea waves.
The 12-scale measure was created by Sir Francis Beaufort, a British naval officer some 200 years ago. The scale has often been referred to as a relatively easy method to indicate strength of wind also for non-naval use, and is also loosely related to the mean wind speed.

Boundary Layer Wind Tunnel. A kind of wind tunnel which has a long test section to develop turbulent boundary layer flow as the simulated natural wind.
It is a means to simulate micrometeorological characteristics of natural wind at the model scale. The idea was originally developed by Martin Jensen, a Danish engineer, who experimentally arrived at the conclusion that "the phenomena induced by natural wind can be reproduced only when the model tests are performed in a boundary layer which was created in a similar way as the case of natural wind and also when the linear scale of its turbulence coincides with the linear scaling of other models placed in it." This scaling factor, h/z_0, is now called the **Jensen number**, where h is a linear dimension of the model, and z_0 is the roughness length of the flow. The idea of the boundary layer wind tunnel was extensively developed by A.G. Davenport and J.E. Cermak in the North America for large-scale industrial applications.

Bridge Management. A process and decision-making framework that covers maintenance and operation of bridges, in order to maintain the structural safety and serviceability of them in cost-effective ways, considering the changes in environment, public expectation and technological advances.

Buckling. An instability failure mode of a structure or its members under compression.
The critical buckling load is decided by the stiffness against deformation and the effective length of the member. Buckling can also take place in torsion, a combination of torsion and bending or only locally in a small portion of a structural member. The infamous collapse of the Quebec Bridge in 1907 is a well-known example of buckling failure in bridge engineering.

Buffeting. Dynamic excitation or induced structural response caused by wind turbulence, which inherently exists in natural wind. Buffeting also can be caused by turbulence that was created by the existence of upstream objects.
It is usually considered and analyzed as a forced vibration caused by time-dependent aerodynamic forces due to velocity fluctuation. Buffeting is a stochastic vibration that has a range of frequency components and vibration amplitude that is randomly fluctuating. The magnitude of vibration is generally greater at higher wind speed. Buffeting vibration is usually not catastrophic but long-time influence of it, such as fatigue damage, can be a serious engineering concern.

Bulk Modulus of Elasticity. Ratio of stress to change in volume of a material subjected to axial loading.
If the material is homogeneous and isotropic, the bulk modulus of elasticity K is related to Young's modulus E and Poisson's ratio ν by

$$K = \frac{E}{3(1 - 2\nu)} \tag{2}$$

Cable. A flexible metal wire or group of wires, used as a structural member against tensile force.
Cables employed as structural tension members are usually one of the following five types:

(a) a single piano wire;
(b) seven-wire strand;
(c) multiwire helical strand;
(d) parallel-wire strand;
(e) locked-coil strand.

A piano wire has much higher tensile strength in the order of 3000 MPa and much smaller ductility than ordinary structural steel. Young's modulus is around 205 GPa. It is usually assembled to form wire strands. The seven-wire strand consists of a single core wire and a single layer of six wires, all having the same pitch and direction of helix. The multiwire helical strands are fabricated by successive spinning of layers generally in opposite direction of a helix, whereas the parallel-wire strand has all wires straight and parallel so that there is no reduction of Young's modulus due to twist of wires. The parallel wires are suitable for main cables of long-span suspension bridges, for example. In the locked-coil strand there are a number of different shapes of wires to form a strand with a smoother and tighter surface. This type is considered to be appropriate for relatively short span cable-stayed bridges.

Cables are primarily assumed to be perfectly flexible and resist only against tension. However, in reality, there is a bending stiffness. Because of its large flexibility, cable often exhibits a relatively large static deflection and, as a result, nonlinear structural characteristics. Once in vibration, cable has very low structural damping, usually an order of magnitude less than other types of structural members.

Cepstral Analysis. A nonlinear filtering technique in signal processing, often applied in speech analysis, when it is difficult to do deconvolution by applying linear filtering.
In many of the **inverse problems** it is experienced that the transfer function constructed by applying Fourier transfer technique for deconvolution is possible only with some ambiguity. However, it has been recognized sometimes that the cepstral analysis technique is useful in smoothing or desensitizing the transfer function such that the reconstructed input/output data become clearer with minimum signal pollution. Mathematically, when the Z-transform of a signal x_n is given by $X(z)$, the cepstrum of x_n is defined by the inverse Z-transform of the logarithm of $X(z)$.

Charpy Test. A method to test the material toughness by applying an impact to break a standard specimen.
The specimen is broken by the impact of a heavy pendulum hammer, which strikes the specimen at a fixed speed. The amount of energy absorbed by the high strain rate fracture is indicated by the remaining potential energy, i.e. by the maximum height of the swing after the specimen is broken.

Comfort. A state of being relaxed and feeling no pain or worry.
The term is used in several different levels in the present context, such as in the **comfort criteria**, when the structural serviceability and/or pedestrian level wind environment are to be considered. Stages of categories are usually based on the clients' freedom from concerns regarding their safety, danger of bodily harm, sickness, inconvenience and nuisance. They are also influenced by the position people are taking, such as walking, standing, sitting, etc., personal health conditions and what they are wearing.

Convolution. A mathematical operator defined as the integral over all space of one function at x and another function at $x - u$.
The dynamic response of a structure due to external force is given by a convolution integral of the **impulse response function** and the force function. In many problems in physics, when there is a linear system with the principle of superposition a convolution appears.

Correlation. A measure to indicate how much two variables are statistically related.
Correlation is one of the most fundamental concepts in describing the statistical relationship between two signals. It can be the relationship between the excitation force and dynamic response, or between dynamic deflections at two different locations of the same structure, for example. As a special case, the correlation of a signal with the same signal itself, but with a given time interval in between, can be taken and in this case it is called the autocorrelation as opposed to the cross-correlation between two different signals. The correlation between two signals can be described in various forms with different degrees of sophistication, including the correlation functions, cross-spectral density functions and coherence.

Corrosion. Chemically induced damage to a material that results in deterioration of the material properties.
It is difficult to prevent corrosion totally but it needs to be minimized by proper choice of material and design, coatings and/or environmental control, since corrosion will eventually result in failure of the component. **Stress corrosion** is a type of failure mechanism that takes place particularly when material is under tensile stresses, which are often **residual stresses** in the material, above a certain threshold value, together with the environmental conditions. It is particularly sensitive to the temperature environment. The collapse of the Silver Bridge, a 680 m long eye-bar chain suspension bridge over the Ohio River, in 1967 is known to be the result of stress corrosion.

Coulomb Damping. Nonlinear damping as a result of dry frictional rubbing that often exists due to sliding at structural joints and supports.

Creep. A slow flow of metal under large normal stresses or high temperature.
As a transient stress-strain status, it is called **creep** if strain is increasing under the same stress, whereas if the stress is decreasing under the same strain, it is called **relaxation**. The method for determining creep or stress relaxation behavior is called creep test. Standard creep testing procedures are defined by the American Society for Testing and Materials standards.

Critical Damping. Magnitude of damping beyond which the energy dissipation is so large that vibratory motion no longer exists.
In ordinary structural vibration, the magnitude of damping is indicated by the fraction of critical. For many of the civil engineering structures, the overall structural damping is the order of 1% of critical.

Critical Reynolds Number. The Reynolds number at which the transition of separation shear flow takes place, resulting in sudden drop of drag force and less organized wake flow.
For the case of a circular cylinder against transverse flow, the transition is known to take place at $(Re)_{cr} \approx 3 \times 10^5$. However, in reality, the critical Reynolds number can only be indicated as a range of Re rather than a definite number, since it is easily influenced by many factors including the surface roughness of the cylinder and the level of flow turbulence. Experimentally, it is evidenced by a sudden drop of drag force on the body and loss of more regular flow pattern in its wake.

***Damage. Unfavorable change in the condition of a structure that can affect structural performance.**

Damage Detection. On-site, nondestructive identification of structural damage.
Periodic visual inspections provide a generally economical means for assessing the structural condition. However, they are inherently subjective so that the reliability of outcome is often less than desired. Also, most of the nonvisible degradation of the bridge would remain undetected by visual inspection alone. The SHM system should be inexpensive, noninvasive and automated, so that subjective differences by the operator can be avoided. In particular, it must be able to detect all the significant structural damage without any exceptions.

The detection of damage is expected to be at four levels:

(a) Is any damage present?
(b) Where is it located?
(c) How severe is the damage?
(d) What is the remaining service life of the structure?

Damping. The capacity of structures to dissipate energy imparted by the external forces.
The dissipation of dynamic energy during vibration results from many different sources, such as the imperfect elasticity and internal friction of structural materials, friction of structural members at their joints and support mechanisms, aerodynamic and hydrodynamic damping due to surrounding environment, the nonlinear structural characteristics, energy dissipation through foundation and substructures, and so on. In any of these, the theoretical evaluation of damping capacity is generally limited. For this reason, it is essentially important to consult the results of field experience as references.

Although the mechanism of damping is quite diverse, their overall effect on vibration is usually characterized by considering an equivalent viscous damping, represented in the single number of a **damping ratio** (ζ) as a fraction of critical. If the overall damping of the system is 1% of critical, for example, the free vibration amplitude will be reduced to a half after 11 cycles, whereas 10% damping will reduce the amplitude to a half at each cycle. When damping is at or beyond critical, there is no vibration.

Data Acquisition (DAQ). Sampling and processing of signals, usually manipulated by a computer, to obtain the desired information.
The components of data acquisition systems include appropriate sensors that convert any measurement parameters to electrical signals, which are acquired, displayed, analyzed and stored on a PC by interactive control software and hardware.

Data Mining. Extraction of potentially useful, previously unknown, information from large databases.
It is the practice of automatically searching large stores of data for patterns, which would not be recognized otherwise. In this sense, it is also known as **knowledge-discovery in databases**. Data mining grew as a result of very rapid developments in storing massive amounts of data and the necessity for applying statistical analyses and search techniques to them for **artificial intelligence**.

Decision Support System (DSS). An interactive computer-based information system that helps decision makers by compiling useful information from raw data, documents and other knowledge.
Started about a 50 years ago, DSS has developed as a powerful interactive concept with applications in any knowledge domain. It is now based on multidisciplinary foundations, including database research, artificial intelligence, human–computer interaction, simulation methods, software engineering and telecommunications.

Degrees-of-Freedom (DOF). The number of displacements for describing the characteristics of a given vibration.
The concept of DOF is applicable only in terms of mathematical modeling of vibration. The same vibration of a structure can be considered as multi-degree-of-freedom (MDOF) or it may be approximated as a single-degree-of-freedom (SDOF) system, depending on how the structure is conceptually modeled.

***Deterioration. Process that adversely affects the structural performance, including the reliability over time.**
Deterioration of structural performance can be caused by various reasons, such as:

(a) naturally occurring chemical, physical and biological actions;
(b) repeated actions such as those causing fatigue;
(c) normal or severe environmental influences;

(d) wear due to use;
(e) improper operation and maintenance of the structure.

Discrete Fourier Transform (DFT). Fourier transform of a discretely indexed series. Usually executed by the use of the fast Fourier transform algorithm, which is extremely efficient in computation.

Ductility. The ability of a material to plastically deform without rupture.
Ductility is usually defined by tension tests but may also be considered in bending. More ductile materials show larger deformation before fracture, and hence, given the same strength and hardness, the material with the higher ductility would be more desirable. The lack of ductility is often termed **brittleness**. The ductility of material may change if conditions are altered. A decrease in temperature tends to make the same material more brittle.

Dynamic Excitation. The extraneous sources which cause dynamic response of structures.
Civil engineering structures are usually designed to withstand the static loading, including dead load. However, in reality, the structures are often exposed to dynamic load as well. The main sources of dynamic excitation for bridge structures are: moving vehicles and pedestrians, wind, earthquakes, and possibly blast loading. It is standard design practice to cover these anticipated dynamic effects by either considering equivalent static forces or dynamic amplification factors, although sometimes more elaborate dynamic analyses would be required.

Earthquakes. Sudden movements of a part of the Earth's crust. They are often caused due to accumulated stress along the boundaries of crustal plates or geological faults, which are slowly moving at the geological timescale. Earthquakes release a large amount of strain energy, which radiates as seismic waves.
The Earth's crust, is up to about 50 km thick and a geological fracture may extend deep below ground level. The origin of the fracture is called the hypocenter and the point on the Earth's surface directly above the hypocenter is called the epicenter. In addition to movements, at plate boundaries, earthquakes are also caused by volcanic activity. Large earthquakes often result in catastrophic consequences to the infrastructure.

The magnitude of an earthquake is related to the amount of energy released by the geological rupture causing it, and is measured by the **Richter scale**. The intensity of an earthquake, on the other hand, is a measure of the observed damage at a particular location, and is often given in the **Modified Mercalli Intensity scale (MMI)**.

Eigenfrequency. Also called the natural frequency.

Eigenmode. Characteristic shape of amplitude distribution when a structure is freely vibrating at one of its natural frequencies. Also simply called the vibration mode.

Elastic Hysteresis. Difference between strain energy required to generate a given stress in a material and elastic energy at that stress.
Hysteresis is caused when the dynamic strain of the system does not instantly follow the applied stress, resulting in the stress–strain curve making a loop for each stress cycle. Since the area under the curve corresponds to the **strain energy**, the area surrounded by the clock-wise loop is the energy dissipated as heat in a material in one cycle of vibration. Elastic hysteresis divided by elastic deformation energy is equal to damping capacity.

Elasticity. Ability of a material to return to its original shape when the applied load that caused deformation is removed.
Elasticity is a concept opposed to **plasticity**, which is a tendency to remain deformed. When a material is linearly elastic, the slope of the straight line portion of a stress–strain diagram is called the modulus,

or coefficient, of elasticity. Since both stress and strain have the normal and shearing components in all three directions, in general, there can be 21 moduli of elasticity for any linearly elastic material. When the material is homogeneous and isotropic, because of symmetry, the number of elastic moduli is reduced to two.

Elongation. A ratio defined in a tensile test by the increase in gauge length measured after rupture to the original gauge length.
It is an important measure of the material's ductility and expected to be 35% or so for the case of ordinary mild steel. Elongation cannot be used to predict behavior of materials subjected to sudden or repeated loading.

Environmental Noise. Unwanted sound that is loud, unpleasant, or unexpected.
It is disturbing, annoying, and even causes health or psychological problems to different extents, such as hearing damage, sleep disorder, high blood pressure or devated sense of frustration. It can come from a variety of sources such as factories, construction projects, vehicle traffic and aviation noise. Philosophically, one difficult aspect of acoustic noise control is that the definition of noise itself is quite subjective; some sounds are considered noise by some but not by others. Even in music, there is rarely a consensus regarding what should be called noise amongst those who are involved. On the other hand, noise control technology in reality is becoming a more and more serious engineering topic. Reduction of noise levels or controlling the airborne and structure-borne noise by the use of curtains and barriers, damping with absorbent materials, enclosure of sources or by isolation of vibrating structural elements are possible means often considered.

Expert System. A software-based artificial intelligence system which analyzes information and upgrades the quality and quantity of databases for specified purposes.
The primary goal of expert systems is to make knowledge-based artificial intelligence that is available to decision makers. A major feature of the system is a reliance on the database comparable to that of human experts, whose knowledge is based on a theoretical understanding of the problem and a collection of heuristic problem-solving rules obtained through professional experience.

Extensometer. A strain gauge. An instrument for measuring changes in linear dimensions.

Extreme Value Distributions. The limiting distributions for the extreme values, such as the maxima or minima, of a large collection of random observations.
In many civil engineering applications, concern often lies with the largest values of many events. This means that our attention is focused upon the upper tail of the parent distribution of actual observations. Fisher and Tippett (1928) proved that there are only three forms of extreme value distributions. Extreme value analysis was largely developed and elaborated by Emil J. Gumbel (1891–1966), a German statistician.

Failure. The state or condition of a structure or its component that becomes unable to function for expected services.
Structural failure could occur because of various reasons, such as: (a) yielding; (b) fatigue failure; (c) corrosion failure; (d) ductile or brittle fracture; and (e) creep rupture; or even too much deflection elastically, if the design was not appropriate or external loads exceed the expected magnitude.

Fast Fourier Transform (FFT). A highly efficient algorithm to compute the discrete Fourier transform (DFT) in high speed.
The method developed by Cooley and Tukey (in 1965) has been commonly used. However, some other algorithms are also known.

Fatigue. The adverse effect on metal of repeated cycles of stress.
Fatigue fracture is said to start with microcracks or defects, which cause a localized **stress concentration**, resulting in growth or propagation of them but without any appreciable deformation of the structure. Repetition of this process would cause decreased toughness, impact strength and tensile elongation and eventually failure of the material at considerably lower stress level than the original tensile strength.

When the number of cycles-to-failure (N) is tested and plotted against the given constant stress level (S), it is called the **S–N curve** or **Wöhler curve**, which allows designers to make a basic estimate of the expected life of the structural part against expected stresses.

Fatigue failure can be influenced by a number of factors including the level of stresses, number of cycles, size and shape of the structural component, condition of the surface and operating environment. There have been some infamous accidents where the cause was attributed to fatigue failures, including the 1842 railway disaster in Versailles, crashes of three de Havilland Comet jets in 1954, and the loss of Japan Airline's flight 123 in 1985.

Fiber Optic Sensors. Highly sensitive sensors by the use of optical fibers.
When an optical fiber is bent, the light in the core no longer meets the cladding at an angle equal to or greater than the critical angle. This means that light escapes into the cladding and does not reach the other end of the fiber. It is called the **microbending loss** and the more the fiber is bent, the more loss takes place. The optical fiber thus works as a transducer by converting a measured quantity into a corresponding change in the optical radiation. Since light is characterized by intensity, phase, frequency and polarization, a change of any one or more of these parameters can be used for the detection of various quantities, such as temperature, stress and strain, angle of rotation or electromagnetic currents. Some of the advantages of fiber optic sensors, on top of high sensitivity, are freedom from electromagnetic interference, wide bandwidth, compactness, geometric versatility and economy.

Fiber Stress. Stress through a point in a part in which stress distribution is not uniform, such as the maximum stress in both tension and compression at extreme surfaces for the case of beam bending.

Filter. An electronic device or mathematical algorithm to process a data stream by means of separating the frequency components of signals.
There are various types of filters, such as low-pass filters, high-pass filters and band-pass filters. For the monitored data, low-pass filters are used to cut-off high frequency noise and to prevent aliasing, whereas the high-pass filters are used to reject low-frequency noise such as the shift of zero reading.

Finite Element Method. A numerical method for solving static and dynamic problems of structures. It has been extensively applied to a variety of other engineering problems including fluid dynamics, heat transfer and material sciences.
The method was originally developed for working on dynamic analysis of aircraft structures. The basic idea is to subdivide a structure of any shape into a large number of simple elements. It is found that if the load-displacement equations for a single element are derived in matrix form, it is possible to use matrix algebra to combine the interacting effects of all the elements in a systematic and conceptually straightforward manner. Taking advantage of the rapid development of large capacity computers, the structural analysis thus became extremely simple and efficient.

Force Balance Accelerometer. A type of accelerometer which is widely used for the measurement of strong earthquake motions. It is based on the force-balance principle.
Instead of directly measuring the inertia force exerted upon the mass by detecting its displacement, the force-balance system measures the compensated inertia force, which is generated by an electromagnetic force transducer, so that the mass of the accelerometer moves as little as possible. The system is particularly suitable for the measurement in low-frequency range. The available frequency bandwidth is, however, somewhat limited because of the phase delays caused by the feedback servo loop.

Fourier Series. An infinite series of sine and cosine functions that can, if convergent, approximate a variety of periodic functions.
Fourier series can be further transformed to a series of exponential functions by the use of the Euler's formula. The study of functions given by Fourier series is called ***Fourier analysis*** or ***harmonic analysis***.

Fourier Transform. An integral transform particularly useful in relating the time domain and frequency domain variables for random vibration analysis.
Concept of the Fourier transform developed from a Fourier series, which is the decomposition of a periodic function. Fourier transform expands this concept to make it further applicable to the nonperiodic functions. The analysis of measured data using the **discrete Fourier transform (DFT)**, particularly in the form of the **fast Fourier transform (FFT)**, is one of the most important techniques for the frequency domain vibration analysis.

Fracture Toughness. Ability of a material to resist crack propagation when subjected to shock load in an impact test.

Frequency Domain Analysis (FDA). Dynamic analysis processed by taking frequency, rather than time, as an independent variable.
Vibration analysis was originally established by taking both the external forces and resultant displacements as functions of time as an independent variable. However, when the random vibrations came into the scope and the statistical treatment of the problem became imminent, another way to look at the problem by taking frequency as a variable was found to be an attractive choice.
Frequency domain analysis, as opposed to the **time domain analysis (TDA)**, is a way of processing dynamic analysis by decomposing the external forces into frequency components by applying Fourier transform and evaluating the structural response by superposing its frequency components. Even a nonperiodic force can be included by pretending its period to be infinity.

Frequency Response Function (FRF). The ratio of the induced response to the excitation force, when an idealized SDOF system is subjected to a simple harmonic fluctuation force.
The FRF, sometimes called the **mechanical admittance function**, shows the sensitivity of a structural system to the excitation frequency. The peak response is reached when the excitation frequency coincides with the natural frequency and its magnitude is inversely proportional to the total damping of the system.

Froude Number. A dimensionless parameter that is decided by the ratio of fluid inertia force to vertical force due to gravity and/or buoyancy.
Froude number is defined by

$$Fr = \frac{V}{\sqrt{gL}} \tag{3}$$

where V, L are the flow speed and a representative length, respectively, and g is the acceleration due to gravity. Hence, it is the square-root of the ratio of the fluid inertia force to the gravity force. If the change of fluid density ρ is involved in the problem, the densimetric Froude number is defined by using the reduced gravity,

$$g' = g \cdot \left(1 - \frac{\Delta\rho}{\rho}\right) \tag{4}$$

in lieu of g. It becomes an important parameter for describing the cases, such as dissipation of airborne particles or wind-induced response of cable-supported structures, where gravity is a dominant factor.

Fuzzy Logic. A problem-solving control system by introducing the concept of partial truth rather than expressing everything in binary terms.
The concept was introduced by L. Zadeh in 1965. Degrees of truth are often confused with being imprecise or probable but they are not. It allows for set membership values between and including 0 and 1, based on vague or even missing information, but arrives at a definite conclusion. Fuzzy logic is, by some engineers, still considered to be controversial but has been applied to many practical purposes including the fields of artificial intelligence, neural network and pattern recognition.

Global Positioning System (GPS). A world-wide radio-navigation system formed from a constellation of satellites and their ground stations. The system uses the network of reference points to calculate any position on the ground accurately.
The GPS provides a means of measuring position or displacement that does not require the component that is being tested to be physically connected to a fixed reference location. The working principle of the GPS involves triangulation of the location using radio signals from satellites as reference points. Depending upon the direction of measurement, GPS accuracy ranges from metres to centimetres.

Gust Factor. The ratio of the peak to the mean wind speed.
The same term is often used for the ratio of the peak to the mean of wind-induced dynamic response as well. Wind-induced buffeting motion often has a **peak factor** of 3.5 to 4 and a coefficient of variation of about 0.3, resulting in a gust factor of approximately 2 or slightly higher than 2.

Hardness. Measure of a material's resistance to localized plastic deformation.
Most hardness tests involve indentation, but hardness also may be reported as resistance to scratching or rebound of a projectile bounced off the material. It is a good indication of tensile and wear properties of a material.

Health Monitoring. Tracking of various aspects of a structure's performance and integrity in relation to the system's expected safety and serviceability.
It is desirable if the SHM system is inexpensive, noninvasive and also automated, so that subjective input by an operator can be avoided. In particular, neither the implementation nor operation of the system should involve closure of the bridge. Carden and Fanning (2004) list the attempted levels of structural identification as follows:

(a) presence of damage in the structure;
(b) location of the damage;
(c) severity of the damage;
(d) prediction of the remaining service life of the structure.

Impact Energy. Energy required to fracture a material that is subjected to shock loading as in an impact test.
Alternate terms are impact value, impact strength and impact resistance, and is usually measured by the energy absorbed in breaking the specimen in a single blow, as in the **Charpy** test. It is an indication of the toughness of the material.

Impulse Response Function (IRF). The ratio of the induced response to the excitation force, when an idealized SDOF system is subjected to a unit impulse load.
The **convolution** integral of IRF with an external force function, if available analytically, is called the **Duhamel Integral**, which gives the induced response of the system by this force. Frequency response function and impulse response function make a Fourier transform pair.

Indicial Response Function. The response of an idealized quiescent SDOF system when it is subjected to a unit step load. Impulse response is obtained by differentiating the indicial response.

Infrastructure. A set of interconnected structural elements that provide the framework for supporting the entire structure.
The term is often used in a very broad range, as the **social infrastructure**, including any life-sustaining social facilities required for municipal or public services, particularly for transportation systems such as roads, railways, airports and water surface transportation, public utilities such as flood control, fire services and waste management, emergency and security services, and even public education, health systems and social welfare.

***Inspection. On-site, nondestructive examination to establish the present conditions of the structure.**

A visible inspection performed on a regular basis is called a routine inspection and a more detailed inspection is usually performed as a follow-up to a routine inspection to identify any deficiencies discovered is called the in-depth inspection.

Instability. A failure mode of a structure by losing its structural integrity in static or dynamic equilibrium.

Buckling is a static instability. There are also dynamic instability phenomena such as parametric excitation and/or self-excited vibrations.

Intensity. The relative strength of a physical quantity, such as electricity, light, heat or sound, usually per unit area or volume of the space over which it is exposed.

It is a measure of the time-averaged energy flux transmitted. Intensity of signal fluctuation is often expressed by the coefficient of variation, which is the ratio of the root-mean-square to the mean value.

Inverse Problem. A task to identify system parameters based on observed output data.

Estimation of structural parameters from the measured vibration record is an example. It is exactly the inverse process compared to conventional response calculations in structural dynamics. Inverse problems exist in many disciplines such as remote sensing, medical imaging and nondestructive testing of materials. A linear inverse problem is essentially expressed in the form of a Fredholm first kind integral equation and at least the idea is straightforward, but nonlinear problems are considerably more complex.

***Investigation. Collection and evaluation of information through inspection, document research, load testing and other experimental methods.**

Jensen Number. See the Boundary Layer Wind Tunnel.

Kalman Filter. An efficient recursive filter which estimates the state of a dynamic system from a series of incomplete measurements with noise.

It means that only the estimated state from the previous time-step and the current measurement are needed for making an estimate of the current state. One example is to provide accurate, continuously updated information about the position and velocity of an object, given only a sequence of observations about its position, each of which includes some error. Kalman filtering is an important topic in control systems engineering. A variety of Kalman filters has now been developed, including Kalman's original simple filter, the extended filter, the information filter, and a variety of square-root filters.

Kármán Vortices. A street of alternating vortices existing behind a circular cylinder which is exposed to the transverse flow. Also called the Kármán–Bénard vortex street.

Existence of a street of alternating vortices behind a circular cylinder which is exposed to the transverse flow was known for quite some time and studied by many including Strouhal, Lord Rayleigh and Bénard. However, the name of von Kármán has been frequently remembered because of his work on the stability of vortex arrangement. He considered two rows of alternating vortices behind a cylinder and mathematically considered a stability condition in terms of their geometric locations.

A very clear vortex street is usually recognizable at the Reynolds number range of 100–300 and these are the original Kármán vortices. However, even at much higher flow speed, in the range beyond the critical Reynolds number, the existence of alternating vortex shedding behind a non-streamlined object has been widely recognized and they are also rather loosely called Kármán vortices. An engineering significance of this vortex shedding is the **vortex shedding excitation** of structures.

Laser Doppler Vibrometer (LDV). A highly sensitive optical instrument to measure displacement and velocity of a moving object.

It consists of an optical head that emits laser light and a converter that processes the Doppler frequency of the reflected laser light. The voltage signal from the converter is proportional to the velocity at which the object moves. There are different types of LTD for: (a) out-of-plane vibration; (b) in-plane vibration; and (c) three-dimensional vibration. A non-contact way of measurement is a distinct advantage of the system, but care should be taken for the fact that the measured results are easily affected by the noise caused by the surface roughness of the target.

Least-Square Method. A process to quantitatively obtain a regression for a set of data by minimizing the sum of the deviations squared.
Polynomial functions are most commonly used for least-square curve-fitting. If more than one control parameters is involved, the multiple regression method can be applied.

Life-Cycle. The total phases through which a structure passes from its birth to the time it ceases to exist. It involves all levels of engineering work, including design, construction, inspection, management, repair, improvement and demolition.
The concept has been developed from the need to consider the overall performance of a structure, both in-service and associated costs. The life-cycle assessment of a structure would include its cost-effectiveness, serviceability, environmental impacts and sustainability.

Limit State. A limit state is a set of performance criteria defined in terms of deflection, vibration levels, etc., as a reference state that must be met by a structure under factored loading.
It is a concept used for design of structures in lieu of the older concept of **allowable stress design**. The most common limit states are the serviceability limit state and the ultimate limit state.

Linear Variable Differential Transformer (LVDT). A type of electrical transformer used for measuring linear displacement without contacts.
The LVDT is a commonly used, reliable position meter. It consists of a hollow metallic casing with solenoid coils around a tube and a ferromagnetic core shaft, which is attached to the object whose position is to be measured and moves freely back and forth along the axis of measurement. An alternating current is driven through the primary coil, causing a voltage to be induced in the solenoid coils and thus measures the traveling distance of the core.

Load. The applied forces which a structure is subjected to and expected to resist against.
It generally includes the weight of the structure itself, traffic to be carried, effects of wind, earthquakes, temperature change, rain, snow, ice, etc., and dynamic amplification of these loads due to their motion, possible collision and/or accidents.

***Load Testing. Test of the structure or its part by loading to evaluate its behavior or properties, or to predict its load-bearing capacity.**

***Maintenance. Routine intervention to preserve appropriate structural performance.**

Markov Process. A statistical process whose relationship to the past does not extend beyond the immediately preceding observation, or its most recent value.
It also could be called the random events whose likelihood depends on what happened last in this process. The stream of events with this character is called the **Markov chain**. A game of Monopoly or Snakes and Ladders, whose moves are determined entirely by dice, is a Markov chain, in contrast to the Poker game, where the cards represent a memory of the past moves.

Mass Parameter. A dimensionless parameter to indicate the significance of structural mass.
Mass parameter can be defined in different ways depending on the nature of engineering problems. However, in the case of the wind-tunnel tests of a bridge model, for example, it is defined by $\mu = m/(\rho B^2)$, where m is structural mass per unit length of the bridge, ρ is the air density, and B stands for the width of the bridge. If the issue is the bridge motion in torsion, another mass parameter, $\nu = J/(\rho B^4)$, would be more important, where J is equal to the polar mass moment of inertia per unit length of the bridge.

***Material Properties. Mechanical, physical or chemical properties of structural materials.**
Engineering properties sometimes required in the present context include the material density, yield strength, elongation, tensile strength, Young's modulus, Poisson's ratio, melting point, thermal expansion coefficient, specific heat capacity, electrical conductivity, etc.

MATLAB. A numerical computing environment, which is probably most widely used as a modern operating system. The same name can also mean its core language.
Short form of **MATrix LABoratory**. It was developed by Cleve Moler in the late 1970s and quickly spread and became popular through the community of applied mathematics. It allows easy matrix manipulation,

plotting of functions and data, implementation of algorithms, creation of user interfaces, and interfacing with programs in other languages.

Maximum Likelihood Method. A procedure of finding the value of one or more parameters for the hypothetical probability distribution, or likelihood, to make it a maximum.
It is said to be a statistically robust method, versatile and applicable to most models and to different types of data. In addition, it provides efficient methods for quantifying uncertainty through confidence bounds. The background theory is relatively simple. Suppose x is a random variable with PDF given by $p(x;\theta)$, where θ is an unknown parameter which needs to be estimated. After obtaining N independent observations $\{x_r\}$ $(r = 1, 2, \ldots, N)$, the likely function is defined by

$$L = \prod_{r=1}^{N} p(x_r;\theta) \tag{5}$$

and the maximum likelihood estimator of θ is obtained by maximizing $L(x_1, x_2, \ldots, x_N \mid \theta)$, or $\partial \ln(L)/\partial\theta = 0$. There can be more than one θ involved in this process.

Melting Point. The temperature at which a solid changes its state to liquid.
The temperature of the reverse change, liquid to solid, is referred to as the **freezing point**. Unlike the boiling point, the melting point is relatively insensitive to pressure. The melting point of steel is approximately 1700 K.

Miner's Rule. A method to predict structural failure due to cumulative fatigue damage.
Fatigue loading is seldom of constant amplitude and hence the method of its assessment for the cumulative damage needs to consider the mean rate of crack propagation under variable-amplitude loading. This approach yields to an empirically derived expression by Palmgren (in 1924) and Miner (in 1945), which is given by $\sum_j n_j/N_j \geq 1$ as a failure condition, where n_j is the number of stress cycles with stress range σ_j and N_j is the number of stress cycles necessary to cause failure at stress range σ_j. It is a simple and convenient criterion but the usefulness of the Miner's rule is admittedly questionable.

Modal Mass. Participating mass.

Modal Parameters. The most fundamental information regarding vibration modes, namely the frequency of vibration, corresponding mode shape and damping.
The ultimate goal of **vibration-based structural health monitoring** is to determine the existence, location and extent of structural damage by identifying these modal parameters. Two key areas where the research efforts are required are related to the following:

(a) whether the measurement of these parameters yields consistently reliable results;
(b) whether the observed parameters are a sensible reflection of structural damage that needs to be identified.

Often the measured **modal data** are influenced not only by possible structural damage but also other environmental factors such as the live load conditions, thermally induced variations and amplitude dependence. Sometimes the results are limited by the availability of instrumentation, and error and noise in measurement and data analysis. Thus the measured modal data are either incomplete or with remaining uncertainties. It becomes necessary then to address this uncertainty by applying probabilistic approaches.

Modified Mercalli Intensity Scale. A widely accepted scale to indicate the intensity of an earthquake, which is a measure of the observed damage at a particular location.
It is based on a subjective assessment of the severity of an earthquake. The intensity varies with distance from the epicenter and local ground conditions. It should be remembered that the evaluated intensity is not necessarily proportional to the magnitude scale. It is analogous to the **Beaufort scale** for wind in that sense.

Moment Magnitude Scale. A method to indicate the magnitude of earthquakes as a possible replacement of the Richter scale. It is a measure of the seismic moment, which is relatable to the dimension of earthquake rupture and released energy.
The method was proposed by Hanks and Kanamori (1979) and defines the magnitude by

$$M = \frac{2}{3} \cdot \left[\log_{10}(M_0) - 9.1\right] \tag{6}$$

where M_0 is the seismic moment in Nm. The seismic moment is a measure of the total energy that is transformed during an earthquake and only a very small fraction, typically 1.6×10^{-5}, is said to be converted into radiated seismic energy and registered on seismographs.

***Monitoring. Frequent or continuous observation or measurement of structural conditions or actions.**

Monte Carlo Method. A method to simulate a large variety of qualitative processes or to provide approximate solutions to mathematical problems by performing statistical sampling experiments on a computer by the use of random number generation.
The Monte Carlo method developed out of nuclear science, when the probabilistic problem of random neutron diffusion in fissile materials was a concern. The method is often envisaged as a statistical sampling technique to solve inherently probabilistic problems but it also has been applied to deterministic issues. In the present context, it is referred to as a method to make use of random generation for the numerical simulation of functional relationships between random variables. It is known to be a useful and effective approach when the theoretical analysis of the input–output relationship is too complicated, particularly in nonlinear problems.

Natural Frequency. The frequency at which a structure is most easily excited.
Natural frequencies are uniquely decided by mass and stiffness of the structure and are rightfully termed the **eigenfrequencies**. However, their magnitude also depends upon the way the structure vibrates, which is called the **mode** of vibration. For example, a bridge can vibrate in vertical bending, horizontal bending or in torsion. Even in vertical bending mode alone, the bridge can vibrate with its maximum dynamic deflection at the span centre, or with no deflection at the midspan but the significant movement at the quarter-span points. Each vibration mode has its own natural frequency corresponding to it.

The mode of vibration is largely influenced by the way the structure is supported. The supporting conditions are called the **boundary conditions**. For the design of the structure, they are usually assumed to be hinges, rollers, rigidly fixed, or sometimes elastic spring supports. However, these conditions, including the conditions of substructure and the ground, in reality are often somewhat different from their mathematical assumptions.

Natural frequencies are usually evaluated under the standard design conditions, in which the structure is free of live loads and extreme temperature effects. Since the instantaneous mass and stiffness of the structure in reality could be different from the design assumptions, the natural frequencies of the bridge in service can be different from the values calculated earlier.

Natural frequency is reduced a little with the increase of the system's **damping**. However, in reality, this effect can be disregarded for the case of civil engineering structures. For example, even if the damping is as high as 10% of critical, the natural frequency is reduced by only 0.5% compared to the case with no damping.

The structure can be most easily excited into vibration if the excitation frequency coincides with, or is very close to, one of the natural frequencies. This phenomenon is called **resonance**. If the excitation force contains many frequencies, the one close to the natural frequency would most effectively excite the structure. The sensitivity of a structure to different excitation frequencies is typically represented by the **frequency response function (FRF)**, or the **mechanical admittance function**.

When the structural system is *nonlinear*, the natural frequencies are amplitude-dependent. For this case, the frequency observed at very small amplitudes is usually defined as the natural frequency.

Neural Network. An analytical technique to explore the relationship between variables by creating a network system conceptually similar to the biological cognitive system.
The neural networks method is one of the data mining techniques, which is an analytical process to explore data in search of consistent patterns and/or systematic relationships between variables. The name obviously came from its resemblance to the biological cognitive system of the brain and layers of neurons. The method consists basically of three steps, the design of a specific object-oriented network architecture, the training of the system by the use of existing data, and the generation of predictions once the network is ready.

A distinct advantage of neural networks is that the method can be applied to any continuous input–output relationship without assuming hypotheses particular to the underlying model. On the other hand, there is an important disadvantage that the final outcome of the work depends upon the initial conditions of the network and the experience-based solution does not really give any insight to theoretical explanation of the physics.

Noise. A random signal that does not convey any useful information.
The analysis of noise signals has been developed in the telecommunication field and became one of the most important tools in dynamics. A signal whose intensity is the same at all frequencies is called the **white noise**, although an infinite-bandwidth white noise is a purely theoretical concept and, in reality, its frequency band has to be limited. For an acoustic aspect of noise, see **environmental noise**.

Nonlinear Vibration. When the structural restoring force and/or damping force are not linearly proportional to the displacement, or the external forces are amplitude dependent, the vibration becomes nonlinear.
Nonlinear stiffness can be caused by material nonlinearity, such as plasticity, or structural nonlinearity, such as the case of cables. The structural damping ratio is usually higher when vibration amplitude is greater. In contrast, the aerodynamic damping is sometimes found drastically reduced, or can even become negative, with greater vibration amplitude. This is because the excitation force is largely nonlinear with the displacement and/or displacement rate. All of these factors indicate that the structural vibration is likely to be nonlinear, unless the dynamic amplitude is very small. However, we try to handle them with linear approximation so long as it is acceptable for practical purposes. Generally speaking, the nonlinear vibrations can be analyzed only by applying numerical methods without closed-form solutions.

Nyquist Frequency. The highest frequency beyond which the signal contents cannot be properly represented by the data in discrete form.
Any time signal can be measured usually only for a limited time period. Consequently, the analyzed results of the sampled data from the original signal would be different from the expected results for which the whole infinite signal is intended. This difference is known as **leakage**. When the original continuous signal is sampled as a discrete time-series usually with a constant sampling time increment Δt, the signal contents with any frequencies higher than $f_N = 1/(2\Delta t)$ cannot be accurately represented because of the sampling resolution. f_N is called the Nyquist (or folding) frequency.

The problem caused by the choice of sampling time increment for digitizing the data is called **aliasing**. In order to reduce these errors, it is considered prudent to low-pass the original signal at, or preferably even below, the Nyquist frequency, before analysis.

Offset. A reading that is other than zero for a zero condition. Every reading thereafter is inaccurate by this amount, for which compensation is required.

Parameter Estimation. The process of finding parameter values that fit a mathematical model to experimental data.
Other terms such as **system identification** can be used for the present context. There are various heuristics developed for parametric estimation particularly in the field of control engineering, such as **maximum likelihood, predictive estimation** and **Kalman filtering**.

Parametric Excitation. A type of self-excited vibration where the stiffness term is time-dependent, resulting in the instability of the system depending on the combination of the system parameters.
Lateral vibration of struts and cables due to periodic fluctuation of axial force are two of the examples often referred to. The equation of motion takes the form of Mathieu's equation, whose solutions are known to become unstable under certain combinations of parameters.

Participating Mass. A part of the physical mass which is actually contributing to vibration. Also called the modal mass, since its magnitude depends upon the vibration mode.
Consider the sway vibration of an elevated mass, supported by an elastic column. If the structure vibrates in the fundamental mode of vibration, the heavy mass will vibrate with a large amplitude. However, when the structure vibrates in its second mode, the large amplitude is experienced by the thin column rather than the top mass. Hence, the effective mass contribution will be more substantial in the fundamental mode rather than the second mode, even if the physical mass distribution of the structure remains the same. The modal mass is calculated by the integration of mass per unit length times the mode function squared, over the whole structure, and it is unique to each mode.

Pattern Recognition. The act of taking in raw data, extracting meaningful information to form the feature vectors out of them, and classifying the measured data into categories based on characterization of patterns.
It is a field within the area of machine learning or **artificial intelligence**. It has been applied typically to automatic speech recognition, classification of spam and nonspam email messages, and machine reading of hand-written postal codes on postal envelopes. Effective application of these techniques to structural parameter estimation is highly desired. The most fundamental tasks in pattern recognition are (a) to learn the probabilistic relationship between the feature vectors and presumed categories; and (b) to make an inference as to which categories the recognized data should belong to, based on the Bayesian decision theory. Also developed in recent years is the application of **neural network** to pattern recognition as an effective tool for learning process.

Peak Counting. A method of cycle counting to produce a statistical summary out of irregular time-histories by counting the number of peaks and valleys.
The valuable outcomes of field observation of structural behavior are usually given as lengthy and irregularly fluctuating time-histories of acceleration, stress, deflection and so on. An absolutely essential matter for engineers then is to produce a small amount of useful information out of them for fatigue analysis, for example. Counting the number of cycles of the record fluctuation is one of them. There are a number of methods to perform this operation, such as counting of **peaks, level crossings** and **ranges**, which are defined by the difference between two successive extremes. All of these are called **one-parameter methods** whereas a two-parameter method called **rainflow analysis** is known to be a state-of-the-art counting method successfully applied to fatigue analysis. Note that the cycle counting yields amplitude distribution with no regard to frequency information.

Peak Factor. The ratio of dynamic peak to its root-mean-square value.
Not to be confused with the **gust factor**. If the dynamic signal is a simple harmonic fluctuation, the peak factor should be $\sqrt{2}$. If the signal has the normal distribution of its magnitude, the peak factor is known to be approximately 3.6. Some peculiar random signals have very high peak factors. Wind-induced suction, for example, on a tall building sometimes shows a peak factor of even higher than 10.

Piezoelectric Accelerometer. A type of accelerometer that uses solid-state strain gauge elements that are physically attached to cantilever beams and electrically connected to a Wheatstone bridge circuit.

Plasticity. Tendency of a material to remain deformed, after reduction of the deforming stress to a value equal to or less than its yield strength.

Poisson's ratio. The ratio of transverse contraction strain to longitudinal extension strain in the direction of stretching force.

Virtually all common materials have Poisson's ratios in the range from 0 (such as cork) to 0.5 (such as rubber). Poisson's ratio of steel is approximately 0.29, resulting in the ratio of Young's modulus to shear modulus to be 1.55.

Power Spectral Density (PSD). A function to represent a random process by the distribution of dynamic energy in terms of its frequency components.
Power Spectral Density is a statistical function consisting of the average squared moduli of the Fourier transform at each frequency. It represents the random characteristics of the process in the frequency domain. The ordinate of the function corresponds to the intensity of energy at that particular frequency and integration of PSD over the whole frequency range gives the variance of the process. Hence, the PSD function divided by the variance is called the **normalized spectral density**.

When the structural system and excitation force are both linear and the principle of superposition is applicable, the PSD functions of input (force) and output (displacement) are related by the **transfer function**, which is given by the square of the frequency response function (FRF).

Predictive Estimation. Parameter estimation consistent with the Bayesian probability theory.
It seeks to minimize the average divergence between the estimated and true distributions. The divergence is measured by Kullback and Leibler's formula. The distribution which achieves minimum divergence corresponds to integrating out the unknown parameter. Hence, predictive estimation can be approximated by averaging over several different parameter choices.

Probability Distribution. A function to represent that the probability of an event is less than or equal to certain value.
The probability that the instantaneous value $X(t)$ is less than or equal to a certain value x is defined by a function $P_X(\leq x)$, which is called the **probability distribution function**, or **cumulative distribution function (CDF)**, not to be confused with its derivative, which is the **probability density function (PDF)**, $p_X(x)$.

Proper Orthogonal Decomposition (POD). A statistical method for system identification to provide modal decomposition.
It is essentially an attempt to extract characteristic information of a multivariate data set into an optimal set of uncorrelated variables called POD modes. The method was originally developed in analyzing the spatial coherent structure in turbulent flow as a useful tool but it has been now applied to dynamics of structures, materials processing and many more fields of pattern recognition, including feedback control design for smart material structures.

Proportional Limit. Highest stress at which stress is directly proportional to strain.
It is the highest stress at which the stress–strain diagram is a straight line. It is usually a little lower than the yield stress and is equal to the elastic limit for many metals.

Rainfall Analysis. A cycle counting method to define an equivalent series of peaks and troughs for determining fatigue life prediction.
The general approach in fatigue life prediction needs to relate a random load fluctuation in real life situation to the **Wöhler curves**, which are based on laboratory experiments of simple specimens subjected to constant amplitude load. The rainflow cycle counting analysis is a method proposed to overcome this difficulty, originally by Endo et al. (in 1968) but has been developed by many researchers, including Downing (in 1972), Rychlik (in 1987), etc., to a state-of-the-art method in fatigue analysis for reducing lengthy irregular time-histories to a small amount of useful knowledge.

Random Decrement (RD) Technique. A technique to identify structural parameters by averaging out the random noise components.
The RD technique was developed by H.A. Cole at NASA in the late 1960s as an alternative to FFT, for identification of dynamic parameters and in-service damage detection of space structures. Its principle can be best explained by a simple example as follows. The random response of a structure at time $t_0 + t$ is composed of the following three parts:

(a) the component due to the initial displacement $x(t_0)$;
(b) the impulse response due to initial velocity $\dot{x}(t_0)$;
(c) the zero-mean, random component due to random loading or noise in the period, t_0 to $t_0 + t$.

If a time segment of $x(t)$ is picked up every time the triggering condition, such as $x(t) = a$, is satisfied, the average of these segments will be a free decay response from the initial displacement, a. It is because (c) above will be eventually averaged out and become negligible, and the sign of the initial velocity is expected to vary randomly with time so the resulting initial velocity will be zero, and so is (b).

The method has merit in the sense that it requires a relatively short data length, although it is also suitable for transforming long-term observations into a small amount of data. It requires a high rate of digitization, usually an order of magnitude higher than the representative frequency. Also, the method has to be carefully applied when there are two or more outstanding modes and frequencies coexisting. It is desirable to band-pass filter the data before processing in order to isolate any outstanding mode should there be one. Particularly when co-existing frequencies are closely spaced, the application of least-square curve fitting to obtain a multifrequency signature has been recommended.

Random Process. The process which generates a sequence of indexed random variables, or the sequence itself.
It is also called the **stochastic process**. The set of random variables is often a time sequence, $X(t)$, sampled from continuous analog signals. The entire collection of all possible sets of sequences is called the **ensemble** and an individual set is called a **sample function** or **realized function**. When the probability distribution of the random process does not evolve appreciably over a time of interest to the engineer, the process is called **stationary**. When the temporal and ensemble averages of a random process are equal, the process is called **ergodic**.

Random Vibrations. A type of vibration where the nondeterministic nature of excitation and/or of the structural system needs to be accounted for.
There are a number of examples. Wind-induced vibration of tall buildings, earthquake excitation of buildings and dams, vibration of offshore oil drilling platforms by action of ocean waves and currents, aircraft vibration during flight and taxiing, and traffic-induced vibration of highway bridges are all random vibrations. Some of them are almost periodic but others are not. Some of them are stationary but others are not, transient or even impulsive. However, generally speaking, because of the nondeterministic nature of their processes, random vibrations are handled statistically.

Reduced Frequency. A dimensionless expression of frequency often employed in wind engineering and aerodynamics.
Usually it is defined as

$$K = \frac{\omega B}{U} \tag{7}$$

where

$$\omega = 2\pi f \tag{8}$$

is the circular frequency, B is a linear structural dimension, and U is the mean wind speed. Sometimes, instead of the circular frequency ω, the frequency itself f is used for the definition. The inverse of the reduced frequency is often called the **reduced velocity**.

Reduced Velocity. Inverse of the reduced frequency.

***Reference Period. Chosen period of time, which is used as a basis for assessing values of variables.**

***Rehabilitation. The work required to repair or upgrade an existing structure.**

Reliability. The probability that a structure will perform its intended functions during a specified period of time under stated conditions.

***Remaining Working Life. Remaining period of the expected life of an existing structure, which is intended or expected to operate with planned maintenance.**

***Repair. Improvement of the conditions of a structure by restoring or replacing existing components that have been damaged.**

Residual Stresses. Internal stress state of a structure or its components, as a result of the preceding thermal and/or mechanical processes such as pre-stressing or welding.

Reynolds Number. A dimensionless parameter that indicates the dynamic flow configuration in relation to the development of turbulence.
It is defined by

$$Re = \frac{VL}{\nu} \tag{9}$$

where V, L and ν are the mean flow speed, a characteristic length and the kinematic fluid viscosity. It is essentially the ratio of fluid inertia force to the viscous force. When Re is very small, the viscous force is large enough relative to the inertia force to suppress the development of turbulence and flow stays laminar, whereas when Re is large, it becomes turbulent. The transition between laminar and turbulent flow is often indicated by a critical Reynolds number, which depends on the exact flow configuration. For example, for the flow around a circular cylinder with a very smooth surface, the **critical Reynolds number** is known to be about 2300.

Richter Scale. A definition of earthquake magnitude proposed by Richter, who defined the magnitude by the logarithm of the largest displacement recorded by a standard seismograph.
The Richter scale M and the energy E released by an earthquake is approximately related by

$$\log_{10}(E) = 4.8 + 1.5M \tag{10}$$

where E is measured in joules. Events with magnitude 5.0 or above can cause major damage to structures. The energy released by a 1 Mt hydrogen bomb is roughly equivalent to a magnitude 7.4. It has been recognized that this scaling method has a saturation effect near 8.5 and it becomes difficult to differentiate the magnitude of events even when they are clearly different in size. Some seismologists now want to replace it with the **moment magnitude scale**, which has been established more recently.

Risk Rating. A measure to classify the risk levels in different categories.
The risk levels are defined by assessing (a) the potential threat; (b) vulnerability of the existing or projected system; and (c) the potential impact that could result.

Robustness. A desirable characteristic of a regulatory network to generate a certain qualitative response over a broad range of parameter values.
In the context of computer software or network system, it is the resilience of the system, especially when under stress or when confronted with invalid inputs. For example, the operating system is considered robust if it operates properly and correctly even when it is starved of memory space, or confronted with an illegitimate application or bugs.

Safety. The condition of a structure being protected against failure, damage, error, accidents, or harm, in both causing and exposure.

Safety is the most overwhelming factor in structural design, construction and maintenance. However, in reality, it is a probabilistic concept and there is a need of a reliability method to decide if the condition is adequately acceptable. Reliability analysis based on appropriate modeling of parameters with reliability indices is a topic of **Safety Engineering**.

***Safety Plan. A plan specifying the performance objectives, the scenarios to be considered for the structure, and all present and future measures such as design, construction, or operation such as monitoring, to ensure the safety of the structure.**

Scruton Number. A dimensionless parameter which indicates a combined effect of mass and damping on the vortex-induced structural response.
Named after Kit Scruton, who was a pioneering British engineer in industrial aerodynamics, it is defined by

$$Sc = \frac{m\zeta}{\rho D^2} \tag{11}$$

in which m is mass per unit length, ζ is the structural damping ratio, ρ is fluid density, and D is a representative linear dimension of the structure. Note that sometimes the Scruton number is defined by

$$Sc = \frac{2m\delta}{\rho D^2} \tag{12}$$

which is 4π times greater than the above definition. If the vortex excitation can be regarded as a resonance to a simply fluctuating excitation force, the induced peak response will be inversely proportional to the Scruton number.

Seismic Waves. Elastic waves that are caused by earthquakes and travel through the Earth. There are different types of waves – body waves and surface waves.
Seismic waves can be also caused by explosions on or under the ground surface. The mechanics of wave motion in solid media, particularly with geological strata, are very complex. There are two different body waves, P and S waves. P waves are longitudinal or compressive waves and travel at the speed of sound, which is about 1.5 km/s in water and 5 to 13 km/s in hard rocks. S waves are transverse or shear waves. They travel at about 60% of the speed of P waves, only through solid. With S waves, the ground moves perpendicular to the direction of wave propagation. The body-wave amplitudes decay at the rate that is inversely proportional to the square of radial distance from the hypocenter. Two kinds of surface waves, Rayleigh and Love waves, are known that travel along the ground surface or intersurface of media. Surface waves are often the direct cause of severe catastrophic consequences of earthquakes. They produce ground surface motion in vertical and horizontal directions, respectively. Surface waves travel with a speed a little slower than S waves and their amplitudes decay much more slowly than the body-wave amplitude, with the rate proportional to the square-root of radial distance.

Sensor. A device that is designed to acquire information from an object and transform it into an electrical signal.
It usually consists of three parts:

(a) the sensing element, such as resisters, capacitor, transistor, piezoelectric materials, photodiode, etc.;
(b) equipment for signal conditioning and processing, such as amplification, linearization, compensation and filtering;
(c) sensor interface to connect with other electronic components.

Service Life. The number of years a structure is intended to be in service.

Serviceability. The ability of a structure to be serving or capable of serving its intended purposes to the user's satisfaction.

Shear Modulus. The ratio of the increase in stress to that of strain of a material subjected to shear loading.
For an isotropic, homogeneous elastic material, there are only two independent elastic moduli. The shear modulus (G), for this case, is related to Young's modulus (E) and Poisson's ratio (ν) by

$$G = \frac{E}{2(1+\nu)} \tag{13}$$

For structural steels, $G \approx 80\,\text{GPa}$. The shear modulus of structural materials is determined by a twisting test, which is regulated in ASTM E-143.

SI Units. The Systeme Internationale d'Unities. An international system of units that was established in an attempt to simplify the language of science.
The system was an outcome of a resolution adopted at the 9th General Conference of Weights and Measures, 1948, and has been gradually accepted by many countries over the world, replacing the traditional local unit systems. It is based on seven standard base units: length (m), mass (kg), time (s), electric current (A), temperature (K), luminous intensity (cd), and the amount of substance (mol), and all other units are derived from these. The system also specifies the standard prefixes to express multiples and submultiples, such as kilo (k), mega (M) and milli (m).

Signal Processing. A data analyzing system which includes data filtering, frequency domain transformation and statistical analysis.

Similitude Requirements. Scaling requirements in engineering model tests so that the scale model test results could be interpreted with proper physical meaning.
The **Buckingham Π theorem** states that a set of dimensionless parameters which consist of suitable combinations of the reference quantities are required to be invariant in model and prototype and with them the governing equations are also rendered dimensionless. However, the complete satisfaction of this requirement for all conceivable dimensionless numbers is possible only when the model and prototype are identical. It means that in any scale model tests, one or more of the similitude requirements need to be relaxed to make the model test possible. The real issue of experimental mechanics, therefore, becomes a matter of interpreting the test results, knowing which of the requirements are actually distorted or disregarded in the tests. The dimensionless numbers often treated in wind-tunnel tests for bridge aerodynamics, for example, are **Reynolds number, Froude number, Jensen number, reduced frequency, mass parameter, damping ratio**, etc.

Simulation. The physical, mathematical or software-based modeling of a system and also of its behavior as an interaction with the environment.

Smart Materials. Materials that have one or more property that can be significantly altered in a controlled fashion by external stimuli.
The external inputs can be stresses, temperature, moisture, pH, and electromagnetic fields, which are sensed and the response of the materials is actuated. There are many types of smart materials including piezoelectric materials and thermo-responsive materials.

Smart Structure. A structure that has the ability to alter its configuration or properties as a response to changes in environment.
The concept has been developed in the field of aerospace engineering but many possible applications in civil engineering structures have been explored and implemented. The use of piezoelectric materials and embedded or surface-mounted fiber optic sensor systems for monitoring the dynamic behavior of structures to actuate the active damper system is one example. A smart structure needs to

have three integral components besides the load-carrying function – the sensors, the processor and the actuators.

S-N Curve. See Wöhler curve.

Spectral Analysis. Dynamic analysis in terms of the frequency-based characteristics of the processes by the use of the auto- and cross-power spectral density functions (PSD).

Spectral Windows. Weighing functions to be applied in spectral analysis for processing only band-pass filtered data in the frequency domain.
Windows are applied to obtain smoother spectra so that the physical interpretation of them would become easier. Basic requirements for the windows are: (a) the integration of a window function over the whole frequency range should be unity; and (b) the window function should be symmetric with respect to zero frequency. There are various types of window functions employed in engineering signal processing including the **digital filters** such as Hanning and Hamming windows. Inverse Fourier transform of the spectral windows are called **lag windows**, which are the windows in time domain and applicable to the autocorrelation functions.

Stationary Process. A stochastic process in which the probability density function (PDF) of a random variable does not change over time or position.

Stochastic Subspace Identification. A modal parameter identification method developed in a MATLAB environment, applicable to ambient vibration survey (AVS).
The method starts with a stochastic state–space representation of the dynamic behaviour of a structure under white noise excitation. The numerical procedures of matrix decomposition for identifying the state–space model are all handled by built-in functions of MATLAB.

Strain Energy. Measure of energy absorption characteristics of a material under load up to fracture. It is equal to the area under the stress–strain curve, and is a measure of the toughness of the material.

Stress Concentration. A phenomenon where an object under load has higher-than-average local stresses due to its shape.
The types of shape that cause these concentrations are: cracks, sharp corners, holes and narrowing of the cross-section. Ratio of the greatest stress in the area to the corresponding average stress is called the **stress concentration factor**.

Strouhal Number. A dimensionless parameter to indicate the frequency of alternating wake vortices.
Formulated by a Czech physicist, V. Strouhal, it is defined as $St = fD/U$, where f is the frequency of vortex formation, U is the flow speed, and D is a representative structural dimension, which is usually taken normal to the flow. The Strouhal number is generally a function of the **Reynolds number**, $Re = UD/\nu$. The Strouhal number of a two-dimensional circular cylinder, for example, is approximately 0.2 when the Reynolds number is in the subcritical range. For ordinary plate-girder or box-girder bridge decks, the Strouhal number is typically in the range of 0.07–0.14.

Symmetry. A relationship of a form, pattern or style to its mirror image, expressed by exact correspondence of the original to the opposite side of a dividing line or plane.
There is also a case of **point-symmetry**, where the same pattern correspondingly exists at a position that is 180° rotated about an axis. **Symmetry** is an extremely important element of thought in any artistic works. As an extended concept, particularly in physics, **symmetry** sometimes means **invariance**. Note, however, that the mirror image may have different characteristics from the original, such as the case of **enantiomers**.
The autocorrelation function of a stationary process $x(t)$ is an even function, which is symmetric with respect to zero, or $R_X(-\tau) = R_X(\tau)$. It follows that, as a result, the power spectral density $S_X(f)$ of the same process becomes an even function of frequency. However, since the negative values of frequency

do not make good physical sense, the one-sided power spectral density,

$$G_X(f) = \begin{cases} 2S_X(f) & 0 \leq f < \infty \\ 0 & \text{otherwise} \end{cases} \tag{14}$$

is often defined only for the positive range of frequency. Note that the area under $G_X(f)$ for the positive frequency is equal to the whole area under the original $S_X(f)$.

System Identification. Technical determination of structural properties from the known response of the system.
Other terms such as **modal identification** or **parameter estimation** have been used for the same procedure in the present context. The goal of system identification is the opposite of classic dynamic analysis, where usually the structural properties are known and response of the system is to be determined under various excitations. The approaches in system identification techniques are broadly classified into two groups; the time-domain procedure, which includes the least-squares fitting and the random decrement method, deterministic as well as stochastic subspace identification method, etc., and the frequency-domain procedure by applying Fourier transform, which includes the half-power bandwidth method, the maximum likelihood method and so on.

***Target Reliability Level. The level of reliability required to ensure acceptable safety and serviceability.**

Thermal Expansion. Material characteristics specified by a linear expansion due to a unit increase in temperature.
A material constant defined by a strain corresponding to a unit increase in temperature is called the coefficient of thermal expansion. It is approximately $12 \times 10^{-6}/^{\circ}C$ for structural steel and concrete.

Time Domain Analysis (TDA). Analysis of signals and their functions with respect to time as opposed to their handling in terms of frequency.
See also the **Frequency Domain Analysis (FDA)**.

Time Series. An ordered sequence of values of a variable at equally spaced time intervals.
Time-series analysis is used for analyzing and understanding the characteristics of discrete data systems sampled from continuous observations and also for fitting of time-series models for forecasting of future values. There are many techniques of model fitting including Box–Jenkins ARIMA models and multivariate models.

Toughness. The ability of a metal to withstand shock loading.
The concept is the exact opposite of brittleness. Toughness can be explained as the ability of a metal to distribute within itself both the stress and strain caused by a suddenly applied load. It is usually measured by the **Charpy test** but its indication is relative, since toughness is also governed, in addition to the material composition, by the shape of the metal.

Trend removal. Removal of the data characteristics defined as any frequency component whose period is longer than the record length. Least square procedures or the average slope method are often used.

Ultimate Strength. Highest engineering stress developed in material before rupture. Normally, changes in area due to changing load and necking are disregarded in determining ultimate strength.

***Upgrading. Modifications to an existing structure to improve its structural performance.**

***Utilization Plan. A plan containing the intended uses of the structure, and listing the operational conditions of the structure including maintenance requirements, and the corresponding performance requirements.**

Vibrations. The periodic to-and-fro motion of a structure or its members.

Vibration is characterized by three basic parameters: how quickly the motion is repeated **(frequency)**, the magnitude of the motion is **(amplitude)**, and how soon it dies out without new supply of excitation energy **(damping)**. Depending upon the sources of excitation, structural vibration in reality may be a regular periodic motion but is more likely stochastic in terms of both frequency and amplitude. The latter case is called the **random vibration**, and it is often more convenient to handle the characteristics of vibration in statistical terms.

The **adverse effects** of structural vibration are examined from various aspects. The structure should not collapse and must maintain its structural integrity so that it does not lose its serviceability. Even if the structure is safe and able to serve, if it produces any discomfort to the users and/or causes any mechanical problems such as overstressing, malfunctioning or misalignment, it is not acceptable. Also, even if there is no immediate problem, any troubles in future such as structural fatigue damage, possibly even compound with material corrosion, must be avoided as much as possible. The **acceptance criteria** are often defined by the combination of amplitude and frequencies but they are related to how often and how long the structure is exposed to dynamic excitation, too. There are various sources of dynamic excitation for bridge vibrations, including earthquakes, wind, moving vehicles and pedestrians, and sometimes even unexpected impact loads such as ship collision on piers or machine operation and blasting.

Vortex Shedding Excitation. Vibration of a structure or its members excited by vortices of air flow created by the interaction of wind and the structure.

When an aerodynamically bluff body is exposed to wind, a trail of alternating vortices (the **Kármán vortices**) is often found in its wake, formed by the flow separated from the body. There is also a fluctuating lift force acting on the body corresponding to the formation of vortices. As a result, when the frequency of vortex formation is close to the structure's eigenfrequency, there will be a resonant vibration. This is the most fundamental concept of vortex excitation. However, once the vibration starts, the body motion itself will influence the flow behavior, which results in the more complicated interaction of flow and structure. Unlike the case of buffeting, the vortex excitation is usually observed in a limited range of wind speed and its amplitude decreases once it hits the peak value. The vibration is usually characterized by a narrow-band frequency spectrum and a somewhat regular amplitude.

Wave. A vibratory motion or disturbance that propagates and yet is not usually associated with mass transport.

Mechanical waves propagate through continuous media, including air, liquid or solids, which are recognized as sound, ocean waves and seismic waves, for example. There are also electromagnetic radiations, including visible lights, infrared or ultraviolet rays, gamma rays, etc., which can propagate through a vacuum. All waves have common characteristics that are experienced as reflection, refraction, diffraction, interference, dispersion and rectilinear propagation, although possession of these characteristics is only a necessary condition to be waves. Waves can be described using standard parameters such as frequency, wavelength, amplitude and period. Waves remaining in one place are called standing waves and waves moving are called traveling waves. Material particles in mechanical waves can be vibrating in the direction of wave propagation or perpendicular to it. They are termed longitudinal and transverse waves, respectively. In the case of **seismic waves**, they are also called the P wave and S wave, respectively.

Wavelet Transform. A tool for decomposing a signal into its time- and scale-dependent components, in terms of so-called wavelet coefficients. It is suitable for the analysis of nonstationary data.

Fourier transform is a very versatile tool in signal analysis, but it is not suitable for identifying nonstationary aspects of the signal. For example, since Fourier transform is applied to the entire signal length, the result cannot indicate at what time in the signal a specific frequency existed. It is really a tool for frequency resolution but not for time resolution. It means that if the method is applied to structural health monitoring, the method may recognize damage occurrence, location, or even its severity, but not exactly when the damage happened. Wavelet transform is basically an extended application of a windowing technique with variable-sized windows. It allows the use of long time intervals where low-frequency information is needed and shorter intervals for high-frequency information.

Wind Power Input. An expression for the magnitude of energy input due to wind. This expression is often used in the field of power line vibration.
The specific power input $\widetilde{P}$ is defined by

$$\widetilde{P} = \frac{P}{f^3 D^4 L} = 2\pi^2 \frac{m}{D^2} \left(\frac{a}{D}\right)^2 \delta \tag{15}$$

where P is the power imparted by wind, D is the cylinder diameter, L is cable length, m is the effective cable mass per unit length, f is vibration frequency, a is vibration amplitude, and δ is the net logarithmic decrement.

Wind Tunnel. An experimental facility to examine the interaction between wind flow and solid objects that are exposed to it by placing a model of the object in an artificially created air flow.
Wind-tunnel tests are carried out for the measurement of wind flow around the body, wind-induced forces or pressure on the body and/or static and dynamic behavior of the structure induced by wind. Since the determination of wind-structure interaction is not easily done by any analytical means, particularly for many of the civil engineering applications, wind-tunnel tests are attractive alternatives. However, care should be taken for proper simulation of both structural models and wind-flow conditions in carrying out these tests so that the test results would be properly interpreted to the situation in reality. A set of rules required for carrying out the tests is called the **similitude requirements**. A wind tunnel in which an artificially developed turbulent boundary layer flow is used as a simulation of natural wind is called the **boundary layer wind tunnel** (BLWT).

Windows. Weighting functions to be applied for required operations, such as band-pass filtering, smoothing and/or distortion of a given set of data.
Measurement of any data is a kind of windowing, too, since the measuring period cannot be infinitely long to cover the whole length of the original data, and the observed results are inevitably influenced by the characteristics of sensing devices such as the frequency response, resolution and precision of the measuring system. The windowing operation can be carried out in time domain as well as the frequency domain. Windowing in time domain becomes convolution in frequency domain. See also the **spectral windows**.

Wöhler Curve. A graph of the magnitude of a cyclic stress (S) against the number of cycles (N) to the material's fatigue failure.
It is also called the ***S–N* curve**. A Wöhler curve is derived from the coupon tests in the laboratory environment, where an ideal sinusoidal stress of constant amplitude is applied to failure. Each coupon test generates a point on the plot, although in some cases the time to failure exceeds the anticipated timeframe. The Wöhler curves allow designers to make a basic estimate of the expected life of the structural part against expected stresses.

Yield Point. The stress at which the metal changes from elastic to plastic in behavior.
Structural steels usually exhibit a yield strength of 300–400 MPa. **Offset yield strength** is determined from a stress–strain diagram. It is the stress corresponding to the intersection of the stress–strain curve, and a line parallel to its straight line portion offset by a specified strain. Offset for metals is usually specified as 0.2%.

Young's Modulus. A ratio of increase in stress to that of strain, when stress and strain has a linear relationship.
Young's modulus of steel is usually taken as 200-210 GPa. When a material exhibits a nonlinear stress–strain relationship, the definition of an equivalent Young's modulus is sometimes convenient. Young's modulus in a dynamic situation may not be always equal to the static value.

12.2 Mathematical Formulations in Dynamics

12.2.1 Elements of Structural Dynamics

12.2.1.1 Dynamics of Single-Degree-of-Freedom Systems

Depending on the purpose of the analysis, many structures can be simulated as a simple Single-degree-of-freedom (SDOF) system, where DOF is the number of displacements for describing the characteristics of a given vibration. However, the concept of DOF is applicable only in terms of mathematical modelling of vibration. The same vibration of a structure can be considered as multi-degree-of-freedom (MDOF), depending on how the structure is conceptually modeled.

When a structure is modeled by a SDOF system of mass m, stiffness k and viscous damping c, subjected to an excitation force $F(t)$, the analysis of its dynamic displacement $z(t)$ can be formulated by the **equation of motion** as follows:

$$m\ddot{z}(t) + c\dot{z}(t) + kz(t) = F(t) \tag{16}$$

or

$$m(\ddot{z}(t) + 2\zeta\omega_0\dot{z}(t) + \omega_0^2 z(t)) = F(t) \tag{17}$$

where

$$\omega_0 = 2\pi f_0 = \sqrt{\frac{k}{m}} \tag{18}$$

is the natural circular frequency and

$$\zeta = \frac{c}{2\sqrt{mk}} \tag{19}$$

is the damping ratio. Natural frequency is a characteristic of a particular structure since it is uniquely decided only by mass and stiffness. Natural frequencies of structures are supposed to be evaluated under the standard design conditions of m and k, in which the structure is free of live loads and extreme temperature effects. However, since the instantaneous mass and stiffness of a structure in reality could be different from the design assumptions, it is worthwhile to note that the natural frequencies of the structure in service condition can be different to the values calculated by Equation (18).

Free Vibration

When $F(t) = 0$, the general solution of Equation (16) is given by the following three cases:

(a) $\zeta < 1$ (decaying vibration)

$$z_H(t) = \mathrm{e}^{-\zeta\omega_0 t}\left[A\sin(\omega_D t) + B\cos(\omega_D t)\right] \tag{20}$$

where

$$\omega_D = \omega_0\sqrt{1-\zeta^2} \tag{21}$$

is the frequency with damping.

(b) $\zeta = 1$ (critical damping)

$$z_H(t) = (At + B)\sin(\omega_0 t) \tag{22}$$

(c) $\zeta > 1$ (overdamped)

$$z_H(t) = \mathrm{e}^{-\zeta\omega_0 t}\left[A\sinh(\omega_N t) + B\cosh(\omega_N t)\right] \tag{23}$$

where

$$\omega_N = \omega_0\sqrt{\zeta^2 - 1}. \tag{24}$$

The integral constants A and B are decided from the **initial conditions**, $z_H(0)$ and $\dot{z}_H(0)$.

Damping can be defined as the capacity of structures to dissipate energy imparted by the external forces. The dissipation of dynamic energy during vibration results from many different sources, such as the imperfect elasticity and internal friction of structural materials, friction of structural members at their joints and support mechanisms, aerodynamic and hydrodynamic damping due to the surrounding environment, the nonlinear structural characteristics, energy dissipation through foundation and substructures, and so on. Theoretical evaluation of damping capacity is generally limited for all these sources and it is therefore important to consult to the results of the field experience as references. Some practical notes on damping are added later in Section 12.2.4.

Although the mechanism of damping is quite diverse, their overall effects on vibration is usually characterized by considering an equivalent viscous damping, represent by a single number in the **damping ratio** (ζ) as a fraction of critical. If the overall damping of the system is 1% of critical, for example, the free vibration amplitude will be reduced to a half after 11 cycles, whereas 10% damping will reduce the amplitude to a half at each cycle. When damping is at or beyond critical, there is no vibration, as in Equations (22) or (23) above.

Natural frequency is reduced a little with an increase in a system's damping, as in Equation (21). However, in reality, this effect can be disregarded for the case of civil engineering structures. For example, even if the damping is as high as 10% of critical, the natural frequency is reduced by only 0.5% compared to the case with no damping.

Forced Vibration

Simple Harmonic Excitation

By a simple harmonic excitation: $F(t) = F_0\sin(\omega t) \quad (\omega = 2\pi f)$

$$z_P(t) = M(\Omega, \zeta)\frac{F_0}{k}\sin(\omega t - \beta) \tag{25}$$

where

$$\Omega = \frac{\omega}{\omega_0} = \frac{f}{f_0} \tag{26}$$

is the frequency ratio,

$$\beta = \arctan\left(\frac{2\zeta\Omega}{1 - \Omega^2}\right) \tag{27}$$

is called the phase lag and the **dynamic load factor** is given by

$$M(\Omega, \zeta) = \frac{1}{\sqrt{(1-\Omega^2)^2 + (2\zeta\Omega)^2}} \tag{28}$$

The dynamic excitation a structure is subjected to may not be exactly simple harmonic. However, by decomposing the external excitation function $F(t)$ to various frequency components by applying Fourier analysis, the structural response to each frequency component can be decided by Equation (25) and hence the total response is given by a linear combination of all frequency response components, unless the system is nonlinear.

Generally the response of the system under excitation is given by $z(t) = z_P(t) + z_H(t)$.

- The dynamic load factor (or magnification factor) takes its peak value

$$M_{max} = \frac{1}{2\zeta\sqrt{1-\zeta^2}} \approx \frac{1}{2\zeta} \tag{29}$$

when

$$\Omega = \sqrt{1-2\zeta^2} \approx 1 \tag{30}$$

or $\omega \approx \omega_0$. This case is called the **resonance**.

- More generally, when a simple harmonic excitation force is expressed by $F(t) = F_0 e^{i\omega t}$, the induced response of the system is given by

$$z_P(t) = \frac{H(\omega)}{k} F(t) \tag{31}$$

where

$$\frac{H(\omega)}{k} = \frac{1}{k} \frac{1}{(1-\Omega^2) + i(2\zeta\Omega)} \tag{32}$$

is called the **frequency response function (FRF)**. Note that the dynamic load factor is the magnitude of FRF, or

$$\frac{|H(\omega)|}{k} = M(\Omega, \zeta) \tag{33}$$

The FRF is sometimes called the **mechanical admittance function**. It shows the sensitivity of a structural system to the excitation frequency. The peak response is reached when the excitation frequency coincides with the natural frequency and its magnitude is inversely proportional to the total damping of the system.

Impulse Load

Forced vibration induced by an impulse load $F(t) = F_0$ $(0 \le t \le \Delta t)$ is given by $z_P(t) = F_0 h(t)\Delta t$, where

$$h(t) = \frac{1}{m\omega_D} e^{-\zeta\omega_0 t} \sin(\omega_D t) \tag{34}$$

is called the **impulse response function (IRF)**. Together FRF and IRF make a **Fourier transform** pair, or

$$\frac{H(\omega)}{k} = \int_{-\infty}^{\infty} h(t)\mathrm{e}^{-i\omega t}\,dt \qquad \text{and} \qquad h(t) = \frac{1}{2\pi}\int_{-\infty}^{\infty} \frac{H(\omega)}{k}\mathrm{e}^{i\omega t}\,d\omega \tag{35}$$

Fourier transform is an integral transform, which is particularly useful in relating the time-domain and frequency-domain variables for random vibration analysis. The concept of the Fourier transform developed from a Fourier series, which is the decomposition of a periodic function. Fourier transform expands this concept to make it further applicable to the nonperiodic functions. The analysis of measured data using the **discrete Fourier transform (DFT)**, particularly in the form of the **fast Fourier transform (FFT)**, is one of the most important techniques for the frequency-domain vibration analysis.

Duhamel Integral

When the system is excited by an external load that can be expressed by a known function $F(t)$ $(t \geq 0)$, the response is given by the Duhamel Integral:

$$z_P(t) = \int_0^t h(\tau)F(t-\tau)\,d\tau \tag{36}$$

Mathematically, Equation (36) is the general solution of Equation (16), when $z(0) = \dot{z}(0) = 0$.

Example 12.1 When a SDOF system of mass m, stiffness k with no damping, is subjected to an external force $F(t) = F_0$ $(t \geq 0)$, the induced response is given by

$$z_P(t) = \frac{F_0}{m\omega_0}\int_0^t \sin\left(\omega_0(t-\tau)\right)d\tau = \frac{F_0}{k}\left[1-\cos(\omega_0 t)\right] \tag{37}$$

Example 12.2 When $F(t) = F_0\sin(\omega t)$ is applied to the same system, the response is

$$z_P(t) = \frac{F_0}{m\omega_0}\int_0^t \sin(\omega\tau)\sin\left(\omega_0(t-\tau)\right)d\tau = \frac{F_0}{k}\frac{\sin(\omega t)-\Omega\sin(\omega_0 t)}{1-\Omega^2} \tag{38}$$

12.2.1.2 Nonlinear Vibration

Equation of motion Equation (16) describes the dynamics of a linear system. When the structural restoring and/or damping forces are not linearly proportional to the displacement or its time derivative, however, or the external forces are amplitude dependent, the vibration becomes nonlinear. Nonlinear stiffness can be caused by material nonlinearity, or structural nonlinearity, such as the case of cables or strain hardening. Structural damping ratio is usually higher when vibration amplitude is greater. In contrast, the aerodynamic damping is sometimes found drastically reduced, or even becomes negative, with greater vibration amplitude. This is because the excitation force is largely nonlinear due to structural interaction with the surrounding fluid.

All these factors indicate that very often the structural vibration is likely to be nonlinear in reality, unless the dynamic amplitude is very small, and the nonlinear vibrations can be analyzed only by applying numerical methods without closed-form solutions. However, we try to handle them with linear approximation so long as it is acceptable for practical purposes. An important discussion to be added to this topic, therefore, is to cover the characteristics of vibration that cannot be fully explained, or not explained at all, unless the linear approximation is removed.

Free Vibration

Let us consider, as an example, a typical nonlinear equation of motion, where the restoring force has a slight strain hardening effect as follows:

$$m\ddot{z} + c\dot{z} + k(1 + \varepsilon z^2)z = 0 \qquad (\varepsilon \ll 1) \tag{39}$$

or

$$\ddot{z} + 2\zeta\omega_0\dot{z} + \omega_0^2(1 + \varepsilon z^2)z = 0 \tag{40}$$

Equation (39) is called the Duffing's equation and its approximate solution is, by neglecting the damping, given by

$$z(t) \approx A \sin\left(\left(1 + \frac{3\varepsilon}{8}A^3\right)\omega_0 t\right) \tag{41}$$

which indicates that the frequency of vibration, $\omega = \omega_0(1 + 3\varepsilon A^3/8)$, is amplitude dependent. As it is clear here, unlike the case of linear vibration, the eigenfrequency cannot be defined for a nonlinear system.

Forced Vibration

Consider the same nonlinear system as before, which is now subjected to a simple harmonic excitation force. It is more natural to include the damping term for this case. The equation of motion becomes

$$\ddot{z} + 2\zeta\omega_0\dot{z} + \omega_0^2(1 + \varepsilon z^2)z = \frac{F_0}{m}\sin(\omega_e t) \tag{42}$$

If $\varepsilon = 0$, the case is linear and the steady-state response is given by

$$z(t) = A\sin(\omega_e t + \beta) \tag{43}$$

When $\varepsilon \neq 0$, because of the nonlinearity of the system, the resultant response is not limited only to the component at the excitation frequency but there are also components of higher harmonics at the frequencies, such as

$$z(t) = \sum_{r=1,3,5,\ldots} A_r \sin(r\omega_e t + \beta_r) \tag{44}$$

This is quite characteristic of nonlinear vibrations.

Also, the response at the excitation frequency behaves differently from the linear case. Let us assume that the solution to this case is somewhat similar to Equation (43) except both A and β are functions of time. Since the steady-state vibration is determined by considering $\dot{A}(t) = \dot{\beta}(t) = 0$, it results that

$$\left(1 - \Omega^2 + \frac{3\varepsilon}{4}A^2\right)^2 A^2 + (2\zeta\Omega)^2 A^2 = \left(\frac{F_0}{k}\right)^2 \tag{45}$$

$$\beta = -\arctan\left(\frac{2\zeta\Omega}{1 - \Omega^2 + (3\varepsilon/4)A^2}\right) \tag{46}$$

where $\Omega = \omega_e/\omega_0$.

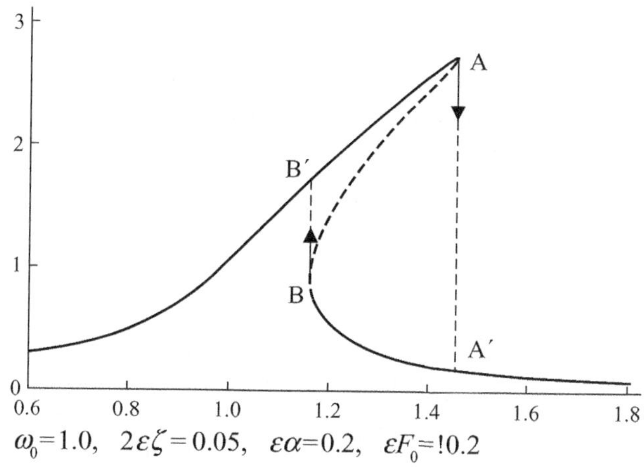

Figure 12.1 Hysteresis phenomenon of a nonlinear system

Equation (44) indicates that, depending on the excitation frequency ω_e, the steady-state amplitude can take three different values. The A–ω diagram (see Figure 12.1) shows that one of them actually is an unstable limit cycle. Depending on ω_e, there is a sudden jump of amplitude **(jumping phenomenon)** and also the magnitude of steady state amplitude can be different for increasing and decreasing cases of ω_e, which is called the **hysteresis phenomenon**. Generally speaking, when $\Omega \approx 1$, which is equivalent to the resonance condition for a linear system, there are somewhat similar phenomena for a nonlinear system too, which is called the **harmonic resonance**. However, unlike the case of linear systems, the harmonic resonance is accompanied by jumps and hysteresis, which are characteristic of nonlinear vibrations.

Chaotic Vibration

Another characteristic aspect of nonlinear vibration is that, even if the system is deterministic, the system's response to a regular simple harmonic excitation can be quite random with a continuous frequency distribution. This is caused by the fact that the response of a strongly nonlinear system can be extremely sensitive to the difference of initial conditions for each cycle of vibration. An example is shown in Figure 12.2. This response vibration looks random but actually there is some regularity within it, known as *chaotic vibration*. The state of this vibration can be fully described by the locus of $(x, \dot{x}, t)$, which is called the **phase portrait** and this 3D space $(x, \dot{x}, t)$ is called the **phase space**. When the points are periodically taken with regular intervals along the phase portrait and a 2D projection of these points is made on the x–$\dot{x}$ plane, it is called the **Poincaré map**. If the nonlinear motion of the system becomes stable and its motion is confined in a limited range, the Poincaré map will form a limited pattern, which is called an **attractor**.

12.2.1.3 Dynamics of Elastic Structures – Modal Analysis

Once the dynamic finite element analysis is carried out for a structure and its natural frequencies and corresponding mode shapes are obtained as ω_r and $\phi_r(x)$ $(r = 1, 2, \ldots, N)$, the dynamic response of the structure under a **known** external force $F(x, t)$ can be calculated as follows. For the rth mode of vibration of the structure, the equation of motion is

$$M_r(\ddot{q}_r(t) + 2\zeta_r\omega_r\dot{q}_r(t) + \omega_r^2 q_r(t)) = F_r(t) \tag{47}$$

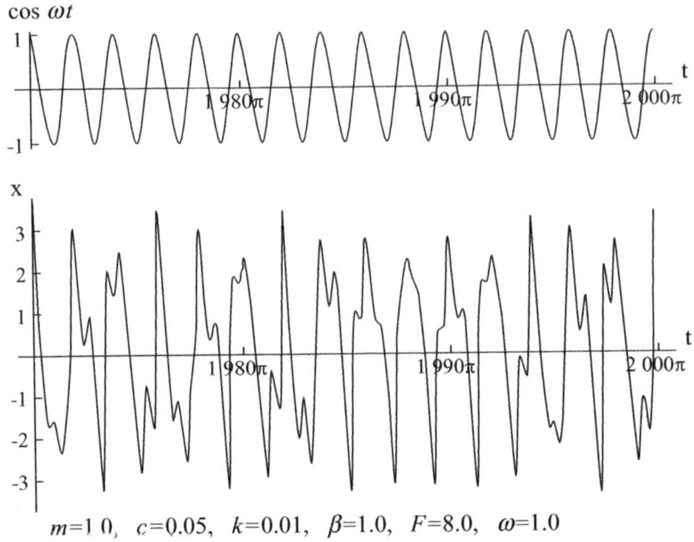

Figure 12.2 Example of chaotic vibration

where

$$M_r = \int_L m(x)\phi_r^2(x)\,dx \tag{48}$$

is the generalized mass,

$$F_r(t) = \int_L F(x,t)\phi_r(x)\,dx \tag{49}$$

is the generalized force and the resultant response is given by

$$z(x,t) = \sum_{r=1}^{N} q_r(t)\phi_r(x) \tag{50}$$

ζ_r is the assumed damping for the rth mode of vibration. The integrations $\int_L [.....]\,dx$ are to be carried out over the whole structure.

Example 12.3 If the structure is a simply supported beam of span L with a uniformly distributed mass m and uniform bending stiffness EI, and the beam is subjected to the axial force T, the natural circular frequencies and corresponding mode shapes are given by

$$\omega_r = \left(\frac{r\pi}{L}\right)^2 \sqrt{\frac{EI}{m}} \sqrt{1 + \left(\frac{\gamma}{r\pi}\right)^2}, \qquad \gamma^2 = \frac{TL^2}{EI} \tag{51}$$

and

$$q_r(x) = \sin\left(\frac{r\pi x}{L}\right) \qquad (r = 1, 2, 3, \ldots) \tag{52}$$

12.2.1.4 Dynamics of MDOF Systems – Complex Eigenvalue Problems

Considering a discrete model of N-degrees-of-freedom for a structure, the equation of motion can be written in a similar way as a SDOF system as follows:

$$\mathbf{M}\ddot{z}(t) + \mathbf{C}\dot{z}(t) + \mathbf{K}z(t) = \boldsymbol{F}(t), \tag{53}$$

where **M**, **C** and **K** are the mass, damping and stiffness matrices, and z and $\boldsymbol{F}(t)$ are the displacement and the external force vectors, respectively.

Free Vibration

The effect of damping can be ignored in free vibration analysis, or

$$\mathbf{M}\ddot{z}(t) + \mathbf{K}z(t) = 0 \tag{54}$$

Hence, by assuming $z\ (t) = \phi e^{i\omega t}$, the natural frequencies ω_r and associated eigenvector ϕ_r, $(r = 1, 2, \ldots, N)$ are decided from

$$|\mathbf{K} - \omega^2\mathbf{M}| = 0 \tag{55}$$

and

$$(\mathbf{K} - \omega_r^2\mathbf{M})\phi_r = 0 \qquad (r = 1, 2, \ldots, N) \tag{56}$$

has to be satisfied by the mode vector ϕ_r, where one of the vector components, ϕ_{1r} for example, needs to be given arbitrarily and other components are decided relative to this value.

In case where the external force is given as a linear function of structural displacement, $\mathbf{F}(t) = \mathbf{F}_1 z(t) + \mathbf{F}_2\dot{z}(t)$, by including these terms, the equation of motion has the homogeneous form as in the case of free vibration, however with damping:

$$\mathbf{M}\ddot{z}(t) + \mathbf{C}\dot{z}(t) + \mathbf{K}z(t) = 0 \tag{57}$$

For this case, the problem can be handled in the same way as before but the frequency equation will become more complicated, including the complex coefficients:

$$|\mathbf{K} + i\omega\mathbf{C} - \omega^2\mathbf{M}| = 0 \tag{58}$$

It is more convenient to rewrite equation (57) as a **state equation set** as follows:

$$\dot{\mathbf{X}}(t) = \mathbf{A}\mathbf{X}(t) \tag{59}$$

where

$$\mathbf{X} = \begin{bmatrix} z \\ \dot{z} \end{bmatrix} \quad \text{and} \quad \mathbf{A} = \begin{bmatrix} \mathbf{0} & \mathbf{I} \\ -\mathbf{M}^{-1}\mathbf{K} & -\mathbf{M}^{-1}\mathbf{C} \end{bmatrix}. \tag{60}$$

X and **A** are a $2N \times 1$ state vector and a $2N \times 2N$ state matrix, respectively, and **I** is a $N \times N$ unit matrix. By assuming a general solution as $\mathbf{X} = \phi e^{st}$, the natural frequencies and damping are obtained as the complex eigenvalues of

$$(s\mathbf{I} - \mathbf{A})\phi = 0 \tag{61}$$

where s and ϕ are the eigenvalue and eigenvector, respectively, and the natural circular frequency ω_r and the modal damping ratio ζ_r in rth mode are related to the eigenvalue $s_r = \alpha_r \pm i\beta_r$ ($r = 1, 2, \ldots, N$) by

$$\omega_r = \sqrt{\alpha_r^2 + \beta_r^2} \quad \text{and} \quad \zeta_n = -\frac{\alpha_r}{\sqrt{\alpha_r^2 + \beta_r^2}} \tag{62}$$

The rth eigenmode of vibration $z_r(t)$ is expressed as

$$z_r(t) = 2e^{-\zeta_r\omega_r t}\left[c_{1r}\phi_r^R \cos(\omega_{Dr}t) - c_{2r}\phi_r^I \sin(\omega_{Dr}t)\right] \tag{63}$$

in which $\omega_{Dr} = \omega_r\sqrt{1 - \zeta_r^2}$, the complex eigenmode is $\phi_r = \phi_r^R \pm i\phi_r^I$, and c_{1r}, c_{2r} are decided from the initial conditions. Note that the nth component of the displacement vector $z_r(t)$ is

$$\begin{aligned} z_{rn}(t) &= 2e^{-\zeta_r\omega_r t}\left[c_{1r}\phi_{rn}^R \cos(\omega_{Dr}t) - c_{2r}\phi_{rn}^I \sin(\omega_{Dr}t)\right] \\ &= 2e^{-\zeta_r\omega_r t}\sqrt{(c_{1r}\phi_{rn}^R)^2 + (c_{2r}\phi_{rn}^I)^2}\cos(\omega_{Dr}t + \theta_{rn}) \end{aligned} \tag{64}$$

where the phase lag $\theta_{rn} = \arctan\left(\frac{c_{2r}\phi_{rn}^I}{c_{1r}\phi_{rn}^R}\right)$ is found to be different for each n, which means that the maximum displacement for each mode (n) does not necessarily take place at the same time, or, unlike the case without damping, the mode of vibration is time-dependent or unsteady.

Forced Vibration

Once the modal quantities ω_r and ϕ_r ($r = 1, 2, \ldots, N$) are found, the modal equation of motion is established as

$$M_r\left(\ddot{q}_r(t) + 2\zeta_r\omega_r\dot{q}_r(t) + \omega_r^2 q_r(t)\right) = F_r^*(t) \tag{65}$$

where $M_r = \phi_r^T\mathbf{M}\phi_r$ and $F_r^*(t) = \phi_r^T\mathbf{F}(t)$ are the modal mass and modal excitation force, and the modal response is calculated as

$$q_r(t) = A_r e^{-\zeta_r\omega_r t}\cos(\omega_{Dr}t - \gamma_r) + \int_0^t F_r^*(\tau)h_r(t-\tau)\,d\tau \tag{66}$$

where

$$h_r(t) = \frac{1}{M_r\omega_{Dr}}e^{-\zeta_r\omega_r t}\sin(\omega_{Dr}t) \tag{67}$$

is the impulse response function (IRF) for rth mode. A_r and γ_r are decided from the modal initial conditions as

$$A_r = \sqrt{\left(\frac{\dot{q}_{r0} + q_{r0}\zeta_r\omega_r}{\omega_{Dr}}\right)^2 + q_{r0}^2} \quad \text{and} \quad \gamma_r = \arctan\left(\frac{\dot{q}_{r0} + q_{r0}\zeta_r\omega_r}{\omega_{Dr}q_{r0}}\right) \tag{68}$$

in which

$$q_{r0} = q_r(0) = \frac{\phi_r \mathbf{M}}{M_r} z(0) \qquad \text{and} \qquad \dot{q}_{r0} = \dot{q}_r(0) = \frac{\phi_r \mathbf{M}}{M_r} \dot{z}(0) \tag{69}$$

Then the total response is obtained as

$$z(t) = \sum_{r=1}^{L} \phi_r q_r(t), \qquad (L \leq N) \tag{70}$$

12.2.1.5 Modal Parameter Identification Techniques

Structural health monitoring is the target, which is to track various aspects of a structure's performance and integrity in relation to the system's expected safety and serviceability. The attempted levels of structural identification are as follows:

1. presence of damage in the structure;
2. location of the damage;
3. severity of the damage;
4. prediction of the remaining service life of the structure.

It is desirable if the structural health monitoring system is inexpensive, noninvasive and also automated, so that subjective assessments by an operator can be avoided. In particular, neither the implementation nor operation of the system should involve closure of the bridge.

Vibration-based structural monitoring is considered in this context. Ideally we are trying to identify structural damage and deterioration from the measured modal parameters. Therefore, the goal of system identification is the opposite of classic dynamic analysis, where the structural properties are known and response of the system under assumed excitation is to be determined, as described in the previous section. For the present case, on the other hand, the structural parameters are to be identified from the measured response of the system.

The conventional modal identifications are based on structural response induced by known excitation such as a simple harmonic vibration imposed by an artificial excitation. For this case, the structural parameters are identified either from the free vibration decay traces, after the excitation is removed, or simply as a transfer function between input and output of the system. A difficulty of this method is to excite huge civil engineering structures for the measurement. The output-only modal identification, as opposed to the procedure based on input–output relationship, has been developed recently in order to avoid this difficulty, supported by technological development in accurate measurements of very low levels of dynamic response induced by ambient excitations.

Various techniques have been developed for determining the structural parameters. The most fundamental idea, however, is to assume mathematical models for the system and minimize the prediction errors by fitting models against the obtained output data. The prediction-error identification approach contains some fundamental procedures, such as the least squares fitting and the maximum-likelihood procedure. It is also closely related to other well-known methods in control engineering, such as the Bayesian maximum a posteriori estimation and Akaike's information criterion. The so-called correlation approach is to carry out a similar procedure by taking correlation between the prediction errors and the output data. It contains the instrumental-variable technique, as well as several methods for rational transfer function models. There is also the subspace approach, which is to identify state–space models. It consists of three steps.

1. Estimating the k-step ahead predictors using the least squares algorithm.
2. Selecting the state vector from these.
3. Estimating the state–space matrices using these states and the least squares method.

Conventional Method Applicable to Free Vibration Records

In terms of data reliability, the forced vibration test using shakers is probably the best method for the evaluation of dynamic characteristics of bridges. However, it usually requires a large-scale operation, which is naturally costly, and could also mean an interruption of services. If this approach is available, the modal parameters are to be obtained from the free vibration record. Identification of natural frequencies is usually less problematic compared to the evaluation of damping. Emphasis is placed on the identification of damping value in this section.

Log-decrement method is probably the classic and most straightforward method of identifying the system's damping. The damping ratio is obtained from the ratio of the vibration amplitudes A_j over N consecutive waves:

$$\zeta = \frac{1}{2\pi N} \ln\left(\frac{A_j}{A_{j+N}}\right) \quad (71)$$

The reading is naturally more accurate with greater N. If the damping is very high, it becomes more difficult to expect good accuracy, since N cannot be too great. With $\zeta = 0.15$, for example, the amplitude becomes 94% less after only three cycles. Even if the damping is not too high, damping often differs at different amplitude levels. It may be better to formulate, hence, the calculation as follows:

$$\zeta(A_j) = \frac{1}{8\pi} \ln\left(\frac{A_{j-2}}{A_{j+2}}\right) \quad (72)$$

A more elaborate way of determining ζ is to find an envelope of $x(t)$, which is given by

$$A(t) = \sqrt{x(t)^2 + \tilde{x}(t)^2} \quad (73)$$

where $\tilde{x}(t)$ is the **Hilbert transform** of $x(t)$, defined by

$$\tilde{x}(t) = \int_{-\infty}^{\infty} \frac{x(u)}{\pi(t-u)}\, du \quad (74)$$

The log-decrement is an ideal method for laboratory tests, where the free vibration decay traces are available relatively easily. However, when it comes to the full scale on-site measurements, it is often difficult to obtain a good decay curve.

Conventional methods applicable to ambient vibration tests

Ambient vibration survey, without any control on the input, is an attractive alternative to a large-scale forced vibration operation. It is a method to determine the dynamic characteristics of a structure by measurement of small vibrations, mostly microtremors, caused by existing disturbances such as earthquakes, wind and traffic, while the structure is in service. This method is based on a few basic assumptions as follows.

1. The input excitation is a broadband stochastic process which is adequately modelled by white-noise.
2. The system characteristics are therefore well represented by the power spectral density function of dynamic response.

3. The technique for measuring the dynamic response is sufficiently reliable.
4. Data acquisition and analysis are also sufficiently reliable.

Hence, the reliability of this method is largely decided by these factors. Errors for this case are caused by various sources such as noise contamination, coexistence of more than one predominant frequency and nonlinearity or amplitude dependence of damping.

Half-power bandwidth method

This widely practiced method is based on the fact that the width of the power spectral density of a SDOF system is proportional to the system's damping ratio. If the original data are given by $x(t) \approx \mathrm{e}^{-\alpha t}\sin(\omega_0 t)$, where $\alpha = \zeta\omega_0$, the power spectral density (one-sided) becomes

$$G_x(\omega) = 2\alpha\left(\frac{1}{\alpha^2 + (\omega + \omega_0)^2} + \frac{1}{\alpha^2 + (\omega - \omega_0)^2}\right) \tag{75}$$

with its peak reading at

$$G_P = G_x(\omega_0) \approx \frac{2}{\alpha} \tag{76}$$

Since the half-peak readings, $G_P/2$, are given at $\omega = (1 \pm \zeta)\omega_0$, the peak spectral width at half-power is given by $\Delta\omega = 2\zeta\omega_0$.

This method cannot be free of all statistical errors associated with the spectral analyses. A possible problem of this method is when the spectral peak is missing. In fact, there is a good chance not to have the reading of the exact spectral peak. If the peak is missing, the half-peak band width tends to be greater than the value in reality and the damping is likely to be overestimated.

Note also that if the peak is not of the power spectral density but the response amplitude itself, or the **dynamic amplification factor**, the frequency width $\Delta\omega$ should be taken at the peak divided by $\sqrt{2}$ rather than the half-power.

Autocorrelation decay method

This long-practiced method was developed to cover the cases where the vibration data are contaminated by random noise. By taking the autocorrelation function of the given data, the outstanding SDOF will give a decaying harmonic function simply superposed on a quickly decaying nonoscillatory function caused by random noise. When the original data are given by

$$x(t) = A\mathrm{e}^{-\zeta\omega_0 t}\sin(\omega_0 t) + n(t) \tag{77}$$

where $n(t)$ is the noise, the autocorrelation function is

$$R_x(\tau) = \frac{A^2}{2}\mathrm{e}^{-\zeta\omega_0\tau}\cos(\omega_0\tau) + C(\tau) \tag{78}$$

where $C(\tau)$ is a quickly decaying function and the frequency and damping can be identified from the first term. This is a good method if a fairly long record is available, and also there is only one outstanding system frequency to be observed. If there are two or more closely spaced system frequencies involved in the data, it becomes increasingly difficult to obtain an exact reading.

Random Decrement Method

Random excitation by noise is a fact of life and cannot be excluded anyway. The random decrement approach is somewhat similar to the autocorrelation decay method. This is also a time-domain method, developed at NASA for use in the aerospace industry (Asmussen 1997). The technique is straightforward. It consists of adding and averaging the segments of the data for predetermined length, whenever the time series exceeds a certain triggering level with the slopes alternating positive and negative. The process can be described as follows:

$$D_x(\tau) = \frac{1}{N}\sum_j x(t_j + \tau) \tag{79}$$

where

$$x(t_j) \equiv x_s \quad (const) \qquad (j = 1, 2, 3, \ldots) \tag{80}$$

$$\dot{x}(t_j) \geq 0 \qquad (j = 1, 3, 5, \ldots) \qquad and \qquad \dot{x}(t_j) \leq 0 \qquad (j = 2, 4, 6, \ldots) \tag{81}$$

$D_x(\tau)$ is called the **random decrement signature** and, in fact, it is proved to be proportional to the autocorrelation coefficient $\widetilde{R}_x(\tau)$. Generally speaking, these segments can be divided into the following three groups.

1. Deterministic step response: $x_1(t) = x_0 e^{-\alpha t}\cos(\omega_0 t)$.
2. Deterministic impulse response: $x_2(t) = \frac{\dot{x}_0 + \alpha x_0}{\omega_0} e^{-\alpha t}\sin(\omega_0 t)$.
3. Transient random response: $x_3(t) = \frac{e^{-\alpha t}}{\omega_0}\int_0^t e^{-\alpha\tau} q(\tau)\sin\left(\omega_0(t-\tau)\right) d\tau$.

The step response is due to an initial displacement x_0, the impulse response results from an initial velocity $\dot{x}_0$, and the random response is with any random component $q(t)$ in the forcing function. When the complete time-history has been searched for these segments and they are overlain and averaged together, (2) and (3) tend to average out to zero and only (1) will remain, from which the damping and frequency are decided, or

$$x(t) = \frac{1}{N}\sum_{j=1}^{N}\left(x_1(t_j) + x_2(t_j) + x_3(t_j)\right) \rightarrow x_0 e^{-\alpha t}\cos(\omega_0 t) \tag{82}$$

The method has a merit in the sense that it requires relatively short data length. However, it also requires a high rate of digitization, usually 16 f_0 or more. Also, the method is not very effective when there are two or more outstanding system frequencies coexisting. It is highly recommended to band-pass filter the data before processing if it is the case. Particularly when co-existing frequencies are closely spaced, the application of least-square curve fitting to obtain a multifrequency signature has been recommended.

Maximum Likelihood Method

This method applies a technique to fit an assumed spectral peak function by maximizing the joint probability of the sampled spectrum with it. In this process, the spectral peak is assumed to correspond to the mechanical admittance of an idealized SDOF system, which is given by

$$|H(f)| = \frac{1}{\sqrt{[1-(f/f_0)^2]^2 + (2\xi f/f_0)^2}} \tag{83}$$

and the response spectrum of the system under a white-noise excitation is given by

$$G_R(f) = |H(f)|^2 G_0 \tag{84}$$

Hence,

$$A_R = \int_0^{\infty} G_0 |H(f)|^2 \, df \approx \frac{\pi}{4} \frac{f_0 G_0}{\zeta} \tag{85}$$

represents the area under the response spectrum. Now the idea of the method is to fit a spectral density function obtained from the original data with this idealized spectrum. After normalizing the fitting function as

$$F(f_r) = \frac{4\zeta}{\pi} \frac{1}{[1-(f_r/f_0)^2]^2 + (2\zeta f_r/f_0)^2} \tag{86}$$

the **likelihood** function $L(f_0, \zeta)$ is defined as the product of $\left(F(f_r)\right)^{\alpha}$, or, by taking the logarithm of it,

$$Q(f_0, \zeta) = \ln(L) = \sum_{r=1}^{N} \left[\alpha_r \ln(F(f_r))\right] \tag{87}$$

The proper f_0 and ζ are decided by maximizing the function Q. This condition is given by

$$\frac{\partial Q}{\partial f_0} = 0 \quad \text{and} \quad \frac{\partial Q}{\partial \zeta} = 0 \tag{88}$$

However, since Q is to take its maximum when the above conditions are satisfied, as an actual operation, it is in fact easier to change the parameters f_0 and ζ within a certain range and find out where Q becomes maximum rather than directly solving the above conditions in Equation (88) as a set of simultaneous equations. A distinct advantage of this method is in the fact that it requires a relatively short record for analysis. It is also said to be relatively insensitive to variance errors. However, it requires larger computational effort compared to other simple methods.

12.2.2 Basic Tools in Statistics

12.2.2.1 Statistical Analysis of Random Processes

Essential probability

The probability of an event E is written as $P(E)$, or $P(E) = \Pr[E]$. There are three fundamental **axioms of probability** as follows. When A is a random event

$$0 \leq P(A) \leq 1 \tag{89}$$

When C is a certain event

$$P(C) = 1 \tag{90}$$

Axiom of additivity

$$P(A \cup B) = P(A) + P(B) - P(A \cap B) \tag{91}$$

The probability of an event A given the condition B is called the **conditional probability** and is written as $P(A \mid B)$. **Bayes' theorem** states that

$$P(A \mid B)P(B) = P(A \cap B) = P(B \mid A)P(A) \tag{92}$$

Statistics of a Random Variable

A sequence that consists of indexed random variables is called a **random process**. A set of random variables is often a time sequence, $X(t)$, sampled from continuous analog signals. The entire collection of all possible sets of sequences is called the **ensemble** and an individual set is called a **sample function**. When the probability distribution of the random process does not change appreciably over a time of interest, the process is called **stationary**. When the temporal and ensemble averages of a random process are equal, the process is called **ergodic**.

When a random signal does not convey any useful information, it is called **noise**. The analysis of noise signal has been developed in the telecommunication field and became one of the most important tools in dynamics. A signal whose intensity is the same at all frequencies is called the **white noise**, although an infinite-bandwidth white noise is a purely theoretical concept and, in reality, its frequency band has to be limited.

For a continuous random variable, $X(t)$ $(-\infty < t < \infty)$, the following functions are defined. **Cumulative distribution function (CDF)**

$$P_X(x) = \Pr[X(t) \leq x] \tag{93}$$

Probability density function (PDF)

$$p_X(x)\,dx = P_X(x + dx) - P_X(x) \tag{94}$$

The probability distribution represents that the probability of an event is less than or equal to a certain value x. These functions satisfy the following:

$$P_X(-\infty) = 0 \tag{95}$$

$$P_X(+\infty) = 1 \tag{96}$$

$$dP_X(x) = \Pr[X(t) \leq x + dx] - \Pr[X(t) \leq x] = P_X(x + dx) - P_X(x), \tag{97}$$

$$P_X(x) = \int_{-\infty}^{x} p_X(\xi)\,d\xi \qquad p_X(x) \geq 0 \tag{98}$$

The **mean**, or average value of $X(t)$, is defined by

$$a_X = E[X(t)] = \int_{-\infty}^{\infty} x p_X(x)\,dx \tag{99}$$

When more than one simultaneous random processes, $X_k(t)$ $(k = 1, 2, \ldots, N)$, are considered, the average value can be taken in two different ways as follows: either by using the ensemble average

$$a_X(t) = \frac{1}{N}\sum_{k=1}^{N} X_k(t) \tag{100}$$

or the time average

$$a_X(k) = \frac{1}{T}\int_0^T X_k(t)\,dt \tag{101}$$

The processes are called ergodic when $a_X(t) = a_X(k) = a_X$.

The **moments** of $X(t)$ are defined as follows: nth moments

$$m_X^{(n)} = E[x^n] = \int_{-\infty}^{\infty} x^n p_X(x)\,dx \tag{102}$$

nth central moments

$$\mu_X^{(n)} = E[(x - a_X)^n] \tag{103}$$

The following magnitudes are defined using the moment functions:

$$m_X^{(0)} = 1 \quad m_X^{(1)} = a_X \tag{104}$$

$$m_X^{(2)} = \Psi_X^2 \ \ldots \text{ mean square} \tag{105}$$

$$\mu_X^{(0)} = 1 \quad \mu_X^{(1)} = 0 \tag{106}$$

$$\mu_X^{(2)} = \sigma_X^2 \ \ldots \text{ variance} \tag{107}$$

$$\sigma_X \ \ldots \text{ standard deviation} \tag{108}$$

$$c_X = \frac{\sigma_X}{a_X} \ \ldots \text{ coefficient of variation} \tag{109}$$

$$\gamma_X^{(1)} = \frac{\mu_X^{(3)}}{\sigma_X^3} \ \ldots \text{ coefficient of skewness} \tag{110}$$

$$\gamma_X^{(2)} = \frac{\mu_X^{(4)}}{\sigma_X^4} \ \ldots \text{ coefficient of kurtosis.} \tag{111}$$

It is sometimes convenient to transform the data $X(t)$ as

$$Z(t) = \frac{X(t) - a_X}{\sigma_X} \tag{112}$$

which results in $a_Z = 0$ and $\sigma_Z = 1$.

The ratio of dynamic peak to its root-mean-square value is called the **peak factor**. If the dynamic signal is a simple harmonic fluctuation, the peak factor should be $\sqrt{2}$. If the signal has the normal distribution of its magnitude, the peak factor is known to be approximately 3.6. Some peculiar random signals have very high peak factors. Wind-induced suction, for example, on a tall building sometimes shows a peak factor of even higher than 10.

Standard Fourier Series Expansion

An infinite series of sine and cosine functions that can, if convergent, approximate a variety of periodic functions is called the **Fourier series**. Fourier series can be further transformed to a series of exponential functions by the use of Euler's formula. The study of functions given by Fourier series is called **Fourier analysis** or **harmonic analysis**.

If $x(t)$ is periodic with a period T_p, $x(t)$ can be represented by the Fourier series as follows:

$$x(t) = \frac{A_0}{2} + \sum_{k=1}^{\infty}\left[A_k \cos\left(\frac{2\pi kt}{T_p}\right) + B_k \sin\left(\frac{2\pi kt}{T_p}\right)\right] \tag{113}$$

where

$$A_k = \frac{2}{T_p}\int_0^{T_p} x(t)\cos\left(\frac{2\pi kt}{T_p}\right) dt \qquad (k = 0, 1, 2 \ldots) \tag{114}$$

$$B_k = \frac{2}{T_p}\int_0^{T_p} x(t)\sin\left(\frac{2\pi kt}{T_p}\right) dt \qquad (k = 1, 2, 3 \ldots) \tag{115}$$

Correlation Functions

Correlation is a measure to indicate how much two variables are statistically related. It is one of the most fundamental concepts in describing the statistical relationship between two signals. It can be the relationship between the excitation force and dynamic response, or between dynamic deflections at two different locations of the same structure, for example. As a special case, the correlation of a signal with the same signal itself, but with a given time interval in between, can be taken and in this case it is called the **autocorrelation** as opposed to the **cross-correlation** between two different signals. The correlation between two signals can be described in various forms with different degrees of sophistication, including the correlation functions, cross-spectral density functions and coherence.

Autocorrelation function is defined as

$$R_X(\tau) = E[x(t)x(t+\tau)] \tag{116}$$

and is normalized as

$$\widetilde{R}_X(\tau) = \frac{R_X(\tau)}{\sigma_X^2} \tag{117}$$

which is referred to as the **autocorrelation coefficient.**

Spectral Analysis

Dynamic analysis based on frequency characteristics of the processes by the use of the auto- and cross-power spectral density functions is known as **spectral analysis. Power spectral density (PSD)** is a function to represent a random process by the distribution of dynamic energy in terms of its frequency components. It is a statistical function consisting of the average squared moduli of the Fourier transform at each frequency. It represents the random characteristics of the process in the frequency domain. The ordinate of the function corresponds to the intensity of energy at that particular frequency and integration of PSD over the whole frequency range gives the variance of the process. When the structural system and excitation force are both linear and the principle of superposition is applicable, the PSD functions of input (force) and output (displacement) are related by the **transfer function**, which is given by the square of the **frequency response function (FRF)**.

The **spectral density function (SDF)** is defined by the Fourier transform of the autocorrelation as

$$G_X(f) = 2\int_{-\infty}^{\infty} R_X(\tau)e^{-i\omega\tau}\, d\tau = 4\int_0^{\infty} R_X(f)\cos(\omega\tau)\, d\tau \tag{118}$$

where $f = \omega/(2\pi)$. Note that

$$R_X(\tau) = \frac{1}{2\pi}\int_0^{\infty} G_X(f)\cos(\omega\tau)\,d\omega \tag{119}$$

and

$$R_X(0) = \int_0^{\infty} G_X(f)\,df = E[x^2] \tag{120}$$

The spectral density defined in Equation (118) is called **one-sided**, because it is defined only in the positive range of frequency. The simple application of Fourier transform on the autocorrelation function will result in the **two-sided** spectral density $S_X(f) = S_X(-f)$ $(-\infty < f < +\infty)$. Since the negative frequency does not make any physical sense, the spectral density is often redefined as follows:

$$G_X(f) = \begin{cases} 2S_X(f) & (0 \le f < \infty) \\ 0 & \text{otherwise} \end{cases} \tag{121}$$

As it is given above, the mathematical definition of spectral density function involves the **Fourier transform** (see Section 12.2.1.1). Two functions related by the Fourier transform, such as the autocorrelation function and spectral density function, make a **Fourier transform pair**, as follows:

$$F(\omega) = \int_{-\infty}^{\infty} f(t)\mathrm{e}^{-i\omega t}\,dt \tag{122}$$

$$f(t) = \frac{1}{2\pi}\int_{-\infty}^{\infty} F(\omega)\mathrm{e}^{i\omega t}\,d\omega \tag{123}$$

It should be noted that the Fourier transform is sometimes defined in somewhat different way:

$$F(\omega) = \frac{1}{\sqrt{2\pi}}\int_{-\infty}^{\infty} f(t)\mathrm{e}^{i\omega t}\,dt \tag{124}$$

$$f(t) = \frac{1}{\sqrt{2\pi}}\int_{-\infty}^{\infty} F(\omega)\mathrm{e}^{-i\omega t}\,d\omega \tag{125}$$

The resultant functions are naturally different by constant factors compared to the previous definition, but their physical meaning is the same as before.

Statistics of Two Time Series

Consider $X(t)$ and $Y(t)$ $(-\infty < t < \infty)$. In addition to the probability functions for each of X and Y, now the following joint functions are defined: **joint probability density function**

$$p_{XY}(x, y) = \frac{\Pr[x \le x(t) \le x + dx,\ y \le y(t) \le y + dy]}{dx\,dy} \tag{126}$$

joint cumulative distribution function

$$P_{XY}(x, y) = \Pr[x(t) \le x,\ y(t) \le y] \tag{127}$$

Note that $p_{XY}(x, y) = p(x)p(y)$ and $P_{XY}(x, y) = P(x)P(y)$, if $X(t)$ and $Y(t)$ are statistically independent of each other.

Transformation of Variables
When the variable $X(t)$ has a known probability density function $p(x)$, $Y = f(X)$ has the probability density function of $p(y) = p(x) \times |dx/dy|$. When $x = f_1(r, s)$ and $y = f_2(r, s)$,

$$p(r, s) = p(x, y)\frac{\partial(x, y)}{\partial(r, s)} = p(x, y) \times \begin{vmatrix} \partial x/\partial r & \partial x/\partial s \\ \partial y/\partial r & \partial y/\partial s \end{vmatrix} \tag{128}$$

Correlation functions and spectra are defined below.

Covariance

$$C_{XY} = E[(x - a_X)(y - a_Y)] = E[xy] - a_X a_Y \tag{129}$$

and

$$\rho_{XY} = \frac{C_{XY}}{\sigma_X \sigma_Y} \quad \text{(correlation coefficient)} \quad \leq 1 \tag{130}$$

Cross-Correlation Function

$$R_{XY}(\tau) = E[x(t)y(t + \tau)] = R_{YX}(-\tau) \tag{131}$$

Cross-Correlation Coefficient

$$\widetilde{R}_{XY}(\tau) = \frac{R_{XY}(\tau)}{\sigma_X \sigma_Y} \tag{132}$$

Cross-Spectral Density Function

$$G_{XY}(f) = 2\int_{-\infty}^{\infty} R_{XY}(\tau)e^{-i\omega\tau}\,d\tau = C_{XY}(f) - iQ_{XY}(f) \tag{133}$$

Note that

$$R_{XY}(\tau) = \int_0^{\infty} C_{XY}(f)\cos(\omega\tau) + Q_{XY}(f)\sin(\omega\tau)\,df \tag{134}$$

and

$$R_{XY}(0) = E[x(t)y(t)] = \int_0^{\infty} C_{XY}(f)\,df \tag{135}$$

Also

$$C_{XY}(f) = 2\int_0^{\infty} \left[R_{XY}(\tau) + R_{YX}(\tau)\right]\cos(\omega\tau)\,d\tau = C_{XY}(-f) \tag{136}$$

$$Q_{XY}(f) = 2\int_0^{\infty} \left[R_{XY}(\tau) - R_{YX}(\tau)\right]\sin(\omega\tau)\,d\tau = -Q_{XY}(-f). \tag{137}$$

Coherence Function.

$$\mathrm{coh_{XY}(f)} = \gamma^2_{XY}(f) = \frac{|G_{XY}(f)|^2}{G_X(f)G_Y(f)} = \frac{C_{XY}(f)^2 + Q_{XY}(f)^2}{G_X(f)G_Y(f)}. \tag{138}$$

Sometimes the following definitions are used instead of Equation (138):
Root-coherence

$$\sqrt{\mathrm{coh_{XY}(f)}} = \gamma_{XY}(f) \tag{139}$$

Co-coherence

$$\mathrm{cocoh_{XY}(f)} = \frac{|C_{XY}(f)|}{\sqrt{G_X(f)G_Y(f)}} \tag{140}$$

12.2.2.2 Statistical Treatment of Discrete Time Series

A time series is an ordered sequence of values of a variable at equally spaced time intervals. Time-series analysis is used for analyzing and understanding the characteristics of discrete data systems sampled from continuous observations and also for fitting of time-series models for forecasting of future values. A continuous signal $X(t)$ $(-\infty < t < \infty)$ is usually measured as a time series for only a limited time period, $0 \le t \le T$, and given by

$$x_j = X(t_0 + j\Delta t) \quad (j = 0, 1, 2, \ldots, N) \tag{141}$$

where $T = N\Delta t$ is the total sampling time, $f_s = 1/\Delta t$ is the sampling frequency, and the starting time can be chosen as $t_0 = 0$. Note that $f_c = 1/(2\Delta t) = f_s/2$ is the **Nyquist frequency**. N is usually taken as an even number. The meaning of the Nyquist frequency will be explained later.

The statistical terms required are, similar to those for a continuous signal, as follows:

Mean

$$a_X = \frac{1}{N}\sum_{j=1}^{N} x_j \tag{142}$$

***n*th central moments**

$$\mu_n = \frac{1}{N}\sum_{j=1}^{N}(x_j - a_X)^n \tag{143}$$

Standard deviation

$$\sigma_X = \sqrt{\mu_2} \tag{144}$$

Data transformation can be applied in the same way as the case for continuous variables:

$$z_j = z(j\Delta t) = \frac{x_j - a_X}{\sigma_X} \tag{145}$$

where $\Delta t = T/N$, resulting in $a_Z = 0$ and $\sigma_Z = 1$.

Trend Removal

A correction is sometimes required to remove a trend in data, defined as any frequency component whose period is longer than the record length. One of the following methods is often applied. **Average slope method:** for $T = N\Delta t$

$$\hat{z}_j = z_j - \alpha_Z\left(t - \frac{T}{2}\right) \qquad (j = 1, 2, \ldots, N) \tag{146}$$

where

$$\alpha_Z = \frac{1}{\Delta t \cdot \nu(N - \nu)}\left[\sum_{j=N-\nu}^{N} z_j - \sum_{j=1}^{\nu} z_j\right] \tag{147}$$

with ν being the largest integer $\leq N/3$.

Least square method: the desired fit is given by

$$\hat{z}_j = \sum_{k=0}^{K} b_k (j\Delta t)^k \quad (j = 1, 2, \ldots, N) \tag{148}$$

where b_k are to be chosen to minimize $Q(b) = \sum_{j=1}^{N}(z_j - \hat{z}_j)^2$. Hence $\partial Q/\partial b_r = 0$, or

$$\sum_{k=0}^{K} b_k \sum_{j=1}^{N} (j\Delta t)^{k+r} = \sum_{j=1}^{N} z_j (j\Delta t)^r \quad (r = 0, 1, \ldots, K) \tag{149}$$

For $K = 1$, for example,

$$b_0 = \frac{2(2N + 1)\sum_{j=1}^{N} z_j - 6\sum_{j=1}^{N}(jz_j)}{N(N - 1)} \tag{150}$$

and

$$b_1 = \frac{12\sum_{j=1}^{N}(jz_j) - 6(N + 1)\sum_{j=1}^{N} z_j}{\Delta t \cdot N(N - 1)(N + 1)} \tag{151}$$

Values of $K \geq 4$ are not recommended.

Statistical Functions for a Discrete Process

The following definitions are made in parallel to the continuous processes.

Probability Mass Function (PMF)

When the random variable Z is discrete

$$p_Z(z) = \Pr[Z = z] \tag{152}$$

Cumulative Distribution Function (CDF)

$$P_Z(z) = \Pr[Z \leq z] \tag{153}$$

Autocorrelation Function
When the maximum lag number is m,

$$R_Z(k\Delta\tau) = \frac{1}{N-k}\sum_{j=1}^{N-k} z_j z_{j+k} \approx \frac{1}{N}\sum_{j=1}^{N-k} z_j z_{j+k} \tag{154}$$

for $k = 0, 1, \ldots, m \ll N$. The resolution bandwidth is decided by m as

$$B_e = \frac{1}{m\Delta t} = \frac{2f_c}{m} \tag{155}$$

Autocorrelation coefficient

$$\widetilde{R}_Z(k\Delta\tau) = \frac{R_Z(k\Delta\tau)}{\sigma_Z^2} \tag{156}$$

Standard Fourier Series Expansion
If the sampled record $z(j\Delta t)$ $(j = 1, 2, \ldots, N)$ is considered as a periodic function of a period $T_p = N\Delta t$, which is the total record length,

$$z_j = z(j\Delta t) = A_0 + \sum_{k=1}^{N/2} A_k \cos\left(\frac{2\pi kj}{N}\right) + \sum_{k=1}^{(N/2)-1} B_k \sin\left(\frac{2\pi kj}{N}\right) \tag{157}$$

where $A_0 = a_Z = 0$, and for $k = 1, 2, \ldots, N/2 - 1$

$$A_k = \frac{2}{N}\sum_{j=1}^{N} z_j \cos\left(\frac{2\pi kj}{N}\right) \tag{158}$$

$$A_{N/2} = \frac{1}{N}\sum_{j=1}^{N} \cos(j\pi) \tag{159}$$

$$B_k = \frac{1}{N}\sum_{j=1}^{N} z_j \sin\left(\frac{2\pi kj}{N}\right) \tag{160}$$

Discrete Fourier Transform
Fourier transform of a discretely indexed series is called the discrete Fourier transform (DFT). The mathematical formulation is defined here but it is usually executed by the use of the fast Fourier transform algorithm, which is extremely efficient in computation

$$Z_k = \sum_{j=0}^{N-1} z_j e^{-\frac{2\pi kj}{N}i} \qquad (k = 0, 1, \ldots, N-1) \tag{161}$$

$$z_j = \frac{1}{N}\sum_{k=0}^{N-1} Z_k e^{\frac{2\pi jk}{N}i} \qquad (j = 0, 1, \ldots, N-1) \tag{162}$$

Spectral Density Function (SDF)

$$G_Z(f_k) = \frac{2\Delta t}{N}|\zeta_k|^2 = \left(\sum_{j=0}^{N-1} z_j \cos\left(\frac{2\pi kj}{N}\right)\right)^2 + \left(\sum_{j=0}^{N-1} z_j \sin\left(\frac{2\pi kj}{N}\right)\right)^2 \tag{163}$$

The SDF is usually obtained by applying the FFT routine program. **Fast Fourier Transform (FFT)** is a highly efficient algorithm for computing the DFT in high speed. The method developed by Cooley and Tukey (in 1965) has been commonly used. However, some other algorithms are also known.

Statistical Functions for Two Discrete Processes

Assume that two standardized time series, x_j and y_j, are given, where $t_k = j\Delta t$, $\Delta t = T/N$. Note that $a_X = a_Y = 0$ is assumed.

Cross-Correlation Function

When the maximum lag number is m,

$$R_{XY}(k\Delta t) = \frac{1}{N-k}\sum_{j=1}^{N-k} x_j y_{j+k} \approx \frac{1}{N}\sum_{j=1}^{N-k} x_j y_{j+k} \tag{164}$$

for $k = 0, 1, \ldots, m \ll N$. The cross-correlation coefficient equals

$$\widetilde{R}_{XY}(k\Delta t) = \frac{R_{XY}(k\Delta t)}{\sigma_X \sigma_Y} \tag{165}$$

Cross-Spectral Density

Or, by considering the DFT of x_j and y_j as Equation (161), the cross-spectral density function can be defined as follows:

$$G_{XY}(f_k) = C_{XY}(f_k) - iQ_{XY}(f_k) \qquad k = 0, 1, \ldots, m \ll N \tag{166}$$

where

$$C_{XY}(f_k) = 2\Delta t\left[A_0 + 2\sum_{j=1}^{m-1} A_j \cos\left(\frac{\pi kj}{m}\right) + (-1)^k A_m\right] \tag{167}$$

is called the **co-spectrum**, and

$$Q_{XY}(f_k) = 4\Delta t\sum_{j=1}^{m-1} B_j \sin\left(\frac{\pi kj}{m}\right) \tag{168}$$

is referred to as the **quad-spectrum**. Furthermore

$$A_j = \frac{R_{XY}(j\Delta t) + R_{YX}(j\Delta t)}{2} \quad \text{and} \quad B_j = \frac{R_{XY}(j\Delta t) - R_{YX}(j\Delta t)}{2} \tag{169}$$

The cross-spectral density is also usually directly computed through the FFT routine.

12.2.2.3 Commonly Used Probability Distributions

Models Based on Simple Discrete Random Trials – Binomial Distribution

A series of random trials in which each trial is independent and yields the binary results, S (success) or F (failure), for example, is called the **Bernoulli trials**. When the Bernoulli trials with the success rate p are repeated n times, the total number of successes X has a **Binomial distribution**, of which the PMF is given by

$$p_X(x) = \binom{n}{p} p^x(1-p)^{n-x} \qquad (x = 0, 1, 2, \ldots, n) \tag{170}$$

The CDF, for this case, can be expressed as follows:

$$P_X(x) = \sum_{r=0}^{x} \binom{n}{r} p^r(1-p)^{n-r} = 1 - \frac{B_p(x+1, n-x)}{B(x+1, n-x)} \tag{171}$$

where $B(a, b)$ is the **beta function** and $B_p(a, b)$ is the **incomplete beta function** defined by

$$B(a, b) = \int_0^1 t^{a-1}(1-t)^{b-1}\,dt = \frac{\Gamma(a)\Gamma(b)}{\Gamma(a+b)} \qquad (a, b > 0) \tag{172}$$

$$B_z(a, b) = \int_0^z t^{a-1}(1-t)^{b-1}\,dt \qquad (0 < z < 1). \tag{173}$$

The mean and variance of this distribution are given as follows:

$$a_X = E[X] = \sum_{x=-\infty}^{\infty} x p_X(x) = \sum_{x=0}^{n} x \binom{n}{x} p^x(1-p)^{n-x} = np \tag{174}$$

$$\sigma_X^2 = E[X^2] - E^2[X] = np(1-p) \tag{175}$$

N, the trial number at which the first S (success) occurs, has a **geometric distribution**, given by

$$p_N(n) = p(1-p)^{n-1} \qquad (n = 1, 2, \ldots) \tag{176}$$

with mean $a_N = 1/p$ and variance $\sigma_N^2 = (1-p)/p^2$; a_N is also called the mean return period.

Models Based on Random Occurrences – Poisson Distribution

Suppose that vehicles are randomly arriving at a specific location. Let $N(t)$ be a process that counts the number of arrivals occurring in $(0, t]$, which satisfies the following conditions.

1. The number of arrivals in nonoverlapping intervals are independent of each other.
2. The mean rate of arrivals λ for a small Δt is defined as follows:
   (a) The probability of having exactly one arrival in Δt is proportional to Δt, or

$$P(1, \Delta t) = \Pr[N + (t + \Delta t) = n + 1 \mid N(t) = n] = \lambda \Delta t. \tag{177}$$

(b) The probability of more than one arrival is negligible and vanishes when $\Delta t \to 0$. Hence, the probability of no arrivals in Δt is given by $1 - P(1, \Delta t)$, or

$$P(0, \Delta t) = \Pr[N + (t + \Delta t) = n \mid N(t) = n] = 1 - \lambda \Delta t \tag{178}$$

3. The process starts at time zero with a count of zero, or $N(0) = 0$.

When these conditions are satisfied, the process is called a **Poisson process**.

The PDF of $N(t)$, $p_N(n, t)$, is now derived. The probability of having the count n at time $t + \Delta t$ is either (a) having the count n at time t and no new arrival in the following Δt; or (b) having the count $n - 1$ at time t and one new arrival in Δt. Therefore

$$p_N(n, t + \Delta t) = P(0, \Delta t) \cdot p_N(n, t) + P(1, \Delta t) \cdot p_N(n - 1, t) \tag{179}$$

$$= (1 - \lambda \Delta t) \cdot p_N(n, t) + \lambda \Delta t \cdot p_N(n - 1, t) \tag{180}$$

or

$$\frac{p_N(n, t + \Delta t) - p_N(n, t)}{\Delta t} = -\lambda \cdot p_N(n, t) + \lambda \cdot p_N(n - 1, t) \tag{181}$$

Considering $\Delta t \to 0$, the following equation is established for $n = 1, 2, \ldots$:

$$\dot{p}_N(n, t) + \lambda p_N(n, t) = \lambda \cdot p_N(n - 1, t) \tag{182}$$

By applying the induction method

$$p_N(n, t) = \frac{(\lambda t)^n e^{-\lambda t}}{n!} \qquad (n \geq 0) \tag{183}$$

This is a **Poisson distribution**, where both the mean and variance are

$$a_N(t) = \lambda t \qquad \text{and} \qquad \sigma_N^2(t) = \lambda t \tag{184}$$

Exponential Distribution

When T is the random time to the first arrival, the probability of $T > t$, which should be $1 - P_T(t)$, is the same as the probability of having no arrivals up to time t. Hence

$$1 - P_T(t) = \Pr[N(t) = 0] = \frac{(\lambda t)^0 e^{-\lambda t}}{0!} = e^{-\lambda t} \tag{185}$$

or, the cumulative distribution of T is hence given by the **exponential distribution**

$$P_T(t) = \Pr[T \leq t] = 1 - e^{-\lambda t} \tag{186}$$

and the density function is

$$p_T(t) = \lambda e^{-\lambda t} \qquad (t \geq 0) \tag{187}$$

The mean and variance are

$$a_T = \frac{1}{\lambda}, \qquad \sigma_T^2 = \frac{1}{\lambda^2} \tag{188}$$

Gamma Distribution

It is also of interest to ask for the distribution of the time S_k to the kth arrival of a Poisson process. Now, the times between arrivals, T_i $(i = 1, 2, \ldots, k)$ are independent and have exponential distribution with common parameter λ, and $X_k = T_1 + T_2 + \cdots + T_k$. From repeated application of the convolution integral, for any $k = 1, 2, \ldots; s \geq 0$:

$$p_S(s) = \frac{\lambda(\lambda s)^{k-1} e^{-\lambda s}}{\Gamma(k)} \tag{189}$$

The mean and variance are

$$a_S = \frac{k}{\lambda} \quad \text{and} \quad \sigma_S^2 = \frac{k}{\lambda^2} \tag{190}$$

The cumulative distribution is expressed as

$$P_S(s) = \int_0^S p_S(s)\, ds = \frac{\Gamma(k, \lambda s)}{\Gamma(k)} \tag{191}$$

where $\Gamma(k, x)$ is the **incomplete gamma function** defined by

$$\Gamma(k, x) = \int_0^x e^{-u} u^{k-1}\, du \tag{192}$$

Normal or Gaussian Distribution

The **central limit theorem** states that when the random variables $Y_j(k)$ $(j = 1, 2, \ldots, N)$ are statistically uncorrelated to each other, $X(k)$ are asymptotically normally distributed when $N \to \infty$, where $X(k)$ are the total sums of these variables defined by

$$X(k) = \sum_{j=1}^{N} Y_j(k) \tag{193}$$

The **normal** or **Gaussian distribution**, $N(a_X, \sigma_X^2)$, is given by

$$p_X(x) = \frac{1}{\sqrt{2\pi}\sigma_X} \exp\left(-\frac{(x - a_X)^2}{2\sigma_X^2} \right) \tag{194}$$

or, by putting $Z = (X - a_X)/\sigma_X$

$$p_Z(z) = \frac{1}{\sqrt{2\pi}} e^{-\frac{z^2}{2}} = N(0, 1) \tag{195}$$

The CDF is given by

$$P_Z(z) = \frac{1}{\sqrt{2\pi}} \int_{-\infty}^{z} e^{-\frac{\xi^2}{2}} \, d\xi = \frac{1}{2}\left[1 + \mathrm{erf}\left(\frac{z}{\sqrt{2}}\right)\right] \tag{196}$$

where

$$\mathrm{erf}(x) = \frac{2}{\sqrt{\pi}} \int_0^x e^{-t^2} \, dt = \frac{2}{\sqrt{\pi}}\left(x - \frac{x^3}{3} + \frac{x^5}{10} - \ldots\right) \tag{197}$$

is the **Gauss error function**.

The nth central moment for even n is

$$\mu_Z^{(n)} = E[z^{2n}] = \frac{n!}{2^{n/2}(n/2)!} \tag{198}$$

resulting in

$$\gamma_Z^{(2)} = \frac{\mu_X^{(4)}}{\sigma_X^4} = 3 \tag{199}$$

which is so-called 'standard' kurtosis or flatness.

When the joint behavior of two or more variables is of interest, the multivariate normal distribution is most commonly considered. For two variables, a **bivariate normal distribution** is given as follows:

$$p_{XY}(x, y) = \frac{1}{2\pi\sigma_X\sigma_Y\sqrt{1-\rho_{XY}^2}} \exp\left(-\frac{Z_X^2 + Z_Y^2 - 2\rho_{XY}Z_X Z_Y}{2(1-\rho_{XY}^2)}\right) \tag{200}$$

where $Z_X = (x - a_X)/\sigma_X$ and $Z_Y = (y - a_Y)/\sigma_Y$ are standardized variables and $\rho_{XY} = C_{XY}/\sigma_X\sigma_Y$ is the correlation coefficient of x and y.

More generally, consider N random variables, $x_j(k)$ $(k = 1, 2, \ldots, N)$. These variables may be correlated and their means, variances and covariances are defined by

$$a_j = E\left[x_j(k)\right] \tag{201}$$

$$\sigma_j^2 = E\left[(x_j(k) - a_j)^2\right] = C_{jj} \tag{202}$$

$$C_{ij} = E\left[(x_i(k) - a_i)(x_j(k) - a_j)\right] \tag{203}$$

The joint distribution is called an ***N*-dimensional normal distribution**, whose density function is

$$p(x_1, x_2, \ldots, x_N) = \frac{1}{(2\pi)^{N/2}\sqrt{|\mathbf{C}|}} \exp\left(\left(-\frac{1}{2|\mathbf{C}|}\right)\sum_{i=1}^{N}\sum_{j=1}^{N}|C_{ij}|(x_i - a_i)(x_j - a_j)\right) \tag{204}$$

where $\mathbf{C}$ is the covariance matrix of the C_{ij}, $|\mathbf{C}|$ is the determinant of $\mathbf{C}$, and $|C_{ij}|$ is the cofactor of C_{ij} in determinant $|\mathbf{C}|$, defined by the determinant of order $N - 1$, formed by omitting the ith row and jth column of $\mathbf{C}$ and multiplied by $(-1)^{i+j}$.

Other Normal-Related Distributions

Chi-Square Distribution

When Y_n is the sum of n squared independent normal random variables, its distributions for PDF and CDF are respectively:

$$p_Y(y) = \frac{(y)^{n/2-1}}{2^{n/2}\Gamma(n/2)} e^{-\frac{y}{2}} \qquad y \geq 0 \tag{205}$$

$$P_Y(y) = \frac{\gamma(n/2, y/2)}{\Gamma(n/2)}. \tag{206}$$

In the above expression, $\gamma(z, p) = \int_0^P e^{-t} t^{z-1}\, dt$, where $Re(z) > 0$, is the **incomplete gamma function**. The mean and variance are n and $2n$, respectively, where $n = 1, 2, \ldots$ is called the degree of freedom.

Rayleigh Distribution

This is a distribution of $Y = \sqrt{X_1^2 + X_2^2}$, where both X_1 and X_2 are normal variables. If, for example, X_1 and X_2 are the N–S and E–W components of wind velocity vector and both of them are normally distributed, Y represents the magnitude of wind speed, regardless of the wind direction, whose distribution for PDF and CDF is characterized by, respectively:

$$p_Y(y) = \frac{y}{\alpha^2} e^{-\frac{y^2}{2\alpha^2}} \tag{207}$$

$$P_Y(> y) = e^{-\frac{y^2}{2\alpha^2}} \tag{208}$$

The mean and variance are

$$a_X = \alpha\sqrt{\frac{\pi}{2}} \approx 1.253\alpha \qquad \text{and} \qquad \sigma_X^2 = \alpha^2\left(2 - \frac{\pi}{2}\right) \tag{209}$$

Weibull Distribution

The Rayleigh distribution has only one parameter and sometimes it does not fit the data so well because of the lack of flexibility. The Weibull distribution tries to cover this deficit for PDF and CDF, respectively, as follows:

$$p_X(x) = \frac{k}{c}\left(\frac{x}{c}\right)^{k-1} \exp\left(-\left(\frac{x}{c}\right)^k\right) \tag{210}$$

$$P_X(> x) = \exp\left(-\left(\frac{x}{c}\right)^k\right) \tag{211}$$

The mean and variance are

$$a_X = c\Gamma\left(1 + \frac{1}{k}\right) \qquad \text{and} \qquad \sigma_X^2 = c^2\left[\Gamma\left(1 + \frac{2}{k}\right) - \Gamma^2\left(1 + \frac{1}{k}\right)\right] \tag{212}$$

The Weibull distribution coincides with the Rayleigh distribution when $c = \alpha\sqrt{2}$ and $k = 2$.

Example 12.4 [Distribution of hourly mean wind speed] The N–S and E–W components of mean wind velocity vector at any location are expressed by

$$X = U\cos(\theta) \qquad \text{and} \qquad Y = U\sin(\theta) \tag{213}$$

where U is the mean wind speed at this location, regardless of its direction, and θ is the wind azimuth angle taken from north. Unless there are strong topographical reasons, both X and Y are expected to have more or less the normal distribution as follows:

$$p_X(x) = \frac{1}{\sigma_X\sqrt{2\pi}} e^{-\frac{x^2}{2\sigma_X^2}}, \qquad p_Y(y) = \frac{1}{\sigma_Y\sqrt{2\pi}} e^{-\frac{y^2}{2\sigma_Y^2}} \tag{214}$$

The joint probability of U and θ are, by definition, given by

$$p(U,\theta) = p(X,Y)\frac{\partial(X,Y)}{\partial(U,\theta)}. \tag{215}$$

However, since X and Y are statistically independent, $p(X,Y) = p_X(x)p_Y(y)$ and

$$\frac{\partial(X,Y)}{\partial(U,\theta)} = \begin{vmatrix} \cos(\theta) & -U\sin(\theta) \\ \sin(\theta) & U\cos(\theta) \end{vmatrix} = U \tag{216}$$

Hence

$$p(U,\theta) = \frac{U}{2\pi\sigma_X\sigma_Y}\exp\left(-\left(\frac{x^2}{2\sigma_X^2} + \frac{y^2}{2\sigma_Y^2}\right)\right) \approx \frac{U}{2\pi\sigma^2} e^{-\frac{U^2}{2\sigma^2}} \tag{217}$$

where $\sigma_X \approx \sigma_Y \approx \sigma$ (const) is assumed. The PDF for both U and θ are given by

$$p(U) = p(U, 0 \le \theta < 2\pi) = \frac{U}{\sigma^2} e^{-\frac{U^2}{2\sigma^2}} \qquad \sigma_U \approx 0.655\sigma \tag{218}$$

$$p(\theta) = p(0 \le U < \infty, \theta) = \frac{1}{2\pi}\int_0^\infty \frac{U}{\sigma^2} e^{-\frac{U^2}{2\sigma^2}}\, dU = \frac{1}{2\pi} \tag{219}$$

This calculation suggests the Rayleigh distribution $P(U) = e^{-\frac{U^2}{2\sigma^2}}$. However, in reality, the meteorological observation has indicated that better agreement is expected with the Weibull distribution $P(U) = e^{-\left(\frac{U}{C}\right)^K}$.

12.2.2.4 Errors Associated with Spectral Analysis

There are several sources of statistical errors when the procedure involves spectral analyses. Principal causes are (a) bias error, (b) variance error and (c) aliasing. There are also possibilities of having some intricacies caused by the presence of nonlinearities and nonstationarities in the data (see Bendat and Piersol 1986).

Bias Error

The **bias term**, or the error caused by frequency resolution, of the estimated spectral density is, in terms of the normalized error, generally expressed by

$$\varepsilon_B[G_X(f)] = \frac{B_e^2}{24} G_X''(f) \approx -\frac{B_e^2}{3B_r^2} \tag{220}$$

where B_e is the narrow-band frequency resolution bandwidth for filtering, and B_r is the half-power bandwidth. This error can be decreased by decreasing the frequency resolution, or by increasing the number of points in the Fourier transform.

Variance Error

The **variance error** is a direct result of estimating each value of an averaged power spectrum. Since the magnitude of each individual power spectral value is random, the average value of them also should be random. The distribution of these values is similar to a χ^2 distribution. The random error, or the standard error, is given by

$$\varepsilon_V[G_X(f)] = \frac{1}{\sqrt{N}} \tag{221}$$

where N is the number of blocks used in the spectral estimates.

Aliasing and Nyquist frequency

Aliasing is a potential source of error associated with sampling of digital data. Sampling of data is usually performed at equally spaced interval Δt. If Δt is too small, the data will be highly redundant. If, on the other hand, Δt is too far apart, the data cannot recognize the frequency components properly. For example, at least two samples per cycle are required to define a frequency component in the original data. Hence, the highest frequency that can be defined by sampling at a rate of $1/\Delta t$ (Hz) is $f_N = 1/(2\Delta t)$, which is called the **Nyquist** (or **folding**) **frequency**.

Nyquist frequency is the highest frequency beyond which the signal contents cannot be properly represented by the data in discrete form. Any time signal can be measured usually only for a limited time period. Consequently, the analyzed results of the sampled data from the original signal would be different from the expected results for which the whole infinite signal is intended. This difference is called **leakage**. The problem caused by the choice of sampling time increment for digitizing the data is called **aliasing**. In order to reduce these errors, it is considered prudent to low-pass the original signal at, or preferably even below, the Nyquist frequency, before analyzing. It is difficult to make a general statement regarding the error caused by nonstationarity and/or nonlinearity which could exist in the data. Nonstationarity, for example, almost always exists even in wind-induced structural responses, which is often analyzed assuming the local stationarity. Some sort of engineering judgement has to be made before the measured data are adopted for analysis.

Leakage

Leakage caused by digitizing can be considered in the following way, too. Suppose there is a continuous time signal $X(t)$ $(-\infty < t < \infty)$ and its Fourier transform is defined by $\Psi(f)$ $(-\infty < f < \infty)$. The measurement of $X(t)$ can be done only for a limited period $(0 \leq t \leq T)$, which is mathematically equivalent to applying a window function $w(t)$ to the signal as $x(t) = w(t)X(t)(-\infty < t < \infty)$. $w(t)$ is a temporal weighting function defined by

$$w(t) = \begin{cases} 1 & \text{for} \quad |t| \leq T/2 \\ 0 & \text{for} \quad |t| > T/2 \end{cases} \tag{222}$$

Thus, the Fourier transform of $x(t)$ is given by the convolution integral

$$\xi(f) = \int_{-\infty}^{\infty} \Psi(\alpha) W(f-\alpha)\, d\alpha \qquad -\infty < f < \infty \tag{223}$$

where

$$W(f) = \frac{\sin(2\pi f T)}{2\pi f} \qquad -\infty < f < \infty \tag{224}$$

is the Fourier transform of $w(t)$. The difference of $\xi(f)$ from $\Psi(f)$ is the leakage.

12.2.2.5 Spectral Windows

Spectral windows are weighing functions to be applied in spectral analysis for processing only band-pass filtered data in the frequency domain. Windows are applied to obtain smoother spectra so that the physical interpretation of them would become easier. Basic requirements for the windows are: (a) the integration of a window function over the whole frequency range should be unity; and (b) the window function should be symmetric with respect to zero frequency. Inverse Fourier transform of the spectral windows are called **lag windows**, which are the windows in time domain and applicable to the autocorrelation functions.

In general, the sample spectra have many fluctuations and smoothing of data by the use of a weighted average is often applied to reduce these fluctuations. A very simple data window in the time domain is to apply the **moving average method**, which is to move the center point of a window and keep taking the average of data over its width b, or

$$f(t) \Rightarrow \overline{f}_b(t) = \frac{1}{b}\int_{t-b/2}^{t+b/2} f(\tau)\, d\tau \tag{225}$$

Rectangular Pulse Window

$$w(t) = 1/b \quad (|t| \leq b/2) \tag{226}$$

Since

$$w(t-\tau) = \begin{cases} 1/b & \text{for } t-b/2 \leq \tau \leq t+b/2 \\ 0 & \text{otherwise} \end{cases} \tag{227}$$

the convolution is given by

$$\overline{f}_b(t) = \int_{-\infty}^{\infty} f(\tau) w(t-\tau)\, d\tau \tag{228}$$

and its Fourier transform is equal to the product $F(f)W(f)$, where $F(f)$ and $W(f)$ are the Fourier transforms of $f(t)$ and $w(t)$, respectively. $W(f)$ is given by

$$W(f) = \int_{-\infty}^{\infty} w(t) e^{-2\pi i f t}\, dt = \frac{\sin(\pi b f)}{\pi b f} = \mathrm{dif(bf)} \tag{229}$$

where

$$\text{dif}(x) = \frac{\sin(\pi x)}{\pi x} \tag{230}$$

is called the **diffraction function**.

Spectral Window

Smoothing of spectra themselves can be done by applying a window as

$$\overline{S}(f) = \int_{-\infty}^{\infty} S(g)W(f-g)\,dg \tag{231}$$

It is required that $\int_{-\infty}^{\infty} W(f)\,df = 1$ and $W(f) = W(-f)$ so that the area $\int_{-\infty}^{\infty} S(f)\,df$ is constant. The shape of the window can be rectangle but there are many other types. However, what matters most is the effective width of the window over which the averaging takes place. It is often defined by

$$B_e = \left[\int_{-\infty}^{\infty} W^2(f)\,df\right]^{-1} \tag{232}$$

which is b for the rectangular window. $W(f)$ has its peak value of unity at $f = 0$ and reduces to zero at $f = 1/b$. Hence it is practically a low-pass filter. However, in the range of $f > 1/b$, there are small ups and downs of this function and these fluctuations are called the **side lobes**.

Filters

Filter is a term used for an electronic device or mathematical algorithm to process a data stream by means of separating the frequency components of signals. There are various types of filter functions employed in engineering signal processing including the **digital filters** such as Hanning and Hamming windows.

Hanning Window

When the raw estimate of a power spectral density is given as $G_k(f)$ by applying Equation (169), where $k = 1, 2, \ldots, m$, a smooth estimate of G_k by the use of a Hanning window is given as follows:

$$\overline{G}_0 = 0.5 \cdot (G_0 + G_1) \tag{233}$$

$$\overline{G}_k = 0.25 \cdot G_{k-1} + 0.50 \cdot G_k + 0.25 \cdot G_{k+1} \quad k = 1, 2, \ldots, m-1 \tag{234}$$

$$\overline{G}_m = 0.5 \cdot (G_{m-1} + G_m) \tag{235}$$

This window is frequently used before applying a fast Fourier transform to avoid aliasing artifacts. It was developed by Julius von Hann and known widely because of good leakage characteristics such as low peak side-lobe height and a rapid decay rate for side lobes far from the centre lobe.

Hamming Window

Another type of digital filter is named after R.W. Hamming, and also is widely employed. It is defined by the following equation in lieu of Equation (234):

$$\overline{G}_k = 0.23 \cdot G_{k-1} + 0.54 \cdot G_k + 0.23 \cdot G_{k+1} \qquad k = 1, 2, \ldots, m-1 \tag{236}$$

12.2.2.6 Wavelet Analysis

Wavelet transform is a tool for decomposing a signal into its time- and scale-dependent components, in terms of so-called wavelet coefficients. It is an extended idea from the Fourier transform so that it is more suitable for the analysis of nonstationary data.

Fourier transform is a very versatile tool in signal analysis, but it is not suitable for identifying nonstationary aspects of the signal. For example, since Fourier transform is applied to the entire signal length, the result cannot indicate at what time in the signal a specific frequency existed. It is really a tool for frequency resolution but not for time resolution. It means that if the method is applied to structural health monitoring, the method may recognize damage occurrence, location, or even its severity, but not exactly when the damage happened. Wavelet transform does not have a single set of basis functions like the Fourier transform, which utilizes only the sine and cosine functions. Instead, wavelet transform has infinite possibility of choosing basis functions. Thus the wavelet analysis provides immediate access to information that can be obscured by other time–frequency methods such as Fourier analysis.

Wavelet transform is basically an extended application of a windowing technique with variable-sized windows. It allows the use of long time intervals where low-frequency information is needed and shorter intervals for high-frequency information.

Wavelet coefficients are calculated by applying a chosen wavelet function to the signal as

$$W_X(a,b) = \frac{1}{\sqrt{|a|}}\int_{-\infty}^{\infty} x(t)\psi^*\left(\frac{t-b}{a}\right)dt \tag{237}$$

Here, $W_X(a,b)$ are wavelet coefficients of the signal $x(t)$, at position (or shift) b and scale a. Scale a is inversely proportional to frequency f. ψ^* is the complex conjugate of the wavelet function $y(t)$. The inverse transform is expressed as

$$x(t) = \frac{1}{C_\psi}\int_{-\infty}^{\infty}\int_{-\infty}^{\infty} W_X(a,b)\frac{1}{\sqrt{a}}\psi\left(\frac{t-b}{a}\right)\frac{da\,db}{a^2} \tag{238}$$

C_ψ is called the admissibility constant, defined by

$$C_\psi = \int_{-\infty}^{\infty}\frac{|\Psi(\omega)|^2}{|\omega|}d\omega < \infty \tag{239}$$

$\Psi(\omega)$ is the Fourier transform of $\psi(t)$.

The discrete wavelet transform of discrete time sequence $x(n)$ is defined by

$$C_{j,k} = \frac{1}{2^{j/2}}\sum_n x(n)\psi_{j,k}(n) \qquad \text{where} \qquad \psi_{j,k}(n) = \frac{1}{2^{j/k}}\psi\left(\frac{n}{2^j}-k\right) \tag{240}$$

in which $\psi(n)$ is the wavelet function and j, k are called the scaling coefficient and shifting coefficient, respectively. $C_{j,k}$ represent the corresponding wavelet coefficients. The inverse transform for this case is given by

$$x(n) = \sum_j\sum_k C_{j,k}\psi_{j,k}(n) \tag{241}$$

12.2.3 Random Vibrations

12.2.3.1 Single-Degree-of-Freedom Systems

When a SDOF system, which is characterized by the frequency response function $H(f)/k$, is subjected to a random excitation $F(t)$, the system's response $x(t)$ is calculated as follows for the mean response

$$a_X = \frac{a_F}{k} \tag{242}$$

with the response spectrum

$$G_X(f) = \frac{|H(f)|^2}{k^2} G_F(f) \tag{243}$$

where $G_F(f)$ is the excitation spectrum (one-sided).

The root-mean-square response σ_X is evaluated by the following integration:

$$\sigma_X^2 = \int_0^\infty G_X(f)\,df = \frac{1}{k^2}\int_0^\infty |H(f)|^2 G_F(f)\,df \tag{244}$$

The peak response is usually estimated as

$$\hat{x} = a_X + g_X \sigma_X \tag{245}$$

where

$$g_X = \sqrt{2\ln(\nu T)} + \frac{\gamma}{\sqrt{2\ln(\nu T)}} \tag{246}$$

is the peak factor. $\gamma = 0.5772\ldots$ is called the 'Euler–Mascheroni constant', T is the evaluation period, and ν is the cycling rate, defined by

$$\nu = \frac{1}{\sigma_X}\sqrt{\int_0^\infty f^2 G_X(f)\,df} \tag{247}$$

Example 12.5 [A random point load $F(t)$ located at $z = a$ on an elastic beam] By the modal analysis, the beam response is expressed as $x(z,t) = \sum_r \phi_r(z) q_r(t)$. Hence the variance of the response is approximately $\sigma_X^2(z) \approx \sum_r \phi_r^2(z)\overline{q_r^2}$, in which the modal variance is

$$\overline{q_r^2} = \frac{\phi_r^2(a)}{M_r \omega_r^2}\int_0^\infty |H_r(f)|^2 G_F(f)\,df \tag{248}$$

where

$$|H_r(f)|^2 = \frac{1}{(1-\Omega_r^2)^2 + (2\zeta_r\Omega_r)^2} \quad \text{and} \quad \Omega_r = \frac{\omega}{\omega_r} \tag{249}$$

ω_r and ζ_r represent the circular frequency and damping ratio for the rth mode of vibration.

$G_F(f)$ is the power spectral density of the random force $F(t)$. If, for example, the excitation spectrum is given by a *white noise* G_0 $(0 < f < \infty)$, Equation (248) reduces to the following:

$$\overline{q_r^2} = \frac{\phi_r^2(a)}{M_r\omega_r^2} G_0 \frac{\pi f_r}{4\zeta_r} \approx \frac{0.7854 f_r}{K_r\zeta_r} \phi_r^2(a) G_0 \tag{250}$$

If there is more than one loading, $F(a_j, t)$ $(j = 1, 2, \ldots, N)$, being applied to the system, the response spectrum is given by

$$\overline{q_r^2} = \sum_{i=1}^{N}\sum_{j=1}^{N} \frac{\phi_r(a_i)\phi_r(a_j)}{M_r\omega_r^2} \int_0^\infty |H_r(f)|^2 G_F(a_i, a_j; f)\, df \tag{251}$$

where the cross-spectrum $G_F(a_i, a_j; f)$ can be approximated by

$$G_F(a_i, a_j; f) = \sqrt{G_F(a_i, f) G_F(a_j, f)} \gamma_F^2(a_i, a_j; f) \tag{252}$$

12.2.3.2 Distributed Random Excitation

When an elastic structure is subjected to a distributed random excitation force $F(z, t)$, the response spectrum in rth mode of vibration is, by extending the concept of Equation (251), as follows:

$$\overline{q_r^2} = \frac{1}{M_r\omega_r^2} \int_0^\infty |H_r(f)|^2 \int_L \int_L G_F(z_1, z_2; f)\phi_r(z_1)\phi_r(z_2)\, dz_1\, dz_2\, df \tag{253}$$

For the case of distributed random load on an elastic beam, often $G_F(z; f)$ does not vary with the space coordinate and the cross-correlation can be simplified as

$$G_F(z_i, z_j; f) = G_F(f)\gamma_F^2(\Delta_{ij}, f) \tag{254}$$

where $\Delta_{ij} = |z_i - z_j|$ is the distance between z_i and z_j. Then

$$\sigma_X^2(z) \approx \sum_r \frac{\Phi_r^2(z)}{M_r\omega_r^2} \int_0^\infty |H_r(f)|^2 |J_r(f)|^2 G_F(f)\, df \tag{255}$$

in which

$$|J_r(f)|^2 = \int_L \int_L \gamma_F^2(\Delta_{12}, f)\phi_r(z_1)\phi_r(z_2)\, dz_1\, dz_2 \tag{256}$$

is called the **joint acceptance function**.

12.2.3.3 Extreme Value Analysis

Extreme value distributions are the limiting distributions for the extreme values, such as the maxima or minima, of a large collection of random observations. In many civil engineering applications, concern often lies with the largest values of many events. This means that our attention is focused upon the upper tail of the parent distribution of actual observations. Fisher and Tippett (1928) proved that there are only three forms of extreme value distributions. Extreme value analysis was largely developed and elaborated by Emil J. Gumbel (1891–1966), a German statistician.

Estimation of Extremes from the Parent Distribution
Consider the random variables X_j ($j = 1, 2, \ldots, N$) and the maximum of them, $Y = \max(X_j)$. The probability that Y does not exceed a certain value y in n discrete events is expressed by using the cumulative distribution function of the parent population as

$$P_Y(< y) = [P_X(< y)]^n \tag{257}$$

For a large value of y, $P_X(> y)$ is generally very small. Hence,

$$1 - P_Y(> y) = [1 - P_X(> y)]^n \approx 1 - nP_X(> y) \tag{258}$$

or

$$P_Y(> y) \approx nP_X(> y) \tag{259}$$

Example 12.6 [Annual maximum wind speed] If $P_Y(> y)$ is the annual maximum of the observed hourly mean wind speed exceeding y, the inverse of it is the *return period* R. The distribution of hourly mean wind is often expressed by the Weibull distribution as

$$P_X(> y) = \mathrm{e}^{-(y/C)^K} \tag{260}$$

Hence, the annual maximum wind can be expressed as

$$y(R) = C(\ln(nR))^{1/K} \tag{261}$$

where $n \approx 10^3$ in our experience.

Type I Distribution
Suppose n samples are randomly taken from the parent group. The parent size is so large that its distribution will not be influenced by taking samples away. At each sampling, the maximum value of the case is retained while the others are returned to the distribution. The sampled maxima will make their own distribution after repeating this process and the resulted distribution will be increasingly removed from the parent, when n increases. This corresponds to the extreme value distribution. Fisher and Tippett (1928) proved that there are only three limiting functional forms of extreme value distributions, which are similar but with a difference caused by the properties of the upper tail of the parent distribution. Gumbel called them **asymptotes** because they are approached asymptotically when $n \to \infty$.

The first of the three, often called the **Type I distribution**, is derived when the distribution of X is (a) unlimited in the positive direction; and (b) the upper tail falls off in an exponential manner as $P_X(< x) \approx 1 - \mathrm{e}^{-g(x)}$, where $g(x)$ is an increasing function of x. The cumulative function is given by

$$P_{\hat{X}}(< \hat{x}) = \mathrm{e}^{-\mathrm{e}^{-\alpha(\hat{x}-u)}} \tag{262}$$

for $-\infty < \hat{x} < \infty$, where u is the mode, or the most probable value, and α is the dispersion. The density function is

$$p_{\hat{X}}(\hat{x}) = \frac{dP_{\hat{X}}(< \hat{x})}{d\hat{x}} = \alpha\mathrm{e}^{-\alpha(\hat{x}-u)-\mathrm{e}^{-\alpha(\hat{x}-u)}} \tag{263}$$

with the mean

$$a_{\hat{X}} = u + \frac{\gamma}{\alpha} \tag{264}$$

where γ is again Euler's constant, and the variance is

$$\sigma_{\hat{X}}^2 = \frac{\pi^2}{6\alpha^2} \tag{265}$$

Hence, the two parameters can be decided by using:

$$u = a_{\hat{X}} - \frac{\gamma}{\alpha} \approx a_{\hat{X}} - 0.450\sigma_{\hat{X}} \tag{266}$$

$$\frac{1}{\alpha} = \frac{\sqrt{6}}{\pi}\sigma_{\hat{X}} \approx 0.7797\sigma_{\hat{X}} \tag{267}$$

Type II and III Distributions

Type II distribution can be applied when the parent distribution of variables is limited on the left at zero, but unlimited to the right in the tail of interest. The probability functions of extremes $\hat{x}$, $\hat{x} \geq 0$, are then defined by

$$P_{\hat{X}}(< \hat{x}) = \mathrm{e}^{-(u/\hat{x})^k} \tag{268}$$

$$p_{\hat{X}}(\hat{x}) = \frac{k}{u}\left(\frac{u}{\hat{x}}\right)^{k+1} \mathrm{e}^{-(u/\hat{x})^k} \tag{269}$$

The jth moment is given by

$$E[\hat{x}^j] = \int_{-\infty}^{\infty} \hat{x}^j p_{\hat{X}}(\hat{x})\, d\hat{x} = u^j \Gamma\left(1 - \frac{j}{k}\right) \tag{270}$$

resulting in

$$a_{\hat{X}} = u\Gamma\left(1 - \frac{1}{k}\right) \qquad (k > 1) \tag{271}$$

$$\sigma_{\hat{X}}^2 = u^2\left[\Gamma\left(1 - \frac{2}{k}\right) - \Gamma^2\left(1 - \frac{1}{k}\right)\right] \qquad (k > 2) \tag{272}$$

Note that if X has Type II distribution with parameters u_x and k, $Y = \ln(X)$ has the Type I distribution with parameters $u_y = \ln(u_x)$ and $\alpha = k$.

Type III distribution is usually applied to a model dealing with smallest values. Assuming the variables as $X \geq 0$, the probability distribution of the minima is actually identical to the Weibull distribution.

Probability of Threshold Crossing

The expected number of up-crossing the level $x = a$ per unit time, a stationary random process $X(t)$ is expected to make, is, according to Rice (1954), given by

$$N_X^+(a) = \int_0^\infty \dot{x} p(a, \dot{x})\, d\dot{x} \tag{273}$$

where $p(x, \dot{x})$ is the joint probability density function of x and $\dot{x}$. For a Gaussian stationary random process with zero-mean $p(x, \dot{x}) = p(x)p(\dot{x})$ where the probability density function is given by

$$p_X(x) = \frac{1}{\sqrt{2\pi}\sigma_X} \mathrm{e}^{\left(-\frac{x^2}{2\sigma_X^2}\right)} \tag{274}$$

Hence,

$$N_X^+(a) = p_X(a) \int_0^\infty \dot{x} p(\dot{x})\, d\dot{x} = \sqrt{2\pi} \nu \sigma_X p_X(a) \tag{275}$$

where

$$\nu = \frac{1}{2\pi} \frac{\sigma_{\dot{X}}}{\sigma_X} = \frac{1}{\sigma_X} \left(\int_0^\infty f^2 G_X(f)\, df \right)^{1/2} = N_X^+(0) \tag{276}$$

is the zero up-crossing frequency and is called the **cycling rate**, which is often considered as the expected frequency of the process. $G_X(f)$ is the power spectral density of $X(t)$.

By substituting Equation (276) into Equation (275), the number of up-crossing can be expressed as

$$N_X^+(a) = \frac{1}{2\pi} \frac{\sigma_{\dot{X}}}{\sigma_X} \mathrm{e}^{-\frac{a^2}{2\sigma_X^2}} \tag{277}$$

Note that $N_X^+(a)$ decreases when the threshold a goes higher. For a given σ_X, the crossing rate increases as the root-mean-square velocity $\sigma_{\dot{X}}$ increases.

Concept of Narrow-Band Envelope

Experience tells us that, if the process is narrow-band, with the mid-band frequency of f_m, most arrivals occur in clusters approximately one cycle period apart, but there may be a substantial length of time between clusters. The waiting times are obviously not exponentially distributed for this case. For engineering design, we are often interested in the first-time crossing and subsequent crossings in the same cluster are less important. To address this situation, Rice introduced the concept of **envelope function** $R(t)$. The up-crossing rate of the envelope is obtained in a similar way as before and

$$N_R^+(a) = \sqrt{2\pi} \frac{a}{\sigma_X} \sqrt{\nu^2 - f_m^2} e^{-\frac{a^2}{2\sigma_X^2}} \tag{278}$$

Note that the envelope up-crossing is zero if ν and f_m coincide.

Concept of First-Passage Failure

Basic Formulation

Consider a stationary random stress process $S(t)$ and the strength of material defined by $R(t)$. Note that $R(t)$ is generally a deterministic function of time. The up-crossing of $S(t)$ past $R(t)$ in a small time increment Δt forms a point process, whose rate of arrival is described by a Poisson process. Then, it follows that the probability of not having an up-crossing in Δt is

$$P(Y < R) = \exp\left(-N_R^+(t)\Delta t\right) \tag{279}$$

Considering the service lifespan of the structure to be T, and $\Delta t = T/k$, where $k \to \infty$ and $\Delta t \to 0$, the probability of not having an up-crossing in T is given by

$$\begin{aligned} P(Y < R) &= \prod_{j=1}^{k} \exp\left(-N_R^+(t_j)\Delta t\right) \\ &= \exp\left(-\sum_{j=1}^{k} N_R^+(t)\Delta t\right) \to \exp\left(-\int_0^T N_R^+(t)\,dt\right) \end{aligned} \tag{280}$$

Hence, the probability of failure is defined by

$$P_f = 1 - \exp\left(-\int_0^T N_R^+(t)\,dt\right) \tag{281}$$

where, if $X(t)$ is a Gaussian process, $N_R^+(t)$ is given by Equation (277) or

$$N_R^+(t) = \nu \exp\left(-\frac{1}{2}\left(\frac{R(t) - a_S}{\sigma_S}\right)^2\right) \tag{282}$$

Example 12.7 If R is constant, and therefore also N_R^+ is constant, then $p_f \approx N_R^+ T$.

Example 12.8 If $X(t)$ is a narrow-band Gaussian process with the central frequency f_m, since $\nu = f_m$, it results in

$$p(f) \approx f_m T \exp\left(-\frac{1}{2}\left(\frac{R - a_S}{\sigma_S}\right)^2\right) \tag{283}$$

Cycle Counting

The valuable outcomes of the field observation of structural behavior are usually given as lengthy and irregularly fluctuating time-histories of acceleration, stress, deflection and so on. An absolutely essential matter for engineers then is to reduce this information to a small amount that is useful for fatigue analysis, for example. Counting the number of cycles of the record fluctuation is one method of data reduction.

There are various ways of performing this operation, such as counting peaks, level crossings and ranges, which are defined by the difference between two successive extremes. All of these are called

one parameter methods, and the two-parameter method **rainflow analysis** is known to be a state-of-the-art counting method successfully applied to fatigue analysis. Note that the cycle counting yields amplitude distribution with no regard to frequency information.

Peak Counting

A relatively simple method where the local maxima above mean are identified, each maximum is paired with a local minimum of the same magnitude, and thus an equivalent time-history is obtained.

Range cycling

A method where the range, defined by the difference between two successive local extremes, is counted as a half cycle.

Rainflow Cycle Counting

The general approach in fatigue life prediction needs to relate a random load fluctuation in a real life situation to the Wöhler curves, which are based on laboratory experiments of simple specimens subjected to constant amplitude load. The rainflow cycle counting analysis is a method proposed to overcome this difficulty, first proposed by Endo et al. (1968) and has been developed by many researchers including Downing (1972) and Rychlik (1987). It is now regarded as a state-of-the-art estimator in fatigue analysis. The basic way of counting is explained in Figure 12.3, which is a strain time-history plotted against time, where the time is vertically downward and the line connecting the strain peaks are imagined to be a series of pagoda roofs, on which rainwater drips and falls.

Several rules are imposed on defining the half-cycles that are counted.

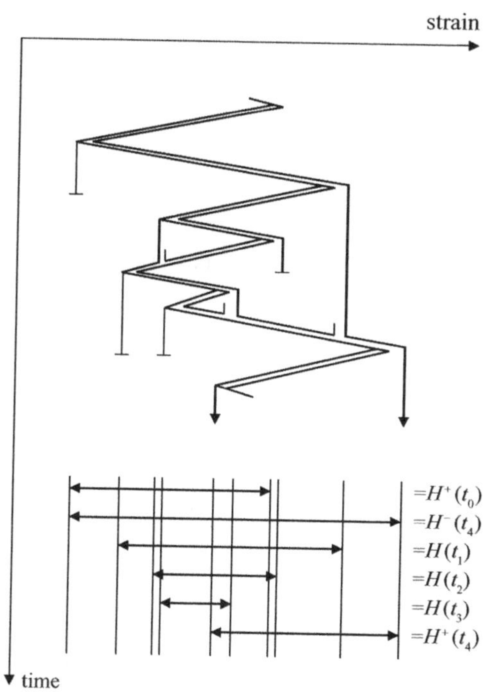

Figure 12.3 Rainflow cycle counting (after Rychlik 1987)

1. Rainflow begins at the inside of each strain peak and strain trough.
2. A rainflow must stop if it meets another flow from the roof above.
3. A rainflow goes on, unless it is stopped, to the tip of the roof and drips down.
4. The rainflow initiating at each trough is allowed to drip down and continue, but stops when it comes opposite to a trough that is more negative than where it originated.
5. The rainflow initiating at each peak is allowed to drip down and continue, but stops when it comes opposite to a peak that is more positive than where it originated.

Thus a random time series is now reduced to an equivalent pair of reversals. A counting matrix, as a result, is obtained where the number of occurrences of closed cycles from a certain level of displacement to another is established for convenience of fatigue life prediction.

12.2.3.4 Simulation Techniques

Once the characteristics of a random process, the observed dynamic excitation for example, are identified, it is useful to simulate the process by a mathematical function or a time series for the future prediction of the process. There have been various simulation techniques developed and investigated by many researchers and the following is a typical example.

Application of Fourier Series

When $x(t)$ is a stationary Gaussian random process with $a_X = 0$, and its power spectral density function is given by $G_X(\omega)$, a sample function of $x(t)$ can be approximately simulated by the following expression:

$$x^d(t) = \sum_{k=1}^{N} \sqrt{2G_X(\omega_k)\Delta\omega}\cos(\omega_k t + \phi_k) \tag{284}$$

where

$$\omega_k = \omega_L + \left(k - \frac{1}{2}\right)\Delta\omega \qquad (k = 1, 2, \ldots, N) \tag{285}$$

with

$$\Delta\omega = \frac{\omega_U - \omega_L}{N} \tag{286}$$

ω_U and ω_L are the upper and lower bounds of $G_X(\omega)$, the phase angles ϕ_k are randomly produced mutually independent variables with a flat distribution in the range of $0 \leq \phi_k < 2\pi$, and N is a sufficiently large integer. For the production of random variables X that follows the cumulative distribution function $P_X(x)$, the Monte Carlo method can be applied.

If the random process to be simulated is nonstationary, such as the case of earthquake waves, the model function can be assumed as $f(t) = g(t)x(t)$, where $g(t)$ is a known deterministic function of time. For this case, a simulated sample function can be simply expressed by $f^d(t) = g(t)x^d(t)$.

Simulation of the process can be also made by applying the ARIMA time series (e.g. Box and Jenkins 1976).

12.2.4 Some Notes on Damping

12.2.4.1 Nature of Vibration Damping in General

The capacity of structures to dissipate energy imparted by the external forces is one of the most basic knowledge requirements for the dynamic analysis of buildings, towers, bridges and other civil engineering structures. For built-up structures, energy dissipation may result from various sources. Bleich and Teller (1952) and Van Koten (1974) listed the following sources.

1. The imperfect elasticity of the structural materials.
2. Plastic yielding and friction due to small relative displacements in the structural joints.
3. Internal friction of the materials such as concrete.
4. Friction in the expansion joints of the floor structure or at the pedestals
5. Friction at the end bearings of the trusses and girders.
6. Aerodynamic and hydrodynamic damping.
7. The nonlinear structural characteristics such as of cables.
8. Energy dissipation through foundation on the ground and any other substructures.
9. Artificial dampers installed on the structures, etc.

Theoretical approaches are possible to some extent for estimating the damping effects on dynamic behavior of structures from each origin based on some basic experimental information. Nevertheless, because of the complexity and interactions of these mechanisms, actual measurements for the structures in reality are essentially important for finding the overall damping capacity of compound structures.

12.2.4.2 Sources of Damping

A brief discussion on various sources of damping follows. References such as Richart et al. (1970), Nashif et al. (1985) and Soong (1990) can be recommended for further study if there is a need. The magnitude of damping in this article is given as a damping ratio (ζ), or the fraction of critical.

Material Damping

Intrinsic material damping always exists due to plastic and viscoelastic behavior of structural materials. This appears in conventional viscous type and the following values, as the damping ratio (ζ) or the fraction of critical damping, are generally accepted from a number of experimental results: steel, $\zeta = 0.001 - 0.004$; concrete $\zeta = 0.010 - 0.020$; and wood $\zeta = 0.020 - 0.025$. At increasing stresses, if the limit of proportionality is exceeded, this deviation will cause extra damping. Damping capacity in material can be defined as the ratio of the energy ΔW dissipated on one cycle of oscillation to the maximum amount of energy W accumulated in that cycle (see Figure 12.4), or

$$\Delta W = \oint F\,dx = \int_0^T F\frac{dx}{dt}\,dt \tag{287}$$

In the case where a hysteresis loop occurs because of the material nonlinearity, as shown in Figure 12.5, the area (A) enclosed by the loop indicates the amount of energy dissipated during each complete cycle. Note that when the stress is related to the strain and strain rate by

$$\sigma = E'\varepsilon + \frac{E''}{\omega}\frac{d\varepsilon}{dt} \tag{288}$$

the stress–strain **hysteresis loop** becomes an ellipse as shown in Figure 12.5 and $A = \oint \sigma\,d\varepsilon$.

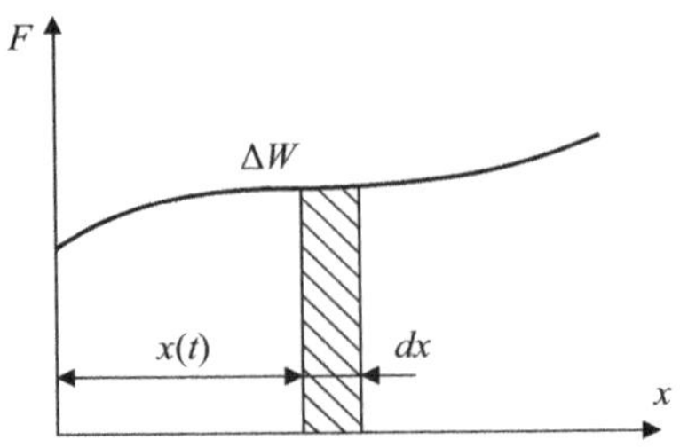

Figure 12.4 Force–displacement relationship

There is a group of special metal alloys called **high damping alloys** developed to have very high damping characteristics. In these special alloys, the gain in damping is often at the expense of stiffness, strength, durability, corrosion resistance, cost, machinability, or long-term stability and they are usually not suitable for construction purposes. They are often highly nonlinear and also temperature sensitive. One example is a material called **Sonoston**, which is commercially available (e.g. Jones and Trapp 1971).

It is also worthwhile mentioning **composite materials**. These are combinations of two or more materials at a macroscopically homogeneous level. Fibers of a material may be embedded uniformly in either single or multiple directions, or short fibers are randomly embedded in a matrix, etc. Boron fibers in an aluminum or titanium matrix and carbon fibers in an epoxy matrix are typical examples. Often the aim of making composite materials is to increase the stiffness and/or reduce the overall density of the material and the change of damping characteristics is not a consideration. There are also disadvantages, such as low natural resistance to erosion and impact damage, high costs, difficulty in repairing, etc. Damping is not necessarily high but often highly nonlinear.

There are also materials categorized as the **viscoelastic materials**. Many polymeric and glassy materials exhibit high viscoelastic damping see e.g. Oberst (1986) and Grootenhuis (1969).

Coulomb Frictional Damping

Significant amount of energy can be dissipated by friction at structural joints, such as bolted, riveted or simply placed with their surfaces contacting each other, between components of the structure, cladding joints, masonry facades, composite floors and so on. The magnitude of damping force for these cases is directly proportional to the coefficient of friction (μ), the unit pressure (P_N) between surfaces and the

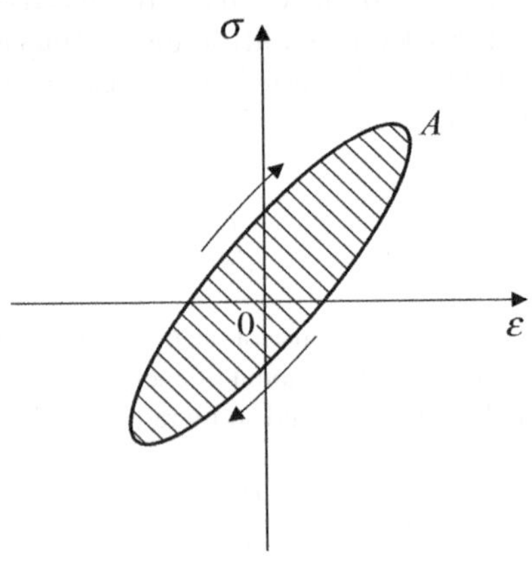

Figure 12.5 Stress–strain hysteresis

area of contact (S), or

$$F_f = \mu S P_N. \tag{289}$$

Typical values for coefficient of friction (μ) are as follows: brass versus steel, 0.15; steel versus steel, 0.15; leather versus steel, 0.35; nylon versus metal, 0.30; and teflon versus metal, 0.05.

In general, the static coefficient tends to be greater than its dynamic counterpart. In the case of steady-state vibration, the relative velocity becomes equal to zero twice during each cycle and thus the effective coefficient of friction falls between two extremes. The values listed above reflect this condition to an extent.

Free vibration with Coulomb damping is completely damped out at time T_e, which is given by

$$\frac{T_e}{T_0} = \frac{y_0 k}{4F_f} \tag{290}$$

where T_0, is the natural period of vibration, k is stiffness and y_0 is the initial amplitude.

Radiation Damping through Foundation

This is due to the propagation of energy away from the structure into the infinite soil mass. The magnitude of radiation damping depends on the soil properties and also on the size of the substructure. Several examples of soil stiffness in terms of an equivalent spring constant (k) and damping ratio (ζ) associated with various modes of vibration of a rigid circular footing on an idealized elastic half-space are given as follows, see Richart et al. (1970) and Figure 12.6.

Vertical	$k_z = \dfrac{4Ga}{1-\nu}$	$\zeta_z = \dfrac{0.425}{\sqrt{B_z}}$	$B_z = \dfrac{1-\nu}{4}\dfrac{m}{\rho a^3}$
Rocking	$k_\phi = \dfrac{8Ga^3}{3(1-\nu)}$	$\zeta_\phi = \dfrac{0.15}{(1+B_\phi)\sqrt{B_\phi}}$	$B_\phi = \dfrac{3(1-\nu)}{8}\dfrac{I_\phi}{\rho a^5}$
Horizontal	$k_X = \dfrac{32(1-\nu)Ga}{7-8\nu}$	$\zeta_X = \dfrac{0.288}{\sqrt{B_X}}$	$B_X = \dfrac{7-8\nu}{32(1-\nu)}\dfrac{m}{\rho a^3}$
Rotational	$k_\theta = \dfrac{16}{3}Ga^3$	$\zeta_\theta = \dfrac{0.50}{1+2B_\theta}$	$B_\theta = \dfrac{I_\theta}{\rho a^5}$

In these examples, m is mass on the half space, I is mass moment of inertia on the half space, a is the radius of the footing, G is the shear modulus of the elastic media, ρ is the density of the elastic media and ν is Poisson's ratio of the elastic media.

As an approximation for various other forms of footings, an equivalent radius a of the footing can be taken by equating the actual contact area to πa^2.

However, the dynamic stiffness should be, generally speaking, a function of the excitation frequency. Hence, it should be more generally given in the form of a complex function

$$K(\omega) = k\left[f_1(\omega) + if_2(\omega)\right] \tag{291}$$

and $C = 1/K$ is called the **dynamic compliance**. f_1 and f_2 are generally given as functions of $\omega a/V_S$ as typically shown in Figure 12.7, where $V_S = \sqrt{G/\rho}$, see Hsieh (1962). The damping ratio is then given by $\zeta = f_2/2f_1$. There have been a number of studies carried out for the derivation of dynamic stiffness, including radiational damping, in the form of Equation (291) for different forms of footings, pile foundations and effects of embeding. They belong to the category called soil-structure interaction problems. The analytical solutions are often complicated, involving so-called **mixed boundary condition**

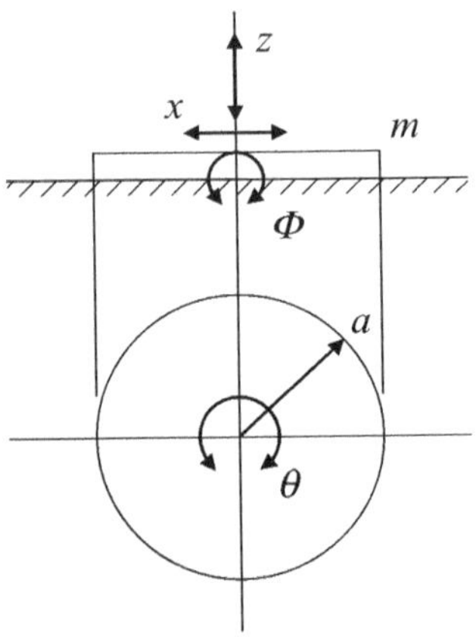

Figure 12.6 Four directions of displacement (after Richart et al. 1970)

problems, which can be simplified by assuming the contact pressure distribution under the footing. It should be noted that the static case corresponds to $\omega \to 0$, resulting in $f_1 \to 1$ and $f_2 \to 0$.

Aerodynamic Damping

The aerodynamic damping is typically caused by a 'fan' action of a structure as it moves through wind. The damping in still air is generally negligible whereas it is a major factor for the case of **hydrodynamic damping**. The magnitude of aerodynamic damping is more or less proportional to wind speed but is also dependent upon the vibration amplitude. It significantly varies with the geometrical shape of the structure. The amount of this damping is often of the order of 1% and 5% of critical for buildings and bridges, respectively, but it could be even negative in some instances, leading to aerodynamic instability.

Consider the linear aerodynamic force acting on a SDOF system. The equation of motion is given by

$$m\frac{d^2x}{dt^2} + c\frac{dx}{dt} + kx = \frac{\rho V^2}{2}A\left(F_R x + F_I\frac{dx}{dt}\right) \tag{292}$$

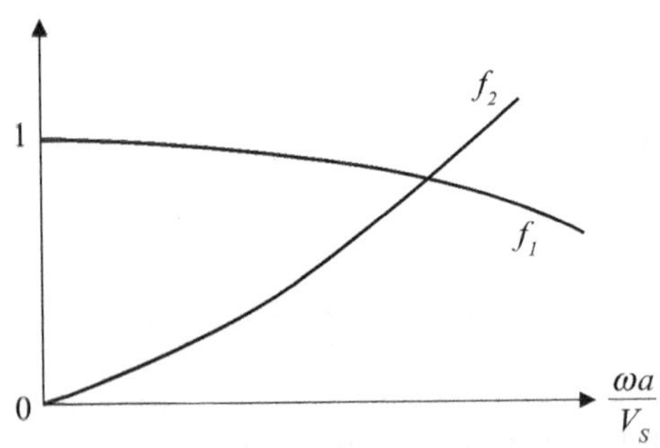

Figure 12.7 General tendency of functions defined in Equation (291)

or

$$m\frac{d^2x}{dt^2}+\left(c-\frac{\rho V^2 A F_I}{2}\right)\frac{dx}{dt}+\left(k-\frac{\rho V^2 A F_R}{2}\right)x=0 \qquad (293)$$

where m, c, k are the mass, damping and stiffness of the system, respectively, A is the area of the body exposed to wind, ρ and V are air density and wind speed, and F_R and F_I are the aerodynamic force coefficients.

From Equation (293), if $F_I > 2c/(\rho V^2 A)$, the total damping becomes negative. It means that once vibration of this structure starts its amplitude will grow endlessly leading to its destruction. This is a case of aerodynamic instability or **flutter**. Similarly, if $F_R > 2k/(\rho V^2 A)$, the total stiffness becomes negative and the structure becomes unstable and also will be destroyed. This phenomenon is called **divergence**. Generally speaking, F_R and F_I are given as functions of the reduced wind speed $V_r = V/(f\sqrt{A})$, where f is the frequency of the vibration and, for this case, is likely to be very close to the natural frequency. Generally speaking, aerodynamic force coefficients have to be obtained experimentally. However, when the wind speed is much higher than the moving speed of the body itself, so-called *quasi-steady* assumption can be introduced, which is to say that the aerodynamic forces at any instant depends only upon the instantaneous position of the body at that particular moment and the temporal memory effect or the history of the motion can be ignored. If this assumption is justified, the aerodynamic damping can be given as follows.

$$\text{Lift} \quad \zeta_{z,aero} = \frac{\rho VA}{4m\omega_z}\frac{dC_L}{d\alpha}$$

$$\text{Drag} \quad \zeta_{x,aero} = \frac{\rho VA}{2m\omega_x}C_D$$

$$\text{Pitch} \quad \zeta_{\theta,aero} \propto \frac{\rho VA^{3/2}}{4\Theta\omega_\theta}\frac{dC_M}{d\alpha}$$

Variables ω_Z, ω_X, ω_θ are the natural circular frequencies in lift, drag and pitching motions, respectively, m and Θ are mass and mass moment of inertia of the body, and C_L, C_D, C_M are the coefficients of the aerodynamic lift force, drag force and pitching moment, respectively.

12.2.4.3 Artificial Dampers

Concept of Vibration Control

Vibration control can be achieved by either (a) reducing the external disturbances applied to the structure; or (b) increasing the damping force to dissipate dynamic energy out of the system. Improvement of aerodynamic performance by choosing and adjusting the geometrical details or adding edge fairings (Figure 12.8), corner vanes and vortex killers often have been employed to avoid adverse dynamic effects of wind, see Wardlaw (1992). These actions are typical of category (a). Base isolation to reduce the seismic response of structures also belongs to this category.

It is obviously not possible to completely isolate the structure from the ground motion. However, the idea is to allow having the relative displacement between the base and the ground to an extent so that the transmission of the base excitation from the ground would be reduced (Figure 12.9). At the same time, it is a usual practice to expect some sort of damping with the isolation mechanism so that a part of the dynamic energy, which is usually proportional to the relative velocity, could be dissipated. If this is achieved by inserting any damping system with the isolation mechanism, the addition of artificially increased damping belongs to category (b).

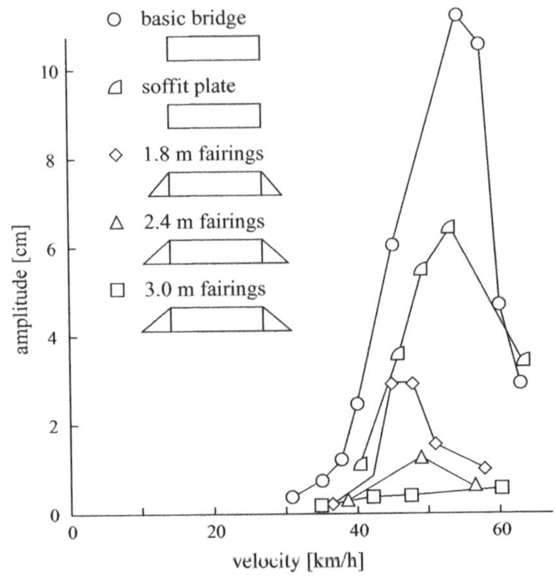

Figure 12.8 Fairings of the Longs Creek Bridge (after Wardlaw (1994))

More typical of category (b) is the installation of artificial dampers. They can be simply an addition of viscous materials or frictional mechanisms at the movable joints. The World Trade Center Towers, New York, had nearly 10,000 viscoelastic damper elements installed in their frames in each building. Artificial dampers also can be a fairly complicated mechanical system, which may need to be operated by supplying additional energy from outside. The damper system which requires the additional energy for controlling structural vibrations is called an **active control** system, see Soong (1990) and Hirsch (1994). Gyro-stabilizer, a system to stabilize the rolling motion of a ship by rotating a gyroscope, is an example of an active control system. Many of the artificial dampers often listed in references such as shock absorbers, hanging chains, water sloshing basins, etc. (Figure 12.10), on the other hand, usually do not require any energy supply from outside. Their action is triggered by the vibration of the primary structure itself and the induced action accordingly works in dissipating energy. This type of damper is called a **passive control** system. Tuned mass dampers can be either active or passive.

Passive Dampers

The simplest kinds of passive control dampers include **hydraulic shock absorbers** which are often used for the bridge stay-cables and installation of either viscoelastic- or hysteresis-type **shear dampers**

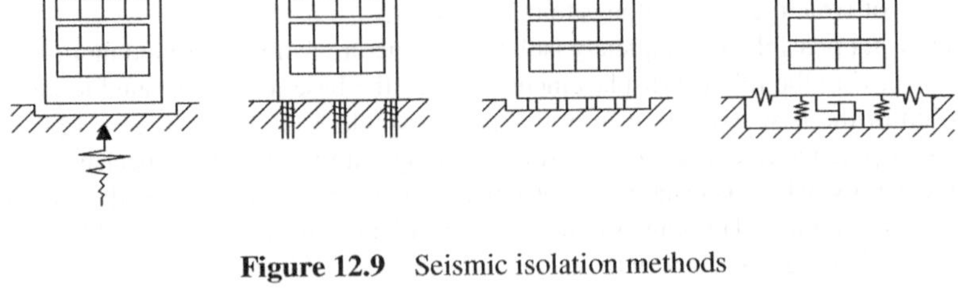

Figure 12.9 Seismic isolation methods

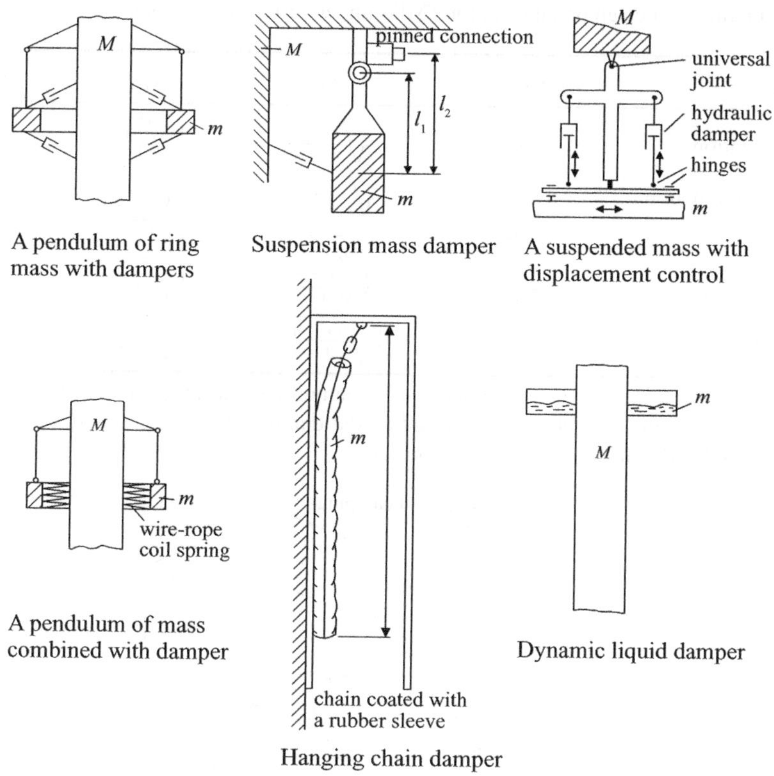

Figure 12.10 Various types of passive dampers (after Ruscheweyh 1982)

applied to the building frames to augment seismic resistance capacity. There is also a group of **impact dampers**. These dampers are not particularly complicated in principle. Rather special dampers that are worthwhile noting are the tuned mass dampers (TMDs) and liquid sloshing dampers (LSDs). Both of them work on the same mechanical principle which involves an auxiliary vibratory system, using a solid mass (TMD) or liquid mass (LSD).

Tuned Mass Dampers

The idea of the TMD is to have a small auxiliary system of a mass-spring damper whose natural frequency is tuned to the frequency of the primary system or the structure so that the vibration energy of the primary structure can be absorbed and dissipated by the auxiliary system. The original idea was extensively discussed by Den Hartog (1956) for the case of a simple harmonic excitation. In civil engineering applications, however, the excitation forces are usually better defined as a random process with a certain bandwidth and hence the choice of physical parameters for the design of TMDs need to reflect this reality to make them effective. In the past 30 years or so, TMDs have been applied to some major structures, including the Centrepoint Tower, Sydney, the CN Tower, Toronto, the John Hancock Tower, Boston, and the Citicorp Center, New York, and also for stabilizing various structures particularly during erection before the structure has reached its full design stiffness (Figure 12.11).

Considering the difference in nature of external excitations, the optimized physical parameters for the design of TMDs can be summarized as seen in Table 12.1, see Fujino and Abé (1993). The optimization for the design of the damper referred to in Table 12.1 is:

Table 12.1 Optimized design parameters for TMDs under various excitations

Excitation	$\beta = \omega_T/\omega_S$	ζ_T	ζ_{eff}	Optimization
Harmonic excitation	$\frac{1}{1+\mu}$	$\sqrt{\frac{3\mu}{8(1+\mu)}}$	$\frac{1}{2}\sqrt{\frac{\mu/2}{1+\mu/2}}$	(1)
Harmonic ground motion	$\frac{1}{\sqrt{1+\mu}}$	$\sqrt{\frac{3\mu}{8(1+\mu/2)}}$	$\frac{1}{2}\sqrt{\frac{\mu(1+\mu)}{2}}$	(1)
Free vibration	$\frac{1}{1+\mu}$	$\frac{u}{\sqrt{1+\mu}}$	$\frac{1}{2}\sqrt{\frac{\mu}{1-\mu/4}}$	(2)
Self-excited vibration	$\frac{1}{\sqrt{1+\mu}}$	$\frac{1}{2}\sqrt{\frac{\mu}{1+\mu/2}}$	$\frac{1}{2}\sqrt{\frac{\mu(1+\mu)}{1+\mu/2}}$	(3)
Random excitation	$\frac{\sqrt{1+\mu/2}}{1+\mu}$	$\frac{1}{2}\sqrt{\frac{\mu(1+3\mu/4)}{1+3\mu/2}}$	$\frac{1}{4}\sqrt{\frac{\mu(1+\mu)}{1+3\mu/4}}$	(4)

ω_T = natural circular frequency of the damper; ω_S = natural circular frequency of the structure; ζ_T = damping ratio of the damper itself; ζ_{eff} = overall modal damping including the effect of the damper; μ = damper mass/modal mass of the structure ($\ll 1$).

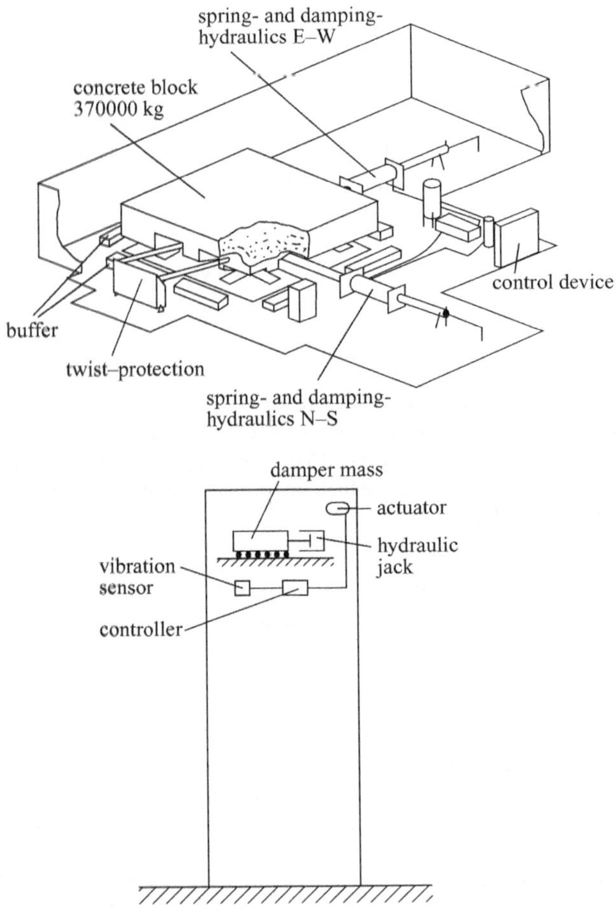

Figure 12.11 A tuned mass damper system, the Citycorp Center, New York (after Ruscheweyh 1982)

1. by maintaining the response as low as possible for any excitation frequencies (in case of harmonic excitation and ground motion);
2. by maximizing the modal damping (in case of free vibration);
3. by maximizing the acceptable negative damping (in case of self-excited vibration);
4. by minimizing the root-mean-square response (in case of random excitation).

There are a few important aspects to be remembered for the application of TMDs.

Tuned mass dampers are most effective when structures are under steady excitations since they require some time to transfer vibration energy before they start dissipating it. They are less effective, therefore, under unsteady excitations such as impulsive loading and earthquakes. Tuned mass dampers tend to be very effective when their frequency is exactly synchronized. However, a slight deviation in such design parameters could significantly reduce their effectiveness. Since a degree of uncertainty in estimation of natural frequencies, for example, is inevitable, the lack of robust performance is a weakness of this type of dampers.

For many civil engineering structures where the modal mass is quite large, a required auxiliary mass for the damper system may become enormous. Also, in order to have the damper effectively functioning, a large stroke of damper movement may be required. By restricting this stroke, the damper will not be fully effective. In order to cover these deficiencies, the TMD system designed and installed in the Citicorp Center Building, New York, was made to be an active control system. The damper is essentially a hydraulic servo system which controls the motion of a mass that weighs 400 t, supported hydrostatically in two mutually perpendicular directions. The damper-system mass is equal to about 1% of the building mass. The spring stiffness is pneumatic and the provided damping is of hydraulic type, see Petersen (1980).

Tuned Liquid Dampers

Instead of using a solid mass, liquid can be used in the same concept as TMDs and, for this case, it is called the tuned liquid damper (TLD). For this case, the gravity field provides the restoring force on the water mass which performs a sloshing motion in a container and hence it is also called the **tuned sloshing damper**. Tuned liquid dampers have been used in the aerospace industry and in marine vessels for quite some time. The use of liquid could be even more advantageous in the following ways.

- It works even for a very slight disturbance, for which TMDs are sometimes ineffective because of the frictional resistance at its solid surface.
- The damper system itself is not complicated and usually less expensive.
- One damper can be effective for the structural movement in two or more directions.
- Easy to install and easy to relocate.
- Easy to maintain and unlikely to have any aging problems such as fatigue failures.

Just like in TMDs, TLDs also require proper frequency tuning and appropriate choice of the damper's own damping to have the system work effectively. There are basically two types of TLDs as follows.

- Shallow liquid in a relatively small container, see Fujino et al. (1988). Energy dissipation is due mainly to surface wave breaking with large vibration amplitude for this case and the synchronization of the damper's frequency does not have too much influence on the overall effectiveness. However, with smaller amplitude, damping effect is quite sensitive to the frequency tuning. The nonlinear surface wave theory is known to predict natural frequencies with good accuracy unless wave breaking is involved.
- Deep water contained in a large tank, see Ueda et al. (1991). The motion of water becomes a relatively smooth sloshing, for which Housner's sloshing model can be applied. However, in order to increase the damper's damping, various grids, nets and barriers are usually inserted to create turbulence, which

also somewhat increases the sloshing frequency. Since theoretical prediction of damping is difficult at any rate, its effectiveness largely relies on experimental evaluation. Another application of the same principle is TLCD, the **tuned liquid column damper**, which is basically to have a liquid column such as a U-tube and utilizes it as an auxiliary mass. Its frequency can be controlled by adjusting pressure of the air room above the water surface and damping is changed by inserting an orifice in the tube.

Active Dampers

Active control of structures has been applied to aircraft and spacecraft for sometime but its application to civil engineering structures has only a short history so far, see Soong (1990) and Hirsch (1994). However, considering the fact that they are generally more capable of controling vibrations with different frequencies and vibration modes compared to the passive type of dampers, it can be expected that they will be used more extensively in future, see François et al. (2000). Some of the successful applications so far include the First National City Corporation Building, New York, see Petersen (1980), as mentioned earlier, and the Crystal Tower, Osaka, see Nagase and Hisatoku (1992). Also, for the past 10 years or so, the active vibration dampers often have been applied to control vortex shedding excitation of suspension bridge pylons during erection in Japan. It was found to be particularly advantageous to use active dampers for these cases since the natural frequencies are almost continuously changing and also because it was desirable to install relatively small dampers which are effective in two or more vibration modes simultaneously.

There is another category of active dampers, which is to install the aerodynamic active control devices to reduce wind-induced motion of buildings and bridges. Mechanically controlled additional flaps are known to be quite effective in controlling the occurrence of flutter instability of wings and streamlined shallow box girders, see Kobayashi and Nagaoka (1992).

12.2.4.4 Effects of Vibrations on Human Bodies

Human Sensitivity to Motion and Stresses

The motion and mechanical stresses resulting from the exposure of the human body to vibrations can have several different effects such as:

- direct interference with physical activities or functions;
- mechanical damage or destruction;
- secondary effects such as changes in an organism, see Harris (1976).

Mechanical Interference

Some types of displacement, velocity and/or acceleration are known to be very disturbing to sensory and neuromuscular activities such as reading, speaking and position controls. For example, the disturbance of visual acuity is recognized in certain frequency ranges and is also found to become proportionally serious to vibration amplitude.

Mechanical Damage

Mechanical damage of the human body arises under application of the accelerative forces in the form of either shock or vibration. Among these are bone fracture, lung damage, injury to the inner wall of the intestine, brain injury, cardiac damage, ear damage, tearing or crushing of soft tissues, and some types of chronic injury such as tendon or joint strains and interruption of circulatory system. The response of humans to mechanical vibration is often frequency dependent, particularly at or near the resonant frequency with the visceral organs. It is also likely that there is heating of the body when it is shaken hard.

Biological Responses

Mechanical stresses and motions may stimulate various receptor organs or may excite parts of the nervous system and hormonal activity. These changes are difficult to measure since they can be somewhat subjective. Nevertheless, considerable indirect evidence exists for the reality of these response patterns such as fatigue, changes in work capacity, ability to maintain attentiveness, etc. Also, there are acute emotional reactions such as fear or unpleasantness which lead to automatic or deliberate compensation or protective behavior.

Human Tolerance Criteria based on ISO Standard

A general guideline for the evaluation of human exposure to whole-body vibration has been developed as ISO 2631 (ISO 1980). Three different levels of human discomfort are distinguished as follows.

- Reduced comfort boundary – applies to tolerable disturbance during ordinary life activities such as eating, reading or writing.
- Fatigue-decreased proficiency boundary – gives the level at which recurrent vibrations would cause fatigue to personnel and reduction of efficiency as a consequence.
- Exposure limit – defines the maximum vibration tolerable with respect to human health and safety and is much higher than the other two.

The bounds given in Figure 12.12 are for the **fatigue-decreased proficiency boundaries**. The **exposure limits** are obtained roughly by multiplying by two and the **reduced-comfort boundaries** are given approximately by dividing by three. These bounds also depend on the direction of incidence to the human body. The reference coordinates are shown in Figure 12.13. The criteria given here are in terms of an effective acceleration, which is the root-mean-square value over the exposure time T.

Design Criteria for Acceptable Comfort

Obviously the knowledge of physical tolerance or human perception itself does not automatically decide any **acceptance** level applicable to everybody. For example, Melbourne and Cheung (1988) have pointed out that 'the least tolerant appears to be those people doing office work, with apartment dwellers seemingly more tolerant and tower-top diners even more.' It also depends on visual and auditory cues that heighten occupant awareness. It is the level and frequency of unacceptable motion that is critical to the development of structural design criteria.

In reality, the criteria for body injuries or clear disturbances for physical functions are far beyond our considerations. Needless to say, with regard to their safety the civil engineering structures are expected to serve people without giving any serious discomfort. What is at stake, therefore, is not only an individual's perception but also the resultant state of mind. This is not a very easy condition to define since it is not only a function of acceleration and frequency but also there is a variety of physiological and psychological factors influencing it. However, the line has to be drawn somewhere.

Influencing factors include the individual's position (moving, standing, sitting or lying), preoccupation and knowledge, existence of any visual or auditory cues, health conditions on that particular day, and so on. There have been a number of early experimental investigations carried out regarding the tolerable peak accelerations for people as functions of frequency, see Harris (1976) and Wiss and Parmelee (1974). Figure 12.14 is a crude summary of them. However, these studies cover only the frequency range down to 1 Hz. Only a few studies have explored the region below 1 Hz and a good summary of them has been presented by Melbourne (1998), which is reproduced in Figure 12.15. Note that the acceleration of Figure 12.15 is given in root-mean-square (RMS), which should be multiplied by $\sqrt{2}$ in theory to be compared with peak values of Figure 12.14, since the motions referred to in Figure 12.15 are all presumably sinusoidal. However, Melbourne suggests the use of 2–2.5 instead as an appropriate factor to be multiplied.

In particular, it is worth noting the results produced by Chen and Robertson (1973), which were the work carried out for the design of the World Trade Center Towers, New York. As indicated in Figure 12.15,

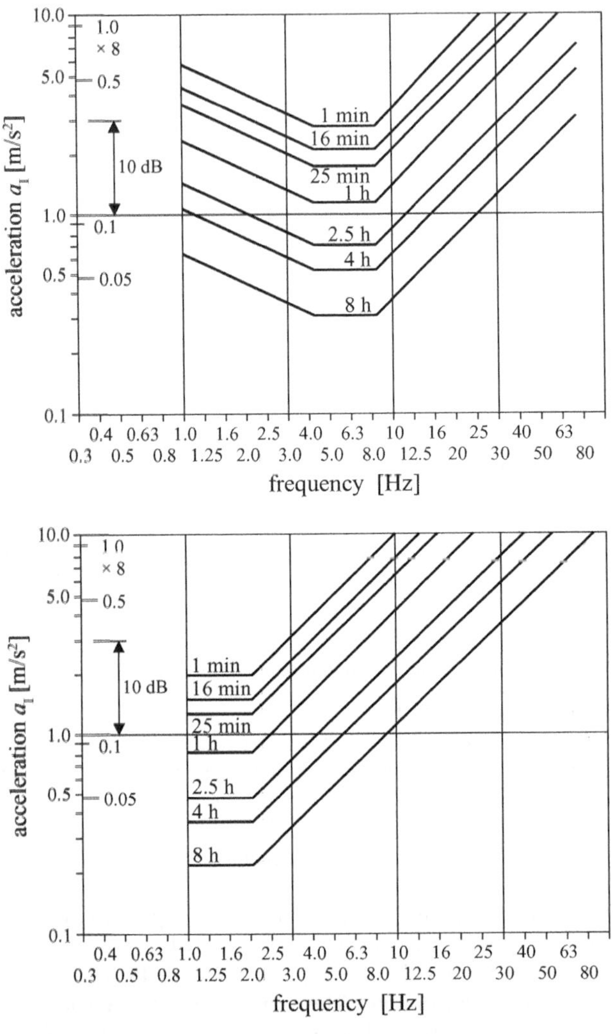

Figure 12.12 Fatigue-decreased proficiency boundaries (after Harris 1976)

the results show a substantial range of people's difference in perception – the perception range of the whole population spans about a decade of accelerations. Also the 'lower threshold' indicated by Irwin (1979) appears to present a good lower limit to motion perception. Chang (1973) has given a similar comparison of results from various sources as shown in Figure 12.16, although the outcome is a little less conclusive. Khan and Parmelee (1971), in connection to the design of the John Hancock Center, Chicago, carried out an experimental programme to investigate accelerations on subjects in various body positions and found that the difference between individuals' perception tends to be more significant than the effects of body positions. It indicates 4 mg for a perception level and 2 mg as a disturbing level for the frequency of 0.13 Hz, which fit well with other findings.

Regarding, in particular, wind-induced motion of tall buildings, the prediction method of structural response was basically established by the end of the 1970s, but their acceptability criteria had not yet

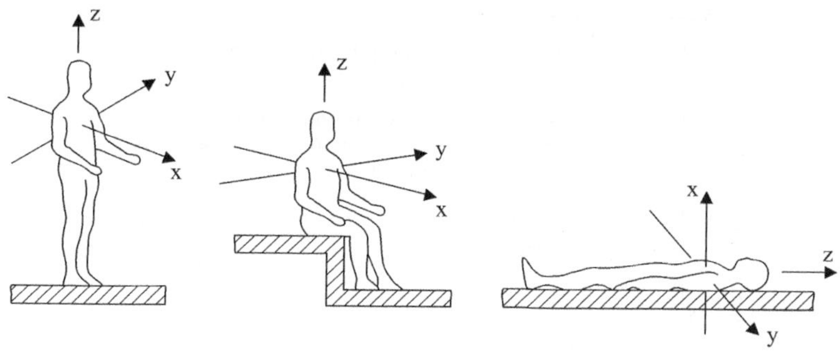

Figure 12.13 Reference coordinates in relation to the position of the body

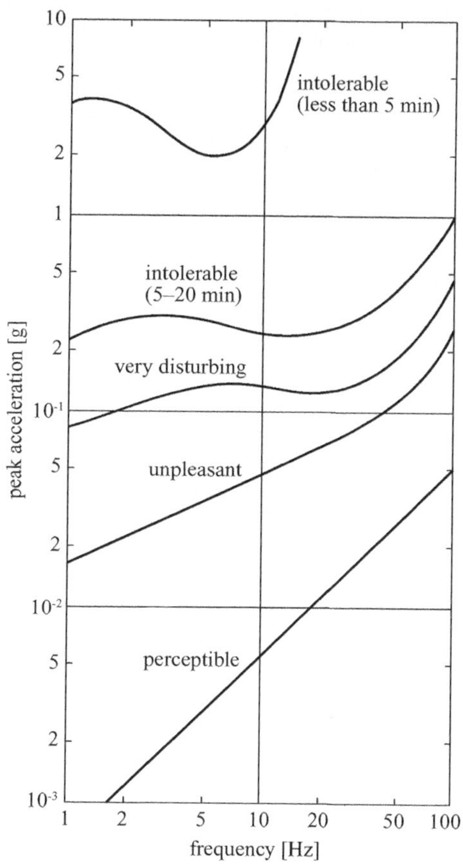

Figure 12.14 Tolerable peak accelerations based on many studies

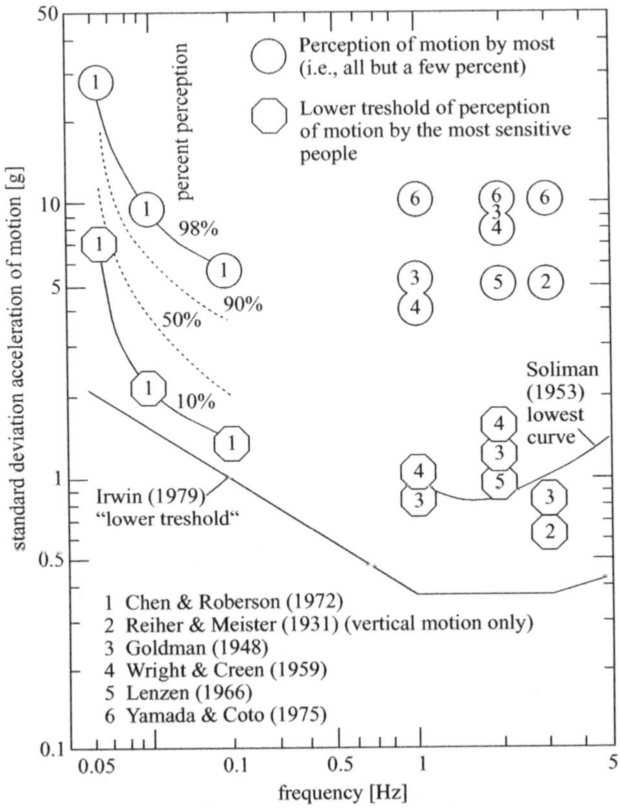

Figure 12.15 Summary of studies of human exposures to frequencies below 1 *rm Hz* (Melbourne 1998)

been established. Hansen et al. (1974) were probably the first to work on this issue systematically. Their study suggested the RMS acceleration should not exceed 5 mg for a return period of 6 years to avoid significant vacancy of an office building.

Based on a number of full-scale observations and experiments, Chen and Robertson (1973) and Irwin (1986) developed a general expression to indicate the level of human occupancy comfort as follows:

$$\sigma_a = \mathrm{e}^{-3.56-0.41\ln(n)} \tag{294}$$

where σ_a is the root-mean-square acceleration [$\mathrm{m/s^2}$] and n is the frequency with an approximately normal distribution [Hz].

Equation (294) has been quoted by ISO 6897 (ISO 1984a) as a satisfactory magnitude indicating a level of acceleration at which 'about 2% of the occupants will comment adversely'. Melbourne and Cheung (1988) further developed the peak acceleration criteria for the return period of less than 10 years as follows:

$$\hat{a} = \sqrt{2\ln(nT)}\left[0.68 + \sqrt{\ln(R)/5}\right]\mathrm{e}^{-3.56-0.41\ln(n)}, \tag{295}$$

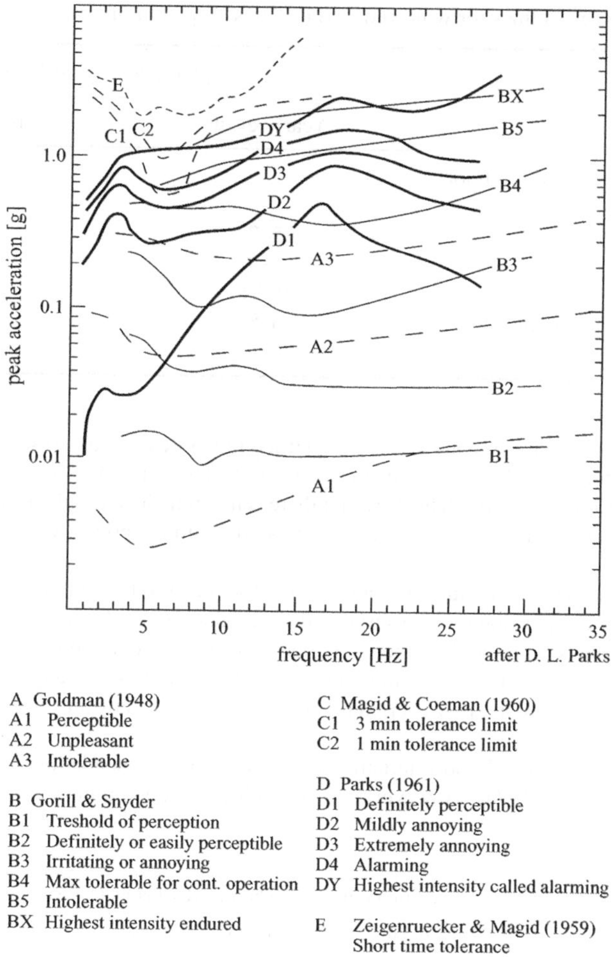

Figure 12.16 Summary of studies of human exposure to acceleration frequencies (from Chang 1973)

where $\hat{a}$ is the peak acceleration in the horizontal plane [m/s^2] R is the return period [years] and T is the evaluation period of wind storm (= 600 s).

Both Equations (294) and (295) are presented in Figure 12.17. Further to these, Isyumov et al. (1995) have come up with tentative guidelines as shown in Table 12.2 considering the importance of visual and auditory cues. These guidelines do not recognize the importance of any frequency dependence. Nevertheless, Melbourne (1998) regards that it would at least avoid any building motion perception problems, although admittedly more work is needed to evaluate these effects exactly.

12.2.4.5 Other Socio-Economic Impacts

Production–Quality Criteria

The acceptance criteria for vibrations of civil engineering structures need to be considered from various aspects. First of all, the structure should not collapse and must maintain its structural integrity so that it

Table 12.2 Tentative guidelines for wind-induced motions in tall buildings Isyumov et al. (1995)

Category of structures	Acceptable hourly peak values (10 Mg)	
	Yearly event	10-year event
Peak resultant accelerations (top floor) should be at or below:		
residential	5–7	10–25
hotels	7–9	15–20
office	9–12	20–25
Peak torsional velocity (top floor) should be at or below	1.5	3.0

does not lose its serviceability. Second, even if the structure is safe and able to serve, if it produces any discomfort to the customers and/or causes any mechanical problems such as overstressing, malfunctioning or misalignment, it is not acceptable. Third, even if there is no immediate problem, any troubles in future, such as structural fatigue damage, possibly even compound with material corrosion, must be carefully avoided.

The acceptance criteria for structural vibrations therefore need to be established in three categories. (a) structural criteria; (b) physiological criteria; and (c) production-quality criteria. The criteria related to human tolerance and perception described in the previous section correspond to the category (b). For industrial or scientific work in particular, more purpose-oriented criteria need to be established and this is the reason of the third category. However, in reality, there cannot be any universally acceptable criteria in this aspect.

There are number of national and international codes and recommendations in the literature, such as ISO 2372 and 2373 (in 1974). However, it is difficult to produce a unified regulation from them.

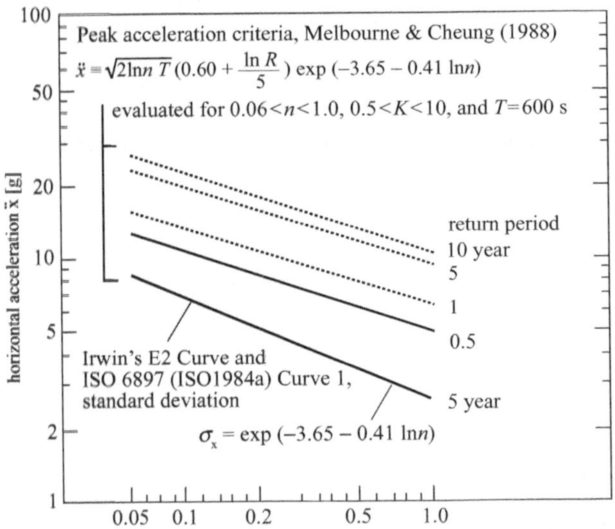

Figure 12.17 Peak acceleration criteria determined by Melbourne and Palmer (1992)

Table 12.3 Acceptance criteria for the use of machinery and equipment

Category	Apparatus	Peak acceleration (mm/s^2) Frequency range: 1 – 10 Hz	Maximum velocity Frequency range: 10 – 100 Hz (mm/s)
I	Highly sensitive optical instruments, mechanical measuring instruments, etc.	6.3	0.1
II	Normally sensitive machinery for grinding, milling, etc.	63	1
III	Little sensitive machinery for metal-working to usual precision	250	4
IV	Insensitive machinery such as blowers, vibratory machines, etc.	> 250	> 4

Specific data on individual types of machinery are also available in references such as Major (1980) and Harris (1976). Acceptance criteria, particularly those in relation to the use of machinery and equipment developed by Korenev and Rabinovič (in 1980), have been listed by Bachmann and Ammann (1987) and reproduced in Table 12.3 as an example.

Structural Aspects

Structural criteria need to be considered with several different levels of consequences in mind.

1. The level of damage that could accumulate to eventual fatigue failure.
2. Development of permanent damage such as plastic deformation and cracks.
3. Serious damage to make the structure unserviceable.
4. Collapse of the structure.

Bachmann and Ammann (1987) list the following parameters as the elements to be counted for structural criteria:

1. type and quality of the structural materials, ductility in particular;
2. type of construction;
3. properties of the foundation;
4. main dimensions of the principal load-bearing members;
5. age of the structure;
6. duration of the vibration effects;
7. characteristics of vibration, namely frequency, amplitude and damping.

There are no standard criteria which can be uniformly applied over the world. The proposed draft of ISO Standard 4866 (ISO 1984b), for example, divides the structures into four different categories and gives the criteria to be considered in the form of pseudospectra, or the response velocity versus frequency. Figure 12.18, for example, shows the criteria for vibration limits for structures subjected to blasting and adopted by the US Bureau of Mines, see Richart et al. (1970). The diagram shows simultaneous values of displacement, velocity and acceleration and the limiting condition for each of these quantities forms an envelope on this diagram. Points falling above this envelope violate the failure conditions. Two shaded

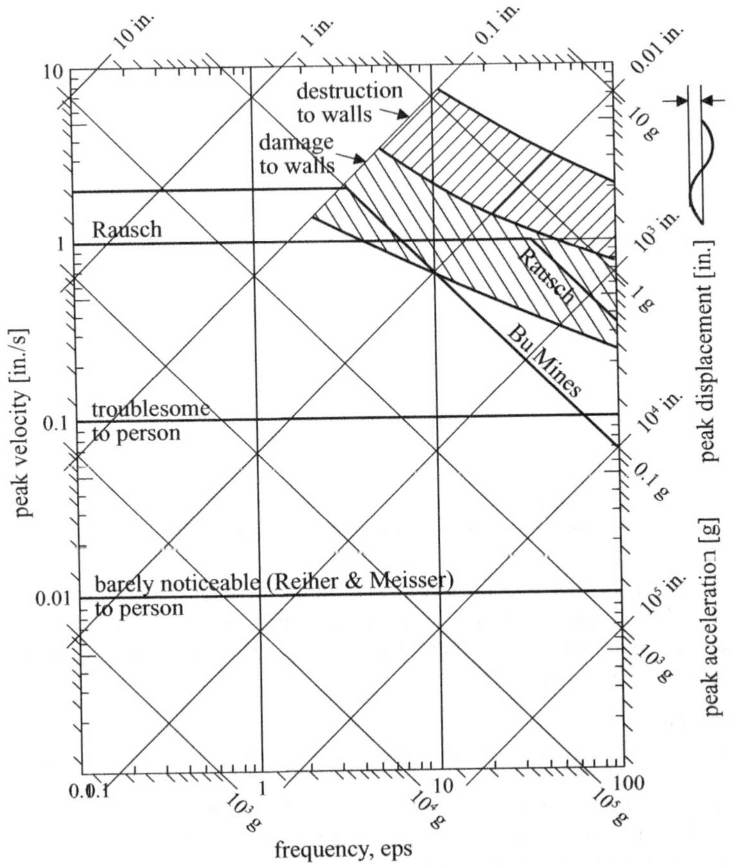

Figure 12.18 Criteria for structures under blast loading (after Richart et al. 1970)

zones describe the possibility of structural damage, particularly to walls for this case, which may be caused by steady-state vibrations.

Another chart showing the safety limits of structural vibration (Figure 12.19) for the frequency range of up to 50 Hz is presented from Leet (1960). It should be noted that the vibration limits are often given in terms of velocity but it is more appropriate to give it in accelerations when the structures are in higher frequency range.

More Contemporary Criteria

Bachmann and Ammann (1987) has made a concise survey of various regulations and codes of practice regarding the structural criteria. German and Swiss codes, as typical examples, are presented in Tables 12.4 and 12.5, respectively.

Overall Acceptance Levels

Having reviewed various existing criteria, Bachmann and Ammann (1987) suggest the overall acceptance levels presented in Table 12.6. There are some examples included in Table 12.6, such as footbridges and some public buildings. As it is clearly stated by Bachmann and Ammann (1987), their intention was

Table 12.4 Standard DIN 4150 (1983)

Structural Category	Peak velocity limits	
	Frequency range (Hz)	Velocity (mm/s)
Industrial buildings	$f \leq 10$	20
Residential buildings	$10 < f \leq 50$	$15 + f/2$
Vulnerable buildings	$50 < f \leq 100$	$30 + f/5$

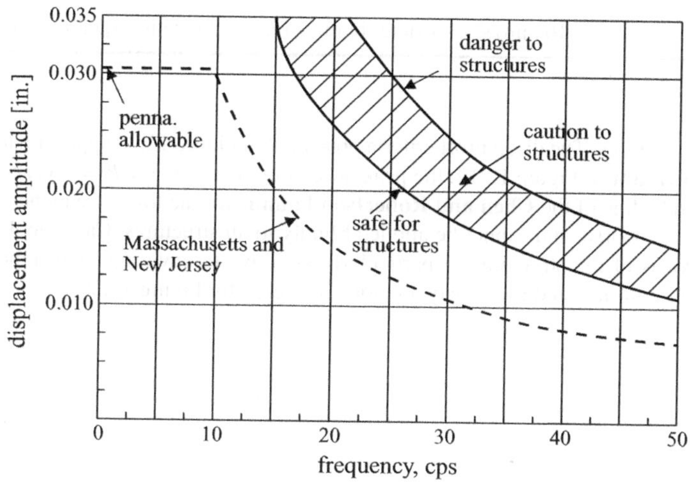

Figure 12.19 Structural criteria by (Leet 1960)

to provide a 'crude but simple global criteria', which of course should be applied with caution, but nevertheless it is quite helpful, particularly for provisional studies.

Figure 12.20 is a tentative proposal as an overall summary of the present study. Generally speaking, the discomfort criteria are much stricter than the structural criteria and hence become more critical for

Table 12.5 Standard SN 640312 (1978)

Structural Category	Machinery / Traffic loads		Blasting load	
	Frequency (Hz)	Velocity limits (m/s)	Frequency	Velocity limit
I (Industrial structures)	$10 \rightarrow 30$	12	$10 \rightarrow 60$	30
	$30 \rightarrow 60$	$12 \rightarrow 18$	$60 \rightarrow 90$	$30 \rightarrow 40$
II	$10 \rightarrow 30$	8	$10 \rightarrow 60$	18
	$30 \rightarrow 60$	$8 \rightarrow 12$	$60 \rightarrow 90$	$18 \rightarrow 25$
III	$10 \rightarrow 30$	5	$10 \rightarrow 60$	18
	$30 \rightarrow 60$	$5 \rightarrow 8$	$60 \rightarrow 90$	$12 \rightarrow 18$
IV (Vulnerable structures)	$10 \rightarrow 30$	3	$10 \rightarrow 60$	8
	$30 \rightarrow 60$	$3 \rightarrow 5$	$60 \rightarrow 90$	$8 \rightarrow 12$

Table 12.6 Overall acceptance levels as structural criteria From Bachmann and Ammann (1987)

Structures	Acceptance level	Comments
Pedestrian structures	$a \leq 5 - 10\%$ g	Normally the lower value does not produce discomfort
Office buildings	$a \leq 2\%$ g	DIN and BS may yield quite different values
Gymnasia (sports halls)	$a \leq 5 - 10\%$ g	The higher value recommended only if (i) the acoustic effect is small, and (ii) only participants are on or near the vibrating floor
Dancing and concert halls	$a \leq 5 - 10\%$ g	The same as for gymnasia
Factory floors	$v \leq 10$ mm/s	Stricter bounds required for high-quality production factories

structural design purposes. The discomfort criteria presented in Figure 12.20 approximately corresponds to the curve **Unpleasant** in Figure 12.14, the peak acceleration criteria for $R = 5$ years in Figure 12.17 and about the middle line of the **Chen and Robertson bounds** for the lower frequency range. It is only a crude indicator, of course, and cannot be applied blindly to all structures. Care should be taken when the required services of the structure are particularly sensitive to environmental tranquility since the production-quality criteria listed previously are not considered for Figure 12.20.

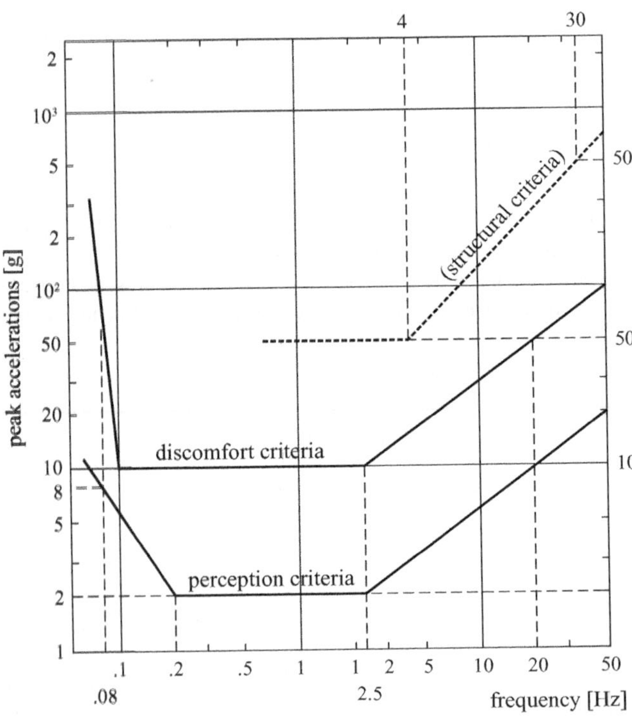

Figure 12.20 Proposed peak acceleration criteria

12.3 Wind-Induced Vibration of Bridges

12.3.1 General Background

12.3.1.1 Early History of Bridge Aerodynamics

On this topic, many people may have in mind the collapse of Tacoma Narrows suspension bridge 67 years ago. All the large bridges, long-span cable-supported bridges in particular, designed and built since then bear the memory of Tacoma. Meanwhile there has been remarkable technical progress, both in experimental mechanics and in analytical methods in the subject. A brief review of its history is provided here, and attention is also focused upon the problems still to be addressed.

Static action of wind

Wind has been an important factor since ancient time for the design of house buildings and town planning. However, it was only in the 18th century that wind force was first considered in a more scientific way in relation to structural design. John Smeaton, known as the first Civil Engineer in England, presented the *Table of Wind Forces* to the Royal Society of London in 1759. Smeaton's table was ahead of its time and, despite of its good quality, its significance was not much appreciated by engineers until in the 19th century, when wrought iron became a popular new material, the size of structures grew to an extent that wind force on them became a more serious issue. The failure of the Firth of Tay Bridge in 1879 provided a reminder to all engineers regarding the recognition of wind load. Clearly, this was a turning point in the history of bridge aerodynamics.

Failure of the first Tacoma Narrows Suspension Bridge

The second turning point in the history of bridge aerodynamics was the collapse of the Tacoma Narrows Bridge in 1940. With a centre span of 854 m, Tacoma was one of the largest bridges of the day. The bridge was designed for the lateral wind load of 2.4 kPa, which is equivalent to roughly 50 m/s of wind speed. What happened in reality, however, was that the bridge kept vibrating in vertical bending modes in much lighter wind, even during its construction. On November 7, 1940, the usual bending vibration suddenly changed to a violent motion in torsion and the bridge collapsed in a wind of less than 20 m/s. It was not a matter of static wind load but a matter of dynamic instability.

It is fair to state that no bridge before Tacoma was designed specifically against the dynamic action of wind. That is not to say that no bridge had collapsed due to dynamic action of wind. In the gale of November 30, 1836, one of the four 78 m spans of the Brighton Chain Pier in England collapsed. A report of this incident included remarkable illustration of severe torsional motion and ultimate deck failure. Surprisingly there is a striking resemblance between these sketches and photographs of the Tacoma failure, which took place a century later.

Application of Deflection Theory

In fact, the progress of modern suspension bridges throughout the first 200 years of history was a continuous struggle against wind action. Many of the major suspension bridges constructed in the 19th century, the Menai Straits Bridge by Thomas Telford to start with, were either destroyed or severely damaged by wind. John Röbling was one of the successful engineers who managed to overcome this difficulty by adding a heavy girder and numerous stay cables to ensure high overall rigidity of the structure. The Niagara Railway Bridge (1855) and the Brooklyn Bridge (1883) were his monumental achievements.

The challenge of bridge building is often indicated by span length. It is interesting to observe how the maximum bridge span length increased in the past 200 years. Most of them were achieved by suspension bridges. A rather sudden increase of maximum span length appeared in early 1930s, supported by the application of the **deflection theory**. When Professor Melan developed a new theory taking the bridge deck deflection into stress analysis, his idea was perhaps simply to carry out more exact stress calculation. However, once this was applied to the design of suspension bridges, its interpretation could have been quite different: 'The more flexible the stiffening girder is, the less stress it has to carry'.

Faithful application of this philosophy to an extreme end in a rather simplistic manner led to the design of the George Washington Bridge (1931), which practically had no stiffening girder at all. This was the first bridge with a clear span of over 1000 m. The achievement of Golden Gate (1937) was possible only with this philosophy. Emergence of suspension bridges with plate girder stiffening such as Bronx-Whitestone (in 1939) and Tacoma (in 1940) was also along the same line. At this point, at the same time, the precious insight of Röbling on the 'stability due to heavy weight' was not paid much attention. The unfortunate incident of Tacoma thus happened.

There has been another very rapid development of maximum bridge span in recent years. This has been made possible by many factors but what should be emphasized in particular are the progress of high-strength steels and welding techniques together with the development of electronic computers and calculation techniques. At the same time, the number of engineers working on the design and construction of these bridges is no comparison to a half century ago.

12.3.1.2 General Characteristics of Wind-Induced Bridge Vibration

There are various types of wind-induced dynamic response of bridges. They have been revealed through difficult experience, measurements and experimental investigations. The difference of response is caused by different mechanisms of aerodynamic action or its interaction with the structure, and it very much depends upon the type of structure. They can be classified typically as shown in Table 12.7. Although the listed examples in Table 12.7 are not necessarily of bridge structures, the difference between three types of dynamic response is of particular importance. They can be identified in two different ways: one is in terms of a response curve, which is the change in magnitude of response amplitude versus wind speed, and another is their dynamic characteristics. Effects of damping can be clearly indicated in the former way, whereas amongst the statistical treatment of the latter, the spectral analysis is generally most informative.

12.3.1.3 Development of Bridge Aerodynamics

From Tacoma to Present

Immediately following the collapse of Tacoma, an intensive investigation into the aerodynamic stability of suspension bridges was initiated by Farquharson in the USA, later joined by von Kármán and Vincent. Their pioneering activities using bridge models in wind tunnels were not only productive (including the reconstruction of the more successful second Tacoma Narrows Bridge and improved aerodynamic

Table 12.7 Various types of wind-induced structural response

Type of response		Examples
Static behaviour	Lateral deflection / lateral buckling	Suspension bridges
	Overturning	Towers, tall buildings
	Negative pressure	Flat roofs, window panes
	Instability/buckling	Shell structures
	Instability/divergence	Light beams
Dynamic response	Buffeting	Tall buildings, bridge decks
	Vortex excitation	Chimneys, bridge decks
	Instability/galloping	Ice-covered cables
	Instability in torsion	Bridge decks
	Instability/flutter	Aerofoils

performance of other major bridges such as Golden Gate, George Washington and Bronx-Whitestone) but also became a precedent for many other research activities on wind resistant design of large suspension bridges all over the world. Somewhat similar activities were carried out by Frazer, Scruton and others in the UK, aiming at the building of the Severn and Forth crossings after World War II.

In this early period, the most important matter was not to repeat the same mistake as at Tacoma. Naturally, the emphasis was placed on finding out the **critical wind speed** where the dynamic instability may occur for a given deck cross-section and to make sure that this critical threshold is higher than the anticipated wind speed in design.

There were also theoretical analyses, including the application of flutter theory by Bleich (1949), lateral buckling analysis by Hirai (in 1947), buffeting analysis by Davenport (in 1958), etc. Theoretical analyses would naturally require knowledge of aerodynamic forces on bridge decks. There has been a significant development of measuring techniques associated with these activities. Rapid development of electronic computers and computing technology is very much hand-in-hand with them for more sophisticated analyses. Application of statistical reliability theory is also a significant development in this field.

Simulation of Natural Wind

Wind tunnels were originally developed by civil engineers at the end of the 19th century. However, following the successful flight experience by Wright Brothers, it became a major experimental tool particularly in the field of aeronautics. In the post-war era, when civil engineers started using wind tunnels, they naturally tested bridge models in conventional aeronautical wind tunnels with idealized uniform smooth air flow rather than with simulated natural wind. It was a contribution of Danish engineers, Nøkkentved and Jensen, who brought in the concept of simulating natural wind, that is necessary in order to properly reproduce wind-induced response in model tests.

Following Jensen's pioneering work (1958), Davenport in Canada further developed the concept of the boundary layer wind tunnel in the 1960s. Davenport applied a large-scale simulation of strong natural wind to many major structures including the World Trade Center Towers, New York, Sears Building, Chicago, and CN Tower, Toronto. Bridges were not exceptions. The Thomas MacKay suspension bridge in Halifax, Nova Scotia, was probably the first bridge that was tested in properly scaled turbulent boundary layer wind, including its deck construction stage.

12.3.1.4 Recent Trends in Bridge Aerodynamics

It is probably acceptable to state now that bridges can be designed without repeating the problem of the former Tacome Narrows suspension bridge. However, bridge engineers today have their own issues to deal with, where the focal points of bridge aerodynamics have somewhat shifted from before. One of the most remarkable developments in long-span bridges in the past four decades or so is the fact that cable-stayed bridges are becoming more competitive in their span length with classic suspension bridges, while the latter have also increased the maximum span length dramatically. When the span is increased, naturally the wind stability becomes a more serious issue. This is not only for the completed bridges but also, if not more so, through their erection stages. General trends in the field of bridge aerodynamics in recent years include the following topics.

1. Security of serviceability conditions – the engineering attention is not only for ensuring the survival of the bridge as a structure but also the consideration of better serviceability of the structure as a part of the social infrastructure, including the safety for transportation and prevention of long-term problems such as fatigue damages.
2. Aerodynamics of bridges during erection – conditions during construction are generally less favorable and yet this issue has been traditionally less focused in research. It includes the consideration of rational design conditions and vibration control measures.

3. Another focal point of wind study for bridges is a dynamic problem of structural members, cables in particular. There have been a number of reports on stay-cable vibrations. Cables are extremely flexible and their structural damping is generally very low compared to other structural components. They are, as a result, much more vulnerable to dynamic wind actions. Accumulated structural damage due to vibration needs to be identified and properly looked after.
4. Advances in cost-effective design tend to impose more stringent requirements not only for qualitative but also for quantitative wind study results. It is also preferable to have response predictions not only conservatively for design purposes but as realistic as possible so that they make better comparison with the full-scale observations.
5. Contemporary research topics include:
    - comparative benchmark study of various wind tunnel testing methods;
    - full-scale measurement and verification of predictions;
    - numerical simulation of wind for response calculation in time domain, including the question regarding the applicability of computational fluid dynamics;
    - identification of flutter derivatives for commonly used road deck configurations and effects of turbulence on them;
    - refinement of buffeting theory for more accurate response prediction;
    - consideration of wind yaw angles;
    - dynamics of cables, particularly on galloping of inclined stay cables;
    - fundamental mechanism of vortex-induced oscillations;
    - Further development of vibration control methods, etc.

12.3.2 Formulation of Aerodynamic Forces

12.3.2.1 Definition of Force Components

When a 3D structure is exposed to an air flow, three force components, the lift, drag and side forces, and three moment components, the pitching, yawing and rolling moments, can be generally considered. However, in many wind engineering problems, it is not necessary to consider all six of these components, and a 2D alternative is considered as a convenient mathematical model.

The **strip theory** assumption, originally introduced for aerofoils, is usually also the method applied to bridges. Since a bridge is usually extended in one direction along its span and our primary concern is its behavior when wind comes perpendicular to the span axis, instead of looking at the whole structure, often a 2D slice (or strip) of unit bridge length, cut off by two planes in the mean wind direction, is to be considered. The idea is the same as the plane strain analysis in the 2D theory of elasticity.

For this case, there are only three components needed to be considered; lift force, drag force and pitching moment, since each spanwise station is considered as though it were a portion of an infinite span bridge with uniform spanwise properties. Corresponding to these, the displacements taken into consideration are (h, p, α) in the lift, drag and pitching directions, respectively. In the following section, only these three directions are counted for the formulation of aerodynamic force components, see Figure 12.21.

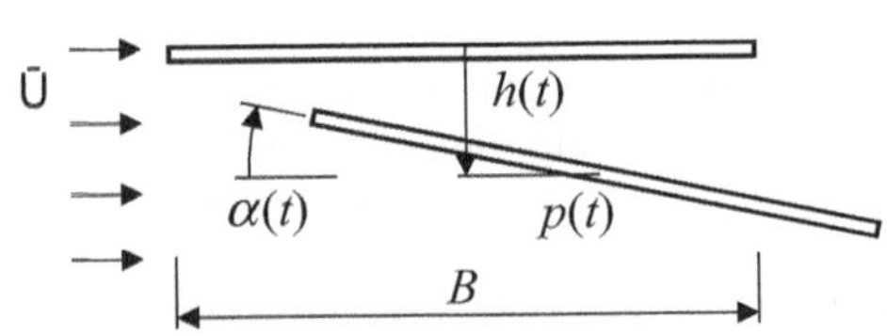

Figure 12.21 Three displacements considered in the analysis

Corresponding to these force components, the following dimensionless coefficients are often introduced and called the lift force, drag force and pitching moment coefficients, respectively:

$$C_L = \frac{L}{\rho \overline{U}^2/2 \cdot B} \qquad C_D = \frac{D}{\rho \overline{U}^2/2 \cdot B} \qquad C_M = \frac{M}{\rho \overline{U}^2/2 \cdot B^2} \tag{296}$$

where $\overline{U}$ is the mean wind speed taken normal to the bridge axis, ρ is the air density, B is the width of the bridge deck, and L, D, M are the 2D lift force, drag force and pitching moment for the deck of unit length exposed to wind, respectively.

12.3.2.2 Quasi-Steady Aerodynamics

Another assumption often introduced is the **quasi-steady** approximation, which is to say that the aerodynamic forces at any time are dependent only on the instantaneous position of the body relative to wind at that particular moment. In other words, the temporal memory effect or the history of the motion in the aerodynamic model is to be ignored.

The strip theory approximation discussed above is unambiguous and its application is generally accepted. However, this is not true for the quasi-steady approximation. It is an acceptable assumption for the case of relatively high wind speed, but clearly unacceptable, for example, in the case of vortex excitations. When the quasi-steady approximation is applied, three aerodynamic force components are given by

$$L = \frac{\rho \overline{U}_{rel}^2}{2} BC_L \qquad D = \frac{\rho \overline{U}_{rel}^2}{2} BC_D \qquad \text{and} \qquad M = \frac{\rho \overline{U}_{rel}^2}{2} B^2 C_M \tag{297}$$

where $\overline{U}_{rel}$ is the relative wind speed given by $\overline{U}_{rel} = \overline{U} + u - \dot{p}$. The velocity vector is given by $(\overline{U} + u, v, w)$. The force and moment coefficients are, as their first approximation, given by

$$C_F = C_F(\alpha) + \frac{dC_F}{d\alpha}\alpha_{rel} \qquad (F = L, D, M) \tag{298}$$

where α_{rel} is the relative angle of attack as follows:

$$\alpha_{rel} = \alpha - \frac{\dot{h} + nB\dot{\alpha} - w}{\overline{U} + u - \dot{p}} \approx \alpha - \frac{\dot{h} + nB\dot{\alpha} - w}{\overline{U}} \tag{299}$$

where n is a dimensionless factor representing the position where the aerodynamic lift force acts upon. For a thin aerofoil at low reduced frequency, for example, $n \approx 0.25$. The force components can be expressed as follows:

$$F = \overline{F} + F_f + F_b \qquad (F = L, D, M) \tag{300}$$

The first term is the static component and is given by

$$\overline{L} = \frac{\rho \overline{U}^2}{2} BC_L(\alpha) \tag{301}$$

$$\overline{D} = \frac{\rho \overline{U}^2}{2} BC_D(\alpha) \tag{302}$$

$$\overline{M} = \frac{\rho\overline{U}^2}{2}B^2 C_M(\alpha) \tag{303}$$

The second term is the motion-dependent component, which is

$$L_f = \frac{\rho\overline{U}^2}{2}B\left(\frac{dC_L}{d\alpha}\alpha - 2C_L\frac{\dot{p}}{\overline{U}} + \frac{dC_L}{d\alpha}\frac{\dot{h} - nB\dot{\alpha}}{\overline{U}}\right) \tag{304}$$

$$D_f = \frac{\rho\overline{U}^2}{2}B\left(\frac{dC_D}{d\alpha}\alpha - 2C_D\frac{\dot{p}}{\overline{U}} + \frac{dC_D}{d\alpha}\frac{\dot{h} - nB\dot{\alpha}}{\overline{U}}\right) \tag{305}$$

$$M_f = \frac{\rho\overline{U}^2}{2}B\left(\frac{dC_M}{d\alpha}\alpha - 2C_M\frac{\dot{p}}{\overline{U}} + \frac{dC_M}{d\alpha}\frac{\dot{h} - nB\dot{\alpha}}{\overline{U}}\right) \tag{306}$$

and the last term, buffeting force term, is given by

$$L_b = \frac{\rho\overline{U}B}{2}\left(2C_L u(t) + \frac{dC_L}{d\alpha}w(t)\right) \tag{307}$$

$$D_b = \frac{\rho\overline{U}B}{2}\left(2C_D u(t) + \frac{dC_D}{d\alpha}w(t)\right) \tag{308}$$

$$M_b = \frac{\rho\overline{U}B^2}{2}\left(2C_M u(t) + \frac{dC_M}{d\alpha}w(t)\right) \tag{309}$$

The first terms relate only to the static displacement and can be set aside for the dynamic analysis. The slope of force coefficients are taken at the vicinity of the static angle. If the cross-section is symmetric with respect to the horizontal plane, C_L, C_M and $dC_D/d\alpha$ are close to zero, and hence

$$L(t) \approx \frac{\rho\overline{U}B}{2}B\frac{dC_L}{d\alpha}\left(\alpha - \frac{\dot{h} - nB\dot{\alpha}}{\overline{U}}\right) + \frac{\rho\overline{U}B}{2}\frac{dC_L}{d\alpha}w(t) \tag{310}$$

$$D(t) \approx -\rho\overline{U}^2 BC_D\dot{p} + \rho\overline{U}BC_D u(t) \tag{311}$$

$$M(t) \approx \frac{\rho\overline{U}^2}{2}B^2\frac{dC_M}{d\alpha}\left(\alpha - \frac{\dot{h} - nB\dot{\alpha}}{\overline{U}}\right) + \frac{\rho\overline{U}B^2}{2}\frac{dC_M}{d\alpha}w(t). \tag{312}$$

12.3.2.3 Unsteady Aerodynamic Force Coefficients

The motion-dependent force components can be expressed more generally, including the virtual mass effects, as follows:

$$\begin{bmatrix} L_f \\ D_f \\ M_f \end{bmatrix} = \begin{bmatrix} L_{\dot{h}} & L_{\ddot{h}} & L_{\dot{\alpha}} & L_{\ddot{\alpha}} & L_{\dot{p}} & L_{\ddot{p}} \\ D_{\dot{h}} & D_{\ddot{h}} & D_{\dot{\alpha}} & D_{\ddot{\alpha}} & D_{\dot{p}} & D_{\ddot{p}} \\ M_{\dot{h}} & M_{\ddot{h}} & M_{\dot{\alpha}} & M_{\ddot{\alpha}} & M_{\dot{p}} & M_{\ddot{p}} \end{bmatrix} \times \begin{bmatrix} \dot{h} \\ \ddot{h} \\ \dot{\alpha} \\ \ddot{\alpha} \\ \dot{p} \\ \ddot{p} \end{bmatrix} \tag{313}$$

The components of the coefficient matrix are generally called the aerodynamic derivatives and are given as functions of the reduced velocity

$$V_r = \frac{\overline{U}}{\omega B} \tag{314}$$

or the reduced frequency

$$K = \frac{1}{V_r} = \frac{\omega B}{\overline{U}} \tag{315}$$

The analytical expressions for the derivatives have been obtained only for a few very limited cases: one is the case of an idealized flat plate performing a simple harmonic motion in coupled heave–pitch mode with infinitesimal amplitudes and zero mean angle of attack. Another simple case analytically resolved is the lift force on a flat plate, induced by a sudden stepwise position change, $h = 1$ and $a = 1$, respectively, when it is exposed to a uniform air flow with zero angle of attack.

For other cases, the aerodynamic derivatives need to be determined experimentally. Usually a simple harmonic oscillatory motion is assumed. For this case, although the velocity term is 90° out-of-phase, the acceleration and displacement terms are put together, since they are in the same phase. The following expression by Scanlan has been widely accepted:

$$\begin{bmatrix} L_f \\ D_f \\ M \end{bmatrix} = \frac{\rho \overline{U}^2 B}{2} \begin{bmatrix} KH_1^* & KH_2^* & K^2H_3^* & \ldots \\ KP_5^* & KP_2^* & K^2P_3^* & \ldots \\ BKA_1^* & BKA_2^* & BK^2A_3^* & \ldots \end{bmatrix}$$

$$\begin{bmatrix} \ldots & K^2H_4^* & KH_5^* & K^2H_6^* \\ \ldots & K^2P_6^* & KP_1^* & K^2P_4^* \\ \ldots & BK^2A_4^* & BKA_5^* & BK^2A_6^* \end{bmatrix} \times \begin{bmatrix} \dot{h}/\overline{U} \\ B\dot{\alpha}/\overline{U} \\ \alpha \\ h/B \\ \dot{p}/\overline{U} \\ h/B \end{bmatrix} \tag{316}$$

where eighteen aerodynamic derivatives, H_1^*, H_2^*, …, A_6^*, are functions of the reduced frequency, $K = \omega B/\overline{U}$. For the case of the flat plate aerodynamics, the aerodynamic derivatives are given as follows:

$$H_1^* = -\frac{\pi}{k}F(k) \qquad H_2^* = -\frac{\pi}{4k}\left(1 + F(k) + \frac{2}{k}G(k)\right)$$

$$H_3^* = -\frac{\pi}{4k^2}\left(2F(k) - kG(k)\right) \qquad H_4^* = \frac{\pi}{2}\left(1 + \frac{2}{k}G(k)\right)$$

$$A_1^* = \frac{\pi}{4k}F(k) \qquad A_2^* = \frac{\pi}{16k}\left(-1 + F(k) + \frac{2}{k}G(k)\right)$$

$$A_3^* = \frac{\pi}{8k^2}\left(\frac{k^2}{8} + F(k) - \frac{k}{2}G(k)\right) \qquad A_4^* = -\frac{\pi}{4k}G(k)$$

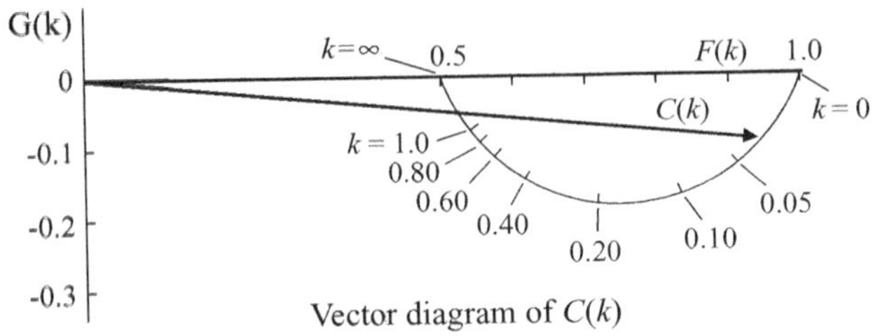

Figure 12.22 Vector diagram of $C(k)$ (after Fung 1955)

Furthermore

$$H_5^* = H_6^* = A_5^* = A_6^* = 0 \qquad \text{and} \qquad P_j^* = 0 \quad (j = 1, 2, \ldots, 6)$$

in which $F(k)$ and $G(k)$ are the real and imaginary parts of the Theodorsen function, defined by using the Hänkel functions:

$$H_\nu^{(2)}(k) = J_\nu(k) - iY_\nu(k) \qquad (\nu = 0, 1) \tag{317}$$

as follows:

$$C(k) = F(k) + iG(k) = \frac{H_1^{(2)}(k)}{H_1^{(2)}(k) + iH_0^{(2)}(k)} \tag{318}$$

or

$$F(k) = \frac{J_1(k)\left[J_1(k) + Y_0(k)\right] + Y_1(k)\left[Y_1(k) - J_0(k)\right]}{\left[J_1(k) + Y_0(k)\right]^2 + \left[Y_1(k) - J_0(k)\right]^2} \tag{319}$$

$$G(k) = \frac{Y_1(k)Y_0(k) + J_1(k)J_0(k)}{\left[J_1(k) + Y_0(k)\right]^2 + \left[Y_1(k) - J_0(k)\right]^2} \tag{320}$$

by using the Bessel functions, $J_\nu(k)$ and $Y_\nu(k)$ ($\nu = 0, 1$), where $k = K/2$ is traditionally used as the reduced frequency (Figure 12.22).

12.3.2.4 Transient Forces

The definition of these aerodynamic derivatives, however, assumes that the body in motion is performing a simple harmonic coupled oscillation, just like the case of an aerofoil flutter analysis, with infinitesimal amplitudes. Hence they are not directly applicable to the description of any transient motion, which can be obtained by a convolution of any forcing function with indicial admittance.

When a thin two-dimensional aerofoil of the chord length B is placed in a uniform flow $\overline{U}$ and the angle of attack α is suddenly given as a step-wise increase from zero, only a half of the final lift force would become effective immediately and another half of the lift force will be gained asymptotically (Figure 12.23)

$$L(\tau) = \pi\rho\overline{U}^2 B\alpha\Phi(\tau) \tag{321}$$

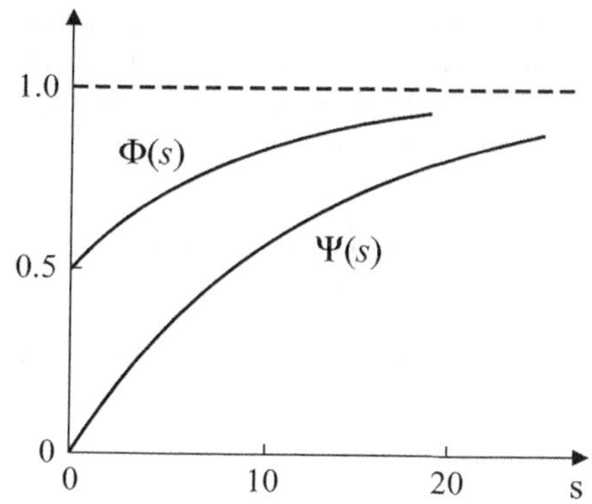

Figure 12.23 Wagner and Küssner functions (after Fung 1955)

where $\tau = 2\overline{U}t/B > 0$ is the dimensionless time. $\Phi(\tau)$ is called the Wagner function and is approximately given by

$$\Phi(\tau) = 1 - 0.165e^{-0.0455\tau} - 0.335e^{-0.300\tau}. \tag{322}$$

The Wagner function is related to the Theodorsen function by the inverse Laplace transform as

$$\Phi(\tau) = L^{-1}\left[\frac{C(-is)}{s}\right] \tag{323}$$

This relationship implies that the indicial lift and steady-state sinusoidal lift term form a Laplace transform pair. Since the Laplace transform of an exponential function is given by a rational form, or

$$L\left[e^{at}\right] = \frac{1}{s-a} \tag{324}$$

it has been established that the unsteady aerodynamic force terms in general can be approximated by a series of partial function forms of the Laplace variable (Karpel 1981). The most common form of the approximate functions, currently used for each unsteady generalized force coefficient, is the Roger's formulation given as follows (Roger (1977)):

$$\mathbf{A}_P = \mathbf{Q}_1 + \mathbf{Q}_2 p + \mathbf{Q}_3 p^2 + \sum_{j=1}^{N} \frac{p}{p+\gamma_j}\mathbf{R}_j \tag{325}$$

Another fundamental formulation is when the lift force is induced by a suddenly applied constant upward gust $w_0(t)$, which is given by

$$L(\tau) = \pi\rho\overline{U}^2 B\frac{w_0}{\overline{U}}\Psi(\tau) \tag{326}$$

where $\tau = 2\overline{U}t/B$ is again the dimensionless time and $\Psi(\tau)$ is the indicial admittance and is called the Küssner function for this particular case (Figure 12.23). The Küssner function is somewhat similar to the

Wagner function, except it starts from zero at time zero and approaches asymptotically its steady-state value of unity when $\tau \to \infty$. If the gust is of an arbitrary vertical velocity distribution $w(t)$, the lift becomes a convolution, or the Duhamel integral, involving the derivative of the Küssner function, which is equivalent to the impulse response function, as

$$L(\tau) = \pi\rho\overline{U}^2 B \int_0^\infty w(\tau - s)\Psi'(s)\, ds \tag{327}$$

In particular, if the gust variation takes a sinusoidal form, $w(s) = w_0 e^{iks}$, where $k = \omega B/2\overline{U}$, the lift force becomes

$$L(s) = \pi\rho\overline{U}^2 B \frac{w(s)}{\overline{U}}\Theta(k) \tag{328}$$

where $\Theta(k)$ is the Sears function, which is related to the Küssner function by

$$\begin{aligned} \Theta(k) &= \int_0^\infty \Psi'(s)e^{-iks}\, ds \\ &= ik \int_0^\infty \Psi(s)e^{-iks}\, ds \end{aligned} \tag{329}$$

The concept of the convolution integral explained above becomes an important tool for the time-domain analysis.

12.3.3 Aerodynamic Instability

12.3.3.1 Concept of Aerodynamic Instability

Consider, as an example, a twisting motion of a beam. If the external excitation is a pitching moment defined in Section 12.3.2.2, the 2D equation of motion is as follows:

$$J\ddot{\phi} + C\dot{\phi} + K\phi = \frac{\rho\overline{U}^2}{2}B^2 C_M(\phi, \dot{\phi}, \ldots) \tag{330}$$

where J is the polar mass moment and C and K are appropriate damping and stiffness terms of the structure. C_M is not known. However, as a simple expression, if $C_M = a\phi + b\dot{\phi}$,

$$\ddot{\phi} + 2\zeta_S\omega_T\dot{\phi} + \omega_T^2\phi = \frac{\rho\overline{U}^2 B^2}{2J}(a\phi + b\dot{\phi}) \tag{331}$$

where $\omega_T = \sqrt{K/J}$ and $\zeta_S = C/(2J\omega_T)$, or

$$\ddot{\phi} + 2(\zeta_S + \zeta_a)\omega_T\dot{\phi} + \omega_T^2(1 - S_a)\phi = 0 \tag{332}$$

where

$$\zeta_a = \frac{\rho\overline{U}^2 B^2}{4\omega_T J}b(U_r), \qquad S_a = \frac{\rho\overline{U}^2 B^2}{2J\omega_T^2}a(U_r), \qquad U_r = \frac{\overline{U}}{B\omega_T} \tag{333}$$

Since the aerodynamic derivatives a and b are not known, Equation (332) indicates that there is a possibility of having two types of aerodynamic instability as follows:

$$\zeta_S + \zeta_a < 0 \rightarrow \text{negative damping} \rightarrow \text{flutter} \quad (334)$$

$$S_a > 1 \rightarrow \text{negative stiffness} \rightarrow \text{divergence.} \quad (335)$$

12.3.3.2 Historical Development of Bridge Flutter Analysis

Flutter and Unsteady Aerodynamics

Although dynamic failures of aircraft wings caused by aeroelastic phenomena were observed in the early days of flight, a real development of the nonstationary airfoil theory did not occur till the 1920s. An analytical expression of the aerodynamic lift force on a harmonically oscillating flat plate was first given by Birnbaum in 1922 as an application of the Prandtl's theory of bound vortices. Through the following decade, the analysis of unsteady aerodynamic forces on an oscillating 2D plate attracted significant interest from aerodynamicists such as Wagner, Glauert, Küssner, Duncan and Collar and the most complete solution to this problem was presented by Theodorsen in 1935. Similar solutions were also given in Europe by Küssner and Schwartz, Cicala, Schmieden, Ellenberger, etc., but the solution by Theodorsen (1930) has been most extensively used.

The collapse of the original Tacoma Narrows Bridge was, from the time immediately following the incident, frequently compared to galloping of iced cables or flutter of aircraft wings. Bleich (see Bleich 1949) tried to analyze the incident as a flutter by applying Theodorsen's aerodynamic formulation to the bridge and found that the critical flutter speed thus calculated was considerably higher than that of Tacoma Narrows. It was obvious that the airfoil flutter coefficients calculated by the potential flow theory were not directly applicable to much more aerodynamically bluff sections such as those of this bridge.

Bleich tried to cover this defect by modifying Theodorsen's expression to consider an additional lift force term corresponding to the effects of vortex formation from the leading edge of the deck but was not very successful. Pugsley commented at this point that experimentally determined aerodynamic coefficients rather than Theodorsen's would be of more help in future. He was right. In later days, the use of streamlined shallow box girders as a suspension bridge stiffening girder became quite popular, having been inspired by successful application of them for both the Severn and Lillebælt crossings in the late 1960s. In these cases, the flow separation is much less than the sections such as Tacoma, and ironically Bleich's original calculation with Theodorsen's force can actually yield a reasonably good approximation.

Experimental Determination of Unsteady Aerodynamic Forces

Theodorsen's analysis was based on the potential flow theory, which assumes that flow is to follow the solid surface of the body. However, there is another category of flutter instability in which the flow separation is involved as an essential feature of it. The phenomenon of stall flutter became an issue particularly as instability of propellers and turbine blades. Severe drop of critical flutter speeds, predominance of pitching or torsional vibration and strong nonlinearity in response are known to be its characteristics.

In the absence of any analytical means to determine unsteady aerodynamic forces for stalled wings, where the flow separation is involved, an extensive experimental evaluation of them has been attempted since 1930s. There are basically two ways to do it. One is to make a direct measurement of aerodynamic force components by dynamometers, strain gauges and so on when the body is given a specific motion, and the other way is to calculate the force indirectly from the induced motion of the body. The same principles have been applied to both airfoils and bridge deck sections.

Försching applied the direct method for the measurement of unsteady aerodynamic forces on various prisms, but et al. Ukeguchi et al. (1966) were probably the first to apply it to bridge deck sections. Rigid bridge deck models were mechanically driven into a simple harmonic motion with a range of specific frequency and amplitude in a 2D air stream and the reactions at the model supports were detected. This is a method which was extensively applied by Halfman (1952) to airfoil aerodynamics. The forced vibration technique was further extended particularly in Japan to cover various aerodynamically bluff sections and also to investigate nonlinear characteristics of them. A recent development of high-speed pressure scanning techniques has made it possible to do practically a simultaneous measurement of dynamic pressure signals at many ports and real-time integration of them. An application of this technique to the forced vibration method has opened up another promising avenue for effective measurement of unsteady aerodynamic forces , (King et al. 1991).

As opposed to the direct measurement, the indirect measurement of aerodynamic forces by detecting the induced response of models in air flow generally requires less-complicated experimental set-ups but more careful conditioning of them. Application of this method in bridge aerodynamics was initiated by Scanlan and Sabzevari (1968) and has been widely practiced over the world. A more sophisticated development of this method is an application of the system identification (SI) technique, which has been developed almost simultaneously in Europe, USA and Japan in recent years (e.g. Poulsen et al. 1991).

Most of these measurements have been done in 2D, smooth air flow and the force coefficients are assumed to be given by linear combinations of displacements and their first derivatives with respect to time. Some attempts have been made to investigate the effects of turbulence and non-linearity on the derivatives but the process and outcomes are still too complicated for practical applications.

Formulation of 2D flutter analysis

Once the unsteady aerodynamic forces are established, the critical conditions for the onset of aerodynamic instability can be determined. The most traditional way for this analysis is by applying the strip assumption, where the interaction between air stream and the body is to be decided in a 2D section perpendicular to the longitudinal axis of the structure. Consequently, any 3D effects along the longitudinal axis of the structure are assumed to be negligible. The equation of motion is given as follows:

$$\begin{bmatrix} m & 0 \\ 0 & J \end{bmatrix} \begin{bmatrix} \ddot{h} \\ \ddot{\alpha} \end{bmatrix} + \begin{bmatrix} C_h & 0 \\ 0 & C_\alpha \end{bmatrix} \begin{bmatrix} \dot{h} \\ \dot{\alpha} \end{bmatrix} + \begin{bmatrix} K_h & 0 \\ 0 & K_\alpha \end{bmatrix} \begin{bmatrix} h \\ \alpha \end{bmatrix} = \begin{bmatrix} L_h \\ M_\alpha \end{bmatrix} \tag{336}$$

The drag-ward motion was considered to be insignificant following the tradition of airfoils. Another restriction usually assumed in a conventional flutter analysis is the fact that the unsteady aerodynamic forces are all given as functions of the reduced frequency $K = \omega B/\overline{U}$ assuming that the body is performing a simple harmonic motion both in heaving and pitching simultaneously with the same frequency (f) and infinitesimal amplitudes. Hence, in Equation (336) above, the left-hand side is all established with respect to time and yet the aerodynamic force terms are actually given in the frequency domain. It means that the equation is applicable only when the body is performing this particular motion as follows:

$$h(t) = h_0 e^{i\omega t}, \qquad \alpha(t) = \alpha_0 e^{i\omega t} \tag{337}$$

By substituting Equation (337) into (336), the flutter conditions, $\overline{U}_F$ and ω_F, are decided and this process is called the flutter analysis.

Simple Formulae for 2D Flutter

It required a great deal of effort to carry out the numerical calculation necessary for flutter analysis in pre-computer days. Because of the tedious process required for its calculation, various simplifications of the flutter calculation were considered. For the case of a thin airfoil, the calculation needs to be done just once for a range of structural parameters, since the aerodynamic tools are established uniquely. An extensive parametric study was carried out and an approximate equation to give the flutter speed was proposed by Theodorsen and Garrick (1940) for airfoils. For the case of bridge decks, similar attempts were made by several researchers and, amongst them, Selberg's formulation has been most frequently referred to:

$$\overline{U}_F = \kappa B f_T \sqrt{\frac{r^*}{\mu}\left[1 - \left(\frac{f_V}{f_T}\right)^2\right]} \tag{338}$$

where

$$r^* = \frac{r}{B}, \qquad r = \sqrt{\frac{J}{m}} \qquad \text{and} \qquad \mu = \frac{\rho B^2}{m} \tag{339}$$

m, J are the mass and mass moment per unit deck length and f_V, f_T are the fundamental natural frequencies in bending and torsion, respectively. The coefficient κ was introduced to accommodate the difference of flutter speed due to the cross-sectional shape of the bridge deck and is unity for a flat plate. Selberg (1963) suggested the κ-values for some aerodynamically bluff sections and different angles of attack. Similar approaches were taken by Klöppel and Weber, Rocard, Frandsen and others in this period.

Time-Domain Versus Frequency Domain Analyses

Bridge flutter analysis was traditionally carried out in the frequency domain but there have been some attempts to do it in the time domain as well. Scanlan et al. (1974) first worked on solving the problem entirely in the time domain, introducing the indicial functions which were put forward earlier in the aeronautical field. The idea was also extended to the coupled mode flutter by Bucher and Lin (1988). One of the difficulties then was to make a proper functional fit to define indicial functions corresponding to the experimentally obtained aerodynamic derivatives, particularly when the cross-sections are far from streamlined. It is only recently that much effort has been invested in developing efficient time-domain formulations of the unsteady aerodynamic forces that could be combined with FE models of the structure and could include all the nonlinearities that have been disregarded in the past. This development has been associated with the planning of very long bridge projects such as Akashi, Storebælt and Messina.

Miyata et al. (1995) clearly presented the advantages of the time-domain approach, particularly when it is combined with a FE modeling of the structure, for the prediction of bridge behavior under wind action. The formulation used by them was conventional quasi-steady aerodynamics with the strip assumption. Much the same approach was also taken by Kovács et al. (1992). Diana et al. (1992), on the other hand, developed a corrected quasi-steady theory by introducing the concept of an equivalent linearization of force coefficients for each reduced velocity. This formulation has been proved to be adequate in many ways, except that it does not deal with the aerodynamic memory effects and the span-wise coherence of the lift forces. Another model for the self-excited forces is a method to approximate the unsteady aerodynamic forces by rational functions, which can be regarded as the Laplace transformed functions (Tiffany and Adams 1988). The idea, in principle, is the same as the use of indicial functions. This method was extensively applied as a state–space method by Xie (1985) to analyze the 3D multimode bridge flutter. Similar approaches also have been taken by Lin and Li (1993), Boonyapingyo et al. (1994) and Fujino et al. (1995), amongst others.

12.3.3.3 3D Flutter Analyses

Direct Method and Modal Method

Application of flutter analysis for 3D structures, without the use of strip assumption, has a relatively short history. The calculation can be performed in two different ways: one is to apply the unsteady aerodynamic forces, either in the frequency domain or in the time domain, directly to a three-dimensional FE model of the structure (direct method); another is to consider the structural response separately in various vibration modes and assemble them (mode superposition method).

The direct method was formulated by Miyata and Yamada (in 1990) leading to a complex eigenvalue problem by the use of aerodynamic derivatives in the frequency domain. The method has a straightforward philosophy but drawbacks are that it requires a large computer capacity and solving a complex eigenvalue problem tends to be a time-consuming process. The mode superposition method, on the other hand, has been employed by many researchers. There have been several methods developed to analyze the multimode flutter in the frequency domain. Agar (1989, 1991) and Chen (1994) developed modal techniques to solve the linearized quadratic eigenequations. As a development of the p-K method which has been used in the aircraft industry, Namini et al. (1992) and Cheng (1995) presented a more general numerical procedure called the p-K-F method to determine the pre- and post-flutter behavior by solution to the modal equations. Further to these, Lin and Yang (1983), Jones and Scanlan (1991), Tanaka et al. (1992), Jain et al. (1996) and others directly utilize the determinant search method to calculate the complex eigenvalues in a general term of the impedance matrix.

Formulation of the Flutter Equation

The actual calculation procedure in detail is explained elsewhere, for example by Miyata and Yamada (in 1990) and Ge and Tanaka (1999). However, the general formulation of 3D flutter analysis can be briefly outlined as follows. The equations of motion of a bridge that is discretized as a n-degree-of-freedom structure can be formulated as follows:

$$\mathbf{M}_S\ddot{\boldsymbol{\delta}} + \mathbf{C}_S\dot{\boldsymbol{\delta}} + \mathbf{K}_S\boldsymbol{\delta} = \mathbf{F}_v\dot{\boldsymbol{\delta}} + \mathbf{F}_d\boldsymbol{\delta} \tag{340}$$

where $\mathbf{M}_S$, $\mathbf{C}_S$ and $\mathbf{K}_S$ are the mass, damping and stiffness matrices of the structure, $\boldsymbol{\delta}$ is a structural deflection vector, $\mathbf{F}_v$ and $\mathbf{F}_d$ are the velocity and displacement parts of the unsteady aerodynamic force matrix, respectively. The unsteady aerodynamic force components were defined earlier by Equation (313). The static action of wind and the aerodynamic gust forces on the right-hand side of the equation can be excluded for the flutter analysis. From Equation (340), the system equation can be very simply

$$\mathbf{M}\ddot{\boldsymbol{\delta}} + \mathbf{C}\dot{\boldsymbol{\delta}} + \mathbf{K}\boldsymbol{\delta} = 0 \tag{341}$$

where $\mathbf{M} = \mathbf{M}_S$, $\mathbf{C} = \mathbf{C}_S - \mathbf{F}_v$ and $\mathbf{K} = \mathbf{K}_S - \mathbf{F}_d$. If the mode superposition method is considered, Equation (341) can be rewritten as follows:

$$\widetilde{\mathbf{M}}\ddot{\boldsymbol{\Phi}} + \widetilde{\mathbf{C}}\dot{\boldsymbol{\Phi}} + \widetilde{\mathbf{K}}\boldsymbol{\Phi} = 0 \tag{342}$$

where $\widetilde{\mathbf{M}} = \mathbf{X}^T\mathbf{M}\mathbf{X}$, $\widetilde{\mathbf{C}} = \mathbf{X}^T\mathbf{C}\mathbf{X}$ and $\widetilde{\mathbf{K}} = \mathbf{X}^T\mathbf{K}\mathbf{X}$ are the generalized mass, damping and stiffness matrices, respectively. The displacement vector is now expressed by

$$\boldsymbol{\delta} = \mathbf{X}\boldsymbol{\Phi} \tag{343}$$

where $\mathbf{X}$ and $\boldsymbol{\Phi}$ are called the mode matrix and general coordinate vector, respectively. By assuming a simple harmonic oscillation for the general coordinate as

$$\boldsymbol{\Phi}(t) = \boldsymbol{\Phi}e^{\lambda t} \tag{344}$$

where $\lambda = \lambda_R + i\lambda_I$, and the flutter conditions are given by

$$\left|\lambda^2\widetilde{\mathbf{M}} + \lambda\widetilde{\mathbf{C}} + \widetilde{\mathbf{K}}\right| = 0 \tag{345}$$

Corresponding circular frequency ω and the damping ratio ζ are defined from the complex eigenvalue as

$$\omega = \sqrt{\lambda_R^2 + \lambda_I^2}, \qquad \zeta = \frac{\lambda_R}{\sqrt{\lambda_R^2 + \lambda_I^2}} \tag{346}$$

Flutter critical speed $\overline{U}_F$ is decided from $\zeta = 0$ and ω_F is decided from the corresponding λ_I.

Mode Participation

Almost all 3D flutter analyses are carried out in the frequency domain and are based on the idea of mode superposition. The assumption is that a dynamic coupling between the natural modes takes place through the self-excited aerodynamic forces. However, it should be noted that there are some fundamental questions regarding this assumption. First of all, there is a question of how many and which natural modes are participating to the instability. Particularly when the structure is very large and/or when the structure is under construction and its full stiffness is not yet reached, there can be more than two vibration modes contributing to the instability. Second, this mode combination is only an approximate expression of the flutter mode anyway and there is no reason why it has to be always true. Particularly when the contributing modes are lacking in their geometrical affinity, the flutter vibration mode could be quite complicated. In view of this, it is desirable to develop a more comprehensive and accurate procedure to do flutter mode analysis and improve the understanding of the aerodynamic instability of cable-supported bridges.

Ge and Tanaka (1999) investigated the issue of mode participation by considering the participation of both multiple natural modes and full natural modes of vibration, which are particularly important for long-span cable-supported bridges. Both the multimode and full-mode flutter analysis methods employed the general expression of linear unsteady aerodynamic forces, which include 18 dimensionless aerodynamic derivatives considering vertical bending, pitching torsion as well as lateral bending and their mode interactions. The multimode approach uses a multidegree-of-freedom model technique and includes the self-excited aerodynamic force interaction in the structural equations of motion to produce an asymmetrical eigenvalue problem with the order of the number of natural modes participating in flutter oscillation. The eigenvalue solution of the characteristic matrix involved is indicative of the nature of preselected possible oscillations that can exist at a given wind speed. For the full-mode approach, it is unnecessary to assume the flutter oscillation mode as a combination of a few natural modes of the target structure. Indeed, this approach formulates the asymmetric eigenvalue problem including the entire natural modes and representing all possible system oscillations, but only solves the first several characteristic roots which contain the most important one corresponding to the lowest flutter speed.

Comparison of numerical results based on the two methods indicated that there may be a significant conservatism in the two-mode method that assumes only the fundamental flexural and torsional natural modes to participate in the flutter response, and that the multimode analysis including a sufficient number of natural modes can approach an approximate solution to the full-mode analysis with good accuracy.

12.3.3.4 Galloping

Aerodynamic instability in translational motion normal to wind is sometimes observed in structures with aerodynamically bluff cross-sections such as towers, cranes and ice-covered cables, and is called galloping because of its violent feature. It is a self-excited motion in which the structural motion itself is the cause of creating the negative aerodynamic damping. The main difference in characteristics of this phenomenon from the previously described flutter instability is in the fact that

1. the response amplitude of galloping tends to grow to a much greater magnitude than the linear dimension of the cross section, often leading to the structural destruction;
2. the vibration tends to be strongly nonlinear as a result;

3. the response is often strongly influenced by the existence of flow turbulence, in the sense that turbulence could trigger the instability which does not exist otherwise.

Since the galloping motion is usually with very large amplitude at high wind speed, it is known that the response is predictable by applying the quasisteady analysis only if the aerodynamic forces on the given structural section are clearly identified. A well-known criterion for instability, introduced by Den Hartog (1932), is given by

$$\left[\frac{dC_L}{d\alpha} + C_D\right]_{\alpha=0} < 0 \tag{347}$$

For the bridge girders, particularly for a deck-on-box section, this condition is sometimes satisfied when the depth of the box is substantial, such as more than a quarter of the bridge deck width.

One of the characteristic points of galloping that is different from flutter type of instability is its strong nonlinearity. Since the dynamic displacement for this case is considered to be large, the direction of wind against the bridge deck will have a relative angle of attack defined by

$$\alpha = \arctan\left(\frac{\dot{z}}{\overline{U}}\right) \tag{348}$$

The aerodynamic lift for this case is then replaced by a lateral force, taken to be positive in the direction of $-z$, which is defined by

$$F_z(\alpha) = -(L\cos(\alpha) + D\sin(\alpha)) = \frac{\rho\overline{U}^2}{2} d \cdot C_{Fz}(\alpha) \tag{349}$$

where, the lift and drag forces are defined by

$$L(\alpha) = \frac{\rho\overline{U}_{rel}^2}{2} d \cdot C_L(\alpha), \qquad D(\alpha) = \frac{\rho\overline{U}_{rel}^2}{2} d \cdot C_D(\alpha), \qquad \overline{U}_{rel} = \overline{U}\sec(\alpha) \tag{350}$$

Hence, the lateral force coefficient is given by

$$C_{Fz}(\alpha) = -\big(C_L + C_D\tan(\alpha)\big)\sec(\alpha) \tag{351}$$

and

$$\frac{dC_{Fz}}{d\alpha} = -\left[\frac{dC_L}{d\alpha} + C_D\left(1 + 2\tan^2(\alpha)\right)\right]\sec(\alpha) - \left(C_L + \frac{dC_D}{d\alpha}\right)\tan(\alpha)\sec(\alpha) \tag{352}$$

The lateral force excites the vibration when $dC_{Fz}/d\alpha$ at $\alpha = 0$ is positive, which gives the Den Hartog's criterion given in Equation (347) as a necessary condition for instability. By substituting Equations (348–351), it follows

$$C_{Fz}(\alpha) = \sum_{i=odd} A_i\left(\frac{\dot{z}}{\overline{U}}\right)^i + \sum_{j=even} A_j\left(\frac{\dot{z}}{\overline{U}}\right)^j \frac{\dot{z}}{|\dot{z}|} \tag{353}$$

In many cases, the first mode of vibration would be predominant in response and hence

$$Q_i(t) = \int_L L(y,t)\psi_i(y)\,dy \approx \frac{\rho \overline{U} d}{2} A_1 \dot{q}_1 \int_L \psi_1^2(y)\,dy \tag{354}$$

Hence, as a first-order equation,

$$\ddot{q}_1 + 2\omega_1 \left(\zeta_1 - \frac{\rho \overline{U} d}{4\omega_1 m_e} A_1 \right) \dot{q}_1 + \omega_1^2 q_1 = 0 \tag{355}$$

which leads to a refined form of the Den Hartog criterion as follows:

$$\left[\frac{dC_L}{d\alpha} + C_D \right]_{\alpha=0} < \frac{4\omega_1 m_e}{\rho \overline{U} d} \zeta_1 \tag{356}$$

m_e above is an equivalent mass. Further, the nonlinear equation of motion is expressed by

$$\ddot{z} + \omega_1^2 z + F(\dot{z}) = 0 \tag{357}$$

where

$$F(\dot{z}) = 2\zeta_1\omega_1\dot{z} - \frac{\rho \overline{U}^2 d}{2m} \left[\sum_{i=odd} A_i \left(\frac{\dot{z}}{\overline{U}} \right)^i + \sum_{j=even} A_j \left(\frac{\dot{z}}{\overline{U}} \right)^j \frac{\dot{z}}{|\dot{z}|} \right] \tag{358}$$

From the energy principle

$$W = \oint F(\dot{z})\,dz(t) = 0 \tag{359}$$

which will decide the nonlinear limit cycle for a range of wind speed.

12.3.3.5 Instability in Torsion

Analysis of torsional instability, considering a SDOF system, should be parallel to SDOF galloping. However, its quasi-steady analysis is not so straightforward because of the difficulty in taking the effective angle of attack. Analyses were attempted by Modi and later by Nakamura and Mizota, which are well summarized by Blevins (1990).

In order to avoid any possible ambiguity, it makes more sense to use unsteady aerodynamic forces. The only relevant terms are A_2^* and A_3^* of the pitching moment. Hence, the equation of motion is

$$\ddot{\alpha} + 2\zeta_\alpha\omega_\alpha\dot{\alpha} + \omega_\alpha^2\alpha = \frac{\rho \overline{U}^2}{2J} B^2 \left(K A_2^* \frac{B\dot{\alpha}}{\overline{U}} + K^2 A_3^* \alpha \right) \tag{360}$$

where, as before, the aerodynamic coefficients A_2^* and A_3^* are assumed to be functions of reduced frequency $K = \omega B/\overline{U}$. If no coupling with any other modes are assumed, $\omega = \omega_\alpha$ and hence the critical condition for the onset of instability is defined by

$$A_2^* = \frac{4J\zeta_\alpha}{\rho B^4} \tag{361}$$

12.3.4 Buffeting

12.3.4.1 General Characteristics

Buffeting is a dynamic structural response caused by wind turbulence, which either inherently exists in natural wind or was created by existence of upstream objects. It is usually considered and analyzed as a forced vibration caused by time-dependent aerodynamic forces due to velocity fluctuation. Buffeting is a stochastic vibration, consisting of a wide range of frequency components, and its amplitude is randomly fluctuating. However, the buffeting of bridge decks often appears as a narrow-band response only in the first couple of modes and quite random in its amplitude. It generally increases in parabolic manner with the mean wind speed. The peak response amplitude is usually four times or so greater than the root-mean-square response. The consequence of having buffeting vibration is usually not catastrophic to the structure but a long-duration influence from it, such as fatigue damage, can be a serious engineering concern.

12.3.4.2 Analytical Prediction

Two-Dimensional Quasi-Steady Assumption

By applying conventional quasi-steady aerodynamics and the concept of strip theory, the 2D aerodynamic lift force per unit deck length can be approximately expressed as

$$L(y,t) = \frac{\rho \overline{U}^2}{2} B \frac{dC_L}{d\alpha} \frac{w - \dot{z}}{\overline{U}} \tag{362}$$

in which, $dC_L/d\alpha$ is the lift slope and $w(t)$ is the vertical component of the velocity vector. Equation (362) is obtained by replacing the lift coefficient by the product of the lift slope and the instantaneous angle of attack, $(w - \dot{z})/\overline{U}$. Equation (362) results in the aerodynamic damping of

$$\zeta_{ai} = \frac{\rho \overline{U} B}{4\omega_i M_i} \frac{dC_L}{d\alpha} \int_L \psi_i^2(y)\, dy \tag{363}$$

and the buffeting force corresponding to the vertical bending of the bridge deck as

$$Q_i(t) = \frac{\rho \overline{U} B}{2} \frac{dC_L}{d\alpha} \int_L w(y,t)\psi_i(y)\, dy \tag{364}$$

By taking the autocorrelation of Equation (364) and applying Fourier transform, the spectral density function of $Q_i(t)$ can be obtained as follows:

$$G_{Qi}(f) = \left(\frac{\rho \overline{U} B}{2} \frac{dC_L}{d\alpha}\right)^2 \int_L \int_L |X_a(y_1, y_2; f)|^2 G_w(y_1, y_2; f)\psi_i(y_1)\psi_i(y_2)\, dy_1\, dy_2 \tag{365}$$

Aerodynamic Admittance and Joint Acceptance

In Equation (365), there appears a new function X_a, which represents the overall effectiveness of the aerodynamic lift force in exciting the structure. It is primarily a function of frequency and the span-wise correlation of lift forces assumed in the strip theory and can be put together with the lateral correlation of velocity as

$$|X_a(y_1, y_2; f)|^2 G_w(y_1, y_2; f) \approx |X_a(f)|^2 G_w(f) \widetilde{R}_L(y_1, y_2; f) \tag{366}$$

Hence

$$G_{Qi}(f) \approx \left(\frac{\rho \overline{U} B}{2} \frac{dC_L}{d\alpha} \right)^2 |X_a(f)|^2 G_w(f) |J_i(f)|^2 \tag{367}$$

where

$$|J_i(f)|^2 = \int_L \int_L \widetilde{R}_L(y_1, y_2; f) \psi_i(y_1) \psi_i(y_2) \, dy_1 \, dy_2 \tag{368}$$

is referred to as the joint acceptance function. $\widetilde{R}_L(y_1, y_2; f)$ is the correlation coefficient of the lift force, and is usually approximated as

$$\widetilde{R}_L(y_1, y_2; f) \approx \exp\left(-c \left(\frac{f \Delta y}{\overline{U}} \right)^k \right) \tag{369}$$

where $\Delta y = |y_1 - y_2|$. The force correlation has been often assumed to be represented by the velocity correlation. However, in reality, it has been observed that the lift forces tend to be better correlated than the velocity components, which means that the calculated response based on this assumption can be somewhat less than in reality. $|X_a(f)|^2$ is called the **aerodynamic admittance function**.

Example 12.9 [Horizontal line-like structures under drag excitation] Consider the buffeting response of a horizontal beam of length L, mass per unit length m and lateral bending stiffness EI. The beam is exposed to a turbulent wind approaching normal to the beam axis, inducing horizontal bending motion due to fluctuating drag. The equation of motion is

$$m\ddot{x} + (EIx'')'' = p(y, t) \tag{370}$$

By expressing the response as $x(y, t) = \sum_r \phi_r(y) q_r(t)$ the modal quantities are introduced:

$$M_r = \int_L m(y) \phi_r^2(y) \, dy \tag{371}$$

$$K_r = (2\pi f_r)^2 M_r \tag{372}$$

$$C_r = 2\zeta_r \sqrt{M_r K_r} \tag{373}$$

$$Q_r(t) = \int_L p(y, t) \phi_r(y) \, dy \tag{374}$$

The modal equation of motion becomes

$$M_r(\ddot{q}_r + 2\zeta_r\omega_r\dot{q}_r + \omega_r^2 q_r) = Q_r(t) \tag{375}$$

The mean-square response is given by

$$\sigma_x^2(y) = \sum_i \sum_j \overline{q_i q_j}\phi_i(y)\phi_j(y) \approx \sum_r \overline{q_r^2}\phi_r^2(y) \tag{376}$$

where

$$\overline{q_r^2} = \frac{1}{K_r^2}\int_0^\infty |H_r(f)|^2 G_{Q_r}(f)\,df \tag{377}$$

$$G_{Q_r}(f) = \int_L \int_L G_P(y_1, y_2; f)\phi_r(y_1)\phi_r(y_2)\,dy_1\,dy_2 \tag{378}$$

and

$$G_P(y_1, y_2; f) = G_P(f)\widetilde{R}_P(y_1, y_2; f) \tag{379}$$

The force coherence $\widetilde{R}_P(y_1, y_2; f)$ is unknown but is assumed to be the same as the velocity coherence $\widetilde{R}_u(y_1, y_2; f)$ by Davenport and the same assumption has been often made. Then

$$\widetilde{R}_P(y_1, y_2; f) = \widetilde{R}(y_1, y_2; f) = e^{-f|y_1 - y_2|/\overline{U}} \tag{380}$$

Hence,

$$G_{Q_r}(f) = G_P(f)|J_r(f)|^2 \tag{381}$$

in which

$$|J_r(f)|^2 = \int_L \int_L \widetilde{R}_P(y_1, y_2; f)\phi_r(y_1)\phi_r(y_2)\,dy_1\,dy_2 \tag{382}$$

is the joint acceptance function. Substituting these expressions, the mean-square modal response becomes as follows:

$$\overline{q_r^2} = \frac{1}{K_r^2}\int_0^\infty |H_r(f)|^2 G_{Q_r}(f)\,df = \frac{1}{K_r^2}\int_0^\infty |H_r(f)|^2 |J_r(f)|^2 G_P(f)\,df \tag{383}$$

where

$$|H_r(f)|^2 = \frac{1}{\left(1 - (f/f_r)^2\right)^2 + \left(2(\zeta_{S,r} + \zeta_{a,r})f/f_r\right)^2} \tag{384}$$

$$G_P(f) = \left(\frac{2\overline{P}}{\overline{U}}\right)^2 |X_r(f)|^2 G_u(f) \tag{385}$$

$\zeta_{S,r}$ and $\zeta_{a,r}$ are the structural and aerodynamic damping, respectively. The aerodynamic damping for the r-th mode is approximately given, based on the quasi-steady theory, by

$$\zeta_{a,r} = \frac{\rho A C_D \overline{U}}{2\omega_r m_r} \tag{386}$$

where

$$m_r = \frac{\int_L m(y)\phi_r^2(y)\,dy}{\int_L \phi_r^2(y)\,dy} \tag{387}$$

The aerodynamic admittance function $|X_r(f)|^2$ needs to be defined in one way or another. A few important comments should be remembered for this analysis as follows.

1. Mode-coupling terms are ignored, though this should not have too much influence.
2. Spanwise force coherence would be probably higher than assumed here.
3. Aerodynamic admittance: is this a scape goat?
4. For the unsteady aerodynamic forces, it is better to use the measured derivatives, if possible.

Example 12.10 [Horizontal line-like structure under lift excitation] Consider this time the same beam as the previous example, but buffeting excitation due to lift force. By expressing the response as

$$z(y,t) = \sum_r q_r(t)\phi_r(y)B \tag{388}$$

the modal equation of motion is

$$\ddot{q}_r + 2\left(\zeta_{S,r} + \zeta_{a,r}\right)\omega_r\dot{q}_r + \omega_r^2 q_r = \frac{Q_r(t)}{M_r} \tag{389}$$

in which

$$M_r = \int_L m(y)\phi_r^2(y)\,dy \tag{390}$$

By using the Scanlan type aerodynamic derivatives, the aerodynamic damping is expressed as

$$\zeta_{a,r} = -\frac{H_1^*}{2\mu_r}\int_L \phi_r^2(y)\,dy \tag{391}$$

in which $\mu_r = M_r/L\rho B^2$ is the modal mass ratio and the derivative H_1^* is a function of reduced frequency $K = B\omega_r/\overline{U}$. If the quasi-steady approach is taken as before,

$$\zeta_{a,r} = \frac{1}{4\mu_r L}\frac{1}{K}\frac{dC_L}{d\alpha}\int_L \phi_r^2(y)\,dy \tag{392}$$

The generalized buffeting force can be formulated as

$$Q_r(t) = \frac{\rho \overline{U}}{2} \frac{dC_L}{d\alpha} \int_L w(y, t)\phi_r(y)\, dy \tag{393}$$

Auto-correlation of $Q_r(t)$ is

$$R_{Q_r}(\tau) = \left(\frac{\rho \overline{U}}{2} \frac{dC_L}{d\alpha}\right)^2 \int_L \int_L \overline{w(y_1, t)w(y_2, t+\tau)\phi_r(y_1)\phi(y_2)\, dy_1\, dy_2} \tag{394}$$

By applying Fourier transform, the spectral density function of $Q_r(t)$ can be obtained. Considering the concept of aerodynamic admittance again,

$$G_{Q_r}(f) = \left(\frac{\rho \overline{U}}{2} \frac{dC_L}{d\alpha}\right)^2 |X_r(f)|^2 \int_L \int_L C_w(y_1, y_2; f)\phi_r(y_1)\phi_r(y_2)\, dy_1\, dy_2 \tag{395}$$

in which $C_w(y_1, y_2; f) = G_w(f)R_w(y_1, y_2; f)$ is the co-spectrum of velocity component $w(t)$. Again, the use of velocity correlation instead of the force correlation is usually the practice here. It is an approximation which is probably not right and will, most likely, lead to the underestimation of the structural response. The modal response spectrum is

$$G_{qr}(f) = \frac{|H(f)|^2}{K_r^2} G_{Q_r}(f) \tag{396}$$

in which

$$|H_r(f)|^2 = \frac{1}{\left(1 - (f/f_r)^2\right)^2 + \left(2(\zeta_{S,r} + \zeta_{a,r})f/f_r\right)^2} \tag{397}$$

hence

$$\sigma_{qr}^2 = \int_0^\infty G_{qr}(f)\, df = \left(\frac{\rho B \overline{U}}{2K_r} \frac{dC_L}{d\alpha}\right)^2 \int_0^\infty G_w(f)|J_r(f)|^2 |H_r(f)|^2\, df \tag{398}$$

where the joint acceptance function is defined as

$$|J_r(f)|^2 = \int_L \int_L \widetilde{R}_w(y_1, y_2; f)\phi_r(y_1)\phi(y_2)\, dy_1\, dy_2 \tag{399}$$

and the mean-square response is given by

$$\sigma_Z^2(y) \approx \sum_r \sigma_{qr}^2 \phi_r^2(y) \tag{400}$$

Example 12.11 [Along-wind response of vertical line-like structure] Consider now a thin vertical structure, such as a tower, of height H, exposed to a boundary layer wind. The mean speed distribution is generally given by

$$\overline{U}(z) = \overline{U}_H \left(\frac{z}{H} \right)^{\alpha} \tag{401}$$

Applying the modal analysis as before,

$$x(y,t) = \sum_r \phi_r(z) q_r(t) \qquad \text{and} \qquad \sigma_x^2(z) \approx \sum_r \overline{q_r^2} \phi_r^2(z) \tag{402}$$

The expressions leading to $\overline{q_r^2}$ are the same as before, except the aerodynamic damping can be approximately given by

$$\zeta_{a,r} \approx \frac{1}{M_r \omega_r} \int_0^H \frac{\overline{P}(z)}{\overline{U}(z)} \phi_r^2(z)\, dz \tag{403}$$

with

$$\overline{P}(z) = \frac{\rho \overline{D} C_D}{2} \overline{U}^2(z) \tag{404}$$

Spectral density for the generalized force $Q_r(t)$ is given by

$$G_{Qr}(f) = \int_0^H \int_0^H G_P(z_1, z_2; f) \phi_r(z_1) \phi_2(z_2)\, dz_1\, dz_2 \tag{405}$$

where

$$G_P(z_1, z_2; f) = \sqrt{G_P(z_1, f) G_P(z_2, f)} \cdot \widetilde{R}_P(z_1, z_2; f) \tag{406}$$

in which

$$G_P(z, f) = \left(\frac{2\overline{P}(z)}{\overline{U}(z)} \right)^2 |X_a(f)|^2 G_u(z, f) \tag{407}$$

where

$$G_u(z, f) \approx G_u(H, f) \tag{408}$$

and

$$\widetilde{R}_P(z_1, z_2; f) \approx \widetilde{R}_u(z_1, z_2; f) \approx \exp\left(-c_V \frac{f|z_1 - z_2|}{\overline{U}(z_m)} \right) \tag{409}$$

with $z_m = \frac{z_1 + z_2}{2}$. Hence

$$\sqrt{G_P(z_1, f) G_P(z_2, f)} = 4 \frac{\overline{P}(z_1)}{\overline{U}(z_1)} \frac{\overline{P}(z_2)}{\overline{U}(z_2)} |X_a(f)|^2 G_u(f) \tag{410}$$

and

$$\frac{\overline{P}(z)}{\overline{U}(z)} = \frac{\rho \overline{D} C_D}{2}\overline{U}(z) = \frac{\overline{P}_H}{\overline{U}_H}\frac{\overline{U}(z)}{\overline{U}_H} \qquad \text{with} \qquad \overline{P}_H = \frac{\rho \overline{D} C_D}{2}\overline{U}_H^2 \tag{411}$$

Substituting these expressions

$$G_{Qr}(f) \approx \left(\frac{2\overline{P}_H}{\overline{U}_H}\right)^2 G_u(f)|X_a(f)|^2 \times \times \int_0^H \int_0^H \widetilde{R}_u(z_1, z_2; f)\phi_r(z_1)\phi_r(z_2)\frac{\overline{U}(z_1)\overline{U}(z_2)}{\overline{U}_H^2}\, dz_1\, dz_2 \tag{412}$$

$$|J_r(f)|^2 = \int_0^H \int_0^H \frac{\overline{U}(z_1)\overline{U}(z_2)}{\overline{U}_H^2} \exp\left(-c_V\frac{f|z_1 - z_2|}{\overline{U}_m}\right)\phi_r(z_1)\phi_r(z_2)\, dz_1\, dz_2 \approx \frac{1}{(1+\alpha)^2}\frac{1}{1 + \frac{c_V}{3}\frac{fH}{\overline{U}_H}} \tag{413}$$

when $\phi(z) = \frac{z}{H}$. It can be assumed as $c_V = 8$, for example.

Example 12.12 [Vertical structure with horizontal extent] In reality, it becomes probably necessary to consider a substantial width of the structure in the above analysis. As a result, the joint acceptance function is now considered in horizontal direction as well as the vertical direction as follows:

$$G_Q(f) = \left(\frac{2\overline{P}_H}{\overline{U}_H}\right)^2 |J_V(f)|^2|J_H(f)|^2 G_u(f) \tag{414}$$

where

$$|J_V(f)|^2 \approx \frac{1}{(1+\alpha)^2}\frac{1}{1 + \frac{8}{3}\frac{fH}{\overline{U}_H}} \qquad \text{and} \qquad |J_H(f)|^2 \approx \frac{1}{1 + 10\frac{fB}{\overline{U}_H}} \tag{415}$$

Furthermore,

$$\left(\frac{\sigma_X}{\overline{X}}\right)^2 = 16(1+\alpha)^2 I_U^2 \int_0^\infty |J_V(f)|^2|J_H(f)|^2 \frac{G_u(f)}{\sigma_u^2}\, df \tag{416}$$

Hence,

$$\left(\frac{\sigma_X}{\overline{X}}\right)^2 \approx 16(1+\alpha)^2 I_u^2 \left[|J_V(f)|^2|J_H(f)|^2 \frac{G_u(f)}{\sigma_u^2} \int_0^\infty |H(f)|^2\, df + \int_0^\infty |J_V(f)|^2|J_H(f)|^2 \frac{G_u(f)}{\sigma_u^2}\right] \tag{417}$$

and

$$\int_0^\infty |H(f)|^2\, df = \frac{\pi f_0}{4\zeta_S} \tag{418}$$

where f_0 is the eigen-frequency, and ζ_S is the structural damping. Finally

$$\int_0^\infty |J_V(f_0)|^2 |J_H(f_0)|^2 \frac{G_u(f)}{\sigma_u^2}\, df \approx \int_{-\infty}^{\ln(f)} \frac{f G_u(f)}{\sigma_u^2}\, d\ln(f) \tag{419}$$

with $f = 3\overline{U}_H/(8H)$. This formulation has been adopted by the National Building Code of Canada for the consideration of wind load on tall buildings.

12.3.4.3 Aerodynamic Admittance

Aerodynamic admittance is a transfer function to express how effectively the frequency characteristics of velocity fluctuation are picked up by the aerodynamic force components. Naturally, it depends upon the geometrical configuration of the body exposed to the flow turbulence. Even if an ideally 2D strip assumption is considered, since there are two velocity components (u, w) and each of them has its own effects on three force components (L, D, M_P), six different admittance functions are conceivable for each cross section of the body, although probably only three of them would have practical significance. It is very difficult to make analytical estimation of these functions, although there are a few cases considered as follows.

Sears Function

A classic example is an aerofoil that is facing a sinusoidal change of the w component. Sears (1941) formulated this case as follows:

$$L = \frac{\rho \overline{U}^2}{2} B \frac{dC_L}{d\alpha} \frac{w(t)}{\overline{U}} \Theta(k) \tag{420}$$

where $k = \frac{\pi f B}{\overline{U}}$ and $w(t) = w_0 \sin(\omega t)$. The Sears function $\Theta(k)$ is analytically given by

$$\Theta(k) = \big(J_0(k) - iJ_1(k)\big)C(k) + iJ_1(k) \tag{421}$$

using $C(k) = F(k) + iG(k)$. It can be also written as

$$\begin{aligned} |\Theta(k)|^2 &= (J_0^2 + J_1^2)(F^2 + G^2) + J_1^2 + 2J_0 J_1 G - 2J_1^2 F \\ &\approx \frac{1}{1 + 2\pi k} \qquad \text{(Liepmann 1952)} \end{aligned} \tag{422}$$

Davenport's Formulation

The admittance function between $u(t)$ and drag force $D(t)$ was considered by Davenport (1964) by integrating the velocity correlation over the surface. When the pressure correlation over a flat surface placed normal to the flow is approximated by

$$\widetilde{R}_u(f) \approx \exp\left(- k\frac{f\Delta}{\overline{U}}\right) \tag{423}$$

with $k \approx 7$, the admittance function is

$$|X(f)|^2 = \frac{1}{A^2}\int_0^D \int_0^D \int_0^B \int_0^B \widetilde{R}_u(\Delta y)\widetilde{R}_u(\Delta z)\, dy_1\, dy_2\, dz_1\, dz_2 \tag{424}$$

where $\Delta y = |y_1 - y_2|$ and $\Delta z = |z_1 - z_2|$. When $B = D = \sqrt{A}$ (a square plate), this formulation results in

$$|X(f)|^2 = \left[\frac{2}{(k\xi)^2}(k\xi - 1 + \mathrm{e}^{-k\xi})\right]^2, \qquad \xi = \frac{f\sqrt{A}}{\overline{U}} \tag{425}$$

Vickery's Expression

Following a similar concept as before, Vickery (1965) came up with an expression as follows:

$$|X(f)|^2 = \frac{1}{\left[1 + \left(2f\sqrt{A}/\overline{U}\right)^{4/3}\right]^2} \tag{426}$$

Measured Results

There have been a number of attempts to measure the aerodynamic admittance functions experimentally. The method in general is to take the ratio of the total force spectrum on a section model to the velocity spectrum detected simultaneously. However, since the testing conditions cannot be perfectly two-dimensional, often there is a problem of having two different effects on the results mixed up: 2D frequency transfer effects of the sectional force at any particular section and the spanwise coherence of it. In order to avoid this problem, the pressure distribution around the sections along very narrow strips has to be detected to measure both of these effects separately. The development of a fast pressure-scanning device has made it possible only recently. There are indications that actual 2D admittance may be much less than has been believed but, at the same time, the spanwise force correlation seems to be much higher than that of velocity components.

12.3.4.4 Peak Factor

The largest instantaneous value of a stationary random function within a specific sampling period can be estimated from the statistical characteristics of the process as a ratio to the value of the standard deviation. Consider first a stationary random process $x(t)$ having a normal probability distribution with the mean a_x and standard deviation σ_x. It is convenient to define the reduced variate $\eta = (x - a_x)/\sigma_x$, in which case the probability density of the process is

$$P(\eta) = \frac{1}{\sqrt{2\pi}}\mathrm{e}^{-\eta^2} \tag{427}$$

The expression for the cumulative distribution function for the maxima of the stationary random process was first derived by Cartwright and Longuet-Higgins. They have shown that, for a large value of η, and

$\varepsilon \neq 1$, the distribution function $q(\eta)$ can be approximately given by

$$q(\eta) \approx \sqrt{1-\varepsilon^2}\mathrm{e}^{-\eta^2/2} \tag{428}$$

in which

$$\varepsilon = \sqrt{1 - \frac{m_2^2}{m_0 m_4}} \quad \textit{and} \quad m_r = \int_0^{\infty} n^r G(n)\, dn \tag{429}$$

where $G(n)$ is the power spectrum of the random process. Davenport used this expression and also by applying the Rice's theory to give the number of maxima N, $N \to \infty$, during a given period, T, as

$$N = \sqrt{m_4/m_2} \tag{430}$$

He reached the probability density for the largest maxima as

$$P_{max}(\eta)\, d\eta = \mathrm{e}^{-\xi}\, d\xi \tag{431}$$

where

$$\xi = Nq(\eta) = \nu T \exp\left(-\frac{\eta^2}{2}\right) \quad \text{and} \quad \nu = \sqrt{\frac{m_2}{m_0}} \tag{432}$$

or, inversely,

$$\begin{aligned} \eta &= \sqrt{2\ln(\nu T) - 2\ln(\xi)} \\ &= \sqrt{2\ln(\nu T)} - \frac{\ln(\xi)}{\sqrt{2\ln(\nu T)}} - \frac{(\ln(\xi))^2}{2(2\ln(\nu T))^{3/2}} + \ldots \end{aligned} \tag{433}$$

Using these expressions, certain properties of the distribution can be decided. The mean, for example, is

$$\overline{\eta_{max}} = \int_{-\infty}^{\infty} \eta P_{max}(\eta)\, d\eta = \int_{-\infty}^{\infty} \eta \mathrm{e}^{-\xi}\, d\xi = \sqrt{2\ln(\nu T)} + \frac{\gamma}{\sqrt{2\ln(\nu T)}} \tag{434}$$

where $\gamma = 0.5772$ is the Euler's constant and T is the sampling period. Since the mean square value is expressed as

$$\overline{\eta_{max}^2} = \int_{-\infty}^{\infty} \eta^2 \mathrm{e}^{-\xi}\, d\xi \tag{435}$$

the standard deviation is given by

$$\sigma(\eta_{max}) = \overline{\eta_{max}^2} - (\overline{\eta_{max}})^2 = \frac{\pi}{\sqrt{6}}\sqrt{2\ln(\nu T)} \tag{436}$$

The mode, which is found by the maximum of Equation (431), is given by

$$\mathrm{mode}(\eta_{max}) = \sqrt{2\ln(\nu T)} \tag{437}$$

and the probability density is $\sqrt{2\ln(\nu T)}/e$.

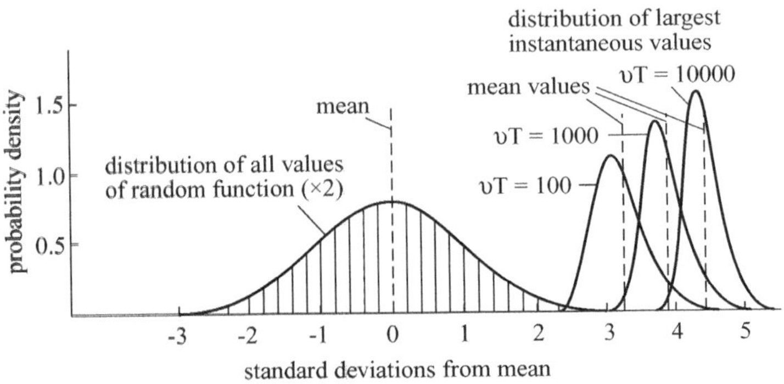

Figure 12.24 Largest value distributions for various νT (after Davenport 1964)

Davenport (1964) has commented on this theory as follows: "The quantity, ν, can be interpreted physically as the frequency at which most of the energy in the spectrum is concentrated. Thus, with lightly damped systems, this will generally be close to the natural frequency when we are talking about the gust induced structural response, for example. Furthermore, in wind-loading problems, we are generally concerned with predicting the largest value likely to occur within a period such as an hour. These conditions suggest that, in the main, the values of νT of practical interest lie between 10^2 and 10^4." The probability density functions for $\nu T = 10^2$, 10^3 and 10^4 are shown in Figure 12.24. For comparison, the distribution of the population is also shown in this figure. Figure 12.24 emphasizes the narrowness of the distribution of largest values, particularly for larger values of νT. It points to the fact that, in many problems, it is probably sufficient to assume the largest value equal to the mean largest value and ignore the variability. From this figure, it is seen that in practical cases the peak value during an hour is likely to be 3.5–4.5 times standard deviations in excess of the mean value.

12.3.4.5 Time-Domain Analysis

The formulation of buffeting analysis was first developed in the time domain just like any other dynamic analysis. However, its development has been done by linearizing equations and solving them in the frequency domain and this practice is still continuing. In more recent years, the development of computer-aided flying and the active flutter control of aircraft wings have pushed the need to express the unsteady aerodynamic forces in a time-variant manner. Very efficient algorithms have been developed in the time domain to perform the servo-control of flaps based on aerodynamic data obtained experimentally in the frequency domain.

In the field of bridge aerodynamics, too, the frequency domain approach has been predominantly applied to solve the problem of wind effects even if the original formulation was in the time domain. As briefly stated in Section 12.3.3.2, in the 1970s Scanlan et al. first worked on solving the problem entirely in the time domain, introducing the indicial functions, which were put forward earlier in the aeronautical field. It is, however, only recently that much effort has been invested in developing efficient time-domain formulation of the unsteady aerodynamic forces that could be combined with FE models of the structure and could include all the nonlinearities that have been omitted in the past. This development has been justified with the planning of very long bridges such as the bridge over the Stretto di Messina and the Akashi Kaikyo.

The problem of the formulation of time varying wind load in the time domain can be summarized as follows.

1. The bridge deck is excited by the instantaneous wind turbulence at a certain time t but is also influenced by what happened between the eddies in the flow and the deck at time $t - \tau$. It is to say that the wind loading involves a memory effect or that there exists a phase lag between the excitations and actual aerodynamic forces themselves. In the frequency domain this memory effect is expressed by aerodynamic admittance functions.
2. A moving fluid induces the initial forcing on a deck. At the same time, however, the motion of a deck interacts with the fluid media and induces additional forces, which are not in phase with the body motion. These motion-dependent, self-excited forces can be expressed in one or several degrees of freedom, either independent of or coupling with each other.
3. The aerodynamic forces are far from being fully correlated along the bridge span. The spanwise correlation of the approaching wind fluctuations, combined with the mode shapes of the deck, governs the spanwise distribution of the aerodynamic forcing. Also, the spanwise coherence of the aerodynamic forces is not necessarily identical to that of the approaching flow turbulence, as opposed to what has been often implied by the strip assumption. It appears that the spanwise coherence of the forces also depends upon the deck geometry, its aspect ratio in particular, length scales of turbulence and turbulence intensity.
4. The simulation of the approaching wind has also its own share of difficulties. An adequate simulation should model all the one-point and two-point statistics of the wind fluctuations. Locally, it should respect the first and second moments for each wind component as well as autospectra and cross-spectra of all velocity components. A discussion of this aspect is found in the literature. Needless to say this is a subject of primary importance because, no matter how good the wind loading algorithm is, if the input wind field is not adequate the output will suffer significantly.

A brief literature review regarding time-domain analyses in bridge aerodynamics is given below. The approaches found in the literature can be classified into five categories:

1. quasi-steady aerodynamics;
2. corrected quasi-steady aerodynamics;
3. indicial functions (Fourier transform);
4. rational function approximations (Laplace transform);
5. equivalent oscillator (neural network and black box).

Miyata et al. (1995) clearly presented the advantages of the time-domain approach for the prediction of wind action on long span bridges when combined with a FE modeling of the structure. The formulation used here is fairly simple and very much along the line of the original formulation by Davenport. Quasi-steady aerodynamics are assumed as well as the strip assumption. Kovács et al. (1992) has a similar approach except that the FE model of the structure includes structural nonlinearities for large deflections in order to evaluate the ultimate state conditions. Although the limitations of these approaches with quasi-steady aerodynamics are numerous, Miyata et al. (1995) obtained an impressively good agreement of their results with the wind-tunnel test results for a very large suspension bridge. The limitations are:

1. noninclusion of the motion-induced forces besides the quasi-steady expression of the aerodynamic damping in the wind load model;
2. memory function expressed as an aerodynamic admittance filtering the wind field simulation;
3. inadequate representation of the spanwise coherence of the forces;
4. the assumption that the center of the aerodynamic lift is fixed at a quarter of the deck width from the deck centerline.

This last assumption allows the inclusion of the torsional motion in the definition of the apparent angle of attack or in the relative wind velocity as shown in Section 12.3.2.2. The position of the aerodynamic center is known to vary with reduced velocity.

The work by Diana et al. (1992, 1995) in relation to the studies for the proposed bridge over the Stretto di Messina has taken a similar approach, with the exception that some of the above limitations have been removed by the development of a "corrected" quasi-steady theory. Here, the motion-induced forces are fully included in the formulation of the aerodynamic forces via the experimentally determined aerodynamic derivatives. The location of the aerodynamic center is also determined from the aerodynamic derivatives. Note that Diana et al. have proposed a different normalization of the aerodynamic derivatives so that they can be interpreted as a deviation from the quasi-steady expression. At high reduced velocities, the process is almost quasi-steady and the aerodynamic derivatives, by their definition, tend to be unity.

To include the motion-induced forces, Diana et al. (1992) proposed an equivalent linearization, for each reduced velocity, of the following type:

$$C_L^*(\alpha_e) = C_{Ls}(\alpha_0) + \int_{\alpha 0}^{\alpha e} K_L^* \, d\alpha \tag{438}$$

where $C_L^*(\alpha_e)$ is the corrected aerodynamic force coefficient, $C_{Ls}(\alpha_0)$ is the static coefficient at angle α_0 of the equilibrium position and K_L^* are the aerodynamic derivatives (i.e., lift slope varying with reduced velocity). Here α_e is defined as the apparent angle of attack based on the relationship

$$\alpha_e = \alpha - \frac{\dot{z}}{\overline{U}} - n\frac{B\dot{\alpha}}{\overline{U}} \tag{439}$$

where z and α represent respectively a vertical displacement and rotation of the deck, B the deck width, $\overline{U}$ the mean wind speed, and n is the position of the aerodynamic center, which is a function of reduced velocity. The formulation has been proved to be adequate in many ways, except when it deals with the response in higher vibration modes. The limitations attributed to the memory effects and the spanwise coherence of the forces have not been dealt with by Diana et al.

In a series of papers spanning several years, Scanlan et al. (1972, 1974) described the use of indicial functions for the inclusion of the motion-induced forces in the time-domain formulation of the aerodynamic forces. The formulation of self-excited force F_{Ls} can be taken as follows:

$$F_{Ls} = \frac{\rho\overline{U}^2 B}{2}\frac{dC_L}{d\alpha}\int_0^S \left(\Phi_L(s_\tau)\alpha'(\tau) + \Psi_L(s-\tau)\frac{z''(\tau)}{B}\right) d\tau \tag{440}$$

where $s = \overline{U}t/B$ and the $(\ldots)'$ means $\frac{d}{ds}$. Furthermore $\Phi_L(s)$ is the indicial response function due to a step change in angle of attack, and

$\Psi_L(s)$ is the indicial response function associated with a step change in vertical velocity. $\Phi_L(s)$ and $\Psi_L(s)$ could be in the form of

$$\Phi_L(s) = c_0 + c_1 e^{c_2 s} + c_3 e^{c_4 s} + \ldots \tag{441}$$

where the coefficients c_i are extracted by nonlinear least-square fitting of the experimentally obtained aerodynamic derivatives. Even though this approach appears to be efficient, very few examples of its application can be found in the literature.

More recently, Lin and Li (1993) proposed another model for the self-excited forces on a bridge deck based on the idea of indicial functions. The self-excited loads are expressed in terms of convolution integrals of response functions due to a unit impulse displacement – vertical and angular. These impulse response functions are obtained from the inverse Fourier transform of the frequency response functions of the form for the moment due to the change of the angle of attack α, for example:

$$\widetilde{F}_{Ms}(\omega) = \rho B^2 \overline{U}^2 \left[c_1 + i c_2 \frac{\omega B}{\overline{U}} + \sum_j \sum_k \frac{i c_j \frac{\omega B}{\overline{U}}}{c_k + i \frac{\omega B}{\overline{U}}} \right] \tag{442}$$

$$F_{Ms}(\omega) = \rho B^2 \overline{U}^4 \omega^2 \left[A_3^*(K) + i A_2^*(K) \right] \tag{443}$$

where the coefficients c_j are obtained from nonlinear fitting of the experimentally determined aerodynamic derivatives $A_j^*(K)$ and $H_j^*(K)$. Li and He (1995) have presented a method to experimentally determine the coefficients c_j and consequently the impulse response functions. A complete example of the use of this formulation is given by Xiang et al. (1995) for the motion-induced forces. Comparisons are made between time-domain predictions and wind-tunnel test results for the Shanton Bay Bridge. The agreement is remarkable, even though here also the buffeting forces are calculated using quasi-steady aerodynamics.

Fujino et al. (1995), in dealing with the problem of active control of flutter for bridges, have formulated a method to express the motion-induced forces in the time domain. This method is directly inspired by the work in the aerospace industry. Rational function approximations are used to express the equation of motion of the deck in a linear time invariant state–space form where the unsteady aerodynamic forces do not depend explicitly on reduced velocity. The aerodynamic derivatives obtained from experiments are stored in tabular form in the reduced frequency domain and are approximated in the Laplace domain by rational functions and a series of coefficients. The rational functions are then inverse Laplace transformed to the time domain to solve the state–space equation of motion, see Tiffany and Adams (1988). The level of accuracy is a function of the number of aerodynamic states used for the approximation, but once the numerical problems are solved this method appears to be effective and is currently used in aerospace applications, for example by Tewari and Brink-Spulink (1992).

Li and He (1995) have also presented a formulation of the unsteady forces that includes rational approximations for the motion-induced forces and the buffeting forces are expressed by convolution integral of impulse aerodynamic transfer functions determined experimentally from wind-tunnel tests. An example of the calculations is given but no comparisons with measured aeroelastic response of a physical model are made. This method appears to be the most complete of the above-described approach. Yagi (1997) analysis is along the same line although the work was apparently carried out independently of any of the above.

Finally, Diana et al. (1995) have initiated research on new methods to express the aerodynamic forces in the time domain including buffeting, motion-induced forces and vortex-shedding forces using numerical models comprised of the bridge deck and an equivalent oscillator. The new methods include 'black-box model', a neural network model and sophisticated parameter identification algorithms using an extended Kalman filter. Preliminary evaluations of the methods have shown satisfactory results, especially for nonlinear phenomena such as vortex-shedding-induced oscillations.

12.3.5 Vortex-Induced Oscillation

12.3.5.1 General Characteristics

Vortices of air flow created by the interaction of wind and the structure induce vibrations in the structure. When an aerodynamically bluff body is exposed to wind, a trail of alternating vortices, the Kármán vortices, is often found in its wake, formed by the flow separated from the body. There is also a fluctuating lift force acting on the body corresponding to the formation of vortices. As a result, when the frequency of vortex formation is close to the structure's eigenfrequency, there will be a resonant vibration. This is the most fundamental concept of vortex excitation. However, once the vibration starts, the body motion itself will influence the flow behavior, which results in the more complicated interaction of flow and structure.

Unlike the case of buffeting, the vortex excitation is usually observed in one or more limited wind-speed ranges and its amplitude is limited without divergence. The vibration is usually characterized by a narrow-band frequency spectrum and somewhat regular amplitude. The bridge deck usually vibrates in vertical bending but also sometimes in torsion. The phenomenon is known to be sensitive to the magnitude of structural damping and also to the existence of wind turbulence. Truss-stiffened bridges are, unless they are equipped with high-solidity railings or parapets, expected to be free of any serious vortex excitation.

Vortex-induced oscillation of bridge decks has been observed so frequently that it is usually of primary concern when the wind stability of bridges is considered. It is often regarded as a relatively easy matter that can be determined by carrying out simple section model wind-tunnel tests. It is mainly because the design criteria applied to this phenomenon often simply ask if a section is "good" or "bad", meaning whether or not any appreciable vortex-induced response would be expected in a possible wind-speed range for a given structural damping. However, for more exact assessment, the prediction has to indicate the level of actual response quantitatively in relation to damping and wind characteristics.

Obviously the most reliable method of prediction at this point is to use an aeroelastic model of the whole bridge with properly simulated natural wind conditions. A possible problem for this case is the wind speed. Often the available wind-speed range of the wind tunnel is too high for proper measurement. On the other hand, if the section model tests are projected, a question still remains of whether a reliable response prediction is obtainable from the test results.

12.3.5.2 Analytical Prediction

Forced Vibration Model

The most classic treatment of vortex shedding excitation is to consider it as a resonance with the aerodynamic forces caused by the periodic shedding of vortices alternately from the upper and lower surfaces of the deck. From this understanding, the following two most fundamental dimensionless parameters are derived:

$$St = \frac{fd}{\overline{U}} \qquad \text{Strouhal number} \tag{444}$$

$$Sc = \frac{m\zeta}{\rho d^2} \qquad \text{Scruton number} \tag{445}$$

where f is the frequency of vortex formation, which coincides with the natural frequency, d is a representative linear dimension of the deck cross-section, usually the depth, $\overline{U}$ is the mean wind speed, m is the deck mass per unit deck length, ζ is the critical damping ratio of the structure, and ρ is the air density. Considering ordinary plate-girder or box-girder bridge decks, the Strouhal number is typically in the range 0.07–0.14, see Figures 12.25 and 12.26. The simplest analysis is to assume a simple harmonic fluctuating force

$$\widehat{L}(y,t) = \frac{\rho\overline{U}^2}{2} d \cdot \widehat{C}_L \sin(2\pi f_S t) \tag{446}$$

which works on a 2D section of the bridge all through the span. Also, the lift forces are assumed to be fully correlated along the bridge span. The peak response amplitude $\hat{z}$, if it is the case, is simply given by

$$\frac{\hat{z}}{d} = \left(\frac{U_R}{4\pi}\right)^2 \frac{\widehat{C}_L}{Sc} \qquad \text{where} \qquad U_R = \frac{1}{St} = \frac{\overline{U}}{fd} \tag{447}$$

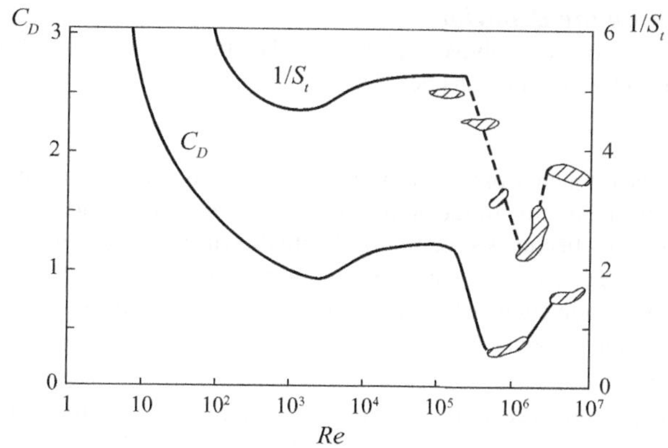

Figure 12.25 Strouhal number versus Reynolds number of a circular cylinder (after Miyata 1997)

$\widehat{C}_L$ is a function of the cross-sectional shape, the Reynolds number, the scale of turbulence and the structural aspect ratio, but is typically in the range 0.2–0.4 and is independent of the response amplitude as a first approximation. Note that the deck mass may not be constant along the span in reality. Generally speaking, there is also the effect of the vibration mode, $\psi_i(y)$. Considering these factors, m in Equation (445) should be replaced by an effective mass defined as follows:

$$m_e = \frac{\int_L m(y)\psi_i^2(y)\,dy}{\int_L \psi_i^2(y)\,dy} \tag{448}$$

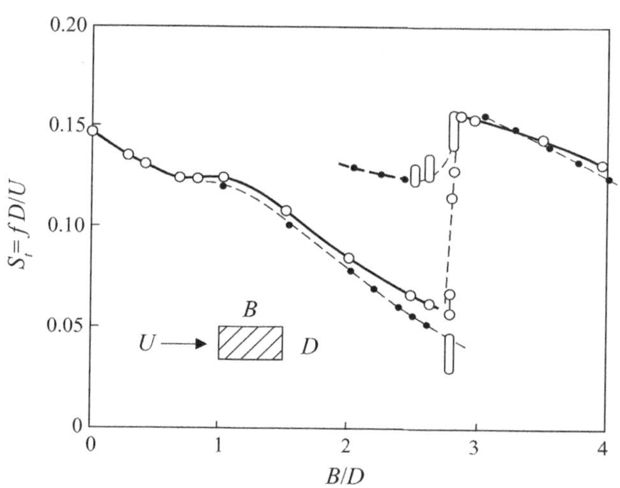

Figure 12.26 Strouhal number versus Reynolds number of rectangular prisms (after Miyata 1997)

Factors to Influence on the Behavior

There are some deviations from the above explanation when the vortex excitation is observed in reality. The three most significant issues are considered below.

Frequency Locking

From Equation (444), the frequency of vortex formation is proportional to the mean wind speed. However, once the vibration starts, the motion of the body tends to control the flow pattern around the body and, as a result, the vortex shedding becomes synchronized with the vibration frequency, which does not follow the Strouhal law. The vortex frequency stays constant over a specific range of wind speed. This situation is called the locked-in phenomenon, which indicates the aspect of self-excited vibration rather than a simple forced vibration and resonance.

Correlation Length

The vortex excitation mechanism is not really uniformly distributed along the whole bridge span; i.e., the cross-correlation of the exciting forces decreases along the bridge axis. Because of the lock-in effects, the excitation force has its maximum at the point of the antinode of the mode shape. The general tendency is that the correlation increases with the increasing vibration amplitude and decreases with the flow turbulence.

Flow Turbulence

The flow turbulence existing in the approaching wind tends to disorganize the regular pattern of excitation forces and thus decreases the response level. This is generally observed in all wind-tunnel tests for vortex excitation and, as a result, there is almost always a high expectation of engineers that the vortex shedding excitation observed in wind-tunnel tests could be averted, or at least reduced in magnitude, in reality because of the wind turbulence that inherently exists in natural wind. Past experience is that the flow turbulence at the bridge site is not necessarily always so much as it is estimated for buffeting prediction, and the magnitude of vortex excitation can be just as high as it is predicted by wind-tunnel tests with smooth flow.

More Refined Method

Experience has shown that even when the flow is smooth and d is uniform, the fluctuating lift force is not strictly periodical but has a narrow-band spectrum over frequencies adjacent to the Strouhal frequency, f_S, where f_S is again defined by $St \cdot \overline{U}/d$. Large-scale turbulence in approaching flow can be considered as a slowly varying mean speed, $\overline{U}$, which affects the central frequency f_S. Thus, if the force fluctuation in smooth flow is sinusoidal, Gaussian turbulence would cause the Gaussian form spectrum of the lift force. Hence,

$$\frac{f \cdot G_L(f)}{\sigma_L^2} = \frac{f/f_S}{b\sqrt{\pi}} \exp\left(-\left(\frac{1 - f/f_S}{b}\right)^2\right) \tag{449}$$

where σ_L is the RMS lift force and b is a bandwidth parameter.

The bandwidth parameter b also depends primarily on the large-scale turbulence. The full-scale experience indicates the approximate expression of $b = 0.10 + 2\sigma_u/\overline{U}$. It also suggests a fairly narrow band, typically 80% of the variance lying within a frequency range of $\pm 20\%$ of f_S. The above spectrum is expected to fit over the frequency range of $f_S(1 \pm b)$ (Vickery 1994).

The spectrum of $Q_i(t)$ can be given in terms of the spectrum of $L(y, t)$ and its normalized co-spectrum as follows:

$$G_{Qi}(f) = \int_L \int_L \sqrt{G_L(y_1, f) G_L(y_2, f)} \widetilde{R}_L(y_1, y_2) \psi_i(y_1) \psi_i(y_2)\, dy_1\, dy_2 \tag{450}$$

in which, $G_L(y, f)$ is given by Equation (449). $\widetilde{R}_L(y_1, y_2)$ is the spanwise correlation of the fluctuating lift forces acting on a stationary structure. The peak amplitude is typically given by two times RMS for a well-developed vortex shedding excitation.

Spanwise Correlation

The spanwise correlation decays with turbulence and improves significantly with the increase of vibration amplitudes. However, where the amplitudes of structural motion are large enough to have significant effects of them, the effect of motion-dependent aerodynamic damping becomes even more predominant in determining the response level. The measured correlation for the case of vortex excitation of bridge decks is much needed but not yet available, and if any new experimental results become available, reference to them is essential. With the absence of any reliable information, it can be taken as unity for the given span length, as a conservative assumption.

Aspect Ratio, Turbulence, Etc.

The Strouhal number is dependent upon the surface roughness, Reynolds number, flow turbulence and the aspect ratio $\Lambda = l/d$ of the structure. A wishful idea here is that very small effects are expected of the Reynolds number at its practical range. The effect of aspect ratio is indicated in some references. The Strouhal number for typical highway bridge sections is usually in the range of 1/6.5 to 1/13 and tends to decrease slightly with roughness and increase with turbulence intensity.

The lift force coefficient is considered to be strongly influenced by the intensity of turbulence and also by the scales of turbulence when they are of the same order or less than the width of the structure. Not much is known on these points for typical bridge deck sections. Equivalent knowledge for chimneys is better available. The proposed British code suggests formulating a characteristic excitation parameter, $C_L/(St)^2$. The order of magnitude of the RMS lift coefficient, σ_{CL}, is known to be 0.1 to 0.2.

Motion-Induced Forces

The motion-dependent forces can be represented by

$$L_Z(y, t) = \frac{\rho \overline{U}^2}{2}\left(\frac{\omega d}{\overline{U}}\right)^2 \left[H_4^* z + H_1^* \frac{\dot{z}}{\omega}\right] \tag{451}$$

where, generally speaking, the aerodynamic coefficients or derivatives H_1^*, H_4^* are amplitude dependent. They can be decided only by experimental means. H_1^* would decide the magnitude of aerodynamic damping. The application of this idea particularly for cylindrical chimney stacks is more elaborated, for example by Vickery (1994) and Ruscheweyh (1994).

12.3.5.3 Typical Box Sections Applicable to Bridges

The critical reduced speed to give the maximum response for typical box sections has been suggested as follows (Wyatt and Scruton (1981)):

B/D	$(U/fD)_{cr}$
< 1.25	6.5
$1.25 - 10$	$5.5 + 0.8 \cdot B/D$
> 10	13.5

where B and D indicate the width and depth of the box girders, and $\overline{U}$ is the mean wind speed. Exact prediction of dynamic response amplitude is difficult unless the aerodynamic lift force is known exactly. However, based on past experience, a simple analysis has been proposed by the Hanshin Expressway Authority, Japan, as follows (Miyata 1997):

Consider a typical box-girder highway bridge. Applying 2D analysis,

$$m\ddot{y} + c\dot{y} + ky = \frac{\rho\overline{U}^2}{2} BC_L \frac{\dot{y}}{\overline{U}} \tag{452}$$

where

$$C_L \approx 0.625\left(\frac{D}{B}\right)^2 \frac{B}{y} \tag{453}$$

This leads to the peak response amplitude of

$$\hat{y} = \frac{\rho\overline{U}B^2C_L}{8\pi mf\zeta} \tag{454}$$

Since $(\overline{U}/fD)_{cr} \approx 1.67\alpha$, Equation (454) can be written as

$$\hat{y} = K_R \frac{\alpha\rho D^2 B}{8\pi m\zeta} \tag{455}$$

where K_R is a factor to consider the effect of flow turbulence (≤ 1) and

$$\alpha = \begin{cases} 1.8 - 0.125Sc & (Sc < 4.8) \\ 1.2 & (Sc \geq 4.8) \end{cases} \tag{456}$$

Further to this analysis, wind-tunnel experience has indicated the following range of maximum response amplitude:

$$\frac{y}{B}\frac{m\zeta}{\rho BD} \leq \begin{cases} 0.024 & (B/D \leq 5) \\ \frac{0.12}{B/D} & (5 < B/D < 13) \end{cases} \tag{457}$$

12.3.5.4 Control and Suppression

Structural vibration could be reduced by installation of vibration dampers, for example. However, in the evolution of the design of modern large bridges, there has been considerable development of aerodynamic performance of road decks, such as the use of shallower and/or closed sections and various edge treatments, to reduce, if not completely suppress, the dynamic bridge response. It is difficult to come up with a comprehensive theory regarding what would be the aerodynamically better cross-sections, but a practically important technique is to learn through past experiences. Wardlaw (1992) has given a good review of successful experience, which is really a treasure box in this respect.

Generally speaking, the following statements can be made.

1. Instability of the bridge deck is often induced corresponding to the vortex formation at the sharp corners of the deck cross section. If the deck has an open cross section, therefore, it is usually a good idea to at least partially close it, or extend the sleeves or fairings from the sharp edges.

2. Another possibility is to install spanwise vertical baffles longitudinally inside the cavity, such as the space between the edge plates or boxes. They help in reducing the formation of vortices.
3. Edge fairings are often found to be effective in reducing vortex excitation. They can be of a triangle shape pointing outside, or more streamlined.
4. The tower legs often have rectangular cross sections, which tend to be subjected to vortex excitation. Installation of small vanes around the corner or removal of a sharp corner edge by including a small cut-out is known to be effective in improving the aerodynamic performance.
5. In any of these provisions, the wind tunnel is an indispensable tool for determining which method is effective and what their size should be.

12.3.6 Cable Aerodynamics

12.3.6.1 Classification of Cable Vibrations due to Wind

The development of cable-stayed bridges as a structural choice for medium to long span bridges was remarkable through the closing decades of the last century. As an essential component of the bridge superstructure, stay-cables play an important role in the dynamic behavior of cable-stayed bridges. Cables are extremely vulnerable to wind excitation mainly due to its low mechanical damping. Many efforts have been made during the past years to clarify the mechanisms of, and find solutions to, various types of wind-induced cable vibrations to alleviate engineering problems. Furthermore, with a rapid development of span length in cable-stayed bridges, even new types of instability, particularly of the inclined cables, have been identified, such as the rain–wind vibration, high-speed vortex excitation, and the dry inclined cable galloping, which have all been new challenges to bridge engineers. The purpose of the present chapter is to attempt a comprehensive state-of-the-art review of various types of wind-induced cable vibrations.

The wind-induced cable vibrations can be categorized into several groups as follows, depending mainly upon their excitation mechanisms but also on their historical context.

1. Vortex-induced vibration, or aeolian oscillation.
2. Buffeting due to wind gust.
3. Classic galloping typically observed in cables with ice accretion.
4. Wake interference, or wake galloping and resonant buffeting.
5. Parametric excitation due to support motion.
6. Reynolds number related drag instability.
7. Rain–wind induced vibration.
8. High-speed vortex excitation.
9. Dry inclined cable galloping.

The first two types of vibration are generally small in amplitude and possible fatigue failure would be the only engineering concern associated with them. The last three types are mainly related to inclined cables, such as bridge stay-cables, and the rain–wind induced vibration is the one most frequently observed in reality. The excitation mechanism of rain–wind vibration has become clearer in recent years, and some effective control methods have been successfully applied in practice. However, the last two types of motion are still much less understood and require more intensive research attention. The dry inclined cable galloping is of particular concern since it results in undesirably large-amplitude motion and yet it is not fully understood precisely under what conditions this would occur and why.

12.3.6.2 Vortex-Induced Vibration, or Aeolian Oscillation

This is a small-amplitude vibration caused by vortex shedding and takes its name from the aeolian harp, an ancient Greek instrument functioned on the same mechanical principle. The basic mechanism of excitation is a resonance to the frequency of vortex shedding f_V, which is given by

$$f_V = St\frac{\overline{U}}{D} \tag{458}$$

where $\overline{U}$ is the mean wind speed, D is the outer diameter of the cable. St is the Strouhal number, which is approximately $0.19 - 0.20$ for a circular cross section and the Reynolds number of $\sim 10^5$ or less. As a typical example, if $D = 0.16\,m$ and $St = 0.19$ are assumed, for the mean wind speed of $\overline{U} = 5 - 25\,\text{m/s}$, corresponding vortex shedding frequency is in the range of $f_V = 10 - 50\,\text{Hz}$. Since the cable's fundamental frequency is often in the range of $0.2 - 2\,\text{Hz}$, the resonance occurs only with much higher harmonic modes with vibration frequencies of $10 - 50\,\text{Hz}$, where the mechanical self-damping is likely to be quite high. The maximum vibration amplitude usually does not exceed one cable diameter, peak–peak. Wind turbulence generally tends to reduce the response amplitude, even down to a half compared to the exposure to smooth air flow (Ehsan et al. 1990).

The same mechanism of wind excitation can cause much more serious engineering problems for towers, chimneys and bridge road decks. Both experimental and analytical investigation have been devoted to this topic and their outcomes have been reflected in the design codes in practice. However, vortex shedding excitation of cables is not a major concern for engineers except that it may cause fatigue damage near the cable clamps. For the case of power transmission lines, it is a common practice to install the Stockbridge-type dampers (Leblond and Hardy 1999) or helical surface rods around the conductor. It is noteworthy that the Stockbridge dampers also contribute in reducing the twist of cables but they also have serious problems of failure due to fatigue. For bridge cables, the viscous dampers are more commonly used. A novel device, the passive damper cable, has been proposed recently (Sauter et al. 2001). The slack nature of the damper cable exhibits large static hysteresis caused by interstrand friction of the cable during bending motion, which is said to dissipate energy effectively.

12.3.6.3 Buffeting due to Wind Gust

Gust-induced random vibration is generally not a very serious concern for structural cables except for the power transmission lines. There are a few fundamental references such as the one by the ASCE Committee in this respect (ASCE 1984).The basic idea of its recommendation is to come up with a gust response factor for the prediction of dynamic response by a conventional buffeting analysis in the frequency domain. The procedure is based upon an earlier publication by Davenport (1979). Further improvement of its content should include particularly the spanwise force correlation, wind yaw angle effects, and the longitudinal loads along the cable span. For bridge cables, buffeting is generally a less serious problem. High tension of bridge-stay cables generally helps to limit the amplitude of buffeting. Sometimes buffeting is observed as a result of wake interference, where the existence of upstream objects is the cause of disturbance for the downstream cables. Electric wires along the main cable of the Golden Gate Bridge exhibited vibration a couple of decades ago, which was probably buffeting of this kind. It has been pointed out (Virlogeux 1998) that wake buffeting could produce aerodynamic instability in bridges when there are two parallel cable planes. If the time required for the wind flow to travel from the upstream plane to the downstream plane of cables is equal to a half of a cycle of the torsional vibration of the deck, it is conceivable that cable buffeting can enhance the deck vibration and cause the bridge instability, although, to the author's knowledge, this has never been an issue in reality. This mechanism is touched upon again in Section 12.3.6.5.

12.3.6.4 Galloping of Iced Cables

The accretion of ice on a conductor modifies its cross-sectional geometry and hence its aerodynamic characteristics. This may result in an aerodynamically induced instability called galloping. The motion is principally in the vertical direction with a low frequency, typically less than 1 Hz, and a large amplitude such as 10 – 20 m, large enough to cause serious design and operational problems. The primary reason of this mechanism is a significant negative slope in the lift force curve against the angle of attack, α, which gives the exciting lift force in the same direction as the cable motion. Thus, energy from the surrounding airflow is fed continuously to the system, leading to an unstable motion similar to flutter. The instability criterion given by Den Hartog (1932) is, as discussed in Section 12.3.3.4,

$$\frac{\partial C_L}{\partial \alpha} + C_D < 0 \tag{459}$$

where C_L and C_D are the lift and drag force coefficients, respectively.

However, strictly speaking, since the drag force also slightly depends on the angle of attack, the path of the cable motion tends to follow an elliptical trace. Furthermore, the instability is known to be triggered by aerodynamic coupling with torsion (Lilien 1997). On the other hand, the static deflection due to drag force and pitching moment under high wind speed can effectively change the aerodynamic characteristics of the cross section, which could even stabilize the cable (Novak and Tanaka 1974). A similar behavior for the case of noncircular cable sections has been also discussed by Fujino et al. (1997). Since the dynamic amplitude of galloping tends to be so large, the cable behavior becomes strongly nonlinear. Unlike torsional flutter, wind turbulence does not necessarily work as a stabilizing factor for galloping (Novak and Tanaka 1974). Engineering problems are not restricted to the power conductors. Some studies (Novak et al. 1978) have shown that the same mechanism of excitation may have resulted in galloping of heavy guy cables, leading to the collapse of a guyed tower. It has been reported that the violent vibration of bridge stay-cables observed at the Øresund link between Denmark and Sweden in February 2004 was likely to be galloping caused by snow and sleet accumulation on cable surface (Larsen and Lafrenière 2005). A comprehensive list of the references related to the research on galloping has been presented by Lilien (1997). The analysis of galloping, as described in Section 12.3.3.4, does not involve too much theoretical difficulty in principle. However, it needs to be supported by reliable aerodynamic data for a range of realistic natural ice shapes formed under various weather conditions of interest. Collection of these data will greatly assist the theoretical study of galloping.

12.3.6.5 Vibrations Caused by Wake Interference

The dynamic behavior of structures excited by wind can be drastically altered by their proximity to neighboring structures. These mechanisms that lead to aerodynamic excitation do not of course exist for a single isolated structure. Much research on this category of fluid–structure interaction has been motivated by problems encountered with the closely spaced tubes exposed to internal flow of heat exchangers and also with bundled conductors used in high voltage electric power transmission lines. There are, however, some other examples of proximity effects also related to spaced cables, slender towers and chimneys. Due to the complexity of flow patterns around closely spaced structures, several different driving mechanisms can arise. Zdravkovich (1997), for example, has studied various cases of aerodynamic interaction between two circular cylinders (Figure 12.27) located close to each other quite extensively. Of particular importance, in the present context, is the case when two circular cylinders are placed in staggered arrangements. The predominant response is (a) resonant buffeting, where the vortex wake of an upstream body is resonant with a natural frequency of a structure submerged in the wake; and (b) wake galloping, where lift and drag forces in the wake shear flow lead to a coupled 2DOF instability of a submerged cylinder.

Wake-interference galloping occurs when two cylinders are either closely or widely spaced. For close spacings, the flow around two cylinders is significantly altered by aerodynamic interference between two cylinders. Stay-cable vibrations of the Yobuko Bridge (Yoshimura et al. 1995) were of this kind. Extensive studies of instability of closely spaced cylinders have been conducted, particularly for heat exchanger

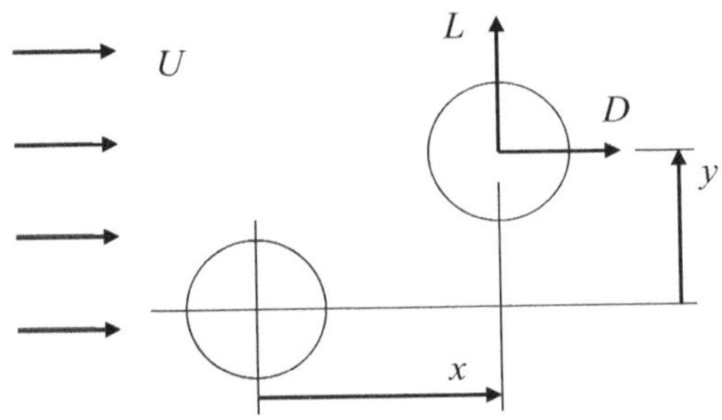

Figure 12.27 Two cylinders in tandem configuration

bundles (Chen 1987) and bridge stay-cables (e.g. Yoshimura et al. 1995). Instability is observed in the range of $-2 < y/D < 2$ and $1 < x/D < 4$. It starts at the critical reduced velocity of $(\overline{U}/fD)_{cr} \approx 40$, which typically corresponds to the flow speed of $5 - 20\,\mathrm{m/s}$, and follows generally an elliptical trajectory with the maximum amplitude of less than $3D$. The motion is known to be sensitive to the Scruton number, $Sc = m\zeta/(\rho D^2)$ and becomes hardly recognizable at $Sc > 50$. To reduce the vibration amplitude, the installation of viscous dampers at cable ends or connecting the adjacent cables by cross-ties have been practiced. However, these methods are not effective for longer cables. As spacing increases, the interference effects diminish until the next 'large spacing' instability range, $8 < y/D < 20$, is reached, where the interference effects are only on the downstream cylinder and the flow around the upstream structure is no longer affected by the second cable. The interference effects of widely spaced structures have attracted less research attention except for the wake galloping of bundled conductors and a comprehensive account both analyses and test results is presented by Wardlaw (1994). The transmission of electric power in very high voltage requires suspension of conductors in bundles to avoid a corona discharge to ground. The separation of parallel conductors, typically in the range of $10 - 20$ conductor diameters, is maintained by the use of spacers that usually divide the span into $50 - 60\,\mathrm{m}$ subspans. This is the reason for having a distinctively different type of cable vibration called subspan oscillation caused by wake interference.

If the upstream cable is fixed, the downstream conductor would move in an elliptic path with the long axis nearly horizontal. Amplitudes become large enough that conductors could clash. Although it is the downstream conductor that becomes aerodynamically unstable, the field observations report the existence of antiphase motion of a conductor pair with both cables having similar amplitudes and with the frequency in the range of $1 - 4\,\mathrm{Hz}$. This is in contrast to the vortex shedding excitation which takes place at higher frequencies of $10 - 50\,\mathrm{Hz}$, and the large-amplitude galloping of iced conductors at the frequency less than 1 Hz. The aerodynamic mechanism of this vibration was extensively investigated in the 1970s and is now fairly well understood. Research activities included aerodynamic force measurements (Wardlaw and Cooper 1974) and mathematical analyses (e.g. Simpson and Flowers 1977). It has been found that the presence of wind velocity gradient across the wake of the windward object would induce position-dependent lift and drag forces on the downstream cable. The most sensitive region for this excitation is approximately at a quarter of the width of the wake from its outer edge shear layer, where the lift force reaches its maximum.

A conventional practice of suppressing this type of motion is with spacer-dampers that are flexible and often installed at unequal intervals. However, the impact of the damper provided by the spacers on the instability is still not clear. Opposite conclusions were drawn from different studies. The use of low-level damping has been recommended by a Hydro-Quebec study (Hardy and Bourdon 1979), whereas Price and Paidoussis (1984) has advised the use of very high level of damping.

The unexpected wake-induced flutter was observed recently (Toriumi et al. 1999) on the hanger ropes of the Akashi Kaikyo Bridge in Japan. Cable distance was $9D$ for this case, normal to the bridge axis. Spiral ropes of 10 mm in diameter were used to wind up the hanger ropes to suppress both the wake-induced flutter and vortex shedding excitation effectively.

12.3.6.6 Parametric Excitation

Vibrations of stay cables can be excited by motion of the anchorage points, as first pointed out by Kovács and Leonhardt (1982). This is sometimes observed in cable-supported structures such as cable-stayed bridges and guyed towers. One of the important points in particular is the fact that a cable can be excited not only with its natural frequency but also with the excitation frequency which is two times its own natural frequency.

The most likely cases are when the excitation frequency is approximately twice or equal to the first natural frequency of the cable. This is only an indirect aerodynamic excitation of cables. Nevertheless, it can be a serious issue and should be briefly mentioned here. Because of its low structural damping, the cable motion can develop to large amplitude even if the end excitation is small. Also, since the bridge stay-cables present a wide range of natural frequencies, there is a good possibility of some of the cables resonating with the movements of the deck or pylons due to wind or traffic loads. Parametric excitation of bridge stay-cables including nonlinear characteristics and varieties of parameter changes is an interesting research topic but is not fully investigated yet (Gani and Tanaka 2005). However, it is also said that the stay-cable vibration is likely to stay linear because of very small sag and small end motions (Liu et al. 2005). It has been reported that the stay cables of the second Severn Crossing experienced severe vibration apparently for this reason. When the cross-ties were installed to alleviate the cable vibration, the bridge deck started vibrating, since now it has lost an effective TMD (Stubler et al. 1999). For the case of the Normandy Bridge, the problem was removed by adding cross-ties between stay-cables to avoid resonant vibrations (Virlogeux 1995).

12.3.6.7 Reynolds Number Related Drag Instability

This phenomenon was observed about four decades ago on a stranded wire conductor crossing the Severn River in England. The vibration was severe enough to cause numerous electrical faults due to the clashing between conductors. There was apparently no vertical conductor motion associated with it. A very interesting fact was that the wind speed and direction at which the instability occurred was confined to rather narrow bands. The physical dimensions involved in the case are as follows: main span length $= 1619$ m, the sag $= 80.5$ m, cable diameter $= 43$ mm and cable mass $= 6.40$ kg/m. Vibration frequency was found in the range of $0.128 - 0.130$, which approximately corresponds to the first asymmetric mode in lateral sway. The vibration was observed at the mean wind speed $13 - 15$ m/s and most of the time the mean wind direction was 10° to 25° deviation from normal to the span. Results from extensive studies (Davis et al. 1963; Richards 1965) pointed out that this unusual instability was due to the sensitivity of the drag force to the Reynolds number.

As it has been well-recognized, the aerodynamics of cylindrical bodies is highly sensitive to the change of the flow speed near the critical Reynolds number. Although the critical Reynolds number is influenced by the roughness and texture of the body surface as well as the flow turbulence, it is characterized by a sudden drop of the drag force with the increase in wind speed. If the cable is swinging back and forth parallel to the wind direction, and if the change in the relative wind speed takes place at this sensitive speed range, it is possible to have the induced aerodynamic force acting in the same direction as the body motion, and thus generate the negative aerodynamic damping. In the case of a smooth circular section, this 'drag crisis' occurs at the Reynolds number range of $2 - 5 \times 10^5$. Considering the cable diameter to be of the order of $50 - 150$ mm, the possibility of having this instability is at the wind speed of $20 - 60$ m/s, which is possible to take place but has never been reported as a problem in reality.

For the case of the Severn crossing, the helical stranding of conductors featured two influences on the aerodynamic characteristics. First, the critical Reynolds number for the conductor was about an order of magnitude lower than that of the smooth circular cylinder, down to $2.6 - 4.0 \times 10^4$ because of the characteristic roughness. Second, in oblique winds, the lay of the strands streamlined the flow on the surface in which the wind was more parallel to the stranding and roughened the opposite surface. This deflected the flow streamlines and generated a mechanism for lift. The effect was further amplified at the critical Reynolds number when the mechanism of drag damping was also active. This mechanism caused a large amplitude motion and clashing. As a countermeasure, the conductor was wrapped in PVC tape to eliminate the unfavorable impact of the stranded surface.

Considering the range of Reynolds number and possible cable dimensions, similar instability is conceivable for bridge stay-cables. However, so far no particular case has been attributed to this mechanism.

12.3.6.8 Rain–Wind-Induced Vibration

Rain–wind-induced cable vibration has been most frequently reported for bridge cable vibration. It was first addressed by Hikami (1986) when the stay-cables of the Meiko-Nishi Bridge in Japan experienced annoying vibrations. The most curious part was that it was observed under certain wind conditions, but only when it was raining. The observed vibration amplitude was up to approximately $2D$, D being the cable diameter, which was typically 14 cm, with a windspeed of $8 - 14$ m/s. A simple analysis led to the conclusion that the observed vibration was not any of the known types. The observed frequency was $1 - 3$ Hz, which was well below the Strouhal frequency for the vortex shedding excitation. The cables were too far apart to cause any aerodynamic interference. The observed cable vibrations were hence assumed to be a new type of instability caused by the combined action of rain and wind.

After this new type of excitation mechanism was reported it became clear that, in fact, there had been some other cable vibrations reported earlier for other bridges, which could have been classified into the same category. Some of the reported cases of this type of cable vibration are given in Table 12.8. Other

Table 12.8 Reported cases of rain–wind-induced cable vibration

Bridge name	Country	Observation year	Maximum double amplitude (m)	References
Köhlbrand	Germany	1974	~ 1	Ruscheweyh and Hirsch (1974)
Brottonne	France	1977	~ 0.6	Wianecki (1979), Hikami (1986)
Meiko-Nishi	Japan	1984	0.55	Hikami and Shiraishi (1988)
Farø	Denmark	1985	~ 2	Langsø and Larsen (1987), Yoshimura et al. (1989)
Aratsu	Japan	1988	~ 0.6	Yoshimura et al. (1995), Miyasaka et al. (1987)
Tenpozan	Japan	-	≈ 2	Ohshima and Nanjo (1987), Cremer et al. (1995)
Ben Ahin	Belgium	1988	~ 1	Lilien and Pinto da Costa (1994)
Burlington	Vermont	1990s	–	Virlogeux (1998)
Glebe Island	Australia	1990s	–	Virlogeux (1998)
Nampu	China	1992	–	Cheng (pers. comm.)
Yangpu	China	1995	–	Gu et al. (1998)
Erasmus	Holland	1996	~ 1.4	Geurts et al. (1998)
Øresund	Denmark-Sweden	2001	Large	Larsen and Lafrenière (2005)
Cochrane	Alabama	2002-04	~ 1.5	Irwin et al. (1999, 2005)

reported cases include oscillation of overhead conductors (Hardy and Bourdon 1979), inclined hangers of the Humber Suspension Bridge (Zasso et al. 1992) and even vertical hangers of two arch bridges (Ruscheweyh and Verwiebe 1995; Verwiebe 1998) all observed under rain and wind conditions. In the case of the arch bridges, the violent vibration of the vertical hangers caused fatigue damage at its welded connection to the gusset plate.

The characteristics and some specific conditions of this type of vibrations through field observations and wind-tunnel tests can be summarized as follows (Hikami and Shiraishi 1988; Yoshimura et al. 1989; Main et al. 2001; Matsumoto 1998; Stubler et al. 1999; Matsumoto et al. 2001).

- Moderate rain – neither light drizzle nor a downpour is conductive to such vibrations.
- Wind speed of 6 – 18 m/s, with the majority of the cases at 8 – 12 m/s.
- Cable frequency of 0.6 – 3.0 Hz.
- Cable diameters ranging from 140 mm to 225 mm.
- Reynolds number of 6×10^4 to 2×10^5, which is the transition range from subcritical to critical.
- Cables located on the leeward side of the pylons in most of the cases.
- Cable inclination of 20° – 45° from horizontal in many cases.
- Wind direction of 20° – 60° relative to the plane of the cable.

The cause of this vibration is considered to be in two steps. First, the formation of an upper water rivulet on the cable surface seems to be a key factor (Hikami 1986; Yamada et al. 1991). It is formed as a result of a sensitive equilibrium between gravity, capillary and aerodynamic forces. The water rivulet effectively alters the geometrical cross section of the cable and hence the aerodynamic forces on it. Depending on the location and size of the rivulet, it tends to give a negative slope of lift curve against the small change in the angle of attack and also significantly reduces the drag force, which results in the Den Hartog type galloping instability (Yamaguchi 1990). Also, once the cable is set to motion, the upper rivulet tends to oscillate along the cable surface in a circumferential direction and this motion can be aerodynamically coupled with the flexural oscillation of the cable, making the modal aerodynamic damping negative. Naturally, it is expected to intensify the vibration (Hikami 1986; Yamaguchi 1990; Ruscheweyh 1999). Contrary to this explanation, Bosdogianni and Olivari (1996) do not believe the motion of liquid rivulets has any influence on the instability. More research on the second triggering factor, in fact, identified three fundamentally different excitation mechanisms associated with alongwind, crosswind and mainly across-wind cable vibration under rain and wind conditions (Verwiebe 1998; Verwiebe and Ruscheweyh 1998). They essentially depend on the cable orientation and wind speed. An approximate method to estimate the amplitude of rain–wind-induced vibration is suggested. A recent finding by Flamand (2001) reveals the dependence of excitation on the thickness of a thin water film moving on the cable surface, and the link between the thickness and surface speed of the water.

There are many other studies on this topic. Yamaguchi (1990) work indicates that this is essentially a galloping instability. Geurts et al. (1998) and Geurts and Staalduinen (1999) developed a numerical model based on SDOF galloping theory to predict the rain–wind-induced cable response for the Erasmus Bridge. The analytical results of vibration amplitude thus obtained approximately agreed with the field data. In the model proposed by Xu and Wang (2001), the interaction between wind, cable and rivulet was considered. The predicted steady-state response showed that the main characteristics of an inclined cable with moving rivulet, such as velocity-restricted and amplitude-restricted, could be captured. The analytical study by Gu and Lu (2001) pointed out the importance of initial rivulet position in generating the cable instability. The 'unstable zone' of initial rivulet position and the 'dangerous zone' of instantaneous rivulet position were identified for cables of different natural frequencies. (Matsumoto et al. 1995a) on the other hand find that the air flow component along the cable is the essential cause of this vibration.

Various kinds of structural and aerodynamic means have been developed to suppress and prevent the vibration. Increase of the system damping by installation of oil dampers (Yoshimura et al. 1989), hydraulic dampers (Geurts et al. 1998; Geurts and Staalduinen 1999), viscous friction dampers

(Kovács et al. 1999), or connecting some of the longest cables by using cross-ties (Langsø and Larsen 1987; Hikami and Shiraishi 1988; Kusakabe et al. 1995; Virlogeux 1998) are found to be effective structural means. The installation of TMD on the vertical hanger of an arch bridge (Ruscheweyh and Verwiebe 1995) was also proven to be effective. As for developing the aerodynamic measures, the main idea is to break up the formation of the upper water rivulet. Thus, various methods have been proposed and applied to the bridge cables, such as the use of helical wire whirling on the cable surface (Flamand 1994; Bosdogianni and Olivari 1996), the adoption of a dimpled cable surface (Kobayashi et al. 1995; Virlogeux 1998), or the use of an axially protuberated surface (Saito et al. 1994). All these have proven to be effective and successful in the field to various extents. The effect of cable surface roughness also has been investigated by the wind-tunnel tests (Miyata et al. 1994). A more recent proposal (Verwiebe and Ruscheweyh 1998) of deflecting the water on the cable surface to control the motion needs further development.

Regarding the amount of damping required to control it, Irwin (1997) has suggested that the vibration can be reduced to a harmless level if the Scruton number is greater than 10, i.e.,

$$Sc = \frac{m\zeta}{\rho D^2} > 10 \tag{460}$$

where m is cable mass per unit length, ζ is the critical damping ratio, ρ is air density and D is cable diameter. This statement is also supported by experience in Japan (Yamada 1997). Considering $D = 15 - 20\,\text{cm}$ and $m = 100\,\text{kg/m}$ for a bridge stay-cable, for example, this is equivalent to the structural damping of approximately $\zeta = 0.5\%$ or more.

12.3.6.9 High-Speed Vortex Excitation of Dry, Inclined Cables

Although majority of the observed stay cable vibration belongs to the rain-wind induced type, it has been found both in field and wind tunnel tests that dry inclined cables can also undergo large-amplitude oscillation without precipitation. Matsumoto et al. (1989) reported the observations of cable vibrations without rain but with characteristics of rain–wind vibrations, up to the maximum amplitude of 23 cm at the wind speed 40 m/s during a typhoon. Matsumoto et al. (1995b) further explained that these vibrations occurred for the cables of the Higashi-Kobe Bridge, too. These phenomena were not properly explained by any known mechanisms and were attributed to the high-speed vortex shedding excitation because, compared to the normal Kármán vortex-induced vibration, the observed instability occurred at much higher reduced wind velocity ranges in multiples of 20, i.e., $U_R = 20, 40, 60, \ldots$.

Matsumoto (1998) tried to explain the mechanism of this vibration with the concept of 3D Kármán vortex interaction. When wind goes over an inclined cable, the vortices are generally shed in the wake. Besides, because wind is oblique to the cable, an axial airflow also exists, which may form and shed axial vortices along the cable. The axial flow, which interrupts the fluid interaction between the two separated shear layers in the wake, plays a role similar to a splitter plate, and behaves like an "air-curtain" or "base-bleed" (Matsumoto et al. 1990). They explained that it was the fluid interaction between the axial and the Kármán vortices, as well as the cable motion that caused this high-speed vortex excitation. In the experimental studies carried out by them, the frequency of the axial vortex shedding was found to be equal to 1/3 of the Kármán vortex shedding. Consequently, the Kármán vortex was amplified with the every third vortex, which correlates well with the observed fact that the instability occurred only at the discrete reduced wind velocity of 20, 40, 60, ... The intermittently amplified Kármán vortices also well explained the beating phenomenon observed in the tests (Matsumoto 1998; Cheng and Tanaka 2002) and on Meiko-Nishi Bridge (Matsumoto 1998).

As a means to suppress the vibration, Matsumoto has suggested the use of discrete elliptical plates attached on the cable surface (Matsumoto et al. 1995b). By doing so, not only the formation of the axial flow, but also the formation of an upper rivulet is prevented. Thus, this aerodynamic countermeasure

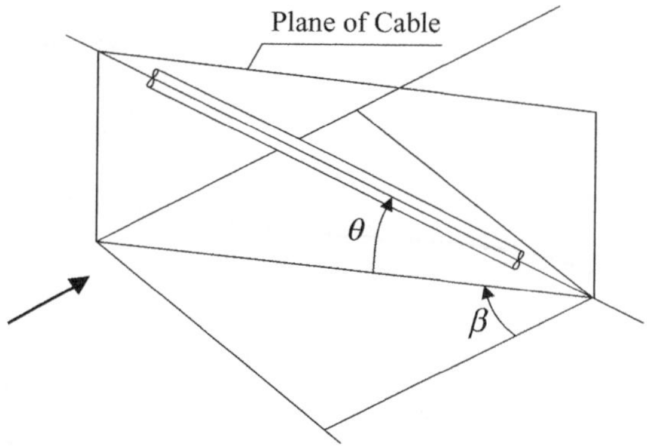

Figure 12.28 A cylinder configuration with angles against wind

helps to control both rain–wind-induced vibration and high-speed vortex shedding. However, any of the damping devices would also effectively work, similar to any other wind excitation of cables.

Although studies have been carried out to investigate this type of vibration for several years, and progress seems to be encouraging as the excitation mechanism gradually becomes clear, there are still lots of unknowns to satisfactorily explain the phenomenon and make proper predictions. The exact damping level required to suppress or eliminate the motion should be quantitatively predicted. Since the suggested 3D vortex shedding mechanism strongly depends on end conditions at the wind-tunnel tests and also the yaw angle of the cable against wind, it is important to compare the conditions between the cable model tests and the full-scale observation of real bridge cables.

12.3.6.10 Dry Inclined Cable Galloping

Beginning

Galloping of dry inclined cable is a relatively new term and its concept has become clearer only recently. First it appeared as a side product of the study on rain–wind vibration. There have been some experimental studies carried out (Saito et al. 1994; Miyata et al. 1994; Matsumoto et al. 1995a, 2005; Cheng and Tanaka 2002; Cheng et al. 2003a,b) to investigate this phenomenon and, so far, it has been observed only in wind-tunnel tests. Results obtained from these studies show that if the wind is oblique, the instability of the cables, which are inclined to wind (Figure 12.28), could have similar response characteristics as galloping. The results are found to be very sensitive to the model end conditions in the wind-tunnel tests and thus sometimes it is difficult even to reproduce it. However, if it takes place in reality as predicted, it would be a very serious engineering problem. Although this type of cable motion has never been clearly observed on real bridges, some field observations in fact are said to be better explained with galloping than calling them rain–wind-induced vibration (Virlogeux 1998; Irwin et al. 1999).

An instability criterion to indicate the critical wind speed originally suggested by Saito et al. (1994) for this instability is approximately given by (Figure 12.29)

$$(U_R)_{cr} \approx 35\sqrt{Sc} \tag{461}$$

where $U_R = \overline{U}/(fD)$ is the reduced wind speed, and $Sc = m\zeta/(\rho D^2)$ is the Scruton number. According to Saito, this criterion is applicable to the cases where the angle between cable axis and wind direction is 30° to 60°. It imposed a difficult design condition for bridge stay-cables with a typical diameter of 150

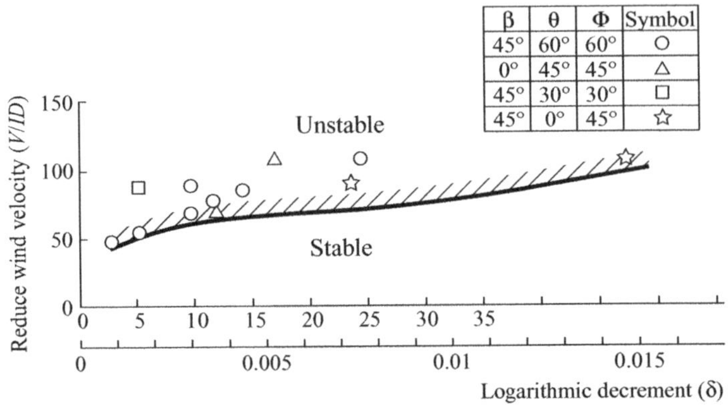

Figure 12.29 Stability criteria proposed by Saito et al. (1994)

to 200 mm, since it would place so many bridge stay-cables into a category of 'prone to galloping'. The reality, however, is that many existing stay-cables seem to be surviving without suffering this instability. Further investigation is urgently required under these circumstances.

It is now required that a more refined stability criterion for the dry inclined cable galloping, and the range of the physical parameters associated with the suggested criterion, should be established. In order to evaluate if the Den Hartog galloping criterion is applicable to explain the phenomenon, aerodynamic forces acting on the inclined cable need to be measured.

There are a few interesting issues raised in relation to the excitation mechanism of this instability. One of them is the role of the air flow along the cable axis and possible interaction of it with the Kármán vortices behind the cable. It has been suggested by Matsumoto et al. (1995a), who indicate that the axial flow in fact also has a significant role in rain–wind vibration. By introducing an artificial axial flow in the experiment, Matsumoto et al. (1995b) showed that the existence of axial flow could induce negative slope of the lift force, and galloping of dry inclined cable would occur when the velocity of the axial flow was 30% more than that of the approaching flow. A question then is whether or not this artificially imposed axial flow well represents the air flow situation behind an inclined cable in reality.

Another significant fact is, as pointed out by Larose and Zan (2001), that the instability is observed clearly in the critical Reynolds number range. It is certainly interesting to note that all of observed rain–wind vibration has occurred almost exactly at the critical Reynolds number. A question is then why and how it relates to the instability, including the dry cable galloping.

Two test series – Ottawa and Milano

Ottawa Project by NRCC/UO/RWDI

A series of wind-tunnel investigations on inclined cable vibrations took place in Ottawa recently (Cheng and Tanaka 2002; Cheng et al. 2003a,b) as a collaborative effort between the University of Ottawa, RWDI Inc. and the National Research Council Canada. The study consisted of two phases: the dynamic model test for response measurement and the static pressure model to investigate the aerodynamic forces on the cable. The dynamic model was a section of a full-size stay-cable with a total length of 6.7 m, a diameter of 0.16 m, and a linear mass of 60.8 kg/m. The model was inclined against wind with equivalent angles of $\phi = 35°$ to $60°$ and $\alpha = 0°$ to $60°$, where ϕ is the cable-wind angle, which is the angle of wind plane relative to the cylinder axis, and α is the angle between wind plane and the vibration plane. They successfully reproduced high-speed vortex excitation and one particular case ($\phi = 60°$ and $\alpha = 54.7°$) in which the dry inclined cable galloping appeared at the wind speed of 32 m/s. The geometrical set-up

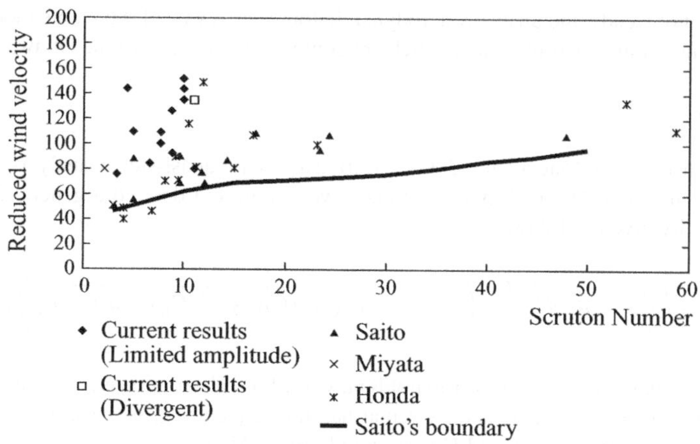

Figure 12.30 Various dynamic test results of inclined cables

of the model is equivalent to a real bridge cable inclined and yawed both 45° to the mean wind direction and the Scruton number was $m\delta/(\rho D^2) = 1.8$.

In Figure 12.30, their test results are compared with the stability criterion suggested earlier by Saito et al. (1994) and also some other experimental results obtained in Japan (Miyata et al. 1994; Honda et al. 1995). It is evident that Saito's criterion for the onset of instability is much more conservative than other experimental results.

The force measurement results clearly indicate that the tendency toward instability, or galloping, takes place at particular combinations of vertical inclination and horizontal yaw angle, and also at the wind speed corresponding to the critical Reynolds number, where the Den Hartog instability criterion is satisfied as a result of a negative slope of the lift force with the angle of incidence and significant reduction of drag force (Cheng et al. 2003a). Further to this, an observed indication toward instability with and without the spanwise force correlation is discussed in conjunction with the appearance of amplitude-limited, high-speed vortex excitation phenomena (Cheng and Tanaka 2005). Further to these results, Macdonald (2005) introduced a more generalized 2D instability criterion, which was applied to this particular case and showed an excellent agreement with the test results. His theoretical criteria are given below in more detail.

Milano Campaign

In 2004, by using a very large wind tunnel at the Politecnico di Milano, a 6 m long, 316 mm diameter cylinder was rigidly supported horizontally with a variable horizontal yaw angle for the measurement of cross-sectional aerodynamic forces and their spanwise coherence. A particular emphasis was placed on finding the aerodynamic behavior in the critical Reynolds number range. There were a number of extremely interesting findings from this series of tests but some of the findings particularly related to inclined cable galloping are as follows (Larose et al. 2005).

1. The existence of the asymmetric, single bubble regime at $Re = 1.6 \times 10^5$ on an inclined cylinder, which was previously observed in the Ottawa study in smooth flow, was confirmed in 2.5% turbulence, which is a good representation of the situation of bridge cables in the field.
2. As was the case in the Ottawa study, the cross-sectional lift forces were found distinctly different at different locations along the cylinder, even if they were only one diameter apart.
3. In this Reynolds number range, a nonlinear change of lift force due to change in wind speed, which could explain partly a quasi-steady cause of instability.

4. A complete set of steady and unsteady aerodynamic forces on a yawed stay-cable mode in the critical Reynolds number range was obtained, which are complementary tests of the Ottawa study.

Critical Issues

Onset Criteria

As it was mentioned earlier, Macdonald (2005) has tried a general quasi-steady analysis of inclined cable galloping, including the effect of Reynolds number variation, and derived a general 2D criterion for galloping instability given as follows:

$$Z_s = \frac{\zeta_s m f_n}{\nu/\rho} > Real\left(\frac{Re}{16\pi}\left(-h(C_D) + \sqrt{g^2(C_D) + g^2(C_L) - h^2(C_L)}\right)\right), \tag{462}$$

where ρ is the air density, $Re = \frac{DU}{\nu}$ is the Reynolds number, D is cable diameter, C_D is the drag coefficient, C_L is the lift coefficient, $\omega_n = 2\pi f_n$ is the circular natural frequency, ζ_s is the structural damping ratio, m is the cable mass per unit length, and U is the wind speed. Also

$$g(C_F) = C_F\left(2\sin(\phi) - \frac{1}{\sin(\phi)}\right) + \frac{\partial C_F}{\partial Re} Re \sin(\phi) + \frac{\partial C_F}{\partial \phi}\cos(\phi) \tag{463}$$

and

$$h(C_F) = g(C_F) + \frac{2C_F}{\sin(\phi)} \tag{464}$$

in which C_F is either C_D or C_L and ϕ is the cable-wind angle, defined as the angle between a flow velocity relative to the cylinder axis. The definition of aerodynamic forces and their coefficients follow Equation (465). The angle relationship is explained in Figures 12.31 and 12.32

$$F_x = \frac{\rho U_R^2 D}{2}\left(C_D\cos(\alpha_R) - C_L\sin(\alpha_R)\right) \tag{465}$$

where U_R is the magnitude of the relative velocity, and α_R is the relative angle of attack.

It is important to note that the right-hand side of the criterion (462) is a function of Re and ϕ only and independent of the direction of cable motion. Note also that the expression Equation (462) is simplified for a circular cylinder. A more general expression and its derivation are given in Macdonald and Larose (2006). Detailed measurement of both lift and drag force components for an inclined cylinder in the

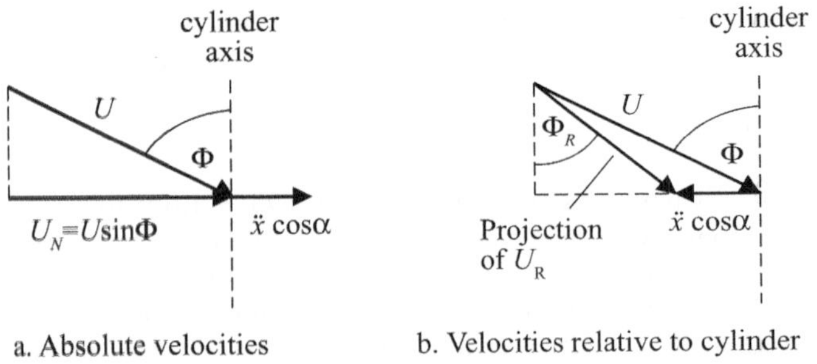

Figure 12.31 Cable–wind plane (after Macdonald 2005)

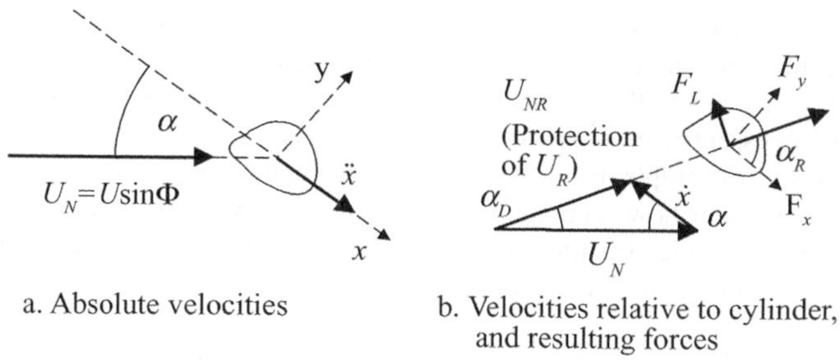

Figure 12.32 The plane normal to the cylinder (after Macdonald 2005)

critical Reynolds number range was carried out by Larose et al. (2003a). It is interesting to note that Macdonald has predicted, based on this analysis and available aerodynamic data, that another instability area, which exists when the cable–wind angle is between 75° and 90°. A practical importance of this analysis is in fact that the magnitude of aerodynamic negative damping actually can be calculated, and hence, the additional damping to the system to suppress any instability can be predicted.

Critical Reynolds Number

Larose and Zan (2001) particularly emphasized the sensitivity of cable vibrations to Reynolds number. It is certainly interesting to note that all of the observed rain–wind vibration has occurred almost exclusively in the critical Reynolds number range. The case of dry inclined cable galloping investigated by the Ottawa group was again at the critical Reynolds number. It is also extremely interesting to note that the conventional Den Hartog criterion would indicate instability corresponding to the drag crisis in this range, but by considering the large influence of Reynolds number on force coefficients, the governing reason of inclined cable galloping is actually found to be the difference in lift force due to *Re* – the $\partial C_L/\partial Re$ term (Larose and Macdonald).

The critical Reynolds number is sensitive to the presence of surface roughness, flow turbulence, motion of the cable and a flow angle not perpendicular to the cylinder axis. More recent study further indicates the fact that the effects of the Reynolds number on aerodynamic force distribution and resultant cable motion are highly dependent upon the orientation of the body to the mean flow direction ; (Larose et al. 2003a; Macdonald 2005).

The state-of-the-art regarding this particular instability, as of 2005, is as follows.

1. Galloping of dry inclined cable does exist as a possible instability.
2. It is a unique aerodynamic phenomenon for a cable that is inclined against wind.
3. Instability takes place in the critical Reynolds number range.
4. There are a few specific geometrical positions where the cable could become unstable.
5. Additional damping required to suppress the instability could be predicted by Equation (462).

12.3.7 Wind-Tunnel Tests

12.3.7.1 Similitude Requirements

General

When the likely effects of wind upon a bridge need to be examined, it is often a practice to undertake wind-tunnel tests. Modelling techniques employed in these testings may vary but the similitude requirements for testing are considered to be well established and have been practiced for quite some time. An alternative to this is to make a prediction of wind effects based on quasi-theoretical mathematical models which are often based on empirical knowledge. Depending on the problems, it is also possible to apply the techniques of computational fluid dynamics although the scope of its application is still somewhat limited. Whichever the employed modeling technique is, it is important to remember that an experimental or theoretical model can never perfectly satisfy the similitude requirements. It means that an appropriate interpretation of the results is always essential for the prediction of the results in reality.

Structural analyses, for example, using large capacity computers are generally regarded to be fairly accurate. However, no matter how accurate the calculation is, obviously the adequacy of the outcome largely depends on the way the structure and its boundary conditions are mathematically formulated to prepare inputs for the computers. Inclusion of aerodynamic factors in this process never makes the problem easier.

The experimental simulation is even more complicated. The general requirement for modelling a physical phenomenon is essentially the same in any wind engineering problem. The concern is typically the behavior of wind flow in a certain space and its interaction with the geometrical and/or mechanical characteristics of the boundaries of the field of concern. For correct modeling in these problems, based on Buckingham Π-theorem, it is required that a set of dimensionless parameters which consist of suitable combinations of the reference quantities are invariant in model and prototype and with them the governing equations are also rendered dimensionless. Various boundary conditions have to be also maintained in dimensionless form. There is no danger of overemphasizing the importance of this principle since most of the wind engineering problems cannot be solved with theoretical approaches alone.

For correct modelling, all of the dimensionless parameters in the prototype must be duplicated in the model. However, almost invariably, complete duplication of the parameters is impracticable. As a matter of fact, all the requirements can be satisfied only when model and prototype are identical. Hence, the decision must be made as to which parameters could be relaxed, or the extent to be distorted, for each testing based on the understanding of the phenomenon and the knowledge of dominant parameters. The weak laws should be relaxed. Perhaps only a part of the entire process can be simulated to clarify the unknown mechanism. Or, analytical means may be able to substitute the deficit in physical modeling.

Similitude Requirements

Geometrical Consistency

First of all it is required that the model and prototype are to be geometrically similar, which means that all linear dimensions must have the same scaling factor applied. The linear dimensions involved are as follows:

- linear dimensions of the structure to be modeled and other structures;
- linear dimensions of the surrounding topography;
- surface roughness of the structures involved;
- linear dimensions of wind flow, including the ground surface roughness, scales of turbulence, depth of the atmospheric boundary layer, etc.

Often in reality, it is difficult to reproduce the structural details in model scale, when the linear scaling is exaggerated and this is one outstanding cause of inaccuracy in physical model study.

Kinematic Consistency

The flow field is defined by six variables; three velocity components (u, v, w), air density ρ_a (kg/m^3), pressure p (N/m^2) and temperature T (°K), all given as functions of space and time variables (x, y, z, t). However, in case of bridge dynamics, ρ and T are generally deemed constants. Hence, the flow conditions are decided by four equations: three components of momentum conservation (Navier-Stokes equations 466) and the mass conservation (continuity 467) equation as follows:

$$\frac{D}{Dt}\mathbf{V} = -\frac{\Delta p}{\rho_a} + \nu\nabla^2\mathbf{V} \tag{466}$$

$$\nabla \mathbf{V} = 0 \tag{467}$$

in which $\mathbf{V} = (u, v, w)$ is a velocity vector (m/s) and ν is the kinematic viscosity (m^2/s).

The structural behavior is influenced by many variables but, ignoring the temperature effects, major parameters are as follows: L is the linear dimensions of the structure (m), ρ_m the material density (kg/m^3), g is the acceleration due to gravity ($\approx 9.8\,m/s^2$), E is Young's modulus (N/m^2) and ζ is the structural damping ratio. Considering these properties, the dimensionless parameters to be considered for physical simulation are given as follows:

Reynolds number	$\Re = \frac{VL}{\nu}$
Froude number	$Fr = \frac{V}{\sqrt{gL}}$
density ratio	$\frac{\rho_m}{\rho_a}$
Cauchy number	$Ca = \frac{E}{\rho_a V^2}$
structural damping	ζ

Boundary Conditions

As the boundary conditions, the following factors need to be considered: pressure gradient

$$p(x, y, z_g) \tag{468}$$

Approaching flow: mean flow speed

$$\overline{U}(y, z \leq z_g) \tag{469}$$

spectral quantities

$$G_{jj}(y, z; f) \quad (j = u, v, w) \tag{470}$$

cross-correlations

$$R_{ij}(\Delta x, \Delta y, \Delta z; f) \quad (i, j = u, v, w) \tag{471}$$

Geometrical shape of the test section boundaries.

Simulation of these factors particularly with wind-tunnel facilities is discussed in Section 12.3.7.2.

Discussion on Dimensionless Parameters

It is generally not possible to satisfy all of the requirements. Just as an example, consider $Re = VL/\nu$ and $Fr = V/\sqrt{gL}$. If ν and g are common between the model and prototype, both of these requirements are simultaneously satisfied when, and only when, two systems have exactly the same V and L between them. In other words, these systems have to be identical. If the requirements are not perfectly satisfied, it is important to know the limit of availability of each similitude and also the possible influence on outcomes when a condition is relaxed or distorted.

Reynolds Number

The Reynolds number can be defined as the ratio of the fluid inertia force to the fluid viscous force. In most wind-tunnel tests it is impracticable to satisfy the Reynolds number similitude. Indeed the viscous forces are usually at least an order of magnitude smaller and relatively unimportant compared to the inertia forces. However, the consequence of the distortion of this requirement must be examined carefully. The following three points are of particular importance.

1. It is well known that the flow pattern around a circular cylinder is very sensitive to the change of *Re* because of the shift of flow separation points with it. There is a corresponding change of the wake width and the drag force as well as the frequency of vortex formation. It is important that in modeling a structure with smoothly curved surface geometry these effects are properly taken into consideration. It should be noted that the critical Reynolds number is also dependent on the surface roughness of the solid boundary as well as the turbulence level in the approaching air flow.
2. In the case of flow around sections with sharp corners, the flow separation points do not move and the flow pattern is less sensitive to a change of Reynolds number. However, a broad wake after separation from the upstream corners may reattach to the body surface, depending on the aspect ratio of the body cross section or the length of the after-body. The flow reattachment, of course, results in a reduction of drag force and increase of the Strouhal number, $St = fd/V$, in general. The critical aspect ratio of the body at which this change occurs depends on the Reynolds number as well as the corner radius and the air stream turbulence level. It also should be noted that this factor is influenced by the wind-tunnel blockage ratio. The effects of Reynolds number on the bridge test results have been taken more seriously in the past few years (see Matsuda et al. 2003).
3. In problems involving the effects of wind turbulence it is essential to simulate the velocity spectra correctly. Townsend (1976) has pointed out that 'geometrically similar flows are expected to be dynamically and structurally similar, if their Reynolds numbers are large enough to allow turbulent flow'. However, it should be remembered that the Reynolds number does play a part in the existence of the inertia subrange of its energy spectra. As the Reynolds number increases, the high-frequency end of the distribution will be extended so that the total dissipation of the turbulence energy remains unchanged. On the other hand, when the Reynolds number is small, the ratio of the size of the dissipating eddies to the representative size of the predominant eddies becomes highly dependent on viscosity. A result of this is an inaccurate simulation of turbulence structure due to a narrower-than-necessary inertia subrange.

Froude Number

Froude number is the ratio of fluid inertia force to vertical force due to gravity and/or buoyancy. Consequently, the Froude similitude becomes important for cases such as dissipation of airborne particles or wind-induced response of cable-supported structures where gravity is a dominant factor. For aeroelastic testing, the Froude scaling often becomes the only available matching parameter for the flow itself. Whenever the influence of gravitational field needs to be considered, this requirement cannot be waived. However, depending on the nature of the problem, when the restoring force of the structure is strictly provided by the elastic force, even this similitude can be relaxed. As a result, the timescale would most likely be decided by the frequency scaling.

Density Ratio

The ratio of structural density ρ_m to air density ρ_a has to be consistent. Here, ρ_m is not necessarily the density of the materials used but the apparent density of the structure as a whole. Therefore the requirement is often satisfied in terms of mass ratio, μ, as follows:

$$\frac{\rho_m}{\rho_a} = \frac{\rho_m L^2}{\rho_a L^2} \Rightarrow \frac{m}{\rho_a B^2} = \mu \tag{472}$$

For the case of a bridge model, m is the mass of the bridge per unit length, and B is the deck width. If it is a problem in torsion, the mass parameter becomes $J/\rho_a B^4 = \mu(r/B)^2$ instead of μ, where J represents the polar mass moment of inertia per unit length of the structure.

Cauchy Number
The Cauchy number, which is the ratio of elastic force to the inertia force of the fluid, can be interpreted in the following way:

$$Ca = \frac{E}{\rho_a V^2} = \frac{EL^4/L^2}{\rho_a V^2 L^2} \Rightarrow \frac{mL^4\omega^2/L^2}{\rho_a V^2 L^2} = \left(\frac{\omega L}{V}\right)^2 \frac{m}{\rho_a L^2} \Rightarrow \frac{\mu}{V_R^2} \tag{473}$$

Since the mass ratio is already a requirement, this implies the consistency of the reduced velocity

$$V_R = \frac{V}{fL} \tag{474}$$

or the inverse of it, reduced frequency, $f_R = fL/V$, as a similitude parameter. This becomes one of the fundamental requirements in aeroelastic testing of structures. Usually it is not difficult to do testing over a wide range of reduced velocity to cover the equivalent range in reality.

Critical Damping Ratio
The magnitude of structural damping is obviously an important parameter for the prediction of structural dynamic response. The problem, however, is that its magnitude is uncertain even for the existing structures. Naturally the wind-tunnel tests are usually carried out for a range of damping ratio. Mass and damping are two most important factors in the design and construction of wind-tunnel models and these two requirements sometimes have been combined together as the Scruton number, or the mass-damping parameter, defined by

$$Sc = \frac{m\zeta}{\rho_a d^2} \tag{475}$$

The concept of this parameter was originally introduced by a simple analytical model to demonstrate the peak amplitude of vortex-induced response, and it should be remembered that the Scruton number does not necessarily always work as a good single parameter to describe the response characteristics. Mass and damping parameters have to be, generally speaking, examined separately (Tanaka and Yamada 1987). Note that the Scruton number is sometimes defined as $Sc = 2m\delta/(\rho_a d^2)$, which is 4π times the number defined by Equation (475).

12.3.7.2 Wind-Tunnel Simulation of Natural Wind

General Principle and its History
Nobody would do model testing of a structure without considering the external loading conditions as realistic as possible together with those of the structure itself. For the testing of bridges, however, the wind action was often considered to be simply a uniformly smooth air flow, with or without a small angle of attack deviated from parallel to the ground, and always normal to the longitudinal bridge axis. This simplified loading condition is something like a consideration of earthquake excitation by a simple harmonic ground motion, which may be a fairly conservative assumption but not without exceptions.

Early wind-tunnel studies of model buildings in the 1930s and 1940s indicated significant influence of wind turbulence on test results such as the pressure distribution patterns. It was about the same time when Prandtl's boundary layer theory was applied by meteorologists to explain the structure of the lower atmosphere. However, the most fundamental principle for wind-tunnel tests in this regard was not clearly recognized until the following statement was made by Martin Jensen:

> "The natural wind is turbulent, and the phenomena [...] take place in the boundary layer of the wind, and, as should be emphasized, are highly dependent on the nature of this boundary layer [...]. The correct model test with phenomena in the wind must be carried out in a turbulent boundary layer, and the model-law requires that this boundary layer be to scale as regards the velocity profile." (Jensen (1958))

Following the collapse of the Tacoma Narrows suspension bridge there was a significant contribution by aeronautical engineers towards the experimental aerodynamics with civil engineering applications. At the same time, it became a general practice to do testing of structures such as bridges in a conventional aeronautical wind tunnel where the air flow is smooth and uniform rather than simulated natural winds. It could be said that it was a 'side effect' of the contribution by aeronautical prejudices. Jensen's micrometeorological consideration was followed up by Davenport's formulation of codified natural wind characteristics and physical simulation of it in the early 1960s, which had a significant impact on practice in this engineering field (Davenport and Isyumov 1967).

Characteristics of Natural Wind

The similitude requirements of wind characteristics for physical wind tunnels and mathematical simulation models have been discussed extensively by both meteorologists and wind engineers (e.g. Plate 1982). The simulation of wind can be considered in two categories: (a) the average characteristics of the turbulent boundary layer wind approaching the site of each project; and (b) the wind structure at the immediate proximity of the structure, which is largely influenced by the particular topographical conditions including surrounding structures. These two are sometimes referred to as the far-field and near-field simulation, respectively.

For convenience, let us take the x-coordinate horizontally along the mean wind, which is assumed normal to the longitudinal bridge axis, and the z-coordinate vertically upward. The y-coordinate is hence along the bridge. The wind-field characteristics for testing of a bridge are generally defined by the following parameters: mean speed distribution

$$\overline{U}(y, z) \tag{476}$$

turbulence intensities

$$I_u(y, z), \qquad I_v(y, z), \qquad I_w(y, z) \tag{477}$$

velocity spectra

$$G_u(f), \qquad G_v(f), \qquad G_w(f) \tag{478}$$

and velocity correlations

$$R_{uu}(\Delta y, \Delta z), \qquad R_{ww}(\Delta y, \Delta z) \tag{479}$$

For all of these factors, the general homogeneity of the flow field is required over some length in the x-direction when the model is rotated to observe the effect of horizontal skew angle.

There have been a number of studies that present good summary of these natural wind characteristics (Counihan 1973; ESDU 1986). A simple mathematical model of the atmospheric turbulence at elevation z, assuming neutral stability, can be given typically as follows.

Turbulence intensities:

$$I_u = \frac{1}{\ln(z/z_0)}, \qquad I_v = 0.8I_u, \qquad I_w = 0.5I_u \tag{480}$$

Scales of turbulence:

$$L_u^x = 2.93z, \qquad L_u^y = L_u^z = 0.98z, \qquad L_v^x = L_v^y = L_v^z = 0.73z \tag{481}$$

and

$$L_w^x = L_w^y = L_w^z = 0.37z \tag{482}$$

Velocity spectra:

$$\frac{fG_u(f)}{\sigma_u^2} = \frac{22n}{(1+33n)^{5/3}}, \qquad \frac{fG_v(f)}{\sigma_v^2} = \frac{6.3n}{(1+9.5n)^{5/3}} \tag{483}$$

and

$$\frac{fG_w(f)}{\sigma_w^2} = \frac{1.3n}{1+5.3n^{5/3}} \tag{484}$$

in which $n = fz/\overline{U}(z)$.
Coherence functions:

$$\gamma_{ii}(f, \Delta r_j) \approx \exp\left(-k_{ij}\frac{f\Delta r_j}{\overline{U}(z)}\right) \qquad (i = u, v, w;\ j = x, y, z) \tag{485}$$

A problem of so-called Davenport-type coherence above is the fact that the coherence function of this model goes to unity at zero frequency, whereas the field measurements indicate that they should be less than one. ESDU (1986) is also following this functional form. As opposed to them, the historically well-known model proposed by von Kármán (1948) is known to agree well with the field measurement, particularly in the low-frequency range.

Boundary Layer Wind Tunnels

Jensen's Experiment

Using a long wind tunnel ($L = 5.5$ m, width 0.6 m) with a floor coated by various roughness, Jensen produced the turbulent boundary layer flows of various roughness length z_0 as shown in Table 12.9, together with the boundary layer thickness, which was found to be given by approximately $\delta(x) = z_0^{0.2} 0.341 x^{0.8}$ when $2 \times 10^3 < Re < 5 \times 10^5$.

Jensen did a measurement of wind-induced pressure around a building ($h = 160$ cm) in the field where $z_0 = 0.95$ cm. Then he placed a 1:20 scaled model of it in his wind tunnel and tried to reproduce the wind-induced pressure distribution. What he found was when h/z_0 is consistent to the full-scale value, the correct pressure distribution was observed. Pressure patterns were found to be significantly different when h/z_0 did not agree.

From these results, Jensen concluded that the phenomena induced by natural wind can be reproduced only when the model tests are performed in a boundary layer which was created in a similar way as the case of natural wind and also when the linear scale of its turbulence coincides with the linear scaling of other models placed in it. h/z_0 is now called the Jensen number after his name. Jensen's idea on the

Table 12.9 Various types of floor coating

Floor coating	z_o (cm)	δ (cm)
Glazed cardboard sheets	1.5×10^{-3}	10
Smooth masonite plates	$0.9 - 1.8 \times 10^{-3}$	10
Sandpaper	2.5×10^{-2}	12
Corrugated paper ($h = 0.35$ cm, $\lambda = 0.90$ cm)	$4.1 - 6.7 \times 10^{-2}$	14
Small broken stones (1.5–2 cm)	0.37	15
Wooden fillets, 2.5 cm(h) ×2.0 cm(w) 15–20 cm spacing, angled ≤ 20°	0.41	20
Large broken stones (3–6 cm)	0.86	22
'Model of a city' 7.0 cm(h) ×2.9 cm(w) rows across the tunnel, angled ≤ 20°	3.50	30

experimental simulation of the atmospheric boundary layer in a wind tunnel was largely extended by Davenport in Canada and Cermak in the USA in the 1960s.

Development of Boundary Layer Wind-Tunnel Simulation

Since then, the simulation of natural wind and research in various wind engineering fields have made significant progress. The idea of boundary layer simulation is now firmly established as one of the most fundamental requirements that must be satisfied in any experimental or analytical simulation of the wind engineering problems.

The boundary layer wind tunnel Davenport first established at the University of Western Ontario had a working section of about 24 m long, 2.4 m wide and an adjustable height, variable from 1.7 m at the entrance to 2.3 m at the end. The adjustment of the roof height allowed control over pressure gradients along the tunnel length. The adjustable stable wind speed range was 0.5 to 15 m/s. The simulation of atmospheric boundary layer flow under neutral condition was made by placing various roughness coverages on the wind-tunnel floor. The representative power-law exponents were in the range up to 0.34 and the available linear scale compared to natural wind was typically 1:400 to 1:500.

Other methods of producing velocity profiles

There have been a number of attempts to artificially create a thick boundary layer without losing the required characteristics. They include the use of graded gauze, grid of rods, plates and other types of vortex generators, fences and other roughnesses in the upstream section. More sophisticated methods are, for example, the use of counter-jets and pulsating grids, which tend to be very expensive. Amongst these, one of the most successful is the use of triangular spires at the entrance of wind-tunnel test section. The particular technique was developed mostly at the Low Speed Aerodynamics Laboratory of the National Research Council of Canada. The work was initiated in 1968 by Templin with the object to generate a thick boundary layer without having a long upwind fetch, thereby enabling the full potential of aeronautical wind tunnels to be realized for wind engineering purposes. It was further developed by Standen and by Campbell and co-workers, and more or less completed by Irwin (1981).

The original idea was to design the spire shape to obtain the desired power-law wind profile with an acceptable lateral uniformity at a distance approximately six spire heights downstream. No considerations of the turbulence characteristics produced as a result entered into the calculation. Standen and Campbell and co-workers were successful in producing a simulated mean flow profile but their calculation needed further improvement. Furthermore the desired shape of spires was still disputable. After many experiments, Irwin concluded that a straight triangle should be a reasonable choice and indicated the calculation for the design of them (see Irwin 1981).

Distorsion of Similitude Requirements and its Consequences
Distorsion of boundary layer simulation, even if all the general characteristics of the atmospheric boundary layer flow are assumed to be well defined, can enter the modeling procedure in two ways. First, because the required flow cannot be defined very well due to inhomogeneities in the terrain surrounding the site to be modeled. Second, because the flow simulation itself may have distorsions inherent in its methodology.

In order to discuss this effectively, it is convenient to limit the issue mostly to structural problems, and the subject of modeling pollutant dispersion, for example, is not being considered. For the establishment of the wind environment, the general practice now is to develop the background boundary layer over a relatively homogeneous terrain (far-field simulation) and then modify it by the detailed surroundings of the site in question, typically for a radius of 300 to 500 m (near-field simulation). The far-field simulation is to produce the overall characteristics of the boundary at the site as they have been discussed through this chapter. The near field, on the other hand, is to provide the intimate complex interactions, which are largely inhomogeneous and intractable to precise definition. A useful classification of structures and a detailed discussion on various aspects of simulation problems within this framework have been attempted by Surry (1982).

12.3.7.3 Mathematical Simulations

The application of numerical simulation techniques in the field of wind engineering or related fields has taken various steps through its history and it is probably fair to say that only recently has it come to a rather realistic stage. The first application of numerical analysis to fluid mechanics was to obtain solution of inviscid, or potential, flow equations described in terms of complex potential $w(x, y) = \varphi + i\psi$, which was possible only for laminar flow. The next development was to solve the Navier–Stokes equation directly by applying various numerical integration techniques. Of course for this case, a proper turbulence model has to be introduced for the expression of Reynolds stresses.

Fundamental Equations
Description of incompressible flows, whatever they are, can be given by the following equations:

$$\frac{\partial u_i}{\partial t} + u_j \frac{\partial u_i}{\partial x_j} = -\frac{1}{\rho}\frac{\partial p}{\partial x_i} + \frac{\partial}{\partial x_j}\left(\nu \frac{\partial u_i}{\partial x_j}\right) \qquad (486)$$

$$\frac{\partial u_i}{\partial x_i} = 0 \qquad (487)$$

With p and T not far from the standard status, the minimum scale of turbulence, typically the depth of the viscous sublayer, is far greater than the average of the molecular free path. This means that it is very unlikely that the flow turbulence cannot be described by the Navier–Stokes equation. Turbulence can be considered essentially as a phenomenon of continuous media. Note that there are microworlds that do not experience turbulence, such as the motion of bacteria or spermatozoa. The mechanics of them is completely described as viscous laminar flow.

The combination of the above equations describes all the possible turbulent phenomena but numerical analysis of these equations does not necessarily provide all the correct solutions, because in numerical analyses, continuous physical quantities are approximated by discrete quantities and hence give an equivalent cut-off frequency of a low-pass filter corresponding to the mesh-size of different grids adopted for the calculation. In general, it is necessary to consider very high-frequency fluctuations of velocity to grasp the characteristics of turbulence, which therefore requires very high-quality computers.

Introducing $p \to \overline{P} + p(t)$ and $u_i \to \overline{U}_i + u_i(t)$, the following Reynolds equation is obtained:

$$\frac{\partial \overline{U}_i}{\partial t} + \overline{U}_j \frac{\partial \overline{U}_i}{\partial x_j} = -\frac{1}{\rho}\frac{\partial \overline{P}}{\partial x_i} + \frac{\partial}{\partial x_j}\left(\nu \frac{\partial \overline{U}_i}{\partial x_j} - \overline{u_i u_j}\right) \tag{488}$$

The last term, corresponding to the Reynolds stress terms, expresses dispersion of momentum caused by velocity fluctuations. Equation (488) describes the averaged field of turbulence; meaning that fine fluctuations are not given by this equation and the spatial gradients of wind velocity and pressure are relatively smooth. Hence it is easier to use Equation (488) than (486) for application of the finite difference method. Many attempts of numerical simulation starts with this equation.

Discussion of the Viscosity Term

By introducing the representative length scale L and the velocity scale V and making

$$u_i^* = \frac{u_i}{V}, \qquad x_i^* = \frac{x_i}{L}, \qquad t^* = \frac{tV}{L}, \qquad p^* = \frac{p}{\rho_a V^2} \tag{489}$$

and $Re = VL/\nu$, Equation (486) becomes

$$\frac{\partial u_i^*}{\partial t^*} + u_j^* \frac{\partial u_i^*}{\partial x_j^*} = -\frac{\partial p^*}{\partial x_i^*} + \frac{1}{Re}\frac{\partial^2 u_i^*}{\partial x_j \partial x_j} \tag{490}$$

The last term of Equation (490), dispersion due to molecular viscosity, is inversely proportional to Re. It means that when $Re \to \infty$, the effect of molecular viscosity cannot be observed well.

When Equation (488) is used, usually eddy viscosity K_m is introduced to model the Reynolds stresses. For this case, as $K_m \sim VL$,

$$Re = \frac{VL}{K_m} \sim 1 \tag{491}$$

One of the most important characteristics of turbulence is the existence of various scales of eddies and their cascading process. The minimum scale of eddies is called 'Kolmogorov's micro-scale η' and is given approximately by

$$\eta \sim \left(\frac{\nu^3}{\varepsilon}\right)^{1/4} \tag{492}$$

where ε is the rate of viscous dissipation per unit time and unit mass, or defined by

$$\varepsilon = \nu \overline{\left(\frac{\partial u_i}{\partial x_j}\right) \cdot \left(\frac{\partial u_i}{\partial x_j}\right)} \tag{493}$$

and hence its unit is m^2/s^3. Eddies dissipate as heat energy at this scale due to viscosity. ε is related to the energy transfer mechanism at a larger scale than η and is approximately valued by

$$\varepsilon \sim V^3/L \tag{494}$$

where $V \sim \sigma_u$ and $L \sim L_u^x$, for example. Therefore $\frac{\eta}{L} \sim (Re)^{-3/4}$, where $Re = \frac{VL}{\nu}$. η and L are the most important linear scales in the numerical simulation of turbulent flows.

Problem of Direct Simulation

If the turbulent flow field is solved, it means that the mechanism of energy dissipation needs to be included as a part of the solution and this can be done only at the scale of η. If the mesh size h (say) of finite difference is much greater than η, this mechanism will not be discussed in this solution. It needs to be at least $h \sim \eta$, and hence, $h/L \sim (Re)^{-3/4}$. If L is the linear size of the space numerically simulated, $L/h = N$ is the number of mesh (grid) in one direction. In a 3D field, the number of grid is $N = (L/h)^3$, or

Re	10^4	10^6	10^8
N	10^9	10^{13}	10^{18}

In practical wind engineering problems, $Re > 10^6$. Therefore, $N > 10^{13}$, which means that turbulence simulation models are needed.

If Navier–Stokes and continuity equations are considered under proper initial conditions and boundary conditions at each grid point in order to calculate velocity and pressure, and the required mathematical operation for this process is 10^2, and this is to be repeated 10^2 times, meaning 10^2 steps going ahead in time, and the required time for each mathematical operation is about 10^{-6} s, for example, the total calculation time is

$$10^{13} \times 10^2 \times 10^2 \times 10^{-6} = 10^{11}\text{s} = 1.16 \times 10^6 \text{ days} \approx 3200 \text{ years!} \qquad (495)$$

Reynolds stresses are simulated by various assumptions, such as $K_m \frac{\partial u_i}{\partial x_j}$, where $K_m \propto u_{ij}$, etc. In order to consider the space average of quantities for a certain mesh area, coarse graining methods such as the **large eddy simulation** have been considered.

12.3.7.4 Wind-Tunnel Testing Techniques

Full-Bridge Model Tests

Full-bridge model testing started immediately following the failure of the Tacoma Narrows suspension bridge. When engineers are faced by an unexpected incident such as this, they naturally attempt to reproduce what happened in reality at the model scale to find out the reason why the bridge failed. It involved construction of a large structural model and a large wind tunnel to accommodate it and, as a result, it became an expensive operation.

Professor Farquharson's experimental study using a full-bridge model was successful in (a) reproducing the failure mechanism of the original Tacoma Narrows Bridge; (b) designing a new bridge which was aerodynamically stable; and (c) establishing a pioneering method of bridge tests. This initial research was followed up by other researchers, including Scruton and co-workers at NPL, England. However, their test results were not quantitatively reproducing what happened in reality in the sense that the critical wind speed for the bridge's final destruction was found to be much lower than in reality. The reason of this could be at least partially attributed to the fact that the bridge model was mechanically equivalent to the prototype bridge following the Froude similitude, but the applied wind was a smooth uniform air flow and no attempt was made to simulate the characteristics of natural wind. Simulation of natural wind, of course, was introduced only in the 1960s.

The overall stiffness of the bridge girder is provided by a few spinal stiffening bars and the geometrical configuration of the girder is made up by short modules of nonstructural material. The distribution of mass and mass moment of inertia can be easily adjusted by attaching additional weights at any conveniently hidden location. The towers can be made in much the same way as the girder. The axial stiffness of the main cables has to be properly simulated. The result is usually the use of very thin piano wires and the cable mass needs to be adjusted by discrete masses attached to them. The scale ratio for a long span bridge is often very much limited because of the availability of large wind tunnels and the linear scale of turbulence created in the wind tunnel. Regarding the latter requirement, if the Kaimal type spectrum is assumed for u-component turbulence, the perfect matching of this component may not be practically achievable at the present.

The model design procedure is as follows.

1. Dynamic FE modeling analysis to decide eigenfrequencies ω_j and mode shapes $\phi_j(x)$.
2. Dynamic analysis at the model scale to decide the boundary conditions, degree of simplifications, etc.
3. Froude similitude ($\lambda_V = \sqrt{\lambda_L}$) is required, particularly for suspension bridges.
4. Simulation of mass, mass moment of inertia and stiffness.
5. Simulation of cables for axial stiffness, mass and drag force.

Sectional Model Tests

The idea is to make use of a rigid, shape-wise representative segment of the bridge as an analog device for extracting wind loads (Hjorth-Hansen 1992). By definition, any spanwise variations of the structure and wind characteristics are ignored in this testing method. The model can be almost immobile or given a forced motion for measuring the reactions at its support, or freely suspended in a given air flow for the response measurement. The tests usually consider only vertical bending and torsional motion. However, it is possible to include the drag-wise sway as well. The wind flow can be either smooth or with turbulence, but it is difficult to include the turbulence effects with good accuracy, particularly in the low-frequency range.

The required similitude usually includes the following items:

- geometrical configuration of the bridge deck;
- mass parameters $\mu = m/\rho B^2$ and $\nu = \mu(r/B)^2$;
- structural damping ζ_T and ζ_V;
- reduced velocity $\overline{U}/f_T B$ or $\overline{U}/f_V B$;
- frequency ratio f_T/f_V;
- location of the center of rotation;

where m, r, B are the linear mass, radius of gyration and width of the bridge deck, ρ is air density, ζ_T, ζ_V are the critical damping ratio in torsion and in vertical bending and f_T, f_V are eigenfrequencies in torsion and vertical bending. The Reynolds number effects will have to be ignored but in this particular issue it is not any worse than in other testing methods . It should be noted, however, that the Reynolds number does have an influence, generally speaking, on bridge response and measured aerodynamic forces (Matsuda et al. 2003).

The aerodynamic force coefficients in lift, drag and pitching moment, and all 18 motion-dependent flutter derivatives can be measured effectively by using a section model. The dynamic tests are usually effective in making a prediction of aerodynamic instability, both in torsion and galloping, and exploring the possibility of vortex-shedding excitation. However, considerable interpretation efforts would be required for buffeting prediction.

Some errors pertaining to the oncoming flow can be expected because it is difficult for the flow to be perfectly two-dimensional. Because the model needs to be supported from outside the wind-tunnel walls, there is a possibility of air flow leaking through these holes. There have been various attempts to prevent or reduce the effects of the end leakage. It is also necessary to give a certain aspect ratio to the model in order to maintain the 2D characteristics of testing. Hjorth-Hansen (1992) provides an excellent review regarding the design of suspension rigs, drag wires, vibrational mode in torsion, devices to control the system's damping, and other practical hints for maintaining good accuracy in measurement (Figure 12.33).

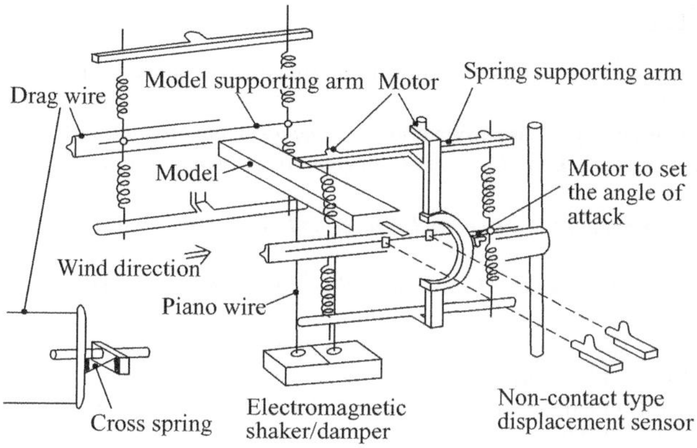

Figure 12.33 Typical set-up of a dynamic rig (after Hjorth-Hansen 1992)

Taut-strip model tests

The use of a taut-strip model was proposed to fill in a gap between two conventional testing methods, the sectional-model method and full-bridge model method. The idea is to have a simulated bridge deck in terms of its geometrical shape and mass distribution but its stiffness being provided only by stretched wires between anchorages. The geometrical scale of the deck has to be chosen in conformity with the linear scaling of wind turbulence. The simulation of main cables, which is often difficult with large scaling, is not included. Since the stretched wires would vibrate with a half sine wave mode shape, the obtained test results have to be treated properly for the prediction of actual bridge response, considering the anticipated mode shapes of the bridge. Frequency in vertical bending is tuned by wire tension, whereas its ratio to the torsional frequency is controlled largely by the separation of two wires and also by addition of an elastic tube located at the centre of twist.

This technique was first introduced by Davenport (in 1972) as a means of taking a second look at the experiments carried out at the early stages of suspension bridge aerodynamics, including this time their 3D response characteristics to simulated turbulent wind. A chief objective of the taut-strip model method is a simulation of the dynamic characteristics of the bridge road deck with the consistent linear scaling factor as the simulated natural wind, and yet not going to the complexity of manufacturing a full-bridge model. Inevitably there are inaccuracies in structural simulation. Since the main cables are not simulated at all, any gravitational acceleration effects and associated structural nonlinearity are altogether ignored. An ironical advantage is an almost free choice of length scale and timescale independent of each other. It means that both the Reynolds and Froude similitudes are violated.

The fundamental response characteristics of taut-strip models have been examined to confirm that the responses in general are in good agreement with buffeting and flutter theories for which the aerodynamic derivatives and aerodynamic admittance functions are assumed to be more or less known. In fact it is one of the advantages offered by this method that when the taut-strip model results are used as input to the buffeting theory, the complication of defining the aerodynamic admittance function is avoided. Also a possible complication due to the nonlinearity of aerodynamic derivatives need not be considered either (Tanaka and Davenport 1982).

More recent development of a high-speed scanning technique for the measurement of wind-induced pressure fluctuations made it possible to apply this testing method to more versatile objectives. In relation to the aerodynamic study of the Storebælt Bridge, a taut-strip model was used for the measurement of unsteady aerodynamic derivatives, aerodynamic admittance function and space correlation of lift forces (Davenport et al. 1992).

Further Reading

Agar T (1989) Aerodynamic flutter analysis of suspension bridges by a modal technique. *Journal of Engineering Structures* **11**(2), 75–82.

Agar T (1991) Dynamic instability of suspension bridges. *Computers and Structures* **41**(6), 1321–1328.

Ang A and Tang W (1975) *Probability Concepts in Engineering Planning and Design: Basic Principles.* Wiley, Chichester.

Ang A and Tang W (1984) *Probability Concepts in Engineering Planning and Design: Decision, Risk, and Reliability.* Wiley, Chichester.

ASCE (1984) *Guidelines for Transmission Line Structural Loading. Structural Division,* American Society of Civil Engineers.

Asmussen J (1997) *Modal analyis based on the random decrement technique: Application to civil engineering structures.* PhD thesis, Aalborg University.

Ayoride E and Warburton G (1980) Minimizing structural vibrations with absorbers. *Earthquake Engineering and Structural Dynamics* **8**, 219–236.

Bachmann H and Ammann W (1987) Vibrations in structures – induced by man and machines. In *Structural Engineering Documents*, 3rd edn, 1ABSE, Zürich (Switzerland).

Balageas D (ed.) (2002) *Proceedings of the 1st European Workshop on Structural Health Monitoring*, Paris.

Bendat J and Piersol A (1986) *Random Data: Analysis and Measurement Procedures*, 2nd edn. Wiley, Chichester.

Benjamin J and Cornell C (1970) *Probability, Statistics and Decision for Civil Engineers.* McGrw-Hill.

Biggs J (1954) Wind Forces on Structures, Final Report of the Task Committee. *Transactions of the American Society of Civil Engineers.*

Bleich F (1949) Dynamic instability of truss-stiffened suspension bridges under wind action. *Transactions of the American Society of Civil Engineers* **114**, 1177–1222.

Bleich F and Teller L (1952) Structural damping in suspension bridges. *Transactions of the American Society of Civil Engineers, Paper* **2486**.

Blevins R (1990) *Flow-Induced Vibration*, 2nd edn. Van Nostrand Reinhold.

Boonyapingyo V, Yamada H and Miyata T (1994) Wind-induced nonlinear lateral-torsional buckling of cable-stayed bridges. *Proceedings of the American Society of Civil Engineers*, **120** (ST2), 486–506.

Bosdogianni A and Olivari D (1996) Wind- and rain-induced oscillations of cables of stayed bridges. *Journal of Wind Engineering and Industrial Aerodynamics* **64**, 171–185.

Box G and Jenkins G (1976) *Time Series Analysis: Forecasting and Control.* Holden-Day.

Brigham E (1974) *The Fast Fourier Transform.* Prentice-Hall.

Brika D and Laneville A (1997) The power imparted by wind to a flexible circular cylinder in the wake of another stationary cylinder. *Institute of Electrical and Electronic Engineers, Transactions on Power Delivery* **12**(1), 398–405.

Bucher C and Lin Y (1988) Stochastic stability of bridges considering coupled modes. *Proceedings of the American Society of Civil Engineers*, **114** (EM2), 2055–2071.

Bucher C and Lin Y (1989) Stochastic stability of bridges considering coupled modes. *Proceedings of the American Society of Civil Engineers*, **115** (EM2), 384–400.

Bullard F (2001) *A Brief Introduction to Bayesian Statistics.* North Carolina School of Science and Mathematics.

Carden E and Fanning P (2004) Vibration based condition monitoring: a review. *Structural Health Monitoring* **3**(4), 355–377.

Cartwright D and Longuet-Higgins M (1956) Statistical distribution of the maxima of a random function. *Proceedings of the Royal Society, London*, **A237**, 212–232.

Cermak J (ed.) (1975) *Proceedings of the 2nd U.S. National Conference on Wind Engineering*, Colorado State University Paper IV-21, Fort Collins, CO.

Cermak J (1987) Advances in physical modeling for wind engineering. *Proceedings of the American Society of Civil Engineers*, **113** (EM5), 737–756.

CETS (1984) *Guidelines for Transmission Line Structural Loading.* Technical Report, Committee on Electrical Transmission Structures. Structural Division, American Society of Civil Engineers.

Chang F (1973) Human response to motions in tall buildings. *Proceedings of the American Society of Civil Engineers*, **98**, 1259–1272.

Chang F (ed.) (2003) *Proceedings of the 4th International Workshop on Structural Health Monitoring* Stanford, CA.

Chen P and Robertson L (1973) Human perception thresholds of holizontal motion. *Proceedings of the American Society of Civil Engineers*, **98**, 1681–1695.

Chen S (1987) A general theory for dynamic instability of tube arrays in cross flow. *Journal of Fluids and Structures* **1**, 35–53.

Chen X, Matsumoto M and Kareem A (2000a) Aerodynamic coupling effects on flutter and buffeting of bridges. *Proceedings of the American Society of Civil Engineers*, **126** (EM1), 17–26.

Chen X, Matsumoto M and Kareem A (2000b) Time domain flutter and buffeting response analysis of bridges. *Proceedings of the American Society of Civil Engineers*, **126** (EM1), 7–16.

Chen Z (1994) The three dimensional analysis of behavior investigation on the critcal flutter state of bridges. *Proceedings of the International Symposium on Cable-Stayed Bridges*, pp. 302–307, Shanghai.

Cheng S (1995) *A 3D finite element flutter analysis of long span bridges*. Master's thesis, Tongji University, Shanghai. (In Chinese.)

Cheng S and Tanaka H (2002) Aerodynamics of dry inclined cables. *Proceedings of the 2nd International Conference Advances in Wind and Structures*, pp. 361–368, Pusan, Korea.

Cheng S and Tanaka H (2005) Correlation of aerodynamic forces on an inclined circular cylinder. *Wind and Structures, an International Journal* **8**(2), 135–146.

Cheng S, Irwin P, Jakobsen J, Lankin J, Larose G, Savage M, Tanaka H and Zurell C (2003a) Divergent motion of cables exposed to skewed wind. *Proceedings of the 5th International Symposium on Cable Dynamics (ISCD)*, pp. 271–278, Santa Margherita.

Cheng S, Tanaka H, Irwin P and Jakobsen J (2003b) Aerodynamic instability of inclined cables. *Proceedings of the 5th International Symposium on Cable Dynamics (ISCD)*, pp. 69–76, Santa Margherita.

Clough R and Penzien J (1993) *Dynamics of Structures*, 2nd edn. McGraw-Hill.

Cole H (1973) *On-line Failure Detection and Damping Measurements by Random Decrement Signatures*. Nasa-cr-2205, NASA.

Cook N (1985) *The Designer's Guide to Wind Loading of Building Structures, Part 1*. Butterworths.

Cook N (1990) *The Designer's Guide to Wind Loading of Building Structures, Part 2*. Butterworths.

Cremer JM, Cournasse C, Goyet V, Lothaire A and Dumontier A (1995) The stays, their dynamic behaviour, their equipments – bridges at Ben-Ahin, Wandre and upon Alzette. *Proceedings of the International Symposium on Cable Dynamics*, pp. 473–480, Liège, Belgium.

Cunha A and Caetano E (2005) From input–output to output-only modal identification of civil engineering structures. *Proceedings of the 1st International Operational Modal Analysis Conference*, pp. 11–27, Copenhagen.

Davenport A (1962) The buffeting of suspension bridge by stormy winds. *Proceedings of the American Society of Civil Engineers*, **83** (ST3), 233–268.

Davenport A (1964) Note on the distribution of the largest value of a random function with application to gust loading. *Proceedings of the Institute of Civil Engineers*, **28**, 187–196.

Davenport A (1979) Gust response factors for transmission line loading. *Proceedings of the 5th International Conference on Wind Engineering (ICWE)*, pp. 899–910, Fort Collins, Colorado.

Davenport A and Isyumov N (1967) The application of the boundary layer wind tunnel to the prediction of wind loading. *Proceedings 2nd International Conference on Wind Engineering (ICWE)*, pp. 201–230.

Davenport A and Wardlaw R (1964) LR416, National Research Council Canada.

Davenport A, King J and Larose G (1992) Taut strip model tests. In *Aerodynamics of Large Bridges*, pp. 113–124. Balkema.

Davis D, Richards D and Scriven R (1963) Investigation of conductor oscillation on the 275 kV crossing over the River Severn and Wye. *Proceedings of the Institute of Electrical Engineers*, **110**, 205–219.

De Roeck G and Peeters B (2000) Benchmark study on system identification through ambient vibration measurements. *Proceedings of the 18th International Model Analyis Conference*, pp. 1106–1112, San Antonio, Texas.

Den Hartog J (1932) Transmission line vibration due to sleet. *American Institute of Electrical Engineers Transaction* **51**, 1074–1076.

Den Hartog J (1956) *Mechanical Vibrations*, 4th edn. McGraw-Hill.

Diana G, Cheli F, Zasso A, Collina A and Brownjohn J (1992) Suspension bridge parameter identification in full scale test. *Journal of Wind Engineering and Industrial Aeroynamics* **41**, 165–176.

Diana G, Cigada A, Damsgaard A, Lavose GL and Falco M (1993) Comparison between wind tunnel test on a full aeroelastic model of the proposed Messina Bridge and numerical results. *Proceedings of the 3rd Asia–Pacific Symposium on Wind Engineering, Hong Kong*, Vol. 1, pp. 137–142, Hong Kong.

Diana G, Cheli F and Resta F (1995) Time domain aeroelastic force identification on bridge decks. *Proceedings of the 9th International Conference on Wind Engineering, New Delhi*, Vol. 2, pp. 938–949.

Doebling S, Farrar C and Prime M (1998) A summary review of vibration-based damage identification methods. *The Shock and Vibration Digest* **30**(2), 91–105.

Dowell E (1989) *A Modern Course in Aeroelasticity*, 2nd edn. Kluwer Academic Publishers.

Downing S and Socie D (1982) Simple rainflow counting algorithms. *International Journal on Fatigue* **4**(1), 31–40.

Ehsan F, Scanlan R and Bosch H (1990) Modelling spanwise correction effects in the vortex-induced response of flexible bridges. *Journal of Wind Engineering and Industrial Aerodynamics*, **36**(1–3), 1105–1113.

ESDU (1985) *Fluctuating Loads and Dynamic Response of Bodies and Structures in Fluid Flow – Background Information*. Technical Report No. 77032, Engineering Science Data Unit, London.

ESDU (1986) *Characteristics of Atmospheric Turbulence Near the Ground*. Technical Report No. 86010, Engineering Science Data Unit, London.

Falco M, Curami A and Zasso A (1992) Nonlinear effects in sectional model aeroelastic parameters identification. *Journal of Wind Engineering and Industrial Aerodynamics* **42**, 1321–1332.

Farar C, Doebling S and Nix D (2001) Vibration-based structural damage identification. *Philosophical Transactions of the Royal Society, London* **A359**, 131–149.

Felber A (1993) *Development of a hybrid bridge evaluation system*, PhD thesis, University of British Columbia, Vancouver.

Fisher R and Tippett L (1928) Limiting forms of the frequency distribution of the largest or smallest member of a sample. *Proceedings of the Cambridge Philosophical Society*, **24**, 180–190.

Flamand O (1994) Rain–wind induced vibration of cables. *Proceedings of the International Conference on Cable-stayed and Suspension Bridges*, Vol. 2, pp. 523–531, Deauville.

Flamand O (2001) An explanation of the rain–wind induced vibration of inclined stays. *Proceedings of the 4th International Symposium on Cable Dynamics*, pp. 69–76, Montreal, Canada.

François A, De Man P, Bossens F and Preumont A (2000) *State-of-the-Art Review of MR Fluids Technology and Semi-active Control*. Consistent Semiactive System Control Workpackage No.1.

Fujino Y and Abé M (1993) Design formulas for tuned mass dampers based on a perturbation technique. *Earthquake Engineering and Structural Dynamics* **22**, 833–854.

Fujino Y, Pacheco B and Sun L (1988) Fundamental study on tuned liquid damper – A new damper for building vibrations. *Proceedings of the Symposium on Serviceability of Buildings*, Ottawa.

Fujino Y, Wilde K, Masubawa J and Bhartio B (1995) Rational function approximation of aerodynamic forces on bridge deck and its application to active control of flutter. *Proceedings of the 9th International Conference in Wind Engineering*, Vol. 2, pp. 994–1005, New Delhi, India.

Fung Y (1955) *An Introduction to the Theory of Aeroelasticity*. Wiley.

Gani F and Tanaka H (2005) Parametric excitation of stay-cables. *Proceedings of the 6th International Symposium on Cable Dynamics*, Chaleston, South Carolina.

Ge Y and Tanaka H (1999) Aerodynamic flutter analysis of cable supported bridges by multi-mode and full-mode approaches. *Journal of Wind Engineering and Industrial Aerodynamics* **86**, 123–153.

Geurts C and Staalduinen P (1999) Estimation of the effects of rain-wind induced vibration in the design state of inclined stay cables. In *Wind Engineering into the 21th Century* pp. 885–892. Balkema.

Geurts C, Vrouwenvelder T, Staalduinen P and Reusink J (1998) Numerical modeling of rain–wind-induced vibration: Erasmus Bridge, Rotterdam. *Structural Engineering International* **8**(2), 129–135.

Grootenhuis P (1969) Vibration control with viscoelastic materials. *Proceedings of Society for Environmental Engineering*, **38**(3–9).

Gu M and Lu Q (2001) Theoretical Analysis of wind–rain induced vibration of cables of cable-stayed bridges. *Proceedings of the 5th Asia–Pacific Conference on Wind Engineering*, pp. 125–128, Kyoto.

Gumbel E (1958) *Statistics of Extremes*. Columbia University Press.

Hanks T and Kanamori H (1979) A moment magnitude scale. *Journal of Geophysical Research* **84**(B5), 2348–2350.

Hansen R, Reed J and Vannarcke E (1974) Human response to wind-induced motion on buildings. *Proceedings of the American Society of Civil Engineers*, **99**, 1589–1605.

Hardy C and Bourdon P (1979) The influence of spacer dynamic porperties in the control of bundle conductor motion. *Institute of Electrical and Electronics Engineers–Power Engineering Society Summer Meeting*, Vancouver.

Haritos N and Owen J (2004) The use of vibration data for damage detection in bridges: a comparison of system identification and pattern recognition approaches. *Structural Health Monitoring* **3**(2), 141–163.

Harris C (1976) *Shock and Vibration Handbook*, 2nd edn. McGraw-Hill.

Hikami Y (1986) Rain vibrations of cables of cable stayed bridge. *Journal of Wind Engineering*. (In Japanese), **27** 17–28.

Hikami Y and Shiraishi N (1988) Rain–wind induced vibrations of cables in cable-stayed bridges. *Journal of Wind Engineering and Industrial Aerodynamics* **29**, 409–418.

Hirsch G (1994) Damping measures to control wind-induced vibrations. In *Wind-Excited Vibrations of Structures*, pp. 1–50. Springer-Verlag.

Hjorth-Hansen E (1992) Section model test, In *Aerodynamics of Large Bridges*, pp. 95–112. Balkema.

Hodges D and Piercs G (2002) *Introduction to Structural Dynamics and Aeroelasticity*. Cambridge University Press.

Honda A, Yamanaka T, Fujiwara T and Saito T (1995) Wind tunnel test on rain-induced vibration of the stay-cable. *Proceedings of the 1st International Symposium on Cable Dynamics*, pp. 225–262, Liège.

Hsieh T (1962) Foundation vibrations. *Proceedings of the Institute of Civil Engineers*, **22**, 211–226.

Ibrahim S (1977) Random decrement technique for modal identification of structures. *Journal of Spacecrafts and Rockets* **11**(14), 696–700.

Ibrahim S and Miklucik E (1977) A method for the direct identification of vibration parameters from the free response. *Shock and Vibration Bulletin* **47**(4), 183–198.

Iliff K (1987) *Aircraft Parameter Estimation – AIAA Dryden Lecture in Research for 1987*. TM 88281, NASA.

Irwin P (1977) *Wind Tunnel and Analytical Investigations of the Response of Lions Gate Bridge to a Turbulent Wind*. LTR-LA-210, National Research Council, Canada.

Irwin P (1979) Human response to dynamic motion of structures. *The Structural Engineer* **9**(56A), 237–244.

Irwin P (1981) Design and use of spires for natural wind simulation. *Journal of Wind Engineering and Industrial Aerodynamics* **7**, 361–366.

Irwin P (1986) Motion in tall buildings. *Proceedings of the Conference on Tall Bridges and Urban Habitat*, pp. 759–778, Chicago.

Irwin P (1992) Full aeroelastic model test. In *Aerodynamics of Large Bridges*, pp. 125–135. Balkema.

Irwin P (1997) Wind vibrations of cables on cable-stayed bridges, building to last. *Proceedings of the American Society of Civil Engineers Structural Congress XV*, pp. 383–387, Portland, Oregon.

Irwin P, Nedim A and Telang N (1999) Wind induced stay cable vibrations – a case study. *Proceedings of the 3rd International Symposium on Cable Dynamics*, pp. 171–176, Trondheim, Norway.

ISO (1980) *Guidelines for Evaluation of Human Exposure to Whole-body Vibration*. ISO 2631, International Standards Organization.

ISO (1984a) *Guidelines for the Evaluation of the Response Occupants of Fixed Structures, Especially Buildings and Off-shore Structures, to Low Frequency Horizontal Motion (0.063 to 1 Hz)*. ISO 6897, International Standards Organization.

ISO (1984b) *Mechanical Vibrations and Shock Measurement and Evaluation of Vibration Effects on Buidlings*. ISO/DIS 4866, International Standards Organization.

Isyumov N (1995) Motion perception tolerance and mitigation. *Proceedings of the 5th World Congress of Council on Tall Buildings and Urban Habitat*, Amsterdam.

Jaeger J and Newstead G (1949) *An Introduction to the Laplace Transformation with Engineering Applications*. Methuen Young books, London.

Jain A, Jones N and Scanlan R (1996) Coupled flutter and buffeting analysis of long-span bridges. *Proceedings of the American Society of Civil Engineers*, 122 (ST7), 716–725.

Jakobsen J (1995) *Fluctuating wind load and response of a line-like engineering structure with emphasis on motion-induced wind forces*. PhD thesis, Norges Tekniske Høgskole, Trondheim.

Jensen M (1958) The model-law for phenomena in natural winds. *Ingenioren, International edition* **4**(2), 121–128.

Jones D and Trapp W (1971) Influence of additive damping on resonance fatigue of structures. *Journal of Sound and Vibration* **17**(2), 157–185.

Jones N and Scanlan R (1991) Issues in the multi-mode aerodynamic analysis of cable-stayed bridges. *Infrastructure International*, 281–290.

Karpel M (1981) *Design of Active and Passive Flutter Suppression and Gust Alleviation*. Contractor Report 3482, NASA.

Khan F and Parmelee R (1971) Service criteria for tall buildings for wind loading. *Proceedings of the 3rd International Conference Wind Effects on Buildings and Structures*, pp. 401–408, Tokyo.

Kim J and Lyon R (1992) Cepstral analysis as a tool for robust processing, deverberation and detection of transients. *Mechanical Systems and Signal Processing* **6**(1), 1–15.

King J, Davenport A and Larose G (1991) *A Study of Wind Effects for the Storebælt Bridge Tender Design, Denmark*. BLWT-SS31-1991, University of Western Ontario.

Kobayashi H, Minami Y and Miki M (1995) Prevention on rain-wind induced vibration of an inclined cable by surface processing. *Proceedings of the 9th International Conference on Wind Engineering*, pp. 753–758, New Delhi.

Kobayashi Y and Nagaoka H (1992) Active control of flutter of a suspension bridge. *Journal of Wind Engineering and Industrial Aerodynamics* **41**, 143–153.

Korenev B and Reznikov L (1993) *Dynamic Vibration Absorbers: Theory and Technical Applications*. Wiley, Chichester.

Kotz S and Nadarajah S (2000) *Extreme Value Distributions – Theory and Applications*. Imperial College Press, London.

Kovács I and Leonhardt F (1982) Zur Frage der Seilschwingungen und der Seildämpfung. *Die Bautechnik* **10**, 325–332.

Kovács I, Svensson H and Jordet E (1992) Analytical aerodynamic investigation of the cable-stayed Helgeland bridge. *Proceedings of the American Society of Civil Engineers*, **118** (ST1), 147–168.

Krishna P (ed.) (1994) *Advances in Engineering*. Wiley Eastern Ltd.

Kovács I, Strommen E and Hjorth-Hansen E (1999) Damping devices against cable oscillations on Suuingesund Bridge. *Proceedings of the 3rd International Symposium on Cable Dynamics*, pp. 145–150, Trondheim.

Kusakabe T, Yokohama K, Kanazaki T and Sekiya M (1995) The effects of cable cross ties for wind-induced vibration. *Proceedings of the 9th International Conference on Wind Engineering*, pp. 783–792, New Delhi.

Langsø H and Larsen O (1987) Generating mechanisms for cable-stay oscillations at the Farø Bridge. *Proceedings of the International Conference on Cable-stayed Bridges*, pp. 1023–1033, Bankok.

Larose G (1992) *The response of suspension bridge deck to turbulent wind: the taut strip approach*. MESc thesis, University of Western Ontario.

Larose G (1997) *The dynamic action of gusty winds on long-span bridges*. PhD thesis, Technical University of Denmark.

Larose G and Zan S (2001) The aerodynamic forces on the stay cables of cable-stayed bridges in the critical Reynolds number range. *Proceedings 4th International Symposium on Cable Dynamics*, pp. 77–84, Montreal.

Larose G, Jakobsen J and Savage M (2003a) Wind-tunnel experiments on an inclined and yawed stay-cable model in the critical Reynolds number range. *Proceedings of the 5th International Symposium on Cable Dynamics*, pp. 279–286, Santa Margherita.

Larose G, Jakobsen J and Savage M (2003b) Wind-tunnel experiments on an inclined and yawed stay-cable model in the critical Reynolds number range. *Proceedings of the 11th International Conference on Wind Engineering*, Vol. 2, pp. 1705–1712, Lubbock, Texas.

Larose G, Zasso A and Grappino S (2005) Experiments on a yawed stay cable in turbulent flow in the critical reynolds number range. *Proceedings of the 6th International Symposium on Cable Dynamics*, Charleston.

Larsen A (ed.) (1992) *Aerodynamics of Large Bridges: Proceedings of the First International Symposium*. Balkema, Copenhagen, Denmark.

Larsen A and Esdahl S (eds) (1998) *Bridge Aerodynamics: Proceedings of the International Symposium on Advances in Bridge Aerodynamics*. Balkema.

Larsen A and Lafreniere A (2005) Application of a limit cycle oscillator model to bridge cable galloping. *Proceedings of the 6th International Symposium on Cable Dynamics*, Charleston, South Carolina.

Leblond A and Hardy C (1999) On the estimation of a $\vartheta \leq 45°$ complex matrix of symmetric Stockbridge-type dampers. *Proceedings of the 3rd International Symposium on Cable Dynamics*, pp. 139–144, Trondheim.

Leet L (1960) *Vibrations from Blasting Rocks*. Harvard University Press.

Li M (1993) *Buffeting response of large bridges in atmospheric turbulence*. PhD thesis, Southwestern Jiatong University, Chengdu.

Li M and He D (1995) Buffeting response of large bridges. *Proceedings of the 9th International Conference on Wind Engineering*, Vol. 2, pp. 893–904, New Delhi.

Liepmann H (1952) On the application of statistical concepts to the buffeting problem. *Journal of the Aeronautical Sciences* **19**(12), 793–800.

Lilien J (1997) Galloping of overheard electrical lines, mechanisms, wind tunnel experiments – field measurements. *Proceedings of the 2nd International Symposium on Cable Dynamics*, pp. 37–48, Tokyo.

Lilien J and Pinto da Costa A (1994) Amplitudes caused by parametric excitations on cable stayed structures. *Jorunal of Sound and Vibration* **174**(1), 64–90.

Lin Y and Li Q (1993) New stochastic theory for bridge stability in turbulent flow. *Proceedings of the American Society of Civil Engineers*, **119** (EM1), 113–127.

Lin Y and Yang J (1983) Multi-mode bridge response to wind excitation. *Proceedings of the American Society of Civil Engineers*, **109** (EM2), 586–603.

Liu M, Zuo D and Jones N (2005) Deck-induced stay cable vibrations: Field observations and analytical model. *Proceedings of the 6th International Symposium on Cable Dynamics*, Charleston, South Carolina.

Ljung L (1999) *System Identification: Theory for the User*. Prentice Hall.

Macdonald J (2005) Quasi-steady analysis of inclined cable galloping in the critical Reynolds number range. *Proceedings of the 6th International Symposium on Cable Dynamics*, Charleston, South Carolina.

Macdonald J and Larose G (2004) Quasi-steady analysis of dry inclined able galloping in the critical Reynolds number range. *Proceedings of the UK Wind Engineering Society*, Cranfield.

Macdonald J and Larose G (2006) A unified approach to aerodynamic damping and drag/lift instabilities, and its application to dry inclined cable galloping. *Journal of Fluids and Structures*, **22**(2), 229–252.

Madsen H, Krenk S and Lind N (1986) *Methods of Structural Safety*. Prentice-Hall.

Main J, Jones N and Yamaguchi H (2001) Characterization of rain–wind induced stay-cable vibrations from full-scale measurements. *Proceedings of the 4th International Symposium on Cable Dynamics*, pp. 235–242, Montreal.

Major A (1980) *Dynamics in Civil Engineering: Analysis and Design*. Akademial Klado.

Mann J (1994) The spatial structure of neutral atmospheric surface-layer turbulence. *Journal of Fluid Mechanics*, **273**, 141–168.

Matsuda K, Cooper K and Tanaka H (2003) The analysis of wind-induced static displacements and flutter for long-span suspension bridges using steady and unsteady aerodynamic forces measured at high Reynolds numbers. *Proceedings 11th International Conference of Wind Engineering*, Vol. 1, pp. 649–656, Texas.

Matsumoto M (1998) Observed behaviour of prototype cable vibration and its generation mechanism. In *Bridge Aerodynamics*, pp. 189–211. Balkema.

Matsumoto M, Yokohama K, Miyata T, Fujino Y and Yamaguchi H (1989) Wind-induced cable vibration of cable-stayed bridges in Japan. *Proceedings of the Canada–Japan Workshop on Bridge Aerodynamics*, pp. 101–110, Ottawa.

Matsumoto M, Shiraishi N, Kitazawa M, Knisely C, Shirato H, Kim Y and Tsujii M (1990) Aerodynamic behavior of inclined circular cylinders-cable aeroynamics. *Journal of Wind Engineering and Industrial Aerodynamics* **33**, 63–72.

Matsumoto M, Ishizaki H, Kitazawa M, Aoki J and Fujii D (1995a) Cable aerodynamics and its stabilization. *Proceedings of the 1st International Symposium on Cable Dynamics*, pp. 289–296, Liège.

Matsumoto M, Yamagishi M, Aoki J and Shiraishi N (1995b) Various mechanisms of inclined cable aerodynamics. *Proceedings of the 9th International Conference on Wind Engineering*, pp. 759–770, New Delhi.

Matsumoto M, Shirato H, Yagi T, Jones N and Hayashi T (2001) Field observation system of cable aerodynamics in natural wind. *Proceedings of the 4th International Symposium on Cable Dynamics*, pp. 219–225, Montreal.

Matsumoto M, Yagi T, Oishi T and Adachi Y (2005) Effects of axial flow and Kármán vortex interferences on dry-state cable galloping of inclined stay-cables. *Proceedings of the 6th International Symposium on Cable Dynamics*, Charleston.

Melbourne W (1998) Comfort criteria for wind-induced motion in structures. *Structural Engineering International* **8**(1), 40–44.

Melbourne W and Cheung J (1988) Designing for serviceable accelerations in tall buildings. *Proceedings of the 4th International Conference on Tall Buildings*, pp. 148–155, Hong Kong and Shanghai.

Melbourne W and Palmer T (1992) Accelerations and comfort criteria for building underoing comply motions. *Journal of Wind Engineering and Industrial Aerodynamics* **41**, 105–116.

Miyasaka Y, Ohshima K and Nakabayashi S (1987) *Experimental Study on Ajikawa Bridge Cable Vibration*. Report No. 7, Hanshin Expressway Public Corporation Engineering.

Miyata T (ed.) (1997) *Wind Engineering of Structures*. Japanese Society of Steel Construction. (In Japanese.)

Miyata T (ed.) (1999) *Long-Span Bridges and Aerodynamics*. Springer-Verlag.

Miyata T, Yamada H and Hojo T (1994) Aerodynamic response of PE stay cables with pattern-indented surface. *Proceedings of the International Conference on cable-stayed and Suspension Bridges*, Vol. 2, pp. 515–522, Deauville.

Miyata T, Yamada H, Boonypinyo V and Santos J (1995) Analytical investigation on the response of a very long suspension bridge under gusty wind. *Proceedings of the 9th International Conference on Wind Engineering*, Vol. 2, pp. 1006–1017, New Delhi, India.

Modi V and Welt F (1988) Damping of wind induced oscillations through liquid sloshing. *Journal of Wind Engineering and Industrial Aerodynamics* **30**, 85–94.

Mufti A (ed.) (2002) *Proceedings of the 1st International Workshop on Structural Health Monitoring*. ISIS Canada Research Network, Winnipeg.

Nagase T and Hisatoku T (1992) Tuned-pendulum mass damper installed in the Crystal Tower. *The Structural Design of Tall Buildings* **1**, 35–36.

Namini A, Albrecht P and Bosch H (1992) Finite element-based flutter analysis of cable-suspended bridges *Proceedings of the American Society of Civil Engineers*, **118** (ST6), 1509–1526.

Nashif A, Jones D and Henderson J (1985) *Vibration Damping*. Wiley, Chichester.

Newland D (1993) *An Introduction to Random Vibrations, Spectral and Wavelet Analysis*, 3rd edn. Addison Wesley Logman.

Novak M and Tanaka H (1974) Effect of turbulence on galloping instability. *Proceedings of the American Society of Civil Engineers*, **100** (EM1), 27–47.

Novak M, Davenport A and Tanaka H (1978) Vibrations of towers due to galloping of iced guy cables. *Proceedings of the American Society of Civil Engineers*, **104** (EM2), 457–473.

Oberst H (1986) Reduction of noise by the use of damping materials. *Philosophical Transactions of the Royal Society, London* **A263**, 441–453.

Ohshima K and Nanjo M (1987) Aerodynamic stability of the cables of a cable-stayed bridge subject to rain (A case study of the Aji River Bridge). *Proceedings of the US–Japan Joint Seminar on Natural Resources*.

Panofsky H and Dutton J (1984) *Atmospheric Turbulence. Models and Methods for Engineering Application*. Wiley, Chichester.

Paxton R (1990) *100 Years of the Forth Bridge*. Thomas Telford.

Peeters B (2000) *System identification and Damage Detection in Civil Engineering*. PhD thesis, Katholieke Universiteit Leuven.

Peeters B, Maeck J and De Roeck G (2001) Vibration-based damage detention in civil engineering: Excitation sources and temperature effects. *Smart Materials and Structures* **10**, 518–527.

Petersen N (1980) *Design of Large Scale Tuned Mass Dampers – Structural Control*. North-Holland.

Plate E (ed.) (1982) *Engineering Meteorology*. Elsevier Scientific, Amsterdam.

Poulsen N, Damsgaard A and Reinhold T (1991) Determination of flutter derivatives for the Great Belt Bridge. *Proceedings of the 8th International Conference on Wind Engineering, London, Ontario*, Vol. 2, pp. 153–164.

Price S and Paidoussis M (1984) The aerodynamic forces acting on groups of two and three circular cylinders when subject to a cross-flow. *Journal of Wind Engineering and Industrial Aerodynamics* **17**, 329–347.

Reda Tana NM, Nouraldin A, Osman A and El-Sheimy N (2004) Introduction to the use of wavelet multi-resolutions analysis for intelligent structural health monitoring. *Can. Journal of Civil Engineering* **31**, 719–731.

Rice S (1954) Mathematical analysis of random noise. In *Selected Papers on Noise and Stochastic Processes*, Dover, 133–294.

Richards D (1963) Aerodynamic properties of the Severn Crossing conductor. *Proceedings of the 1st International Conference on Wind Engineering*, pp. 699–765, Teddington.

Richart F, Hall J and Woods R (1970) *Vibrations of Soils and Foundations*. Prentice-Hall.

Roger K (1977) *Airplane Math Modeling Methods for Active Control Design*. AGARD-CP-228, NASA.

Ruscheweyh H (1982) *Dynamische Windwirkung an Bauwerken*. Bauverlag (2 Vol).

Ruscheweyh H (1994) Vortex excited vibrations, In *Wind-Excited Vibrations of Structures*, pp. 51–84. Springer-Verlag.

Ruscheweyh H (1999) The mechanism of rain–wind-induced vibration, In *Wind Engineering into the 21st Century*, pp. 1041–1047. Balkema.

Ruscheweyh H and Hirsch G (1974) *Vibration Measurements at the Cable of Koehlbrand Bridge in Hamburg*. Technical Report, Institute for Lightweight Structures, RTWH Aachen. (In German.)

Ruscheweyh H and Verwiebe C (1995) Rain–wind-induced vibrations of steel bars. *Proceedings of the 1st International Symposium on Cable Dynamics*, pp. 469–472, Liège.

Rychlik I (1987) A new definition of the rainflow cycle counting method. *International Journal on Fatigue* **9**(2), 119–121.

Saito T, Matsumoto M and Kitazawa M (1994) Rain–wind excitation of cables on cable-stayed Higashi-Kobe Bridge and cable vibration control. *Proceedings of the International Conference on Cable-Stayed and Suspension Bridges*, Vol. 2, pp. 507–514. Deauville.

Salawu O (1997) Detection of structural damage through changes in frequency: a review. *Engineering Structures* **19**(9), 718–723.

Sauter D, Liu Q, Hagedorn P and Rahman A (2001) On the vortex-excited vibrations of stay-cables. *Proceedings on the 4th International Symposium on Cable Dynamics*, pp. 277–284, Montreal.

Scanlan R and Rosenbaum R (1951) *Aircraft Vibration and Flutter*. Macmillan.

Scanlan R and Sabzevari A (1968) Aerodynamic stability of suspension bridges. *Proceedings of the American Society of Civil Engineers*, **94** (EM2), 489–519.

Scanlan R and Simiu E (1999) *Wind Effects on Structures: Fundamentals and Applications to Design*, 3rd edn. Wiley, Chichester.

Scanlan R, Béliveau J and Budlong K (1972) Flutter and aerodynamic response considerations for bluff objects in a smooth flow. *Proceedings of the International Union of Theoretical and Applied Mechanics–International Association for Hydraulic Research, Symposium on Flow-Induced Structural Vibrations*, pp. 339–354, Karlsruhe.

Scanlan R, Béliveau J and Budlong K (1974) Indicial aerodynamic fuction for bridge decks. *Proceedings of the American Society of Civil Engineers*, **100** (EM4), 657–672.

Scott R (2001) *In the Wake of Tacoma: Suspension Bridges and the Quest for Aerodynamic Stability*. American Society of Civil Engineers Press.

Scruton C (1981) *An Introduction to Wind Effect on Structures*. Oxford University Press.

Selberg A (1963) Aerodynamic effects on suspension bridges. *Proceedings of the 1st International Conference on Wind Engineering, Teddington*, Vol. 2, pp. 462–486, Teddington.

Smith J (1988) *Vibration of Structures. Applications in Civil Engineering Design*. Chapman and Hall.

Soong T (1990) *Active Structural Control: Theory and Practice*. Longman Science and Technical.

Steinman D and Watson S (1957) *Bridges and Their Builders*. Dover Publications.

Stubler J, Ladret P, Domage J and Peltier M (1999) Bridge stay-cable vibration: Phenomena, criteria und damper technology. *Proceedings on the 3rd International Symposium on Cable Dynamics*, pp. 163–170, Trondheim.

Surry D (1982) Consequences of distorsions in the flow including mismatching scales and intensities of turbulence. *Proceedings International Workshop on Wind Tunnel Modelling Criteria and Techniques in Civil Engineering Applications*, pp. 137–185, Gaithersburg, MA.

Tanaka H (1992) Balkema chapter Similitude and modelling in bridge aerodynamics, In *Bridge Aerodynamics*, pp. 83–94.

Tanaka H and Davenport A (1982) Response of taut strip models to turbulent wind. *Proceedings of the American Society of Civil Engineers*, **108** (EM1), 33–49.

Tanaka H and Yamada H (1987) Mass and damping simulation for the modelling of aeroelastic responses. *Proceedings International Conference on Flow Induced Vibrations*, pp. 103–110, Bowness-on-Windermere.

Tanaka H, Yamamura N and Tatumi M (1992) Coupled mode flutter analysis using flutter derivatives. *Journal of Wind Engineering and Industrial Aerodynamics* **42**, 1279–1290.

Tanaka H et al. (1996) *Modelling of Wind Action on Bridges*. Research Report 94869, Danish Maritime Institute.

Tewari A and Brink-Spulink J (1992) Multiple-pole rational function approximations for unsteady aerodynamics. *Journal: Aerocraft* **30**(3), 426–428.

Theodorsen T (1935) *General Theory of Aerodynamic Instability and the Mechanism of Flutter*. Technical Report No.496, National Advisory Committee for Aeronautics.

Theodorsen T and Garrick I (1940) *Mechanism of Flutter: a Theoretical and Experimental Investigation of the Flutter Problem*. TR685, National Advisory Committee for Aeronautics.

Tiffany S and Adams W (1988) Non-linear programming extensions to rational function approximation methods for unsteady aerodynamic forces. Technical Paper 2776, NASA.

Toriumi R, Furuya N, Takeguchi M, Miyazaki M and Saito Z (1999) A study on wind-induced vibration of parallel suspenders observed at the Akashi-Kaikyou Bridge. *Proceedings of the 3rd International Symposium on Cable Dynamics*, pp. 177–182, Trondheim.

Torrence C and Compo G (1998) A practical guide to wavelet analysis. *Bulletin of the American Meteorological Society* **79**(1), 61–78.

Townsend A (1976) *The Structure of Turbulent Shear Flow*, 2nd edn. Cambridge University Press.

Ueda T, Fujiwara T and Nakagaki R (1991) Suppression of wind induced oscillation by tuned sloshing dampers in the free-standing tower of a cable-stayed bridge. *Proceedings of the International Symposium for Innovation in Cable-Stayed Bridges*, pp. 207–216, Fukuoka.

Ukeguchi N, Sakata H and Nishitani H (1966) An investigation of aeroelastic instability of suspension bridges. *Proceedings of the International Symposium on Suspension Bridges, Lisbon*, pp. 273–284.

Van Koten H (1974) *Damping of building structures.* Technical Report No.BI-74-64, Institute TNO voor Bouwmaterialen en Bouwconstructies, Delft.

Veit R and Wenzel H (2004) Measurement data based lifetime estimation of the Europabrücke due to traffic load- a three level aproach. *Proceedings of the 11th International EG-ICE Workshop*, Weimar.

Verwiebe C (1998) Rain–wind-induced vibrations of cables and bars. In *Bridge Aerodynamics*, pp. 255–263. Balkema.

Verwiebe C and Ruscheweyh H (1998) Recent research results concerning the excitation mechanisms of rain-wind-induced vibrations. *Journal of Wind Engineering and Industrial Aerodynamics* **74/76**, 1005–1013.

Vickery B (1965) *On the Flow Behind a Coarse Grid and its Use as a Model of Atmospheric Turbulence in Studies Related to Wind Loads in Buildings.* US National Physics Laboratory Aerodynamics Report 1143.

Vickery B (1994) The response of chimneys and tower-like structures to wind loathing. In *Recent Advances in Wind Engineering*, pp. 205–233. Wiley Eastern.

Virlogeux M (1998) Cable vibrations in cable-stayed bridges. In *Bridge Aerodynamics*, pp. 213–233, Balkema.

Von Kármán T (1948) Progress in the statical theory of turbulence. *Proceedings of the National Academy of Science*, **34**, 530–539.

Wardlaw R (1992) The improvement of aerodynamic performance. In *Aerodynamics of Large Bridges*, pp. 59–70. Balkema.

Wardlaw R (1994) Interference and proximity effects. In *Wind-Excited Vibrations of Structures*, pp. 321–363. Springer-Verlag.

Wardlaw R and Cooper K (1974) Mechanisms and alleviation of wind-induced structural vibrations. *Proceedigs 2nd symposium on Applications of Solid Mechanics*, pp. 369–399, Hamilton, Ontario.

Wenzel H and Pichler D (2005) *Ambient Vibration Monitoring.* Wiley Chichester.

Wianecki J (1979) Cables wind excited vibrations of cable-stayed bridges. *Proceedings of the 5th International Conference of Wind Engineering*, pp. 1381–1393, Colorado.

Wirsching P, Paez T and Ortiz H (1995) *Random Vibrations: Theory and Practice.* Wiley, Chichester.

Wiss F and Parmelee R (1974) Human perception of transient vibrations *Proceedings of the American Society of Civil Engineers*, **100** (ST4), 773–787.

Wu Z and Abe M (ed.) (2003) *Proceedings of the 1st International Conference on Structural Health Monitoring and Intelligent Infrastructure*, Tokyo.

Wyatt T and Scruton C (1981) *Bridge Aerodynamics.* Thomas Telford.

Xiang H, Liu C and Gu M (1995) Time domain analysis for coupled buffeting response of long span bridges. *Proceedings of the 9th International Conference on Wind Engineering*, Vol. 2, pp. 881–892.

Xie J (1985) *Bridge flutter theory and research of flutter characteristics of cable-stayed bridges.* PhD thesis, Tongji University, Shanghai. (In Chinese.)

Xu Y and Wang L (2001) Analytical study of wind-rain-induced cable vibration. *Proceedings of the 5th Asia-Pacific Conference on wind Engineering*, pp. 109–112, Kyoto.

Yagi T (1997) *Wind-induced instabilities of structures.* PhD thesis, Kungl Tekniska Høgskolan, Department of Structural Engineering, Stockholm.

Yamada Y, Shiraishi N, Toki K, Matsumoto M, Matsuhashi K, Kitazawa M and Ishizaki H (1991) Earthquake-resistant and wind-resistant design of the Higashi-Kobe Bridge. In *Cable-Stayed Bridges: Recent Development and their Future*, pp. 397–416. Elsevier.

Yamaguchi H (1990) Analytical study on growth mechanism of rain vibration of cables. *Journal of Wind Engineering and Industrial aerodynamics* **33**, 73–80.

Yang C (1986) *Random Vibration of Structures.* Wiley, Chichester.

Yoshimura T, Inoue A, Kaji K and Savage M (1989) A study on the aerodynamic stability of the Aratsu Bridge. *Proceedings of the Japan–Canada Workshop on Bridge Aerodynamics*, pp. 41–50, Ottawa.

Yoshimura T, Savage M and Tanaka H (1995) Wind-induced vibrations of bridge stay-cables. *Proceedings of the 1st International Symposium on Cable Dynamics*, pp. 437–444, Liege, Belgium.

Zasso A, Bocciolone M and Brownjohn J (1992) Rain–wind aeroelastic instability of the inclined hangers of a suspension bridge. *Proceedings of the Inaugural Conference on Engineering Society, Cambridge.*

Zdravkovich M (1997) *Flow Around Circular Cylinders*, Vol. 1. Oxford University Press, New York.

Abbreviation Index

Person Index

Index

Printed and bound in the UK by
CPI Antony Rowe, Eastbourne

Printed and bound by CPI Group (UK) Ltd, Croydon, CR0 4YY
16/09/2026
14963387-0002